Lecture Notes in Networks and Systems

Volume 177

The series "Lecture Notes in Networks and Systems" publishes the latest developments in Networks and Systems—quickly, informally and with high quality. Original research reported in proceedings and post-proceedings represents the core of LNNS.

Volumes published in LNNS embrace all aspects and subfields of, as well as new challenges in, Networks and Systems.

The series contains proceedings and edited volumes in systems and networks, spanning the areas of Cyber-Physical Systems, Autonomous Systems, Sensor Networks, Control Systems, Energy Systems, Automotive Systems, Biological Systems, Vehicular Networking and Connected Vehicles, Aerospace Systems, Automation, Manufacturing, Smart Grids, Nonlinear Systems, Power Systems, Robotics, Social Systems, Economic Systems and other. Of particular value to both the contributors and the readership are the short publication timeframe and the world-wide distribution and exposure which enable both a wide and rapid dissemination of research output.

The series covers the theory, applications, and perspectives on the state of the art and future developments relevant to systems and networks, decision making, control, complex processes and related areas, as embedded in the fields of interdisciplinary and applied sciences, engineering, computer science, physics, economics, social, and life sciences, as well as the paradigms and methodologies behind them.

Indexed by SCOPUS, INSPEC, WTI Frankfurt eG, zbMATH, SCImago.

All books published in the series are submitted for consideration in Web of Science.

More information about this series at http://www.springer.com/series/15179

Rajendra Prasad Mahapatra · B. K. Panigrahi ·
Brajesh K. Kaushik · Sudip Roy
Editors

Proceedings of 6th International Conference on Recent Trends in Computing

ICRTC 2020

Springer

Editors
Rajendra Prasad Mahapatra
SRM Institute of Science and Technology
Delhi-NCR Campus
Modinagar, Uttar Pradesh, India

Brajesh K. Kaushik
Indian Institute of Technology Roorkee
Roorkee, Uttarakhand, India

B. K. Panigrahi
Indian Institute of Technology Delhi
New Delhi, Delhi, India

Sudip Roy
Indian Institute of Technology Roorkee
Roorkee, Uttarakhand, India

ISSN 2367-3370 ISSN 2367-3389 (electronic)
Lecture Notes in Networks and Systems
ISBN 978-981-33-4500-3 ISBN 978-981-33-4501-0 (eBook)
https://doi.org/10.1007/978-981-33-4501-0

This Springer imprint is published by the registered company Springer Nature Singapore Pte Ltd.
The registered company address is: 152 Beach Road, #21-01/04 Gateway East, Singapore 189721, Singapore

Preface

This Lecture Notes in Networks and Systems (LNNS) volume contains the papers presented at the 6th International Conference on Recent Trends in Computing (ICRTC 2020) held during July 3–4, 2020, at SRM Institute of Science and Technology, Delhi-NCR Campus, Modinagar, Ghaziabad, India.

ICRTC 2020 is aiming at bringing together the researchers from academia and industry to report and review the latest progress in the cutting-edge research on various research areas of image processing, computer vision and pattern recognition, machine learning, data mining, big data and analytics, soft computing, mobile computing and applications, cloud computing, green IT, and finally to create awareness about these domains to a wider audience of practitioners.

ICRTC 2020 received 240 paper submissions including two submissions from foreign countries. All the papers were peer reviewed by the experts in the area from India and abroad, and comments were sent to the authors of accepted papers. Finally, 77 papers were accepted for online Zoom presentation in the conference. This corresponds to an acceptance rate of 34% that is intended to maintain the high standards of the conference proceedings. The papers included in this Lecture Notes in Networks and Systems (LNNS) volume cover a wide range of topics in intelligent computing and algorithms and their real-time applications in problems from diverse domains of science and engineering.

The conference was inaugurated by Prof. Mike Hinchey, University of Limerick, Ireland, on July 3, 2020. The conference featured distinguished keynote speakers as follows: Shri.Aninda Bose, Senior Editor, Springer New Delhi, India; Mr. Avik Dutta, Country Representative, DigitalED, Waterloo, Canada.

We take this opportunity to thank the authors of the submitted papers for their hard work, adherence to the deadlines, and patience with the review process. The quality of a referred volume depends mainly on the expertise and dedication of the reviewers. We are indebted to the Technical Committee members, who produced excellent reviews in short time frames. First, we are indebted to the Hon'ble Dr. T. R. Paari Vendhar, Member of Parliament (Lok Sabha), Founder-Chancellor, SRM Institute of Science and Technology; Shri. Ravi Pachamoothoo, Pro-Chancellor—Administration, SRM Institute of Science and Technology; Dr. P. Sathyanarayanan, Pro-Chancellor—Academics, SRM Institute of Science and

Technology; Dr. R. Shivakumar, Vice President, SRM Institute of Science and Technology; Dr. Sandeep Sancheti, Vice Chancellor, SRM Institute of Science and Technology; Prof. C. Muthamizhchelvan, Pro Vice Chancellor—E&T, SRM Institute of Science and Technology for supporting our cause and encouraging us to organize the conference there. In particular, we would like to express our heartfelt thanks for providing us with the necessary financial support and infrastructural assistance to hold the conference. Our sincere thanks to Dr. D. K. Sharma, Professor and Dean; Dr. S. Viswanathan, Deputy Registrar; Dr. Navin Ahalawat, Professor and Dean (Campus Life) SRM Institute of Science and Technology, Delhi-NCR Campus, Modinagar, Ghaziabad, for their continuous support and guidance. We specially thank Dr. Dambarudhar Seth, Professor; Dr. Pritee Parwekar, Associate Professor; and Dr. Veena Khandelwal, Assistant Professor, Co-conveners-ICRTC 2020, SRM Institute of Science and Technology, Delhi-NCR Campus, of this conference for their excellent support and arrangements. Without them, it is beyond imagination to conduct this conference. We thank the International Advisory Committee members for providing valuable guidelines and inspiration to overcome various difficulties in the process of organizing this conference. We would also like to thank the participants of this conference. The faculty members and students of SRM Institute of Science and Technology, Delhi-NCR Campus, Modinagar, Ghaziabad, deserve special thanks. Without their involvement, we would not have been able to face the challenges of our responsibilities. Finally, we thank all the volunteers who made great efforts in meeting the deadlines and arranging every detail to make sure that the conference could run smoothly. We hope the readers of these proceedings find the papers inspiring and enjoyable.

Ghaziabad, India — Dr. Rajendra Prasad Mahapatra
New Delhi, India — Dr. B. K. Panigrahi
Roorkee, India — Dr. Brajesh K. Kaushik
Roorkee, India — Dr. Sudip Roy
July 2020

Contents

Editors and Contributors

About the Editors

Prof. Rajendra Prasad Mahapatra is B.Tech., M.Tech. and Ph.D. He is currently working at SRM Institute of Science and Technology, NCR Campus, as Professor and Head—Computer Science and Engineering and Dean Admissions. Prof. Mahapatra has vast experience of 18 years as an academician, a researcher and an administrator. During these 18 years, he has worked in India and abroad. He has been associated with Mekelle University, Ethiopia, for more than 2 years. Four candidates have successfully completed their Ph.D. under his supervision, and 7 students are pursuing their research under his guidance. Prof. Mahapatra has authored more than 70 research papers which are published in international journals like Inderscience, Emerald, Elsevier, IEEE and Springer. He is Editor of Taylor & Francis Group; he is Reviewer of Elsevier, Springer Journals; he is a fellow member of I.E. (India), a senior member of IACSIT, Singapore, a life member of ISTE and a member of IEEE and many more reputed bodies.

Dr. B. K. Panigrahi is working as Professor in the Electrical Engineering Department, IIT Delhi, India. Prior to joining IIT Delhi in 2005, he has served as Faculty in Electrical Engineering Department, UCE Burla, Odisha, India, from 1992 to 2005. Dr. Panigrahi is a senior member of IEEE and Fellow of INAE, India. His research interest includes application of soft computing and evolutionary computing techniques to power system planning, operation and control. He has also worked in the field of bio-medical signal processing and image processing. He has served as the editorial board member, Associate Editor and special issue Guest Editor of different international journals. He is also associated with various international conferences in various capacities. Dr. Panigrahi has published more than 100 research papers in various international and national journals.

Dr. Brajesh K. Kaushik received Doctorate of Philosophy (Ph.D.) in 2007 from the Indian Institute of Technology, Roorkee, India. He joined the Department of Electronics and Communication Engineering, Indian Institute of Technology,

Roorkee, as Assistant Professor in December 2009; and since April 2014 he has been Associate Professor. He has served as General Chair, Technical Chair and Keynote Speaker of many reputed international and national conferences. Dr. Kaushik is a senior member of IEEE and a member of many expert committees constituted by government and non-government organizations. He is Editor of IEEE Transactions on Electron Devices; Associate Editor of IET Circuits, Devices and Systems; Editor of Microelectronics Journal, Elsevier; the editorial board member of Journal of Engineering, Design and Technology, Emerald and Circuit World, Emerald. He has received many awards and recognition from the International Biographical Center (IBC), Cambridge. His name has been listed in Marquis Who's Who in Science and Engineering® and Marquis Who's Who in the World®. Dr. Kaushik has been conferred with Distinguished Lecturer award of IEEE Electron Devices Society (EDS) to offer EDS Chapters with quality lectures in his research domain. He has 12 books to his credit published by reputed publishers such as CRC Press, Springer, Artech and Elsevier. One of his books titled "Nanoscale Devices: Physics, Modeling, and Their Application," CRC Press has won 2018 Outstanding Book and Digital Product Awards in the Reference/Monograph Category from Taylor and Francis Group. His research interests are in the areas of high-speed interconnects, low-power VLSI design, memory design, carbon nanotube-based designs, organic electronics, FinFET device circuit co-design, electronic design automation (EDA), spintronics-based devices, circuits and computing, image processing, and optics and photonics-based devices.

Sudip Roy completed his schooling (Class I to XII) from Kalyani University Experimental High School at Kalyani, a town of West Bengal, India, in 1998. He passed Secondary (Class X Board Exam.) in 1996 from WBBSE and Higher Secondary (Class XII Board Exam.) in 1998 from WBCHSE. He studied Bachelor of Science (B.Sc.) with Physics Honors from Rahara Ramakrishna Mission Vivekananda Centenary College and received his B.Sc. degree from the University of Calcutta in 2001. Then, he studied Bachelor of Technology (B.Tech.) in Computer Science and Engineering (a 3-year program) from the University College of Science and Technology (a.k.a. Rajabazar Science College) and received his B.Tech. degree from the University of Calcutta in 2004. He received the M.S. (by research) and Ph.D. degrees in Computer Science and Engineering from the Indian Institute of Technology Kharagpur (IIT KGP), India, in 2009 and 2014, respectively. During 2004–2009, he had research experiences in different intervals as a Project-Linked-Personnel (November 2004–July 2005) at the Machine Intelligence Unit, Indian Statistical Institute Kolkata, India, as Junior Project Assistant (August 2005–June 2008) at the Department of Computer Science and Engineering, Indian Institute of Technology Kharagpur (IIT KGP), India, and as a Project-Linked-Personnel (July 2008–September 2009) at the Advanced Computing and Microelectronics Unit, Indian Statistical Institute Kolkata, India. Since July 2014, Dr. Roy has been working as Assistant Professor at the Department of Computer Science and Engineering, Indian Institute of Technology

Roorkee (IITR), India. Before joining with IITR, he was Research Associate (February 2014–July 2014) at the Electronic Design Automation Laboratory in the Department of Computer Science and Information Engineering (CSIE), National Cheng Kung University, Tainan, Taiwan. His research interest includes electronic design automation (EDA) of microfluidic lab-on-a-chips and digital VLSI integrated circuits. He has published 13 research articles in peer-reviewed journals and 24 research articles in international conference proceedings and holds 2 US patents. He has authored one book and one book chapter. Dr. Roy is a member of the IEEE and ACM.

Contributors

R. Jaswanth Aakash Sri Krishna College of Technology, Coimbatore, Tamilnadu, India

Misganu Tuse Abetu Department of Computer Science, Wollega University, Nekemte, Ethiopia

Ashish Agarwal Time & Frequency Metrology, National Physical Laboratory, New Delhi, India

Isha Agarwal Bharati Vidyapeeth's College of Engineering, New Delhi, India

Mohit Agarwal Computer Science Engineering, Bennett University, Uttar Pradesh, Greater Noida, India

Ritu Agarwal Delhi Technological University, Shahbad Daulatpur, Bawana Road, New Delhi, India

Shivangi Agarwal SRMSCET, Bareilly, India

Ashish Agrawal SRMSCET, Bareilly, India

Nimish Agrawal Division of Instrumentation and Control Engineering, Netaji Subhas University of Technology (Erstwhile NSIT), Dwarka, New Delhi, India

Subhash Chand Agrawal Computer Engineering and Applications, IET, GLA University, Mathura, Uttar Pradesh, India;
Department of Computer Engineering and Applications, GLA University, Mathura, UP, India

Shameem Ahmad Department of Electronics Engineering, Z.H.C.E.T, Aligarh Muslim University, Aligarh, India

Nisha Ahuja Computer Science Engineering, Bennett University, Uttar Pradesh, Greater Noida, India

T. Akash Department of Computer Science and Engineering, Sri Krishna College of Technology, Coimbatore, Tamil Nadu, India

M. Afshar Alam Department of Computer Science and Engineering, School of Engineering Sciences and Technology, New Delhi, India

Supriya Anand Department of Computer Science and Engineering, Punjabi University, Patiala, India

Kushal Anjaria IT and Systems Area, Institute of Rural Management Anand (IRMA), Anand, Gujarat, India

R. Annie Uthra Department of Computer Science and Engineering, SRM Institute of Science and Technology, Chennai, Tamil Nadu, India

Akbar Ansari IIIT Allahabad, Allahabad, Uttar Pradesh, India

Sonia Aribam Computer Science Engineering, NIT Manipur, Imphal, Manipur, India

Chirag Arora KIET Group of Institutions, Ghaziabad, India

Deepak Arora Department of Computer Science and Engineering, Amity School of Engineering and Technology, Amity University, Lucknow Campus, India

Mukesh Arora Swami Keshwanand Institute of Technology, Management and Gramothan, Jaipur, India

Amutha Arunachalam Time & Frequency Metrology, National Physical Laboratory, New Delhi, India

Meenakshi S. Arya Department of Computer Science and Engineering, SRM Institute of Engineering and Technology, Vadapalani, Chennai, India

Ashwin Ashok PES University, Bengaluru, India

Sanchit Bathla Department of Information Technology, Delhi Technological University, Delhi, India

Mohammad Abid Bazaz Department of Electrical Engineering, National Institute of Technology Srinagar, Srinagar, India

Francesco Bellotti DITEN, University of Genova, Genova, Italy

Riccardo Berta DITEN, University of Genova, Genova, Italy

Rajesh Bhadada Department of Electronics and Communication Engineering, M.B.M Engineering College, JNV University, Jodhpur, India

Gayatri Bhandari Department of Computer Science and Engineering, Shri JJT University, Chudela, Jhunjhunu, Rajstan, India

G. Bharathi Department of Information Technology, Anil Neerukonda Institute of Technology and Sciences, Visakhapatnam, India

Ankur Bhardwaj Computer Science and Engineering Department, KIET Group of Institutions, Ghaziabad, India

Geetanjali Bhola Department of Information Technology, Delhi Technological University, Delhi, India

Tejaswini Bhosale Department of Computer Science, Birla Institute of Technology, Mesra, Ranchi, India

Shilpi Birla Manipal University, Jaipur, India

Siddhartha Sankar Biswas Department of Computer Science and Engineering, School of Engineering Sciences and Technology, New Delhi, India

Padma Bonde Department of Computer Science and Engineering, Shri Shankracharya Technical Campus, Bhilai, Chhattisgarh, India

Purbasha Bose Techno International Batanagar, Kolkata, India

Myron Carvalho Department of Information Technology, K. J. Somaiya College of Engineering, Vidyavihar, Mumbai, India

Sushil Chandra Institute of Nuclear Medicine and Allied Sciences, Delhi, India

Dhrati Chandrawat Computer Science Engineering, SRM Institute of Science and Technology, Chennai, Tamil Nadu, India

Naman Chauhan Department of CSE, SRMIST, Chennai, India

Rishi Pal Chauhan National Institute of Technology Kurukshetra, Kurukshetra, India

A. Christy Jeba Malar Department of Information Technology, Sri Krishna College of Technology, Coimbatore, Tamil Nadu, India

Shrey Dabhi Technology and Strategy Department, Innovation & Incubation Business Unit, Robert Bosch Engineering and Business Solutions Private Limited, Bengaluru, Karnataka, India;
Department of Computer Science and Engineering, Institute of Technology, Nirma University, Ahmedabad, Gujarat, India

G. Dakshin Department of Computer Science and Engineering, Sri Krishna College of Technology, Coimbatore, Tamilnadu, India

Ashwini Dalvi Department of Information Technology, K. J. Somaiya College of Engineering, Vidyavihar, Mumbai, India

Ronnie V. Daniel Institute of Nuclear Medicine and Allied Sciences, Delhi, India

Bhaskarjyoti Das PES University, Bengaluru, India

Alessandro De Gloria DITEN, University of Genova, Genova, Italy

R. Deepa Department of Computer Science and Engineering, SRM Institute of Engineering and Technology, Vadapalani, Chennai, India

S. Deepak Sri Krishna College of Technology, Coimbatore, Tamilnadu, India

Tadele Debisa Deressa Department of Computer Science, Wollega University, Nekemte, Ethiopia

M. Deva Priya Department of Computer Science and Engineering, Sri Krishna College of Technology, Coimbatore, Tamil Nadu, India

Anjali Dhamiwal Amity University, Gurgaon, Haryana, India

Hariom Dhungana DITEN, University of Genova, Genova, Italy

Yogesh Mohan Dubey Pranveer Singh Institute of Technology, Uttar Pradesh, Kanpur, India

Anil Joseph D'Costa Department of Computer Science and Engineering, School of Engineering Sciences and Technology, New Delhi, India

Hussain Qudsia Ejaz Department of Computer Science and Engineering, School of Engineering Sciences and Technology, Jamia Hamdard, New Delhi, India

Jayati Gandhi Department of Computer Science and Engineering, G. H. Raisoni College of Engineering, Nagpur, Maharashtra, India

Vaibhav Gaur Malaviya National Institute of Technology, JLN Marg, Jaipur, India

Hima George Department of Information Technology, K. J. Somaiya College of Engineering, Vidyavihar, Mumbai, India

Radha Tamal Goswami Techno International New Town, Kolkata, India

B. Govinda Satyanarayana Computer Science Engineering, SRKR Engineering College, Andhra Pradesh, Bhimavaram, India

Amit Kumar Gupta Department of Production Engineering, Birla Institute of Technology, Mesra, Ranchi, India

Mohit Gupta Department of Information Technology, Delhi Technological University, Delhi, India

Neha Gupta Bharati Vidyapeeth's College of Engineering, New Delhi, India

Shriti Gupta Computer Science Engineering, Mody University of Science and Technology, Sikar, Rajasthan, India

Vinodini Gupta Department of Computer Science and Engineering, Shri Shankracharya Technical Campus, Bhilai, Chhattisgarh, India

Vishesh Gupta GLA University, Mathura, India

J. Muhammad Haaris Department of Computer Science & Engineering, Sri Krishna College of Technology, Coimbatore, Tamil Nadu, India

Saif Haque Department of Information Technology, Delhi Technological University, Delhi, India

Shoeb Hussain Department of Electrical Engineering, University of Kashmir, Srinagar, India

Kalpesh R. Jadav Department of Electronics and Communication Engineering, Parul Institute of Engineering and Technology, Parul University, Vadodara, Gujarat, India

Megha Jadhav Department of Computer Science and Engineering, Shri JJT University, Chudela, Jhunjhunu, Rajstan, India

Rakesh Jain Malviya National Institute of Technology Jaipur, Jaipur, India

Rekh Ram Janghel National Institute of Technology Raipur, Raipur, India

Ranjan Kumar Jha CSIR-Central Scientific Instruments Organisation, Chandigarh, India;
Academy of Scientific & Innovative Research (AcSIR), Ghaziabad, India

Kamal Mohammed Jimalo Department of Information Technology, Wollega University, Nekemte, Ethiopia

Rahul Joshi CSE Department, Parul Institute of Engineering and Technology, Parul University, Limda, India

Rajendra Kachhwaha M.B.M. Engineering College, Jai Narain Vyas University, Jodhpur, India

Rohit Kumar Kaliyar Computer Science Engineering, Bennett University, Uttar Pradesh, Greater Noida, India

M. Lakshmi Kamalanjali Department of CSE, Dadi Institute of Engineering and Technology, Visakhapatnam, India

Sujatha Karimisetty Department of CSE, Dadi Institute of Engineering and Technology, Visakhapatnam, India

Jyotirmoy Karjee Embedded Systems and Robotics, TCS Research and Innovation, Bangalore, India

S. Karthick Computer Science and Engineering, SRMIST, Delhi-NCR Campus, New Delhi, India

S. Karthik SRM Institute of Science and Technology, Ghaziabad, India

M. R. Khan Electronics and Telecommunication Engineering, Government Engineering College, Jagdalpur, India

Namit Khanduja Gurukul Kangri Vishwavidyalaya, Haridwar, India

Prajwal Kotian Department of Information Technology, K. J. Somaiya College of Engineering, Vidyavihar, Mumbai, India

Prakhyat G. Kulkarni PES University, Bengaluru, India

Jagrati Kulshrestha GLA University, Mathura, India

Prasanna Kumar Lakineni Department of CSE, Dadi Institute of Engineering and Technology, Visakhapatnam, India

Hemant Kumar Rath Embedded Systems and Robotics, TCS Research and Innovation, Bangalore, India

Ashwini Kumar Varma Department of Electronics Engineering, Indian Institute of Technology (Indian School of Mines), Dhanbad, India

K. Vijaya Kumar Department of Computer Science and Engineering, Sri Krishna College of Technology, Coimbatore, Tamilnadu, India

Mohit Kumar Computer Science Engineering, UIET Punjab University, Chandigarh, Punjab, India

Nikhil Kumar Delhi Technological University, New Delhi, India

P. Karthik Kumar Information and Communication Engineering, Anna University, Chennai, India

S. Dinesh Kumar Department of Computer Science and Engineering, Sri Krishna College of Technology, Coimbatore, Tamilnadu, India

Sanjeev Kumar CSIR-Central Scientific Instruments Organisation, Chandigarh, India;
Academy of Scientific & Innovative Research (AcSIR), Ghaziabad, India

Shiv Kumar Department of CSE, SRMIST, Chennai, India

Sudesh Kumar National Institute of Technology Raipur, Raipur, India

Suman Kumar Computer Science and Engineering, SRM Institute of Science and Technology, Chennai, India

Vivek Kumar Pranveer Singh Institute of Technology, Uttar Pradesh, Kanpur, India

Yogesh Kumar CSIR-Central Scientific Instruments Organisation, Chandigarh, India;
Academy of Scientific & Innovative Research (AcSIR), Ghaziabad, India

Mrinalini Kumawat Department of Electronics and Communication Engineering, Birla Institute of Technology, Mesra, Ranchi, Mesra, India

Arunita Kundaliya School of Computer and System Sciences, Jawaharlal Nehru University, New Delhi, India

Shivam Latiyan Delhi Technological University, Shahbad Daulatpur, Bawana Road, New Delhi, India

Likhitha Kakinada Department of Electrical Engineering, Delhi Technological University, Rohini, Delhi, India

Yajie Liu Department of Computer Science and Engineering, SRM Institute of Science and Technology, Chennai, Tamil Nadu, India

D. K. Lobiyal School of Computer and System Sciences, Jawaharlal Nehru University, New Delhi, India

Ritika Mahajan National Institute of Technology Kurukshetra, Kurukshetra, India

R. P. Mahapatra Computer Science and Engineering, SRMIST, Delhi-NCR Campus, New Delhi, India

Medhavi Malik Department of CSE, SRMIST, Chennai, India

Prashant Mani Department of Electronics and Communication Engineering, SRM Institute of Science and Technology, Ghaziabad, U.P., India

G. Manju Department of Computer Science and Engineering, SRM Institute of Science and Technology, Kattankulathur, India

Anushree Mathur Department of Electronics and Communication Engineering, Birla Institute of Technology, Mesra, Ranchi, Mesra, India

Neha Mathur Manipal University, Jaipur, India

Priyanka Meel Department of Information Technology, Delhi Technological University, New Delhi, India

Animesh Mishra Computer Science and Engineering, SRM Institute of Science and Technology, Chennai, India

Debjani Mitra Department of Electronics Engineering, Indian Institute of Technology (Indian School of Mines), Dhanbad, India

Avijit Mondal Techno International Batanagar, Kolkata, India

M. Murali Computer Science and Engineering, SRM Institute of Science and Technology, Chennai, India

Mohd Mustaqeem Department of Electronics Engineering, Z.H.C.E.T, Aligarh Muslim University, Aligarh, India

Sameena Naaz Department of Computer Science and Engineering, School of Engineering Sciences and Technology, Jamia Hamdard, New Delhi, India

Md Tabrez Nafis Department of Computer Science and Engineering, School of Engineering Sciences and Technology, New Delhi, India

Linga Reddy Nagulapalli Department of Computer Science & Engineering, Himalayan University, Itanagar, India

Sayan Nath Techno International Batanagar, Kolkata, India

Naveen IIIT Delhi, Delhi, India

Ezhilarasu Umadevi Palani Computer Science and Engineering, SRMIST, Delhi-NCR Campus, New Delhi, India

Vishal Pandey Institute of Nuclear Medicine and Allied Sciences, Delhi, India

Manojkumar Parmar Technology and Strategy Department, Innovation & Incubation Business Unit, Robert Bosch Engineering and Business Solutions Private Limited, Bengaluru, Karnataka, India;
HEC Paris, Jouy-en-Josas, France

Pritee Parwekar Department of CSE, SRMIST, NCR Campus, Delhi-NCR, Ghaziabad, Uttar Pradesh, India

Nirmal Patel Computer Engineering Department, Gujarat Power Engineering and Research Institute (GPERI), Mehsana, Gujarat, India

Amitkumar Patil Malaviya National Institute of Technology Jaipur, Jaipur, Rajasthan, India

L. N. Pattanaik Department of Production Engineering, Birla Institute of Technology, Mesra, Ranchi, India

Dipti Pawade Department of Information Technology, K. J. Somaiya College of Engineering, Vidyavihar, Mumbai, India

P. Anantha Prabha Department of Computer Science and Engineering, Sri Krishna College of Technology, Coimbatore, Tamilnadu, India

Parmanand Prabhat Department of Computer Science and Engineering, Himalayan University, Itanagar, India

S. Y. Praneet PES University, Bengaluru, India

Shashank Pushkar Department of Computer Science, Birla Institute of Technology, Mesra, Ranchi, India

R. B. Raghav PES University, Bengaluru, India

Gaurav Rai IIIT Delhi, Delhi, India

Sumit Kumar Rai Central Scientific Instruments Organisation, Chandigarh, India

D. Rajeswari Computer Science Engineering, SRM Institute of Science and Technology, Chennai, Tamil Nadu, India

R. Ramakrishnan National Institute of Epidemiology, Ayappakkam, Chennai, India

Bhupendra Ramani Department of Computer Science and Engineering, Parul Institute of Engineering & Technology, Vadodara, India

Sukhjeet K. Ranade Department of Computer Science, Punjabi University, Patiala, India

Parul Rani CSIR-Central Scientific Instruments Organisation, Chandigarh, India; Academy of Scientific & Innovative Research (AcSIR), Ghaziabad, India

Akash K. Rao Indian Institute of Technology Mandi, Mandi, India

A. P. S. Rathore Malaviya National Institute of Technology, JLN Marg, Jaipur, India

Shanti Rathore Electronics and Communication Engineering, Dr. C.V. Raman University, Bilaspur, India

Mohd Adil Raushan Department of Electronics Engineering, Z.H.C.E.T, Aligarh Muslim University, Aligarh, India

Rahul Singh Rawat Department of Information Technology, Delhi Technological University, Delhi, India

Taurai George Rebanowako CSE Department, Parul Institute of Engineering and Technology, Parul University, Limda, India

Annem Abhishek Reddy Department of Computer Science and Engineering, SRM Institute of Science and Technology, Kattankulathur, India

M. Rekha Sundari Department of Information Technology, Anil Neerukonda Institute of Technology and Sciences, Visakhapatnam, India

Baratam Renuka Department of CSE, Dadi Institute of Engineering and Technology, Visakhapatnam, India

Ramena Venkata Satya Rohit Computer Science Engineering, SRM Institute of Science and Technology, Chennai, Tamil Nadu, India

R. Ruban Department of Computer Science & Engineering, Sri Krishna College of Technology, Coimbatore, Tamil Nadu, India

B. Sabarinathan Department of Computer Science & Engineering, Sri Krishna College of Technology, Coimbatore, Tamil Nadu, India

Gaurav Sahu Pranveer Singh Institute of Technology, Uttar Pradesh, Kanpur, India

Satya Prakash Sahu National Institute of Technology Raipur, Raipur, India

V. Sai Naveen Department of Information Technology, Anil Neerukonda Institute of Technology and Sciences, Visakhapatnam, India

Baratam Sailaja Department of CSE, Vignan Institute of Engineering for Women, Visakhapatnam, India

T. Saravanan Department of Computer Science and Engineering, Galgotias College of Engineering and Technology, Greater Noida, Uttar Pradesh, India

P. Sasikumar School of Computing Science and Engineering, Galgotias University, Greater Noida, Uttar Pradesh, India

Deboshree Seal Computer Science and Engineering Department, KIET Group of Institutions, Ghaziabad, India

K. Seetharaman Department of Computer & Information Sciences, Annamalai University, Chidambaram, India

C. Y. Poovizhi Selvan Department of Computer Science & Engineering, Sri Krishna College of Technology, Coimbatore, Tamil Nadu, India

J. Sengathir Department of Information Technology, CVR College of Engineering, Hyderabad, Telangana, India

K. Senthil Kumar Computer Science and Engineering, SRM Institute of Science and Technology, Chennai, India

Dambarudhar Seth SRM Institute of Science and Technology, Uttar Pradesh, Ghaziabad, India

Pranit Gopaldas Shah Department of Computer Science and Engineering, Parul Institute of Engineering & Technology, Vadodara, India

Chandra Shakher Tyagi Department of Electronics and Communication Engineering, Noida International University, Noida, U.P., India

Shalini Department of Electrical Engineering, Delhi Technological University, Rohini, Delhi, India

Chetna Sharma Central Scientific Instruments Organisation, Chandigarh, India; Academy of Scientific & Innovative Research (AcSIR), Ghaziabad, India

Harshit Sharma Computer Science and Engineering, SRM Institute of Science and Technology, Chennai, India

Jyoti Sharma Department of Electronics and Communication Engineering, Birla Institute of Technology, Mesra, Ranchi, Mesra, India

Mayank Sharma Delhi Technological University, New Delhi, India

Nitesh Kumar Sharma Electronics and Communication Engineering, Dr. C.V. Raman University, Bilaspur, India

Poonam Sharma Amity University, Gurgaon, Haryana, India

Pratyush Sharma Delhi Technological University, New Delhi, India

Puneet Sharma Department of Computer Science and Engineering, Amity School of Engineering and Technology, Amity University, Lucknow Campus, India

R. L. Sharma Department of Electronics and Communication Engineering, Noida International University, Noida, U.P., India

Shubham Sharma Quantum School of Technology, Roorkee, India

Yogeshkumar Sharma Department of Computer Science and Engineering, Shri JJT University, Chudela, Jhunjhunu, Rajstan, India

Vikash Shaw CSIR-Central Scientific Instruments Organisation, Chandigarh, India;
Academy of Scientific & Innovative Research (AcSIR), Ghaziabad, India

Shivam Kukrety Department of Electrical Engineering, Delhi Technological University, Rohini, Delhi, India

Nitisha Shrivastava Department of EEE, Inderprastha Engineering College, Delhi-NCR, Ghaziabad, India

A. P. Shukla Computer Science and Engineering Department, KIET Group of Institutions, Ghaziabad, India

Irfan Siddavatam Department of Information Technology, K. J. Somaiya College of Engineering, Vidyavihar, Mumbai, India

Mohd Jawaid Siddiqui Department of Electronics Engineering, Z.H.C.E.T, Aligarh Muslim University, Aligarh, India

Aashish Pratap Singh Department of Production Engineering, Birla Institute of Technology, Mesra, Ranchi, India

Anshuman Singh Department of Information Technology, Delhi Technological University, New Delhi, India

Bablu Kumar Singh Department of Electronics and Communication Engineering, M.B.M Engineering College, JNV University, Jodhpur, India

Mrigank Singh Division of Instrumentation and Control Engineering, Netaji Subhas University of Technology (Erstwhile NSIT), Dwarka, New Delhi, India

Nikita Singh Delhi Technological University, New Delhi, India

Raghvendra Singh Pranveer Singh Institute of Technology, Uttar Pradesh, Kanpur, India

Rajat Singh Department of Information Technology, Delhi Technological University, Delhi, India

Nidhi Sinha HKBK College of Engineering, Bangalore, India

Shweta Sinha Amity University, Gurgaon, Haryana, India

G. Siva Rama Krishna Department of Information Technology, Anil Neerukonda Institute of Technology and Sciences, Visakhapatnam, India

Gunjan Soni Malaviya National Institute of Technology Jaipur, Jaipur, Rajasthan, India

Mini Sreejeth Delhi Technological University, New Delhi, India

Avireni Srinivasulu Department of Electronics and Communication Engineering, JECRC University, Jaipur, India

Suhail Ahmad Suhail Department of Electrical Engineering, National Institute of Technology Srinagar, Srinagar, India

Shifa Surin Department of Production Engineering, Birla Institute of Technology, Mesra, Ranchi, India

A. Deva Surya Sri Krishna College of Technology, Coimbatore, Tamilnadu, India

C. S. Surya Department of Computer Science and Engineering, Sri Krishna College of Technology, Coimbatore, Tamilnadu, India

M. Surya Department of Computer Science & Engineering, Sri Krishna College of Technology, Coimbatore, Tamil Nadu, India

Surendra Talari Department of Mathematics, GITAM Institute of Science, Visakhapatnam, India

Ganesh Tanpure Malviya National Institute of Technology Jaipur, Jaipur, India

Aabhash Tiwari Computer Science and Engineering, SRM Institute of Science and Technology, Chennai, India

Shivam Tiwari Department of Computer Science and Engineering, Amity School of Engineering and Technology, Amity University, Lucknow Campus, India

Smita Tiwari Computer Science Engineering, Bennett University, Uttar Pradesh, Greater Noida, India

Rajesh Kumar Tripathi Computer Engineering and Applications, IET, GLA University, Mathura, Uttar Pradesh, India;
Department of Computer Engineering and Applications, GLA University, Mathura, UP, India

Ujjwal Department of Electrical Engineering, Delhi Technological University, Rohini, Delhi, India

Syed Umar Department of Computer Science, Wollega University, Nekemte, Ethiopia

Pragya Varshney Division of Instrumentation and Control Engineering, Netaji Subhas University of Technology (Erstwhile NSIT), Dwarka, New Delhi, India

A. Venmani Department of Mathematics and Statistics, SRM Institute of Science and Technology, Kattankulathur, Chennai, India

S. A. Vighneshbalaji Department of Computer Science and Engineering, Sri Krishna College of Technology, Coimbatore, Tamilnadu, India

Vibhor Vimal Department of Information Technology, Delhi Technological University, New Delhi, India

Maheedhar Vundela Department of Computer Science and Engineering, SRM Institute of Science and Technology, Kattankulathur, India

Ajay P. Vyas M.B.M. Engineering College, Jai Narain Vyas University, Jodhpur, India

Arvind R. Yadav Department of Electronics and Communication Engineering, Parul Institute of Engineering and Technology, Parul University, Vadodara, Gujarat, India

Om Prakash Yadav North Dakota State University, Fargo, USA

Parul Yadav Bharati Vidyapeeth's College of Engineering, New Delhi, India

Sarita Yadav Bharati Vidyapeeth's College of Engineering, New Delhi, India

Tariku Birhanu Yadesa Department of Computer Science, Wollega University, Nekemte, Ethiopia

Sherin Zafar Department of Computer Science and Engineering, School of Engineering Sciences and Technology, New Delhi, India

An Integrator Using Voltage Differencing Transconductance Amplifier

Mrinalini Kumawat, Anushree Mathur, Jyoti Sharma, and Avireni Srinivasulu

Abstract This paper presents a voltage differencing transconductance amplifier (VDTA)-based integrator circuit. The VDTA block is simulated using gpdk 180 nm CMOS innovation. The proposed integrator employs single VDTA block and only one capacitor. No resistor is used; thus, the circuit is suitable for IC fabrication. The proposed circuit is simulated using Cadence Virtuoso Tool, and the results are verified mathematically.

Keywords Voltage differencing transconductance amplifier · VDTA · Integrator circuit · Current mode logic · Voltage mode logic

1 Introduction

The output of the supply can be controlled in two different regulation states, i.e., current and voltage mode. It is progressively helpful to utilize the current mode over voltage mode for its exceptional performance [1–16]. In the prior years, for the development in the microelectronic zone signal processing, some active blocks were introduced, such as current differencing transconductance amplifier (CDTA), differencing difference current conveyor (DDCC) and voltage differencing transconductance amplifier (VDTA). A VDTA block is a compact structure where two different

M. Kumawat · A. Mathur · J. Sharma
Department of Electronics and Communication Engineering, Birla Institute of Technology, Mesra, Ranchi, Jaipur Center, Mesra, India
e-mail: kumawatmrinalini30@gmail.com

A. Mathur
e-mail: anushreemathur01@gmail.com

J. Sharma
e-mail: jyotisharma@bitmesra.ac.in

A. Srinivasulu (✉)
Department of Electronics and Communication Engineering, JECRC University, Jaipur 303905, India
e-mail: avireni@jecrcu.edu.in

R. P. Mahapatra et al. (eds.), *Proceedings of 6th International Conference on Recent Trends in Computing*, Lecture Notes in Networks and Systems 177,
https://doi.org/10.1007/978-981-33-4501-0_1

voltages and a multiple transconductance gain are analyzed. Due to these special features, VDTA is easily tunable with unique circuits. Another noteworthy component of VDTA is its use as an active element in various applications as amplifiers, filters, etc. [16–25].

An integrator circuit is used to measure and analyze the element whose output signals co-ordinate with the time and yield for its data signal [26–31]. This block may be designed in current mode or voltage mode. The input in current mode integrator is in terms of current, while in the voltage mode integrator, it is in terms of voltage. Integrators are also used to measure electric charge and its movement [31–39].

In this paper, we have designed a current mode integrator using one VDTA active block and only one capacitor. Section 1 gives the basic idea behind taking VDTA as an active block. The fundamental related to VDTA element is elaborated in Sect. 2. Section 3 shows the results obtained on simulating the circuit using Cadence Virtuoso tool. Finally, the conclusion is given in Sect. 4.

2 VDTA Fundamentals and Proposed Design

The voltage differencing transconductance amplifier circuit is shown in Fig. 1. The VDTA circuit can be partitioned in two parts. The first half circuit generates the difference of the two input voltages V_p and V_n and then provides current IZ by multiplying that difference with the input transconductance g_{mI}. The second part of the circuit gives the output current in terms of voltage by using the output transconductance g_{mO}. These transconductance g_{mI} and g_{mO} can be electronically tuned with the help of bias currents. Thus, this device can be used for transconductance mode of applications along with other voltage and current mode applications such as rectifiers, waveform generators and oscillators. The input ports p, and n of ideal VDTA are

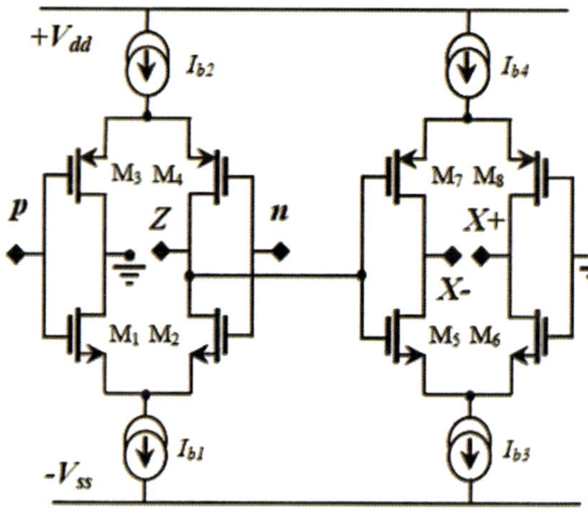

Fig. 1 Device-level circuit diagram of VDTA

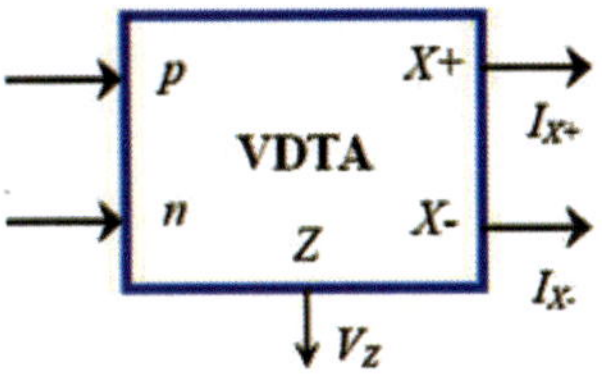

Fig. 2 Block diagram of VDTA

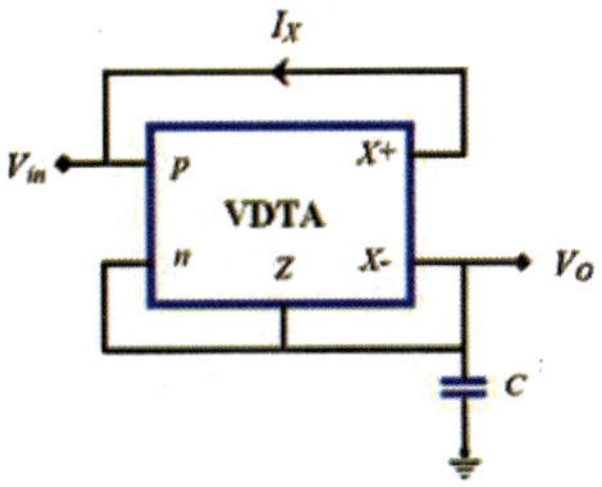

Fig. 3 Proposed VDTA integrator circuit

high impedance nodes and their relationships are characterized by (1). The symbol for VDTA block given in Figs. 2 and 3 shows the circuit of the proposed integrator which consists of single VDTA block and only one capacitor.

$$\begin{aligned} I_p &= I_n = 0 \\ I_Z &= g_{\mathrm{mI}} V_p - g_{\mathrm{mI}} V_n \\ I_{X\pm} &= \pm g_{\mathrm{mO}} V_Z \end{aligned} \tag{1}$$

The transfer function or the output voltage of the VDTA-based integrator circuit of Fig. 3 has been obtained by doing mathematical derivations with the help of (1) and Fig. 3 and can be given as shown in (2).

$$V_0 = \left(\frac{g_{\mathrm{mI}}}{g_{\mathrm{mI}} + g_{\mathrm{mo}} + sC} \right) V_{\mathrm{in}} \tag{2}$$

3 Simulation Results

The VDTA-based integrator circuit of Fig. 3 is simulated in the Cadence Virtuoso Tool using the gpdk CMOS 0.18 μm process parameters. The width and length taken for the MOS transistors in the VDTA transistor-level circuit are shown in Table 1. The power supply voltage of ± 0.9 V and the bias currents of $I_{b1} = I_{b2} = I_{b3} = I_{b4}$ = 60 μA used in Figs. 3 and 4 show the DC transfer characteristics of the currents at $X+$ and $X-$ terminals w.r.t. to V_Z. The square wave input of amplitude ± 1 V

Table 1 Aspect ratios of the transistors in the VDTA circuit

Transistors	W (μm)	L (μm)
M_1–M_2	3.6	0.36
M_3–M_4	16.64	0.36
M_5–M_6	3.6	0.36
M_7–M_8	16.64	0.36

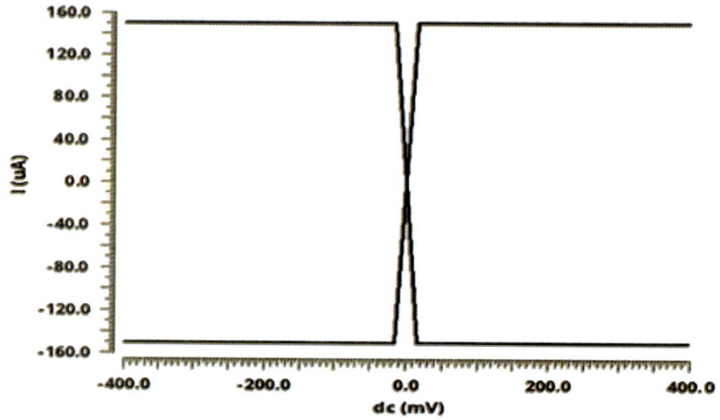

Fig. 4 Input and output characteristics of VDTA block

and the frequency of 1 kHz is forced to the proposed integrator circuit. The input–output relationship of the VDTA block is shown in Fig. 4. The capacitor values are varied to obtain the output of the proposed integrator for short, medium and long time constants. The values taken for short, medium and long time constants are $C = 0.05$ μF, $C = 0.25$ μF and $C = 1$ μF, respectively. The output waveforms of the integrator for short, medium and long time constants are shown in Figs. 5, 6 and 7, respectively.

Table 2 shows the brief comparison of the proposed circuit in this work with the other integrator circuits presented in the literature.

4 Conclusion

In the present work, an integrator circuit utilizing one voltage differencing transconductance amplifier is exhibited. The proposed circuit employs one active element and only one passive element, i.e., capacitor which is grounded thus suitable for IC implementation. The output of the integrator circuit has been investigated by forcing a square wave input and verifying it for short, medium and long time constants. The

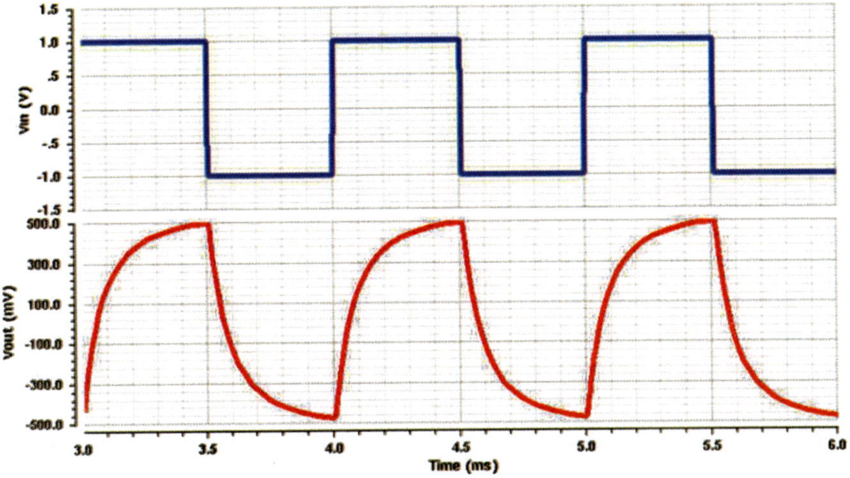

Fig. 5 Output waveform of the proposed integrator for short time constant ($C = 0.05$ μF)

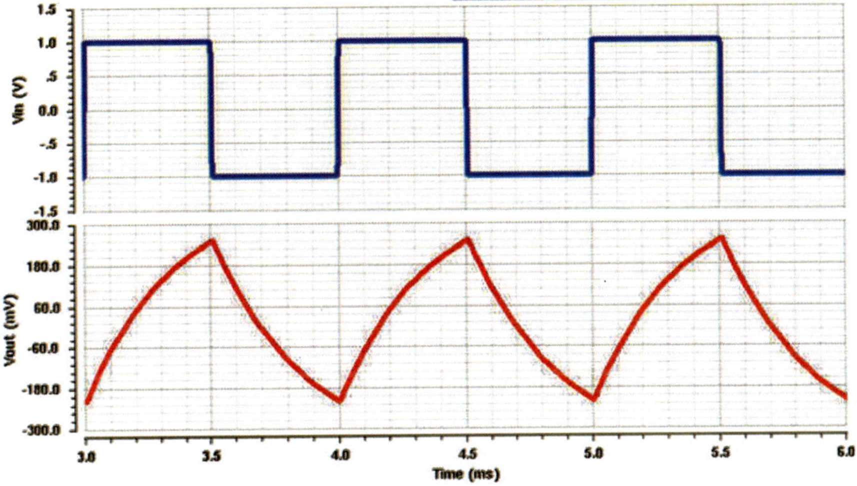

Fig. 6 Output waveform of the proposed integrator for medium time constant ($C = 0.25$ μF)

VDTA integrator can be used for numerous analog signal processing (ASP) applications such as filters, oscillators, instrumentation amplifiers, inductance simulators and rectifiers.

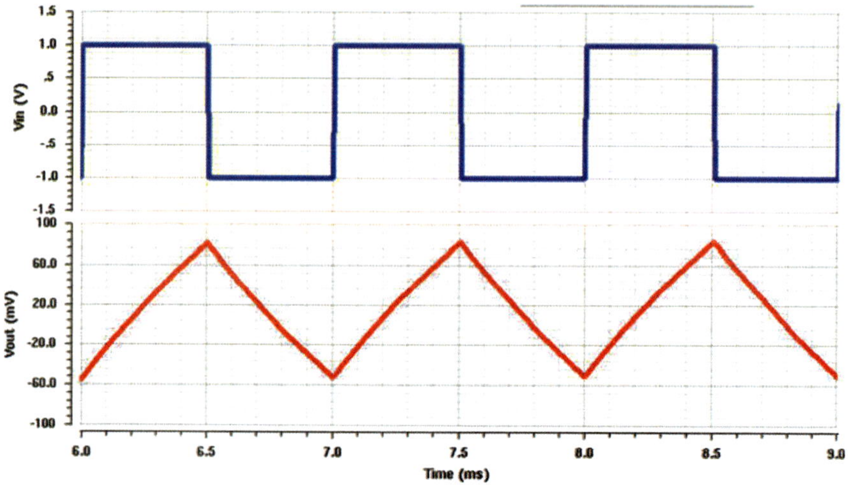

Fig. 7 Output waveform of the proposed integrator for long time constant ($C = 1\ \mu F$)

Table 2 Comparison of designed circuit with conventional circuits

Reference	Active block	Passive blocks used	Active blocks used
[26]	DVCC	3	1
[27]	Op-Amp	2	1
[28]	Op-Amp	2	1
[29]	CCII	2	1
[30]	CCII	2	1
[31]	DDCCTA	1	1
Present work of Fig. 3	VDTA	1	1

References

1. Biolek D, Senani R, Biolková V, Kolka Z (2008) Active elements for analog signal processing: classification, review, and new proposals. Radioengineering 17(4):15–32
2. Siripruchyanun M, Payakkakul K, Pipatthitikorn P, Satthaphol P (2016) A current mode square/triangular wave generator based on multiple-output VDTA's. Int Electr Eng Congr 86:152–155. https://doi.org/10.1016/j.procs.2016.05.040
3. Shankar C, Singh SV (2017) Electronically tunable current mode biquad filter based on single VDTA and grounded passive elements. Int J Eng Technol 9(2):271–279. https://doi.org/10.21817/ijet/2017/v9i1/170902302
4. Mehra R, Kumar V, Islam A (2018) Floating active inductor based Class-C VCO with 8 digitally tuned sub-bands. Int J Electron Commun 83:1–10. https://doi.org/10.1016/j.aeue.2017.08.018
5. Yesil A, Kacar F, Kuntman H (2011) New simple CMOS realization of voltage difference transconductance amplifier and its RF filter application. Radioengineering 20(3):632–637
6. Gupta G, Singh SV, Bhooshan SV (2015) VDTA based electronically tunable voltage- mode and trans-admittance biquad filter. Circ Syst 6:93–102. https://doi.org/10.4236/cs.2015.63010

7. Yesil A, Kacar F (2013) Electronically tunable resistor less mixed-mode biquad filters. Radioenginnering 22(4):1016–1025
8. Alaybeyoglu E, Kuntman H (2016) CMOS implementations of VDTA based frequency agile filters for encrypted communications. Analog Integr Circ Sig Process 89(3):675–684. https://doi.org/10.1007/s10470-016-0760-y
9. Prasad D, Ahmad J, Srivastava M (2018) A novel grounded to floating admittance converter with electronic control. Indian J Phys 92(1). https://doi.org/10.1007/s12648-017-1077-0
10. Chen HP, Hwang YS, Ku YT (2017) A new resistorless and electronic tunable third-order quadrature oscillator with current and voltage outputs. Inst Electron Telecommun Eng Tech Rev 1–13. https://doi.org/10.1080/02564602.2017.1324329
11. Pandey N, Kumar P, Paul SK (2015) Voltage Differencing Transconductance Amplifier based resistor less and electronically tunable wave active filter. Analog Integr Circ Sig Process 84(1):107–117. https://doi.org/10.1007/s10470-015-0546-7
12. Pal D, Srinivasulu A, Pal BB, Demosthenous A, Das BN (2009) Current conveyor-based Square/Triangular wave generators with improved linearity. IEEE Trans Instrum Meas 58(7):2174–2180. https://doi.org/10.1109/TIM.2008.2006729
13. Srinivasulu A (2012) Current conveyor based relaxation oscillator with tunable grounded resistor/capacitor. Int J Des Anal Tools Circ Syst (Hong-Kong) 3(2):1–7
14. Srinivasulu A (2011) A novel current conveyor based-Schmitt trigger and its application as a relaxation oscillator. Int J Circ Theor Appl 39(6):679–686. https://doi.org/10.1002/cta.669
15. Pal D, Srinivasulu A, Goswami M (2009) Novel current-mode waveform generator with independent frequency and amplitude control. In: Proceedings of the IEEE international symposium on circuits and systems, pp 2946–2949. https://doi.org/10.1109/ISCAS.2009.5118420
16. Bhasker DR, Tripati MP, Senani R (1993) A class of three OTA-two capacitor oscillators with non-interacting controls. Int J Electron 74(03):459–463. https://doi.org/10.1080/00207219308925849
17. Chung WS, Kim H, Cha HW, Kim HJ (2005) Triangular/square wave generator with independently controllable frequency and amplitude. IEEE Trans Instrum Meas 54(1):105–109. https://doi.org/10.1109/TIM.2004.840238
18. Srinivasulu A, Shaker PC (2014) Grounded resistance/capacitance-controlled sinusoidal oscillators using operational transresistance amplifier. WSEAS Trans Circ Syst 13:145–152
19. Shaker PC, Srinivasulu A (2014) Quadrature oscillator using operational transresistance amplifier. In: Proceedings of international conference on applied electronics, pp 117–120. https://doi.org/10.1109/AE.2014.7011681
20. Lo YK, Chien HC (2007) Switch controllable OTRA based square/triangular waveform generator. IEEE Trans Circ Syst-II 54(12):1110–1114. https://doi.org/10.1109/TCSII.2007.905879
21. Linitha R, Srinivasulu A, Reddy VV (2015) A Schmitt trigger based on DDCCTA without any passive components. In: Proceedings of IEEE international conference on communications and signal processing, pp 1695–1698. https://doi.org/10.1109/ICCSP.2015.7322808
22. Jaikla W, Siripruchyanum M, Bajer J, Biolek D (2008) A simple current-mode quadrature oscillator using CDTA. Radioengineering 17(4):33–40
23. Tangsrirat W (2011) Synthesis of current differencing transconductance amplifier –based current limiters and its applications. Int J Circ Syst Comput 20:185–206
24. Kumbun J, Siripruchyanun M (2010) MO-CTTA-based electronically controlled current mode square/triangular wavegenerator. In: Proceedings of international conference on technical education, pp 158–162
25. Sedra A, Smith KC (1998) Microelectronic Circuits, 5th edn. Oxford University Press, London, UK, pp 105–112
26. Minaei S (2004) Dual-input current-mode integrator and differentiator using single DVCC and grounded passive elements. In: Proceedings of the 12th IEEE mediterranean electro technical conference, vol 1, pp 123–126. https://doi.org/10.1109/MELCON.2004.1346788
27. Nagaria RK, Goswami A, Venkateswaran P, Sanyal SK, Nandi R (2003) Voltage controlled integrators/differentiators using current feedback amplifier. In: 2003 International symposium

on signals, circuits and systems. SCS 2003, Iasi, Romania, vol 2, pp 573–576. https://doi.org/10.1109/SCS.2003.1227117
28. Patranabis D, Ghosh DK (1984) Integrators and differentiators with current conveyors. IEEE Trans Circ Syst 31(6):567–569. https://doi.org/10.1109/TCS.1984.1085535
29. Lee J-Y, Tsao H-W (1992) True RC integrators based on current conveyors with tunable time constants using active control and modified loop technique. IEEE Trans Instrum Meas 41(5):709–714. https://doi.org/10.1109/19.177348
30. Chiu W, Tsay J-H, Liu S-I, Tsao H-W, Chen J-J (1995) Single-capacitor MOSFET-C integrator using OTRA. Electron Lett 31(21):1796–1797
31. Linitha R, Srinivasulu A, Reddy VV (2017) An integrator circuit using differential difference current conveyor transconductance amplifier. In: Proceedings of IEEE international conference on signal processing, communications and networking, pp 1–4. https://doi.org/10.1109/ICSCN.2017.8085420
32. Santhoshini KM, Sarada M, Srinivasulu A (2019) An integrator circuit using voltage difference transconductance amplifier. Solid State Electron Lett 1(1):10–14. https://doi.org/10.1016/j.ssel.2018.08.001
33. Fragoulis N, Haritantisl I, Constantinides AG (2000) Active filter synthesis based on tunable log-domain lossy integrators. In: Proceedings of IEEE international symposium on circuits and systems, pp 409–412. https://doi.org/10.1109/ISCAS.2000.857458
34. Tsukutani T, Kinugasa Y, Sumi Y, Higashimura M, Fukui Y (2003) Novel current-mode active-only biquad with loss-less and lossy integrators. Int J Electron 90(10):627–633. https://doi.org/10.1080/00141840320001593360
35. Lovchakov VI, Shopin SA (2016) Solution set of time optimal control problem for four series connected integrators. In: Proceedings of international Siberian conference on control and communication, pp 1–4. https://doi.org/10.1109/SIBCON.2016.7491693
36. Xin Z, Zhao R, Wang X, Loh PC, Blaabjerg F (2016) Four new applications of second-order generalized integrator quadrature signal generator In: Proceedings of IEEE applied power electronics conference and exposition, pp 2207–2214. https://doi.org/10.1109/APEC.2016.7468173
37. Sanyal SK, Sarker UC, Nandi R (1990) Increased time-constant dual-input integrators. IEEE Trans Instrum Measur 39:672–673. https://doi.org/10.1109/19.57257
38. Liu S-I, Hwang Y-S (1994) Dual-input differentiators and integrators with tunable time constants using Current Conveyors. IEEE Trans Instrum Meas 43(4):650–654. https://doi.org/10.1109/19.310164
39. Arbel AF, Goldminz L (1992) Output stage for current-mode feedback amplifiers, theory and applications. Analog Integr Circ Sig Process 2(3):243–255

Traffic Prediction Protocol for a Novel Cloud-Based Self-adaptive Cognitive Radio Network Structure

Ashwini Kumar Varma, Jyotirmoy Karjee, Hemant Kumar Rath, and Debjani Mitra

Abstract Constraint of battery, data processing and memory of a mobile cognitive user (CU) limits the implementation of cognitive radio (CR) technology over real-time scenario. The paper claims to eliminate such issues by proposing a novel cloud-based self-adaptive cognitive radio network (CSCRN) structure which possesses to moves the complex computation of sensing a wideband of spectrum from mobile node to the network. The proposed network seems to consists of three essential building blocks, namely prioritization, optimal relay selection and relay monitoring. The paper gives a brief overview of the two initial building blocks while the latter block is the main focus of the paper. A new traffic prediction protocol (TPP) is introduced as a part of relay monitoring block which uses the existing machine learning (ML) techniques to predict network traffic. The results in this paper are presented in two parts where the first part demonstrates the effectiveness of the proposed network structure in terms of energy consumption at CUs while the second part presents TPP result which is demonstrated in the form of a comparative study of the various variants of the prediction models over Web traffic time-series dataset.

Keywords Cognitive radio · Cloud · Network structure · Communication protocol · Traffic prediction protocol

A. Kumar Varma (✉) · D. Mitra
Department of Electronics Engineering, Indian Institute of Technology (Indian School of Mines), Dhanbad 826004, India
e-mail: akvarma.iitism@gmail.com

D. Mitra
e-mail: debjanim@iitism.ac.in

J. Karjee · H. Kumar Rath
Embedded Systems and Robotics, TCS Research and Innovation, Bangalore, India
e-mail: jyotirmoy.karjee@ieee.org

H. Kumar Rath
e-mail: hemant.rath@tcs.com

R. P. Mahapatra et al. (eds.), *Proceedings of 6th International Conference on Recent Trends in Computing*, Lecture Notes in Networks and Systems 177,
https://doi.org/10.1007/978-981-33-4501-0_2

1 Introduction

Cognitive radio (CR) is an upcoming technology to overcome the inefficient utilization of the spectrum by allowing unlicensed users, also referred as cognitive users (CUs), to access the spectrum opportunistically [1]. Over such technology, CUs obtain knowledge about the current state of the spectrum and access the unoccupied spectrum while keeping the occupied spectrum free from interference. The process of obtaining knowledge about the current state of the spectrum is known as spectrum sensing that is an essential block of CUs [2, 3]. However, sensing a wide range of frequency requires a good hardware capability such as a radio frequency (RF) unit, a high-speed analog-to-digital converter (ADC) and a high-speed signal processing board. Since in most of the cases, the CU is considered to be a mobile unit, battery life turns out to be a major constraint. Therefore, embedding CU with such heavy computation will result in the degradation of battery life exponentially. However, the solution to this problem can be achieved in the form of a cognitive radio network (CRN), which performs all such heavy computation and provides the desired output as required.

In literature, CRN structure is basically discussed with infrastructure, ad-hoc, mesh or cluster-based network [4]. Authors in [5] and [6] introduce an essential entity in the CRN, so-called spectrum monitoring devices and "Spectrum Agent," respectively, which performs spectrum sensing for CUs in order to overcome their high energy consumption, design complexity and hardware cost. However, both the papers does not talk about any communication protocol which needs to be followed. Authors of [7] introduce another essential entity, cloud to utilize the network effectively while interacting with other CUs. Although the idea of computing spectrum sensing outside the SU will have a significant impact on CRNs, still there are several issues such as sensor design, network architecture, communication protocol, dynamic path planning/ optimal multi-hop relay selection, realy monitoring and a few others which need to be explored while building it.

For a large-scale outdoor scenario as shown in Fig. 1, where thousands of cognitive users (CUs, such as smart devices, sensors, actuators, robots and drones, etc.,) are deployed. Each CU attempts to establish communication with another CU based on certain tasks assigned to it. However, with the large deployment region, it is usually not possible for the CU to communicate with another CU mainly due to the large distance, small communication range, increased signal interference and obstacles, etc. Hence, a hierarchical multi-hop network structure is required in this deployment scenario, where a CU can establish a connection with another CU through an intermediate entity called cognitive access point (CAP). The CAP can be a gateway or a Wi-Fi access point or an eNodeB, etc. Each CAP is associated with a group of CUs, where the CUs continuously transmit data packets to the CAP. However, due to the continuous transmission over a certain period of time, a huge amount of data packets are captured at the CAP. This increases the communication overhead at the CAP. In this situation, network functionality such as prioritization is considered at the CAP which prioritizes the received data packets from the CUs. Prioritization signifies

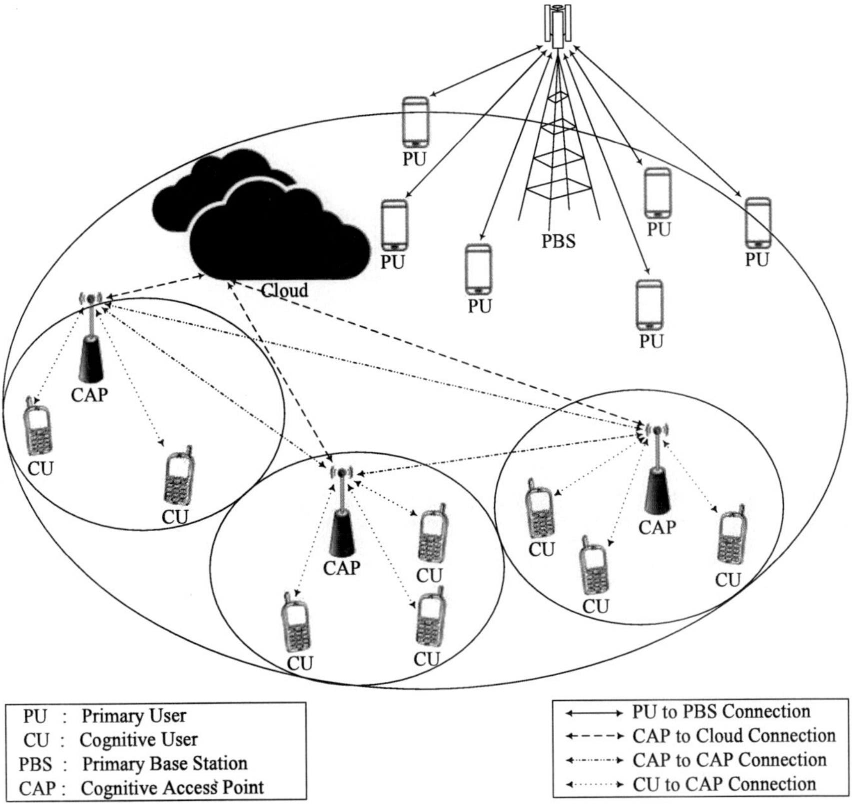

Fig. 1 Cloud-based network structure

the importance of data packets based on attributes of statistical measurements and desired quality of services (QoS).

The data packets transmitted by a prioritized CU needs to reach a destination CU to establish an end-to-end communication. To establish communication in an ad hoc or cluster-based radio system [4], source CU transmits data packets through intermediate CAPs to reach the destination CU. The source CU can consider different paths/ relays to communicate with the destination CU, where CAPs act as relay entities. However, both CU and CAP do not have complete information over which optimal path/ relay [8, 9] must be chosen. In this situation, a central entity needs to be considered which can provide enough information about the network topology. To overcome this problem, "Cloud" [7] is introduced as a central entity which has information about the entire network. Cloud computes the optimal relay/path and establishes communication between the source CU and the destination CU through the intermediate CAPs. Further, it is possible that the radio access network (RAN) gets overloaded or congested due to continuous data packets received at the intermediate CAPs. Therefore, relay monitoring functionality needs to be considered at each intermediate CAP which can monitor the flow of data traffic. Moreover, a traffic prediction protocol (TPP) is also desired at the CAP such that congestion can be predicted apriori. To overcome the above problems, we present a cloud-based self-adaptive cognitive radio network (CSCRN) structure considering prioritization, optimal relay selection and relay monitoring through a TPP.

The paper contributes in terms of a novel cloud-based self-adaptive cognitive radio network structure that eliminates most of the issues discussed above. While describing the proposed network, three essential blocks, namely prioritization, optimal relay selection and relay monitoring were introduced. In our recent work [8–10], a couple of novel algorithms were developed for the two initial blocks of prioritization and optimal relay selection while this paper focuses over the third block of relay monitoring. A new traffic prediction protocol (TPP) incompatible with CR technology is presented which has not been reported in any work. With ML as its core block, various variants of the machine learning (ML) algorithms are evaluated in terms of accuracy and time complexity to determine the effectiveness and reliability of the proposed TPP. Rest of the paper is organized as follows: Sect. 2 presents the related work while Sect. 3 presents the system model of CSCRN structure and TPP protocol. Results and discussion are presented in Sect. 4 and finally, we conclude the work in Sect. 5.

2 Related Work

In any network including CRN, the dynamic changes in traffic load over the network can affect its performance badly [11, 12]. Due to this, we require a model which can predict the occurrence of traffic congestion in advance, so that preventive measures can be taken to avoid congestion. In literature, numerous authors attempt to contribute over such area with prediction models. These models basically work in 2-folds: training and testing. In the training phase, the system builds a prediction model which can classify the new unseen testing dataset. However, the literature groups the traffic prediction algorithms into two categories, namely parametric and nonparametric techniques.

Traffic prediction using parametric techniques are basically approached by time-series model. Authors in [13] proposed a popular time-series method, namely auto-regressive integrated moving average (ARIMA) model to predict the traffic over the network. Further, the accuracy of the prediction model is proposed to improve by variants of ARIMA model [14, 15]. In [16], another approach under statistical time-series method was presented, where the frequency component of variable traffic is analyzed via fast Fourier transform (FFT) and then the temporal traffic model is obtained by data fitting. Since time-series techniques are conventional linear models, prediction of stochastic and nonlinear nature of traffic are perhaps not a good idea. Hence, some researchers have shifted their attention toward non-parametric techniques which are much more suitable for nonlinear data. Support vector machine (SVM) is one of the most accurate prediction models for nonlinear data. Therefore, authors of [17] propose to explore the performance of SVM as a traffic prediction model. The results show the effectiveness of the model but is computationally intensive while dealing with the huge dataset. Sekwatlakwatla et al. in [13] attempt to predict the network traffic using artificial neural network (ANN) due to its ability to process high dimensionality data with strong forecasting capability. However, the performance of the network might decrease with simple ANN model. Hence, researchers applied more advanced and deeper models as in [18]. It is observed that in all the techniques discussed above use a sequence of time-series traffic observation $\{X_t\}$, $t = 1, 2, \ldots, T$ to predict the traffic over a time interval $(t + \Delta)$ for some prediction horizon Δ . However, which does not suit the CR technology, where CUs uses the licensed spectrum opportunistically. Under such circumstances, the model predicts incorrect traffic condition using traffic observation of primary users (PU). Hence, keeping the need for traffic prediction in mind, this paper presents a novel traffic prediction protocol (TPP) to predict the traffic over multi-hop relay in the proposed CRN structure. The proposed TPP uses the existing ML techniques to predict the traffic.

3 System Model and Problem Formulation

In this section, we present the cloud-based self-adaptive cognitive radio network (CSCRN) structure which consists of three essential entities, namely the cloud, the CAP (referred as spectrum monitoring devices/ Spectrum Agent in literature) and the CUs. The CU discussed in this paper differs from the one in the literature. Here, CU does not have the capability to sense the spectrum. However, when it requires accessing the spectrum, it sends a spectrum state request (SSR) to the associated CAP, which in response transmits the current state of the spectrum. Therefore, CU does not have to perform the complex computation to determine the current state of the spectrum. This reduces the cost, complexity and high energy consumption of CUs. In this paper, the CUs are considered as mobile while cloud and CAPs are static over the area of deployment. Static nature of cloud and CAPs enables it with predefined connectivity.

The proposed CSCRN structure divided the macro-cellular area into smaller region known as subnet associated with corresponding CAPs. Asset directory (AD) installed over each CU is as shown in Fig. 2. The AD basically publishes a list of

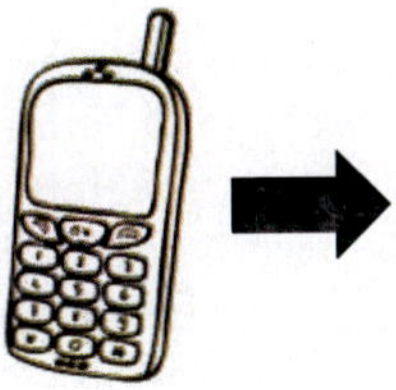

List of neighboring CAP's	
CAP_j	$D_{CAP_j \rightarrow CU_i}$
CAP_1	d_{1i}
CAP_2	d_{2i}
⋮	⋮
CAP_J	d_{Ji}

Fig. 2 Assets directory

CAP

$CAP \rightarrow CU_i$	$CAP \rightarrow CAP_j$
IP_{CU_1}	IP_{CAP_1}
IP_{CU_2}	IP_{CAP_2}
⋮	⋮
IP_{CU_I}	IP_{CAP_J}

(a)

CAP

CU_i	Cost Function	CAP_j	Cost Function
CU_1	CF_1	CAP_1	CF_1
CU_2	CF_2	CAP_2	CF_2
⋮	⋮	⋮	⋮
CU_I	CF_I	CAP_J	CF_J

(b)

Fig. 3 **a** Network Map **b** Cost map

nearby CAPs along with the distance from the corresponding CU. However, as the network is initiated the CU starts retrieving information from the AD which assist them to identify an appropriate CAP to which it can connect. The selection of CAP, CAP_j where $j = \{1, 2, \ldots, J\}$ is based on the distance, $D_{CAP_j \rightarrow CU_i}$ from the corresponding CU, CU_i where $i = \{1, 2, \ldots, I\}$. Once the connection between CUs and CAPs is established, parallel processing starts at CAPs. On the one hand, it senses the spectrum periodically while on the other hand, it publishes subnet information through control data packet, namely cognitive user services (CUS) which is stored in the cloud. CUS associated with each CAP provides with two essential information, namely Network Map and Cost Map. Network Map defines the set of CUs and other CAPs associated to the corresponding CAP. The connectivity is represented by their IP-addresses in the corresponding Network Map (assuming Internet Protocol (IP)-based network). While the Cost Map defines the pair wise cost function, CF of the set of entities associated with the corresponding CAP. Both the maps are shown in Fig. 3. This subnet information from various CAPs in the network helps the cloud to build its own "network view" of the entire network. Hence, this completes the initialization process.

Working procedure of CSCRN introduces three essential blocks, namely prioritization, optimal relay selection and relay monitoring. The flow diagram of the proposed CSCRN structure is depicted in Fig. 4 and is explained broadly as given below:

3.1 Prioritization

In CSCRN structure, if a CU requires accessing the spectrum to reach its destination. It sends spectrum state request (SSR) to the associated CAP, which in response provides with the current state of the spectrum. However, in a situation where the CAP receives multiple SSR from the group of associated CUs at a particular time than CU with the highest priority is preferred over other CUs. Prioritization indicates the importance of CUs associated with each CAP in the network. However, priority can be allocated to CUs based on certain attributes such as transmission deadline or packet type or based on other parameters. For the proposed network structure, a novel

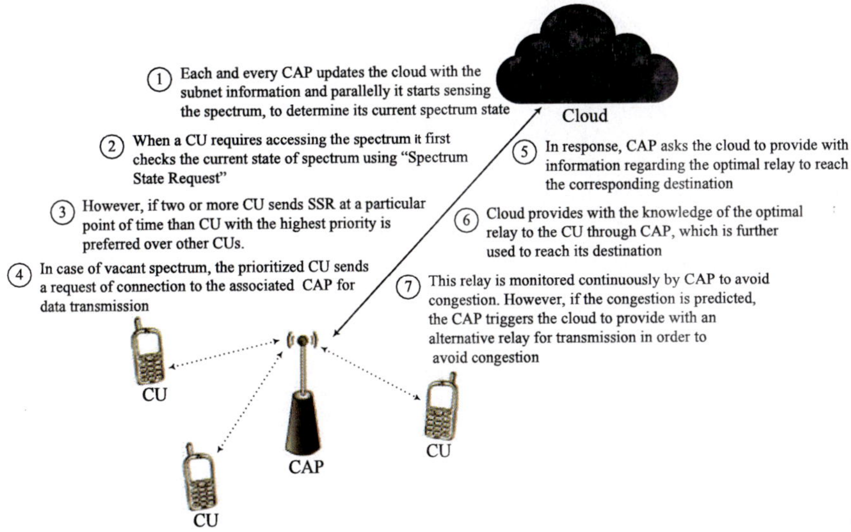

Fig. 4 Flow diagram of cloud-based self-adaptive network structure

methodology of prioritizing CU on the basis of transmission deadline and information contained is explained broadly in [10]. Hence, as the process of prioritization determines the CU with the highest priority, it selects an optimal frequency band for transmission depending on the current state of the spectrum, required data rate and QoS. Finalizing frequency band is followed by initiation of transmission process by sending a connection request message to the associated CAP. In response to the CU request, the CAP asks the cloud to provide information regarding an optimal relay to reach the corresponding destination. Since the cloud has information about the entire network topology, it is the best entity to provide such information.

3.2 *Optimal Relay Selection and Relay Monitoring*

In CSCRN structure, cloud provides the optimal relay information to the CUs through associated CAPs. Using the information of optimal relay provided by the cloud, the prioritized/ source CU (CUs) establishes a communication link through intermediate CAPs to reach its destination CU (CU_D) as shown in Fig. 5; where N represents the number of CAPs combines to form an optimal relay. The optimal relay selection methodology is explained broadly in [8, 9]. It is to be noted that while communicating among CUs, the radio access network (RAN) can get overloaded which leads to packet drop, increasing communication overhead and reduced QoS. To overcome this problem, we need a traffic prediction functionality which can monitor the data traffic of the selected relay. In the proposed model, relay monitoring is considered at the CAP which can perform traffic prediction in CSCRN structure.

Fig. 5 Relay selected for communication among CUs

Relay monitoring using traffic prediction is one of the crucial tasks which manages the traffic smartly and improve the efficiency of the network. In CSCRN structure, we propose a traffic prediction protocol (TPP) to predict the data traffic over the selected multi-hop relay. The traffic is predicted at the CAP connected to the source CU (CUs). Hence, it keeps tracking the data traffic and has the capability to predict the occurrence of congestion. However, as the congestion is predicted, the CAP triggers a request to the cloud to provide information about another optimal relay for establishing connection in order to avoid congestion. The following sub-section illustrates the two phases of TPP, namely accumulation of data traffic information and traffic prediction.

Accumulation of Data Traffic Information. In this phase, the CAP associated with CUs periodically broadcasts a beacon signal indicating other CAPs in the selected relay to provide with the data traffic information (DTI). DTI helps to provide traffic information in the network. The DTI packet is basically divided into two parts where the first part provides with the sender identity (of CAP) while the other part indicates the number of data packets received over time interval T. The time interval T is similar to the time over which the beacon signal is broadcasted. The sender identity is essential information as it indicates the entity to which the DTI packet belongs, while the information regarding the number of data packets received by each CAP assists to predict the traffic over the multi-hop relay. This is because as the number of received data packet increases, the CAP fails to process the data in time due to the limited available resource or computational capability. Hence, the received data packets are enforced to wait for processing by the CAP. This introduces delay in the network and affects the traffic conditions over the selected relay.

In this paper, the CAP associated with CUs is named as Parent CAP while other CAPs in the selected relay are referred as Child CAPs. The Parent CAP (PCAP) is represented by CAP_1 while Child CAPs are denoted by $\text{CCAP} = \{\text{CAP}_n | n = 2, 3, \ldots, N\}$, as shown in Fig. 5. However, upon reception of beacon signal from PCAP, CCAP in the selected relay transmits DTI to PCAP immediately. In cases, where CCAP is not directly connected to the PCAP, the intermediate CCAP must retransmit the traffic information toward the PCAP.

Traffic Prediction. In the proposed TPP, the traffic prediction takes place at PCAP. As PCAP receives DTI from all the CCAPs, the traffic of the corresponding relay is predicted. In the proposed model, the prediction of traffic can be presented in the form of binary hypothesis problem, which is represented as:

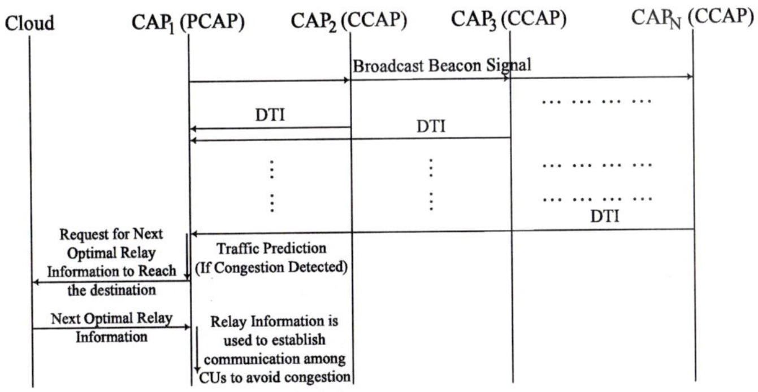

Fig. 6 Timing flow diagram of traffic prediction protocol

$$H = \begin{cases} H_0; & \text{Free flowing data traffic} \\ H_1; & \text{Presence of Congestion} \end{cases} \quad (1)$$

where H_0 and H_1 represent free flowing data traffic (absence of congestion) and presence of congestion, respectively. To predict the traffic over the selected relay, the paper considers various variants of Machine Learning (ML) algorithms. In this paper, we investigate the performance of the popular ML algorithms for this specific CSCRN deployment. The proposed TPP can predict the occurrence of traffic congestion over multi-hop relay by using DTI as an input to the ML algorithms. However, when the prediction model predicts the presence of congestion in response to this, PCAP sends a request to the cloud to provide with the knowledge of another optimal relay for establishing a connection in order to avoid congestion. Otherwise, when free flowing data traffic is predicted, the PCAP continues to predict the traffic over a regular interval of time. Figure 6 illustrates the timing flow diagram of the traffic prediction phase when congestion is detected.

4 Results and Discussion

4.1 Energy Consumption

The section investigates the amount of energy consumed by the processor using existing energy model. As we know, over the proposed network architecture, the CAP performs the computational tasks of spectrum sensing and the results are communicated to the specific entity as required. Hence, both sensing and communication cost have been compared to show that the proposed model eliminates the issue of high energy consumption at CUs. Hence, the total energy consumed while scanning a wide range of frequencies using a powerful spectrum sensing algorithm such as maximum eigenvalue-based detection (MED) [19] is represented as [20]:

$$E_{\text{Sen}} = E_{\text{mem\_ inst}} + E_{\text{mem\_ data}} + E_{\text{cpu\_ inst}} + E_{\text{cpu\_ data}} \tag{2}$$

where $E_{\text{mem\_ inst}}$ and $E_{\text{mem\_ data}}$ represent instruction and data dependent cost at instruction and data memory, respectively, while $E_{\text{cpu\_ inst}}$ and $E_{\text{cpu\_ data}}$ stand for instruction and data dependent cost inside the CPU. In general, MED method is basically based on eigenvalue decomposition of the covariance matrix which is formulated by performing auto-correlation of the input data. Since instruction dependent terms, $E_{\text{mem\_ inst}}$ and $E_{\text{cpu\_ inst}}$ depend on number and nature of instruction in the algorithm. $E_{\text{mem\_ inst}}$ and $E_{\text{cpu\_ inst}}$ are proportional to L^3 where L represent the size of the input signal. While the other two data dependent terms are proportional to L. Hence, (2) can be rewritten as:

$$E_{\text{Sen}} = f\left(L^3\right) + f(L) + f\left(L^3\right) + f(L) = 2\left(f\left(L^3\right) + f(L)\right) \tag{3}$$

The proposed CSCRN structure seems to eliminate this huge energy cost where sensing cost is replaced by communication cost, E_{Comm}. However, communication cost basically depends on the number of transmitted and received bits which can be represented as [21]:

$$E_{\text{Comm}} = E_{\text{Tx}} + E_{\text{Rx}} = f(x) + f(L) \tag{4}$$

where x represents the y-bits SSR signal. Since, $x \ll L$

$$E_{\text{Comm}} = f(L) \tag{5}$$

Hence, from (3) and (5), it can be established that computing complex computation such as spectrum sensing over an independent entity, CAP reduces the energy consumption of CU. Therefore, increases the lifetime of the network.

4.2 *Evaluation of Machine Learning Algorithms for Traffic Prediction*

The section presents a comparative study of various variants of machine learning (ML) algorithms such as support vector machine (SVM), artificial neural network (ANN) and k-nearest neighbors (KNN) to predict traffic over the selected relay as described in the proposed TPP model. These algorithms are compared using Web traffic time-series dataset. Even though a large volume of time-series traffic data is captured at the CAP (transmitted from CUs) for a longer duration, the time computation for these ML algorithms is extremely low. Therefore, these ML algorithms are suitable for predicting data traffic at the CAP. However, the performance of a prediction model cannot be demonstrated through accuracy metric since it misleads the result if the dataset is unbalanced (i.e., when the number of data samples in each class varies). Hence, sensitivity and specificity are the two metrics used in this paper to analyze the performance of the prediction block. Sensitivity measures the ratio of "Presence of congestion" identified correctly, while specificity measures the ratio of "Free flowing data traffic" identified correctly. The values of sensitivity and specificity lie between zero and one with values closer to one represent better performance of the prediction block.

The performance of the prediction block basically depends on the parameters chosen while training the model considered for prediction. These parameters are chosen using a ten-fold cross-validation technique for various prediction algorithms, which can further be initialized in the simulation model for a comparative study in terms of sensitivity, specificity and time complexity. Appropriate parameters boost the efficiency of the prediction model. Various parameters chosen to construct the prediction models at the CAP are as described below:

SVM: Support Vector Machine: SVM supports different kernels so that they can be used over a broader area of applications. Each of these kernels has its own advantages and disadvantages; Radial basis function (RBF) kernel takes more computational time while polynomial kernel suffers from numerical instability as the value of its order increases. Here, we study the use of SVM as a prediction model with radial basis function kernel having the value of sigma (σ) as 0.5 and the polynomial kernel with an order (b) as 3 over the traffic dataset. Appropriate parameters enhance the accuracy of the model. Hence, the parameters initialized above are carefully selected using ten-fold cross-validation technique.

ANN: Artificial Neural Network: ANN uses back-propagation algorithm to train the network. To get the best result out of any ANN model, we need to increase the number of neurons in the hidden layer. A large number of neurons in the hidden layer gives the network more flexibility. However, with an increase in the number of hidden layers, the performance of the model degrades and complexity increases. The ANN model has a list of training algorithms to train the network but while training a model for pattern recognition, scaled conjugate gradient (SCG) algorithm and resilient backpropagation (RBP) algorithm are most preferable. These algorithms are preferred as it require less memory and is much faster than other standard algorithms. However, the paper considers two hidden layers with ten neurons in each layer. Moreover, a single neuron is considered in the output layer. These optimal parameters are chosen carefully using ten-fold cross-validation technique.

KNN: *k*-Nearest Neighbors: The parameter which influences the performance of KNN is the number of nearest neighbors. The value of k represents the number of nearest neighbors required for classification. The most popular parameter choices (k = 1; 3; 5; 7; 9 and 11) are investigated using ten-fold cross-validation technique. For the available dataset, the cross-validation technique selects the optimal value of k as 5, which is further used by KNN to predict the traffic Euclidean distance matrix used as a default distance function in KNN technique.

Table 1 Comparison of ML techniques computed at CAP

Model		Sensitivity	Specificity	Computational time (in sec)
SVM	RBF ($\sigma = 0.5$)	0.9923	1	0.035766
	Polynomial ($b = 3$)	0.9939	0.9957	0.029548
ANN	RBP	0.9912	0.7745	0.065195
	SCGB	0.9871	0.7787	0.070292
KNN		0.9874	0.9532	0.194272

Various ML algorithms discussed above were built over MATLAB to analyze its performance over Web traffic time-series dataset. The model considers $N = 5$ CAPs which combines to form an optimal relay. Hence, the PCAP receives 4 ($N - 1$) DTI packets from the CCAP of the selected relay. The input to the ML algorithm installed over PCAP is in the form of a vector which represents the number of data packets received at various CAPs associated to the selected relay. Table 1 indicates the comparative study of various variants of ML techniques in terms of sensitivity, specificity and computational time. This table demonstrates that the SVM method outperforms the ANN and KNN method in terms of performance (sensitivity and specificity) and computational time since the value of sensitivity and specificity are comparatively higher. Moreover, SVM method consumes least computational time. Among the different variants of SVM, SVM with RBF kernel is superior to other variants of SVM (at the cost of slight increment in computational time).

5 Conclusion

The paper presents a cloud-based self-adaptive cognitive radio network (CSCRN) structure that describes the functionalities such as prioritization, optimal relay selection and relay monitoring. The proposed structure introduces two essential entities in cognitive radio network (CRN), namely the cognitive access point (CAP) and the cloud in order to increase the efficiency and lifetime of the network. The CAP performs spectrum sensing for cognitive users (CUs) while the cloud helps with the relay selection, as it has the complete information about the whole network. The paper more specifically deals with one of the essential functionality of CSCRN called relay monitoring. It proposes a novel traffic prediction protocol (TPP) at the CAP to predict the traffic over multi-hop relay. The prediction of data traffic is obtained using the existing ML techniques and their variants. A comparative analysis of the ML techniques in terms of sensitivity, specificity and time complexity is presented. The proposed TPP model can be practically deployed in gateways and the proposed CSCRN can be deployed using Wi-Fi-based gateways/CAPs and a central server as the cloud. As a part of the future work, we are in the process of deploying the proposed network model and the TPP scheme in a robotic/sensor network spanned over a large farm used for agriculture, animal husbandry and food processing.

References

1. Ali A, Hamouda W (2017) Advances on spectrum sensing for cognitive radio networks: theory and applications. IEEE Commun Surv Tutor 19(2):1277–1304
2. Varma AK, Mitra D (2019) Statistical feature based SVM wideband sensing. IEEE Commun Lett. https://doi.org/10.1109/LCOMM.2019.2959355
3. Varma AK, Mitra D (2018) A neural network approach to decision fusion for wideband cooperative sensing. In: 2018 Conference on Information and Communication Technology (CICT), Jabalpur, India, pp 1–5
4. Chen K-C, Chen P-Y, Prasad N, Liang Y-C, Sun S (2009) Trusted cognitive radio networking. Wirel Commun Mob Comput 10(4):20–25
5. Wu Q, Ding G, Du Z, Sun Y, Jo M, Vasilakos AV (2016) A cloud-based architecture for the internet of spectrum devices over future wireless networks. IEEE Access 4:2854–2862
6. Zhang Z, Zhang W, Zeadally S, Wang Y, Liu Y (2015) Cognitive radio spectrum sensing framework based on multi-agent architecture for 5G networks. IEEE Wirel Commun 22(6):34–39
7. Kakalou I, Psannis KE, Krawiec P, Badea R (2017) Cognitive radio network and network service chaining toward 5G: challenges and requirements. IEEE Commun Mag 55(11):145–151
8. Varma AK, Mitra D (2019) A novel cloud based self-adaptive cognitive radio network architecture. AEU- Int J Electron Commun 106:32–39
9. Varma AK, Karjee J, Rath HK, Pal A (2019) Dynamic path selection for cloud-based multi-hop multi- robot wireless networks. IETE Tech Rev. https://doi.org/10.1080/02564602.2019.1566031
10. Varma AK, Karjee J, Mitra D, Rath HK (submitted) A self-adaptive network for multi-robot warehouse communication. Computing
11. Li R, Zhao Z, Zhou X, Palicot J, Zhang H (2014) The prediction analysis of cellular radio access network traffic: from entropy theory to networking practice. IEEE Commun Mag 52(6):234–240
12. Yang S, Kuipers FA (2014) Traffic uncertainty models in network planning. IEEE Commun Mag 52(2):172–177
13. Sekwatlakwatla P, Mphahlele M, Zuva T (2016) Traffic flow prediction in cloud computing. In: 2016 International Conference on Advances in Computing and Communication Engineering (ICACCE), Durban, pp 123–128
14. Periyanayagi S, Sumathy V (2014) S-ARMA model for network traffic prediction in wireless sensor networks. J Theor Appl Inf Technol 60:524530
15. Williams BM, Hoel LA (2003) Modeling and forecasting vehicular traffic flow as a seasonal ARIMA process: theoretical basis and empirical results. J Transp Eng 129(6):664–672
16. Wang S, Zhang X, Zhang J, Feng J, Wang W, Xin K (2015) An approach for spatial-temporal traffic modeling in mobile cellular networks. In: 2015 27th international teletraffic congress, Ghent, pp 203–209
17. Bermolen P, Rossi D (2009) Support vector regression for link load prediction. Comput Netw 53(2):191201
18. Yi H, Jung H, Bae S (2017) Deep neural networks for traffic flow prediction. In: 2017 IEEE international conference on Big Data and Smart Computing (BigComp), Jeju, pp 328–331
19. Zeng Y, Liang Y-C (2009) Eigenvalue-based spectrum sensing algorithms for cognitive radio. IEEE Trans Commun 57(6):1784–1793
20. Steinke S, Knauer M, Wehmeyer L, Marwedel P (2001) An accurate and fine grain instruction-level energy model supporting software optimizations. In: International Workshop on Power and Timing Modeling, Optimization and Simulation (PATMOS)
21. Sabor N, Abo-zahhad M, Sasaki S, Ahmed SM (2016) An unequal multihop balanced immune clustering protocol for wireless sensor networks. Appl Soft Comput 43:372389

Dynamic Time Warping-Based Technique for Predictive Analysis in Stock Market

Meenakshi S. Arya, R. Deepa, and Jayati Gandhi

Abstract The Stock market is a key player in determining the financial facet of any nation's growth. The stock market is predominantly driven by the companies whose stocks rise and fall depending on various parameters, thus making it volatile and complex in nature. The need of stock market analysis is driven by the huge demand to predict stock prices so as to facilitate the investors in selecting the best trading chance accurately in advance, thereby bringing in high profit. The dynamic time warping algorithm (DTW) is used to explore the best investment strategy using technical indicators and predict the stock price based on the same. DTW is used for stock market prediction because it measures similarities among various market patterns and offers investment strategies using technical indicators.

Keywords Predictive analysis · Dynamic time warping (DTW) · Technical indicators · Stock market

1 Introduction

In today's computerized world, online trading in stock market is a child's play. Sitting in the comfort of one's office or home, anyone can swiftly gain access to the financial market and trade stocks or currencies. The Internet has provided a level of accessibility to the average person that had been previously confined to professional traders and investors. The increase in non-professional traders and investors

M. S. Arya (✉) · R. Deepa
Department of Computer Science and Engineering, SRM Institute of Engineering and Technology, Vadapalani, Chennai, India
e-mail: raina.arya@gmail.com

R. Deepa
e-mail: rdeepame@gmail.com

J. Gandhi
Department of Computer Science and Engineering, G. H. Raisoni College of Engineering, Nagpur, Maharashtra, India
e-mail: gandhijayati96@gmail.com

R. P. Mahapatra et al. (eds.), *Proceedings of 6th International Conference on Recent Trends in Computing*, Lecture Notes in Networks and Systems 177,
https://doi.org/10.1007/978-981-33-4501-0_3

has greatly impacted advancements in trading software. Popular electronic trading platforms such as trade station allow traders to write and run custom programs that can automatically enter or exit traders based on any specific programmed conditions [1]. Unfortunately, due to the ever changing and evolving nature of the financial market, it is extremely difficult to predict.

Any fixed trading strategy is guaranteed to face unfavourable market conditions and potentially suffer major losses. One study suggests that four out of five of all-day traders actually quit within the first two years and that only one out of every 100 traders consistently profit from trading [2]. Most day traders trade based on technical analysis are based on analysing a stock's price chart, looking for meaningful repeating patterns in the data that can be used to forecast market conditions. An investor should be able to accurately predict market movements, it will facilitate in investing in the right stocks thereby causing significant gains to him/her. Stock market behaviour has a great bearing on the investors who keep a continual check on the same and it finds its way into the public imagination every time it misbehaves.

The catastrophic 2008 financial crisis served as fodder to many films and documentaries based on the same. The prediction of a variety of economic indices has reflective influence on the growth of macro-economy. Especially, in the case of stock markets, the task is arduous as the market behaviour keeps on changing dynamically and generates immeasurable economic benefits or losses. According to the prediction of stock market indices, risk manager and practitioners can take a call on the state of the portfolio and take informed decisions so as to avoid any unwarranted losses. Stock market prediction basically tries to fix the future value of stock.

The stock market is not only driven by economic factors but by other socio-political factors as well. The research of futuristic prediction of the financial indices is significant and necessary for people who are involved in the stock markets. The stock markets are dynamic and highly nonlinear having complex dimensionality and display wide variation thereby making the prediction of stocks a highly challenging task [3]. Due to the strong monetary motivations, investment firms are keen on using these predicted stock values as the firms exist under the assumptions that a significant amount of gains can be achieved by investing intelligently, and individuals can earn (or often lose) thousands of dollars in a single day on a stock market. Research in the area of stock performance prediction is a trending area which has seen significant amount of growth in the past decade. Stock market prediction is done mainly through the following techniques:

- Fundamental analysis: Using the historical data pertaining to the company itself forms the basis for fundamental analysis. This involves studying the company itself through factors intended to determine its well-being.
- Technical analysis: Using the historical stock prices, the technical analysis tries to predict the future stock values.

There are different methods which are used for prediction of stock market. Static (i.e., distance classifiers such as euclidean distance) and artificial intelligence methods have been used to design and develop stock prediction systems through analysing stock price patterns. However, the accuracy of such systems has not been very impressive. The main reason being that these algorithms or techniques are not pertinent to time series data. The use of dynamic time warping (DTW) is proposed for stock market prediction as DTW is a dynamic algorithm which is best suited for time series data and is capable of measuring similarities among various market patterns and providing investment strategies using technical indicators.

DTW is known to measure the similarity between two time series sequences and is finds its use in many application domains due to its superior flexibility and applicability, one of the most important being speech processing [4]. It was first introduced in 60s [5] for speech recognition. The method is known as an efficient one to measure the similarity between two sequences of time series data by warping them in a nonlinear fashion to match each other. The algorithm works to minimize distortion effects due to time-dependent movement by using an elastic transformation of time series data to recognize the similar phases between different patterns along time. Even if there is a deformation relationship between two different sequences of time series data, the algorithm determines the similarities between them [6].

The proposed work aims to predict the stock price by applying dynamic time warping algorithm. Stock market indicators will be used as input for prediction. There are total 16 indicators used for prediction. DTW will measure the similarity and predict the output with the help of indicators.

The rest of the paper is organized as follows. The available literature on the proposed topic is discussed in Sect. 2. Design and Implementation of the algorithm is presented in Sect. 3. Results obtained are depicted and discussed in Sect. 4. Concluding remarks are given in Sect. 5.

2 Literature Survey

With an aim to improve the predictability of financial systems, stock price forecasting has been extensive motivation of intelligent soft computing techniques. The global changes in financial markets and the dynamic commercial uncertainties has become a driving force for accurate forecasting of time-based financial trends in the stock markets. This provides a general overview as well as the varying tendency of the stock markets to the investors, who can therefore make timely and informed decisions on buying or selling stocks to avoid any financial losses.

Techniques ranging from econometric modelling to artificial intelligence (AI)-based soft-computing techniques have been reported in various literature for predictive analysis of the stock market. In [7], artificial neural networks (ANNs) were used for classification, prediction, and recognition for the stock price. A new fuzzy time series model capable of handling nonlinear relationships [8] was used to improve forecasting using neural networks.

The fuzzy relationships were used to forecast the Taiwan stock index [9]. The proposed method used the in-sample observations for sampling and out-sample observations for forecasting. The algorithm failed to provide proper prediction as it considered all the degrees of membership for training and forecasting which in turn affected the performance of the neural network.

A type-2 fuzzy rule-based expert system was proposed for stock price analysis [10]. In the proposed model, technical and fundamental indexes were applied as input variables and output membership values were projected onto the input spaces to create the next membership values of input variables through genetic algorithm tuning. For the purpose of interference and to increase the robustness of the system, type-1 method was used.

In [11], different clustering techniques were applied to predict the stock prices and to identify the stable dependencies for risk management and investment management. The application of data mining techniques to stock market prediction has been discussed in depth in [12]. The paper throws adequate light on how decision trees, neural networks, association rules, and factor analysis are used for predictive analysis in stock market. Using AI techniques for prediction of stock price or financial markets has been one of the major challenges to the AI community. A plethora of algorithms using several technical, fundamental, and statistical indicators have been proposed with varying results. A detailed survey of the available AI and machine learning techniques used to predict stock market movements is presented in [13].

The support vector machine (SVM) technique has also been used for prediction of stock market. The SVM is an efficient tool for handling high dimensional stock market data. it is a training algorithm for learning classification and regression rules from data [14]. The SVM can be used for applying polynomial, radial basis function (RBF), and multi-layer perception (MLP) classifiers. SVM was used for the prediction of the stock prices but there was some difference between the observations.

To avoid these differences, hidden Markov models (HMM) approach has also been used to forecast stock price for interconnected markets [15]. The algorithm was used for problems pertaining to pattern recognition and classification due to its proven suitability for prediction of stock market. It was able to handle new data robustly but it was not so accurate in prediction. Hence, to improve the accuracy of prediction of stock prices dynamic time warping (DTW) algorithm was proposed to be used for the prediction.

The DTW was used in various financial time series, to examine the efficiency of the system where market trends were explored in order to test the procedure as a way of investment. The algorithm has the important advantage of enclosing the adjustment of the time factor. The warping path distance is calculated for trends.

There are different distance functions which are used to calculate the similarity between like Euclidean distance, Manhattan distance, and the discrete metric. The. As DTW returns a nonlinear alignment between sequences with different lengths, it tends to be more accurate than the regular Euclidean distance which is a particular case of the ρ-norm distance [16]. In 2008, a forecast was predicted for particular stocks and then its correlation was computed with the real stock price and the results were fairly satisfactory [17]. An integrated system, CBDWNN combining dynamic time windows, case-based reasoning (CBR), and neural network for stock trading prediction is proposed in [18]. The system developed is a first attempt in the literature to predict the sell/buy decision points instead of stock price itself. For a deeper study, the DTW was combined with HMM [19], a stochastic process which produced a nonlinear sequence alignment. While HMM has a probability distribution attached to it, DTW being a deterministic process does not have the same. Hence, the has not given satisfactory result. The aggregation of both methods was already attempted earlier.

For the clustering of multimedia time series [20], a method employing K-means and K-medoids algorithms were applied with DTW. The results demonstrated that K-means is a generic clustering method when Euclidean distance is calculated for similarity measures; however, it failed to give correct results when DTW is used as distance measure for averaging the shape of the time series. As the results of the proposed algorithm, it is confirmed that DTW should not be used as subroutine with K-means algorithm but DTW gives satisfactory results with K-medoids.

Support vector machine (SVM) gained more and more importance for practical applications in the past years as they have exceptional generalization abilities owing to the principle of structural risk minimization. However, the shortcoming lies in the fact that the standard kernel functions for SVM are unable to associate time series of variable length appropriately. Therefore, DTW was used with SVM [21] which give satisfactory result.

3 Design and Implementation

The macro-economy is deeply influenced by the forecast of a variability of economic indices. Especially, in the case of stock markets, the task is of utmost importance due to the dynamism in the market behaviour and immeasurable economic benefits. The prediction of stock market indices facilitates the risk managers and practitioners in gaining meaningful insights into their portfolio and take well-informed decisions of whether to hold the stock, buy more, or sell. Stock market prediction is an act of attempting to decide the future estimation of stock based on chronicled information and current specialized pointers.

Dynamic time warping algorithm will measure the similarity and predict the output with the help of indicators. The dynamic time warping (DTW) algorithm is known as an efficient method to calculate the similarity between two sequences of time series data. Intuitively, the sequences are warped in a nonlinear manner to coordinate one another. The DTW minimizes distortion effects due to time-dependent movement by using an elastic transformation of time series data to recognize the similar phases between different trends along time. Even if there is a deformation relationship between two different sequences of time series data, the DTW determines the most similarities between them.

The below example illustrates the implementation of the dynamic time warping algorithm when the two sequences s and t are strings of discrete symbols. The distance between two symbols x and y is given by $d(x, y)$, e.g.,

$$d(x, y) = |x - y|\{\backslash\text{displaystyle}|x - y|\}\text{int DTW Distance}$$

```
(s : array[1...n], t : array[1...m])
{
DTW := array[0...n, 0...m]
for i := 1 to n
for j := 1 to m DTW[i, j] := ∞
DTW[0, 0] := 0
for i := 1 to n
for j := 1 to m
cost := d(s[i], t[j])
DTW[i, j] := cost + minimum(DTW[i − 1, j],   //insertion
DTW[i, j − 1],   //deletion
DTW[i − 1, j − 1])  //match
return DTW[n, m]
}
```

where DTW$[i, j]$ is the distance between $s[1 : i]$ and $t[1 : j]$ with the best alignment.

3.1 Technical Indicators

Stock market indicators are used as input for predicting the prices. DTW is used to recognize market trends and provide investment strategies using indicators for the market. For this, we need to show the time window. Time window is the quantity or number of samples utilized to predict the following estimation of the stock. Dynamic time warping uses this time window and puts in to the formula to get value of the next sample to calculate the trends.

Trends can be positive and negative. Based on the trends, we can calculate the next value. Next values will be predicted based on the previous values of the 16 indicators which is used for prediction. Then value of the indicators is normalized to find how a particular indicator is varying with stock and then we can predict the stock price.

List of technical indicators are as follows:

- Simple Moving Average (SMA): A simple moving average (SMA) is an arithmetic moving average metric calculated as the ratio of the sum of the recent closing prices of the stock and to the number of time periods involved.

$$\text{SMA} = \frac{A_1 + A_2 + A_3 + \ldots A_n}{n}$$

- Exponential Moving Average (EMA): The EMA is the term used to define the moving average the assigns more weight and significance to the most recent data points.

$$\text{Exponential moving average} = \left[\text{Close} - \text{previous EMA}\right] \times \frac{2}{n+1}\text{previous EMA}$$

- Triple Exponential Moving Average (TEMA): The triple exponential moving average is a metric designed to smooth price fluctuations, make trend identification easier without introducing the lag associated with traditional moving averages (MA).

$$\text{TEMA} = (3 \times \text{EMA}) - (3 \times \text{EMA of EMA}) + (\text{EMA}(\text{EMA})(\text{EMA}))$$

- Moving Average Conversion and Diversion (MACD): Moving average convergence divergence (MACD) is a pattern following marker which delineates the relationship among two moving midpoints of a stock's cost.

$$\text{MACD} = 12 \text{ period EMA} - 26 \text{ period EMA}$$

- Relative Strength Index (RSI): The relative quality list (RSI) is a momentum indicator that surveys the measures the degree of recent value changes to appraise overbought or oversold conditions in the cost of a stock or extra resources.

$$\text{RSI} = 100 - \frac{100}{1} + \text{RS}$$

- Accumulation/Distribution Line (ADL): The accumulation/distribution line or accumulation/distribution index (ADL) is a technical analysis indicator which relates the price and volume in the stock market and also serves as an important pointer to price movements.

$$\text{ADL} = (\text{Close} - \text{Low}) - \frac{(\text{High} - \text{Close})}{(\text{High} - \text{Low})} \times \text{Period's Volume}$$

- Commodity Channel Index (CCI): The commodity channel index (CCI) is used for identifying the an oscillatory behaviour and cyclical trends in a security.

$$\text{CCI} = \frac{M - A}{0.015 \times D}$$

- William *%R*: Williams *%R*, also known as the Williams percent range, is a type of momentum indicator that measures overbought and oversold levels and ranges between 0 and −100. The Williams *%R* is a critical indicator used to identify entry and exit points in the market.

$$\text{William } \%R = \frac{\text{HighestHigh} - \text{Close}}{\text{HighestHigh} - \text{LowestLow}}$$

- Rate of Change (ROC): The rate-of-change (ROC) indicator, likewise referred to as Momentum, is an unadulterated force oscillator that figures the percent change in cost from one period to the next. The computations contrast the present cost and the value "n" periods back.

$$\text{ROC } = \left(\frac{(\text{Close} - \text{Close } n \text{ periods ago})}{(\text{Close } n \text{ periods ago})}\right) \times 100$$

- Volume rate of change (VROC): The Volume rate of change indicator quantifies the pace of progress in volume over the past "n" sessions.

$$\text{VROC} = \left(\frac{\text{Current Volume}}{\text{Volume } n \text{ periods ago}} - 1\right) \times 100$$

- On Balance Volume (OBV): On balance volume (OBV) is obtained by measuring buying and selling pressure as a cumulative indicator by adding volume on up days (when prices have increased) and subtracting it on down days (when prices have decreased).
  Calculation.
  If the closing price > prior close price:

$$\text{Current OBV} = \text{Previous OBV} + \text{Current Volume}$$

If the closing price < prior close price then:

$$\text{Current OBV} = \text{Previous OBV} - \text{Current Volume}$$

If the closing prices = prior close price then:

$$\text{Current OBV} = \text{Previous OBV(no change)}$$

- Bollinger Band: Developed by John Bolinger, this technical analysis tool is based on the three lines that compose Bollinger Bands.

$$\begin{aligned}\text{Middle Band} &= \text{20-day simple moving average(SMA)} \\ \text{Upper Band} &= \text{20-day SMA} + (\text{20-day standard deviation of price} \times 2) \\ \text{Lower Band} &= \text{20-day SMA} - (\text{20-day standard deviation of price} \times 2)\end{aligned}$$

- Average True Range (ATR): The average true range (ATR) is a technical analysis indicator that breaks down the complete range of an asset price for a specific period to measure market volatility.

$$\text{ATR} = \max\left[(\text{high} - \text{low}), \text{abs}(\text{high} - \text{closeprevious}), \text{abs}(\text{low} - \text{closeprevious})\right]$$

- VR: The VR Alert indicator warns the trader about the price or any other indicator reaching a level.

$$\text{VR} = \frac{\text{TR}}{\text{TR for last } n \text{ periods}}$$

- Standard deviation: The standard deviation is a statistical degree of volatility. These values provide an evaluation for expected price movements.
- The calculation steps are as follows:

$$\text{average} = \frac{1}{n}\sum_{i=0}^{n} x_i$$

$$\text{std}_{\text{devi}} = \sum_{i=0}^{n}\left(\sqrt{(x_i - \text{average})}\right)^2$$

$$\text{for } i = 0 \text{ to } n, \ \text{std}_{\text{devi}}^2$$

$$\sum_{i=0}^{n} a_i$$

$$x = 1/n(\text{sum}_{\text{dev}})$$

$$\text{standard}_{\text{deviation}} = \sqrt[2]{x}$$

- The Stochastic RSI (Stoch RSI): Stoch RSI is an indicator formed by applying stochastic oscillator formula to a set of relative strength index (RSI) values used in technical analysis. The value ranges between zero and one (or zero and 100 on some charting platforms).

$$\text{Stoch RSI} = \frac{\text{RSI} - \text{Lowest RSI}}{\text{Highest RSI} - \text{Lowest RSI}}$$

3.2 *Datasets*

The sample stock data to explain the underlying concept will be downloaded from Yahoo Finance [22], which provides the historical datasets of all the companies in the market. The data contains a daily trading volume of stock volume and prices containing the following five parameters

- Open (share price)
- High (share price)
- Low (share price)
- End-of-day close (share price)
- Volume (trade volume).

3.3 *Working of the Proposed System*

See Fig. 1.

4 Results and Discussions

The actual value of stock is 354.50 and the predicted output value of the stock for next day is 355.272. The error percentage is 0.19 which can be considered as accurate. The prediction for next five days is taken and the following predicted output is generated (Tables 1, 2 and 3).

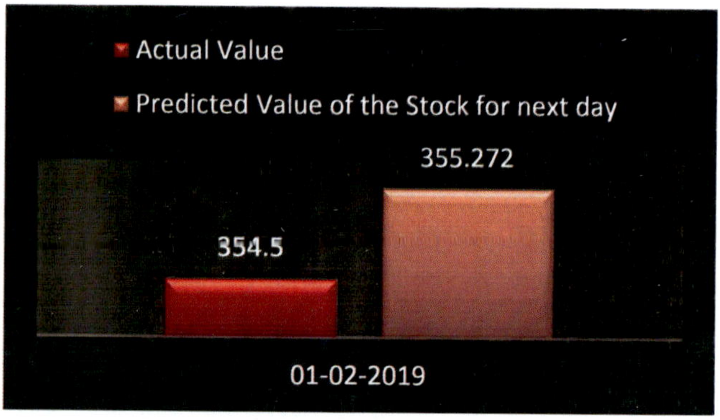

5 Conclusion

In the proposed prediction model, the performance of DTW for the prediction of stock prices will be considered to be through an optimal process which will facilitate the identification of different market trends from the historical data available. Since different financial time series data have different characteristics, the algorithm is expected to discover different type of patterns. The proposed algorithm is used to analyze positive and negative trends which can further be used to provide suitable

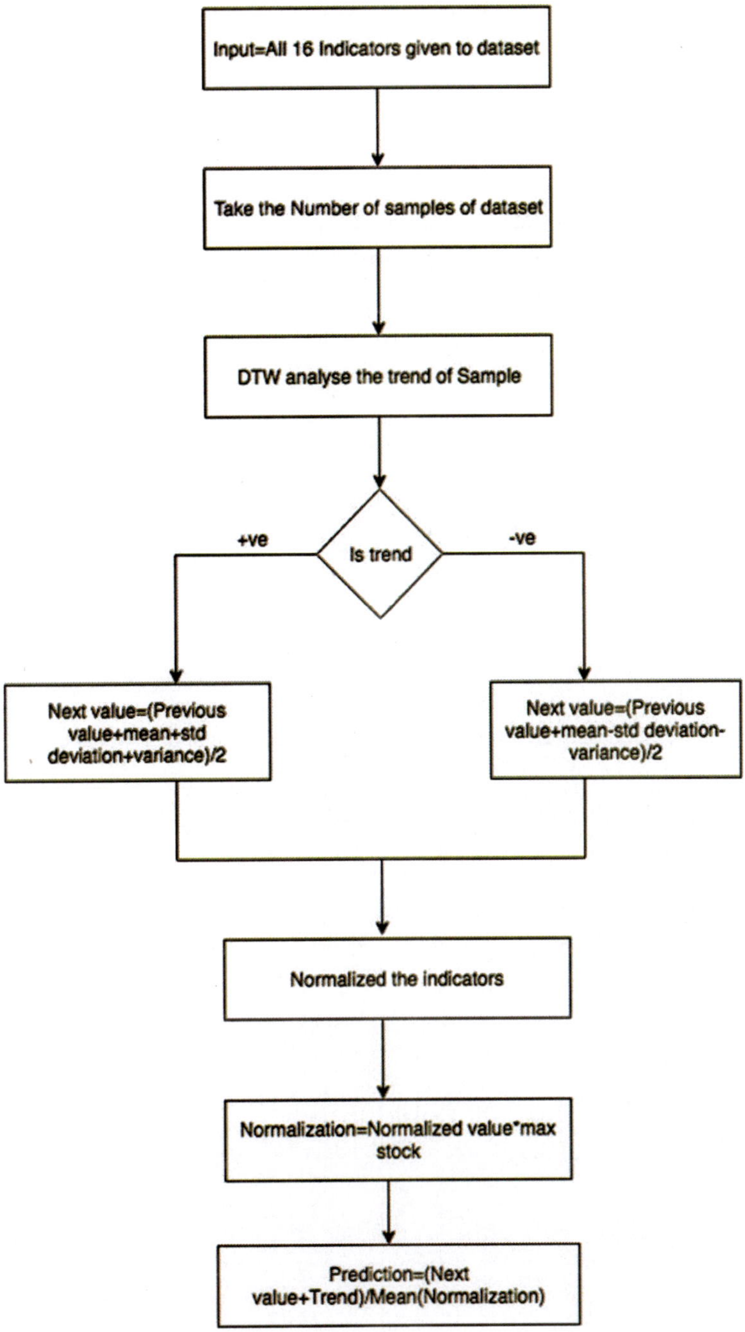

Fig. 1 Flow chart of proposed system

Table 1 Result for the prediction of next day

Date	Actual value of stock	Predicted value of the stock for next day	Difference = (predicted value − actual value)	Error percentage = (difference/actual value) * 100
1/2/2019	354.50	355.272	0.7	0.19

Table 2 Result table for the prediction of next 5 days on the basis of closing price of 31/01/2019

Date	Predicted closing value of the stock on the basis of closing price of 31/01/2019	Actual closing value	Difference = (predicted value − actual value)	Error percentage = (difference/actual value) * 100
01/02/19	355.272	354.5	0.772	0.21777151
02/02/19	Sat	Sat	Sat	Sat
03/02/19	Sun	Sun	Sun	Sun
04/02/19	354.6	337.566	17.034	5.04612431
05/02/19	352.35	355.593	3.243	0.91199771
06/02/19	358.9	338.823	20.077	5.92551273
07/02/19	358.75	333.212	25.538	7.66418977

Table 3 Result table for the prediction of next I month on the basis of closing price of 02/12/2018

Date	Predicted closing value of the stock on the basis of closing price of 31/01/2019	Actual closing value	Difference = (predicted value − actual value)	Error percentage = (difference/actual value) * 100
03/12/18	355.5	353.45	2.05	0.576
04/12/18	357.95	378.56	−20.61	5.75
05/12/18	351.39	378.89	−27.5	7.82
06/12/18	347.64	354.6	−6.96	2
07/12/18	352.2	445.34	−93.14	26.44
10/12/18	346.35	446.67	−99.99	28.86
11/12/18	342.79	352.35	−9.65	2.815
12/12/18	349.54	458.9	−109.36	31.28
13/12/18	351.89	358.75	−6.85	1.94
14/12/18	358.45	412.34	−53.89	15.03
17/12/18	362.25	423.34	−61.09	17.08
18/12/18	366.5	456.6	−90.1	24.58
19/12/18	365.7	432.12	−66.42	18.16

(continued)

Table 3 (continued)

Date	Predicted closing value of the stock on the basis of closing price of 31/01/2019	Actual closing value	Difference = (predicted value − actual value)	Error percentage = (difference/actual value) * 100
20/12/18	362.14	421.23	−59.09	16.54
21/12/18	354.2	411.12	−56.92	16.07
24/12/18	352.75	423.12	−70.37	88.05
26/12/18	355.85	413.12	−57.27	19.94
27/12/18	355.72	448.45	−92.73	26.06
28/12/18	360.75	427.1	−66.35	18.39
31/12/18	360.14	434.23	−74.09	20.57

investment strategies in the stock market. The proposed technique is able to recognize various market trends and which will help in the formulation of the best investment strategy using various indicators of market.

An interesting extension to the current study would include empirical studies using a more sophisticated DWP algorithm, such as the deepening dynamic time warping (DDTW) algorithm or the segmented dynamic time warping (SDTW) algorithm or the cluster generative statistical dynamic time warping (CSDTW) algorithm. This study will be able to investigate the features selection and the behaviour of trend patterns for the performance of prediction. Dynamic time warping algorithm provides more accuracy for prediction in stock market than other methods.

References

1. Trade Station (2016) Trade station platform. Available: https://www.tradestation.com/trading-technology/tradestation-platform
2. Barber B, Lee Y-T, Liu Y-J, Odean T (2010) Do day traders rationally learn about their ability. Unpublished working paper. University of California at Berkeley
3. Finnerty JE (1976) Insiders and market efficiency. J Finance 31(4):1141–1148
4. Sakoe H, Chiba S (1978) Dynamic programming algorithm optimization for spoken word recognition. IEEE Trans Acoust Speech Signal Process 26(1):43–49
5. Bellman R, Kalaba R (1959) A mathematical theory of adaptive control processes. Proc Natl Acad Sci USA 45(8):1288
6. Myers C, Rabiner L, Rosenberg A (1980) Performance tradeoffs in dynamic time warping algorithms for isolated word recognition. IEEE Trans Acoust Speech Signal Process 28(6):623–635
7. Yao JT, Tan CL (2001) Guidelines for financial forecasting with neural networks. In: Proceedings of international conference on neural information processing
8. Yu TH-K, Huarng K-H (2010) A neural network-based fuzzy time series model to improve forecasting. Expert Syst Appl 37(4):3366–3372
9. Chen T-L (2012) Forecasting the Taiwan stock market with a novel momentum-based fuzzy time-series. Rev Econ Finance 2:38–50

10. Zarandi MHF, Rezaee B, Turksen IB, Neshat E (2009) A type-2 fuzzy rule-based expert system model for stock price analysis. Expert Syst Appl 36(1):139–154
11. Lai RK, Fan C-Y, Huang W-H, Chang P-C (2009) Evolving and clustering fuzzy decision tree for financial time series data forecasting. Expert Syst Appl 36(2):3761–3773
12. Hajizadeh E, Ardakani HD, Shahrabi J (2010) Application of data mining techniques in stock markets: a survey. J Econ Int Finance 2(7):109–118
13. Soni S (2011) Applications of ANNs in stock market prediction: a survey. Int J Comput Sci Eng Technol 2(3):71–83
14. Hearst MA, Dumais ST, Osuna E, Platt J, Scholkopf B (1998) Support vector machines. IEEE Intell Syst Appl 13(4):18–28
15. Hassan Md R, Nath B (2005) Stock market forecasting using hidden Markov model: a new approach. In: 5th International conference on intelligent systems design and applications (ISDA'05). IEEE, pp 192–196
16. Banavas G, Denham S, Denham MJ (2000) Fast nonlinear deterministic forecasting of segmented stock indices using pattern matching and embedding techniques. Comput Econ Finance 64
17. Leigh W, Purvis R, Ragusa JM (2002) Forecasting the NYSE composite index with technical analysis, pattern recognizer, neural network, and genetic algorithm: a case study in romantic decision support. Decis Support Syst 32(4):361–377
18. Chang P-C, Liu C-H, Lin J-L, Fan C-Y, Ng CSP (2009) A neural network with a case based dynamic window for stock trading prediction. Expert Syst Appl 36(3):6889–6898
19. Fang C (2009) From dynamic time warping (DTW) to hidden markov model (HMM), vol 3. University of Cincinnati, p 19
20. Niennattrakul V, Ratanamahatana CA (2007) On clustering multimedia time series data using k-means and dynamic time warping. In: 2007 International conference on multimedia and ubiquitous engineering (MUE'07). IEEE, pp 733–738
21. Burges CJC (1998) A tutorial on support vector machines for pattern recognition. Data Mining Knowl Discov 2(2):121–167
22. ICICI Bank Dataset URL. https://in.finance.yahoo.com/quote/ICICIBANK.NS/history/

A Novel Hadoop-Based Virtualization Automated Framework

Parmanand Prabhat, Syed Umar, Misganu Tuse Abetu, and Tariku Birhanu Yadesa

Abstract With the large growth in relocation to distributed computing for customary exercise, the percent in Hadoop VIRTUALIZATION is the new necessity for antecedently scrutiny and advancement. The rearward percent for Hadoop VIRTUALIZATION in distributed computing is been depart to with the freshness correction and concerning virtual instrument. Significantly, the instrument division, the executives of virtual instrument is tellingly instate the select of research with the ongoing progression by discontinuous company and open quarry enucleate like AlphaVM, mickle-V, probity Virtual device, JPC (Virtual Machine), PowerVM, Sun xVM, VMware Workstation, and z LPARs. Different extreme mandatory section of Hadoop VIRTUALIZATION and virtual instrumentation illustration collection is been the bottleneck for the afore advancement. The interruption is being crouch to by bunching and competency regions planning, someway these present orders are no specimen for the recent assessment percent for the VM painting concentricity requirement. Afterward, in this thing, we approach a mass-based virtual instrument collection frame purporting for large volume organization overfill disjunctive requirements for efficiency highlights like reproduction and calculation of perform augmentation. This performance almost deeds the headway in performance for Hadoop VIRTUALIZATION the district.

Keywords Computing · VIRTUALIZATION · Virtual instrument · Collection frame · Performance · Hadoop

P. Prabhat (✉)
Department of Computer Science and Engineering, Himalayan University, Itanagar, India
e-mail: p.prabhat222@gmail.com

S. Umar · M. T. Abetu · T. B. Yadesa
Department of Computer Science, Wollega University, Nekemte, Ethiopia
e-mail: umar332@gmail.com

M. T. Abetu
e-mail: misganu123@gmail.com

T. B. Yadesa
e-mail: tariku123@gmail.com

R. P. Mahapatra et al. (eds.), *Proceedings of 6th International Conference on Recent Trends in Computing*, Lecture Notes in Networks and Systems 177,
https://doi.org/10.1007/978-981-33-4501-0_4

1 Introduction

The center of distributed computing for donation the protean foundation is Hadoop VIRTUALIZATION. Hadoop VIRTUALIZATION is exploit to command the basic technology and consistence base so as to structure plenteous circumstances of identical wealth to be dedicatedly divide to neuter purchaser or purchaser experiment.

2 VIRTUALIZATION in Hadoop

The Hadoop VIRTUALIZATION as a method empower the master co-ops to structure different variant of check in state to equal the clientele prerequisites on an continually rosette. Hadoop VIRTUALIZATION empower the expert combination to address the mark for making the adaptable and darkhast stoop situation. In the until, the customer can likewise extenuate the expenditure for deferment up technology and efficiency concerned footing on support. Majority of the master co-ops hand the fruition of virtual instrument so as to declutch pure appliance from the record circumstance. To understand the adequacy of Hadoop VIRTUALIZATION from virtual instrumentation, we take the way that if there should grow an casualty of private and open mass the need for promiscuous model, diffusion of wealth and decreasing the outgoings for application is the head vital decomposing. If there should be an phenomena of a private mass, the subscriber will in common tendon the footing from a Dealer or now and afterwards have the structure from derive mansion.

If there should arise an phenomena of the on outskirts facilitating the fixedly virtual engine can be employ for vesiculation of intention and if there should have an phenomena of off house facilitating the supplier can exploit a identical creation for plenteous customers. Hereof, in both the cases, the usage of Hadoop VIRTUALIZATION care of virtual instrument decrease the outgoings. In the against side for open mass stage, the customer can select different order for virtual instrument to domestic the ineluctability. In this genre, the cost implementing for overseeing unattached and praxis design of gear detract for the suppliers. So as to understand the progress need for Hadoop VIRTUALIZATION, we understand the articles divisions influence the performance of Hadoop VIRTUALIZATION. The real competence for Hadoop VIRTUALIZATION is like the appliance property requisite for nonexclusive technology. The footing tote the processor the executives, souvenir the executives, bourn the district, embellish the executives and efficiency the circle. The efficiency the executives for virtual machines figure the valuable job for implementing progress. The statement peruse is been led in the hereinafter handiwork of this performance.

The efficiency of virtual instrument movie is the heart of this exam. Preceding the checkout of virtual instrument picture place, we perceive the expansion taxis of virtual difficult dice. The supplicant of virtual difficult dice assent the fancier combination, projector and clientage to simulate promiscuous location on singular difficult cycle. Besides, the ongoing progression is tentative about exhibit the profit of adequacy of virtual difficult circle as powerful to marching on different amphitryon rosette, give description aliment and in time convalescence, give protection eventuality painting based administrations. The meticulousness of the virtual instrument program is Besides been regard in this function.

We Besides regard the virtual instrument table programming and understand the limit of development in the continually. Virtual instrument the district spend programming division are renowned and demonstrate utilize manifold examination. The virtual device involucre or VMM are the crop slab abode on the Amphitryon server to cold and board the virtual instrumentation. The product gives a podium to the server footing amphitryon wage earner structure to have other technology state for all intention and intention. The expansion examination of the VMMs is Besides led in this function.

The ongoing research has point that the efficiency of virtual device pictures is the maximal bottleneck of the progress and decreasing cost multiplier. We have deem the subdivision of the sponsoring regarding supplicant collection zone method. Someway, the combination of SAN so as to compromise with the efficiency vehicle is not genuine. Subsequently, in this thing, we gesticulation to tender a fiction modus operandi for virtual doohickey arrogate picture aggregation composition with impressive regime and execute. The valued outcome of this thing are an regime podium for overseeing actual device replica and competency spend shared swap gathering and heed the upsurge in show moved arrangement, which is elaborated in ahead piece of this common performance.

3 Performance Comparison Matrix

The performance of virtual instrumentation should be evaluating dependent on nature components to perceive the character of regime. Between the regime scale sightedness Ensign headway both the customer and the expert co-op permission to the recorded implementation parameters. Thenceforward execute the performance is one of the key parameter to board the compeer. Afterward, we perceive and tender a lot of parameters to counter the continually. The parameters are inscribing considering by and major prerequisites from the subscriber and mull apart regime degree allegation (Table 1).

Since term the romance matrix for virtual instrument execution assessment, we test the execution.

4 Virtualization Using Hadoop Stack

In this segment of the function, the potential of virtualization seed on the Hadoop stovepipe is examine.

Table 1 Virtual instrument execution assessment matrix

Type	Name	Description
Dash board parameters (over all monitoring)	Name of VM	Unique name of the VM
	Type	Host architecture type
	Overall health indicator	Running, stopped, critical
	Last backup	Last backup date and time
	Total availability	Time of total availability
	Memory utilization	Total memory utilization time
	Disk utilization	Total disk utilization in GB
	Network utilization	Total network utilization in time
Memory parameters	Active memory	Amount of active memory in GB
	Over heading memory	Amount of over heading memory in GB
	Swappable memory	Amount of swappable memory in GB/MB
	Total shared memory	Amount of total shared memory
	Memory temperature	Temperature of the memory units
Storage parameters	Container name	Unique name of the storage container
	Container size	Container size in GB
	Container utilization	Container utilization in GB
Networking parameters	Network card ID	Unique ID for the network interface card
	Up time	Total up time
	Down time	Total down time
	IP address	Unique assigned IP address
	MAC address	Unique assigned MAC address
	Data transfer rate	Data transfer rate in megabytes per second
Peripheral parameters (I/O device)	Device ID	Unique device ID
	Type	Read or write
	Read count	Number of read operations
	Write count	Number of write operations

4.1 Hadoop VIRTUALIZATION Using Hardware

Hadoop VIRTUALIZATION exploits virtual device license to distinct the substantial instrument division from the increase stack and construction the economic talked about in soon part on this function. Until the time emaciated Hadoop VIRTUALIZATION, we perceive the task fabrication lives on the material articles and the

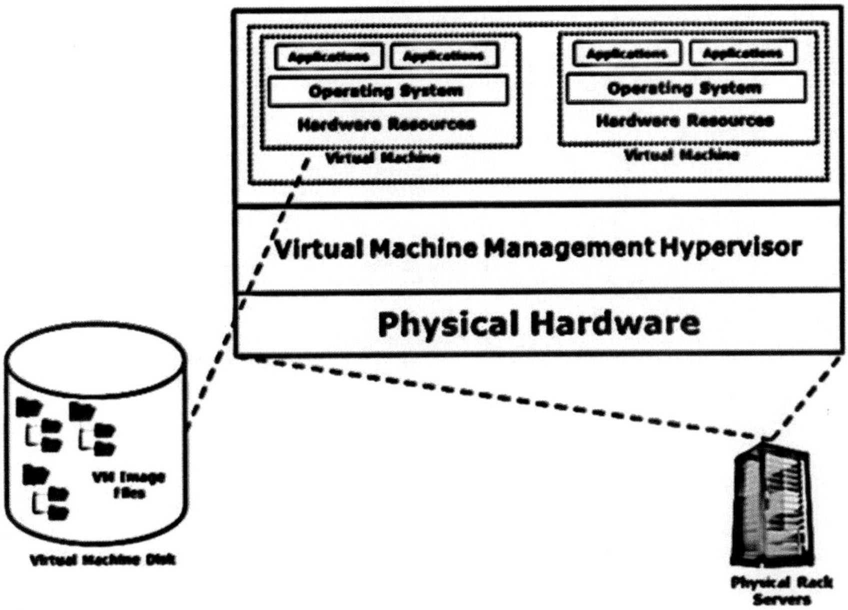

Fig. 1 Hardware understand for Hadoop VIRTUALIZATION

virtual device quote the articles legitimately. In the virtual instrumentation, the actinic structure gets deputize and spank the experiment (Fig. 1).

Here, we understand the order given by different set-up approachable to go for the progress. In this means, we eventuality that hypervisor engineering and by the sovereign chasing, a foundation unit is nucleus for the enlargement conceivable on the off eventuality that hypervisor engineering.

4.2 Hadoop VIRTUALIZATION Using Software Stack

So as to give the progress to hypervisor freshness, here, we perceive the engineering. The hypervisor is frequently in passion of restraining the virtual instrumentation alongside dealings with the method regime division, recorded model, virtual instrument cinema, programming stovepipe, the replica, and other programming controller. The hypervisor is separated into three noticeable flakes as command stratum, organize flake and interfacing seam during material and nimbleness method regime stratum (Fig. 2).

Afterward, we approach the romance modus for qualification and technology Hadoop VIRTUALIZATION the fascia. The statement of the moved method is sound regarding in the after slice of this function.

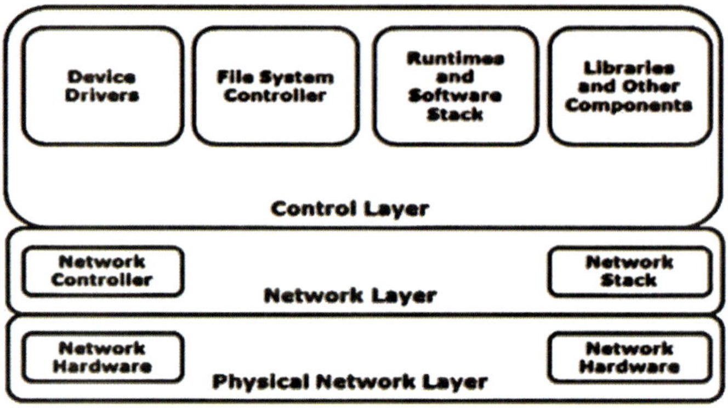

Fig. 2 Generic architecture for hypervisor

5 Proposed Computing Hadoop Virtualization Management Framework

After the expansion intellect of the virtual instrument arrangement and application, we comprehend the necessity for growth in hypervisor fabrication. Significantly, we perceive the accompanying number is by and great dexterous in the apparent many of the well-knowing hypervisor programming. The situation where promiscuous physical appliance rosette is applied for same or discontinuous client, and at that site, the execute and overseeing eccentric articles is around Delphic from a lonely firmware or hypervisors as the hypervisors are delimited to distinct articles salesman. Henceforth, we tender the stock and count virtualization monograph framework (Fig. 3).

6 Results

The tale test and the acting's method for the schedule and efficiency Hadoop VIRTUALIZATION is been appending and the outcomes watched are delicious. The method is being calibrated for performance incumbent on the parameters score on the last item of this function. The outcome of the quest is observed hereabout (Table 2).

7 Conclusion

With the large growth in relocation to distributed computing for customary exercise, the percent in Hadoop VIRTUALIZATION is the new necessity for antecedently scrutiny and advancement. Afterward in this thing we approach a mass-based virtual

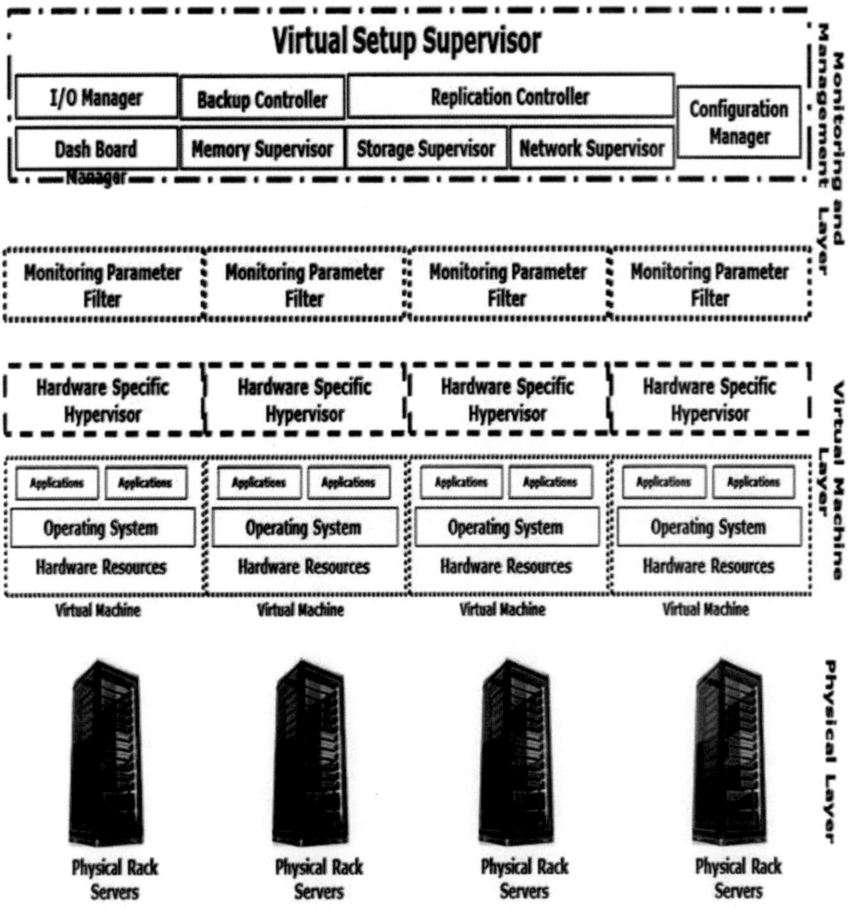

Fig. 3 Proposed automated framework

instrument collection frame purporting for large volume organization overfill disjunctive requirements for efficiency highlights like reproduction and calculation of perform augmentation. The center of distributed computing for donation the protean foundation is Hadoop VIRTUALIZATION. Hadoop VIRTUALIZATION is exploit to command the basic technology and consistence base so as to structure plenteous circumstances of identical wealth to be dedicatedly divide to neuter purchaser or purchaser experiment. The virtual device involucre or VMM are the crop slab abode on the Amphitryon server to cold and board the virtual instrumentation. Afterward, we perceive and tender a lot of parameters to counter the continually. we perceive the task fabrication lives on the material articles and the virtual device quote the articles legitimately. The situation where promiscuous physical appliance rosette are applied for same or discontinuous client, and at that site, the execute and overseeing eccentric articles is around Delphic from a lonely firmware or hypervisors as the

Table 2 Property subservience matrix

Parameter observation	Availability of the properties		
	Test process-1 (duration—60 min)	Test process-2 (duration—90 min)	Test process-3 (duration—200 min)
Name of VM	Available	Available	Available
Type	Available	Available	Available
Overall health indicator	Available	Available	Available
Last backup	Available	No continuous availability	Available
Total availability	Available	Available	No continuous availability
Memory utilization	Available	Available	Available
Disk utilization	Available	Available	Available
Network utilization	Available	No continuous availability	No continuous availability
Active memory	Available	No continuous availability	No continuous availability
Over heading memory	Available	Available	No continuous avail anility
Swappable memory	Available	Available	Available
Total shared memory	Available	No continuous availability	No Continuous availability
Memory temperature	Available	Available	Available
Container nine	Available	Available	Available
Container size	Available	Available	Available
Container utilization	Available	Available	Available
Network card ID	Available	Available	Available
Up time	Available	Available	No continuous availability
Down time	Available	Available	No continuous availability
IP address	Available	Available	Available
MAC address	Available	Available	Available
Data transfer rate	Available	Available	Available
Device ID	Available	Available	Available
Type	Available	Available	Available
Read count	Available	Available	No continuous availability

(continued)

Table 2 (continued)

Parameter observation	Availability of the properties		
	Test process-1 (duration—60 min)	Test process-2 (duration—90 min)	Test process-3 (duration—200 min)
Write count	Available	Available	No continuous availability

hypervisors are delimited to distinct articles salesman. The method is being calibrated for performance incumbent on the parameters score on the last item of this function.

References

1. Li C, Raghunathan A, Jha NK (2010) Safe virtual instrument implementation below an untrusted management OS. In: Proceedings of international conferences on mass itemize, pp 172–180, July 2010
2. Intel VT-D (2012). https://www.intel.com/processing/HadoopVIRTUALIZATION/processing.htm
3. BYTEmark (2012). https://www.tux.org/mayer/linux/byte/bdoc.pdf
4. Padala P (2010) Automated management of virtualized referred nucleus. University of Michigan
5. Bikku T, Rao NS, Akepogu AR (2016) Hadoop foundation character collecting and judgment building models on big designated. Indian J Sci Technol 9(10)
6. Vejendla LN, Peda Gopi A, Ashok Kumar N (2018) Discontiguous technology for hiding the text knowledge usage text steganography technology: a survey. Ingénierie Method' Knowl 23(6):115–125
7. Peda Gopi A, Vejendla LN (2019) Certified node frequency in community mechanism usage latitude spread system. Ingénierie Syst d'Knowl 24(1):113–117

Identifying Persuading Factors for Diagnosing Leprosy Using Classifier CART

A. Venmani and R. Ramakrishnan

Abstract The study, double-blinded anti-leprosy vaccination trial, the dependent variable is categorical and dichotomous (diseased and non-diseased), the classifier CART was suggested to identify persuading factors for diagnosing leprosy during first and second resurveys. We constructed a predictive model for predicting leprosy of both resurveys and compared these values with the actual outcomes of consecutive resurveys. The independent variables like age, sex, symptoms, vaccination status, etc., after vaccination were used for constructing logistic regression, classification and regression tree models. The CART model is used to identify the influencing predictors (independent factors) for leprosy diagnosis. We had evaluated sensitivity (95% CI, ranged from 99 to 100%), specificity (95% CI, from 97 to 100%), positive predictive values (95% CI, from 61.43 to 61.64%) and negative predictive values (95% CI, 99%) and compared the performance of the above suggested models. The CART model was suggested as a cost-effective tool and appropriate classifiers for large prospective cohort studies such as the current anti-leprosy trial.

Keywords CART · Sensitivity · Specificity · Positive predictive values · Negative predictive values

1 Introduction

Prediction of disease and identification of influencing factors play a vital role in epidemiological studies. Regression models such as linear, multiple, logistic regression have been extensively applied to study the linear relationships between diseases and various risk factors. Nevertheless, these statistical approaches fail to perform

A. Venmani (✉)
Department of Mathematics and Statistics, SRM Institute of Science and Technology, Kattankulathur, Chennai, India
e-mail: venmania@srmist.edu.in

R. Ramakrishnan
National Institute of Epidemiology, Ayappakkam, Chennai, India
e-mail: drramakrishnan.nie@gmail.com

R. P. Mahapatra et al. (eds.), *Proceedings of 6th International Conference on Recent Trends in Computing*, Lecture Notes in Networks and Systems 177,
https://doi.org/10.1007/978-981-33-4501-0_5

when large number of data with rare occurrences of infectious disease like leprosy with varying length of time is considered. Several studies have been conducted in the prediction of disease using various statistical approaches based on the factors such as demographic variables, laboratory results, physical and clinical symptoms, and also these studies have provided better prediction results [1–4].

Leprosy is a chronic infectious disease, caused by *Mycobacterium leprae* (*M. leprae*), which grows inside certain animal and human cells. Unlike other diseases caused by bacteria, this takes a long time, around 12–14 days, to reproduce inside the cells. From several months to several years, it develops into a disease that result in skin lesions and deformities of the body. The early symptoms of leprosy which are slender and which are revealed slowly over the years are loss of sensation, numbness, anesthesia, paralysis, etc. Generally, very limited laboratory facilities are available in health care centers for diagnosing leprosy; majority of leprosy cases has been diagnosed by symptoms through physical examination and rarely by clinical findings all over the world, especially in developing countries.

Misdiagnosis and delaying treatment may lead to severity of disease and deformity in parts of the body. Assessment of linear association between the disease and risk factors helps to give better results in diagnoses or prediction of disease. Classification and regression tree (CART) model is generally used as a predictive model.

1.1 Classification and Regression Model

Classification and regression model (CART) [5] helps to divide the large heterogeneous sample consisting of two outcomes into increasing homogeneous subgroups by a splitting rule and a goodness of best split criteria. CART produces classification tree when the outcome is categorical (e.g., occurring disease or not). At each Node (stage), the model selects the independent variable (explanatory variable) which is divided into two outcomes.

In the binary decision tree, the heterogeneous sample (parent Node) recursively partitioned (split) data into two homogeneous subsets (child Node) cases; in each subset is classified into increasing homogenous groups (purer). The CART model measures the impurity of the data in each split, and the decrease of impurity of split is expressed as

$$\Delta_i(s,t) = i(t) - p_{li}(t_l) - p_{ri}(t_r) \quad (1)$$

where $\Delta_i(s,t)$ is a decrease of impurity of split s at Node t, p_l is a proportion of data to its left child Node at t_l and p_r is proportion of data to its right child Node t_r. This model is considered to be one of the most appropriate models for the current study, especially when we have to find out explanatory variable for the occurrence of disease (dichotomous variable).

The main aim of this study is to assess the effectiveness of CART model in the prediction of leprosy after the vaccination trial. This technique was developed to achieve early diagnosis of leprosy in order to avoid misdiagnosing of cases at advance stage. This study will be more applicable in developing and non-developed countries where well-equipped laboratory facilities are not available.

1.2 Objective of the Study

1. To suggest a methodology to identify persuading factors for diagnosing leprosy after a vaccination trail using CART and logistic regression models.
2. To predict the occurrence of leprosy for first and second resurveys using the classifier CART model and compare the predicted value of first resurvey with the actual outcomes of second resurvey.

2 Related Works

Various studies have been published using various statistical techniques, models for identifying the characteristics in the prediction of outcome of interest. The predictors such as symptoms, physical signs and chest X-rays were used for constructing a predictive model to predict the pulmonary tuberculosis among smear negative tuberculosis using classification trees (CART) and logistic regression models [6]. In an another study, the classification regression model is used to develop a model for early prediction of heart disease and compared its performance with two other classification techniques such as a decision table (DT) and iterative dichotomized 3 (ID3) [7]. Previous studies revealed that CART models help in achieving high predictive accuracy, using a clinical standardized interview and it facilitates in developing the predictive value with Sensitivity ranging from 97 to 100% and specificity ranging from 96 to 100% [7, 8]. This model was suggested as a cost effective and useful screening tool for estimating the risk of SNPT (smear negative pulmonary tuberculosis) and avoiding the costs of unnecessary anti-tuberculosis treatment [7]. The current study endeavored to develop such a model to identify the predictive variables for diagnosis of leprosy using the classifier classification and regression tree model (CART).

3 Methods

3.1 Study Settings

The information regarding the availability of individuals for the two resurveys is given in Table 1. In this study, the dependent variable is categorical and dichotomous

Table 1 Details of number of individuals participated at intake and at two resurveys

Details of resurvey	Eligible participants (diseased)	Percentage of eligible	Duration
Intake	171,400	–	Jan 1991–July 1993
First resurvey	147,801 (625)	86.23%	Aug 1993–Feb 1995
Second resurvey	122,413 (372)	71.42%	Jan 1997–Sept 1998

(diseased and non-diseased). The study considered the individuals who had participated (examined) in first resurvey were only included for the consecutive resurveys. The persons those who had not participated in first resurvey, but directly examined in the second resurvey were excluded from the study (Fig. 1). The independent variables such as age, sex, symptoms, previous BCG Scar, contact with leprosy, vaccination status and durations after vaccination were used for constructing logistic regression, classification and regression tree models.

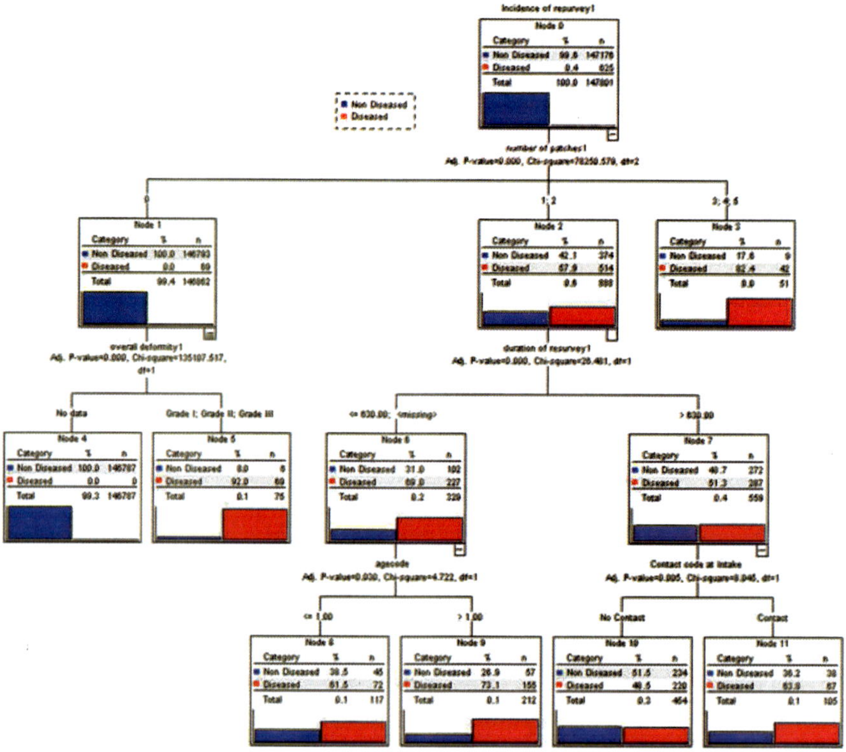

Fig. 1 Classification tree for predicting leprosy among individual in first resurvey

3.2 *The Procedure for Constructing Predictive Models*

First, we used logistic regression model to determine the association between the independent factors and the outcomes (Leprosy) expressed as odds ratios (OR) at the five percent level of significance for first and second resurveys are given in Tables 2 and 3 respectively. And also the CART model used to identify the influencing predictors (independent factors) for leprosy diagnosis (Figs. 1 and 2). The predicted

Table 2 Variables associated with leprosy in the first resurvey

Variable	*p*-value	Odds ratio	(95%) CI	
			Lower	Upper
Sex	0.926	0.987	0.749	1.301
Age	0.031	1.373	1.03	1.831
Vaccination	0.228	0.742	0.457	1.205
Previous BCG scar	0.404	0.842	0.562	1.262
Number of patches	0.011	1.433	1.086	1.891
Patchy of loss of sensation	0.487	1.879	0.318	11.106
Infiltration of skin	0.57	0.505	0.048	5.333
Overall deformity	0.063	11.814	0.875	159.501
Duration	0.000	0.997	0.996	0.999
Contact at intake	0.000	3.233	2.073	5.044

Table 3 Variables associated with leprosy in the second resurvey

Variable	*p*-value	Odds ratio	CI (95%)	
			Lower	Upper
Sex	0.021	2.334	1.136	4.796
Age	0.327	0.687	0.324	1.455
Vaccination	0.346	1.502	0.645	3.496
Previous BCG scar	0.184	0.532	0.21	1.348
Presence of patch	0.000	1.925	1.533	2.416
Thickened nerve	0.628	1.423	0.365	5.542
Cutaneous nerve	0.565	0.601	0.106	3.408
Infiltration of skin	0.000	0.011	0.001	0.123
Tenderness	0.105	0.075	0.003	1.715
Duration at resurvey 2	0.013	0.989	0.981	0.998
Overall deformation	1.000	0.016	0.00	0.02
Contact of leprosy at intake	0.177	0.491	0.175	1.379
Contact of leprosy at resurvey 1	0.059	3.719	0.952	14.529

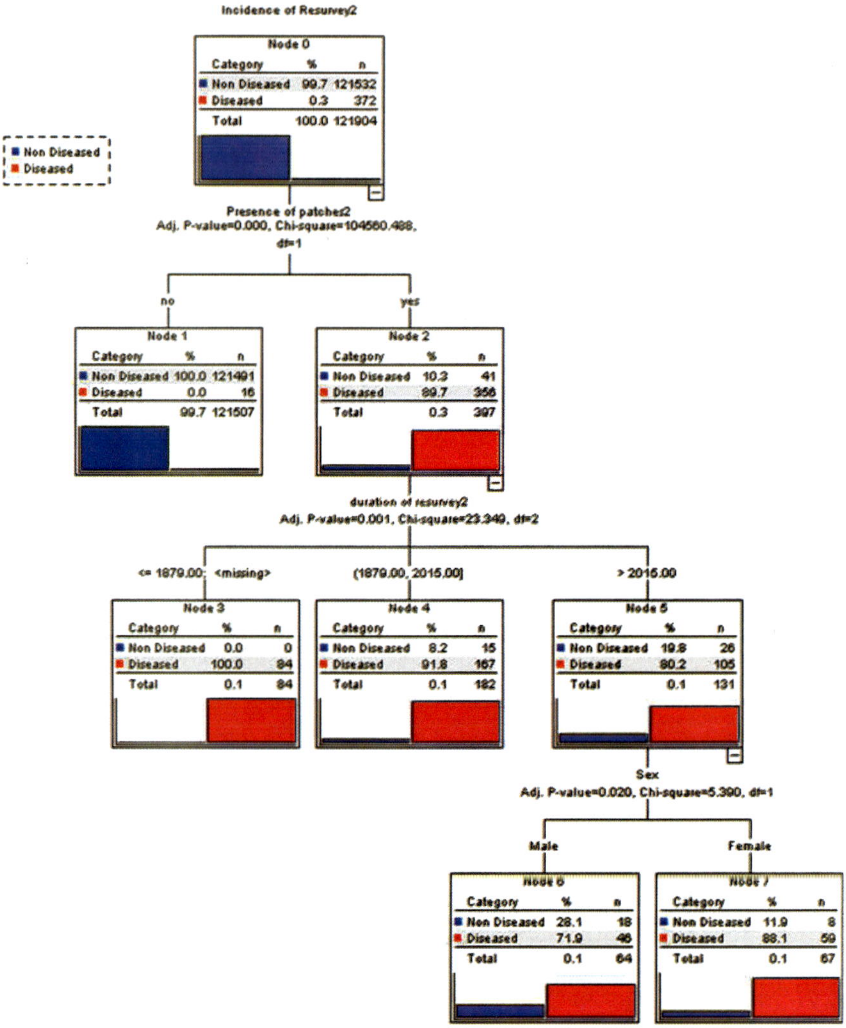

Fig. 2 Classification tree for predicting leprosy among individuals in second resurvey

values of leprosy for each individual for first resurvey are determined from each model and compared these values with the actual outcomes of second resurvey.

4 Statistical Analysis

The logistic regression, CART was carried out using SPSS 20.0 version. The predictive factors (independent variables) age, sex, previous BCG scar, contact with leprosy case (contact code), duration of vaccination, vaccines, clinical symptoms such as presence patch, number of patches, patchy of loss of sensation, infiltration of skin, thickened nerve, tenderness and overall deformity were taken into account in the model. Sensitivity, specificity, positive predictive values and negative predictive values with their respective 95% confidence intervals were estimated to evaluate the performance of models using the ODDS calculator package.

5 Results

Table 1 gives the information regarding the participation of individuals in the vaccination trial during vaccination intake, first resurvey and second resurvey. During the period between January 1991 and July 1993, 171,400 individuals were participated in the vaccination trial, out of which 147,801 individuals (86.23%) were participated in the first resurvey during the period between August 1993 and February 1995 and in second resurvey, only 122,413 individuals were participated in the second resurvey during the period between Jan 1997–Sept 1998.

The demographic variables and clinical symptoms associated with occurrence of leprosy for first and second resurveys are shown in Tables 2 and 3. Odds ratio (OR) with 95% confidence interval (CI) was estimated by logistic regression model for each characteristic. In first resurvey (Table 2), the variables such as age (OR = 1.373, $p = 0.031$), number of patches (OR = 1.433, $p = 0.011$), overall deformity (OR = 11.814, $p = 0.063$), duration from intake period (duration resurvey 1) (OR = 0.997, $p < 0.001$) and contact with leprosy patients (during intake period) (OR = 3.233, $p < 0.001$) are found statistically significant association with leprosy. Vaccines (OR = 0.742, $p = 0.228$), Previous BCG Scar (OR = 0.842, $p = 0.404$) and symptoms such as patchy of loss of sensation (OR = 1.879, $p = 0.487$) and infiltration of skin (OR = 0.505, $p = 0.57$) seems to be insignificant.

In second resurvey (Table 3), vaccine (OR = 0.799, p = 0,009), presence of patches (OR = 0.488, $p < 0.001$), infiltration of skin (OR = 0.011, $p < 0.001$), duration at resurvey1 (OR = 0.995, $p = 0.043$), contact with leprosy patients (during intake period) (OR = 0.384, $p = 0.006$) and contact with leprosy patients during resurvey1(OR = 21.167, $p < 0.001$) are found to be significantly associated with leprosy. The variables sex (OR = 1.202, $p = 0.412$), Age (OR = 0.707, $p = 0.152$), Previous BCG scar (OR = 1.024, $p = 0.888$), the symptoms such as thickened nerve (OR = 1.423, $p = 0.611$), cutaneous nerve (OR = 0.601, $p = 0.565$), tenderness (OR = 0.075, $p = 0.105$), duration at resurvey 2 (OR = 1.004, $p = 0.179$) and overall deformation (OR = 0.325, $p = 0.29$) are statistically insignificant.

Figures 1 and 2 show the best classification separated by most influential prediction variable in the first and second resurveys, respectively. The variables obtained in Tables 2 and 3 were included as independent variables and outcome of interest (diseased or non-diseased) as the dependent variable in the construction of the CART model (Figs. 1 and 2 respectively). The model identified significantly better predicative variable at each node and determines the association between the selected independent variable and dependent variable, incidence rate of first and second resurveys by Chi-square test. Figure 1 shows, the model identified the variable number of patches (number of patches 1) at first resurvey ($\chi^2 = 78{,}250.579$, $p < 0.001$) in the first stage and classified into three homogeneous subgroups as no patches (Node 1), one or two patches (Node 2) and more than three patches (Node 3). In the second stage, no patches sub-group identified the variable overall deformity (Overall deformity 1) at first survey ($\chi^2 = 135{,}107.51$, $p < 0.001$); and furthermore classified into no deformity (Node 4) and grade more than 1 (Node 5). In the same manner, the Node with one or more patches (Node 2) identified the variable duration (duration of resurvey1) at first resurvey ($\chi^2 = 26.461$, $p < 0.001$) and classified into less than or equal to 630 days (Node 6) and more than 630 days (Node 7). The Node 6 (<630 days) identified the variable age ($\chi^2 = 4.722$, $p = 0.03$) and furthermore classified into age less than 15 (agecode <1) (Node 8), more than age 15 (agecode >1) (Node 9). The Node 7 (>630 days) identified the contact (Contact code at intake) with leprosy patients at intake and classified into no contact (Node 10) and contact (Node 11).

Figure 2 shows the model identified the variable presence of patches (presence of patches 2) at second resurveys ($\chi^2 = 104{,}560.48$, $p < 0.001$) in the first stage and classified into two subgroups No (Node 1) and Yes (Node 2). In the second stage, Node 2 identified the variable duration (duration at resurvey 2) at second resurvey ($\chi^2 = 23.35$, $p = 0.001$)and classified into three sub-groups: less than 1879 days (Node 3), between 1879 and 2015 days (Node 4) and more than 2015 days (Node 5). Finally, the Node 5 identified the variable sex ($\chi^2 = 5.39$, $p < 0.02$) and classified into two subgroups male (Node 6 and Node 7).

Table 6 depicts the comparison of the predictive performance of different approaches in terms of sensitivity, specificity, positive predictive values and negative predictive values. The CART model achieved the sensitivity of 100% (95% CI, (99.41–100.00%)), specificity of 99.74% (95% CI, (99.71–99.76%)), positive predictive value of 61.43% (95% CI, (58.56–64.64%)) and negative predictive values 100% (95% CI, 100%). Nevertheless, this study includes the variables like sex, age, presence of BCG scar Contact (leprosy) at home, clinical symptoms, etc., (refer Tables 2 and 3) as the independent variables in both CART model and predicted values determined and portrayed in the Tables 4 and 6. Compared to other diseases, leprosy has a very unique quality that the symptoms may develop after two or more years of infection, so this kind of misdiagnosis may lead to severity of disease and may cause permanent disability. Table 4 shows that the CART model predicted 389 individuals as diseased among those who were observed as non-diseased. We had compared those groups with the actual outcome of the second resurvey in Table 5. Of 389 individuals, 25 predicted as diseased during second resurvey that is the suggested models had predicted these (25 individuals) diseased group nearly two years before (during first resurvey) itself.

As in logistic regression, number of patches, age, overall deformity, duration of resurvey 1 and contact with leprosy (during intake period) which were identified as independent variables were also identified as significant variables by the classification tree in first resurvey (Fig. 1). Table 4 shows the classification between the observed group and the predicted group obtained from the CART model for first resurvey. The

Table 4 Classification of observed and predicted value of first survey through CART Model

Resurvey 1	Observed outcomes		
Predicted value	Diseased	Non-diseased	Total
Diseased	625	**389**	1014
Non-diseased	0	146,787	146,787
Total	625	147,176	147,801

Table 5 Comparison between actual outcome of second resurvey and predicted value of first resurvey

Predicted value of resurvey 1	Outcome of resurvey 2		
	Diseased	Non-diseased	Total
Diseased	**25**	364	389
Non-diseased	347	146,440	146,787
Total	372	146,804	147,176*

*625 diseased from first resurvey has been removed

Table 6 Predictive performance of CART model

Measures	CART model
Sensitivity (95% CI)	100.00% (99.41–100.00%)
Specificity (95% CI)	99.74% (99.71–99.76%)
Positive predictive value (95% CI)	61.43% (58.56–64.64%)
Negative predictive value (95% CI)	100.00% (100.00–100.00%)

CART model predicted 1014 individuals as a diseased category, out of which only 625 individuals were diagnosed as infected with leprosy in the first resurvey and 389 individuals predicted as a diseased category which comes under non-diseased in observed group. Out of these 389 individuals, 25 individuals were observed as disease group in the second resurvey (Table 5). Table 5 shows the comparison between the predicted values of first resurvey (obtained from CART) and actual observed values of second resurvey.

6 Conclusion

Since predicted models are rarely applied in diagnoses of leprosy, current proposed model might be useful in assisting clinicians and public healthcare managers in identifying the suspicious cases and confirm the disease. The CART model could be suggested as a cost-effective tool and appropriate classifiers for large prospective cohort studies such as the current anti-leprosy trial.

References

1. Kanaya AM, Glidden DV, Chambers HF (2001) Identifying pulmonary tuberculosis in patients with negative sputum smear results. Chest 120:349–355
2. Wang C-S, Chen H-C, Chong I-W, Hwang J-J, Huang M-S (2008) Predictors for identifying the most infectious pulmonary tuberculosis patient. J Formos Med Assoc 107(1):13–20
3. Hammermeister KE, De Rouen TA, Dodge HT (1959) Variables predictive of survival in patients with coronary disease selection by univariate and multivariate analyses from the clinical, electrocardiographic, exercise, arteriographic, and quantitative angiographic evaluations. Circulation 59(3):421–430
4. Wilson PWF, D'Agostino RB, Levy D, Belanger AM, Silbershatz H, Kannel WB (1998) Prediction of coronary heart disease using risk factor categories. Circulation 97:1837–1847
5. Breiman L, Friedman J, Olshen R, Stone C (1984) Classification and regression trees. Wadsworth Books, 358
6. de Queiroz Mello FC, do Valle Bastos LG, Soares SLM, Rezende VMC, Conde MB, Chaisson RE, Kritski AL, Ruffino-Netto A, Werneck GL (2006) Predicting smear negative pulmonary tuberculosis with classification trees and Logistic regression: a cross-sectional study. BioMed Central Public Health 6(43):1–8
7. Chaurasia V, Pal S (2013) Early prediction of heart diseases using data mining techniques. Carib J Sci Technol 1:208–217
8. Bhushan P, Kabirsardana S, Koranne RV, Choudhary M, Manjul P(2008) Diagnosing multibacillary leprosy: a comparative evaluation of diagnostic accuracy of slit-skin smear, bacterial index of granuloma and WHO operational classification. Indian J Dermatol Venereol Leprol 74(4):322–326

Metamaterial-Loaded Circularly Polarized Patch Antenna Array for C Band Applications

Chirag Arora

Abstract The article presents a splitted circular ring metamaterial-loaded circularly polarized microstrip patch antenna array. The conventional microstrip patch antenna array possesses the resonant frequency of 5.8 GHz providing the impedance bandwidth of 540 MHz and gain of 9.8 dBi and exhibits linear polarization. However, when this conventional patch antenna array is incorporated with splitted circular ring metamaterial superstrate, then this composite structure exhibits circular polarization at same resonant frequency with three dB axial ratio bandwidth of 390 MHz. Moreover, enhancement in gain by 0.4 dBi is also observed in the proposed metamaterial-loaded patch antenna array possessing circularly polarized.

Keywords Antenna array · Circular polarization · Metamaterial · C band

1 Introduction

Patch antennas exhibiting circular polarization are heavily utilized in many wireless devices, including satellite communication and global positioning system, as they are capable to reduce multipath effect and suppress interference due to rainy and foggy weather [1–4]. They are also capable to improve the channel capacity. Till now, many techniques have been followed to devise circular polarized (CP) antennas [5–18]. Various antennas like helix antennas [5], spiral-shaped antennas [6], magneto-electric dipoles [7], microstrip patch antennas [8–14] and slotted antennas [15–18] have been used as the radiating element for designing the CP antennas. Also, feed networks for the CP antenna arrays are generally designed using coplanar waveguides, microstrip lines, striplines, substrate integrated waveguide (SIW) etc. Thus, the overall structure and weight of a CP antenna results due to the combined effect of radiating element and the feeding network structure. For example, in [5], a 4 × 4 helix radiator with a stripline feed network was designed using LTCC technology. Due to three-dimensional structure of this radiator, the final size of this CP antenna

C. Arora (✉)
KIET Group of Institutions, Delhi-NCR, Ghaziabad, India
e-mail: c_arora2002@yahoo.co.in

R. P. Mahapatra et al. (eds.), *Proceedings of 6th International Conference on Recent Trends in Computing*, Lecture Notes in Networks and Systems 177,
https://doi.org/10.1007/978-981-33-4501-0_6

is about 0.36 times of the wavelength of the smallest frequency of operation and also the cost of fabrication is quite high. In [7], two layers of SIW networks were used to feed 8 by 8 dipole CP antenna array, resulting in high profile structure.

But in modern communication systems, the requirement of low profile antennas is highly essential. For designing the compact CP arrays, planar microstrip patch radiators play important role because of their compact profile. To produce circular polarization, in [19], a square-shaped microstrip patch antenna with circular slots has been proposed. But, the low gain of 0.5 dBic is obtained, along with the thick profile of 4.8 mm. Thus, it is concluded that attainment of circular polarization by using these conventional methods enhances the overall dimensions of antennas and hence make them heavy.

In order to overcome these shortcomings, recently, antenna researchers of entire world are paying attention towards the artificially designed typical structures called metamaterials or zero index materials. Because of their extraordinary properties, they are capable to reduce the size and improve the performance of conventional patch antennas [20–22]. Smith et al., in [23], first time presented the experimental demonstration of these artificially designed double negative materials.

After this, because of their typical characteristics, these artificially designed materials are being used for wide variety of antenna applications. Their unusual electromagnetic behaviour results in reduction of antenna size, performance improvement and most recently, attainment of circular polarization by incorporating them with linearly polarized conventional antennas, without considerable change in overall size of the traditional antenna array [24].

In this article, a metamaterial superstrate based circular polarized microstrip patch antenna array has been explored. The designed antenna array consists of a traditional microstrip patch antenna array and a metamaterial superstrate, consisting of splitted circular rings etched on both sides of the substrate. This metamaterial superstrate is placed on the traditional antenna array at some appropriate height so as to attain the circular polarization. Section 1 of this paper presents the introduction of the article. Section 2 shows the structure of proposed antenna array. Whereas, the simulated results are interpreted in Sect. 3. However, the article is ended in Sect. 4.

2 Design of Metamaterial Superstrate and Traditional & Metamaterial Inspired Microstrip Patch Antenna Array

Figure 1 shows the structural sketch of circular metamaterial unit cell which is used to design the metamaterial superstrate. The thickness of splitted circular metamaterial unit cell is 0.3 mm. Corners of splits subtend the angle of α and β at the centre, the values being 24° and 5°, respectively. The diameter of each circular ring is 6 mm. An array of 3×3 metamaterial unit cells, etched on both sides of the substrate, act as metamaterial superstrate for the traditional microstrip patch antenna array. The sketch of designed metamaterial superstrate is presented in Fig. 2.

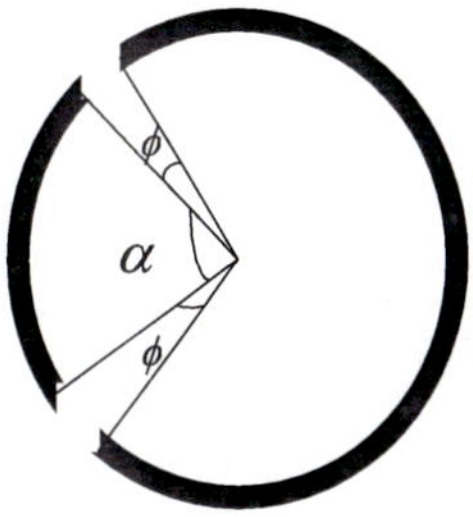

Fig. 1 Top view of splitted circular metamaterial unit cell

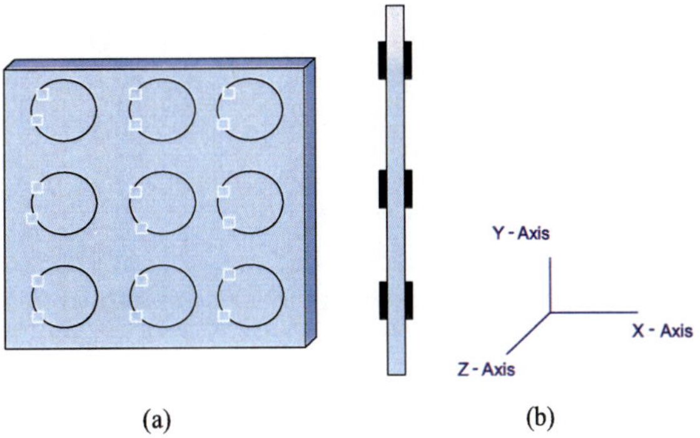

Fig. 2 Proposed splitted circular ring metamaterial superstrate **(a)** top view **(b)** side view

The conventional 4-element linearly polarized microstrip patch antenna array is shown in Fig. 3. The designing of this traditional microstrip patch antenna array is done in such a manner that it resonates at 5.8 GHz. This resonant frequency is generated by a patch of length (*L*) 11.95 mm and width (*W*) 15.88 mm. These dimensions are evaluated using equations of transmission line model [25, 26]. The overall ground plane dimensions are calculated to be 63.88 mm × 63.88 mm. Probe feeding is done to this antenna array by a corporate feed network. The sketch of proposed circularly

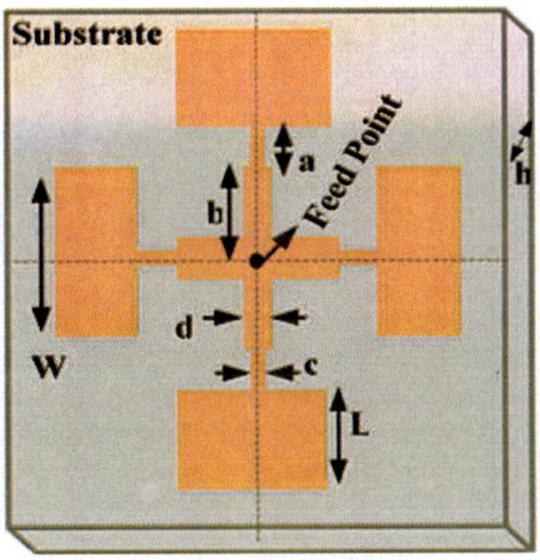

Fig. 3 4-element conventional microstrip patch antenna array

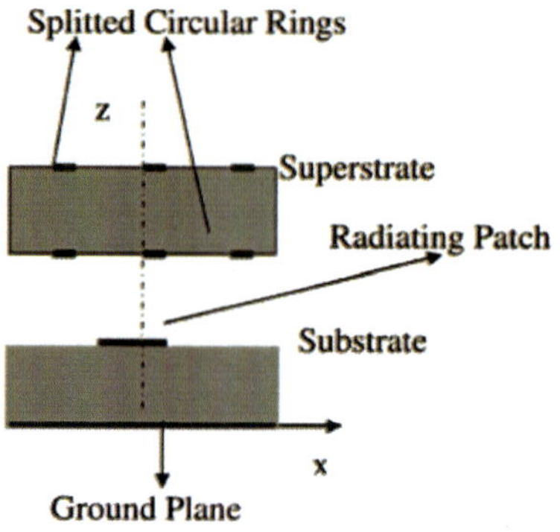

Fig. 4 Circularly polarized microstrip patch antenna array loaded metamaterial superstrate

polarized metamaterial superstrate-loaded microstrip patch antenna array is depicted in Fig. 4.

1.48 mm thick (h) FR-4 substrate is utilized to design this traditional antenna array. The other specifications of the this substrate are as: loss tangent (δ) = 0.01 and dielectric constant (ε_r) = 4.3. These structures are simulated using commercially available electromagnetic software, Ansoft HFSS.

3 Simulation Results and Their Analytical Analysis

Figures 5 and 6, respectively, illustrate the simulated return loss curve (S_{11}) of linearly polarized traditional patch antenna array and the designed circularly polarized splitted circular ring metamaterial superstrate inspired microstrip patch antenna array. From

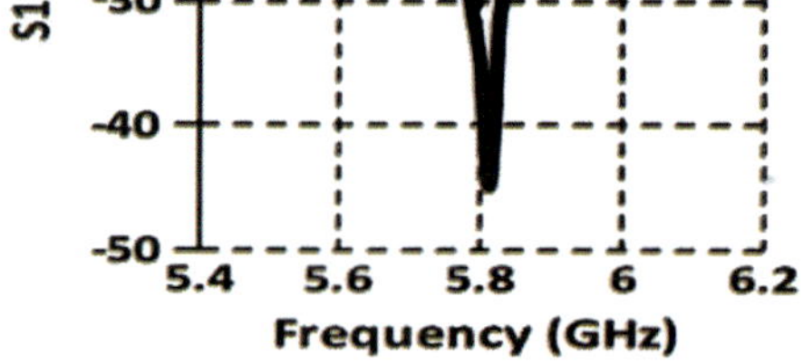

Fig. 5 Simulated return loss curve of traditional linearly polarized microstrip patch antenna array

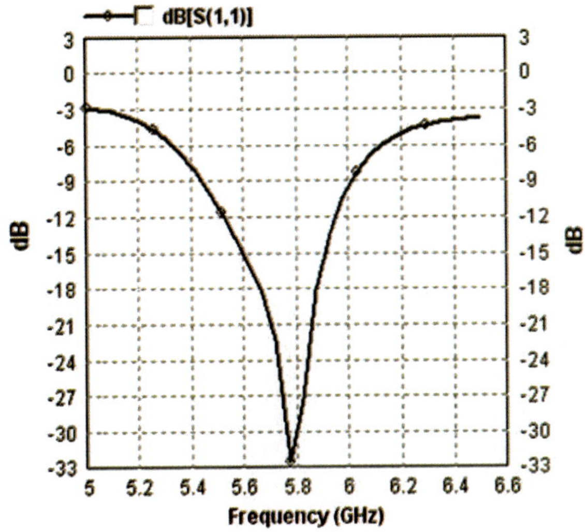

Fig. 6 Return loss curve of proposed circularly polarized patch antenna array

the return loss curves of Figs. 5 and 6, respectively, it is seen that the conventional antenna array has the resonant frequency of 5.8 GHz with impedance bandwidth of 540 MHz and the loaded circularly polarized antenna array resonates at the same resonant frequency with impedance bandwidth of 520 MHz.

Figure 7 depicts the E-plane radiation pattern characteristics of the conventional patch antenna array. It is concluded from this figure that traditional antenna array radiates with overall gain of 9.8 dBi. But, when this conventional patch antenna array is loaded with a metamaterial inspired superstrate, comprising of splitted circular

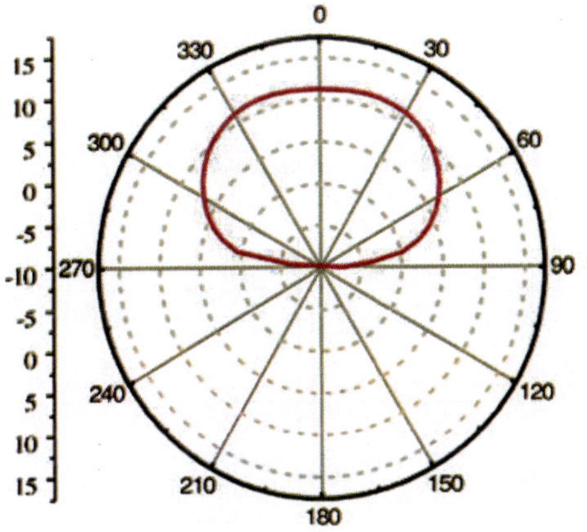

Fig. 7 E-plane radiation pattern of linearly polarized traditional patch antenna array at 5.8 GHz (simulated)

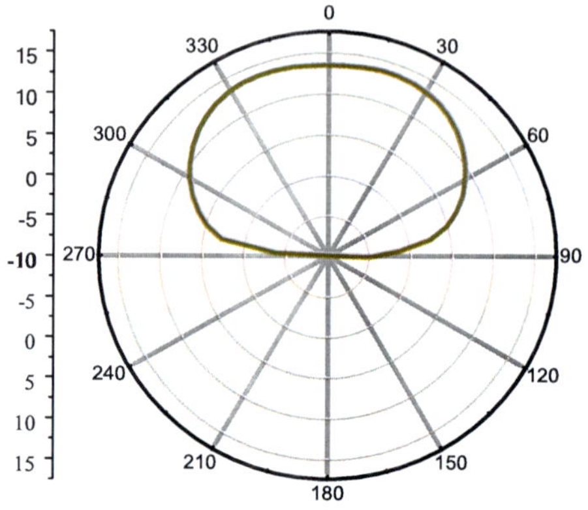

Fig. 8 E-plane radiation pattern of metamaterial inspired circularly polarized patch antenna array at 5.8 GHz (simulated)

ring, the gain of 10.2 dBic is obtained, as observed from the radiation pattern curves of the proposed metamaterial loaded circularly polarized antenna array which is shown in Fig. 8.

To determine the polarization of the traditional and proposed metamaterial inspired patch antenna array, the axial ratio of both the antennas arrays is plotted, as depicted in Fig. 9. As observed from this curve, it is obvious that the designed metamaterial-inspired antenna array possesses the axial ratio below 3 dB, implying that this metamaterial superstrate inspired antenna array is circularly polarized in

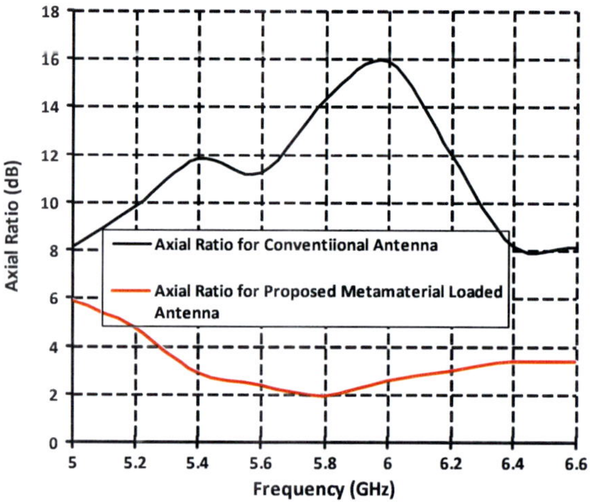

Fig. 9 Axial ratio for conventional and metamaterial superstrate-loaded circular polarized patch antenna array at 5.8 GHz

nature. Also the designed metamaterial-loaded antenna array has the three dB axial ratio bandwidth is 390 MHz.

The improvement in gain and linear to circular polarization conversion can be discussed on the basis of the novel feature of the metamaterials. The metamaterial superstrate is capable to increase aperture efficiency of the traditional antenna array and hence, it focuses all the incoming waves in the broadside direction, and thus enhancing the gain of the designed metamaterial-inspired antenna array. The circular polarization is obtained due to excitation of two orthogonal field components which have same amplitude. These orthogonal components are produced by the two split gaps, present asymmetrically on the circular ring. The split gap of two asymmetrical splits is adjusted to further improve performance of axial ratio.

4 Conclusions

In this article, a typical technique to design a metamaterial inspired circularly polarized antenna array has been presented. This has been achieved by incorporating a metamaterial superstrate, composed of 3×3 splitted circular ring array with a 4-element traditional microstrip patch antenna array. Incorporation of this metamaterial superstrate to the traditional array has been carried out in such a manner that there is no considerable change in the profile and weight of the composite structure of the proposed antenna array. Peculiar nature of metamaterials is utilized to analyse the functioning of this metamaterial loaded antenna array. The designed antenna array will show a good candidature for C band applications. An extended version of this article will contain an equivalent circuit of this composite structure along with the measured results, which will be submitted in an international journal.

References

1. Jeon SI, Kim YW (2000) New active phased array antenna for mobile direct broadcasting satellite reception. IEEE Trans Broadcast 46(1):34–40
2. Weily A, Guo YJ (2009) Circularly polarised ellipse-loaded circular slot array for millimetre-wave WPAN applications. IEEE Trans Antennas Propag 57(10):2862–2870
3. Lin W, Wong H (2014) Circularly polarized conical beam antenna with very thin profile and wide bandwidth. IEEE Trans Antennas Propag 62(12):5974–5982
4. Chu QX, Lin W, Lin WX, Pan ZK (2013) Assembled dual-band broadband quadrifilar helix antennas with compact power divider networks for CNSS application. IEEE Trans Antennas Propag 61(2):516–523
5. Liu C, Guo YX, Bao X, Xiao SQ (2012) 60-GHz LTCC integrated circularly polarized helical antenna array. IEEE Trans Antennas Propag 60(3):1329–1335
6. Zhu Q, Ng K-B, Chan CH (2017) Printed circularly polarized spiral antenna array for millimeter-wave applications. IEEE Trans Antennas Propag 65(2):636–643
7. Luk YLKM (2016) A 60-GHz wideband circularly polarized aperture-coupled magneto-electric dipole antenna array. IEEE Trans Antennas Propag 64(4):1325–1333

8. Guntupalli AB (2014) 60-GHz circularly polarized antenna array made in low-cost fabrication process. IEEE Antennas Wirel Propag Lett 13:864–867
9. Arora C, Pattnaik SS, Baral RN (2018) Dual band microstrip patch antenna array loaded with split ring resonators and via holes. Int J Electron Commun (AEÜ) 93:253–260
10. Li Y, Chen ZN, Qing X, Zhang Z, Xu J, Feng Z (2012) Axial ratio bandwidth enhancement of 60-GHz substrate integrated waveguide-fed circularly polarized LTCC antenna array. IEEE Trans Antennas Propag 60(10):4619–4626
11. Sun H, Guo YX, Wang Z (2013) 60-GHz circularly polarized U slot patch antenna array on LTCC. IEEE Trans Antennas Propag 61(1):430–435
12. Wu J, Cheng YJ, Fan Y (2016) Millimeter-wave wideband high efficiency circularly polarized planar array antenna. IEEE Trans Antennas Propag 64(2):535–542
13. Lai HW, Xue D, Wong H, So KK, Zhang XY (2017) Broadband circularly polarized patch antenna arrays with multiple-layers structure. IEEE Antennas Wirel Propag Lett 16:525–528
14. Arora C, Pattnaik SS, Baral RN (2018) Metamaterial inspired DNG superstrate for performance improvement of microstrip patch antenna array. Int J Microwave Wirel Technol 10(3):317–318
15. Guan DF, Cqian DZP, Zhang YS, Guo YJ, Gong K (2016) Broadband high-gain SIW cavity-backed circular-polarized array antenna. IEEE Trans Antennas Propag 64(4):1493–1497
16. Lang Y, Qu SW, Chen JX (2014) Wideband circularly polarized substrate integrated cavity-backed antenna array. IEEE Antennas Wirel Propag Lett 13:1513–1516
17. González JMF, Padilla P, Domínguez GE, Castañer MS (2011) Lightweight portable planar slot array antenna for satellite communications in X-band. IEEE Antennas Wirel Propag Lett 10:1409–1412
18. Hao ZC, Liu X, Huo X, Fan KK (2015) Planar high-gain circularly polarized element antenna for array applications. IEEE Trans Antennas Propag 63(5):1937–1948
19. Nasimuddin, Chen ZN, Qing X (2010) Asymmetric-circular shaped slotted microstrip antennas for circular polarization and RFID applications. IEEE Trans Antennas Propag 58(12):3821–3828
20. Alu A, Engheta N, Erentok A, Ziolkowski RW (2007) Single negative, double-negative, and low index metamaterials and their electromagnetic applications. IEEE Antennas Propag Mag 49(1):23–36
21. Arora C, Pattnaik SS, Baral RN (2015) SRR inspired microstrip patch antenna array. Prog Electromagn Res C 58(10):89–96
22. Arora C, Pattnaik SS, Baral RN (2015) Microstrip patch antenna array with metamaterial ground plane for Wi-MAX applications. In: Proceedings of Springer second international conference on computer and communication technologies (IC3T-2015), pp 665–671
23. Smith DR et al (2005) A positive future for double negative metamaterials. IEEE Trans Microwave Theory Tech 53(4):1535–1556
24. Rajanna PKT, Rudramuni K, Kandasamy K (2019) A high-gain circularly polarized antenna using zero-index metamaterial. IEEE Antennas Wirel Propag Lett 18(6):1129–1133
25. Garg R, Bhartia P, Bhal I, Ittipiboon A (2001) Microstrip antenna design handbook. Artech House, Boston
26. Balanis CA (2011) Modern antenna handbook. Wiley, New York

Vote Projection Model Based on Past Election Results

Hussain Qudsia Ejaz and Sameena Naaz

Abstract Data analysis is a multi-disciplinary approach for developing mathematical models for analyzing data and doing predictions from it. The aim of this work is to study the past election results from the available data and develop a model that can be used to find the results of upcoming elections. The ups and downs in a particular region or constituency has been analyzed using the proposed algorithm, and the results have been represented by using bar graphs. The steps carried out include data scrapping of the voter list and preprocessing it. The model makes it easy and simple to read every person's data, which includes the caste of the person (caste analysis) and the economic value of the constituency that can be used in support of favorable candidate that can be chosen for a particular constituency.

Keywords Data mining · Predictive analytics · Elections · Caste analysis

H. Q. Ejaz · S. Naaz (✉)
Department of Computer Science and Engineering, School of Engineering Sciences and Technology, Jamia Hamdard, New Delhi 110062, India
e-mail: snaaz@jamiahamdard.ac.in

H. Q. Ejaz
e-mail: rubina986902134@gmail.com

R. P. Mahapatra et al. (eds.), *Proceedings of 6th International Conference on Recent Trends in Computing*, Lecture Notes in Networks and Systems 177,
https://doi.org/10.1007/978-981-33-4501-0_7

1 Introduction

Data science is all about acquiring knowledge and insights from a vast set of data [1]. It consists of various disciplines, including mathematical and statistical modeling, extracting data from its source, and using different data visualization techniques. It often involves incorporating big data techniques to collect both structured and unstructured data.

Data analysis is the process of inspection, modifying, cleaning and modeling data to find information, useful for decision making [2]. The big data tools can be used by political parties for analyzing the data and making their campaigning plans based upon that in order to reach the potential voters. It is important that both candidates and voters alike understand the repercussion of predictive analysis, and big data can have on election results.

In India, the world's largest democracy, studying election is a big challenge due to the size of the country and its population. This work uses the election results of a state, in a particular constituency, and tries to find out which politician's popularity is increasing or decreasing by using the basic algorithm based on predictive analysis. A large number of people connect through social networks like Twitter during the polling season, and their sentiments can also be studied for doing predictions [3].

2 Background and Related Work

A lot of research has been done to study people's voting patterns in predicting the outcome of any election. In [4], the authors formulated a voting pattern based on their closeness to the candidate. The outcome of this study shows that while partisan turnout and incumbent win rates are higher in-house elections, voters do consider ideological contiguity and may penalize candidates who hold positions. The recent work by [5, 6] presents a more hybrid model of voting. While a pure partisan model may predict that an elector will support his or her party candidate, the hybrid model predicts that the partisan bias is not an absolute factor because of spatial proximity and that a candidate may take a position so extreme with respect to his or her opponent that even the voters of his or her own party will not support them.

Election process and caucuses politics of India are different from that of European Parliamentary Democracy [7]. There are several reasons for this: one of them is that the stable and effective alliances at the state level in India are based on the proportionate allocation of seats in the former electoral coalition based on an electoral quota, under which seats are allocated to fellow parties, which they either won, or were the runner-ups, and also when coalition governments share portfolios proportionally [8]. The authors in [9] speak of 2017 assembly elections in Punjab by placing them in the history of state elections and providing a comparative perspective. The election campaign was dominated by issues of development and governance rather than ethnic issues of Punjab, and it saw a 75 percent higher electoral share and was above the national average.

An ethnographic study using big data for analyzing the political discussions in a hybrid media system has been done in [10]. The authors here established that the ethnographic data can be used for the analysis of large datasets, and this can be applied to validate the relation between ethnography and voting pattern. Google trends are also being increasing used in election result and referendum outcome prediction. As suggested by Mavragani and Tsagarakis [11], the use of online data for voting pattern study has increased a lot. The penetration of Internet has increased a lot in our lives [12], and hence, the monitoring and analysis of online data are proving to play a major role in now casting elections.

3 Role of Predictive Analytics in Elections

Big data analytics was used in elections for the first time in 2012 in the US presidential elections [13]. The amount spent for the electoral strategies was approximately $7 billion; out of which 10% was spent on online advertisements of both the parties.

Big data analytics helps the political parties to understand the psyche of their voters on an individual level and not as a mass [14]. With the help of sophisticated tools and superior technology, it is easier to understand the needs of the voters and to approach them based upon it.

In a diverse country like India, most political parties rely on the minority. The voters share their sentiments about the candidates on the social networking sites and show their support for their preferred party and individual. With big data analytics, voters that do not belong to any special minority, can be targeted. The work depicted in this paper is designed to target voters belonging to a particular caste (caste analysis) and the economic value of that region. The constituency is also segregated on the basis of the economic value. Based on the economic value and the caste analysis, the profile of a candidate can be made for that constituency. Therefore, these floating voters can be targeted by a party to gain their votes. Big data analytics could also be used for text and sentiment analysis of data from social network such as Facebook and Twitter [15].

4 Methodology

This work is based upon the nature of the elections held in 2015, 2017, and 2019. The model categorizes each party within a range and depending on that range; the party is declared as strong, favorable, battleground, or weak. This is done using an algorithm which can be modified depending on the nature of the past election. Analysis is done by extracting data and applying an algorithm which consists of a library including various caste (the library is also designed and not taken from any source). Based on the economic value of each constituency and the high number of people of similar caste, the profile of the candidate can be made for that constituency. These two parameters can be considered as foremost apart from the various parameters used in above methods for a win–win election.

4.1 Data Extraction

The data was collected from the election commission of India for the past terms of data spanning from 2015 to 2019 results for the constituency XYZ for the Legislative Assembly Election of 2015, Municipal Corporation election of 2017, and the Lok Sabha election of 2019. The collected data was in the pdf form. The module known as tabula has been used to get it converted to a csv format file.

4.2 Voters List

All the pdfs of a constituency were downloaded from the Delhi electoral website for each area. The pfds included the column names as name of person who has votes, relation, house number, sex, locality details, etc. The data from pdfs was extracted using python libraries like tqdm.

4.3 Caste Analysis

This task does the job of distributing the caste booth-wise and separating out the unidentified surnames. In booth-wise caste distribution, percentage of voters based upon their caste was calculated using previous matching task files and voters list in each area. This task made the use of tqdma and pandas libraries of python.

Voter distribution based on caste for area X and Y has been represented pictorially in Fig. 1.

It can be seen very clearly from this figure that the highest number of voters in area X is general, and the second highest is Brahmin, so the parties can choose a candidate here from the general category specifically Brahmin. Similarly, in area Y, a Muslim candidate can be chosen by the party as he/she has higher chances of winning the election.

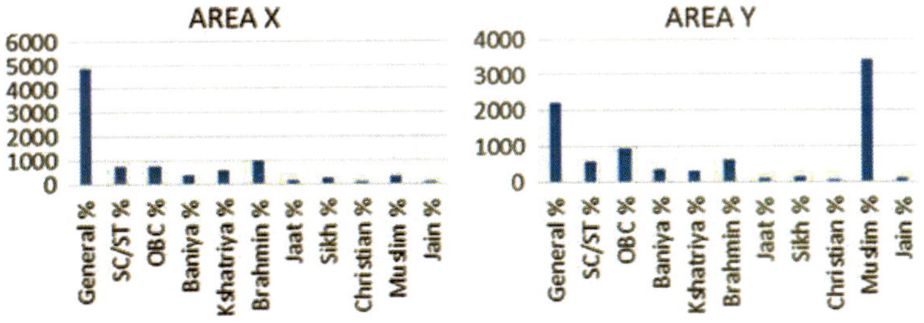

Fig. 1 Voter distribution based on caste for area X and Y

4.4 Booth Matching

The task of booth matching was performed manually as well as with the help of python algorithms. First, the files of 2015 and 2017 were matched on the basis of the address and polling station number, and then, their results were matched with the file of 2019. This combination in different orders was taken in order to get an average value. While booth matching all special characters, abbreviations and spaces were removed.

At the end, three files were obtained which contained the result of elections and polling areas (areas being the same in each file).

4.5 The Economic Status

After matching all the areas for each year, a column known as the economic status column has been added in the file which includes the economic status based on the area and address. The economic status column was classified as: (i) JJ: i.e., slums, (ii) Villa: i.e., village areas, (iii) Govt. flats: i.e., DDA flats, (iv) Society flats: i.e., private flats/CGHS, (v) Kothi: i.e., bungalows/villas.

4.6 Percentage Votes for Each Party

Here, the candidates belonging to each party were recorded. The party-wise columns were added in the result sheets, and the ones with names were deleted. The file included only five party name columns—AAP, BJP, CONGRESS, BSP, and OTHERS.

After this, the percentage of each party votes was calculated. Also, the columns polling number and address from the main sheet were merged together. The Python libraries pandas and numpy were used for this task.

4.7 *Prediction of Election Winner*

In this task, the prediction is made about the party winning the next election area-wise. The pandas, numpy, scikit libraries were used to fetch out the results, and for the visualization, matplotlib library of python was used.

- To make an analysis, there is a score given in total, and each party needs to achieve a score. According to the score associated, each party is then categorized as strong, favorable, battleground, or week.
- To become a strong candidate in any constituency, the party needs to score a maximum score. For, e.g., 60/100.
- Accordingly, a week candidate in any constituency is the one who scores less.
- There exist two conditions that totally depends upon strength and the weakness of our favorable party, that is, A, in that constituency.
- Suppose the scores are taken as 60/100 to become a strong candidate and weak as anywhere between 0 and 10, the areas under battleground are the number of areas in the constituency or booths which lie between strong and weak provided that the weak candidate in that area is our favorable candidate, A.
- The areas under favorable is the number of areas or booths which lie between strong and weak, provided that the strong candidate in those area is our favorable party, A.

5 Algorithm

```
1. Define weights of each year
2015 (0.65), 2017 (0.10), 2019 (0.25)
Len(BJP) = list of BJP numbers
Len(AAP) = list of AAP numbers
Len(INC)= list of INC numbers
2. Start
   AAP
   for i in range(0,len(bjp)):
     if aap[i]>bjp[i] and aap[i]>inc[i]:
       if bjp[i]>inc[i]:
         if (aap[i]-bjp[i])>15:
           'STRONG'

         elif 10<(aap[i]-bjp[i])<=15:
           'FAVOURABLE'

         elif 10>=(aap[i]-bjp[i]):
           'BATTLEGROUND'

         else:
           'CANNOT DECIDE'

       if bjp[i]<inc[i]:
         if (aap[i]-inc[i])>15:
           'STRONG'

         elif 10<(aap[i]-inc[i])<=15:
           'FAVOURABLE'
         elif 10>=(aap[i]-inc[i]):
           'BATTLEGROUND'
         else:
           'CANNOT DECIDE'
     if aap[i]<bjp[i] and aap[i]>inc[i]:
       if (bjp[i]-aap[i])>10:
         'WEAK'
       elif (bjp[i]-aap[i])<=10:
         'BATTLEGROUND'
       else:
         'CANNOT DECIDE'

     if aap[i]>bjp[i] and aap[i]<inc[i]:
       if (inc[i]-aap[i])>10:
         'WEAK'

       elif (inc[i]-aap[i])<=10:
         'BATTLEGROUND'

       else:
         'CANNOT DECIDE'

     if aap[i]<bjp[i] and aap[i]<inc[i]:
       if bjp[i]>inc[i]:
         if (bjp[i]-aap[i])>10:
           'WEAK'

         elif 7<(bjp[i]-aap[i])<=10:
           'BATTLEGROUND'
       else:
           'CANNOT DECIDE'

       if bjp[i]<inc[i]:
         if (inc[i]-aap[i])>10:
           'STRONG'

       elif 7<(inc[i]-aap[i])<=10:
           'BATTLEGROUND'

         else:
           'CANNOT DECIDE'
3. Repeat Same for INC and BJP
```

6 Results and Discussion

The state is divided into 70 constituencies. Each constituency is taken as a sample space. Three different election results are used for making the predictions, and as these are different types of elections, so different weights are associated with each of them.

First, election is the state election, wherein the favorable party, that is, AAP, is the current state party, and hence, the weight associated with state election is taken as maximum. Second is MCD election, wherein the election does not depend on one, single ruling party, be it for the Lok Sabha or Rajya Sabha. Hence, the weight associated with this is the least. Third election is the Rajya Sabha election, wherein AAP wasn't fighting for the seat in Rajya Sabha. Hence, the weight associated with this is an average.

Group analysis was done to synchronize the number of booths of each constituency in all the three election, since the number of booths gets changed in many of the cases. To make an analysis, there is a score given in total, and each party needs to achieve a score. According to the score associated, each party is then categorized as strong, favorable, battleground, or weak.

The results obtained from this algorithm has been plotted for different areas in Figs. 2, 3, 4, and 5. In these figures *X*, axis gives the positions of the three major parties in an area, and *Y* axis is the number of booths in that area. As can be seen from Fig. 2, AAP and BJP are equally strong, whereas INC is the weakest. In area X, AAP is seen to be strongest at most of the polling booths except at a few, where they need to work enough to gain those seats. BJP comes next to it but INC has got no chance to win. This situation is clearly depicted in Fig. 3 for the said area. Figure 4 depicts

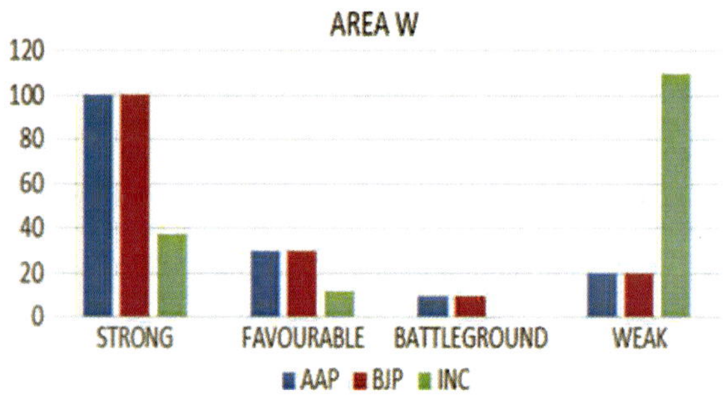

Fig. 2 Prediction of the party position in area W

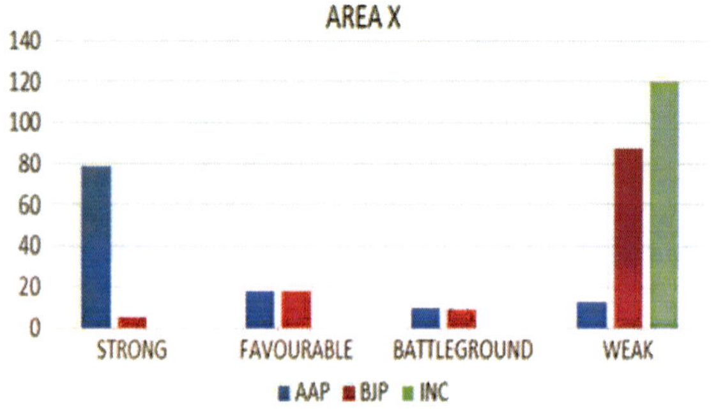

Fig. 3 Prediction of the party position in area X

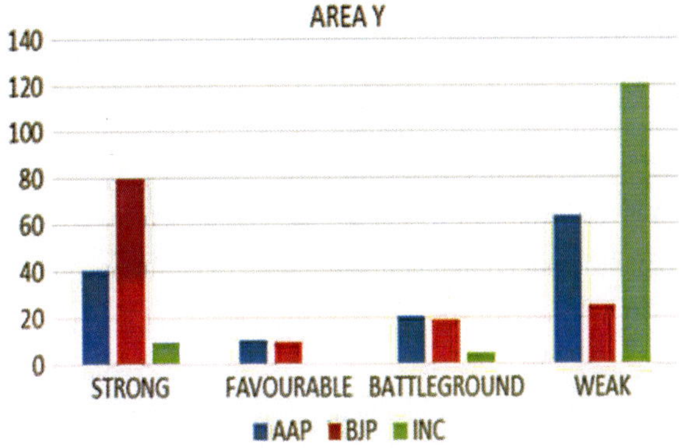

Fig. 4 Prediction of the party position in area W

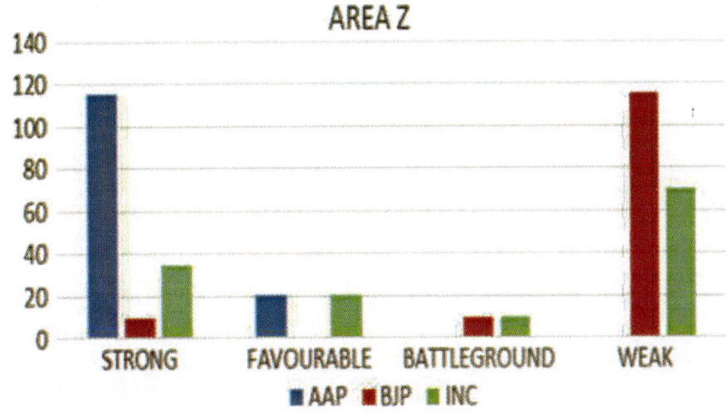

Fig. 5 Prediction of the party position in area W

the situation in area Y. There are 133 booths in this area; out of which, according to the algorithm, BJP has been predicted as the strongest candidate since it holds the highest numbers of votes in 80 booths compared to AAP and INC.

Accordingly, INC is considered to be the weakest. Since AAP got more booths covered than INC, it is in favorable with BJP, and on the other hand, the booths which is in battleground condition, i.e., the number of booths to be considered when AAP is the weaker (considering prediction of AAP since this is the party for which the algorithm is made). The graph shows that INC got the low number votes in almost 90% of the booths, and hence, it is the weakest. So, for this region, if AAP wants to win, then it needs to work harder in this region compared to BJP, and again, INC has negligible chances to win. In area W, as can be seen from Fig. 5 that BJP is weakest, and hence, AAP needs to work more than INC in this region. Also, INC and BJP are seen to be in battleground, and hence, AAP cannot win in such area. Since AAP got higher votes than those in battleground, it still can win in this area.

7 Conclusion

Predictions about the election results can help the parties, and their workers coordinate and campaign in a manner which can help them get closer to a win. Mostly predictions are done on the basis of people's opinion but in this work, it has been done based upon the analysis of the past election results inclusive of the nature of the election. The data has been collected from the election commission of India for the past terms spanning from 2015 to 2019 result and captured it with various features including state, constituency, caste, gender, votes polled, votes majority, and finally party won. Percentage of each caste voted in each year and for each booth was calculated using libraries of python. The visualization part for every topic was covered with the help of python visualization libraries. Finally, after running the entire model, the result was stored in the .csv format. The results obtained can be used to streamline the future course of action by the parties in choosing the candidates and working in a better manner of gain people support for any area in which they are lagging.

References

1. Dhar V (2012) Data science and prediction
2. Wimmer A, Cederman L-E, Min B (2009) Ethnic politics and armed conflict: a configurational analysis of a new global data set. Am Sociol Rev 74(2):316–337
3. Smith A (2011) 22% of online Americans used social networking or Twitter for politics in 2010 campaign. Rep Pew Internet Res Cent
4. Simas EN (2013) Proximity voting in the 2010 US House elections. Electoral Stud 32(4):708–717
5. Jessee SA (2009) Spatial voting in the 2004 presidential election. Am Polit Sci Rev 103(1):59–81
6. Jessee SA (2010) Partisan bias, political information and spatial voting in the 2008 presidential election. J Polit 72(2):327–340
7. Farooqui A, Sridharan E (2017) Uttar Pradesh elections 2017. Econ Polit Wkly 52(15):17
8. Sridharan E (2014) Coalition politics in India: selected issues at the centre and the states. Academic Foundation
9. Kumar A, Mahajan K, Kaur H (2018) The 2017 assembly elections in Punjab: emergence of a triangular electoral system. J Punjab Stud 25(1)
10. Laaksonen S-M, Nelimarkka M, Tuokko M, Marttila M, Kekkonen A, Villi M (2017) Working the fields of big data: using big-data-augmented online ethnography to study candidate–candidate interaction at election time. J Inf Technol Polit 14(2):110–131
11. Mavragani A, Tsagarakis KP (2016) YES or NO: predicting the 2015 GReferendum results using Google Trends. Technol Forecast Soc Change 109:1–5
12. Wagner SA, Vogt S, Kabst R (2016) The future of public participation: empirical analysis from the viewpoint of policy-makers. Technol Forecast Soc Change 106:65–73
13. Wang H, Can D, Kazemzadeh A, Bar F, Narayanan S (2012) A system for real-time twitter sentiment analysis of 2012 us presidential election cycle. In: Proceedings of the ACL 2012 system demonstrations, pp 115–120
14. Tufekci Z (2014) Engineering the public: big data, surveillance and computational politics. First Monday 19(7)
15. Mehndiratta P, Sachdeva S, Sachdeva P, Sehgal Y (2014) Elections again, twitter may help!!! A large scale study for predicting election results using twitter. In: International conference on big data analytics, pp 133–144

Assessment of the Impact of Publication Year and Keywords in Research Manuscript Quality Using Fuzzy Inference System

Shivangi Agarwal and Ashish Agrawal

Abstract The start of good research is always through the good literature obtained in support of it. In today's scenario, the researcher's job of selecting the relevant work is a major complex task. The number of research works in the form of a research paper, review paper, etc., is abundantly available. There are various tools and algorithms available which are used in selecting and ranking the scientific publications. This paper mainly focuses on the impact of publication year and keywords on the selection of a research manuscript. The authors in this paper have proposed a fuzzy logic system for evaluating the impact of these parameters on the quality of the manuscript. The system developed was tested on the primary data collected by downloading the multiple research papers from various journals. The results obtained showed that the proposed system is efficiently supporting the objectives and correctly evaluating the impact of these parameters.

Keywords Scientific publication · Publication year · Keywords · Research manuscript · Quality assessment · Fuzzy logic

1 Introduction

Every year thousands of papers published in many conferences and journals [1]. The topics of the research papers span in almost every area of research. The complexity of research work is increased, so now, the selection of right and important literature has to be done from such a large repository of papers. Selection from such a large pool is always done based on some parameters. These parameters help in extracting the relevant work.

S. Agarwal · A. Agrawal (✉)
SRMSCET, Bareilly, India
e-mail: agarwal.ashish01@gmail.com

S. Agarwal
e-mail: shivangiagarwal4196@gmail.com

R. P. Mahapatra et al. (eds.), *Proceedings of 6th International Conference on Recent Trends in Computing*, Lecture Notes in Networks and Systems 177,
https://doi.org/10.1007/978-981-33-4501-0_8

The quality of a research paper is normally judged by the impact factor of the journal or conference in which it has been published. It sometimes becomes a myth that a journal having a good impact factor will always contain a high-quality paper [2]. Many times the less recognized publications also contain highly conceptual and quality works. Apart from the impact factor, some researcher focuses on the citation count, that is, how many times the paper has been referred by any other author. No doubt it is a quality measure but looking only at the citation value does not tell anything about the work. Suppose any article is having high citation count but it has been published 20 years back, normally it is not advisable to refer such an old paper in the literature review.

In this paper, the authors' emphasize on the importance of publication year and the keywords present in a paper as an important metric in judging its quality. Whenever a researcher searches any research paper, the first question that came to his mind is "What is the publication year of this work?" [3] This question is considered very important in defining the value of any work. If any new researcher is writing a research paper, he is always advised to search the current and latest articles related to his topic of research. This assures him that the work which he is doing has not already been done by anyone else at some other place. The research only gained recognition in society when something new is introduced.

Similarly, the keywords are also very important. Keywords act as a password for getting the information. Good researchers always keep themselves updated by regularly reading the latest articles, but sometimes, the papers get missed because they might not be published in good journal or researcher does not frequently read the articles of that particular journal in which it was published.

Every researcher always wants his work to be read by other researchers and get maximum citation count. Here, the role of keywords arises. Keywords help in searching the relevant papers for researchers. There are multiple situations where titles fail to indicate the relevance of papers. Through the title, the exact idea about the work done in the paper cannot be fully reflected. The selection of the appropriate and a large number of keywords directly relate to the quality of the manuscript. This keyword count helps the indexers and search engines to quickly locate the paper. The selection of good keywords makes the paper appear at the top of the search list. The possibility of achieving higher citation count also increases.

In Sect. 2, the literature available related to this concept is described. Further, in Sect. 3, the research methodology proposed is discussed, and the dataset on which working is performed is also mentioned. Following it in Sect. 4, the implementation is stated, and the results obtained are shown in the form of a table. Finally, the conclusions drawn are mentioned in Sect. 5.

2 Literature Review

The attempt of collecting the literature for any area is a difficult task. Some of the work done by the researchers in this field is as follows:

Authors in paper [4] primarily focus on the use of supervised learning algorithms giving ranking to scientific papers automatically on the basis of key parameters such as keywords, publication type, publication year and citation. Authors in paper [5] discussed the major problems that researchers are facing in India. The disadvantages of the extensive use of impact factor and h-index as a quality research measure were explained.

The methodology proposed by the authors in paper [6] was composed of nine phases. The methodology was used for ranking the scientific papers according to their relevance. They mainly emphasized on changing the values of alpha element and determining the quality. The main reasons affecting the research productivity are described in the paper [7]. The major impact of publication year on the relevance of scientific publication is clearly described in the paper [8]. The authors implemented a Methodi Ordinatio methodology in nine phases for obtaining the results. The increase in research work impacts the reputation of an institute [9].

In paper [10], author has described how a fuzzy logic system can be made while evaluating any manuscript. They proposed a model for obtaining the decision of selection or rejection of the manuscript on behalf of reviewers and editors. In paper [11], authors focused on the increase in quality research over quantity. They evaluated the contributions of researchers through influence rather than the research work they produced. The authors in paper [12] developed a fuzzy logic-based system estimating the quality of drinking water. The fuzzy inference system was made, and inputs are the indicators that affect the quality of drinking water.

The author in [13] focuses on the importance and role of keywords in retrieving the essential information. The correct and accurate use of keywords enhances the retrieval speed of the literature. Reading all the papers completely is a very difficult task. So, the keywords act as a filter in selecting the appropriate literature. Keywords reflect the area of interest.

3 Methodology

The selection of the parameters publication year and number of keywords was done by the authors through conducting a survey. The aim of the survey was to collect the level of importance for different parameters, and eventually, publication year and keywords were turned out to be most important parameter as per the response given by respondents. These are some of the essential parameters which are responsible for estimating the quality of a published research manuscript. The survey was circulated among the researchers through both online/offline modes. The majority of respondents were postgraduates and doctorates. They gave high priority to parameters like publication year and the number of keywords among all other options given for research manuscript quality evaluation.

After the parameter selection, the next step was to decide their range. Here, for this model, the range of publication year was taken as 1–20, and for the number of keywords, it was 1–10. The range for publication year 1–20 describes that users can consider the research manuscript of the last 20 years only for measuring its quality. The value to be entered as input was directly the number of years in which research manuscript was published. For example, if the publication year was 2018, then 18 was the value for the first input.

After the study of various journal templates and conference templates, it was observed that a maximum of ten keywords can be taken by the researcher in his work. That is how the range of 1–10 has been taken. As we know, 11 built-in membership functions are present in the fuzzy logic toolbox. The Gaussian membership function was used in this work because it gives smooth curves [9] and non-zero values at every point. The formula of Gaussian membership function is

$$\mu_{\mathrm{A}}^{i}(x) = \exp\left(-\frac{(c_i - x)^2}{2\sigma_i^2}\right)$$

The next step involves the construction of a rule base. Membership function values for parameters are defined by five linguistic terms. The linguistic variables are taken for each membership function of the input publication year and keywords, and also, for the output, publication quality is tabulated in Table 1 shown along with the categorization of their ranges.

A total of 25 if-then-rules are made following the concept of truth logic. Figure 1 shows the if-then-rule editor of MATLAB Fuzzy logic Toolbox. In if-then-rule editor, we can add new if-then-rules, delete the previously made if-then-rules and could also make changes in the existing ones.

Table 1 Linguistic variables along with the ranges taken

Parameter	Linguistic variable	Range
1. Publication year	• Very old • Old • Average • New • Very new	1–4 5–8 9–12 13–16 17–20
2. Number of keywords	• Very less • Less • Average • Much • Very much	1–2 3–4 5–6 7–8 9–10
3. Publication quality	• Very poor • Poor • Average • Good • Very good	1–5 6–10 11–15 16–20 21–25

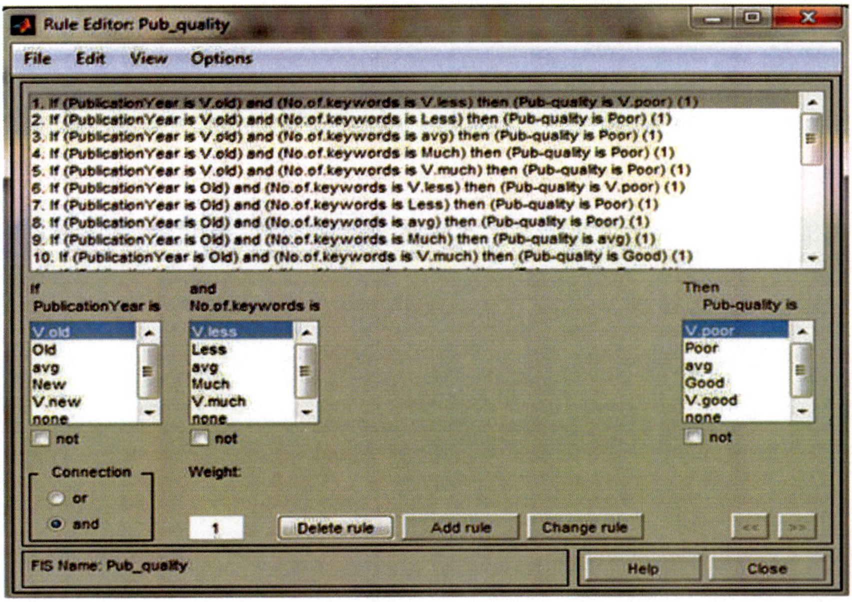

Fig. 1 If-then-rule editor

4 Implementation and Results

The centroid method was used in this study for defuzzifying the output values. The value of publication quality obtained for different inputs was recorded in Table 2. When the value of these two input parameters will be entered by the user, a publication quality will be obtained. The dataset of 15 research papers was taken for this work. These research papers were downloaded from research publication data repositories such as IEEE Explore, Elsevier and Academia. All the papers were published in different Journals/Conferences. Each paper taken belongs to the different fields of engineering.

Table 2 shows the complete list of outputs of the manuscripts taken as a dataset (due to the confidential reasons, the name of the papers taken for dataset is not disclosed, but they are referenced at the end of the paper in Appendix). For example, considering the first case where the input values were [17, 4] (P1) then theoretically AVERAGE should be the answer. The value obtained by the system was 14.9 which lies in the range of AVERAGE according to the membership function which was taken for output. The decision about the quality of these research manuscripts is based on these parameters only. These research manuscripts may have good quality in some other extent.

Table 2 Results obtained

S. No.	Title	Year of publication	No. of keywords	Publication quality
1	P1	2017	4	14.9
2	P2	2018	1	10.6
3	P3	2019	5	17.6
4	P4	2012	4	11.2
5	P5	2018	4	15
6	P6	2017	5	17.4
7	P7	2014	7	17.3
8	P8	2019	8	19
9	P9	2005	5	7.31
10	P10	2016	8	18.7
11	P11	2015	7	18.2
12	P12	2019	5	17.6
13	P13	2008	7	10.9
14	P14	2018	5	17.3
15	P15	2018	3	12.3

5 Conclusions

The results obtained from this work conclude that publication year and keywords affect the quality of a manuscript. The paper published so many years ago may find as less useable for current use. The research manuscripts having the number of keywords more than four result in better quality as compared to research manuscripts that have less than four keywords. As search engine optimization works on the number of keywords, if a research manuscript has less number of relevant keywords, then the chances for being in the search results list for that research manuscript get reduced. The newly published paper will be having more advanced content as compared to the older ones. The results, calculated through the proposed fuzzy inference system, the objectives defined by the authors are fulfilled completely. The details of papers taken for research work (as dataset) are as follows:

6 Appendix: Research Manuscripts Taken for Data Analysis

[P1] Huisman, J., Smits J.: Duration and quality of the peer review process: the author's perspective. Scientometrics.113: 633 (2017) https://doi.org/10.1007/s11192-017-2310-5.

[P2] Muthu Madhan., Subbiah Gunasekaran., Subbiah Arunachalam.: Evaluation of research in India—are we doing it right?. (2018) https://doi.org/10.20529/IJME.2018.024.

[P3] W. Sygocki., Korzeniewska E.: How much is Impact Factor worth? Applications of Electromagnetics in Modern Engineering and Medicine (PTZE). (2019) https://doi.org/10.23919/ptze.2019.8781744.

[P4] Nidhi Jain Kansal., Inderveer Chana.: Cloud Load Balancing Techniques: A Step To-wards Green Computing IJCSI. Vol. 9, Issue 1 (2012).

[P5] Shobha N S., Dr. Subramanya K N.: A Review of Kanban based Manufacturing Systems. (IJRSI) Volume V, Issue IV, April (2018) ISSN 2321–2705.

[P6] Mădalina Manole., Mihai Șerban Avramescu.: A Comparative Analysis of Agile Project Management Tools. Economy Informatics vol.17. (2017).

[P7] Eshetie Berhan., Ajith Abraham.: Hierarchical Fuzzy Logic System for Manuscript Evaluation. ISSN1 990–9233, IDOSI Publications (2014) https://doi.org/10.5829/idosi.mejsr.2014.19.9.11886.

[P8] Sai Kireet Patri., Elena Grigoreva., Wolfgang Kellerer., Carmen Mas Machuca.: Rational Agent Based Decision Algorithm for Strategic Converged Network Migration Planning. Opt. Commun. Network Vol. 11, No. 7(2019).

[P9] Yi Tan., Guo Ji Zhang.: The Application of Machine Learning algorithm In Underwriting Process, Proceedings of the Fourth International Conference on Machine Learning and Cybernetics. Guangzhou, 18–21 (2005).

[P10] Ik Soo Kwon., Jae-Sang Hwang., Jae-Hong Koo., Jin-Yong Na., Bang Wook Lee Member IEEE.: Comparison of the Electrical Conductivity of Polypropylene Laminated Paper (PPLP) and Kraft in LN2 According to the Number of Layers, 1051–8223 (c) (2015).

[P11] Ke Gu., Guangtao Zhai., et.al,.: No Reference Image Sharpness Assessment in Autoregressive Parameter Space. IEEE Transactions on Image Processing, Vol. 24, No. 10, October (2015).

[P12] Pingping Xie., Jizhong Zhu., Jin Zou., Peizheng Xuan.: Uncertainty Analysis of Wind Power Fluctuations in Yunnan Power Grid. IEEE PES Innovative Smart Grid Technologies Asia (2019).

[P13] James P Lyons, Sr. Member IEEE, Michael C. Robinson, Paul Veers, and Robert W. Thresher. Wind Turbine Technology—The Path to 20% US Electrical Energy. IEEE Power and Energy Society General Meeting—Conversion and Delivery of Electrical Energy in the 21st Century (2008).

[P14] Kheder Haddouche, S., Boudia, S. M Feasibility Study of a Wind Farm in El Golea Region in the Algerian Sahara. IRSEC(2018). https://doi.org/10.1109/irsec.2018.8702902.

[P15] Kivimaa, P., Boon, W., Hyysalo, S., Klerkx, L.: Towards a typology of intermediaries in sustainability transitions: A systematic review and a research agenda. Research Policy (2018) https://doi.org/10.1016/j.respol.2018.10.006.

References

1. SandstroÈm U, van den Besselaar P (2016) Quantity and/or quality? The importance of publishing many papers. PLoS ONE 11(11):e0166149. https://doi.org/10.1371/journal.pone.0166149
2. Sygocki W, Korzeniewska E (2019) How much is impact factor worth? Appl Electromagn Mod Eng Med (PTZE). https://doi.org/10.23919/ptze.2019.8781744
3. https://falconediting.com/en/blog/6-tips-for-choosing-keywords-for-your-scientific-manuscript
4. El Mohadab M, Bouikhalene B, Safi S (2018) Predicting rank for scientific research papers using supervised learning. Appl Comput Inf https://doi.org/10.1016/j.aci.2018.02.002
5. Madhan M, Gunasekaran S, Arunachalam S (2018) Evaluation of research in India—are we doing it right? Indian J Med Ethics https://doi.org/10.20529/IJME.2018.024
6. de Campos EAR, Pagani RN et al (2018) Construction and qualitative assessment of a bibliographic portfolio using the methodology Methodi Ordinatio. Scientometrics. https://doi.org/10.1007/s11192-018-2798-3
7. Jung H, Seo I, Kim J, Kim BK (2017) Factors affecting government-funded research quality. Asian J Technol Innov 25(3):447–469. https://doi.org/10.1080/19761597.2018.1436411
8. Pagani RN, Kovaleski JL, Resend LM (2015) Methodi Ordinatio: a proposed methodology to select and rank relevant scientific papers encompassing the impact factor, number of citation, and year of publication. https://doi.org/10.1007/s11192-015-1744-x
9. Wollersheim J, Lenz A, Welpe IM, Spörrle M (2015) Me, myself, and my university: a multilevel analysis of individual and institutional determinants of academic performance. J Bus Econ 85(3):263–291
10. Berhan E, Abraham A (2014) Hierarchical fuzzy logic system for manuscript evaluation. IDOSI Publications. https://doi.org/10.5829/idosi.mejsr.2014.19.9.11886. ISSN 1990-9233
11. Donaldson MR, Cooke SJ (2014) Scientific publications: moving beyond quality and quantity toward influence. University of Waikato Library
12. Tayal T, Prema KV (2013) Drinking water quality estimation using fuzzy logic technique. Int J Res Advent Technol 1(5)
13. Grant MJ (2010) Key words and their role in information retrieval. Health Inf Librar J. https://doi.org/10.1111/j.1471-1842.2010.00904.x
14. https://www.scribbr.com/tips/how-do-you-determine-the-quality-of-a-journal-article/

Adaptive Threshold-Based Approach for Facial Muscle Activity Detection in Silent Speech EMG Recording

Rajendra Kachhwaha, Ajay P. Vyas, and Rajesh Bhadada

Abstract Nature blessed us with one of its best gifts, the human voice, which is an essential tool for vocal communication with others. During silent speech generation, few facial muscles play a vital role. An electromyographic signal is generated around these muscles, which can be captured, processed, and used as communication. The goal of this paper is to automate the signal files reading and the preprocessing module for surface EMG speech signal, which is captured from the facial muscles and saved on disk by LabVIEW software in.tdms file format. An adaptive threshold-based approach is used for detecting facial muscle activity during the silent speech, across random interval recordings in multiple files, where each file has two-channel EMG recordings. With our approach, it is noticed that zygomaticus major muscles show less activity as compared to levator anguli oris muscle for silent speech communication purposes.

Keywords sEMG · Silent speech · TDMS data · Muscle activity detection

1 Introduction

We, humans, are blessed with one of the best gifts of nature, the human voice, which is an essential tool for vocal communication with others and self-expression. The speech production process involves selective control of facial and neck muscles in conjunction with larynx, glottis, vocal cords, tongue, and pharynx [1]. In speech generation, few facial muscles, mainly zygomaticus major, levator anguli oris, masseter, and orbicularis oris, get contracted [1, 2]. As a result of this muscle contraction,

R. Kachhwaha (✉) · A. P. Vyas · R. Bhadada
M.B.M. Engineering College, Jai Narain Vyas University, Jodhpur, India
e-mail: rajendra1983@gmail.com

A. P. Vyas
e-mail: ajaypvyas@gmail.com

R. Bhadada
e-mail: rajesh_bhadada@rediffmail.com

R. P. Mahapatra et al. (eds.), *Proceedings of 6th International Conference on Recent Trends in Computing*, Lecture Notes in Networks and Systems 177,
https://doi.org/10.1007/978-981-33-4501-0_9

an electromyographic signal is induced around these muscles. Using a non-invasive recording approach, this generated signal on facial skin is captured, and this process is known as surface electromyography (sEMG). sEMG plays a vital role in silent speech communication as there is no sound generation. Using sEMG on facial muscles, the EMG signal can be captured and further processed for speech recognition purposes. At the surface of facial skin, the amplitude of this EMG signal is quite low (mV or μV) and also ranges in lower frequency (up to 500 Hz), but the dominant energy concentrates around 50–150 Hz [1–3]. This work is motivated by the requirement of a computer-based automated method for signal preprocessing and speech segment detection in real time, which is identified by Vyas and Bhadada [2], Špulák et al. [4]. This work is the first module of that system, which will be used for silent speech communication purposes. In this work, that automated process is achieved up to a level, which includes preprocessing of the signal captured from facial muscles and detection of facial muscle activity during silent speech using adaptive thresholding-based approach. For the detection of facial muscle activity, we used the 'mean energy' feature of the extracted frame from the signal [5–8]. The paper is organized as follows: The details about methods used are presented in Sect. 2, which includes hardware to acquire EMG signal, its preprocessing at the software level and adaptive threshold-based approach, input data & experimental results using our approach are reported in Sect. 3. Finally, conclusions are given in Sect. 4.

2 Methods Used

In this section, we describe hardware used to acquire EMG signal; it is preprocessing at the software level and adaptive threshold-based approach.

2.1 Hardware Used to Acquire EMG Signal

For data acquisition from facial muscles, we used the electronic hardware developed by Vyas and Bhadada [2]. This hardware is capable of acquiring EMG signals on two channels. A USB-6009 DAQ card from National Instrument is used for signal amplification and conditioning purposes. Further, that hardware processed signal is recorded using LabVIEW biomedical toolkit. The surface EMG signals for five alphabets *i*, *m*, *p*, *u*, *x* are captured from two facial muscles, namely zygomaticus major (channel 1) and levator anguli oris (channel 2) on the left side of the human face. At the same time, the reference electrode is placed at the forehead [1, 2]. The recording of EMG signals is performed at a sampling rate of 2000 Hz using a five lead cables with snap type connectors and pre-gelled surface EMG electrodes [2].

2.2 Signal Preprocessing at a Software Level

To achieve automation at the software level, we tried to automate the EMG signal file reading module and the preprocessing module. The facial muscle activity during silent speech is captured and saved on disk by LabVIEW software in .tdms file format. All such files must be kept at a particular location. At a very first step, we initialize all required parameters like *file_path*, which holds the path of all .tdms files; *sample_freq* = 2000, which is the sampling frequency at which signals recorded; *tdms_group* = 'biosignal', which contain the name of channel groupings; *channel_no* = 2, is the actual number of the channel used; *channel_id* = ['Dev_ai0', 'Dev_ai1'], which is the ids for the two channels used; *window* = 1000, is the size of signal extracted and processed at a time; *stride* = 100, is the value by which we move toward the end of the signal; *factor* = 10.0, is used to increase lower energy values; *thres* = 0.5, is used to detect muscle activity; *signal_no* = 5, is the number of muscle activities recorded. Once these parameters get initialized for the input module, our program can directly read those files. A variable named *signal_index* is used to hold the index values of actual muscle activities identified by our approach. For automation of preprocessing of EMG signal by software, as per [2, 6], we apply a band-pass filter (with order = 3, lowcut = 50.0 Hz and highcut = 150.0 Hz) to remove noise, mean zero for detrending purpose, and rectification to process only positive part of the signal read using a Python-based program.

2.3 Adaptive Threshold-Based Approach

Here, we are detecting facial muscle activity in the read signal files, from pre-defined disk location, using adaptive thresholding-based approach. For this, we require variables like *window*, *stride*, *factor*, *signal_no*, *thres*, which were initialized at the input module. Now, a segment of size equal to the *window* is extracted from the read signal, and its 'mean energy' is computed and stored in variable *eng*. If the calculated mean energy of this extracted segment (*eng*) is above the threshold variable value (*thres*), we make a count. A variable *stride* is used to move toward the end of the signal. It is noticed that sometimes the energy of the extracted segment is quite low, in the case of alphabet '*p*'. For detecting such a lower energy segment, we used another variable *factor*. This variable is multiplied to energy variable *eng* to increase the energy of that segment. We repeat this process until we not identified samples equal to the given sample number (*signal_no*) over the entire range of signal read. If we do not detect samples equal to *signal_no* over the entire signal read, we make an additive increase of 0.5 in *thres* variable, a multiplicative increase of 10 in *factor* variable, and repeat the process. A single file of the EMG signal is processed over a range of 0.5–3.0 for *thres* variable and 10–10,000 for *factor* variable for detecting muscle activity.

3 Input Data and Results

The surface EMG signals for five alphabets *i*, *m*, *p*, *u*, *x* are captured from two facial muscles, namely zygomaticus major on channel one and levator anguli oris on channel two. Table 1 summarizes the input details such as file name, alphabet, sample length, number of channels, and total muscle activity recorded during the silent speech.

Our approach can detect facial muscle activity during the silent speech, from a given signal, with their index value. Table 2 illustrates the values of energy threshold *thres* and *factor* variables used at each channel for each alphabet for detecting muscle activity with its processing time and muscle activity segment.

The testing has been carried on Intel Core i3, 1.8 GHz with 4 GB RAM, having Python, Jupyter notebook under Anaconda environment. We also used various Python packages like numpy, npTDMS, matplotlib, time, scipy [9–13]. For testing, the same alphabet was spoken five times with one recording span of approx 15 s. The same alphabet was mimicked (as if spoken) at approx same interval within 15 s, i.e., nearly every 3 s.

3.1 *Identification of Muscle Activity*

By using our approach, we can detect a muscle activity during silent speech with a segment of size 1000 samples (0.50 s) with a *stride* of 100 samples (0.050 s) in speech EMG signal. Figure 1 illustrates the plots of channel one and channel two for the alphabet '*i*' from file i_01.tdms. Each figure depicts four different plots for (i) RAW EMG signal, which represents signal directly taken from the subject whereas

Table 1 Input files details

File name	Alphabet	Sample length	No. of channels	No. of muscle activities per channel	Total muscle activities
i_01.tdms	*I*	30,400	2	5	10
i_02.tdms	*I*	30,200	2	5	10
m_01.tdms	*M*	30,200	2	5	10
m_02.tdms	*M*	30,600	2	5	10
p_01.tdms	*p*	30,200	2	5	10
p_02.tdms	*p*	30,400	2	5	10
u_01.tdms	*u*	30,200	2	5	10
u_02.tdms	*u*	30,200	2	5	10
x_01.tdms	*x*	30,400	2	5	10
x_02.tdms	*x*	30,600	2	5	10

Table 2 Channel-wise muscle activity detection

File name	Processing time (s)	Value of *eng*	Value of *factor*	Channel no: index range of muscle activity detection
i_01.tdms	0.254	1.0	1000.0	1: [5800–6800,11,800–12,800,19,700–20,700,24,400–25,400,29,100–30100]
		0.5	100.0	2: [6000–7000,11,900–12,900,19,700–20,700,24,300–25,300,29,100–30100]
i_02.tdms	0.144	0.5	1000.0	1: [3900–4900,8800–9800,14,500–15,500,20,300–21,300,25,900–26900]
		0.5	100.0	2: [4100–5100,8800–9800,14,500–15,500,20,300–21,300,26,100–27100]
m_01.tdms	0.136	0.5	100.0	1: [5100–6100,11,200–12,200,17,600–18,600,23,100–24,100,28,200–29200]
		0.5	10.0	2: [5100–6100,11,200–12,200,17,400–18,400,23,200–24,200,28,200–29200]
m_02.tdms	0.232	1.0	1000.0	1: [5500–6500,10,800–11,800,16,800–17,800,22,200–23,200,27,300–28300]
		0.5	100.0	2: [5500–6500,10,800–11,800,16,800–17,800,22,000–23,000,27,300–28300]
p_01.tdms	0.399	1.5	1000.0	1: [4500–5500,10,400–11,400,17,000–18,000,23,000–24,000,27,600–28600]
		1.0	1000.0	2: [4500–5500,10,500–11,500,17,000–18,000,23,000–24,000,27,700–28700]
p_02.tdms	0.180	0.5	1000.0	1: [5600–6600,9900–10,900,15,700–16,700,21,000–22,000,27,200–28200]
		0.5	1000.0	2: [5700–6700,10,100–11,100,15,900–16,900,21,400–22,400,27,200–28200]
u_01.tdms	0.226	1.0	1000.0	1: [3800–4800,10,400–11,400,16,900–17,900,21,800–22,800,26,300–27300]
		0.5	100.0	2: [3800–4800,10,300–11,300,16,900–17,900,22,000–23,000,26,300–27300]
u_02.tdms	0.145	0.5	1000.0	1: [6600–7600,11,100–12,100,16,300–17,300,22,100–23,100,26,400–27400]
		0.5	100.0	2: [6500–7500,11,100–12,100,16,400–17,400,21,800–22,800,26,400–27400]
x_01.tdms	0.136	0.5	100.0	1: [5300–6300,11,400–12,400,18,100–19,100,23,300–24,300,28,200–29200]
		0.5	100.0	2: [5300–6300,11,400–12,400,18,200–19,200,23,200–24,200,28,200–29200]
x_02.tdms	0.467	2.0	1000.0	1: [5700–6700,11,200–12,200,16,700–17,700,22,900–23,900,27,200–28200]
		0.5	100.0	2: [6000–7000,11,100–12,100,16,900–17,900,22,800–23,800,26,800–27800]

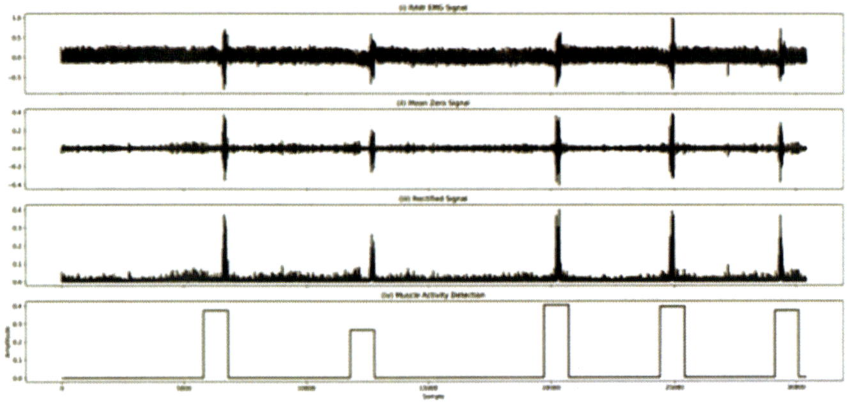

(a) Channel 1

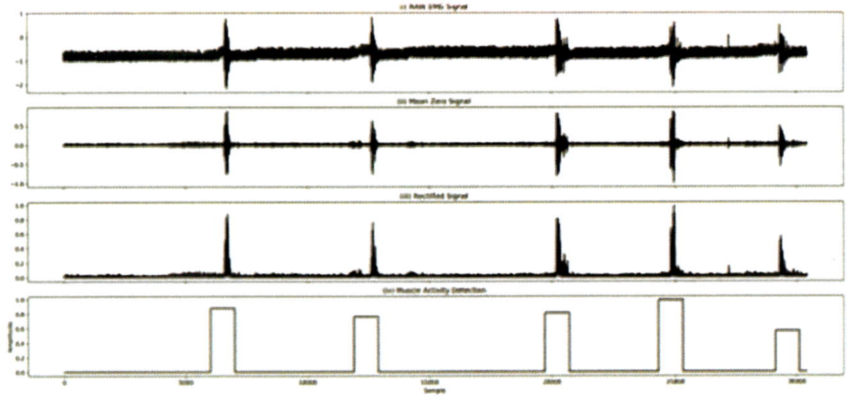

(b) Channel 2

Fig. 1 For alphabet *i* from file i_01.tdms

(ii) mean zero signal, (iii) rectified signal, and (iv) muscle activity detected signal represents output generated by our processing modules, respectively. Here, RAW EMG signal means signal directly read from .tdms file, saved on disk by LabVIEW, mean zero signal means filtered and detrend signal, and rectified signal represents only the positive part of the signal.

3.2 *Test Cases*

Figures 2, 3, 4, 5, 6, 7, 8, 9 and 10 illustrate channel-wise plotting of muscle activity detection by our approach for EMG signal of alphabet *i*, *m*, *p*, *u*, *x*, respectively. Each channel plot contains four different plots, which include RAW EMG, it is mean zero,

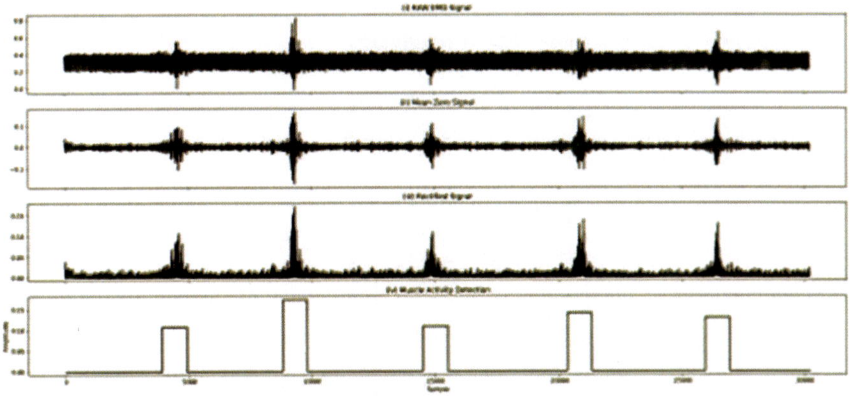

(a) Channel 1

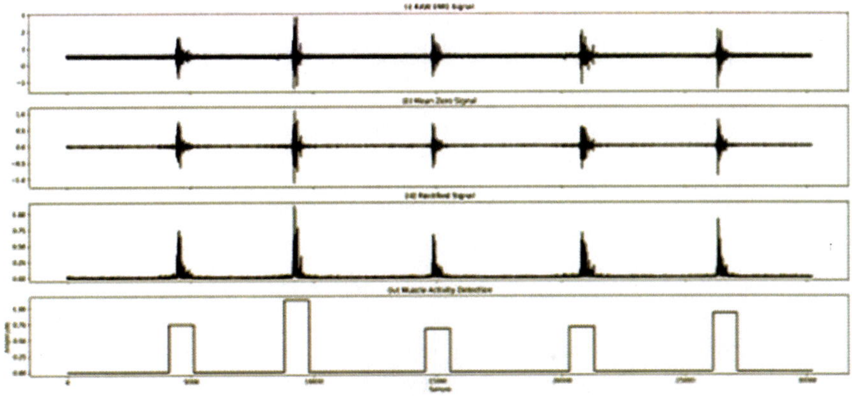

(b) Channel 2

Fig. 2 For alphabet *i* from file i_02.tdms

rectified, and muscle activity detected equivalent with speech segment of size 1000 samples (0.50 s) with a *stride* of 100 samples (0.050 s).

From Table 2, we draw a relationship between threshold variable *thres* and *factor* for zygomaticus major and levator anguli oris muscle for all five spoken alphabets, respectively. This relationship is illustrated in the form of a graph in Figs. 11 and 12. The variables *Thres_01* and *Factor_01* are used for *thres* and *factor* values for channel 1. Similarly, *Thres_02* and *Factor_02* are the variables used for *thres* and *factor* values for channel 2. We also noticed that our approach requires higher values of *thres* and *factor* variables to detect muscle activity in zygomaticus major muscle, which is on channel one as compared to levator anguli oris muscle, which is on channel two.

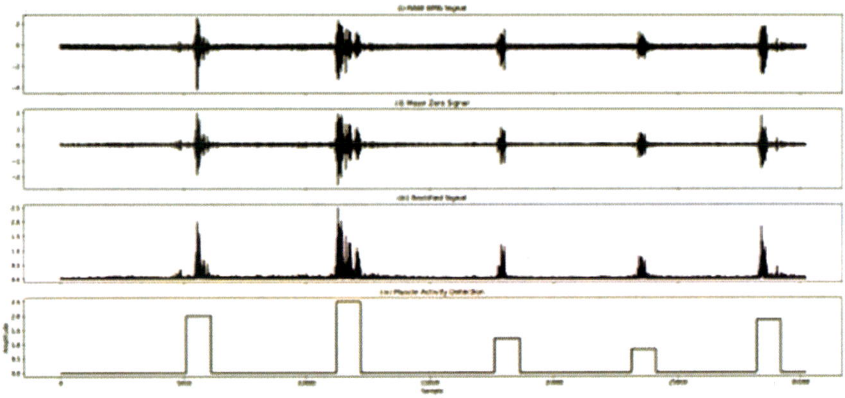

(a) Channel 1

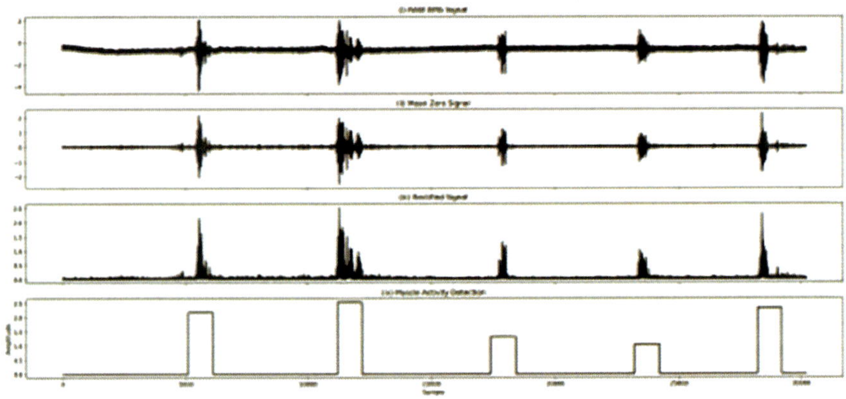

(b) Channel 2

Fig. 3 For alphabet *m* from file m_01.tdms

4 Conclusions

As we know that small automation turns a significant add-on to any manual system. In this work, we tried to achieve the automation at the reading and processing phase of our manual system. By a Python-based process, direct reading of .tdms files becomes automated. With the help of programming logic, we can read & process (i) single file having multiple segments and (ii) multiple files, having multiple segments, one-by-one. With our approach, we achieved 100% muscle activity detection present within 15 s of EMG recordings, over ten EMG signal files for alphabets *i*, *m*, *p*, *u*, *x*, in a few seconds. Due to the adaptive nature of the *thres* and *factor* variable, we can detect shallow muscle activities in a better way. It has been observed that our approach takes less than 0.5 s to detect muscle activities, which can be further reduced with the help of higher processing capability systems. With our approach, it is also noticed

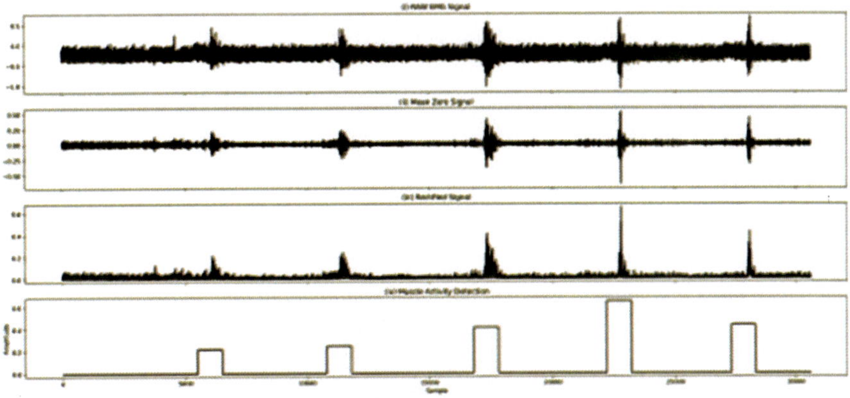

(a) Channel 1

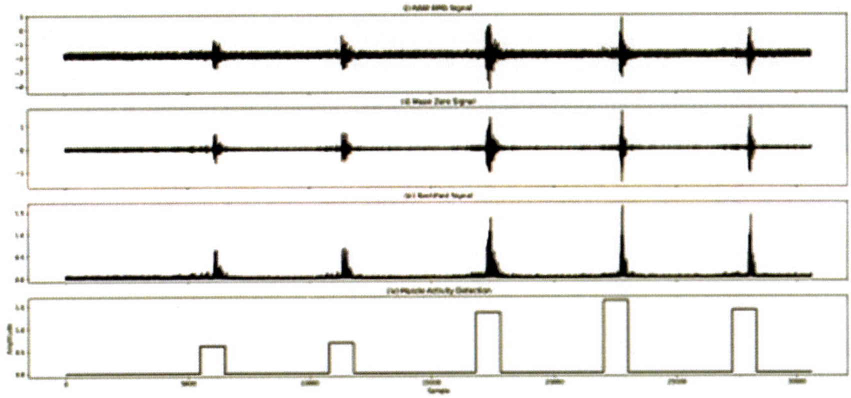

(b) Channel 2

Fig. 4 For alphabet *m* from file m_02.tdms

that to detect muscle activity, we need higher values of the *thres* and *factor* variable on channel one as compared with values of the *thres* and *factor* variable on channel two. It means that the zygomaticus major muscle shows less activity as compared to levator anguli oris muscle for silent speech. So, levator anguli oris muscle is a more critical muscle for silent speech communication as compared to zygomaticus major muscle. Further, the muscle activity detected segment extracted and might be passed as an input to a machine learning model for classification purposes using various classification techniques.

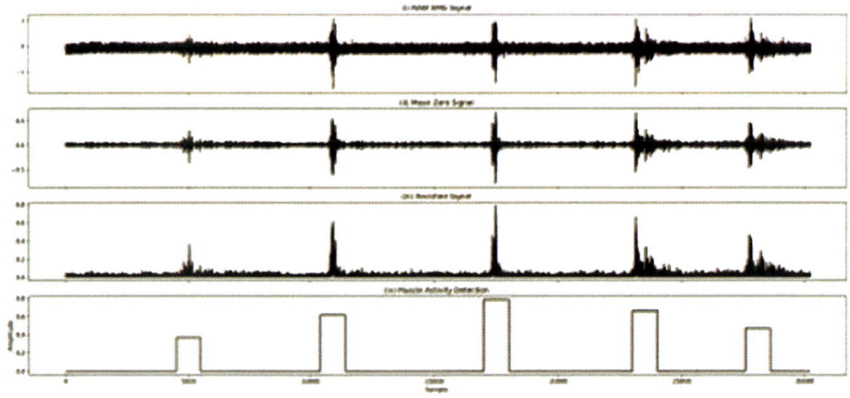

(a) Channel 1

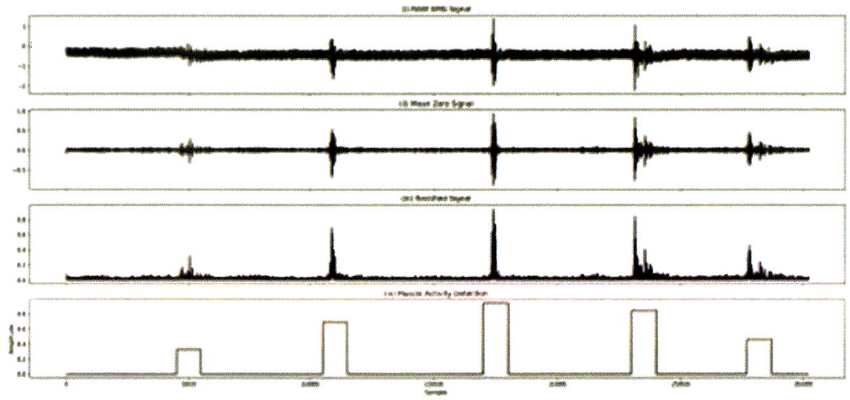

(b) Channel 2

Fig. 5 For alphabet *p* from file p_01.tdms

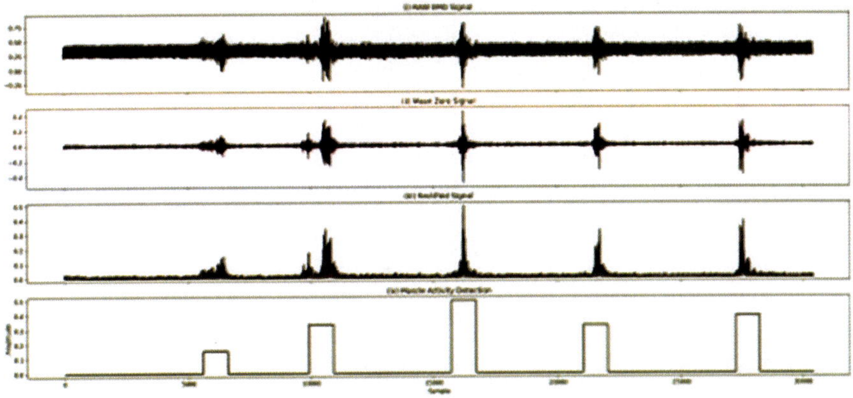

(a) Channel 1

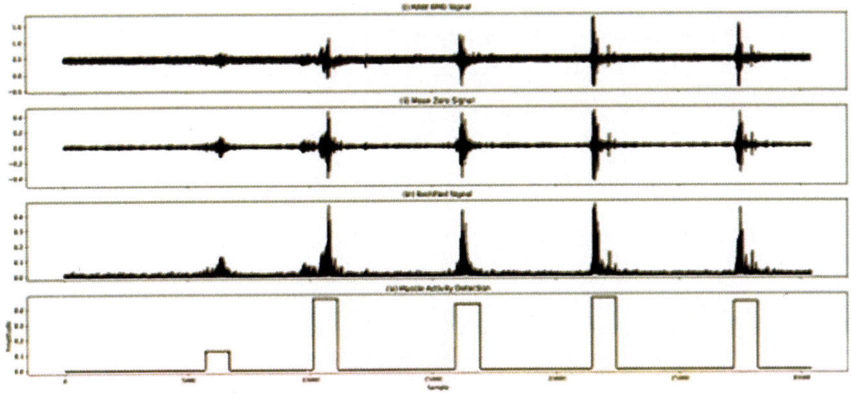

(b) Channel 2

Fig. 6 For alphabet *p* from file p_02.tdms

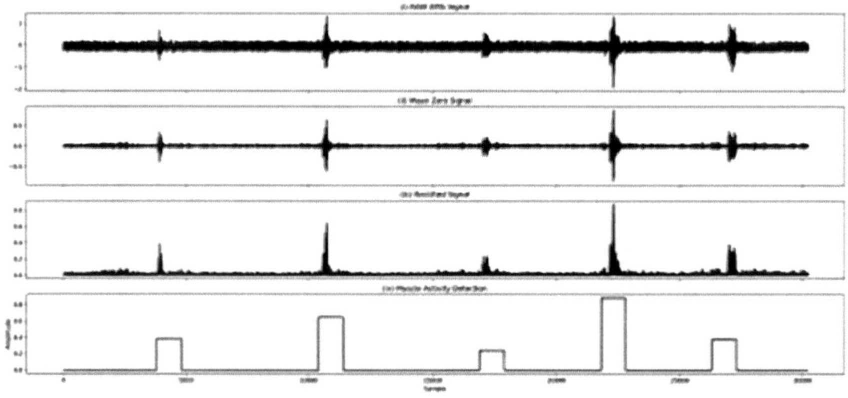

(a) Channel 1

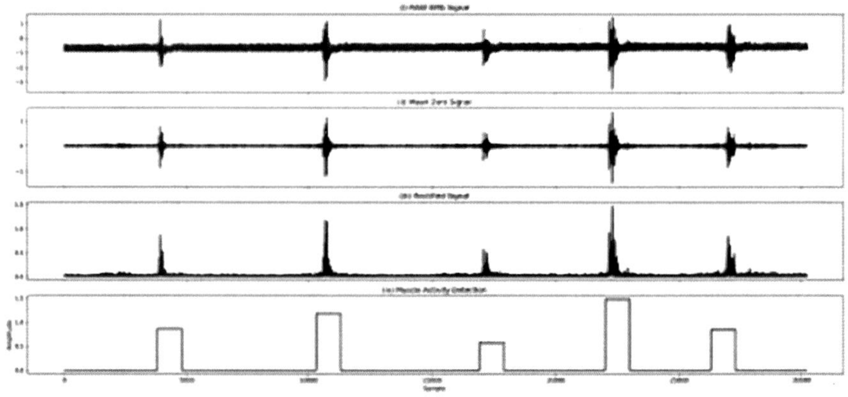

(b) Channel 2

Fig. 7 For alphabet *u* from file u_01.tdms

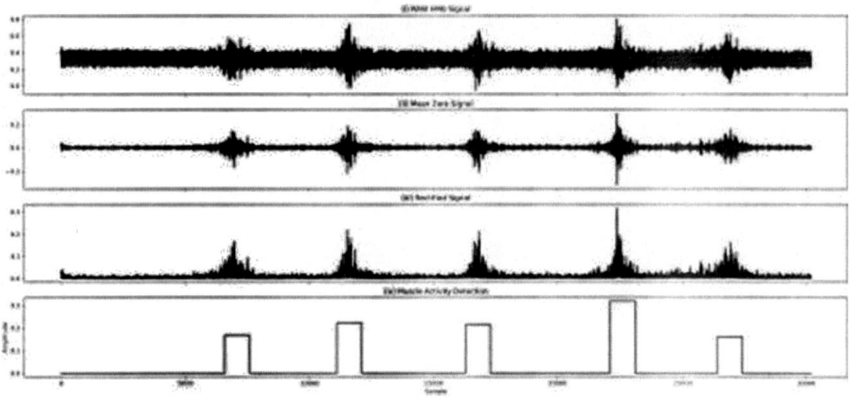

(a) Channel 1

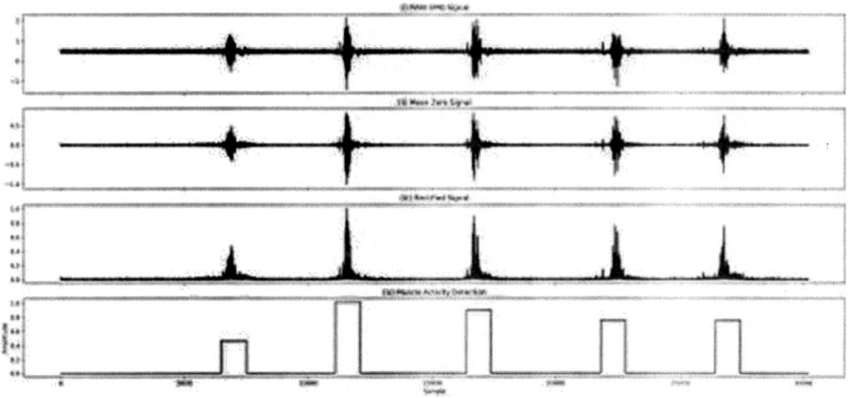

(b) Channel 2

Fig. 8 For alphabet *u* from file u_02.tdms

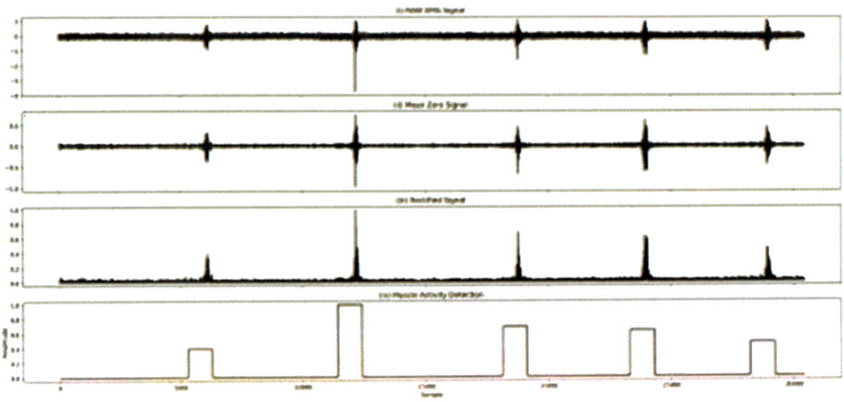

(a) Channel 1

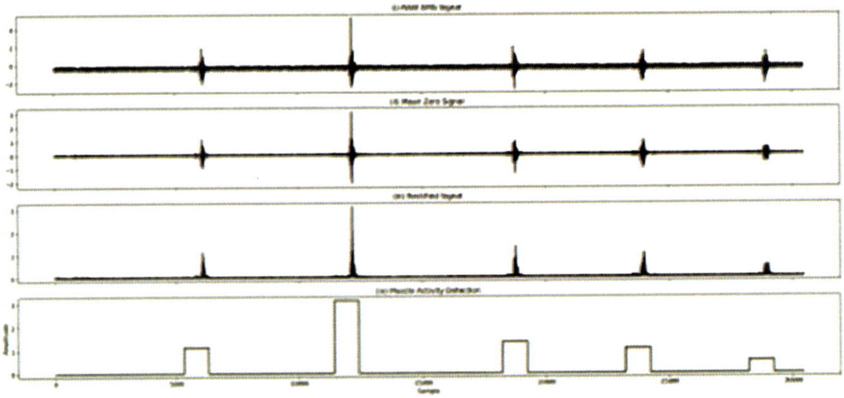

(b) Channel 2

Fig. 9 For alphabet *x* from file x_01.tdms

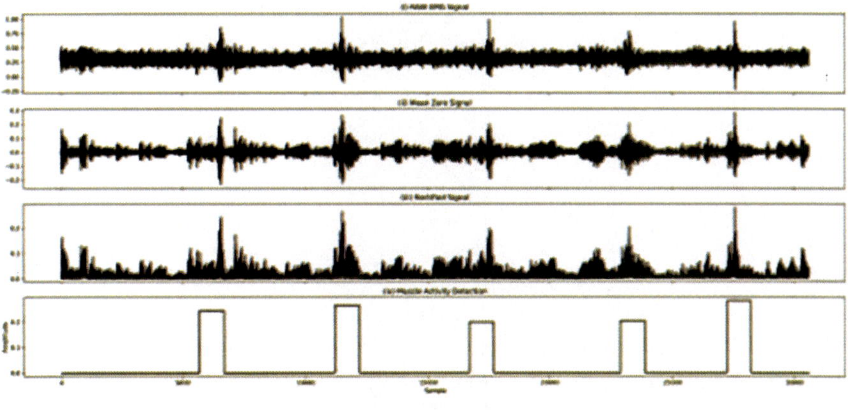

(a) Channel 1

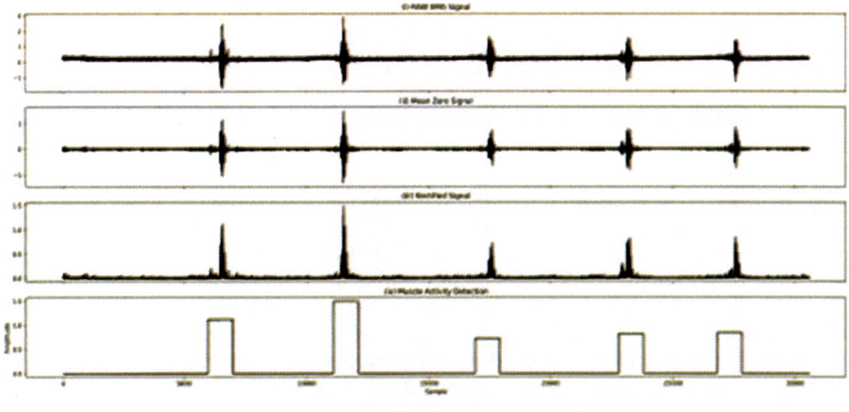

(b) Channel 2

Fig. 10 For alphabet *x* from file x_02.tdms

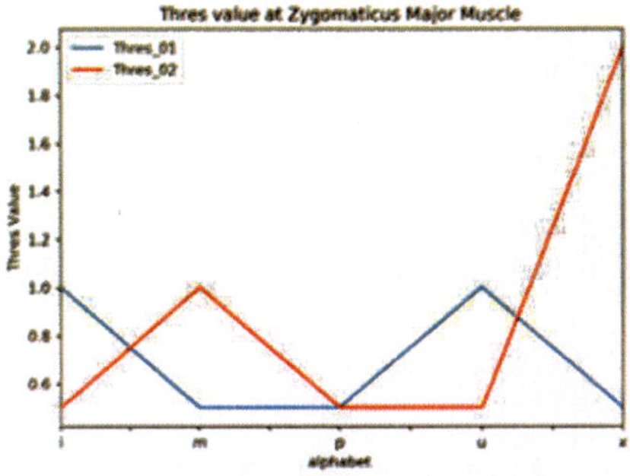

(a) Values of *thres* variable

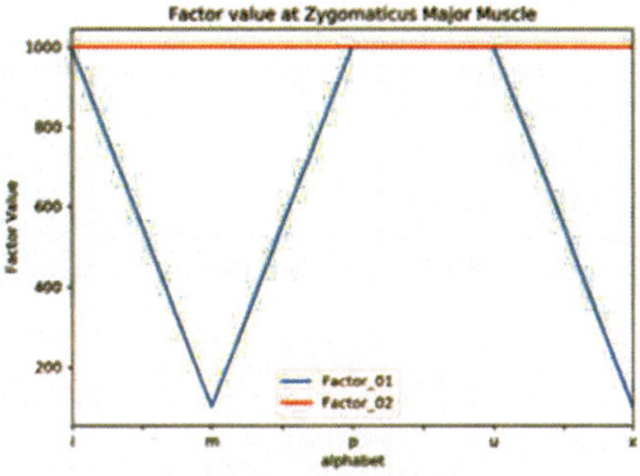

(b) Values of *factor* variable

Fig. 11 For zygomaticus major muscle

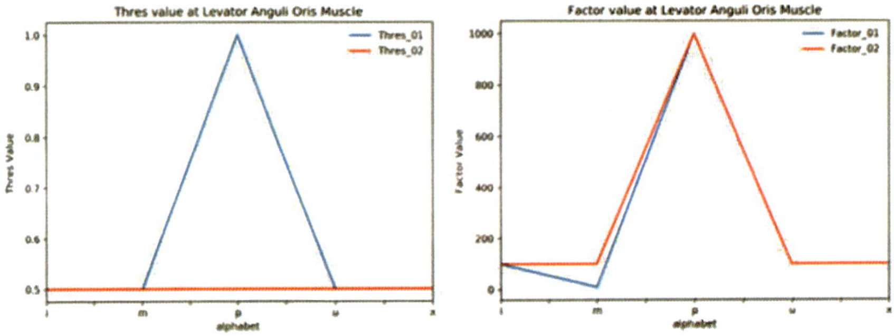

(a) Values of *thres* variable

(b) Values of *factor* variable

Fig. 12 For levator anguli oris muscle

References

1. Fridlund AJ, Cacioppo JT (1986) Guidelines for human electromyographic research. Psychophysiology 23(5):567–589
2. Vyas AP, Bhadada R (2017) Feature extraction cum frequency analysis system for facial surface electromyography signals based human speech recognition. Int J Res Appl Sci Eng Technol 5(12):1998–2006
3. Geryes M, Charara J, Skaiky A, Mcheick A, Girault J (2017) A novel biomedical application for the Myo gesture control armband. In: 29th International conference on microelectronics (ICM), Beirut, pp 1–4
4. Špulák D, Čmejla R, Bačáková R, Kračmar B, Satrapová L, Novotný P (2014) Muscle activity detection in electromyograms recorded during periodic movements. Comput Biol Med 47(1):93–103
5. Chowdhury R, Reaz M, Ali M, Bakar A, Chellappan K, Chang T (2013) Surface electromyography signal processing and classification techniques. Sensors 13(9):12431–12466
6. Wu J et al (2019) sEMG signal processing methods: a review. J Phys Conf Ser 1237:032008
7. Rasoo G, Iqbal K (2012) Muscle activity onset detection using energy detectors. In: Annual international conference of the IEEE Engineering in Medicine and Biology Society, San Diego, CA, pp 3094–3097
8. Caycedo AU, Hurtado DA, Sandoval OLR (2017) Electromyographic analysis for silent speech detection. ARPN J Eng Appl Sci 12(1):107–114
9. Adam Reeve T (2019) npTDMS Documentation Release 0.17.1
10. Python Homepage. https://www.python.org/. Last accessed 23 Jan 2020
11. The NI TDMS File Format. https://www.ni.com/product-documentation/3727/en/\#toc7. Last accessed 23 Jan 2020
12. npTDMS. https://pypi.org/project/npTDMS/. Last accessed 23 Jan 2020
13. npTDMS. https://nptdms.readthedocs.io/en/latest/. Last accessed 23 Jan 2020

A Non-contact Approach for Detection of Sleep Apnea Using Doppler Phenomena

Sumit Kumar Rai, Chetna Sharma, Vikash Shaw, Ranjan Kumar Jha, and Sanjeev Kumar

Abstract Sleep apnea is defined as a disorder in which the patient feels uneasiness during sleep occurring as a consequence of irregularity in breathing. If sleep apnea is not treated or diagnosed on time, it may lead to life-threatening diseases. The current medical practices to detect sleep apnea rely on contact-based approach. Therefore, this paper is an attempt to propose a non-contact approach for the detection of the sleep apnea disorder in which the breathing rate of the patient is detected using a motion detector followed by the processing of the acquired signal and then the analysis of the signal. The non-contact approach if implemented could be a boon for the medical experts and will open new horizons for the real-time monitoring of parameters of the patients.

Keywords Sleep apnea · Polysomnography (PSG) · Doppler radar · Breathing pattern · Real time

---

S. K. Rai · C. Sharma · V. Shaw · R. K. Jha · S. Kumar (✉)
Central Scientific Instruments Organisation, Chandigarh 160030, India
e-mail: virdi205@gmail.com

S. K. Rai
e-mail: raisumit38@gmail.com

C. Sharma
e-mail: chetna.bme@gmail.com

V. Shaw
e-mail: vikashshaw.aec@gmail.com

R. K. Jha
e-mail: ranjan.drmgr@gmail.com

C. Sharma · V. Shaw · R. K. Jha · S. Kumar
Academy of Scientific & Innovative Research (AcSIR), Ghaziabad 201002, India

R. P. Mahapatra et al. (eds.), *Proceedings of 6th International Conference on Recent Trends in Computing*, Lecture Notes in Networks and Systems 177,
https://doi.org/10.1007/978-981-33-4501-0_10

1 Introduction

Sleep apnea is defined as a type of sleep disorder which is characterized by the interruption (apnea) or reduction (hypopnea) in the flow of air, followed by the awakening of the patient while sleeping. When the airflow obstruction exceeds a duration of 10 s, then it is said that the person is suffering from the sleep apnea [1]. Sleep apnea itself is a precursor for a number of other diseases, for instance, hypertension, heart failure, heart attack [2, 3]. According to a study carried in different parts of India, it was found that 2.4–4.96% of men and 1–2% of women are suffering from sleep apnea which amounts to around 35 million people suffering from sleep apnea syndrome (SAS) in India alone [4]. Although sleep apnea has become a common disorder, still a large number of SAS cases (82%) are not even detected [5]. SAS are basically of two types: central sleep apnea and obstructive sleep apnea. In central sleep apnea, the brain is unable to send the right message signals to the muscles that controls breathing, whereas in obstructive sleep apnea, there is an obstruction in the mouth and the pharynx due to which the upper airway passage is blocked [6]. The major difficulty faced by the doctors in the treatment of this type of disorder is its detection which makes it significantly challenging for the healthcare experts to treat of the patient. Presently, nocturnal polysomnography or home sleep tests are the general medical examinations that the clinicians recommend to their patients.

However, these tests are expensive and prone to a number of errors. In nocturnal polysomnography (PSG), a number of sensors are attached to the body of the patient for the measurement of the activity of heart, lung and brain, pattern of breathing, movements of arms and legs, and level of oxygen in the blood during the sleep [7]. Moreover, the acquired physiological parameters for the detection of the disorder are also affected by the body movement of the patient which results in the alteration of the acquired signal [8, 9]. In addition, PSG also requires specific equipment that is available only in a limited number of hospitals and medical institutions [10]. On the other hand, home sleep tests is rather a simple test in which only a few parameters can be measured [11]. Unfortunately, it is not always feasible to go for the contact-based sensors for monitoring the vital signs of the patients, for instance, in case of the patients with burns or infants because the electrode contact, adhesive tape, and various probes utilized for the recording of information can cause damage to the skin of the patients [12].

As a result, a non-contact approach for the recording of the important parameters from the patient which could further lead to the detection of the sleep apnea disorder by the healthcare experts is the need of the hour. Continuous monitoring of the patient through the non-contact approach will be invaluable tool for healthcare professionals, significantly useful when considering nocturnal monitoring [13].

Doppler radar monitoring systems will be a boon for the non-invasive, continuous recording of the patients and allow for the study and detection of SAS [14], as well as other medical conditions where there is a requirement for the measurement of the continuous non-contact vital sign (NCVS).

Following the introduction, this paper is organized into six sections. Section 2 highlights the methodology steps that can be used for the non-contact method for sleep apnea detection. Section 3 discusses the data processing methods used in the proposed technique, while Sect. 4 provides the details and experimental monitoring system setup. The results are discussed in Sect. 5, and the conclusion is detailed in Sect. 6, and future scope is elaborated in the Sect. 7.

2 Methodology

The proposed methodology used in the study is shown in Fig. 1. The radar transmits the signal to the subject and receives the signal through the antenna. A RADAR is an electromagnetic system that is used to transmit the signal and receive the reflected signal from the object (subject) with the help of the antenna. The radar sensor used is a type of motion detection sensor and has a frequency range from 9.41 to 10.5 GHz with a wavelength that varies from 2.8 to 3.2 cm in air. This sensor works on the principle of utilizing the Doppler shift phenomenon to "sense" motion. It is a miniature microwave Doppler radar sensor mainly used for the low power consumption and short range (<30 m). The radar sensor is connected to the hardware module to isolate, amplify, and filter the desired signal. The signal is isolated using a buffer. The buffer is used to transform the signal from one circuit to the succeeding stage with unity gain. After the buffer, the hardware module consists of the amplifier

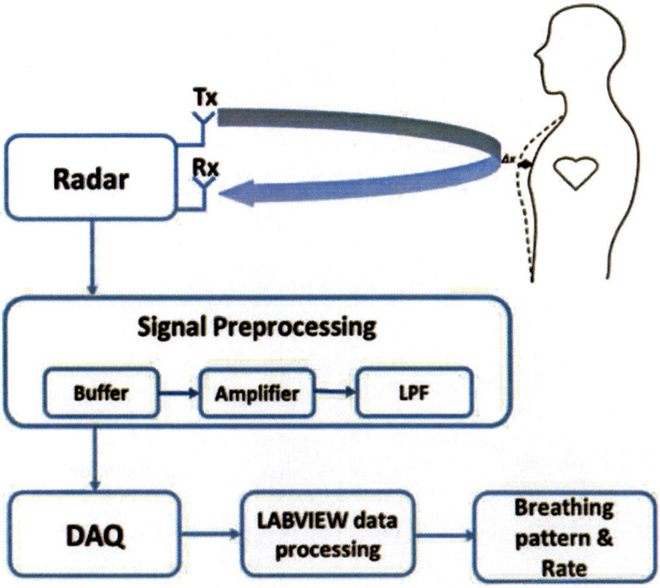

Fig. 1 Methodology

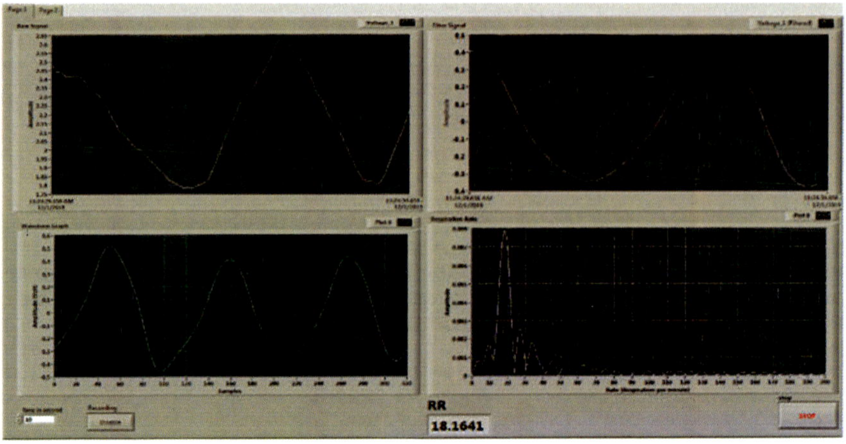

Fig. 2 Front panel for data acquisition and visualization

and the low pass filter. The amplifier is designed using LM358 operational amplifier with a gain of 100 to increase the strength of the acquired signal, and a low pass filter with a cut off frequency of 15 Hz is used. The processing of the acquired signal is done using the NI-LabVIEW software. The NI DAQmx module is employed to connect the hardware module to the computer for acquiring the data from the hardware for further processing to get the results.

3 Data Acquisition

Since the normal range of breathing varies between 12 and 20 (0.2–0.33 Hz) respirations per minute, the data were acquired at 32 samples per second rate with the help of NI USB 6003 DAQ and NI real-time LabVIEW platform. The acquired signals were passed through a 5th order Butterworth bandpass filter having lower and upper cut off frequency of 0.1 Hz and 1 Hz, respectively. Ten seconds of respiration signal were taken to extract the respiration rate. The respiration rate was calculated through the peak detection method, i.e., peak of fast Fourier transform (FFT) of signal with a duration of 10 s (Fig. 2).

4 Experiment

The Doppler radar was placed 70 cm away from the subject. In the first phase of the experiment, the subject was asked to breathe in the various patterns such as normal, slow, fast, and hold with and without sweater. During second phase, the

Fig. 3 Experimental setup

subject was asked to perform the combination of different breathing patterns, i.e., normal→hold→normal→fast→slow→normal with and without sweater (Fig. 3).

5 Results and Discussions

The following graphs obtained for various breathing patterns uphold the fact that the proposed system is able to detect the respiration patterns of the person.

Figure 4a–d denotes the respiration pattern for normal, slow, fast, and combination breathing, respectively. It can be clearly observed from the graphs that in normal breathing includes six breathing cycles in 17 s which corresponds to approximately

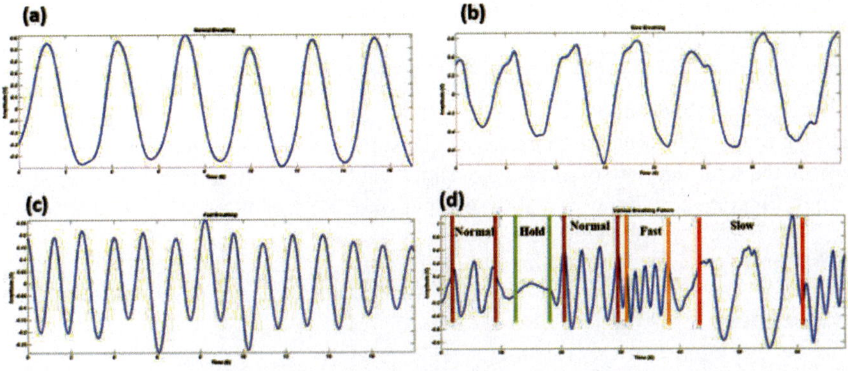

Fig. 4 Respiration signal for **a** normal, **b** slow, **c** fast, **d** combination breathing pattern

21 respiration per minute, whereas there are seven breathing cycles in 37 s and 14 breathing cycles in 17 s for slow and fast breathing patterns that corresponds to approximately 11 and 49 respiration per minute, respectively. The last graph, i.e., (d) shows the breathing signal for the above-mentioned combination of various breathing patterns (normal→hold→normal→fast→slow→normal).

It was also observed that the signal remains significantly unaffected with more number of layers of clothes on the body.

6 Conclusion

Sleep apnea is one of the types of the sleep disorder which is mainly characterized by the obstruction in the flow of air in the respiratory tract resulting in the awakening of the patient during sleep. The traditional method for the diagnosis of this disease is polysomnography includes the attachment of a number of sensors on the body of the patient during examination. However, this technique of the diagnosis is not suitable for the patients with the sensitive skin or with burns. Moreover, it may also hinder with pattern of sleeping of the patient. This paper is an attempt to propose the non-contact approach for the detection of the sleep apnea. The proposed methodology eliminates the need of the placement of any sort of the sensors on the body of the patient. The results obtained significantly prove the fact that the proposed methodology has the capability to detect the different patterns of breathing (slow, fast, normal, hold) very efficiently. The various patterns obtained in the breathing are helpful in the detection of the sleep apnea as the breathing pattern of the patient suffering from this disorder involves several periods of obstruction (hold stage in our case) which can be easily detected. Since the system holds the capability to detect fast and slow breathing pattern of the subject, the application of this system may further be extended to detect tachypnea (fast breathing) and bradypnea (slow breathing).

7 Future Scope

The system performance will validate with the gold standard system (Capnography) of respiration measurement in medical. The signal received by Doppler radar can be correlated with the flow of air. At the same time, the optimum time scale signal for respiration rate extraction in real time can also be explored.

References

1. Guilleminault C, Tilkian A, Dement WC (1976) The sleep apnea syndromes. Ann Rev Med 27:465–484

2. Azagra-Calero E, Espinar-Escalona E, Barrera-Mora J-M, Llamas-Carreras J-M, Solano-Reina E (2012) Obstructive sleep apnea syndrome (OSAS). Review of the literature. Med Oral Patol Oral Cir Bucal 17(6):e925-9
3. Aurora RN, Quan SF (2016) Quality measure for screening for adult obstructive sleep apnea by primary care physicians. J Clin Sleep Med 12(8):1185–1187
4. Sharma SK, Reddy EV, Sharma A, Kadhiravan T, Mishra HK, Sreenivas V et al (2010) Prevalence and risk factors of syndrome Z in urban Indians. Sleep Med 11:562–568
5. Carev M, Karanović N, Dogas Z (2008) Obstructive sleep apnea and anesthesia. Liječ Vjesn 130(3–4):78–86
6. Hall T, Lie DYC, Nguyen TQ et al (2017) Non-contact sensor for long-term continuous vital signs monitoring: a review on ıntelligent phased-array doppler sensor design. Sensors (Basel) 17(11):2632. Published 2017 Nov 15
7. Watanabe T, Watanabe K (2004) Noncontact method for sleep stage estimation. IEEE Trans Biomed Eng 51(10):1735–1748
8. Yang C, Wang X, Mao S (2019) Unsupervised detection of apnea using commodity RFID tags with a recurrent variational autoencoder. IEEE Access 7:67526–67538
9. Lin F et al (2017) SleepSense: a noncontact and cost-effective sleep monitoring system. IEEE Trans Biomed Circ Syst 11(1):189–202
10. Kushida CA et al (2005) Practice parameters for the indications for polysomnography and related procedures: an update for 2005. Sleep 28(4):499523
11. Bloch KE (1997) Polysomnography: a systematic review. Technol Health Care 5(4):285305
12. Boric-Lubecke BKS, Lubecke O (2016) Doppler radar physiological sensing. Wiley, Hoboken
13. Hong H et al (2019) Microwave sensing and sleep: noncontact sleep-monitoring technology with microwave biomedical radar. IEEE Microwave Mag 20(8):18–29
14. Yang X et al.(2018) Sleep apnea syndrome sensing at C-band. IEEE J Transl Eng Health Med 6:1–8, Art no. 2701008

State-of-the-Art Applications of Graph Convolutional Neural Networks

Rajat Singh, Sanchit Bathla, and Priyanka Meel

Abstract Graphs are practiced widely in various real-world applications because of their structure modelling capability. Currently, deep neural networks have been employed to produce excellent outcomes in tasks such as classification. A graph convolutional network (GCN) is a deep learning model that operates on the graphs. We offer the general architecture of a GCN and its utilization in semi-supervised learning. We also investigate the application and performance of GCN in the classification of fake news. Finally, we present different application areas of GCN and open challenges for future research.

Keywords GCN · CNN · Graph · Hidden layer · Semi-supervised learning

1 Introduction

Graphs usually appear in several real-world applications, including social analysis, duplicity detection, traffic prediction, computer vision, and various more. By converting the data that is available to us in the form of graphs (nodes and edges), the fundamental data can be structured in such a way to represent the relationships among different objects, and provide more assuring insights holding the information.

R. Singh (✉) · S. Bathla · P. Meel
Department of Information Technology, Delhi Technological University, Shahbad Daulatpur, Main Bawana Road, Delhi 110042, India
e-mail: rajat.singh1084@gmail.com

S. Bathla
e-mail: sanchitbathla25@gmail.com

P. Meel
e-mail: priyankameel86@gmail.com

R. P. Mahapatra et al. (eds.), *Proceedings of 6th International Conference on Recent Trends in Computing*, Lecture Notes in Networks and Systems 177,
https://doi.org/10.1007/978-981-33-4501-0_11

In context to artificial intelligence(AI), the single most important question is how does AI learn structure where structure is the relationships among data how is one piece of data related to another we have a pretty good way of representing structure and that's with graphs and graphs are everywhere so everything from state machines to molecular networks to social networks all these can be represented with graphs and graphs give us a way to explicitly and mathematically represent this complex information.

1.1 Graph Convolution Networks

A graph neural network, also recognized as a graph convolutional networks (GCN), implements convolution on a graph, rather than on an image composed of pixels.

Convolutional neural network (CNN) is designed to extract the useful information from the image to classify the image similarly, a GCN apply a filter over the graph, scanning for required vertices and edges that can assist the task of classifying nodes within the graph [1].

CNN and GCN have fixed number of parameters that means you have a low memory footprint and a low computational cost at least relative to other deep learning techniques you have a local kernel which lets you build hierarchies of information and you have these very nice spatial and variance properties.

1.2 Architecture of GCN

A graph convolution network's architecture is inspired by the architecture of convolutional neural network (CNN). It consists of layers such as graph convolution layer, dropout layer, pooling which are then stacked collectively to build a deep network [2]. The ith hidden layer in the GCN can be depicted as

$$H^i = f\left(H^{i-1}, A\right). \tag{1}$$

where f is the propagation and the $H^0 = X$. Every layer Hi equals to an $N \times F^i$ feature matrix and the feature of ith node is described by jth row of the matrix. The propagation used here is related to every other neural network techniques where the current layer's features are collected together and uses the input features of the subsequent layers applying function f. In this way, by each layer, the features grow increasingly abstract and thus can be used for classification techniques. Alterations of GCNs can be done by applying various propagation rule f. the basic propagation rule is

$$f\left(H^i, A\right) = \sigma\left(AH^iW^i\right). \tag{2}$$

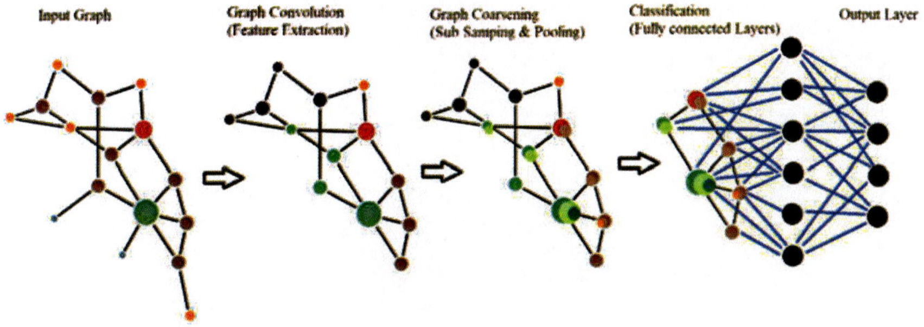

Fig. 1 Architecture of GCN

where W^i is the weight matrix for ith layer and σ is a nonlinear activation function such as the ReLu function. The dimensions of the weight matrix are the number of features in the current layer by the number of features in the following layer (Fig. 1).

2 Semi-supervised Learning with GCN

The foremost difficulty in deep learning over Internet's data is that the number of labelled data is restricted. Very rare articles can be reviewed and interpreted as one thing. To solve this issue, we opted for semi-supervised learning approach. In particular, we can implement graph-based semi-supervised deep learning methods, based on graph convolutional networks (GCN).

Considering everything in this kind of model is differentiable and parameterized, we can add some labels, train the model, and inspect how the embeddings act. We can apply the semi-supervised learning algorithm for GCNs. We can easily label one node per class and begin training for several iterations.

This type of model can immediately produce a two-dimensional latent space with the help of word embeddings of text articles which one can directly visualize. We note that the three-layer GCN model leads to linearly classify the dataset, provided just 1 labelled sample per class. This is a reasonably exceptional result, given that the model took no feature information of the nodes.

3 Semi-supervised Fake News Detection Using GCN

The query for fake news detection has become an important issue in modern studies. To tackle the point that labels are usually very limited and inadequate, we can go with semi-supervised detection methods which are based on graph convolutional networks (GCN). For this purpose, we rely on the fact that graphs are capable of representing

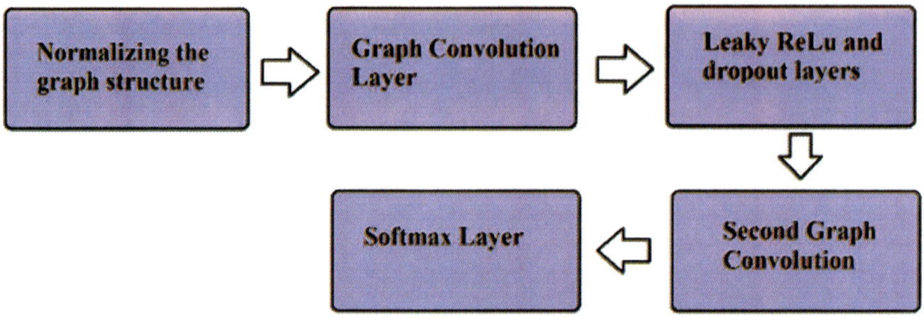

Fig. 2 Stages of a simple GCN architecture

structural and contextual dependencies between its nodes, each representing an object and subsequently can be utilized in cases where the number of labels are limited.

To execute this approach, we can map the news articles onto Euclidean space by using some word embeddings methods, and then by structuring those embeddings into graphs, we can apply graph convolutional neural networks on those graphs to do classification of the news articles (Fig. 2).

4 Related Work

It can be observed that these models cover a wide area of problems and graph convolutional networks have been used in each of those to give good results. Various architectures have been designed and can be designed for such classification tasks. There is also the scope of applying GCN by changing its internal layers (Table 1).

5 Future Scope

Graph convolutional network models, as one section of the graph neural network models, have become a very trendy tool in both machine learning and artificial intelligence, and large number of models have been introduced to solve complex problems.

The areas in which GCN can be utilized is widely immense as well as novel. Text classification is one field where GCNs have been used and has produced better results. In images and videos as well GCNs can be used, by constructing similarity graphs between similar image frames. Other areas where GCNs can be used are biological sciences, materials sciences as well as in fields like social network analysis.

There are various areas in which a graph convolutional network needs to be improved, such as, GCNs expect input graphs to be static, but in real-life scenarios, there are numerous situations where graphs used to model the problems are dynamic.

Table 1 Summary of results of work based on GCN

S. No.	Research paper	Application	Methodology	Figure number	Results
1	Benamira et al. [3]	Fake news detection	Embedding of articles + Graph construction + Classification (GCN, AGNN)	Fig. 3	84.94% ± 2.30%
2	Bian et al. [4]	Rumour detection	Construct propagation and dispersion graphs + Calculate the high-level node representations + Root feature enhancement + Representations of propagation and dispersion for rumour classification	Fig. 4	96.1%
3	Dong et al. [5]	Multiple rumour source detection	GCNSI + LPSI + NetSleuth	Fig. 5	0.63 (error distance)
4	Li and Goldwasser [6]	Political perspective detection	GCN + SkipThought/GCN + HLSTM	Fig. 6	91.74%
5	Wu et al. [7]	Social spammer detection	GCN + Markov random field (MRF)	–	83.9%
6	Aljohani et al. [8]	Bot prediction on social networks	Bot detection using SNA (community detection, degree and triangle, clustering coefficient) + GCN	Fig. 7	71%
7	Ying et al. [9]	Web-scale recommender systems	PinSage (random-walk graph convolutional network (GCN))	–	67% (hit-rate)
8	Yao et al. [10]	Text classification	Text graph convolutional networks (text GCN)	–	86.34%
9	Marcheggiani and Titov [11]	Semantic role labelling	Word embedding + BiLSTM encoder + GCN encoder + Classifier	–	88.0% ($F1$-Score)

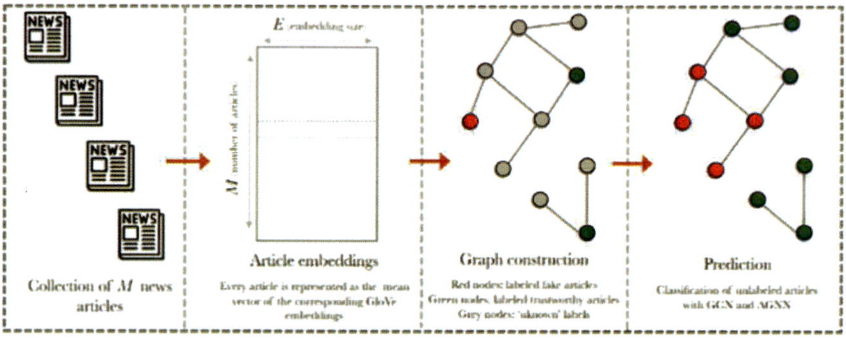

Fig. 3 Architecture of GCN and AGNN for fake news detection [3]

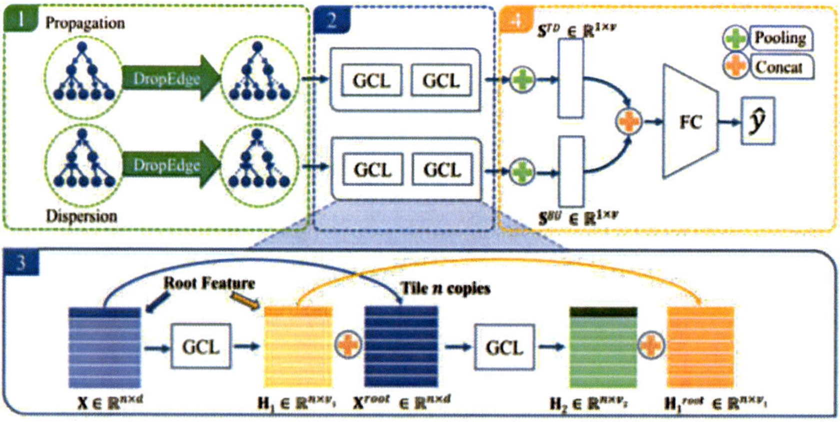

Fig. 4 Architecture of bi-directional GCN for rumour detection [4]

Also, there is scope to use multiple GCNs for improved results which is currently not possible because of the inability of spectral graph convolutional networks to adapt from one graph to another graph. Also, most of the current GCN models utilizes shallow architecture because in deep architectures, even for nodes that are away from each other, their representations become smooth [12], and thus deep graph convolutional network models are still not in use. A possible solution to avoid smoothing in GCN is to use Poison kernel [13].

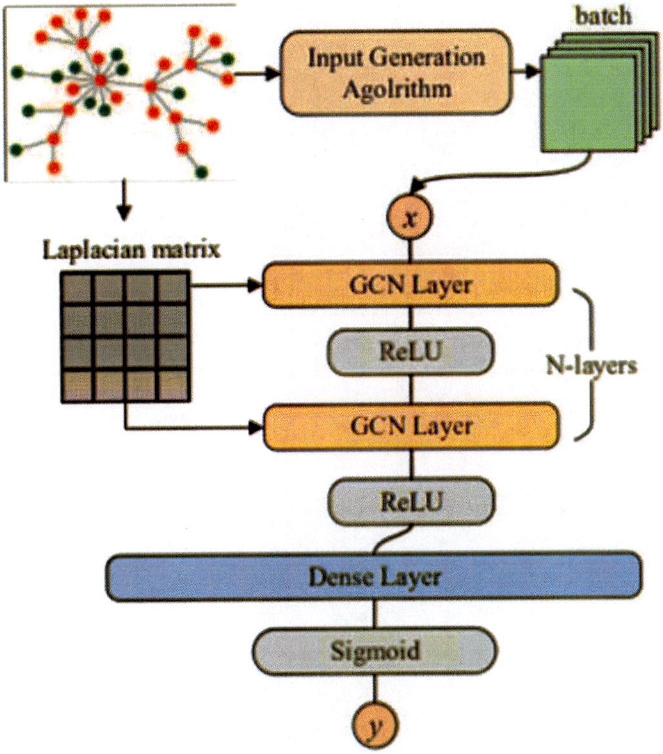

Fig. 5 Architecture of GCNSI for rumour detection [5]

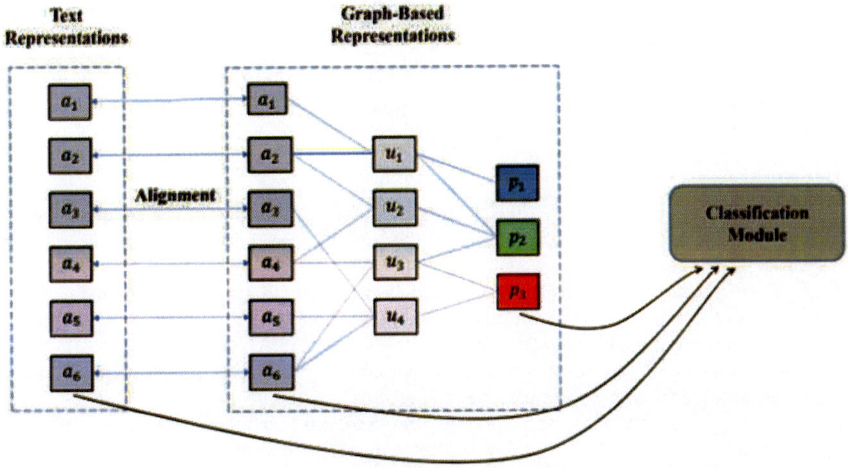

Fig. 6 Architecture of SNA + GCN for bot detection [6]

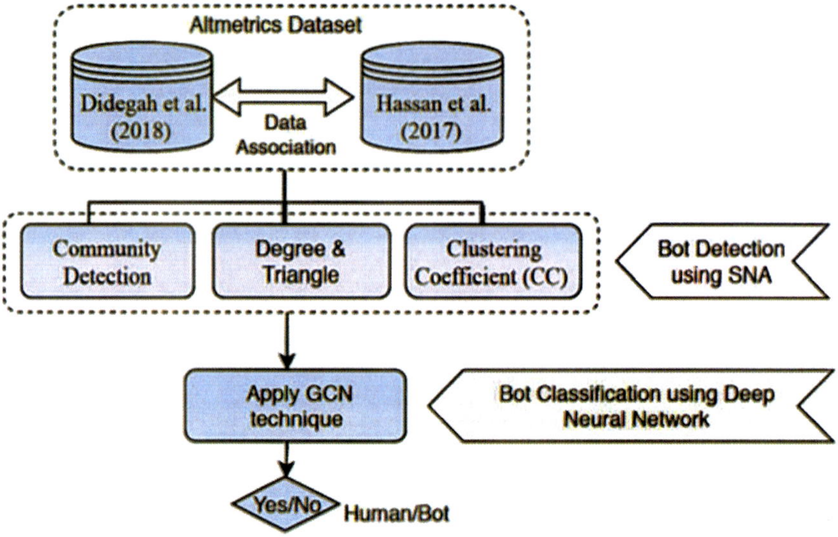

Fig. 7 Model architecture [8]

References

1. Huang K-H (2019) A gentle introduction to graph neural networks (basics, DeepWalk, and GraphSage), 10 Feb 2019. [Online]. Available: https://towardsdatascience.com/a-gentle-introduction-to-graph-neural-network-basics-deepwalk-and-graphsage-db5d540d50b3
2. Jepsen TS (2018) How to do deep learning on graphs with graph convolutional networks, 18 Sept 2018. [Online]. Available: https://towardsdatascience.com/how-to-do-deep-learning-on-graphs-with-graph-convolutional-networks-7d2250723780
3. Benamira A, Devillers B, Lesot E, Ray AK, Saadi M, Malliaros FD (2019) Semi-supervised learning and graph neural networks for fake news detection. In: IEEE/ACM International conference on advances in social networks analysis and mining, Chicago
4. Bian T, Xiao X, Xu T, Zhao P, Huang W, Rong Y, Huang J (2020) Rumor detection on social media with bi-directional graph convolutional networks. arXiv preprint arXiv:2001.06362, Chicago
5. Dong M, Zheng B, Hung NQV, Su H, Li G (2019) Multiple rumor source detection with graph convolutional networks. In: 28th ACM International conference on information and knowledge management, Harvard
6. Li C, Goldwasser D (2019) Encoding social information with graph convolutional networks for political perspective detection in news media. In: 57th Annual meeting of the association for computational linguistics, Harvard
7. Wu Y, Lian D, Xu Y, Wu L, Chen E (2020) Graph convolutional networks with markov random field reasoning for social spammer detection
8. Aljohani N, Fayoumi A, Hassan S (2020) Bot prediction on social networks of Twitter in altmetrics using deep graph convolutional networks. Soft Comput
9. Ying R, He R, Chen K, Eksombatchai P, Hamilton WL, Leskovec J (2018) Graph convolutional neural networks for web-scale. In: 24th ACM SIGKDD international conference on knowledge discovery & data mining, Chicago, pp 974–983
10. Yao L, Mao C, Luo Y (2019) Graph convolutional networks for text classification. In: AAAI Conference on artificial intelligence, Vancouver

11. Marcheggiani D, Titov I (2017) Encoding sentences with graph convolutional networks for semantic role labeling. arXiv preprint arXiv:1703.04826, Harvard
12. Zhang S, Tong H, Xu J, Maciejewski R (2019) Graph convolutional networks: a comprehensive review, 10 Nov 2019. [Online]. Available: https://link.springer.com/article/10.1186/s40649-019-0069-y
13. Yang Z, Han S, Zhao J (2020) Poisson Kernel avoiding self-smoothing in graph convolutional networks. arXiv preprint arXiv:2002.02589, 7 Feb 2020, Vancouver

Performance Evaluation and Comparison of CNTFET-Based 7T SRAM with 6T SRAM

Neha Mathur and Shilpi Birla

Abstract Excellent thermal conductivities, superior current capabilities and ballistic transport operation these prominent properties of carbon nanotube field-effect transistor technology make it superior or alternative approach over traditional CMOS and conventional FETs. A detailed simulation-based analysis of this technology with the evaluation of circuit performance of SRAM is presented here. Performance of CNTFET-based 6T SRAM and 7T SRAM is analyzed and compared in terms of static noise margin and static power dissipation. Variations of parameters for performance evaluation include fermi level, dielectric constant, oxide thickness, work function and chiral vector. The work has modelled in HSPICE software using Stanford CNTFET model for channel length of 32, 22, 16 and 7 nm. Evaluated results show that CNTFET-based 7T-SRAM has high static noise margins when compared with 6T-SRAM.

Keywords CNTFET · CMOS · FET · SNM · SRAM

1 Introduction

Tremendous enhancement in leakage current, reduced reliability and high power consumption are consequence of scaling of devices in order to shrunk the size of devices or to increase the functionality of devices considerably with the demand of electronic industry. With the scaling beyond 32 nm conventional CMOS has limited in terms of short channel effect and leakage current thus the challenging investigation for researcher is to find prominent alternative for nanoscale devices to provide high-performance digital circuits [1–15]. CNTFETs became an attractive alternative by reducing the limitations of conventional CMOS. High value of gate dielectric, low

N. Mathur (✉) · S. Birla
Manipal University, Jaipur, India
e-mail: nmsnehlata@gmail.com

S. Birla
e-mail: Shilpi.Birla@jaipur.manipal.edu

R. P. Mahapatra et al. (eds.), *Proceedings of 6th International Conference on Recent Trends in Computing*, Lecture Notes in Networks and Systems 177,
https://doi.org/10.1007/978-981-33-4501-0_12

voltage operated quasiballistic transport, higher drive current, higher transconductance, lower heat dissipation, higher average carrier velocity and higher ON–OFF current ratio are prominent properties of CNTFETs that became more dominant in comparison to CMOS [2–21]. Its scalable feature and higher bond strength between carbon–carbon make it more suitable for interconnection purpose. CNTFETs are available with diameter range from 1 to 3 nm and length with several micrometres. Excellent electrical properties of CNTFETs not only offer nanoscaling regime but also provide desired performance [24]. Modern system-on-chip (SoC) applications include memory as its main component. Hence, appropriate designing of SRAM is required to achieve efficient performance and high reliability as per requirement of future SoCs. CNTFETs-based SRAM cells are investigated and proposed by researcher to achieve more stable, highly performed and low power circuitry in recent years. Thus, low-power cache memory can be implemented by using CNTFET-based SRAM cell. In this paper, various SRAM cells are analysed and simulated for their performance evaluation and to a proper comparison is observed among them. HSPICE tool is used to design 6T and 7T SRAM cell Stanford CNTFET model file.

2 CNTFET-Based SRAM Cells

6-T SRAM Cell: A pair of inverter is formed using M1, M2, M3 and M4 transistor to store information of a bit while rest two transistors M5 and M6 are access transistors required to access the read and write operation of inverter pair. Word line (WL) is used to control the both M5 and M6 transistors so WL is access transistor controller because it provides memory element access to bit line (BL) and bit line bar (BL bar) [10–18]. Hold operation is performed by setting WL at '0' and BL with BLbar at voltage supply (VDD).When WL is set at '1' with BL and BLbar at voltage supply (VDD) then read operation is performed. Same '1' is set for WL with BL '0' and BLbar '1' for write '0' while for write '1' BL is set at '1' and BLbar is at '0' [12–19] (Fig. 1).

7-T SRAM Cell: 7T SRAM cell is designed to improve static noise margin of SRAM cell. With existing structure of 6T SRAM, one additional transistor M7 is used (Fig. 2).

Thus, a proper amplification of high value of first inverter to low value at second inverter and vice versa takes place by adjusting M7 in feedback form [10–22]. Write select and Read select signals are used to control the access transistors (M5 and M6), respectively, to perform read and write operation from memory cell.

Write select, read select and write bar are set at '0' with write bit, read bit at voltage supply (VDD) for hold operation. Transistor T7 is turned on by setting write bar at '1' during the read and write operation. write select and read select are set at '1' during read operation with write bit and read bit at voltage supply (VDD).To write '0' write bit is set at '0' and read bit is set at '1' while to write '1' write bit is set at '1' and read bit is set at '0' with write select at '0' read select at '1' [6].

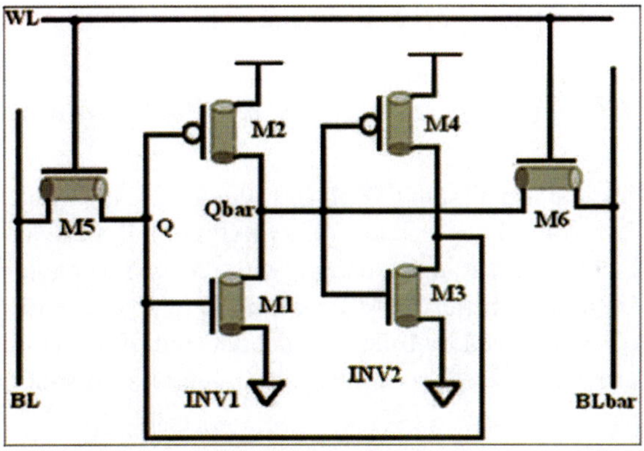

Fig. 1 CNTFET-based 6T SRAM cell [22]

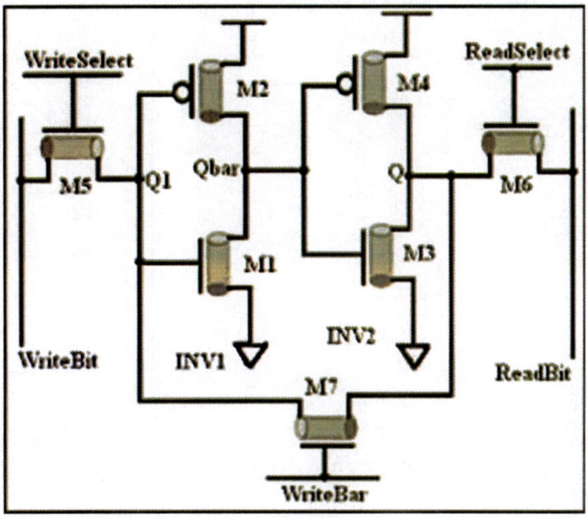

Fig. 2 CNTFET-based 7T SRAM cell [23]

3 Performance Analysis Parameters

There are several design parameters of CNTFETs which play important contribution for the performance evaluation of SRAM in terms of stability, static noise margin, delay, power and leakage current. Extraction of optimum design parameters among them is a necessary investigation [8–22]. This paper deals with performance analysis of 6T and 7T SRAM cells with respect to variations of design parameters of CNTFETs. Some design parameters used here with their variation in values are fermi level, oxide thickness, work function, chiral vector, dielectric constant, etc. [11]. The stability of SRAM cell depends on its static noise margin (SNM) value. Therefore, SNM is evaluated with the variation of each design parameter for both 6T and 7T SRAM cell.

4 Implemented Results

4.1 Performance Analysis with Variations of Channel Length

Static noise margin (SNM) has been evaluated with channel length variation for 32, 22, 16 and 7 nm technology node (Fig. 3). Both 6T and 7T SRAM cells are showing increased static noise margin with reduced technology node [4, 5]. 7T SRAM cell is showing better SNM for all hold, read and write operation in comparison with 6T SRAM cell by following the demand of device scaling. Channel length 7 nm is providing excellent stability for all modes of operations.

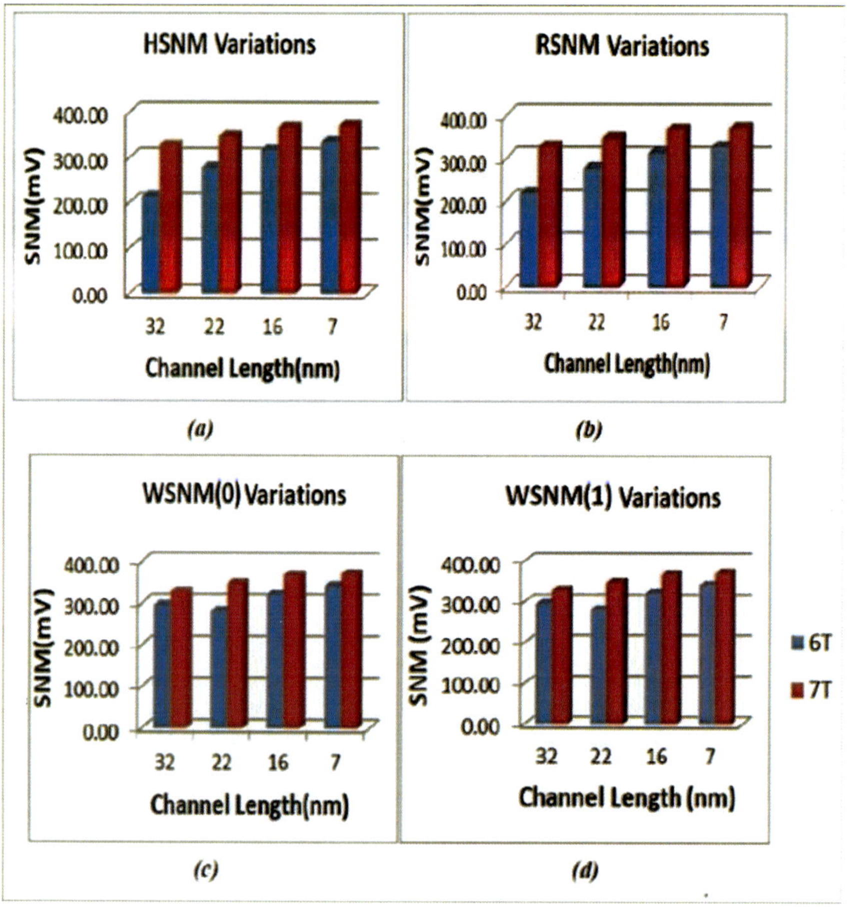

Fig. 3 Variations of **a** HSNM, **b** RSNM, **c** WSNM (0) and **d** WSNM (1) with the variation of channel length (nm)

4.2 Performance Analysis with Variations of Dielectric Material

Silicon dioxide shown excellent thermal and electrical stability interface between substrate and dielectric that is why silicon dioxide with low dielectric constant value, i.e. 3.9, is chosen as a perfect gate dielectric but with technology scaling circuit power dissipation is increased because of gate leakage which is difficult to control thus the alternative solution for this problem is to use dielectric material having high-k dielectric constant [13–20].

Thus, different dielectric materials have been used in this paper having their value from 3.9 to 80. Silicon dioxide (SiO_2), aluminium oxide (Al_2O_3), hafnium oxide (HfO_2), zirconium oxide (ZrO_2), lanthanum oxide (La_2O_3) and titanium oxide (TiO_2) are used with 1 nm dielectric thickness. Performance analysis and a proper comparison are shown in figure by evaluating SNM for 32, 22, 16 and 7 nm technology node. It is observed that for both 6T and 7T SRAM, static noise margin (SNM) is decreased for 32, 22 and 16 nm technology node but excellent stability performance in terms of static noise margin is observed at 7 nm technology node for both 6T and 7T SRAM cell (Fig. 4). In comparison, 7T SRAM cell shown efficient stability than 6T SRAM cell.

4.3 Performance Analysis with Variations of Oxide Thickness

In active state of device transistor, conductivity is increased by reducing the thickness of dielectric constant. Oxide thickness can be reduced at a limit because below 1.5 nm it starts to increase direct tunnelling rapidly consequence degrading the reliability because of high power consumption [16].

In order to reduce tunnelling effect, another efficient approach is to use high-*k* dielectric material with respect to optimum oxide thickness [19]. In this paper oxide, thickness of dielectric material is varied from 1.5 to 9 nm with different dielectric material having dielectric constant 3.9, 9, 15 and 80, respectively. Both 6T and 7T SRAM cells are showing reduced static noise margin for 7 nm technology node (Fig. 5). In comparison 7T SRAM cell is excellent with its stability performance over 6T SRAM.

4.4 Performance Analysis with Variations of Fermi Level

Optimum fermi level is required to obtain efficient performance of CNTFET-based device. Doping level decides the value of fermi level. Barrier height is controlled by increasing the doping level so as to shift fermi level towards the conduction band [6–23]. In this paper, both 6T and 7T SRAM cells are analysed for different values of

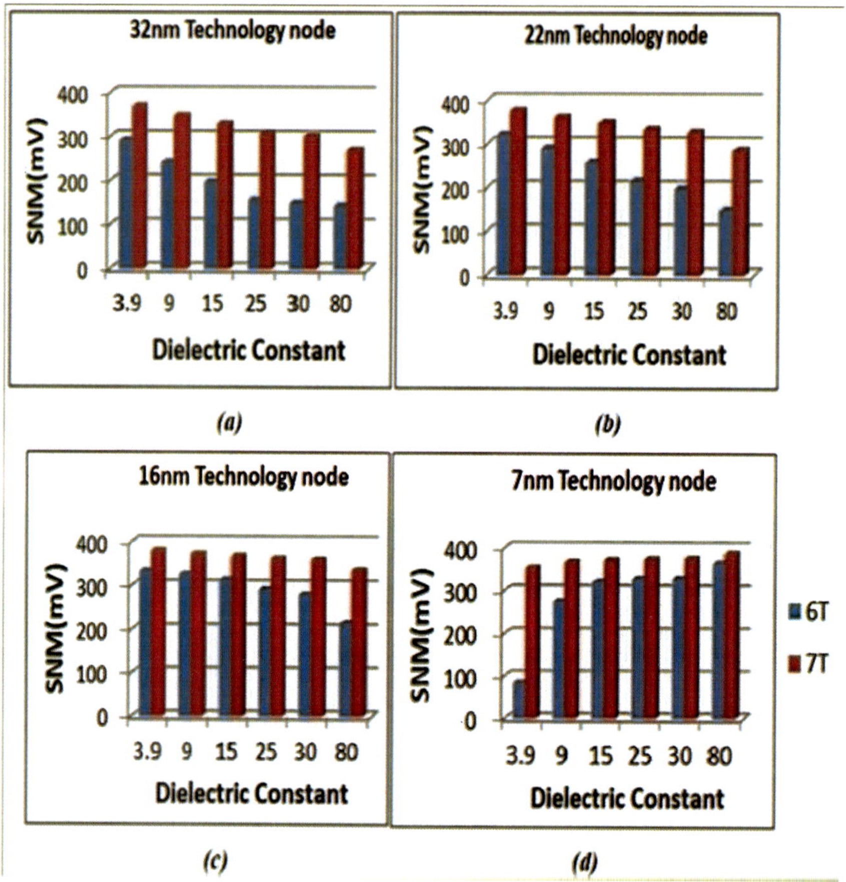

Fig. 4 Variations of SNM for **a** 32 nm, **b** 22 nm, **c** 16 nm and **d** 7 nm technology node with the variation of dielectric constant

fermi level, i.e. 0.4, 0.5, 0.6 and 0.7 eV. Evaluated results show that static noise margin increases rapidly for both 6T and 7T SRAM cell for 32, 22 and 16 nm technology node while it remains constant for 7 nm technology node for both the designs. Again 7T SRAM cell has excellent static noise margin when compared with 6T SRAM cell (Fig. 6).

4.5 Performance Analysis with Variations of Chiral Vector

The structural orientation of CNT depends on chiral vector which defines it behaviour in terms of metal or semiconductor but for CNTFETs its semiconductor behaviour is required. Several parameters like diameter, threshold voltage, shape and size of

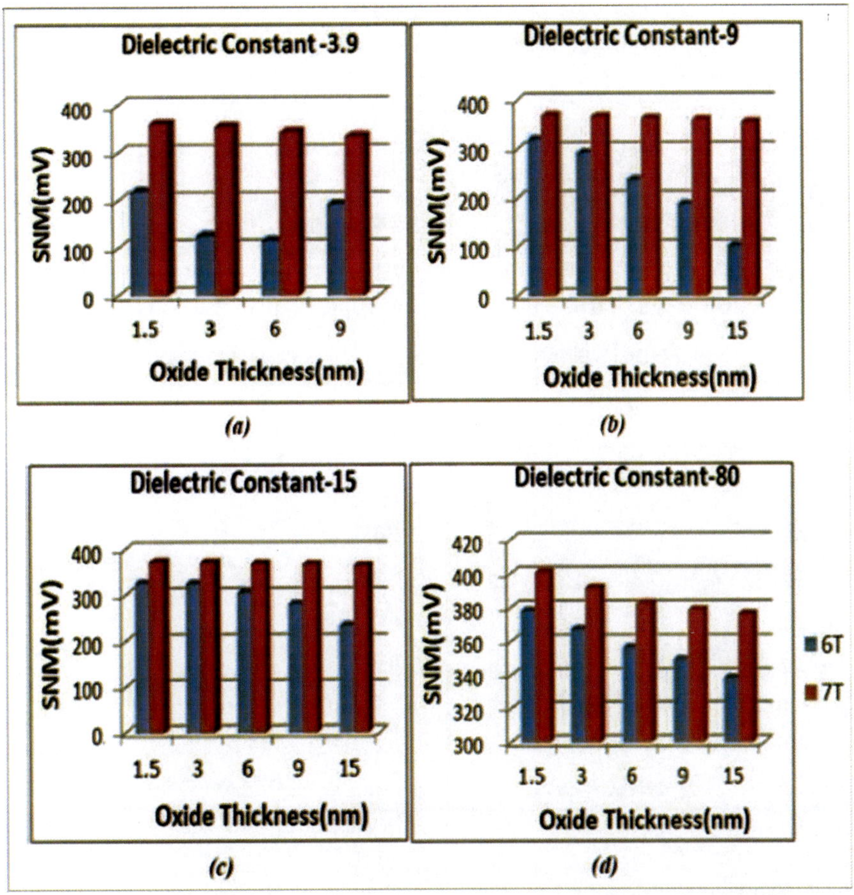

Fig. 5 Variations of SNM for dielectric constant of **a** 3.9, **b** 9, **c** 15 and **d** 80 with the variation of oxide thickness

CNT are determined by chiral vector [7–18]. In this paper, both 6T and 7T SRAM cells are analysed for possible combinations of chiral vector as (10, 0), (13, 0), (16, 0), (19, 0), (22, 0) to evaluate static noise margin to check the stability of cell. Out of all possible combinations, (10, 0) provides more excellent stability for both 6T and 7T SRAM cells for 32, 22, 16 and 7 nm technology node and 7T SRAM cell is more stable than 6T SRAM cell in performance metric (Fig. 7).

5 Conclusion

This paper evaluates stability of both 6T and 7T SRAM cells by evaluating static noise margin for all the necessary operations of SRAM cell and analyses the effect

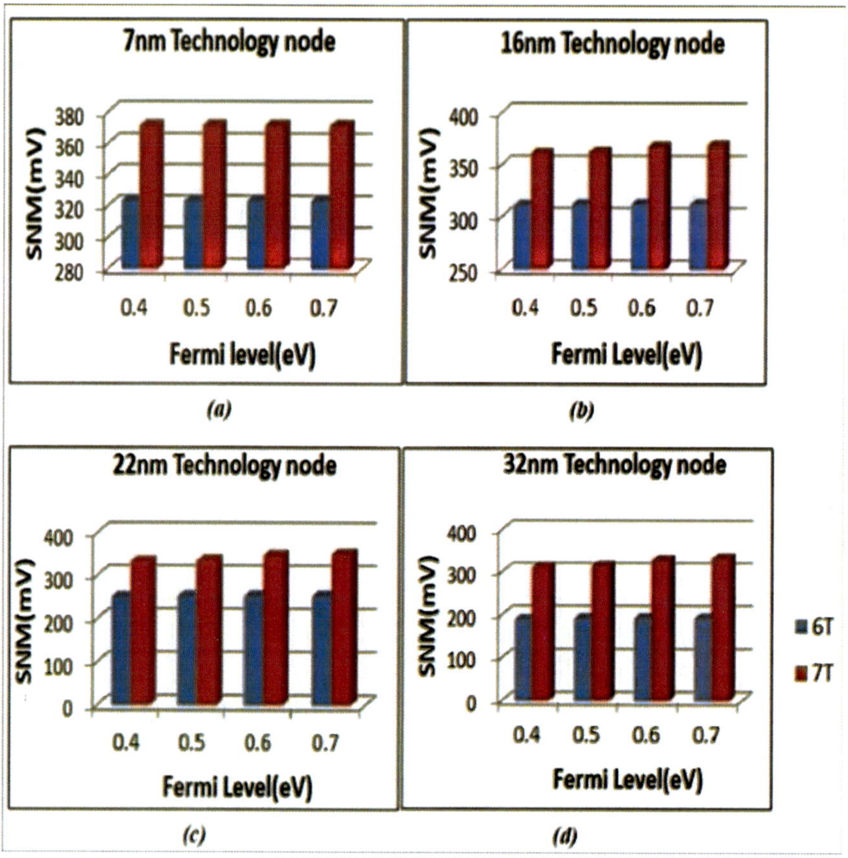

Fig. 6 Variations of SNM for **a** 7 nm, **b** 16 nm, **c** 22 nm and **d** 32 nm technology node with the variation of fermi level

of variation of several design parameters. It is revealed from the results that 7 nm technology node provides efficient stability for both 7T and 6T SRAM design and 7T STAM cell is more stable than 6T SRAM cell. 7 nm technology node yields the best performance with SNM of 371.53 mV. Dielectric materials are varied from 3.9 to 80. The highest value of SNM is obtained for 7 nm technology node with dielectric constant 80 for both 7T- and 6T-based SRAM cell with SNM of 387.77 mV and 363.3 mV, respectively, with oxide thickness 4 nm. Both the cells are analysed for variable values of oxide thickness varied from 1.5 to 9 nm and the efficient results are obtained at 1.5 nm oxide thickness with SNM 401.01 mV and 378.43 mV for 7T and 6T SRAM cells, respectively, for dielectric material having value 80 at 7 nm technology node. The value of fermi level is varied from 0.4 to 0.7 eV for dielectric constant 16 as a dielectric material for oxide thickness 4 nm. The results revealed that HfO_2 yields the optimum SNM of 371.53 eV for 7T SRAM design and 323.37 mV for 6T SRAM design for 7 nm technology node. The evaluated results indicate that

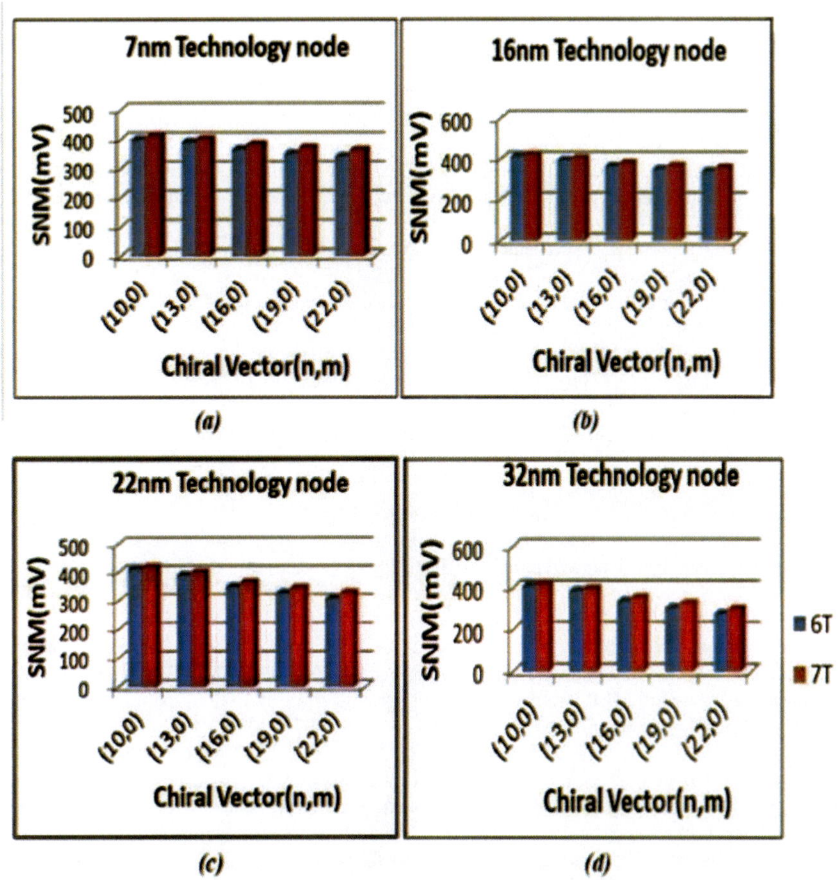

Fig. 7 Variations of SNM for **a** 7 nm, **b** 16 nm, **c** 22 nm and **d** 32 nm technology node with the variation of chiral vector

(10, 0) chiral vector gives the efficient performance from SNM point of view by providing it 410.96 mV and 401.68 mV for both 7T- and 6T-based SRAM cell, respectively. Thus, it is clear from the analysis and evaluation that CNTFET-based 7T SRAM cell is more efficient than CNTFET-based 6T SRAM cell in its stability metrics in terms of static noise margin for hold, read and write operation of SRAM with the variations in the value of different design parameters of CNTFETs to get optimum design parameter for excellent performance.

References

1. Zhang Z, Delgado-Frias JG (2011) Low power and metallic CNT tolerant CNTFET SRAM design. In: 11th IEEE International conference on nanotechnology, Portland Marriott, vol A247, pp 1177–1182
2. Prasad R, Madhavi BK, Lal Kishore K (2011) Design of low write-power consumption SRAM cell based on CNTFET at 32 nm technology. Int J VLSI Design Commun Syst (VLSICS) 2:167–177
3. Prasad R, Madhavi BK, Lal Kishore K (2012) Low leakage-power SRAM cell design using CNTFETs at 32 nm technology. LNICST 108:165–171
4. Murotiya SI, Matta A, Gupta A (2012) Performance evaluation of CNTFET-based SRAM cell design. Int J Electr Electron Eng (IJEEE) 2:78–83
5. Parimala Devi M, Sharmila D, Meenakshi K (2013) Comparative study of 8T SRAM cell using CMOS, FinFET and CNTFET in nanoscale technologies. Int J Electron Commun Technol 4:7–10
6. Sinhal SK, Kumar K, Chaudhurl S (2013) CNTFET: the emerging post-CMOS device. IEEE 8:372–374
7. Sun L, Mathew J, Shafik RA, Pradhan DK, Li Z (2014) A low power and robust carbon nanotube 6T SRAM design with metallic tolerance. EDAA 4:177–182
8. Ahmed SS, Makdey S, Bhoir D (2015) Performance enhancement of VLSI circuits using CNTFETs. Int J Eng Trends Technol (IJETT) 23:1–6
9. Jaiswal M, Singh AR (2015) Design and analysis of CNTFET-based SRAM. Int Res J Eng Technol (IRJET) 2:11–15
10. Nandy T, Dutta A, Mahmood ZH (2015) Analytical distinction between CNTFET based and Mosfet Based SRAMS and logic gates. Electr Electron Eng Int J (ELELIJ) 4:26–37
11. Sayed SI, Abutaleb MM, Nossair ZB (2016) Optimization of CNFET parameters for high performance digital circuits. Adv Mater Sci Eng 2016:1001–1009
12. Mohita TN, Roy T (2016) Design and stability analysis of CNTFET based SRAM cell. In: IEEE Students' conference on electrical, electronics and computer science, vol 2, pp 77–81
13. Sardana N, Ragh LK (2016) Carbon nano tube field effect transistor: a review. Int J Eng Sci Res Technol 5:861–868
14. Rizvi A, Jaiswal P (2016) Study of CNTFET performance over MOSFET using digital logic circuits. IJVDCS 4:363–375
15. Singh A, Khosla M, Raj B (2016) CNTFET modeling and low power SRAM cell design. In: IEEE 5th Global conference on consumer electronics, vol 2, pp 5090–5094
16. Das S, Das D, Rahaman H (2016) Design of content addressable memory cell using carbon nanotube field effect transistors. In: Proceedings of the 2016 IEEE students' technology symposium, vol 2, pp 131–135
17. Bhavar Sonali R, Khule RS (2017) Design and analysis on the Performance of CNTFET Based Inverter in nanometer regime. IJARIIE 3:1779–1784
18. Joshi S, Alabawi U (2017) Comparative analysis of 6T, 7T, 8T, 9T, and 10T realistic CNTFET based SRAM. Hindawi J Nanotechnol 2017:177–185
19. Spasova M, Nikolov D, Angelov G, Radonov R, Hristov M (2017) SRAM design based on carbon nanotube field effect transistor's model with modified parameters, vol 2. IEEE, pp 1–4
20. Patel PK, Malik M, Gupta TK (2018) Design of an ultralow power CNTFET based 9T SRAM with shared BL and half select free techniques, vol 4, pp 1–9. wileyonlinelibrary.com/journal/jnm
21. Prakasha P, Mohana Sundarama K, Anto Bennetb M (2018) A review on carbon nanotube field effect transistors (CNTFETs) for ultra-low power applications. Renew Sustain Energy Rev 89:194–203
22. Divya MK, Singh G (2018) Performance analysis of CNFET based 6T SRAM. COJ Electron Commun 1:1–9

23. Tamil Selvan S, Sundararajan M (2018) Performance evaluation of 3 value logic 7t And 8t CNTFET based SRAM cell design. Int J Pure Appl Math 120:33–47
24. Maheswar Y, Raju BL, Soundara Rajan K (2019) 256K memory bank design with 9T SRAM bit cell and 22 nm CNTFET optimizing for low power and area. Int J Innov Technol Exploring Eng (IJITEE) 8:675–681

Energy Assessment Approach During Context Integration in New-Generation Ubiquitous Networks

Vinodini Gupta and Padma Bonde

Abstract Newer technological advancements in ubiquitous networking have boosted digital dependency of users, thereby facilitating human–computer interaction (HCI). Quality of Experience (QoE) is an important aspect to analyze network performance. However, achieving desired level of user satisfaction is quite challenging which has led to the proliferation of context-aware computing (CAC). The paper focused over various context-gathering mechanisms, system architecture and contemplated on how to improve the Quality of Content (QoC).

Keywords Activity recognition (AR) · Context aware computing (CAC) · Quality of experience (QoE) · Quality of context (QoC) · Network throughput

1 Introduction

With the advent of technology, the lifestyle of humans has become more luxurious and civilized year after year. Evolution of networking from 1G to 5G has increased the digital dependency of users and leads to the integration of different technologies over a single platform to cope up every demand of users.

Context can be defined as the relevant information about user and its surrounding arena based on situational or intentional activities and responses toward external stimuli. Quality of Context (QoC) is a measure to judge the relevancy of the context information. Context integration is unification of parallel occurring activities over different attributes. Context-aware computing (CAC) is intelligent blending of system state, users and the adjoining environment.

To integrate the principles of machine with human understanding and behavior, collaboration of judicious devices with networking devices is necessary. Traditional

V. Gupta (✉) · P. Bonde
Department of Computer Science and Engineering, Shri Shankracharya Technical Campus, Bhilai, Chhattisgarh, India
e-mail: gupta.vinu9@gmail.com

P. Bonde
e-mail: bondepadma@gmail.com

R. P. Mahapatra et al. (eds.), *Proceedings of 6th International Conference on Recent Trends in Computing*, Lecture Notes in Networks and Systems 177,
https://doi.org/10.1007/978-981-33-4501-0_13

systems provide poor HCI. Hence to enhance system performance as well as HCI, context-aware computing is call of hour. CAC can be either network related or user related. Former comprises network-related factors like signal strength, channel capacity, packet size, node mobility, etc. and later includes activities related to user or the mobile host like frame velocity, users' motion, applications installed on host, etc.

The paper summarizes basics of context-enabled networking for correlating network performance and user satisfaction. The further outline comprises description of prominent issues and challenges existing in the field of CAC in Sect. 2. Section 3 briefly described various parameters for computing context information. A brief discussion of context computing mechanism is presented in Sect. 4. A model to calculate energy assessment and battery drainage is discussed in Sect. 5. An overview of context-aware networking techniques is presented in Sect. 6. Section 7 presented brief discussion and analysis of existing trends. Finally, conclusion is presented in Sect. 8.

2 Challenges and Issues in Context Sensing in New-Generation Networks

Context sensing is an intricate activity especially during handoff. In ubiquitous networking, calibration of context information must suit the dynamics of the network. In high mobility scenario, topology changes frequently and therefore managing contextual hierarchy is a complex task. In distributed and scalable scenario, context identification can lead to uneven traffic patterns and loads. So, it must be handled with utmost care. Also, these systems must be persistent and responsive toward real-time requirements of users. Context computing needs additional infrastructure and can cause communication and computational overheads. Due to this, network performance and throughput can be reduced to great extent [1, 2].

Context-aware system requires expensive technology, but it should be implemented in cost-effective manner. Furthermore, it increases infrastructure overhead, so resources must be properly managed. Since wireless devices are battery operated, context gathering becomes tough task. Context fusion and dynamic modeling lead to reliability issues and security threats [3]. To achieve privacy, validation, trust construction and service fairness formulation of authentication protocols is extremely necessary. Similarly, modeling unstructured and inconsistent data like social networking data is tough.

High level of resolution, precision and accuracy is required for fusion and mapping of contextual data into human understandable format. Enhancing visualization, user friendliness and user participation of contextual data is a tedious task [4]. For better performance, conflicts arising in context-aware systems must be resolved properly. QoC management is quite difficult because it gets affected by context attributes, modeling techniques and networking scenario. Context gathering needs additional technological support and therefore degrades the overall QoS standards. Post implementation, maintenance issues like configurability, fault tolerance, accounting, security and overall performance also need to be addressed properly [5].

3 Context Parameters in Ubiquitous Networks

Context parameters describe the relationship between users' activity and the surrounding environment at any given instant of time. These parameters can be categorized as per the computational requirement of space, time and relationship. Based on conceptual grouping [1], context can be classified as either environmental or user-centric (virtual) context, whereas, according to physical representation, the contexts can be either low-level contexts [1] or high-level contexts. The literature in [6] classified context on the basis of source of acquisition and domain of influence. Based on former criteria, the context can be static or primary context, derived or secondary context, intellectual context and profiled context. However, based on latter criteria, it can be grouped as active context and passive context. Similarly according to the perspective of context, it can be of four types, namely physical context, computational context, user context and time context [7]. With the viewpoint of placement site, the context information can be either terminal-side context or network-side context, and on the basis of changes, context can be either static or dynamic [8].

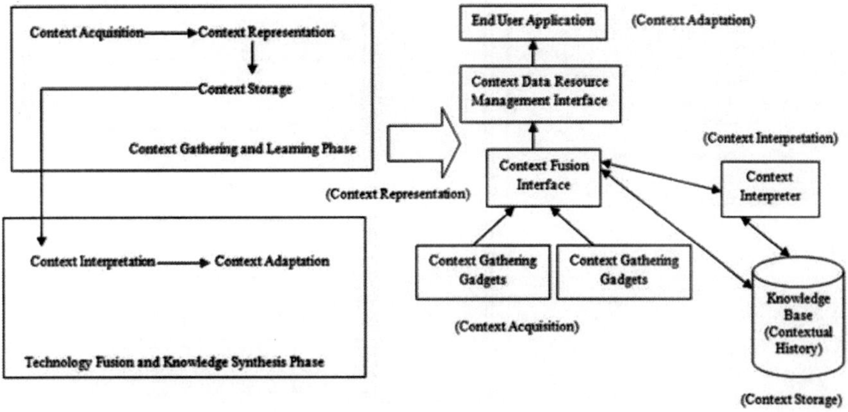

Fig. 1 Context computing mechanism

4 Context Computing Mechanism in Ubiquitous Networks

CAC is a two-phase mechanism [2]. Various steps involved are depicted in Fig. 1. In context learning and gathering phase, the raw data is processed to build contextual hierarchy and formulate metadata. Basic steps involved are context acquisition, in which data is gathered from the vicinity, context representation, where the gathered data is properly modeled, validated and organized into hierarchal structure and context storage where the structured data is stored at knowledge base to form and maintain metadata. In technology fusion and knowledge synthesis phase, accurate matching of context information and human perception is performed. Gathered context is transformed into useful context through various reasoning techniques [1], and context adaptation is performed to enhance user participation in decision making.

Context computing is a layered activity and can be modeled through different kinds of context models [6]. Various modeling techniques [1] can be categorized as object-based, ontology-based, key-value based, graphical, logical or markup scheme modeling. The service layers involve network infrastructure layer, middleware layer, user infrastructure layer and application layer. These are assisted by context data distribution module (CDDM) [7]. The layered organization [3] is diagrammatically represented in Fig. 2.

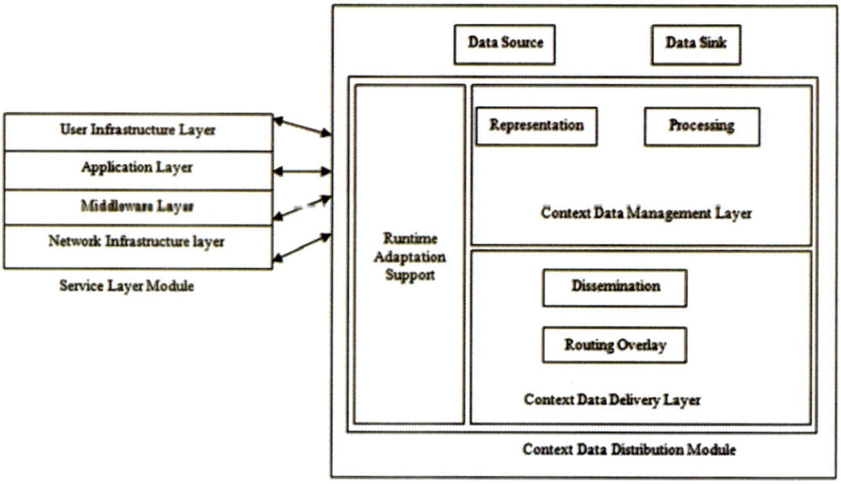

Fig. 2 Context computing layered architecture

5 Generic Assessment of Energy During Context Gathering

Energy consumption during context computing depends on three factors, namely voltage, current ratings and average power consumed by the device. The basic steps followed in energy assessment are represented in Fig. 3.

Mathematically, power can be defined in terms of voltage and current by Eq. 1.

$$\text{Power } (P) = \text{Voltage } (V) \times \text{Current } (I) \tag{1}$$

Similarly, instantaneous power (P_i) (in Watts) can be computed from instantaneous voltage (V_i) (in Volts) and instantaneous current (I_i) (in Ampere) using Eq. 2.

$$\text{Instant Power } (P_i) = \text{Instant Voltage } (V_i) \times \text{Instant Current } (I_i) \tag{2}$$

For n number of samples, average power (P_{avg}) (in Watts) can be calculated by Eq. 3.

$$\text{Average Power } \left(P_{avg}\right) = \frac{1}{n} \times \sum_{i=1}^{n} P_i \tag{3}$$

For time period T (in secs), overall energy (E_i) (in Joules) can be calculated from Eqs. 4 and 5.

$$T = T_{end} - T_{start} \tag{4}$$

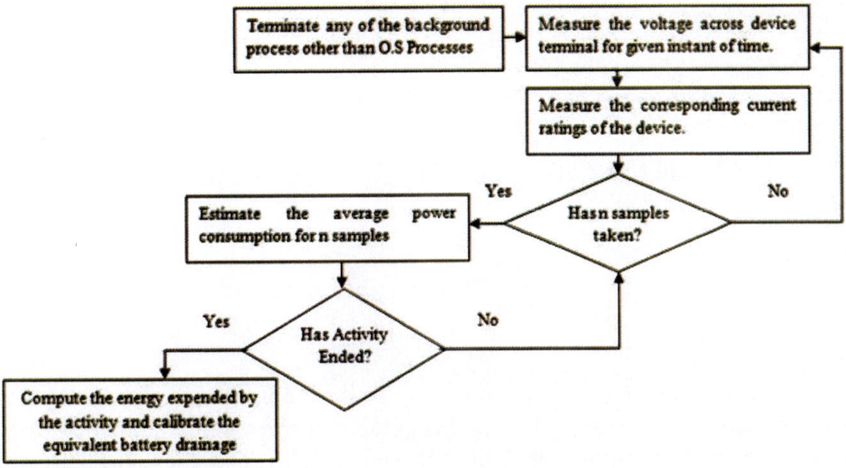

Fig. 3 Energy assessment model during context gathering

where

T Total runtime of context activity
T_{start} Start time of context activity
T_{end} End time of context activity.

$$\text{Energy Consumed } (E_i) = \frac{1}{n} \times \sum_{i=1}^{n} P_i \times t \tag{5}$$

Average energy (E_{avg}) (in Joules) of any ongoing activity is represented by Eq. 6.

$$\text{Average Energy } \left(E_{\text{avg}}\right) = \frac{1}{k} \times \sum_{i=1}^{k} E_i \quad (\text{where } k = \text{no. of iterations}) \tag{6}$$

Considering the standard deviation in time and energy, the above Eqs. 5 and 6 can be represented as Eqs. 7 and 8 respectively.

$$\text{Standard Energy } (E_{\text{std}}) = \sqrt{\frac{1}{k-1} \times \sum_{i=1}^{k} \left(E_i - E_{\text{avg}}\right)^2} \tag{7}$$

$$\text{Standard Time } (T_{\text{std}}) = \sqrt{\frac{1}{k-1} \times \sum_{i=1}^{k} \left(T_i - T_{\text{avg}}\right)^2} \tag{8}$$

Equivalent battery drainage (in Ampere hours) can be computed by Eqs. 9 and 10.

$$\text{Average Battery Drainage } \left((\text{BD})_{\text{avg}}\right) = E_{\text{avg}} / (\text{voltage} \times 3600) \tag{9}$$

$$\text{Standard Battery Drainage } ((\text{BD})_{\text{std}}) = E_{\text{std}} / (\text{voltage} \times 3600) \tag{10}$$

6 Context-Gathering Techniques in Ubiquitous Networks

Context awareness plays an important role in delimiting the service quality and network performance. Korpipaa et al. [9] proposed a software-based framework to enhance context computing capability of mobile devices. The study [10] presented framework for integrating resource management techniques with context awareness to improve interoperability and performance of the networks. Alhazmi and Akkari

[11] presented categorization and brief review of basic behavioral pattern of context-aware schemes. The recommendation system at [12] proposed collaborative filtering for enhancing alliance of users' equipments with ubiquitous networks. Xu et al. [13] applied the concept of context awareness for load balancing in LTE-Advanced Relay Networks.

For enhancing user experience, Quadros et al. [14] proposed extension of Media Independent Handover (MIH) schemes. Orlandi et al. [15] proposed a connection management technique by exploiting the features of Wi-Fi and cellular technology. Similarly, algorithm [16] proposed to enhance Quality of Experience in vehicular networks. A handoff management scheme based on machine learning was proposed in [17]. Adell and Gonzalez [18] proposed a positioning-based handoff algorithm for enhancing user experience by facilitating indoor and outdoor transitions.

For enhancing both service continuity and QoE, context-enabled handoff approach for load balancing [19] was proposed. Sarma et al. [20] proposed context-based mobility management approach for effective traffic offloading. Cherkaoui and Agoulmine [21] proposed a context-based network selection strategy during handover in dynamic and diverse scenario. Sarma et al. [22] presented a context-based handoff scheme to enhance the QoS and QoE standards.

For energy optimization, context-aware bandwidth allocation technique was proposed in [23] to optimize transmit power. Ylianttila et al. [24] presented location-aware mechanism to integrate WLAN hotspot with cellular networks for analyzing handoff. Pawar et al. [25] presented a context-aware handoff mechanism for nomadic mobiles and evaluated its performance in remote patient monitoring system.

To enhance the mobility support, TalebiFard and Leung [26] proposed a context-aware decision algorithm for incorporating the state of user while selecting target network for handoff. A novel context-aggregation mechanism. Miettinen et al. [27] proposed a context-aware framework for enhancing the access control of mobile devices in ubiquitous networking. Lin et al. [28] proposed a prediction algorithm for location detection during context-based handoff in telecom networking. Niutanen [29] proposed a novel idea for effectively managing the dynamic and diverse context-related data. For enhancing performance, Yu et al. [30] proposed location prediction algorithm based on users' activity.

To cope up mobility issues, Guidolin et al. [31] presented context-enabled handover policy to enhance system performance and user satisfaction simultaneously. To enhance the Quality of Service (QoS) and reducing latency and signaling overheads, Al Emam et al. [32] introduced Multi-layer Parallel Handover Optimization (MPHO) for cross-layer triggering. Naeem et al. [33] proposed a user-centric approach for multi-homing of various multi-Radio Access Technologies (multi-RAT) by traffic offloading for load balancing, higher data rates, increased capacity and better QoS. Pihkola et al. [34] discussed the energy consumption pattern during mobile data transfer over cellular networks.

7 Discussion and Analysis

In ubiquitous networks, integration of context-aware techniques is very challenging, especially during handoff. Balasubramaniam and Indulska [8] presented a survey over handoff scenario in context-aware environment. For this, context can be defined on the basis of frequency of change and their placement position. Therefore, context information can be modeled as either static model or dynamic model for both network and mobile host (MH). Also, the communication scenario can be classified as terminal initiated and network controlled, network initiated and terminal controlled, terminal initiated and controlled with network assistance, network initiated and controlled with terminal assistance.

Ahmed et al. [2] briefly explained the context-aware handoff mechanism and categorized the context-aware handover strategies as mobile agent-based schemes, AHP-based schemes, mobility prediction-based schemes, cooperation-based schemes and MIH (Media Independent Handoff)-based schemes. Similarly, Ylianttila et al. [24] compared dwell timer based and power based handoff algorithms over mobility and delay factors in location-aware integrated network. Simulations showed dwell timer performed better and was less sensitive. However, certain adaptive measures need to be taken to further enhance the performance of the algorithm. Study in [25] analyzed the effect of multi-homing over remote patient monitoring system and provided various useful inferences to enhance handoff support by extending middleware functionalities. However, experimentations can be conducted to reduce the deviations.

Similarly, modified WPM (Weighted Product Mean) is proposed at [26] for enhancing decision making during context-aware handoff. The algorithm found to be better in terms of performance and robustness when compared to TOPSIS. Prototype architecture for handoff has also been developed to support seamless video transmission in pervasive computing.

For enhancing user association, Alhazmi and Akkari [12] proposed collaborative filtering technique to enhance recommendations at base stations according to various measurable parameters. It rated users' satisfaction through satisfaction equilibrium. This approach proved advantageous over other techniques. It successfully dealt with the cost-profit trade off and congestion problem.

For resource management, COTS (Channel-Aware Optimized Traffic Shifting) [13] mechanism effectively offloaded traffic at user end thereby reducing both call blocking and dropping probability. Simulation results [23] showed power optimization by 50% during on demand video streaming. Tuysuz et al. [35] presented a brief discussion over the energy gains during handover and concluded that energy efficiency can be achieved by transmitting set of varying information over network.

QoEHand [14] improvised MIH scheme and analyzed subjective and objective impact of various performance metrics to figure out the benefits of the proposed scheme over user perception. Orlandi et al. [15] gathered context information for managing connections and proved fruitful in improving security, service continuity, reducing implementation cost thereby enhancing user satisfaction. However, scalability and latency issues still need to be considered. To cope up mobility issues, the algorithm [15] upgraded the performance and robustness of the system. However, accuracy can be further enhanced and user interference can be allowed to reduce the subjective error.

Simulations [9] showed that the proposed framework facilitated context reusability, improved accuracy and proved to be reliable in processing heterogeneous context data. The algorithm [17] detected factors hindering the performance and QoE standards during handover, which can be enhanced by taking larger and more diverse networking samples into consideration. Marquez-Barja et al. [16] evaluated connectivity requirements of vehicles during handoff and improved connectivity, user satisfaction and network throughput. The context-based model presented in [31] enhanced the handover performance and can further be modeled with exact knowledge of equipment trajectory and machine learning. conXsense [27] proposed probabilistic approach for enhancing user access. However, performance can further be enhanced by considering richer contextual sets. Similarly, energy-efficient handoff mechanism [18] minimized deployment cost and enhanced the positioning accuracy and user experience irrespective of the networking scenario and class of the users. Prediction algorithm [28] enhanced throughput, performance and prediction capability of network within 20% error range. However, the efficiency of the algorithm can still be enhanced. Cherkaoui and Agoulmine [20] proposed context-based traffic scheduling for effective data transmission in e-health services to ensure better service quality and continuity in e-health system. To enhance the QoS standards, Mougy and Mouftah [10] proposed a conceptual framework for resource management in context-enabled networking. The study formulated basic challenges faced in bandwidth allocation and traffic prioritization. The rate adaptive streaming methodology proposed in [32] reduced the handoff failure rate thereby increasing the throughput and overall Quality of Service (QoS) under dynamic network conditions.

The interoperability of Wi-Fi-WiMAX integrated scenario [19] effectively balanced the load to meet both QoS and QoE standards simultaneously. However, adaptability can be further enhanced. Niutanen [29] envisioned a novel research idea to promote interoperability, resource management and efficient use of dynamic context-related data in a widespread networking ecosystem. The context-based handover policy [22] ensured proper bandwidth utilization, load balancing and traffic scheduling and promoted QoE and QoS standards to desired level.

8 Conclusion

This paper presented a brief literature store on context-aware computing paradigm in ubiquitous networking. Context-gathering applications are rapidly gaining roots in world of ubiquitous networking. The paper discussed the attributes affecting the Quality of Context (QoC) and various research challenges prevailing in context-aware computing domain. The paper explained the basic structure of context-aware systems to provide new researchers elementary knowledge of this field. The study also presented a model for energy assessment during context gathering and integration. It also focused over presently existing context-aware techniques and presented a brief analysis of existing techniques to find out the gap still existing in this field. This study serves as the blueprint to novel researchers for enhancing the richness of human–computer interaction (HCI) in an energy-efficient manner which will consequently lead to revolutionary changes in the world of ubiquitous networking leading toward more competent green networks.

References

1. Zhang D, Huang H, Lai CF, Liang X, Zou Q, Guo M (2013) Survey on context-awareness in ubiquitous media. Multimed Tools Appl 67:179–211
2. Ahmed A, Boulahia L, Dominique Gaιti D (2014) Enabling vertical handover decisions in heterogeneous wireless networks: a state-of-the-art and a classification. In: IEEE communications and survey tutorials, vol 16, no 2, Second Quater 2014
3. Hong J, Suh E, Kim S (2009) Context-aware systems: a literature review and classification. Expert Syst Appl 36:8509–8522
4. Nazir Malik N, Mahmud U (2007) Future challenges in context—aware computing. In: IADIS international conference
5. Hegering HG, Kupper A (2003) Management Challenges of context-aware services in ubiquitous environments. In: 14th IFIP/IEEE international workshop on distributed systems: operations and management, DSOM 2003
6. Charith Perera C, Zaslavsky A, Christen P, Georgakopoulos D (2013) Context aware computing for the internet of things: a survey. In: IEEE communications and survey tutorials
7. Bellavista P, Corradi A, Fanelli M, Foschini L (2013) A survey of context data distribution for mobile ubiquitous systems. ACM Surveys
8. Balasubramaniam S, Indulska J (2003) Handovers between heterogeneous networks in pervasive systems. In: IEEE Proceedings of ICCT 2003, April 2003
9. Korpipaa P, Mantyjarvi J, Kela J, Keranen H, Malm EJ (2003) Managing context information in mobile devices. Pervasive Computing, IEEE CS and IEEE ComSoc, 2003
10. Mougy A, Mouftah H (2013) On resource management and context-awareness in LTE-based networks for public safety. In: 13th annual IEEE workshop on wireless local networks. IEEE
11. Alhazmi H, Akkari N (2011) An overview of context-aware vertical handover schemes in heterogeneous networks. Int J Comput Sci Eng Surv (IJCSES) 2(4)
12. Meng Y, Jiang C, Xu L, Ren Y (2015) User association in heterogeneous networks: a social interaction approach. IEEE Trans Vehic Technol
13. Xu L, Luan Y, Chao K, Cheng X, Zhang H, Schormans J (2014) Channel-aware optimised traffic shifting in LTE-advanced relay networks. In: IEEE 25th international symposium on personal, indoor and mobile radio communications
14. Quadros C, Cerqueira E, Neto A, Riker A, Immich R (2012) A mobile QoE architecture for heterogeneous multimedia wireless networks. In: The 4th IEEE international workshop on mobility management in the networks of the future world
15. Orlandi B, Bizouarn E, Taburet F, Scahill F, Lafragette J, Evenden R, Ringland S, Johnson S, Melia T, Twell T, Mghazli Y (2011) Improving the customer experience for heterogeneous wireless access. Alcatel-Lucent
16. Marquez-Barja JM, Ahmadi H, Tornell SM, Calafate CT, Cano JC, Manzoni P, DaSilva LA (2015) Breaking the vehicular wireless communications barriers: vertical handover techniques for heterogeneous networks. IEEE Trans Vehic Technol 64(12)

17. Ali Z, Baldo N, Bafalluy JM, Giupponi L (2016) Machine learning based handover management for improved QoE in LTE. In: IEEE/IFIP network operations and management symposium (NOMS 2016)
18. Adell MC, Gonzalez JP (2013) Smart indoor-outdoor positioning handover for smartphones. In: International conference on indoor positioning and indoor navigation, 28–31 Oct 2013. IEEE
19. Sarma A, Chakraborty S, Nandi S (2016) Deciding handover points based on context-aware load balancing in a WiFi-WiMAX heterogeneous network environment. IEEE Trans Vehic Technol 65(1)
20. Cherkaoui E, Agoulmine N (2014) Context-aware mobility management with WiFi/3G offloading for ehealth WBANs. In: IEEE 16th international conference on e-health networking, applications and services (Healthcom)
21. Lijuan L, Tingjie LV, Xia C (2013) Network selection based on context-awareness services. In: ICT industry convergence, China communications, December 2013
22. Sarma A, Chakraborty S, Nandi S (2014) Context aware handover management: sustaining QoS and QoE in a public IEEE802.11e Hotspot. IEEE Trans Netw Serv Manage 11(4)
23. She C, Yang C (2015) Context aware energy efficient optimization for video on-demand service over wireless networks. In: IEEE/CIC ICCC 2015 symposium on communication and control theory
24. Ylianttila M, Makela J, Pahlavan K (2005) Analysis of handoff in a location-aware vertical multi-access network. Comput Netw 47:185–201
25. Pawar P, Beijnum BJ, Sinderen M, Aggarwal A, Maret P, Clercq F (2008) Performance evaluation of the context-aware handover mechanism for the nomadic mobile services in remote patient monitoring. Comput Commun 31:3831–3842
26. TalebiFard P, Leung VCM (2011) A dynamic context-aware access network selection for handover in heterogeneous network environments. In: IEEE, The mobility management in the networks of the future world (MobiWorld) workshop
27. Miettinen M, Heuser S, Kronz W, Sadeghi A, Asokan N (2014) ConXsense—automated context classification for context-aware access control. In: ACM 2014
28. Lin Y, Huang-Fu C, Alrajeh N (2013) Predicting human movement based on telecom's handoff in mobile networks. IEEE Trans Mobile Comput 12(6)
29. Niutanen J (2013) Improving usability and context awareness over fifth generation (5G) wireless networks. In: Proceedings of 5th conference of Fruct Association
30. Yu C, Liu Y, Yao D, Yang LT, Jin H, Chen H, Ding Q (2015) Modeling user activity patterns for next-place prediction. IEEE Syst J
31. Guidolin F, PappalardoI, Zanella A, Zorzi M (2016) Context-aware handover policies in HetNets. IEEE Trans Wirel Commun 15(3)
32. Al Emam FA, Nasr ME, Kishk SE (2019) Context-aware parallel handover optimization in heterogeneous wireless networks. Ann Telecommun
33. Naeem M, ELAttar HM, Dahab MA (2019) An optimized load balance solution for multi-homed host in heterogeneous wireless networks. Sens MDPI J
34. Pihkola H, Hongisto M, Apilo O, Lasanen M, Vatanen S (2018) Energy consumption of mobile data transfer—increasing or decreasing?—Evaluating the impact of technology development & user behavior. In: ICT4S2018. 5th international conference on information and communication technology for sustainability. EPiC Series in Computing, vol 52, pp 301–315
35. Tuysuz MF, Trestian R (2017) Energy-efficient vertical handover parameters, classification and solutions over wireless heterogeneous networks: a comprehensive survey. Wireless Pers Commun

Modeling and Analysis of Fractal Antenna Using Minkowski Island Technique for Wireless Body-Centric Communication

Raghvendra Singh, Vivek Kumar, Yogesh Mohan Dubey, Gaurav Sahu, Dambarudhar Seth, and Mukesh Arora

Abstract This paper approaches and investigates the study of modeling and development of microstrip patch antennas for wireless body-centric communication. In view of new developments, it exhibits and analyzes the fractal antennas using Minkowski Island technique for on-body communication to miniaturize and make the antenna more efficient. A fractal figure is used multiple times or its down scaled figures to broaden the boundary of it which results in generation of additional current and miniaturization of antenna. The design of fractal antenna using Minkowski Island split ring resonator technique achieves multiband operation (frequency bands 2.4 GHz, 3.5 GHz, and 5.2 GHz) which concentrates on wireless local area network (WLAN) and worldwide interoperability for microwave access (Wi-Max). This fractal antenna using Minkowski Island technique is designed and analyzed on the basis of the parameters: reflection coefficient, voltage standing wave ratio, radiation pattern, and SAR.

R. Singh (✉) · V. Kumar · Y. M. Dubey · G. Sahu
Pranveer Singh Institute of Technology, Uttar Pradesh, Kanpur, India
e-mail: planetraghvendra@gmail.com

V. Kumar
e-mail: rastogi0807@gmail.com

Y. M. Dubey
e-mail: yogesh.ec@psit.ac.in

G. Sahu
e-mail: sahugaurav12@gmail.com

D. Seth
SRM Institute of Science and Technology, Uttar Pradesh, Ghaziabad, India
e-mail: ddgayatri@yahoo.co.in

M. Arora
Swami Keshwanand Institute of Technology, Management and Gramothan, Jaipur, India
e-mail: arora.mukesh02@gmail.com

R. P. Mahapatra et al. (eds.), *Proceedings of 6th International Conference on Recent Trends in Computing*, Lecture Notes in Networks and Systems 177,
https://doi.org/10.1007/978-981-33-4501-0_14

Keywords Wireless body-centric communication · Fractal antennas · Minkowski Island technique · On-body antenna · Layered tissue model · Specific absorption rate

1 Introduction

Nowadays, there has been increasing interest in the field of wearable electronics for medical applications, entertainment industry, and military purpose. The focused aim of wearable electronics is to provide wireless communication from the body of user to external world via antennas. Therefore, wearable antennas are playing the key role of providing communication between the wearable devices and receiving stations for structuring wireless body-centric communication. Due to growing wireless applications, the appropriate antennas requirement is also gaining heights in the field of body-centric wireless communication. The desired properties of antennas for this purpose are: miniaturized size, low battery consumption, conformable, and low cost [1, 2]. Figure 1a is depicting the basic schematic of wireless body-centric communication using on-body sensors on patient's body and its connectivity with physician who is at far away from the patient. Figure 1b is showing the branching of wireless body-centric communication according to the placement of sensors on/around the body of user. The sensors on the body of patient record the vital data and send it to other external wireless environment using efficient body-worn antenna [3–5].

The term body area network (BAN) is currently used to provide a common platform to all types of wireless body-centric communication. Wireless body-centric communication is branched in three parts (i) On-body: Communication is established between different types of wearable sensors, and these sensors are integrated with the antenna to communicate the signals wirelessly which is depicted in schematic. (ii) Off-body: Communication between wearable devices and outer devices (receiver, router, or Internet). (iii) In-body: Antennas and sensors placed inside the human body for therapeutic purposes are called medical implants. Submicron electronics, nanotechnology, and microelectromechanical systems (MEMS) technologies prepared a background of implanted sensing devices [4]. IEEE made standards for body area network and an IEEE task group, IEEE 802.15.6, has been established in November 2007.

Khan et al. [4] designed and investigated a novel dual-band antenna which shows diverse radiation pattern and efficient power transmission. The performance of antenna in proximity of human body is studied. This antenna operates in two bands 2.45 GHz and 1.9 GHz; it is capable of generating sufficient power for on-body and off-body communication. The radiation pattern, gain, detuning, and return loss parameters are analyzed. Roblin et al. [5] investigated an antenna which operates in two bands, 800–900 MHz and 2.4 GHz, and used in wearable applications. This small size and narrow band antenna is designed for body area network. The experimental study is also done in presence of the body tissue model.

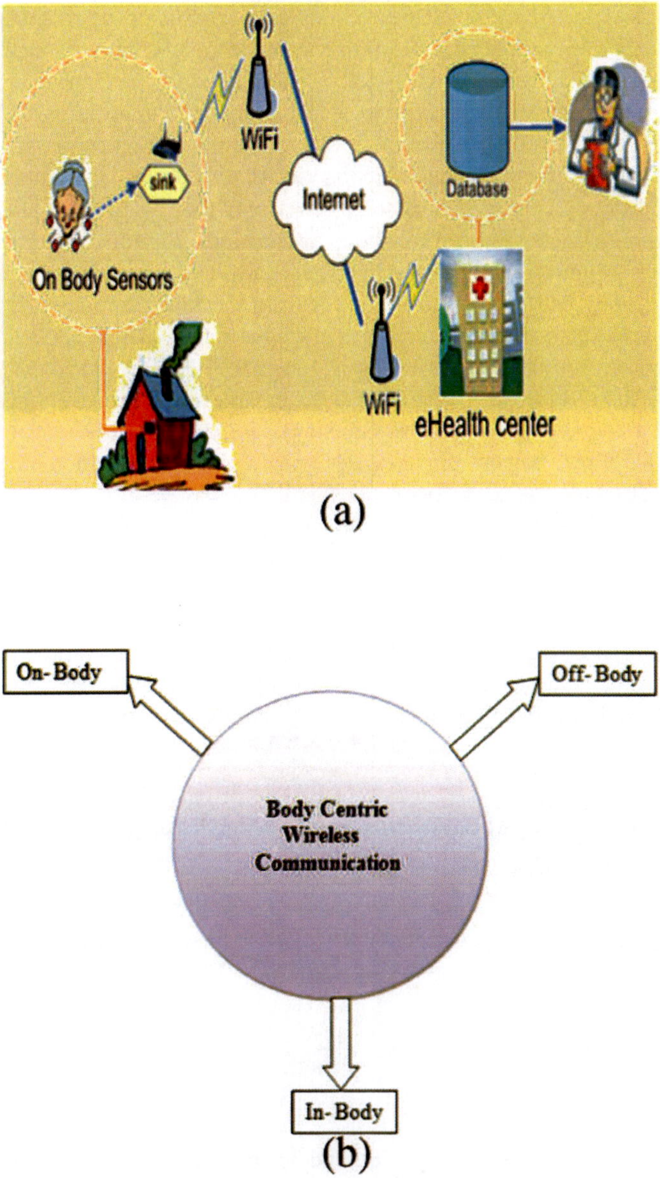

Fig. 1 **a** Schematic of wireless body-centric communication architecture using on-body sensors on patient's body and its connectivity with physician at far distance. **b** Branching of wireless body-centric communication according to the placement of sensors on/around the body [5]

Kanaujia et al. [6] presented a novel design of microstrip patch antenna and its analysis for Wi-Max applications. An equivalent electrical circuit of antenna is drawn for theoretical formulation. Conway and Scanlon [7] studied compact and wearable microstrip patch antennas which are excited at higher modes operating in ISM band. The HMMPA achieved path gain of 11 dB and compared with a classic monopole antenna, and this antenna is studied in the vicinity of a sample muscle tissue.

Mackowiak and Correia [8] statistical behavior of the gain pattern of the proposed antenna is investigated for movable body applications. Motion capture data is used for making the movement of body in this scenario and it seems the realistic view of moving body. Head, chest, arm, and leg are the four positions of antenna for testing at 2.45 GHz. Tanaka et al. [9] analyzed a novel antenna operating at GSM band (1.8 GHz) for wearable applications. Liu et al. [10] proposed an antenna which operates in 2.4 GHz and 5.8 GHz bands for wearable applications. Nornikman et al. used fractal shapes to achieve multifrequency operation for a basic rectangular patch [11], and a multifrequency antenna is designed with miniaturization of dimensions by using self-similarity property of fractal structures.

Minkowski Island split ring resonator technique is applied here in this work to achieve multiband operation for the designing of miniaturized fractal antenna, which is analyzed for wireless body-centric communication applications. Two highly demanded bands wireless local area network (WLAN) and worldwide interoperability for microwave access (Wi-Max) are covered in the operation of proposed fractal antenna. This technique can be used in future for the designing of other efficient and miniaturized antennas in wireless body-centric communication field.

2 Design of Fractal Antenna Using Misnkowski Island Split Ring Resonator Structure

Nowadays fractal structures are popularly used for antenna design to achieve broadband operations. Fractals are characterized by the parameter, fractal dimension D [12].

$$D = \frac{\text{Log}(L)}{\text{Log}(r)} \tag{1}$$

where L and r are one-dimensional total length before fractalization and one-dimensional length after fractalization, respectively. Box dimension DB is another parameter and defined as [12]

$$\text{DB} = \lim_{\delta \to 0}\left(\frac{\text{Log}N\delta(F)}{\text{Log}(1/\delta)}\right) \tag{2}$$

where F is non-empty subset of R^n and $N\delta(F)$ is the smallest number of sets of diameter atmost δ [13] Hausdorff-Besicovich dimension fractal is defined as

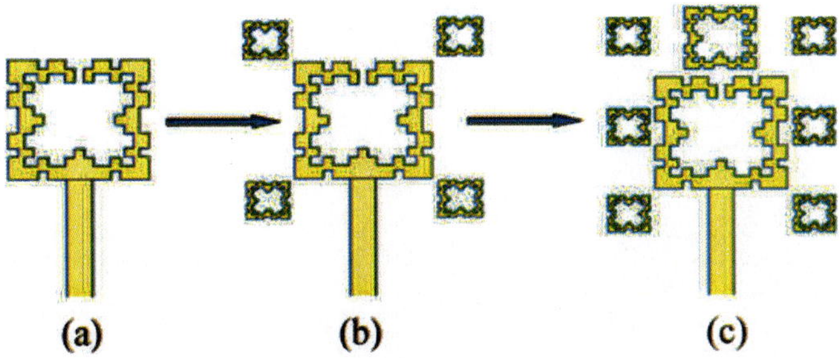

Fig. 2 Steps of designing Minkowski Island split ring resonator structure **a** basic fractal resonator, **b** four Minkowski Island Resonators, **c** improved fractal antenna with Minkowski Island resonators

$$1 = K1\left(\frac{1}{h1}\right)D + K2\left(\frac{1}{h2}\right)D + K3\left(\frac{1}{h3}\right)D \ldots Km\left(\frac{1}{hm}\right)D \tag{3}$$

where *Km* and *hm* are the number of copies of the generator and number of different scale that the fractal containing, respectively [14]. Pfeiffer introduced a new term perimeter compression (PC);

$$\text{PC} = \text{full sized antenna element length/fractal reduced antenna element length} \tag{4}$$

A multifrequency antenna is designed with miniaturization of dimensions by using self-similarity property of fractal structures. Nornikman et al. [11] used fractal shapes to achieve multifrequency operation for a basic rectangular patch. Minkowski Island split ring resonator structure is depicted in Fig. 2.

Improved Minkowski Island split ring resonator structure is used here for WBAN to perform in the three frequency bands 2.4 GHz, 3.5 GHz, and 5.2 GHz. This fractal antenna is designed over a FR4 substrate of dimensions 32 mm × 40 mm × 1.6 mm, dielectric constant (ε) 4.3 and tangent delta 0.002. The dimensions of rectangular patch are 13.73 mm × 13.73 mm, the Minkowski Island I (MI-SRR) 4.06 mm, Minkowski Island II 6.93 mm. Figure 3 is depicting the front view and the back view of fractal antenna with Minkowski Island split ring resonator structure. The dimensions of partial rectangular ground plane are 21 mm × 7 mm (Table 1).

The layered tissue model is designed according to the mathematical model proposed by Gabriel et al. and Kobayashi et al. [15, 16]. The electrical parameters of proposed antenna are investigated in the presence of human body tissue model. The proposed antenna is placed horizontally over the tissue structure [17] and the observations are taken at four positions, 2 mm, 4 mm, 6 mm, and 8 mm, far away from the human tissue model as shown in Fig. 3c, considering the air in between the proposed antenna and tissue model.

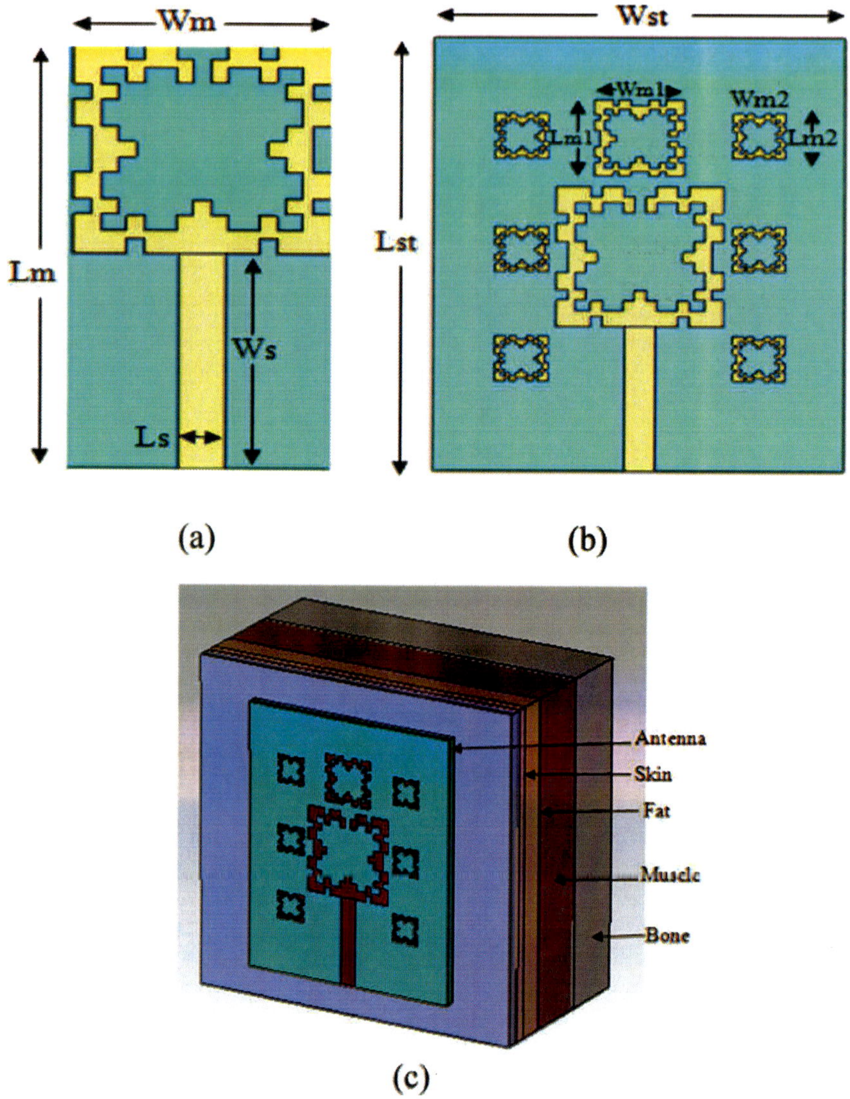

Fig. 3 **a** Fractal antenna: front view. **b** Fractal antenna with Minkowski Islands. **c** Perspective view of Minkowski Island antenna placed horizontally over the tissue model

Table 1 Dimensions of proposed antenna

Dimensions	Value	Dimensions	Value	Dimensions	Value
Wm	13.73	Wst	32.0	Wm2	4.06
Lm	13.73	Lst	40.0	Lm2	4.06
Ws	13.5	Wm1	6.93	Wg	21.0
Ls	2.82	Lm1	6.93	Lg	7.0

3 Results and Discussions

The layered tissue model is designed according to the mathematical model proposed by Gabriel [16]. Fractal Antenna with Minskowski Island split ring resonator structure is designed in CST Microwave studio and placed over the layered tissue model [18, 19] which is the electric equivalent of human body.

3.1 *Reflection Coefficient*

This antenna is investigated for on-body applictaions, hence layered tissue model (electrical equivalent of human tissue called Phantom) is used here for placing this antenna horizontally. In practical applications also, these antennas are placed at some distance from the human tissue as in [20, 21]; the same conditions are mimicked here and observations are taken at 2 mm, 4 mm, 6 mm, and 8 mm from the tissue model and the comparison of reflection coefficient is in Fig. 4.

The effect of human body tissue which is absorptive in nature for radio frequencies is obseved from Fig. 4, which is depicting the loss of reflection coefficient. The reflection coefficient is decreasing in the presence of body tissue model and

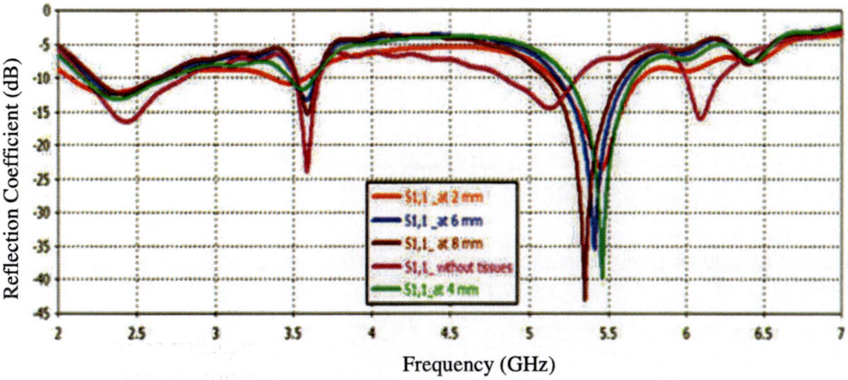

Fig. 4 Comparison of reflection coefficient at various distances from tissue model, i.e., 2 mm, 4 mm, 6 mm, 8 mm and without tissue

Table 2 Reflection coefficient at various distances from tissue model

Distance from tissue (mm)from tissue	Resonant frequency 1 (GHz)	Reflection coefficient (dB)	Resonant frequency 2 (GHz)	Reflection coefficient (dB)	Resonant frequency 3 (GHz)	Reflection coefficient (dB)
2	2.35	−12.33	3.47	−11.0	5.44	−23.5
4	2.35	−13.18	3.55	−11.78	5.45	−39.51
6	2.39	−12.47	3.57	−13.31	5.41	−35.13
8	2.39	−12.39	3.58	−15.36	5.34	−42.9

it is increasing when the distance between tissue model and antenna is increasing. Distance 8 mm from the human tissue model is showing the best results in terms of reflection coefficient as per Table 2.

3.2 *VSWR Parameter*

Graph of voltage standing waves ratio parameter with frequency is shown in Fig. 5. This graph is for various placings of antenna at 2 mm, 4 mm, 6 mm, and 8 mm from the body tissue model. The all values of VSWR at resonant frequencies are below the value 2, which is acceptable for proper operation of antenna. The VSWR parameter graph is depicting the dips at resonant frequency bands and at other frequencies its value is comparatively higher, hence the maximum radiations are propagating into the space efficiently at resonant frequencies.

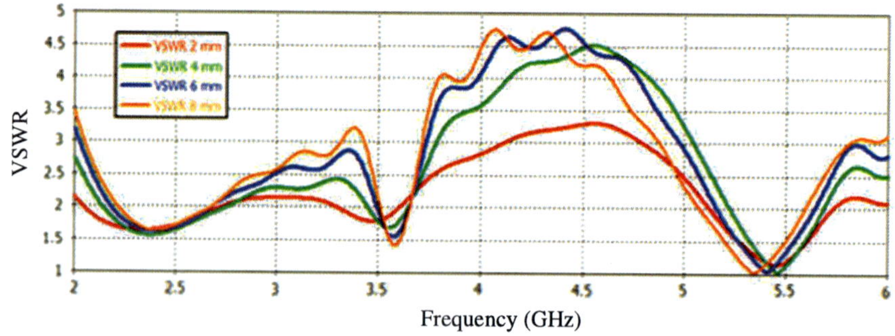

Fig. 5 VSWR graph of proposed antenna at various distances from tissue model, i.e., 2 mm, 4 mm, 6 mm, and 8 mm

3.3 Radiation Pattern

The radiation pattern is a geographical representation of radiation, Fig. 6 is depicting the polar plot and 3D plot of proposed antenna. The 3-dB angular width at 2.4 GHz, 3.5 GHz, and 5.2 GHz is 255.8°, 108.5°, and 64.1°. This radiation pattern is computed at 2 mm from the layered tissue model and it is co-polar pattern.

The effect of frequency over the radiation performance of antenna is interesting; the beamwidth is decreasing with the frequency so antenna is being more directional with the increasing frequency.

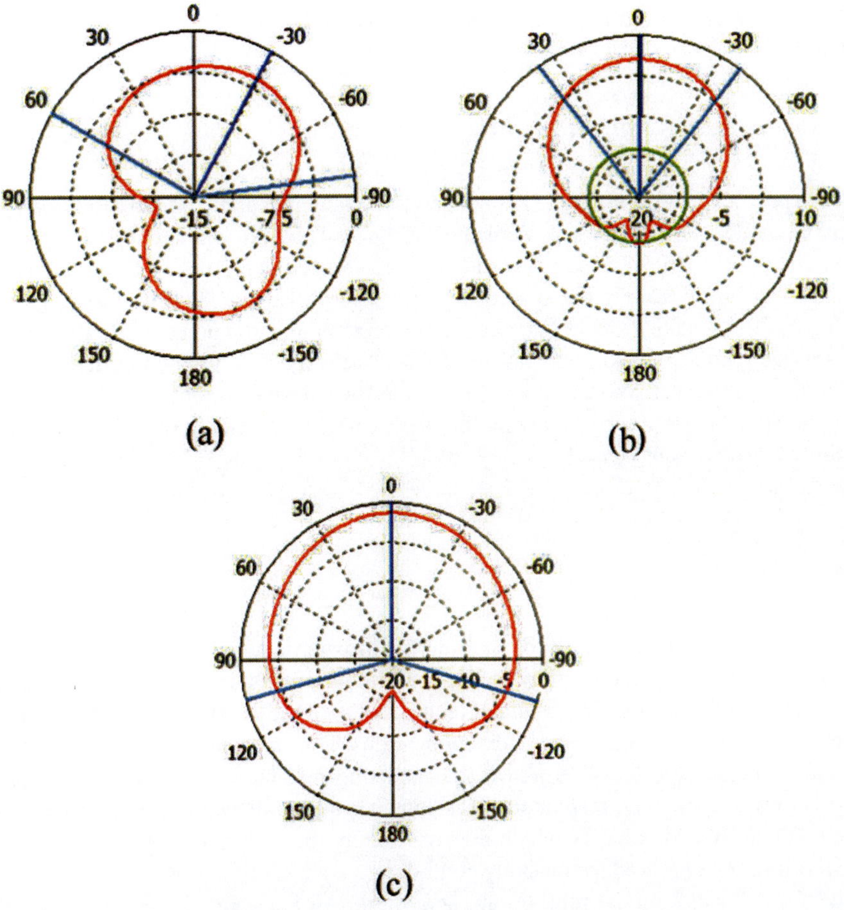

Fig. 6 Polar plot at **a** 2.4 GHz, **b** 3.5 GHz, **c** 5.2 GHz

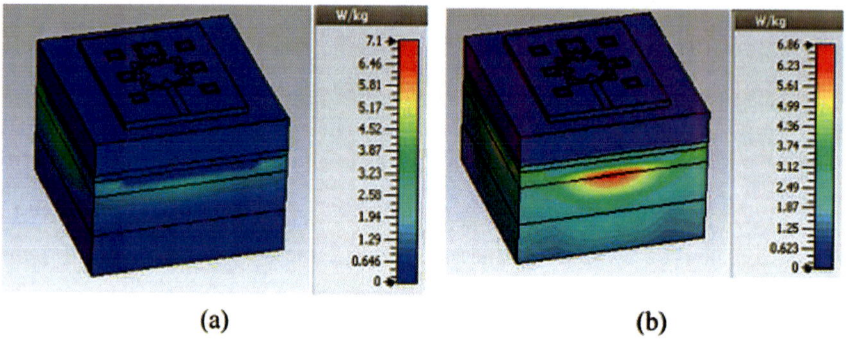

Fig. 7 SAR distribution over the tissue model computed maximum 10 g averaged SAR values, distance from tissue **a** 6 mm, **b** 2 mm

3.4 Specific Absorption Rate

The specific absorption rate (SAR) values are calculated and its distribution on the tissue model is also drawn as depicted in Fig. 7. The SAR valuation is calculated based on IEEE C95.1 which is standard and averaged over 10 g of tissue. Results shown in Fig. 7 are for the case 6 mm and 2 mm distance from the tissue mode 7(a) & (b); to make SMA connector placing of antenna more practical. In the layers of tissue model like skin and fat, the SAR is below 2 W/kg which is depicted in Fig. 7. Nowadays, the SAR calculation for the BAN devices is very much concerned issue. So the SAR values following the IEEE C95.1 standard are required, which is observed by this antenna.

4 Conclusions

This paper exhibits the study of the microstrip patch antennas for wireless body-centric communication. In view of the new developments, the modeling of fractal antenna using the Minkowski Island technique is investigated to make an efficient multifrequency antenna for on-body applications. This antenna is designed by using a main fractal patch, one Minkowski Island ring resonator and other six smaller Minkowski Island ring resonators. It is showing the multifrequencies operation in WLAN and Wi-Max bands which are required by the body area network systems. All results are observed for on-body conditions at various positions of antenna from the body tissue: 2 mm, 4 mm, 6 mm, and 8 mm. All the obtained results of fractal antenna using Minkowski Island are compared in two conditions; without the body tissue and with the body tissue. The moderate on-body performance of antenna is making it outstanding choice for wireless body area network.

References

1. Constantine A (2005) Balanis, antenna theory: analysis and design, 3rd edn. Wiley, Hoboken, New Jersey
2. Balanis CA (1982) Antenna theory, 2nd edn. Wiley, New York
3. Conway GA, Scanlon WG, Linton D (2012) Low-profile microstrip patch antenna for over-body surface communication at 2.45 GHz. IEEE
4. Khan MM, Rahman MA, Talha MA, Mithila T (2014) Wearable antenna for power efficient on-body and off-body communications. J Electromag Anal Appl 6:238–243
5. Roblin C, Laheurte JM, D'Errico R, Gati A, Lautru D, Alvès T, Terchoune H, Bouttout F (2011) Antenna design and channel modeling in the BAN context—part I: antennas. Springer, pp 139–155
6. Kanaujia BK, Khandelwal MK, Dwari S, Kumar S, Gautam AK (2016) Analysis and design of compact high gain microstrip patch antenna with defected ground structure for wireless applications. Springer
7. Conway GA, Scanlon WG (2009) Antenna for over-body-surface communication at GHz. IEEE Trans Antennas Propag 57(4):844–854
8. Mackowiak M, Correia LM (2013) A statistical model for the influence of body dynamics on the gain pattern of wearable antennas in off-body radio channels. Springer, pp 381–399
9. Tanaka M, Jang J-H (2003) Wearable microstrip antenna. IEEE Antennas Propag Int Symp 2:704–707
10. Liu H-W, Ku C-H, Yang C-F (2010) Novel CPW-fed planar monopole antenna for WiMAX/WLAN applications. IEEE Antennas Wirel Propag Lett 9:240–243
11. Nornikman H et al (2015) Design tri-band fractal antenna with Minkowski Island split ring resonator structures. Appl Mech Mater 781:73–76
12. Falconer KJ (2003) Fractal geometry: mathematical foundations and applications, 3rd edn. Wiley, New York, NY
13. Pfeiffer A (1994) The Pfeiffer quad antenna system. QST, 28–32
14. Naji DK, Aziz JS, Fyath RS (2012) Design and simulation of RFID aperture coupled fractal antennas. Int J Eng Bus Manage 4:1–14. https://doi.org/10.5772/50927
15. Kobayashi T, et al (1993) dry phantom composed of ceramics and its application to SAR estimation. IEEE Trans Microwave Theory Tech 41(1):136–140
16. Gabriel C, Gabriel S et al (1996) The dielectric properties of biological tissues: I. Literature Survey. Phys Med Biol 41:2231–2249
17. Alomainy A et al (2005) Comparison between two different antennas for UWB on-body propagation measurements. IEEE Antennas Wirel Propag Lett 4
18. Kalkhambkar G, Khanai R, Chindhi P (2019) Fractals: a novel method in the miniaturization of a patch antenna with bandwidth improvements. In: Information and communication technology for intelligent systems. smart innovation, systems and technologies, vol 106
19. Kumar M, Nath V (2018) Introducing multiband and wideband microstrtip patch antennas using fractal geometries: development is last decade. Wireless Pers Commun 98:2079–2105
20. Reha A, Amri AE, Benhmammouch O (2016) CPW-Fed dragon fractal antenna for UWB applications. In: Advances in ubiquitous networking 2: Proceedings of the UNet'16. Springer, Singapore
21. Manafi S, Deng H (2014) Design of a small modified Minkowski fractal antenna for passive deep brain simulation implants. Int J Antennas Propag Hindawi, Article ID 749043

Pulsar Star Detection Using Ensemble Algorithms and Stack Generalization

M. Murali, Harshit Sharma, and Animesh Mishra

Abstract Pulsar stars are known as 'celestial clocks' and are the most accurate clocks of the universe, giving us a detailed insight into relativistic gravity, cosmology, the structure of interstellar medium and the origin and dynamics of the planetary systems. In the field of machine learning, pulsar star detection has been a topic of interest. Many algorithms have been used in order to detect the pulsar stars, such as fuzzy KNN classifier, artificial neural networks and Gaussian Hellinger very fast decision tree, to name a few. Ensemble algorithms and stack generalization have been widely used in many research areas and have proved quite efficient in increasing the accuracy. In this paper, stack generalization method has been proposed to increase the accuracy of pulsar star detection using the HTRU2 dataset. Ensemble algorithms have been used as individual algorithms for building the stack, and the final accuracy received has been greater than 98%. A comparison was done among several algorithms and stacking performed quite good as compared to other algorithms.

Keywords Stacked generalization · Ensemble learning · Boosting · Pulsar star · KNN · Xgboost · Random forest · HTRU2 · Support vector classifier

1 Introduction

Pulsars are magnetized rotating neutron stars having magnetic poles, from where they emit electromagnetic radiation. This radiation may be observed whenever the beam is pointed towards earth. They are very dense having quite short and regular rotational periods. The rotational period helps in producing a precise interval between

M. Murali (✉) · H. Sharma · A. Mishra
Computer Science and Engineering, SRM Institute of Science and Technology, Chennai, India
e-mail: muralim@srmist.edu.in

H. Sharma
e-mail: harshits1997@gmail.com

A. Mishra
e-mail: animeshmishra.chitta@gmail.com

R. P. Mahapatra et al. (eds.), *Proceedings of 6th International Conference on Recent Trends in Computing*, Lecture Notes in Networks and Systems 177,
https://doi.org/10.1007/978-981-33-4501-0_15

pulses. This interval can be in the range of milliseconds to seconds for any individual pulsar. Since pulsars emit regular pulses of radio waves, the data need not be rectified regularly/daily. It can be used to create a spacecraft navigation system or to make a satellite navigation system. Pulsar stars can be used as probes for interstellar medium and of space time as well. In [11], the authors have given us a detailed study about three pulsars. Moreover, they may be used as gravitational wave detectors and as precise clocks. Some of the notable pulsars include: CP1919 (First Radio Pulsar), PSR 1913 + 16 (First binary pulsar), Vela Pulsar (Brightest Pulsar) and Cen X-3 (First X-ray pulsar).

In this paper, ensemble learning has been used to predict the accuracy of the model. Ensemble learning is a technique in which several algorithms are used, and the individual outputs of each algorithm are combined to get a final result. This helps in selecting the algorithms that work well and improve the accuracy of the final model. The ensemble model will finally generate a result that would be better than the results of the single models that will be used. Variance and bias may be reduced by the model which in turn will improve the predictions. They are generally divided into two groups: sequential ensemble techniques, where models are generated sequentially and the parallel ensemble techniques, where models are generated parallelly. Homogenous ensembles are a type of ensemble technique, where a single type of base learning algorithm is used. On the other hand, in heterogenous ensembles, different types of base learning algorithms are used.

There are various applications covered by ensemble learning that includes, remote sensing such as change detection, computer security such as intrusion detection and malware detection, face recognition, emotion recognition, fraud detection and for financial decision making. On the other hand, in the stacked generalization technique, a few models would be used as first level classifiers. The combined results from this level would be used in the second level classifier models. The results coming out of this second level would be used as the result. Wolpert [19] first introduced this topic. Bagging [3] is also called as bootstrap aggregating and is a meta-algorithm that improves the accuracy and stability of the algorithms. In this, all models run independently, and then, the outputs are aggregated together without any preference to any of the models. Boosting [9] helps in reducing variance and bias in the models. If there are any weak learners in the model, then boosting would help to train these weak learners into strong learners. It will train every weak learner sequentially and improve the next model based on feedback from the predecessor. Every model that undergoes boosting will be deciding upon the features that the next model will be focussing on. Examples of boosting that can be given include adaptive boosting (AdaBoost), XGBoost and gradient boosting.

In Fig. 1, the data has been divided into different parts, and the new data is being fed separately into every classifier. All the classifiers are seeing different types of data and are getting trained accordingly. Hereafter, the output is being given to the meta learner that would separately get trained and give another output, which is always better than the previous individual learners. There are many algorithms that one can use as the first level learner or as the meta learner as well. The algorithms can be simple algorithms like k-nearest neighbour, naive Bayes classifier, logistic

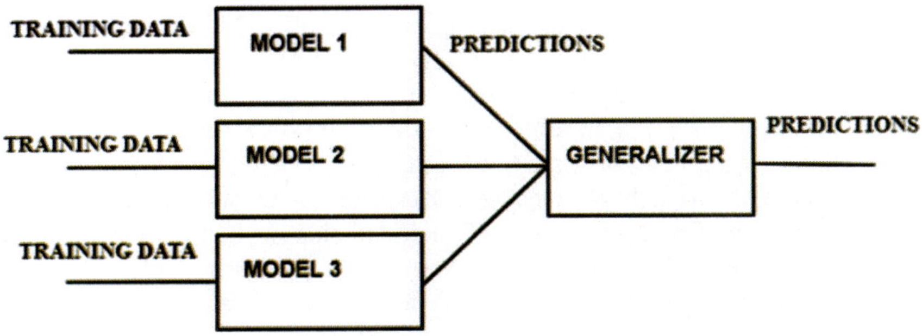

Fig. 1 Concept of Stacking. Three models have been used in the above figure, and the generalizer gives the final accuracy

regression, support vector machines or boosting. In this paper, ensemble methods have been used that would help us in keeping the classification score high. Three different algorithms have been used that would be the first level learners:

(a) K-nearest neighbour
(b) Random forest
(c) Extreme gradient boosting/ XGBoost.

In the second level, XGBoost has been used, so that it can boost the accuracy of the final model. XGBoost is the meta learner for the model. It will learn from the outputs from the first level learner. Vecstack library has been used for helping us perform the stacking process. At the end, the accuracy score, precision score, F1 score, and the recall score are calculated in order to prove the efficiency of the models. The model is performing better than any other previous work on this dataset of HTRU2 [15]. Besides the ensemble model, two more algorithms have been used: Gradient boosting and support vector classifier, for comparison purposes. For hyper-parameter tuning, the GridSearchCV function has been used that would help in choosing the best parameters. The GridSearchCV would then give the best parameters from a list of values, and these values may be given to the parameters in the main model manually.

2 Literature Survey

There have been many attempts to solve the problem of pulsar star detection and choosing the best features to be used in the process. In [14], the authors have helped in calculating a score that would tell whether a score is a real pulsar star or not. It considers six quality factors and then looks to increase the rate of identifying the pulsar stars. In [17], the authors have processed the relatively large data outputs from the pulsar surveys. It also produces continuous and real valued outputs. In this paper, a different technique of stacked generalization is being proposed that will help improve the accuracy of the prediction. Candidates from the Parkes multi-beam pulsar survey were used by authors in [7], and they were able to discover a previously unidentified pulsar star as well. Considering the large volume of data that was involved in the research which was 2.5 million, they used two different ANN structures, one 8:8:2, and another 12:12:2. 12 scores have been used in the input vectors. 75 pulsars were presented in [2] by the authors which were discovered in the mid-latitude portion of the HTRU survey. The paper successfully detected pulsars at all pulse periods but couldn't detect candidates with large pulse duty cycles and millisecond periods with much good accuracy. The ANN detected the stars with an accuracy of approximately 85% and missed around 15 pulsar stars and detected them false.

In [17], the authors have introduced SPINN that would be able to process large data that are received from surveys. Candidates were collected from the southern high time resolution survey and were successfully able to identify all the pulsars. It performed a fivefold cross-validation to find the most optimal network architecture and has used six features. An introduction has been given to an algorithm, the fuzzy KNN classifier in [16]. The author has used the HITRU two dataset. The proposed algorithm has outperformed five other classifiers which also includes the neural network classifiers and has used three evaluation metrics. The F-score, G-mean and accuracy were also enhanced by this algorithm. Pulsar Image-based Classification System (PICS) has been introduced in [20]. The candidates were collected from the pulsar Arecibo L-band feed array survey. The AI used in this has two layers.

In [18], the authors have used a trained random forest classifier for predicting the class of each new candidate. They have used ten features: (a) Minimum observing frequency, (b) estimated dispersion measure, (c) signal to noise ratio, (d) Max asymmetry, (e) mean asymmetry, (f) max outlierness, (g) mean outlierness, (h) max zeros, (i) mean zeros and (g) log ratio of covariance during event versus before and after the event. It has used data collected by the very long baseline array. It has detected fast radio bursts. In [6], the authors have used an approach that consisted of DPG identification and DPG classification. For identification of the DPG, it has used a novel algorithm RAPID. It has used six machine learning algorithms of different types: Artificial neural network, support vector machine, a direct rule learner, a standard tree learner, hybrid rule and tree leaner and an ensemble method that was the random forest classifier. It has used 16 features. Ford J M made a detailed study about the performance of different classifiers, including the ensemble classifiers for detection of pulsars in his PhD thesis [8]. In [15], the authors Gaussian Hellinger Very Fast Decision Tree (GH-VFDT), a tree-based algorithm has been used, which involves the use of eight features.

A full survey of the research that has been done on pulsar detection has been given in [4]. The survey has done a detailed study on pulsar star and given studied all the major pulsar surveys done, the software tools available for pulsar detection and the research works that have been done till date. It has also studied the detection strategies by dividing it into two parts, the offline strategies and the online strategies. In [13], the authors have used artificial neural network in their work and have got a good accuracy of 98.01%. They used five dense layers; and the input had ten neurons; second layer has six neurons; third layer has four neurons; and the fourth layer has a single neuron. Finally, the fourth layer has only one neuron. They have worked with the HTRU2 dataset itself using the eight features as has been mentioned in the dataset. Three ensemble methods have been compared in [1] that includes random forest, boosting and bagging. Then, using R language, they have tested their works on the HTRU2 dataset. They have concluded that AdaBoost and random forest algorithms perform better than the bagged cart classifier. A feed forward and cascade forward backpropagation neural network architecture were proposed by [10]. They used the eight features as input neurons for their architecture. Using all the eight features, they trained their algorithm on the HTRU2 dataset. They achieved a 92.3% accuracy on their FB neural network and a 90.69% accuracy on the other.

Table 1 gives us a detailed knowledge on the experiments done on the HTRU2 dataset that has been used in this paper (Table 2).

3 Dataset

The dataset—'Predicting a Pulsar Star' dataset from Kaggle has been used. The data was collected during the high time resolution survey. The main source is the UCI machine learning repository [15]. The dataset used here contains:

Spurious examples: 16,259 (caused by noise/RFI).
Real pulsar: 1,639 examples.

These examples have all been checked and properly verified and then the dataset made. Each row mentions the class label as the final entry and all other variables initially. The class labels are 0 and 1 that stand for negative and positive, respectively. Every candidate has been described by one single class variable and eight variables that are continuous. The details are given below:

Integrated profile statistics:

(a) Mean
(b) Standard deviation
(c) Excess kurtosis
(d) Skewness

DM-SNR curve statistics:

(e) Mean
(f) Standard deviation
(g) Excess kurtosis
(h) Skewness

Finally, the resultant:

(i) Class.

Integrated pulse profile: A pulsar produces periodic pulsation signals which are generally too weak to detect. But the fold/integrated pulsar signals increase the

Table 1 Research done on HTRU2 dataset only

Paper	Dataset	Number of features	Algorithm used	Final accuracy on HITRU2	Observations
Mohamed [16]	HITRU2	8	Fuzzy KNN classifier	97.8%	Addressed the importance and relationship between the different features
Lahiri et al. [13]	HTRU2	8	Artificial neural network	98.01%	Used 5 dense layers with a different number of neurons at every stage
Lyon et al. [15]	HTRU1, HTRU2, LOTAAS1	8	Gaussian Hellinger very fast decision tree	97.8% on HTRU2	Combined different features and checking if accuracy increases
Azhari et al. [1]	HTRU2	8	Random forest, Adaboost, bagged cart classifier	Greater than 96%	They have compared the three algorithms and concluded finally that the random forest and AdaBoost algorithms are better than the bagged cart classifier
Furat and Ibriki [10]	HTRU2	8	Feed forward backpropagation and cascade forward backpropagation neural network algorithms	FB neural network: For Test Data:92.336% CFB neural network: For test data:90.692%	All the eight features were treated as eight input neurons into the neural network. A future improvement could be to perform feature selection

signal quality and form the integrated pulse profile. All pulsars are distinguished by the integrated pulse profiles. Every pulsar has a unique pulsar profile which can help us distinguish between all the pulsars that have been found till date.

Kurtosis: It is the measure that checks whether data is heavy-tailed or light-tailed that is relative to the normal distribution. If a dataset has many outliers or has a

Table 2 Details on experiments done on other datasets of pulsar stars

Paper	Dataset	Number of features	Algorithm used	Final accuracy	Observations
Eatough et al. [7]	Parkes multi-beam pulsar survey	12	2 Artificial neural network	92% of pulsars present were classified correctly	A few were missed due to: Poor training on MSP, unbalanced training sets
Bates et al. [2]	HTRU survey	22	Artificial neural network	85%:15 pulsars in the dataset weren't detected	Understanding how improving the training process will bring improvements
Morello et al. [17]	Southern HTRU Survey	6	SPINN	False positive rate of 0.64%, ranks 95% among top 0.01%	Characterization of RFI is limited

heavy tail, then it is said to have high kurtosis, and if it has less outliers, or has a light tail, then it is said to have light kurtosis. The metric that has been used here; excess kurtosis is kurtosis minus 1. On the other hand, skewness is the measure by which the distribution differs from a normal distribution (Fig. 2).

A total of 17,898 rows are present that consists of both true and false values, i.e. both the data that indicate pulsar star and not the pulsar star. Out of this 17898, 1639 is the data that is valued 1, the data indicates the Pulsar Star exists. 9.15% of the data shows the presence of pulsar star. 16,259 values represent the non-presence of pulsar star. This can be said to be an imbalanced dataset, where there is minority as well as a majority class; in this case, the true class is said to be minority and false case to

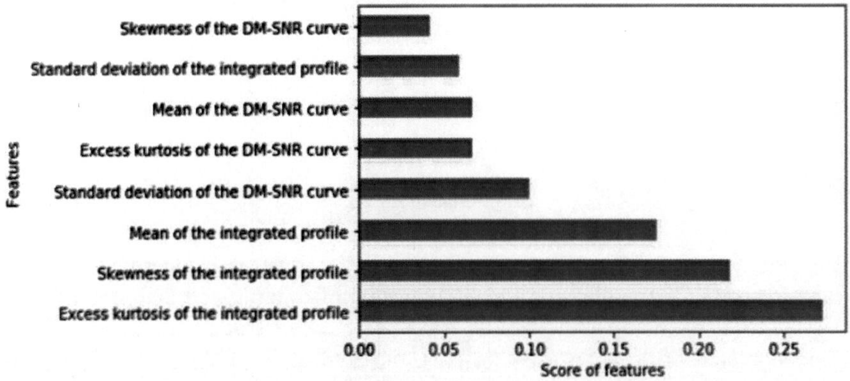

Fig. 2 Importance of features present in the dataset. As can be seen from the figure, excess kurtosis of the integrated profile has the most importance

be majority. Resampling techniques may be used for solving this issue. Some of the techniques that may be used are:

(1) Random undersampling
(2) Random oversampling
(3) Cluster-based Oversampling
(4) Informed over sampling
(5) Modified synthetic minority oversampling.

The other way is to use ensemble algorithms and modify the existing algorithms so that they would give us appropriate results. It helps in improving the performance of single classifiers and helps in boosting the accuracy to a better result. Using the ensemble methods can help us increase the accuracy and help improve the model. The various ensemble techniques that may be used for improving the imbalanced dataset include, adaptive boosting, random forest, gradient boosting, extreme gradient boosting and bagging algorithms. These techniques are highly efficient and will improve the accuracy and solve the problem of an imbalanced dataset.

4 Methods

4.1 K-Nearest Neighbours

K-nearest neighbour algorithm assumes that all the similar things exist together at one place or near each other. One of the most important part of the algorithm is to detect the value of K. Different values of K are taken, and then, the algorithm is run with those values. The K value that gives the least error and gives a good accuracy should be chosen as the final K value. The different hyper-parameters that are covered by the algorithm are as follows:

(a) n_neighbours
(b) weights
(c) metrics.

4.2 Random Forest

They are random decision trees that are an example of ensemble learning methods, which construct many decision trees at the time of training and then finally gives class which is the mode of the classes. They prevent overfitting that sometimes may be encountered in decision trees. In [12], the authors have used a better random forest model for classification of diabetes Mellitus, which helped us understand the concept to improve the model. The different hyper-parameters that are covered by the algorithm are as follows:

(a) maximum depth
(b) maximum features
(c) maximum number of leaf nodes
(d) maximum number of samples at leaf node
(e) number of trees
(f) number of jobs that may run in parallel
(g) verbose
(h) random state.

4.3 XGBoost

It stands for extreme gradient boosting. It can be used when one has to boost an already fitted model. It also has a block structure that helps in parallel tree construction. The execution speed of this algorithm can be said to be extremely fast and is generally very good for classification and regression prediction problems. For proper parameter tuning in this algorithm, the following may be done:

(1) Choosing a higher learning rate
(2) Tuning the regularization parameters
(3) Tuning the tree specific parameters.

In [5], the author gives to us a detailed explanation about the XGBoost metrics. The algorithm automatically learns about the best missing values and handles different sparsity patterns as well. It performs LASSO (L1) and ridge (L2) regularization that prevents overfitting.

4.4 Support Vector Classifier

This hasn't been used in the final ensemble model. But it has been used to compare the accuracy and show that ensemble algorithms perform better than this algorithm. Hyper-parameter optimization hasn't been done in this case, and it has been used with the default parameters only. The SVC basically finds out a hyper-plane that will help in differentiating between two parts that would classify the data. The main aim would be to find out the hyper-plane having the maximum margin. There should be maximum distance between the data of both the classes, so that in the future, if classification of new data points has to be made it can be done with much greater confidence. The hyper-plane construction on the other hand depends upon the number of features that will be used. If the number of features is less, then a thin hyper-plane will be used, but if there are many features, then a thick hyper-plane is constructed to differentiate between the two points. Support vectors are those points that are nearer to the hyper-plane, and these are the points that would be influencing the construction of the hyper-plane. If there are any alternations to these particular points, then the hyper-plane position and orientation and all its characteristics will be changing.

4.5 Gradient Boosting

This is an ensemble model, wherein the weak learners are the decision trees. A loss function is used in gradient boosting that needs to be boosting, and the gradient is used to do so. Every time a decision tree is built, it is done greedily so that the right split is always expected. So, the correct outcome that is expected is compared with the outcome of the decision tree, and the error rate is calculated. Now, the gradient may be calculated using these error rates. The gradient will be describing how steep the loss function is depending upon that the parameters that is being used may be changed, so that the error rate gets reduced. The gradient boosting uses learning rate, number of iterations and a few other values as it is hyper-parameters.

4.6 Stacking

The training data and testing data need to be passed along with the list of bases. It trains every model separately and then makes predictions on it separately. Thereafter, the predictions are made separately on every test set. The total number of predictions are made and stored temporarily. Every model is trained, and the predictions of the data are stored. Finally, the mean of the predictions is found out that was stored earlier. Then, the meta learner is trained, and finally, the outputs are checked. This output is always better than the outputs of the individual base learners. For this reason, stacking is always better than the individual base learners.

5 Metrics

The first thing that needs to be understood is the ROC curve. One may check if it is good or not by checking the Area Under the Curve (AUC) and other parameters as well which make up the confusion matrix.

A confusion matrix is a table that will describe to us the performance of a model that does classification on a particular set of data, which is the test data and for those data for which the true value is known.

There are four parameters that are taken into consideration which are:

(a) True positive: When the algorithm predicts yes, and the actual value is also yes that condition is called as true positive.
(b) True negative: When the algorithm predicts no, and the actual value is also no that condition is called as true negative.
(c) False positive (Type 1 Error): When the algorithm predicts yes, and the actual value is no that condition is called as false positive.
(d) False negative (Type 2 Error): When the algorithm predicts no, and the actual value is yes that condition is called as false positive.

From the above four parameters, another four parameters can be calculated which are:

(a) Accuracy:

$$\text{Accuracy} = \text{TP} + \text{TN}/\text{TP} + \text{TN} + \text{FP} + \text{FN}. \tag{1}$$

(b) Precision score:

$$\text{Precision} = \text{TP}/\text{TP} + \text{FP}. \tag{2}$$

(c) Recall score:

$$\text{Recall} = \text{TP}/\text{TP} + \text{FN}. \tag{3}$$

(d) $F1$ score:

$$F1 = 2 * (\text{Recall} * \text{Precision})/(\text{Recall} + \text{Prec.}). \tag{4}$$

This score is used for information retrieval and for measuring the performance of classification and search. A confusion matrix has a structure which has been depicted in Table 3.

ROC curve is called as the receiver operating characteristic curve. The ability of a binary classifier system is plotted on a graph, while keeping the value of the discrimination threshold different. It is a comparison of two operating characteristic, the TPR and FPR, as the criteria is changed continuously.

GridSearchCV helps in selecting the best hyper-parameters for increasing the accuracy of the models. It helps in constructing many models with all possible combinations of hyper-parameters and will return the best one according to the parameters chosen for that model. Considering that GridSearchCV goes through many combinations and constructs many versions, therefore, it is computationally very expensive and takes time to give back the result. Suppose there are a lot of parameters and a lot of values associated with them, the training time becomes too high, and it is highly expensive to train.

Table 3 Classification report on the other hand measures the quality of the predictions predicted form any classification algorithm

	Predicted YES	Predicted NO
Actual YES	True positive	False negative
Actual NO	False positive	True negative

6 Experiments

Before building the main model, GridSeachCV has been used to find out the best hyper-parameters that would give the accurate result for the stacked generalization. The stack was built using XGBoost, KNN and random forest algorithms. The second level learner was XGBoost. The Vecstack library and the stacking function within it has been used. Four folds have been used for cross-validation purposes. A result of 98.1% accuracy with an error rate of (±)0.2% has been achieved. Overall the stacking process gave a better result than the individual algorithms that have been used. Support vector classifier and gradient boosting were used for comparison purposes only. Hyper-parameter tuning helped find the best parameters that would give a better accuracy.

7 Results

Table 4 gives the details about the metrics for the various algorithms that have been used. The stacking method has produced a better accuracy than the other individual algorithms. The accuracy has been greater than 98% in case of stacking. Ensemble algorithms individually perform on par with the stacking process. In the figures below, an illustration has been given of the various ROC curves that have been obtained for the individual algorithms, including stacking as well (Figs. 3, 4, 5, 6, 7 and 8).

In Fig. 9, a graphical comparison has been done of the prediction accuracies of various individual algorithms including the stacking algorithm. As it can be

Table 4 A summary of the prediction scores, precision scores, recall scores, F1 scores and the confusion matrix

Algorithms	Final prediction score	Final precision score	Final recall score	Final F1 score	Confusion matrix	Area under the curve
Support vector classifier	97.41	97.06	73.98	83.96	[[4867 11] [128 364]]	0.869
Gradient boosting	97.97	91.72	85.56	88.53	[[4840 38] [71 421]]	0.924
K-nearest neighbours	97.35	89.77	80.28	84.70	[[4833 45] [97 395]]	0.901
Random forest	98.00	93.45	84.14	88.55	[[4849 29] [78 414]]	0.918
XGBoost	97.87	91.81	84.34	87.92	[[4841 37] [77 415]]	0.918
Stacking	98.10	92.95	85.77	89.21	[[4846 32] [70 422]]	0.926

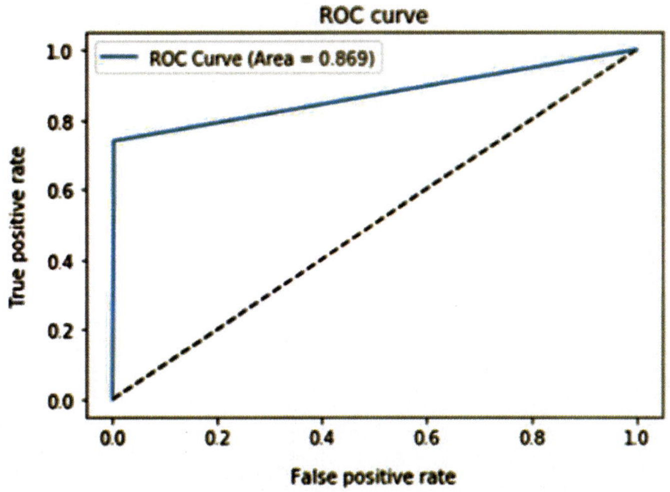

Fig. 3 SVC ROC curve

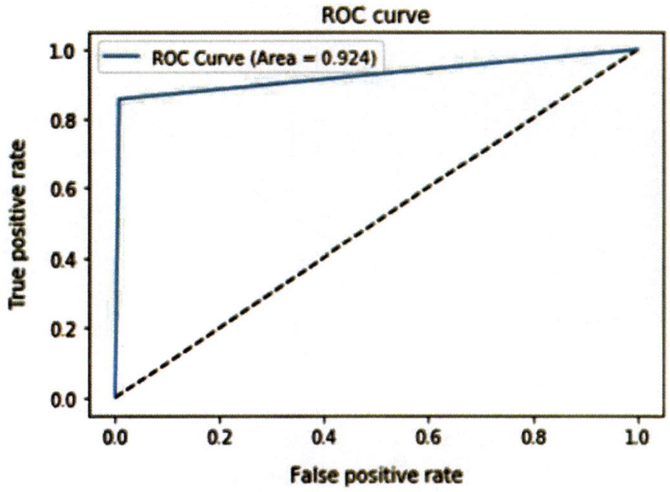

Fig. 4 Gradient boosting ROC curve

seen, stacking has performed better than the individual algorithms, although the ensemble algorithms perform almost on par with the stacking process and produce good accuracies as well.

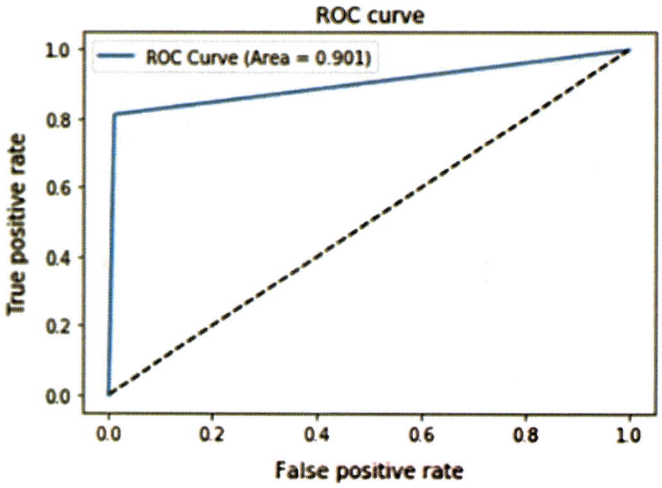

Fig. 5 KNN ROC curve

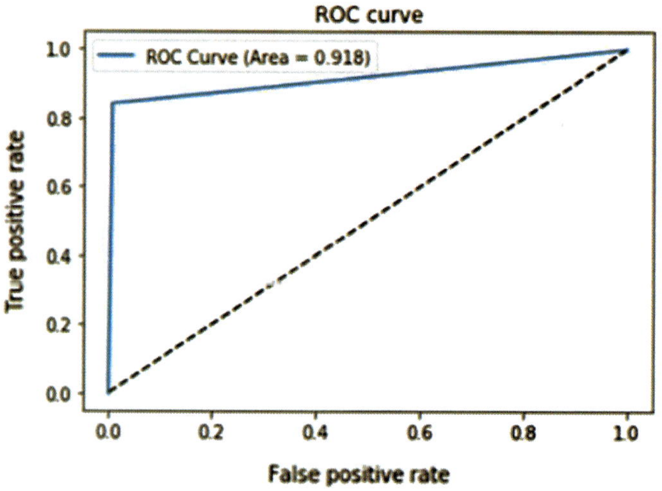

Fig. 6 Random forest ROC

8 Conclusion

The stacked generalization technique gave an accuracy of 98.1% with an error rate of (±)0.2%. This is an improvement over the individual algorithms that have been used as the first level learners. It has performed better than the fuzzy KNN classifier and the Gaussian Hellinger very fast decision tree and equally well as compared to the artificial neural networks. XGBoost as the second level learner has improved the

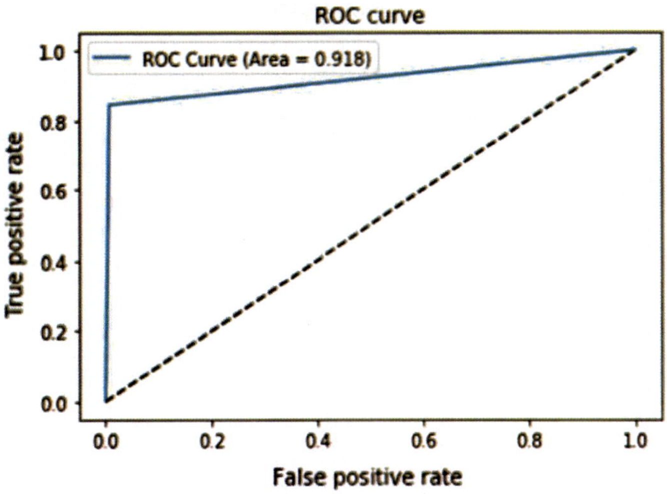

Fig. 7 XGB ROC curve

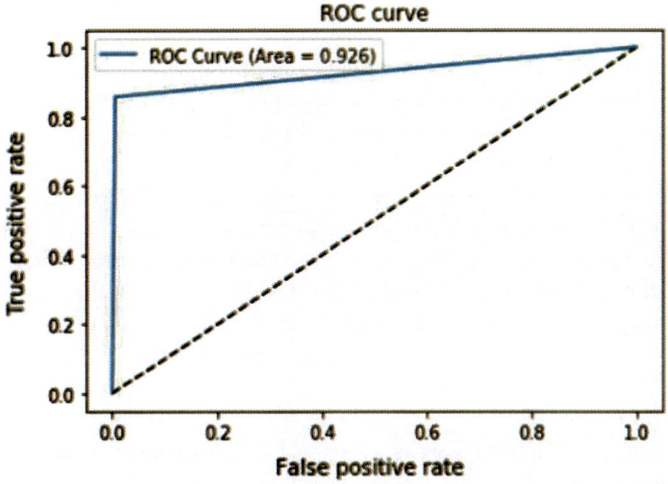

Fig. 8 Stacking ROC curve

score quite considerably. Further, Stacking using ensemble algorithms are a good way of increasing the accuracy of the model. This proves that stacking can be used for improvement of performance/accuracy of the models. Along with that it is also proved that the ensemble algorithms too give an almost equal accuracy as stacking.

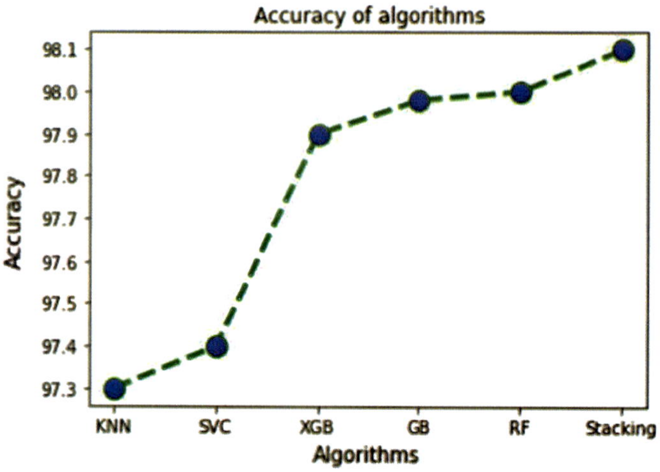

Fig. 9 Performance of individual algorithms as compared to the stacking accuracy

9 Further Works

Various other algorithms that would help in improving the accuracy can be used. More than three algorithms may be used, as the first level learner, with the right hyper parameters. Considering that the field of deep learning has been introduced, neural networks and multi-layer perceptron may be used to improve the accuracy of predictions. The parameters of the neural network may be changed and used as different learners of the first level, and another algorithm may be used as the meta learner of the second level. In this way, the accuracy may increase quite efficiently.

This technique may be used on other datasets as well such as the southern HTRU dataset or any other available dataset thus checking the accuracy of this model. This model has not been tested on any other dataset except HTRU2. A major issue that persists is the selection of features, which may be checked for in further works. The best features that would give more accurate results may be chosen or assigning of weights to every feature and assigning the most weight to the important features and a lesser weight to the ones that are not important, is another way. One can also use ensemble learning and combine it with various other algorithms, using it as base learners that may help in increasing the accuracy. As said earlier one may combine neural networks, multi-layer perceptron to get a good accuracy.

The other work that may be done is feature selection. Identifying the most important features that would help in increasing the accuracy can be done. The number of features may be changed according to the datasets that are present and assign weights to the features as has been mentioned earlier.

References

1. Azhari M, Alaoui A, Abarda A, Ettaki B, Zerouaoui J (2019) Using ensemble methods to solve the problem of pulsar search. In: International conference on big data and networks technologies. Springer, Cham, pp 183–189
2. Bates SD, Bailes M, Barsdell BR, Bhat ND, Burgay M, Burke-Spolaor S, Champion DJ, Coster P, D'Amico N, Jameson A, Johnston S (2012) The high time resolution universe pulsar survey—VI. An artificial neural network and timing of 75 pulsars. Mon Not R Astron Soc 427(2):1052–1065
3. Breiman L (1996) Bagging predictors. Mach Learn 24(2):123–140
4. Chakraborty PS (2019) Detection of pulsars using machine learning algorithms—a study. Artif Intell Internet of Things, 210
5. Chen T, Guestrin C (2016) Xgboost: a scalable tree boosting system. In: Proceedings of the 22nd ACM SIGKDD international conference on knowledge discovery and data mining. ACM, pp 785–794
6. Devine TR, Goseva-Popstojanova K, McLaughlin M (2016) Detection of dispersed radio pulses: a machine learning approach to candidate identification and classification. Mon Not R Astron Soc 459(2):1519–1532
7. Eatough RP, Molkenthin N, Kramer M, Noutsos A, Keith MJ, Stappers BW, Lyne AG (2010) Selection of radio pulsar candidates using artificial neural networks. Mon Not R Astron Soc 407(4):2443–2450
8. Ford JM (2017) Pulsar search using supervised machine learning
9. Freund Y, Schapire R, Abe N (1999) A short introduction to boosting. J-Jpn Soc Artif Intell 14(771–780):1612
10. Furat FG, İbrikçi T (2017) Application of feed forward backpropagation and cascade forward backpropagation neural network algorithms to detect pulsar stars
11. Karuppusamy R, Stappers BW, Serylak M (2011) A low frequency study of PSRs B1133+ 16, B1112+ 50, and B0031–07. Astron Astrophys 525:A55
12. Kumar NK, Vigneswari D, Krishna MV, Reddy GP (2019) An optimized random forest classifier for diabetes mellitus. In: Emerging technologies in data mining and information security. Springer, Singapore, pp 765–773
13. Lahiri R, Dey S, Roy S, Nag S (2020) Detection of pulsars using an artificial neural network. In: Emerging technology in modelling and graphics. Springer, Singapore, pp 147–157
14. Lee KJ, Stovall K, Jenet FA, Martinez J, Dartez LP, Mata A, Lunsford G, Cohen S, Biwer CM, Rohr M, Flanigan J (2013) PEACE: pulsar evaluation algorithm for candidate extraction—a software package for post-analysis processing of pulsar survey candidates. Mon Not R Astron Soc 433(1):688–694
15. Lyon RJ, Stappers BW, Cooper S, Brooke JM, Knowles JD (2016) Fifty years of pulsar candidate selection: from simple filters to a new principled real-time classification approach. Mon Not R Astron Soc 459(1):1104–1123
16. Mohamed TM (2018) Pulsar selection using fuzzy KNN classifier. Future Comput Inf J 3(1):1–6
17. Morello V, Barr ED, Bailes M, Flynn CM, Keane EF, van Straten W (2014) SPINN: a straightforward machine learning solution to the pulsar candidate selection problem. Mon Not R Astron Soc 443(2):1651–1662
18. Wagstaff KL, Tang B, Thompson DR, Khudikyan S, Wyngaard J, Deller AT, Palaniswamy D, Tingay SJ, Wayth RB (2016). A machine learning classifier for fast radio burst detection at the VLBA. Publ Astron Soc Pac 128(966), 084503
19. Wolpert DH (1992) Stacked generalization. Neural Networks 5(2):241–259
20. Zhu WW, Berndsen A, Madsen EC, Tan M, Stairs IH, Brazier A, Lazarus P, Lynch R, Scholz P, Stovall K, Ransom SM (2014) Searching for pulsars using image pattern recognition. Astrophys J 781(2):117

Improving Weed Detection Using Deep Learning Techniques

Shriti Gupta, Mohit Kumar, B. Govinda Satyanarayana, Sonia Aribam, Smita Tiwari, Rohit Kumar Kaliyar, Mohit Agarwal, and Nisha Ahuja

Abstract In recent years, weeds are responsible for agricultural losses. To get rid of this problem, the farmers have to uniformly spray the whole field with the weedicides which require a huge quantity of weedicides. The process of spraying weedicides affects the environment. Weed detection in dense culture is a plant science problem that is important for field robotics where the detection of weed is currently a challenge so that the use of phytochemical products on crops can be reduced. To control and prevent specific weeds, a method of detecting the weed is presented in this paper. By collecting the plants and weeds datasets which are grayscale images, data is divided into training, validation, and testing datasets and then transported to the convolutional

S. Gupta
Computer Science Engineering, Mody University of Science and Technology, Sikar, Rajasthan, India
e-mail: shritigupta6@gmail.com

M. Kumar
Computer Science Engineering, UIET Punjab University, Chandigarh, Punjab, India
e-mail: 1997mohit23@gmail.com

B. Govinda Satyanarayana
Computer Science Engineering, SRKR Engineering College, Andhra Pradesh, Bhimavaram, India
e-mail: b.govin99@gmail.com

S. Aribam
Computer Science Engineering, NIT Manipur, Imphal, Manipur, India
e-mail: aribamsoniasharma@gmail.com

S. Tiwari (✉) · R. K. Kaliyar (✉) · M. Agarwal · N. Ahuja
Computer Science Engineering, Bennett University, Uttar Pradesh, Greater Noida, India
e-mail: st1934@bennett.edu.in

R. K. Kaliyar
e-mail: rk5370@bennett.edu.in

M. Agarwal
e-mail: ma8573@bennett.edu.in

N. Ahuja
e-mail: na6742@bennett.edu.in

R. P. Mahapatra et al. (eds.), *Proceedings of 6th International Conference on Recent Trends in Computing*, Lecture Notes in Networks and Systems 177,
https://doi.org/10.1007/978-981-33-4501-0_16

neural network. Based on the knowledge gained by the model, it can detect the weeds among plants. Utilization of a pre-trained VGG16 model for weed detection in dense cultures demonstrated improved performance compared to state of the art without the need for large datasets and high computational power for training.

Keywords Transfer learning · Convolutional neural network · Computer vision · Convolutional layer · SoftMax layer · Pooling layer

1 Introduction

According to studies [1] carried out by the professors of the Faculty of Agronomy of the University of Buenos Aires (FAUBA) has estimated that 1300 million dollars were spent each year on weed control. Most of the people who are practicing in the agricultural activities are the people who use conventional methods for crop management, land treatment, irrigation of crops, maintaining the quality of the product at an outstanding level. The control of weed is important in the agricultural process because it causes problems in crop production and they are considered as unwanted by the farmers. Weeds give shelter to undesirable insects and diseases and allow the growth of other pests which results in increases in the cost of irrigation. Ecological irrigation is the base of development of modern agriculture and important support for the economic development and agricultural environment protection. To achieve modernization in the area of ecological irrigation, the goal is to establish an effective and convenient barrier to decrease the environmental pollution. At present, the main cause of agricultural pollution is the constant and massive use of pesticides and weedicides. Using a lot of weedicides and pesticides causes agricultural pollution. According to the United Nation's food and agricultural organization, there are approximately 800 types of weeds found worldwide and more than 250 types of weeds are considered to be harmful to the crops. Indian farmers use pesticides to control weeds. Spraying a lot of chemicals on crops affects the environment and also increases the cost of agriculture.

Numerous techniques [10] are applied to classify and eradicate weeds from our vegetations. Professors and several researchers practice different methods by applying different algorithms to get rid of the weeds without harming the environment. According to the research work "Algorithm of Weed Detection in Crops by Computational Vision," which mainly focused on image processing algorithms for the detection of weeds, their main objective is to find the formula in which weed detection system can be developed through binary classification. According to the research [2], work by Dian et al., a fully automatic method for weed detection using drone images was introduced. The method is mainly based on the training data set which is unsupervised and convolutional neural networks (CNNs). Another related approach that aims at the ecological irrigational area, a classification method of weeds based on convolutional neural network (CNN) was introduced by [13], the classification results exhibit the recognition rate of weeds can reach 95.6%. In this paper, we have taken the image dataset of plants and weeds. The first part of the work is done on training the classification model based on CNN to detect about the pants or weeds in the dense culture. Further, the performance of the classification model is evaluated on the unseen image dataset. The second part of the work is to improve the weed detection model using training done by the VGG16 pre-trained model.

2 Related Work

According to the research, [3] done by Irias Tejeda et al. in their research work "Algorithm of Weed Detection in Crops by Computational Vision," this research work is mainly based on the utilization of agricultural equipment for the control of the weeds in the vegetation. Their target is to achieve a technique in which the detection of unwanted and undesirable weeds can be done through a binary classification.

In the study done by Bah et al. [14], the author proposed a technique for the classification of weed using drone images. Their method mainly focused on the set of training data that is unsupervised and the convolutional neural network. The above-mentioned method is practiced using three various steps. Firstly, various rows of crop lines are detected and took advantage of the method to classify the weeds grown within the particular row. Secondly, the weed that is detected within the row is used as the training dataset. A deep learning model is produced using the database which is created using the above-mentioned steps. The benefit of using this technique is that it is more adaptative. This technique is unsupervised, and it is usually independent of the training set. Therefore, in this method, convolutional neural network is used to distinguish between the crops and weeds from a given image.

A study made by Bakhshipour et al. [7] proposed a classification technique using convolutional neural networks and the result exhibits up to 95.6%. In their research work, they applied two convolutional layers, two pooling layers, one fully connected layer. The activation function which is applied in this work is the Rectified Linear Unit (ReLu). The overall accuracy achieved using the 300 * 300-pixel experiment and 100 * 100-pixel experiment is 95.6% and 74% respectively.

One of the recent research papers [9] used supervised learning techniques to detect the weeds present in the culture crops. Here, authors have used the scattering transform which is a multiscale level for the classification of the texture of the crops. The supervised learning has used the training of the crop data which was synthetic. They have tested the trained model on the real-time dataset and obtained an accuracy of 85%.

One of the prominent research papers [11] used encoder and decoder structure in the fully convolutional network for obtaining the spatial information on the images of the crops for the purpose of the classification of weed and crops in culture crop environment and showed their results through simulation.

Deep learning techniques are playing a major role in the field of agricultural research due to the availability of a huge amount of data in today's scenario. The deep learning model can be trained very well with a huge set of data and can provide suitable solutions to the farmer. In [12], authors have used unsupervised techniques for the training of the crop data obtained through unmanned aerial vehicle (UAV) images of crops and used for the classification of weeds and crops through fully connected convolutional neural network (CNN).

3 Methodology

The following section describes the flow diagram of the proposed model, datasets used for the training and testing of the model, techniques used for the classification of weeds and crops.

3.1 Datasets

Data for the task of differentiating plants and weeds from images is available in grayscale format. All the images are taken from the closer distance as the task is to find out the weed from dense culture. All data provided to us is already preprocessed as all the images are of size 34 KB and 325 × 250 pixels in dimensions. This is indeed a challenge because the intensity of the plants in the provided data set is very less. A total amount of 3292 patches containing weed and 3292 patches only with plants was received. The binary classification (weed/plant) is realized on these patches. This balanced dataset serves both for the training and the testing stages to assess the performance of different machine-learning tools. The data sets together with the simulator are proposed as supplementary material under the form of a free executable and a collection of images (Figs. 1, 2 and Table 1).

3.2 Method

In this paper, the performance of the classification model for weed detection is improved using more training of the model and solution is implemented by using the pre-trained model VGG16. The proposed approach will use the convolutional base to extract features, and the extracted features are used to train the classifier to classify the input image as plants or weeds. Therefore, the features extracted from the convolutional base and then later will be fed to the classifier model for the training of the model. A comparison has been made to elaborate on the difference between the proposed classifier and the use of the pre-trained model in combination with a classifier. The project has been done in Python using a Keras package and trained on the NVIDIA supercomputer.

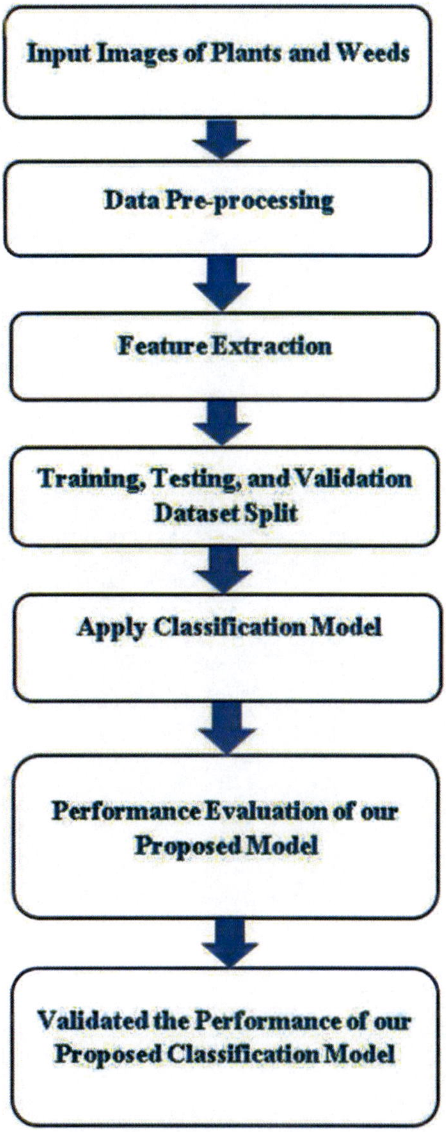

Fig. 1 Flow diagram

3.3 *Transfer Learning*

Transfer learning is a technique which is used to apply the knowledge gained while solving one problem to another but related problem. In this method, first, training of base network on-base datasets is done and then the learned features are transferred to another network which is needed to be trained on the targeted dataset. This method will work only when the features of both the base and target datasets are related but

(a) Plants

(b) Weeds

Fig. 2 Sample dataset

Table 1 Description of dataset

	Train data	Validation data
Plants	3292	274
Weeds	3138	282
Total	6530	556

instead of specific to a base dataset. Transfer Learning is implemented by using two approaches. One is building our model, second is using the pre-trained model. In our project, the pre-trained model approach has been used. This VGG16 is chosen as the model and then the model is fine-tuned (Fig. 3).

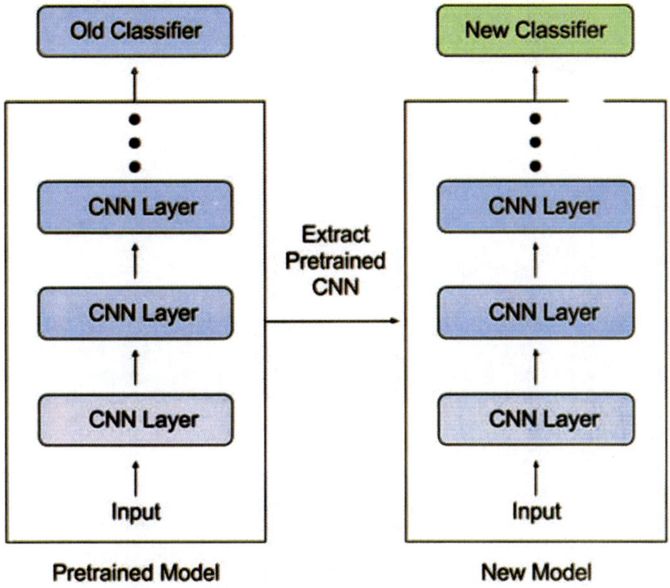

Fig. 3 Example of transfer learning with extracted features

3.4 *Vgg16*

VGG16 was proposed by [9] using convolutional neural networks in their research paper using VGG16, and test accuracy of 92.7% is achieved in ImageNet. The input of the stack of a convolutional layer is fixed; all images should be of the same size that is 224×224 RGB. The images are processed through a stack of convolutional layers. The authors also utilized 1×1 convolutional filters. They fixed the convolutional stride to 1 pixel. Max-pooling is performed by a 2×2 pixel window, with a stride 2. The final layer comprises of SoftMax layer.

In the proposed model, VGG16 has been transformed into binary classifications by not including the fully connected layers on the top of the network. Then new fully connected layers have been added with the random initialization, with the correct number of outputs. The convolutional layers will be using the weights based on training on the ImageNet dataset. Generally, it can be said that the convolutional layers work as feature extractors. The whole network is trained with two classes. This training will train the new fully connected layers (Fig. 4).

4 Experimental Results

Without using any pre-trained model. Building a model from scratch that included five convolution layers and max-pooling layers with three dropout layers, two dense

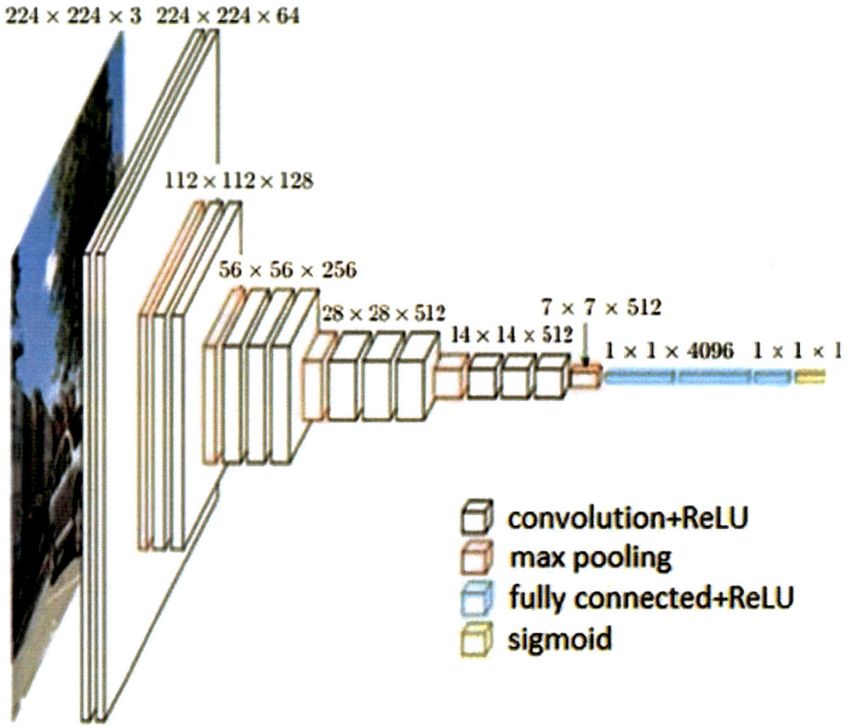

Fig. 4 Visualization of VGG16

layers and one flatten layer. The tabular values in Table 2 show the results with the built model. When we use the VGG16 pre-trained model, the training is done with a large number of images and weights and results are more promising as shown in Table 2 having highest accuracy of 85.6%. The model accuracy versus epoch graph is shown in Fig. 5 and the model loss versus epoch graph is shown in Fig. 6 corresponding to the best result of 86.6% accuracy.

Table 2 Comparison table for different model classification model with VGG16

Input size	Batch size	Optimizer	Learning rate	Validation accuracy
64,64	32	SGD	–	0.7673
64,64	32	SGD	0.1	0.76
64,64	32	SGD	0.01	0.78
64,64	32	SGD	0.001	0.71
64,64	32	Adam	0.1	0.50
64,64	32	Adam	0.01	0.758
64,64	**32**	**SGD**	**0.01**	**0.856**

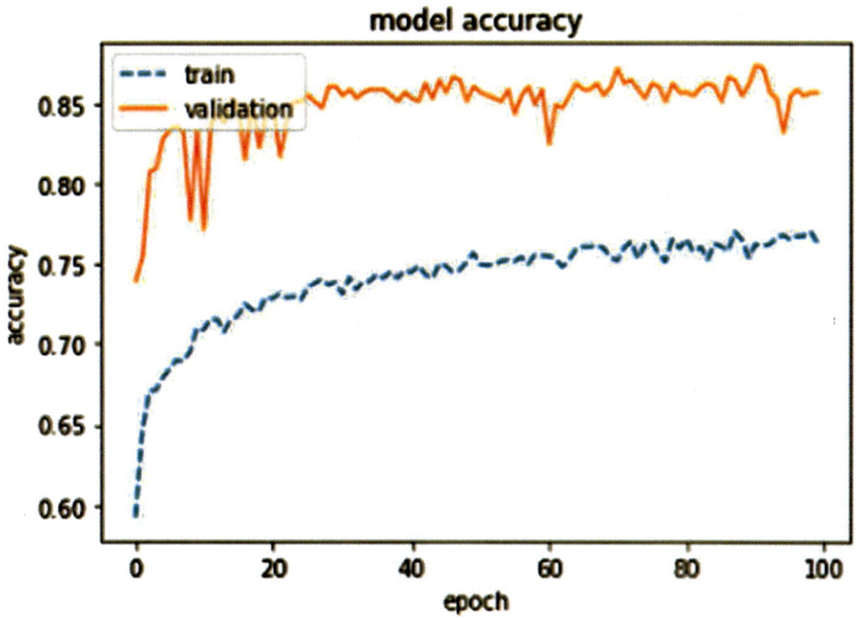

Fig. 5 Model accuracy

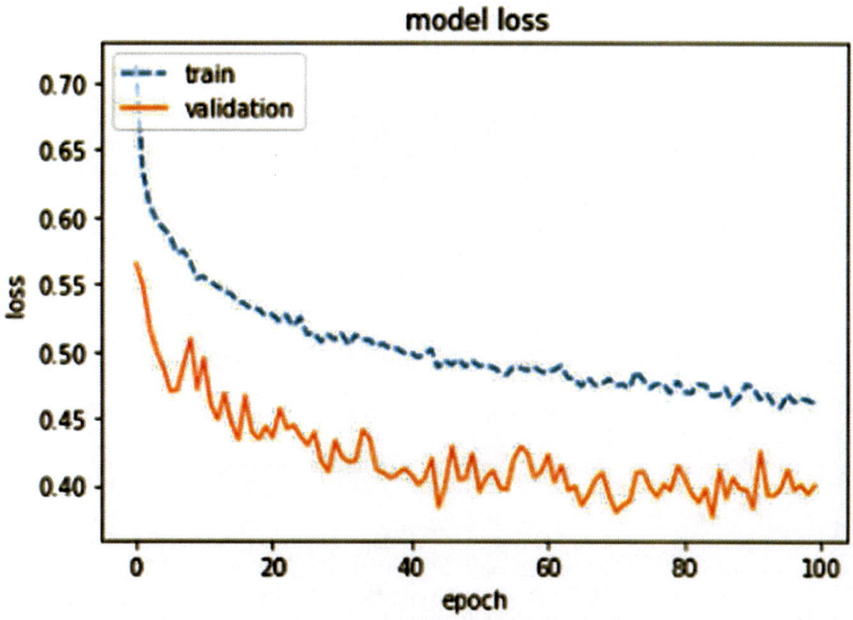

Fig. 6 Model loss

5 Conclusion

In this paper, an effort has been made to detect the weeds in plant images using convolution neural network and pre-trained CNN models. The problem is thus a binary image classification with two classes: with weeds or without weeds. This could help is in run time a field robot could know from an image in front of it is having weed or not. Then selectively it can spay needed fungicide on the plant. The training takes a lot of time, so CNN weights are saved in a file and used by a testing program that needs less time to predict the presence of weed. In the future work, it is intended to solve the location of weeds also on plant images.

References

1. Tejeda AI, Castro RC (2019) Algorithm of weed detection in crops by computational vision. In: 2019 international conference on electronics, communications and computers (CONIELECOMP). IEEE, pp 124–128
2. Chen L, Zhang JG, Su HF, Guo W (2010) Weed identification method based on probabilistic neural network in the corn seedlings field. In: 2010 international conference on machine learning and cybernetics, 11 July 2010, vol 3. IEEE, pp 1528–1531
3. He K, Zhang X, Ren S, Sun J (2016) Deep residual learning for image recognition. In: Proceedings of the IEEE conference on computer vision and pattern recognition, pp 770–778
4. Zhang C, Kovacs JM (2012) The application of small unmanned aerial systems for precision agriculture: a review. Precision Agric 13(6):693–712
5. Lottes P, Khanna R, Pfeifer J, Siegwart R, Stachniss C (2017) UAV-based crop and weed classification for smart farming. In: 2017 IEEE international conference on robotics and automation (ICRA) 29 May 2017. IEEE, pp 3024–3031
6. Ibrahim Z, Sabri N, Mangshor NA (2018) Leaf Recognition using texture features for herbal plant identification. Indonesian J Electr Eng Comput Sci 9(1):152–156
7. Bakhshipour A, Jafari A, Nassiri SM, Zare D (2017) Weed segmentation using texture features extracted from wavelet sub-images. Biosys Eng 1(157):1–2
8. Barrero O, Rojas D, Gonzalez C, Perdomo S (2016) Weed detection in rice fields using aerial images and neural networks. In: 2016 XXI symposium on signal processing, images and artificial vision (STSIVA), Aug 2016. IEEE, pp 1–4
9. Rasti P, Ahmad A, Samiei S, Belin E, Rousseau D (2019) Supervised image classification by scattering transform with application to weed detection in culture crops of high density. Remote Sensing 11(3):249
10. Kamilaris A, Prenafeta-Boldú FX (2018) Deep learning in agriculture: a survey. Comput Electron Agric 1(147):70–90
11. Lottes P, Behley J, Milioto A, Stachniss C (2018) Fully convolutional networks with sequential information for robust crop and weed detection in precision farming. IEEE Robot Autom Lett 3(4):2870–2877
12. Bah MD, Hafiane A, Canals R (2018) Deep Learning with unsupervised data labeling for weeds detection on UAV images. arXiv preprint arXiv:1805.12395. 31 May 2018
13. Wang S, Han Y, Chen J, Pan Y, Cao Y, Meng H, Zheng Y (2018) A deep-learning-based low-altitude remote sensing algorithm for weed classification in ecological irrigation area. In: International conference on intelligent technologies and applications. Springer, Singapore, pp 451–460
14. Bah MD, Hafiane A, Canals R (2018) Deep learning with unsupervised data labeling for weed detection in line crops in UAV images. Remote Sensing 10(11):1690 (2018)

Comparison Between Centrality Measures for a Network Based on Cascading Nature of Nodes

Vaibhav Gaur, Gunjan Soni, Om Prakash Yadav, and A. P. S. Rathore

Abstract Cascading nature of nodes is very important to study the diffusion of information and commodity across networks. Centrality measures are generally used to address cascading behaviors of nodes. This article applies four centrality measures namely, degree, betweenness, closeness, and coreness centrality for modeling of cascading. The results of centrality measures are then compared with an independent simulated cascade model. The simulation result shows coreness centrality to be the best measure for cascade modeling in networks.

Keywords Networks · Cascading · Centrality measures · Coreness

1 Introduction

The research domain has evolved quite significantly in the field of network sciences over the years. The reason behind this is the presence of networks in complex systems ranging from biological, metabolic, social infrastructures to power, communication, and transportation systems [1]. It has been empirically studied that most of the real world complex networks follow the power law degree distribution given by $P(k) \sim k - \gamma$ where the exponent γ ranges from 2 to 3 [2]. This results in some nodes being highly connected whereas some nodes poorly connected within a network. This scale-free property of networks signifies that there are certain nodes playing

V. Gaur (✉) · G. Soni · A. P. S. Rathore
Malaviya National Institute of Technology, JLN Marg, Jaipur, India
e-mail: gaurvaibhav08@gmail.com

G. Soni
e-mail: gunjansoni1980@gmail.com

A. P. S. Rathore
e-mail: apsr100@yahoo.co.in

O. P. Yadav
North Dakota State University, Fargo, USA
e-mail: om.yadav@ndsu.edu

R. P. Mahapatra et al. (eds.), *Proceedings of 6th International Conference on Recent Trends in Computing*, Lecture Notes in Networks and Systems 177,
https://doi.org/10.1007/978-981-33-4501-0_17

vital roles in the structure and functioning of networks. The process of identifying such vital nodes is quite important for studying networks as it could authorize better diffusion of information, control of epidemics, prevention of cascading power outage, maintenance of connectivity in communication networks etc.

Cascading is a concept which is prominently studied in network theory. It has been widely used to analyze power, social, and internet networks. Cascading failure is defined as the sequential failure of interconnected entities in a network triggered from a small failure and resulting in major connectivity loss and breakdown issues in the entire network [3]. The robustness of a network can be addressed by considering the cascading behaviors of nodes. Cascading is also used to examine the diffusion pattern of information in a network [4]. There have been several studies done in the past to analyze cascading in networks. The popular centrality measures used to study cascading includes degree centrality, closeness centrality, betweenness centrality and coreness centrality [5]. To the best knowledge of authors, such a comparative study of these centrality measures has not been fully explored yet. This motivates the authors to investigate the centrality measures and their cascading behavior in a network. This article compares these measures with the results of a simulated cascading model. The comparative study would reveal which of the available measures could be used to address cascading in a network.

Rest of the article follows the structure as: Sect. 2 explains the centrality measures used for studying cascading. Section 3 explains cascading in networks in a laconic manner. An independent simulated cascading model along with centrality measures are discussed in Sect. 4. Section 5 presents the results of this study and Sect. 6 concludes this article.

2 Centrality Measures

In this section, the commonly used centrality measures used to investigate cascading in networks have been discussed.

2.1 Degree Centrality

In case of undirected network $G(V,E)$ where V is the set of nodes and E is the number of edges respectively, the degree for a node i denoted by k_i is defined as the directly connected neighbors of i [6]. Degree centrality is the most simple centrality measure. The normalized degree centrality is calculated as

$$\mathrm{DC}_i = \frac{k_i}{n-1} \tag{1}$$

where n is number of nodes in the graph G. Degree centrality provides a useful measure for the study of networks because of its computational simplicity and wide range of applications. Degree centrality is used for vulnerability analysis, interdependent network analysis and reliability analysis of networks [7].

2.2 Betweenness Centrality

The betweenness of a node i in a network is the number of shortest path of the other nodes that passes through i. Betweenness centrality for a node i can be defined as [8]

$$\mathrm{BC}_i = \sum_{i \neq s, i \neq t, s \neq t} \frac{g_{st}^i}{g_{st}} \tag{2}$$

where g_{st} is the shortest path numbers between node s and t and g_{st}^i is the number of paths that passes through i and from all the shortest path between s and t. Betweenness centrality is extensively used in electric power networks where the load capacity of nodes are modelled [9].

2.3 Closeness Centrality

Closeness centrality represents how efficiently a node can transfer information or commodity to other nodes in the network. The mathematical formulation of closeness centrality involves the inverse of harmonic mean of distances between the nodes in a network and is given by [10]

$$\mathrm{CC}_i = \frac{1}{n-1} \sum_{j \neq i} \frac{1}{d_{ij}} \tag{3}$$

where n is the number of nodes and d_{ij} is the distance between nodes i and j. Closeness centrality has been used primarily to discuss the performance or functioning of a network.

2.4 Coreness Centrality

Coreness centrality is based on the location of a node relative to the entire network. It proposes that the nodes located at core of a network have the maximum influence in spreading effects across the entire network. The process of identifying coreness centrality is also called k-shell decomposition in which the network is decomposed into k-layers where k represents the degree of the respective layer. A modified degree measure based on k-shell decomposition is used to infer the coreness of a particular node in the network structure and for a node i is given by [11]

$$k_i^{\text{mix}} = k_i^r + \varepsilon k_i^e \tag{4}$$

where k_i^r denotes the number of links of remaining nodes and k_i^r is the number of links connected to the removed nodes and ε is an adjustable parameter. Coreness centrality has been used for studying the spreading of information within the networks through vital nodes.

3 Cascading in Networks

Cascading denotes the ability of nodes to diffuse information in a network. Cascading in networks can take many forms such as diffusion of idea in social networks, failure of components in electric networks, contamination of commodity in water networks, spreading of epidemics, etc. Thus, it becomes quite essential to discuss the cascading behavior of nodes in a network.

The structural property of a network offers an insight to scrutinize cascading behaviors of nodes. The identification of nodes which have maximum influence on cascading is of utmost importance in order to take correct actions. In cases where cascading is required such as spreading (diffusion) of important information through a social network, the vital nodes should be infused with the information. In cases where cascading is undesired such as epidemic breakout or failure in electric networks, the vital nodes should be cut off to prevent cascading.

4 Cascading Model and Centrality Measures

For the scope of this study, an independent cascading based on susceptible infected recovered (SIR) model has been simulated. The sample network selected is based on Erdos Renyi (ER) model which has been extensively used for network analysis [12]. It has been proved that ER model provides complete abstraction of many real world networks and has been applied predominantly for understanding the dynamics of networks [1]. For the purpose of examining cascading with the centrality measures a small ER network would provide lucrative results. Therefore, the sample network included in this study consists of 20 nodes having a connection probability of 0.5 between the nodes.

The following steps have been performed in order to carry out cascading on the network:

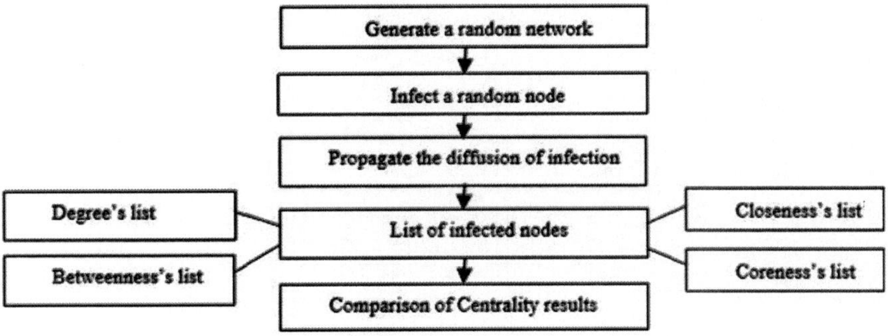

Fig. 1 Flowchart for comparing centrality measures

Step 1—Generate an Erdos Renyi network with 20 nodes and 0.5 connection probability.
Step 2—Infect a node at random in the network.
Step 3—Propagate the diffusion of infection based on topological structure of the network.
Step 4—Generate list of all the infected nodes at the end of simulation.

By following the aforementioned steps, an independent cascading in a network has been simulated. The next task at hand is to compare the results obtained through different centrality measures with the cascading results. For comparing results of the centrality measures, lists are generated based on degree, betweenness, closeness, and coreness centrality. The generated lists are then compared with the cascading list to evaluate which measure provide the best results for cascading. The framework for performing this study is given in Fig. 1.

5 Results and Discussions

Based on the aforementioned framework, cascading along with different centrality measures are simulated on the sample network. The results of the simulated model are given in Table 1.

It can be observed from Table 1 that the list of nodes for cascading and coreness are quite similar. Out of the chosen centrality measures, coreness is well suited for explaining the cascading behavior in a network. The percentage similarity for the first five nodes between cascading list with degree's list, betweenness's list, closeness's list and coreness's list is 60%, 40%, 60%, and 100%, respectively. This demonstrates that for modeling cascading in a network, the coreness centrality measure provides the best result.

Cascading behavior in networks is generally triggered by small changes in attributes of nodes. This dynamicity of nodes might leads to catastrophic results for the entire network. An avalanche of information flow could be prompted by

Table 1 Cascading and centrality measures results

S. No.	Cascading's list	Degree's list	Betweenness's list	Closeness's list	Coreness's list
1	13	5	17	10	13
2	14	13	5	13	14
3	15	14	14	5	15
4	10	15	13	11	16
5	16	11	8	14	11
6	11	10	10	15	10
7	5	16	11	16	12
8	12	12	1	17	1
9	7	1	3	4	2
10	3	2	2	6	3

bringing marginal change in vital node's properties. Such a situation is quite critical for the efficient functioning of networks. Coreness centrality provides a measure to quantify the central tendency of nodes in a network. The effect of bringing change in the central or core region of a network would bring drastic changes in the entire network.

The findings of this work have major implications in the realm of network sciences. Analyzing the cascading behavior through coreness centrality could aid in developing maintenance plan for the core of a network. The elements of critical infrastructures and the potential impacts of their failure can be quantified using the given model. The identification of core in a system could help decision makers in taking ideal actions to prevent catastrophic cascading.

6 Conclusion and Future Scope

This paper compares the popular centrality measures for a network based on their cascading effects. An independent cascading SIR model has been simulated on an Erdos Renyi network for this study and results of the cascading are compared with the results of different centrality measures. The centrality measures compared in this article are degree, betweenness, closeness, and coreness centrality. Among these, coreness centrality provides the best results for scrutinizing cascading behaviors in networks.

Since the dynamicity of nodes in networks is quite high, it is impossible to account for all factors in this paper. We have kept our scope limited to random networks and four centrality measures with the aim to provide readers with an enriched literature in a concise manner. The present work could further be extended by incorporating other type of networks and other available centrality measures.

References

1. Zhao D, Wang L, Xu S, Liu G, Han X, Li S (2017) Vital layer nodes of multiplex networks for immunization and attack. Chaos, Solitons Fractals 105:169–175
2. Buldyrev SV, Parshani R, Paul G, Stanley HE, Havlin S (2010) Catastrophic cascade of failures in interdependent networks. Nature 464:1025–1028
3. Cadini F, Agliardi GL, Zio E (2017) A modeling and simulation framework for the reliability/availability assessment of a power transmission grid subject to cascading failures under extreme weather conditions. Appl Energy 185:267–279
4. Kitsak M, Gallos LK, Havlin S, Liljeros F, Muchnik L, Stanley HE, Makse HA (2010) Identification of influential spreaders in complex networks. Nat Phys 6:888–893
5. Costa LF, Rodrigues FA, Travieso G, Boas PRV, Costa LF, Rodrigues FA, Travieso G, Boas PRV (2007) Advances in Physics Characterization of complex networks : a survey of measurements, vol 8732 (2007)
6. Hu J, Du Y, Mo H, Wei D, Deng Y (2016) A modified weighted TOPSIS to identify influential nodes in complex networks. Phys A Stat Mech Its Appl 444:73–85
7. Chen G, Dong ZY, Hill DJ, Zhang GH, Hua KQ (2010) Attack structural vulnerability of power grids: a hybrid approach based on complex networks. Phys A Stat Mech Its Appl 389:595–603
8. Bompard E, Napoli R, Xue F (2009) Analysis of structural vulnerabilities in power transmission grids. Int J Crit Infrastruct Prot 2:5–12
9. Wang S, Hong L, Ouyang M, Zhang J, Chen X (2013) Vulnerability analysis of interdependent infrastructure systems under edge attack strategies. Saf Sci 51:328–337
10. Guidotti R, Gardoni P, Chen Y (2017) Network reliability analysis with link and nodal weights and auxiliary nodes. Struct Saf 65:12–26
11. Lü L, Chen D, Ren XL, Zhang QM, Zhang YC, Zhou T (2016) Vital nodes identification in complex networks. Phys Rep 650:1–63
12. Li D, Zhang Q, Zio E, Havlin S, Kang R (2015) Network reliability analysis based on percolation theory. Reliab Eng Syst Saf 142:556–562

Impact of Pocket in a Doping-Less Tunnel Field Effect Transistor

Mohd Adil Raushan, Mohd Mustaqeem, Shameem Ahmad, and Mohd Jawaid Siddiqui

Abstract In this paper, we have proposed the use of n-type pocket in a doping-less tunnel field effect transistor to improve the performance of the device. We have proposed a smart strategy to improve the on-state current and sub-threshold slope SS of the device. The main concept is to use the chemical doping at a specific point rather than the whole device. This strategy helps to inherit the doping-less nature of the device up to a large extent. The presence of n-type pocket at gate-source interface reduces the tunnel barrier and increases the electric field in the on-state ($V_{GS} = V_{DS} = 1$ V). This leads to higher I_{ON} (8×), better point SS (~7 mV/dec), and larger I_{ON}/I_{OFF} ratio (10^{13}). We have also investigated other parameters such as g_m (10×), f_T (6×), C_{GG} (1.45x) which have shown an increase in n-type pocket DLTFET. The improved linearity parameters such as g_{m3}, VIP2, VIP3, IIP3 have also been reported. Gate scaling down to 20 nm has also been performed which has yielded significantly better results for pocket DLTFET.

Keywords Doping-less · Tunneling · Impurity pocket · Charge plasma

M. A. Raushan (✉) · M. Mustaqeem · S. Ahmad · M. J. Siddiqui
Department of Electronics Engineering, Z.H.C.E.T, Aligarh Muslim University, Aligarh, India
e-mail: adil.raushan02@gmail.com

M. Mustaqeem
e-mail: mustqeem143@gmail.com

S. Ahmad
e-mail: engineershameem@gmail.com

M. J. Siddiqui
e-mail: mjsatamu@gmail.com

R. P. Mahapatra et al. (eds.), *Proceedings of 6th International Conference on Recent Trends in Computing*, Lecture Notes in Networks and Systems 177,
https://doi.org/10.1007/978-981-33-4501-0_18

1 Introduction

The continuous miniaturization of transistors has been a constant bottleneck for device engineers in this era of rapid scaling. The fabrication of p-n junctions of sharp gradients at sub-100 nm lengths is a very challenging task. The minimum sub-threshold slope "SS" of 60 mV/dec for conventional MOSFET devices are also an inherent problem due to thermionic emission over the barrier in the sub-threshold region [1, 1]. To counter the problem of minimum "SS," tunnel field effect transistors "TFET" have been proposed which are based on the mechanism of tunneling of electrons via the barrier instead of climbing over it [3, 4]. They also have a very low leakage current as compared to MOSFETs due a higher barrier present in the OFF-state. But for smaller channel lengths (<20 nm), these TFETs also suffer from the problem of junction fabrication and random dopant fluctuations [5]. The movement of impurity atoms from highly doped source and drain regions to the intrinsic channel region is also a concern. To overcome these problems, doping-less TFET has been proposed which use electrodes of different work functions to create the n-type, intrinsic, p-type regions for TFET device operation [6, 7]. These doping-less TFETs are free from the problem of random dopant fluctuations, as they use intrinsic substrates. Electrostatic doping is used instead of chemical doping to generate the n^+-i-p TFET. But all types of TFETs suffer from a common problem of low I_{ON} or ON-state current. Many techniques have been proposed to increase the ON-state current of doped and doping-less TFET. They include the use of hetrodielectric [8], heterojunctions [9], strained silicon [10], dual-material gate [11], insertion of metal chip [12], dual-k spacers [3], and different materials [9]. The use of highly doped n-type pocket at the gate edge near the source is also an established technique which has been proposed for doped TFETs [13]. But this technique is proposed for chemically doped TFET. In this paper, we propose the use of highly doped n-type pocket in the doping-less TFET. The basic idea behind this is the use of chemical doping at specific points, rather than the whole device. Thus, in this paper, we analyze the impact of n-type pocket at the gate-source interface on the device behavior of doping-less TFET. Our device is almost doping-less in nature except the presence of n-type pocket in the channel region near the source.

2 Device Structure and Simulation Parameters

Figure 1a, b shows conventional doping-less TFET (DLTFET) without N pocket and doping-less TFET with N pocket (NP-DLTFET). The source, gate, and drain lengths taken for both the devices are 40 nm. The spacers at the gate-source interface and gate-drain interface are 5 nm and 15 nm, respectively. These doping-less devices use electrostatic doping (ED) for the formation of n and p regions [14]. In this ED, the carrier concentration in the semiconductor depends on the electrode besides it. The device thickness should be smaller than the Debye length to generate reasonable and

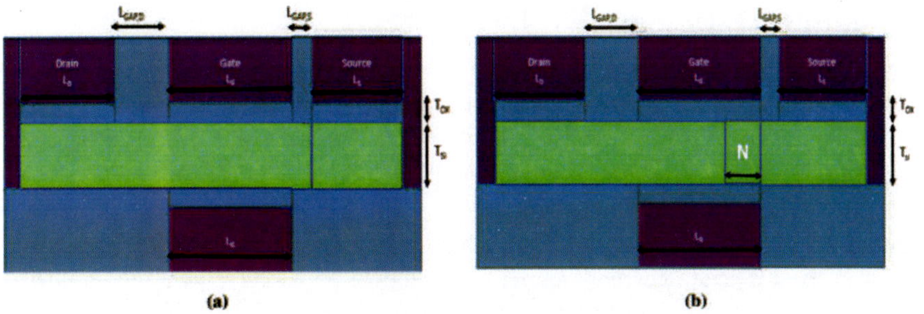

Fig. 1 Cross-sectional view of **a** Conventional doping-less TFET, **b** Doping-less TFET with N pocket

uniform carrier concentration [15]. For p-type region, high work function electrodes (5.93 eV) such as platinum is used and low work function electrodes (3.9 eV) such as hafnium is used for n-type doping. The use of n pocket is proposed in doping-less TFET. The position of n-type pocket is at the gate edge on the source side to increase the electric field at the tunneling junction. The length of the pocket is varied from 2 to 5 nm. The carrier concentration of the pocket is $10^{20}/cm^3$. The device simulations are carried out by Silvaco Atlas 2D device simulator [16]. Lombardi mobility model has been used for mobility calculations. Fermi–Dirac statistics have been used for intrinsic carrier calculations; Shockley–Read–Hall (SRH) recombination model was used for lifetime calculations. The quantum confinement model by Hansch was also used as the device thickness was 10 nm. The non-local band to band tunneling model has been used to capture the tunneling in ON-state. The tunneling mesh was taken from the gate edge on the drain side up till the source to account for accurate tunneling (Table 1).

Table 1 Device parameters

Parameter	Symbol	Values
Gate length	L_g	40 nm
Oxide thickness	T_{ox}	1 nm
Silicon thickness	T_{si}	10 nm
Channel doping	N_{ch}	$10^{15}/cm^3$
Source gap length	$L_{GAP,S}$	5 nm
Drain gap length	$L_{GAP,D}$	15 nm
Gate work function	φ	4.5 eV
N pocket length	L_N	2–5 nm

3 Results and Analysis

The band diagram for both the devices in ON-state ($V_{GS} = V_{DS} = 1$ V) is shown in Fig. 2a. The presence of pocket reduces the tunnel barrier at the tunneling junction. Figure 2b compares the I_D–V_G curves for both the devices DLTFET and NP-DLTFET. The curve with pocket length $L_N = 0$ nm corresponds to the conventional DLTFET. It can be observed that NP-DLTFET shows a remarkable improvement in the ON-state current. The I_{ON} has shown an 8× ($L_N = 2$ nm) improvement for a similar I_{OFF}. The point sub-threshold slope for NP-DLTFET is less than 10 mV/dec which is much smaller than DLTFET. The average sub-threshold slope SS of the NP-DLTFET has also decreased considerably from ~50 mV/dec (DLTFET) to ~25 mV/dec for $L_N = 2$ nm. The I_{ON}/I_{OFF} for NP-DLTFET has also increased by almost an order, thereby enhancing its switching credentials. This increase in ON-state current and decrease in SS is attributed to reduced tunnel barrier and increased electric field and carrier concentration at the tunneling junction due to the presence

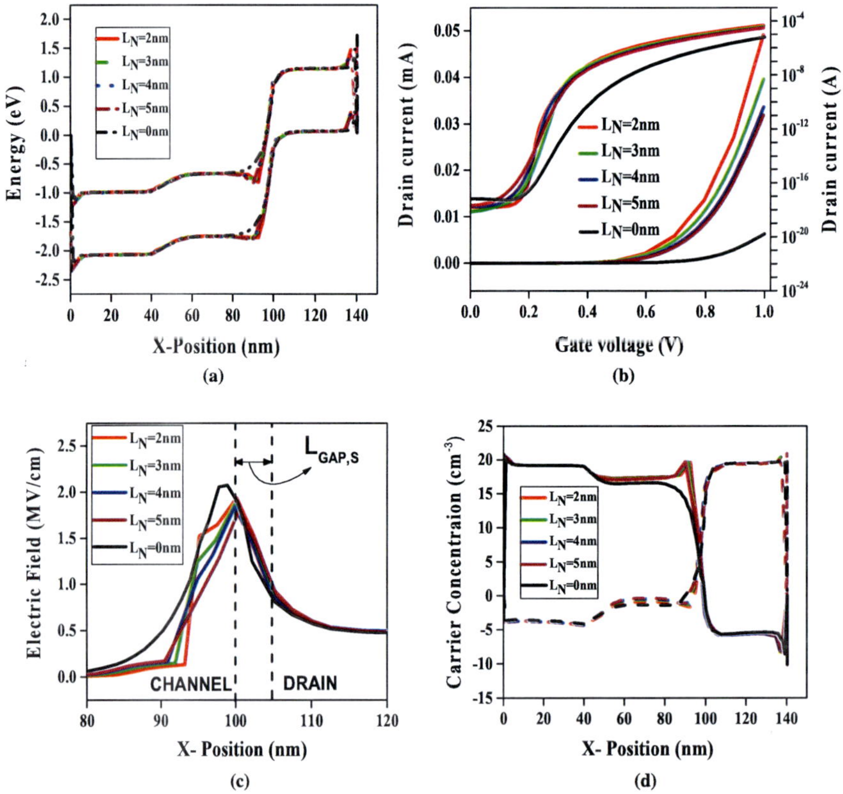

Fig. 2 **a** Band diagram in ON-state ($V_{GS} = V_{DS} = 1$ V), **b** I_D–V_G curves on log and linear scale, **c** Electric field in ON-state, **d** Carrier concentration in ON-state

of n-type pocket (Fig. 2c, d). From Fig. 2c, we can observe that there is a drop in electric field under the spacer $L_{\mathrm{GAP,S}}$ for conventional DLTFET. For NP-DLTFET, electric field is higher under the spacer region which assists tunneling phenomenon. Thus, the pocket improves the electric field and reduces the tunnel barrier which enhances tunneling given by Eq. 1. In the equation E_g and m^* represent the band gap and tunneling mass. F is the electric field, and h and q represent the Planck's constant and charge. These results are in accordance to the results given by Jhaveri et al. [13]. Thus, our device has also shown improved performance by the use of n-type pocket.

$$T_t = \exp\left(-\frac{4\sqrt{2m^*}}{3q\hbar F}E_g^{3/2}\right) \quad (1)$$

To further explore the effect of n pocket, transconductance g_m, gate capacitance C_{gg}, and cut-off frequency f_T were calculated. These parameters are shown in Fig. 3a–c. The transconductance g_m ($\mathrm{d}I_\mathrm{D}/\mathrm{d}V_\mathrm{G}$) for NP-DLTFET shows an almost 10 × increase (Fig. 3a) while the C_{gg} also shows a 1.45 × increase for pocket length L_N = 2 nm (Fig. 3b). The cumulative effect of both the parameters is reflected in

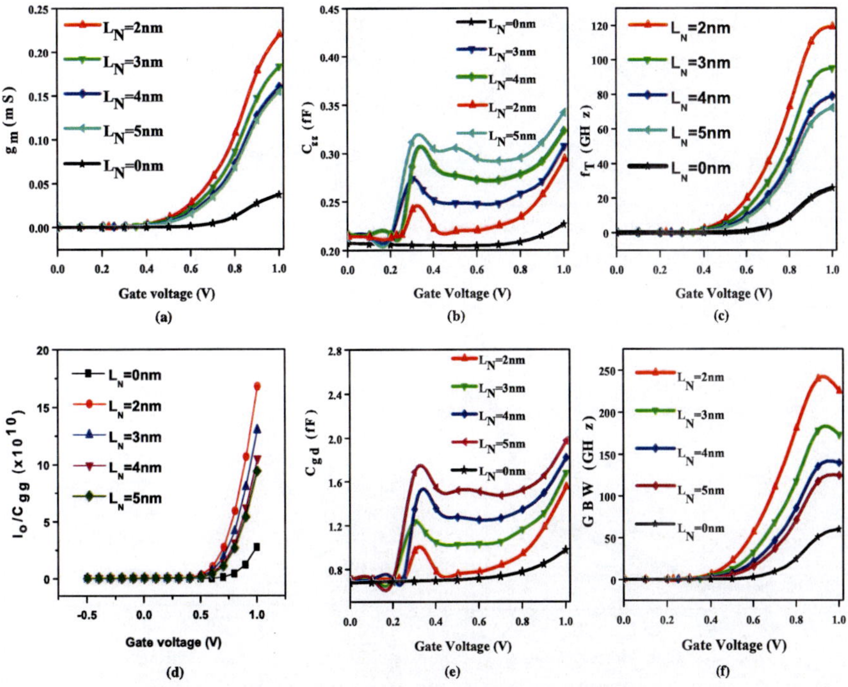

Fig. 3 Comparative plots of **a** Transconductance g_m, **b** Gate capacitance C_{gg}, **c** Cut-off frequency f_T, **d** I_O/C_{GG}, **e** Gate-drain capacitance C_{GD} (f) Gain bandwidth product GBW

$f_T(g_m/2\pi C_{gg})$ shown in Fig. 3c. The f_T, which determines the high frequency performance, shows a larger value for all gate voltages up to 1 V. The cut-off frequency f_T shows a $6\times$ increase for $L_N = 2$ nm at $V_{GS} = 1$ V. Even the I_O/C_{gg} parameter for NP-DLTFET shows a $7\times$ increase for NP-DLTFET ($L_N = 2$ nm) as compared to DLTFET at $V_{GS} = 1$ V (Fig. 3d). I_O/C_{gg} is a measure of transistors drive strength, and higher the value better is the device drive strength. The gate to drain capacitance C_{GD} which plays a dominant role in determining the C_{GG} is shown in Fig. 3e.

C_{GD} increases with increase in L_N, as the coupling between gate and drain improves with increase in L_N. The gain bandwidth product GBW ($g_m/2\pi C_{GD}$) for both the devices are plotted in Fig. 3f. The NP-DLTFET shows higher GBW for all pocket lengths L_N (2–5 nm). For $L_N = 2$ nm, NP-DLTFET shows a value of 250 GHz as compared to DLTFET having 50 GHz at $V_{GS} = 1$ V. All the parameters except C_{GD} and C_{gg} have shown improvement for NP-DLTFET. To investigate the non-linearity issues, we have calculated the linearity and distortion parameters g_{m3}, VIP2, VIP3, and IIP3 of both the devices [17]. Figure 4a shows the g_{m3} coefficient of both the devices.

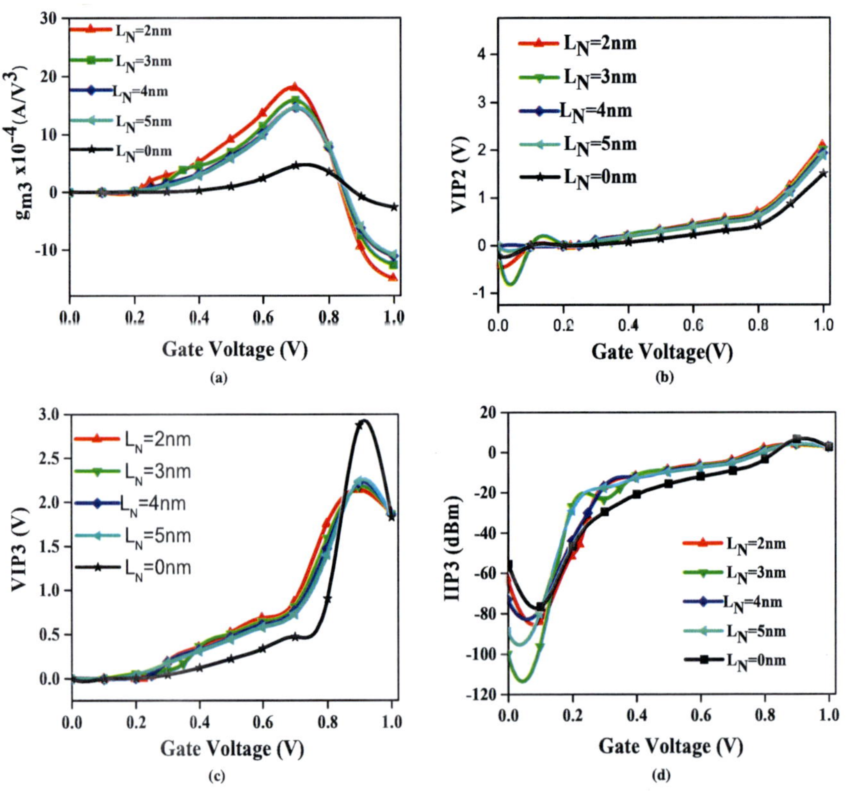

Fig. 4 Comparative plots of **a** Third-order transconductance coefficient g_{m3}, **b** VIP2 parameter, **c** VIP3 parameter, **d** IIP3

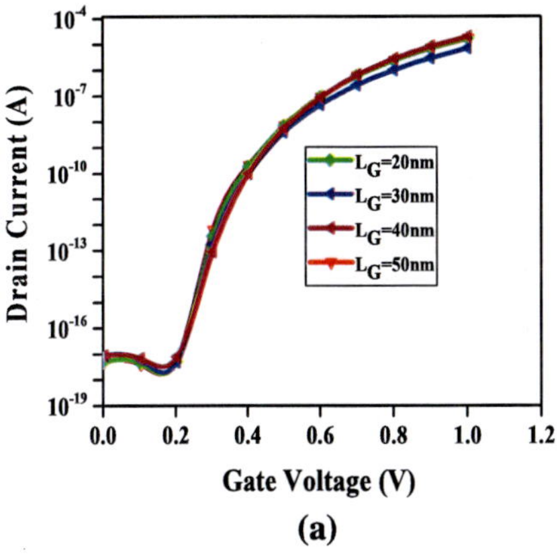

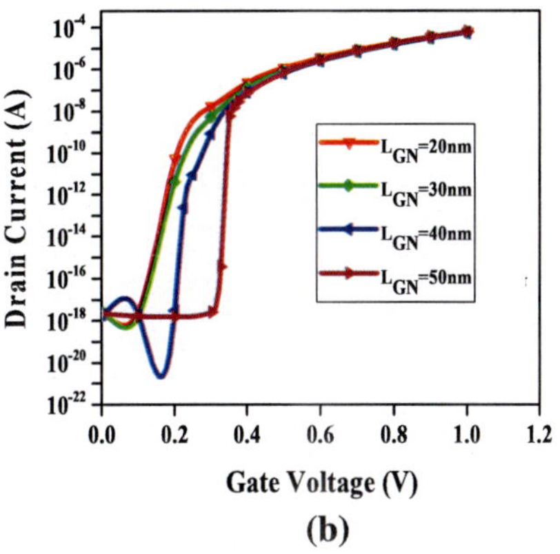

Fig. 5 **a** Gate length (L_G) scaling from 50 to 20 nm for DLTFET with no pocket, **b** Gate length (L_{GN}) scaling from 50 to 20 nm for NP-DLTFET

The proposed NP-DLTFET for different L_N shows the peak and zero crossover point at lower gate voltages as compared to DLTFET. The parameters VIP2 and VIP3 show an increase with gate voltage (Fig. 4b, c). Our proposed device shows a higher value of VIP2 indicating better linearity. The parameter VIP3 shows a slightly decreased value after $V_{GS} = 0.8$ V, indicating reduced cancellation of third-order transconductance coefficient of the device. The parameter IIP3 is also calculated to estimate the linearity of the device characteristics in Fig. 4d. The higher value of IIP3 for lower gate voltages signifies better linear behavior of NP-DLTFET at lower gate voltages [17]. The Fig. 5a, b shows the gate length scaling of both the devics from 50 nm to 20 nm. The NP-DLTFET shows better performance than conventional DLTFET at all gate lengths.

4 Conclusion

In this paper, we have investigated the impact of an n-type pocket on the DLTFET. The introduction of pocket has led to a remarkable improvement in the overall device performance. As expected, the use of pocket has served the purpose of reduction in SS and increased ON-state current. Thus, we have proposed a new strategy to improve the device performance of DLTFET. Although the device is not completely doping-less but considering the negligible pocket length (2–5 nm) as compared to device length ($L_G + L_S + L_D$), it can be considered a doping-less device with improved behavior. Our results can provide an incentive to further explore this device architecture.

Acknowledgements This work was carried out under the project of "Visvesvaraya PhD Scheme for Electronics and IT" at Aligarh Muslim University, by Media Lab Asia (A Section 25 Company of

Department of Electronics and Information Technology, Ministry of Communications and Information Technology, Govt. of India). The revised implementation Order No. is "PhD-MLA/4(39)/2015-16/ Dated 30.05.2016." The authors are also thankful to the University Grant Commission to support this work in part under the DSA-I and start-up grants.

References

1. Raushan MA, Alam N, Siddiqui MJ (2019) Electrostatically doped drain junctionless transistor for low-power applications. J Comput Electron 18(3):864–871
2. Ghosh B, Akram MW (2013) Junctionless tunnel field effect transistor. IEEE Electron Device Lett 34(5):584–586
3. Raushan MA, Alam N, Akram MW, Siddiqui MJ (2018) Impact of asymmetric dual-k spacers on tunnel field effect transistors. J Comput Electron 17(2):756–765
4. Musalgaonkar G, Sahay S, Saxena RS, Kumar MJ (2019) A line tunneling field-effect transistor based on misaligned core—shell gate architecture in emerging nanotube FETs. IEEE Trans Electron Dev, 1–8
5. Damrongplasit N, Kim SH, Liu T-JK (2013) Study of random dopant fluctuation induced variability in the raised-GE-source TFET. IEEE Electron Device Lett 34(2):184–186
6. Raushan MA, Alam N, Siddiqui MJ (2018) Dopingless tunnel field-effect transistor with oversized back gate: proposal and investigation. IEEE Trans Electron Devices 65(10):4701–4708
7. Kumar MJ, Janardhanan S (2013) Doping-less tunnel field effect transistor: design and investigation. IEEE Trans Electron Devices 60(10):3285–3290
8. Venkatesh P, Nigam K, Pandey S, Sharma D, Kondekar PN (2017) Impact of interface trap charges on performance of electrically doped tunnel FET with heterogeneous gate dielectric. IEEE Trans Device Mater Reliab 17(1):245–252
9. Kondekar PN, Nigam K, Pandey S, Sharma D (2017) Design and analysis of polarity controlled electrically doped tunnel FET with bandgap engineering for analog/RF applications. IEEE Trans Electron Devices 64(2):412–418
10. Molaei Imen Abadi R, Sedigh Ziabari SA (2016) Representation of strained gate-all-around junctionless tunneling nanowire filed effect transistor for analog applications. Microelectron Eng 162:12–16
11. Raushan MA, Alam N, Siddiqui MJ (2018) Performance enhancement of junctionless tunnel field effect transistor using dual-k spacers. J Nanoelectron Optoelectron 13(6):912–920
12. Raad BR, Tirkey S, Sharma D, Kondekar P (2017) A new design approach of dopingless tunnel FET for enhancement of device characteristics. IEEE Trans Electron Devices 64(4):1830–1836
13. Jhaveri R, Nagavarapu V, Woo JCS (2011) Effect of pocket doping and annealing schemes on the source-pocket tunnel field-effect transistor. IEEE Trans Electron Devices 58(1):80–86
14. Gupta G, Rajasekharan B, Hueting RJE, Member S (2017) Electrostatic doping in semiconductor devices. IEEE Trans Electron Devices 64(8):3044–3055
15. Rajasekharan B, Hueting RJE, Salm C, Van Hemert T, Wolters RAM, Schmitz J (2010) Fabrication and characterization of the charge-plasma diode. IEEE Electron Device Lett 31(6):528–530
16. Atlas User's Manual (2013) Silvaco, Santa Clara, CA, USA
17. Nigam K, Pandey S, Kondekar PN, Sharma D, Kumar Parte P (2017) A barrier controlled charge plasma-based TFET with gate engineering for ambipolar suppression and RF/linearity performance improvement. IEEE Trans Electron Devices 64(6), 2751–2757

Portfolio Optimization Using Genetic Algorithms with Nelder–Mead Algorithm

Annem Abhishek Reddy, Maheedhar Vundela, and G. Manju

Abstract Portfolio selection has a literal sense that means that the selection of an optimal portfolio which results to an asset that has minimum risk and maximized cost. The process refers to the formulation of an objective function that determines the weights of the portfolio that has to be invested in. In this paper, we have tried to optimize the selection of portfolio using genetic algorithms. Sharpe ratio has been used as a cost function to determine the fitness of stocks, and a local search algorithm, Nelder–Mead, has been used to optimize the selection of the portfolio process and to improve the efficiency.

Keywords Genetic algorithm · Portfolio optimization · Sharpe ratio · Nelder–Mead · Simplex and risk-free rate

1 Introduction

Wall-street is consistently striving toward making smart investments. Selection of portfolios should be optimum so as to yield the maximum profit with minimum risks. They need speed and precision to plan and make decisions regarding the investments that are being made. The primary task is to plan to acquire optimum allotment of the desirable wealth under controlled risk with the best return. These limitations are considered to develop a model to help the decision-making process in the process of portfolio selection. The models developed to optimize the portfolio in the selected companies are investigated using the genetic algorithms. An optimum set of portfolios

A. A. Reddy (✉) · M. Vundela (✉) · G. Manju (✉)
Department of Computer Science and Engineering, SRM Institute of Science and Technology, Kattankulathur, India
e-mail: annemabhishekreddy@gmail.com

M. Vundela
e-mail: maheedharmahi1@gmail.com

G. Manju
e-mail: manjug@srmist.edu.in

R. P. Mahapatra et al. (eds.), *Proceedings of 6th International Conference on Recent Trends in Computing*, Lecture Notes in Networks and Systems 177,
https://doi.org/10.1007/978-981-33-4501-0_19

has been attempted to be determine in this paper using a genetic algorithm coupled with a local maximization search algorithm called Nelder–Mead.

2 Literature Survey

Chen et al. [1] have made a study on how to use island-based genetic algorithms to optimize the portfolio. They have divided the stocks into several subsets, and individual subset has been allotted to pick a random subset from the population. Jin et al. [2] have conducted a study on optimization using matrix inequalities for control problems and robust portfolios using the ellipsoidal algorithm.

Chan et al. [3] have made a study to solve multi-stage portfolio optimization using algorithms of genetic in nature. The actual architecture as represented by them is shown in Fig. 1. The results were evaluated using the outcomes generated by the standard deviation. Sasaki et al. [4], have figured out a method to develop an interactive system for portfolio optimization using genetic algorithms which used the modern portfolio theory (MPT) and evaluated the results using the reviews of domain experts.

Mishra et al. [5] have tried implementing the multi-objective approach to portfolio optimization.

The types of GA that were studied were:

1. Parallel single front
2. Strength Pareto Evolutionary
3. Non-dominated sorting
4. Multi-objective particle swarm.

The results portrayed that the fourth method that was studied outperformed the others. The evaluation results show cased using the evaluation methods are given in Table 1.

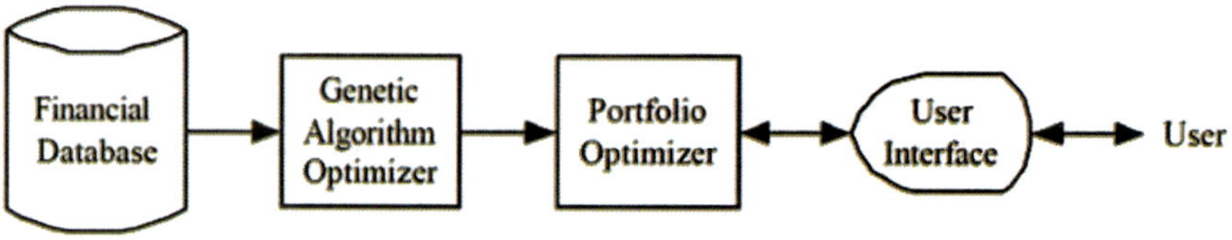

Fig. 1 Architecture of the proposed portfolio optimization system

Table 1 Comparison of the observed algorithms

Algorithm	PSFGA	SPEA 2	NSGA II	MOPSO
Metric S	0.000423124	0.000304616	0.0000067874	0.0000005943
Metric Δ	0.8898852851	0.8337976192	0.5967844252	0.561273589

Rankovic et al. [6] have conducted a study to solve the problem of portfolio allocation that are static based on the historical value at risk (VaR) by using GA. They also showcase the ability to control the computation time and solutions that are diverse in nature. Chou et al. [7] have performed research on the portfolio optimization based on funds standardization and genetic algorithm.

They have combined the Sharpe ratio of the genetic algorithm and the funds standardization to find the find the optimal portfolio. Figure 2 represents the general procedure in which the genetic algorithm works.

Chen and Zhou [8] have created a model that implements the mapping coding scheme (MDM). Further, they have also integrated this scheme in to the evolutionary algorithm with multiple objectives based on the decomposition of the portfolios.

Chan et al. [3] have made a study on solve multi-stage portfolio optimization using different types of genetic algorithm. The actual architecture as represented by

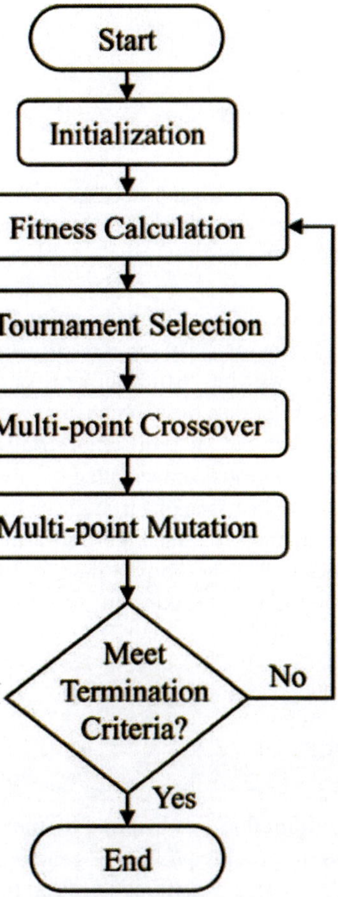

Fig. 2 Flowchart of the genetic algorithm

them is shown in Fig. 2. The results were evaluated using the outcomes generated by the standard deviation. Sasaki et al. [4] have figured out a method to develop an interactive system for portfolio optimization using genetic algorithms which used the modern portfolio theory (MPT) and evaluated the results using the reviews of domain experts.

Xu et al. [9] have conducted a study and their research findings have presented a novel PSO algorithm hybridized with Nelder and Mead procedure (NM-PSO). Test results show that this novel algorithm has a great ability to solve high-dimension optimization, and the performance does not decrease as the dimension increase.

Chang et al. [10] have proposed a new framework SNM-Q that combines stochastic Nelder–Mead method with quantiled based procedure to obtain the ordinal information of solutions instead of requiring gradient information. A global and local search framework further enables SNM-Q to prevent the algorithm from premature converge and guarantees a positive probability of finding a better value in each iteration.

3 Nelder–Mead Algorithm

Since, the genetic algorithms are resource exhaustive, we have tried to use a general-purpose deterministic local search algorithm. The algorithm that has been used is the Nelder–Mead search algorithm.

The Nelder–Mead algorithm also known as a simplex search algorithm is mostly used for multidimensional unconstrained optimization for problems without derivatives. This is used usually to solve the optimization problem with an analytical method. This algorithm follows a heuristic approach for the optimization. This algorithm adds a degree of randomness to the problem that means that the same algorithm could run on the same problem but yield different results each time.

This algorithm uses a geometrical shape called the simplex to search the domain. The simplex is a n-dimensional version of a triangle. Initially, the algorithm starts off with a randomly-generated simplex. This algorithm tries out a set of modifications and selects the optimum simplex that moves the algorithm closer to the optimum value. This basically involves the shrinking of the simplex to the point at which it converges to the optimum value.

4 Genetic Algorithm

It is heuristic search algorithm that was inspired by the Charles Darwin's theory of evolution. This algorithm mimics the process of gene replication and selection. In a similar way, this algorithm performs natural selection based on the inputted fitness or the cost function. Genetic algorithms are used to attain the optimum value in deterministic cases. Since, the stock market involves a very volatile time-series data, this algorithm has been chosen for the optimization.

5 Fitness Function

Fitness function used to determine the exact fit of the selection of the chromosomes that are being generated by the genetic algorithm. Fitness function also known as the cost functions also involves a penalty function that checks on the bounds of the data that is being inputted. The fitness function compromises of the following two functions that are used to determine the fit of the offspring:

- Objective function
- Penalty function.

Objective function has been created using the Sharpe ratio (Eq. 1) that is often used in the field of finance to estimate the risk of any portfolio, bonds, or stocks. In this paper, it has been considered as a stability metric for the set of stocks that is being selected, this is due to the fact that investors usually aim for a higher stability metric rather than the higher profits.

$$\text{Sharpe Ratio} = R_{\mathrm{p}} - R_{\mathrm{f}}/\sigma_{\mathrm{p}} \tag{1}$$

where

R_{p} return of portfolio,
R_{f} risk-free rate,
σp deviation of the portfolio's excess return.

Usually risk-free rate is taken from a portfolio that usually has no risk. Practically, a completely risk-free stock is not possible, and therefore, the prices of the U.S. Treasury bills has been taken into consideration for a risk-free asset. This is due to the fact that government assets are less volatile and are immune to sudden plunges due to any unexpected market scenarios.

The weights have always been kept positive, and they have been constrained to sum up to 1 in an intention to cover the entire capital. Therefore, the constraints determined for this study is as follows (Eq. 2):

$$\sum_{i=1}^{N} x_i = 1 \quad 0 \leq x_i \leq 1 \tag{2}$$

A penalty function has been introduced to transform the problem into an unconstrained problem of optimization. This penalty function introduces a positive penalty to maintain the optimization character of the algorithm. The Sharpe ratio function is multiplied by a negative sign to make the problem as a minimization problem since penalty function is maximizing the algorithm.

6 Optimization

The generic form of genetic algorithms has been used for this study. The data set has undergone preprocessing to reduce the features and to bring the data into the desired form. Everyday profits have been calculated for all the stocks that have been selected to work with. The portfolio's historical returns are a function of the weights. The goal is to find an optimal portfolio by changing the weights. The portfolio returns which are time series in nature are defined as a function of weights that are multiplied by the weight related to it, and then, these weighted time-series returns are summed element by element.

The portfolio return at any time is given by the following equation (Eq. 3):

$$P(t) = \sum_{i=1}^{N} x_i \cdot p_i(t) \tag{3}$$

The **return** of the i-th asset between period t and period $t-1$ is defined as (Eq. 4):

$$P_i(t) = (\text{Price}_i(t) - \text{Price}_i(t-1)) / (\text{Price}_i(t-1)) \tag{4}$$

This study aims at finding the values of the weights that maximize the performance metric of our portfolio according to the weight's constraints, which are (Eq. 5):

$$\sum_{i=1}^{N} x_i = 1 \quad 0 \leq x_i \leq 1 \tag{5}$$

thus making it a constrained optimization problem. Then, the fitness function has been created by using the Sharpe ratio as the objective function which also includes the risk-free rate which is of the treasury bills of the U.S. Government. The penalty function has been created keeping in mind the constraints of our algorithm.

The genetic algorithm, i.e., a heuristic search algorithm, has been equipped with the above-described fitness function that it can apply to its population at its iteration levels. Genetic algorithm heuristically runs to find the global optimum of the given parameters and the input values. At every chromosome generation, it checks the fit of the so mutated chromosome with the fitness function that it has been equipped with. During the selection, the chromosome with the highest fitness would be selected further for optimization.

This algorithm is a variant of the gradient decent problem. This iteratively uses the centroids of the triangles to converge at the local optimum. Therefore, this algorithm has been opted to be used as another level of optimization over the genetic algorithms. Since, this algorithm attempts to find the local optimum at every iteration of the genetic algorithm, the overall execution time and the number of iterations have been effectively reduced, and a better optimum value has been achieved.

7 Data and Research Findings

The New York Stock exchange data (obtained from Kaggle) from the year 2010 to 2016 has been used for this study. A set of six stocks have been used for a more descriptive study of this method of optimization. The U.S. Treasury bills have been used as the risk-free rate. These values are also from the same data as the above stocks so as to maintain the legitimacy. The profits of these bills have been calculated. The calculated values have been used in the Sharpe ratio for determining the fitness of the offsprings.

Initially, the genetic algorithm has been run on the basic dataset without any tweaking and it has generated a portfolio. Then, the risk-free rate has been introduced into the Sharpe ratio. Also, the deterministic algorithm has also been introduced to the optimization algorithm. The results that have been observed showed a change in the percentage of the selection of the portfolio among the stocks. Only six stocks of certain companies have been selected and worked upon considering the resources and the limitations of the system. The number of records were over 1700. These data included the opening and closing rates of the stocks on all the working business days from 2010 to 2016. Fig. 3 represents the allocation of the portfolios of the six companies when the Nelder–Mead was not used.

The results have shown that the use of a general-purpose optimization such as the Nelder–Mead has improved the selection of the overall optimization. These gradient descent algorithms have also helped in increasing the overall efficiency of the optimization. Thus, the iterations required to converge at the optimum values have been evidently reduced. Figure 4 shows the difference between the previous methods and evidently usage of this algorithm has created a mature portfolio that is robust and also has taken into consideration the aspect of the risk-free rate. In Fig. 5, the bold black line shows the returns of the selected portfolio as compared to the individual returns of other companies.

8 Conclusion and Future Work

Thus, the research findings of our study have been presented in this paper. This has shown that the deterministic optimization algorithms if used with the genetic algorithms would increase the efficiency of the process, i.e., increase the speed, utilize less memory, and get better results. The market data of the U.S. Treasury bills

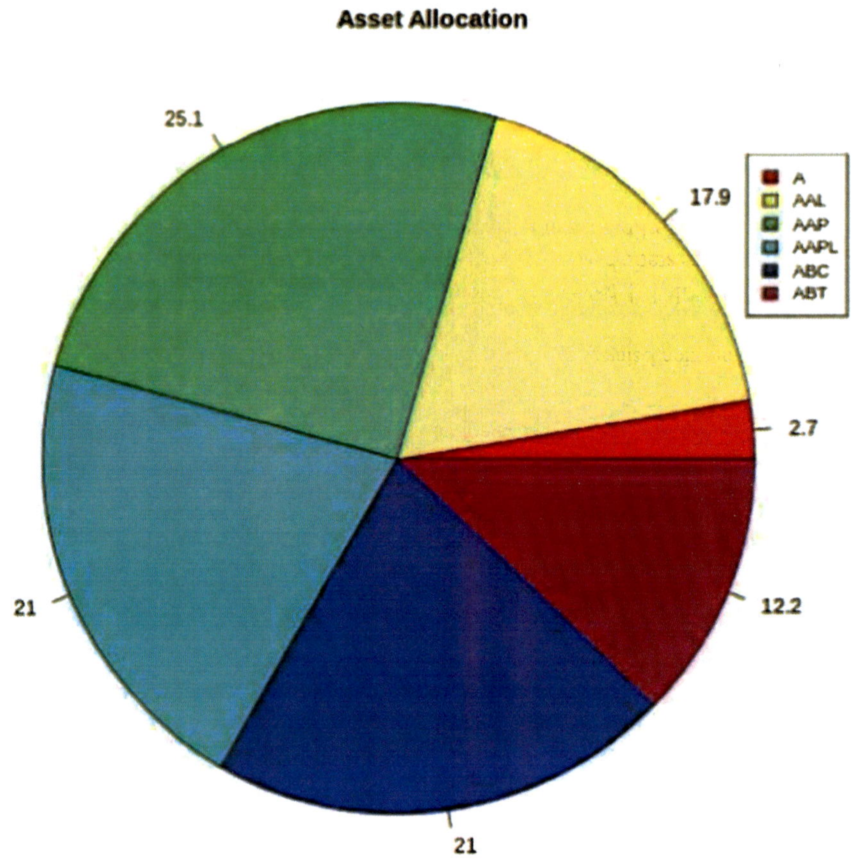

Fig. 3 Allocation without the Nelder–Mead

have also been added for the same time period of the stocks that are being used so as to maintain the legitimacy of the objective function, i.e., the Sharpe ratio.

The Nelder–Mead tends to optimize the function relatively fast usually within a few iterations, but it tends to get stuck in the local optima, and this usually requires a restart of the complete algorithm, but in this case, it does not halt our process as it is checking at each generations of the genetic algorithms, and therefore, it will be receiving different values. But this might cause a problem when there is a more complex generation of a new off-spring. Therefore, this problem has to be addressed in the future work.

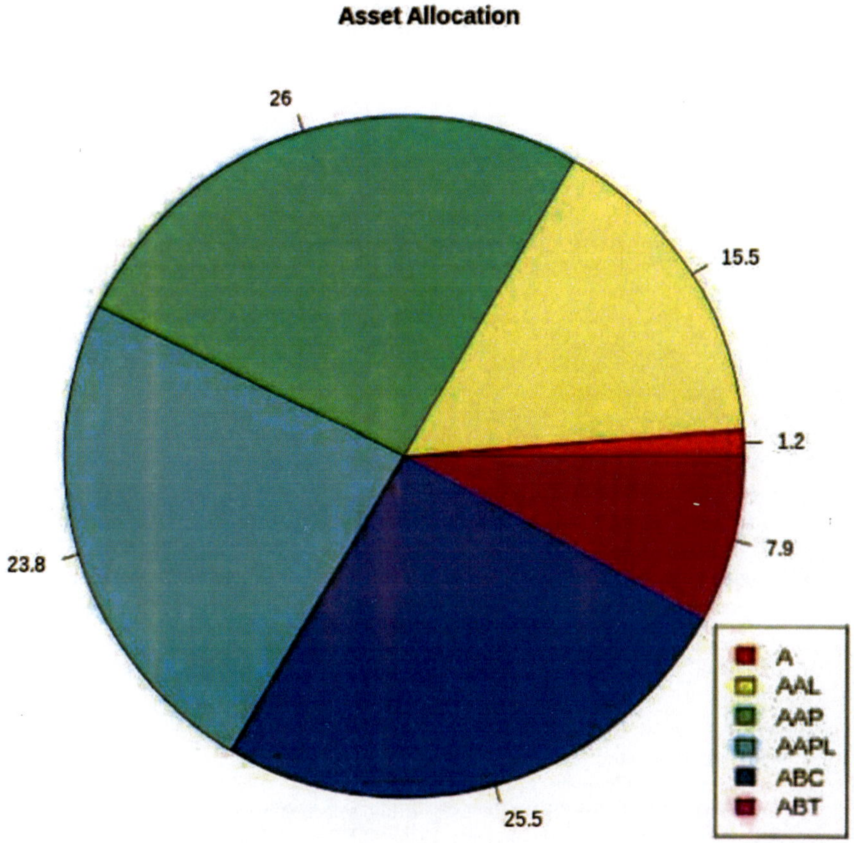

Fig. 4 Allocation with the Nelder–Mead

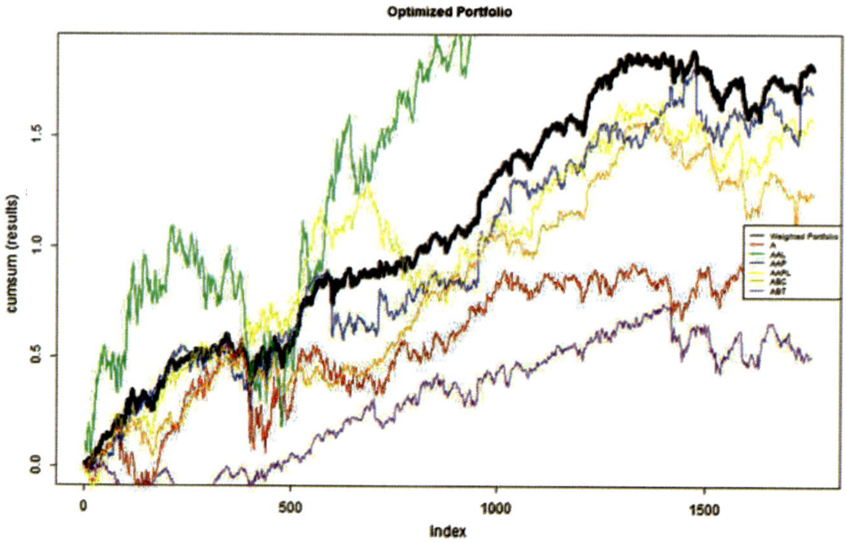

Fig. 5 Comparison of results from the selected portfolio with the individual returns of the company

References

1. Chen C-H, Shen W-Y, Wu M-E, Hong T-P (2019) A divide-and-conquer-based approach for diverse group stock portfolio optimization using Island-BASED genetic algorithms. In: 2019 IEEE congress on evolutionary computation (CEC). IEEE
2. Jin C-X, Chen W-Y, Yu S-J (2018) Robust portfolio selection based on optimization methods. In: 2018 37th Chinese control conference (CCC). IEEE
3. Chan M-C, Wong C-C, Cheung BK-S, Tang GY-N (2002) Genetic algorithms in multi-stage asset allocation system. In: IEEE international conference on systems, man and cybernetics, vol 3. IEEE
4. Sasaki M, Laamrani A, Yamashiro M, Alehegn C, Kamoyedji A (2018) Portfolio optimization by fuzzy interactive genetic algorithm. J Adv Manage Sci 6(3)
5. Mishra SK, Panda G, Meher S (2009) Multi-objective particle swarm optimization approach to portfolio optimization. In: 2009 world congress on nature & biologically inspired computing (NaBIC). IEEE
6. Ranković V, Drenovak M, Stojanović B, Kalinić Z, Arsovski Z (2014) The mean-Value at Risk static portfolio optimization using genetic algorithm. Comput Sci Inf Syst 11(1):89–109
7. Chou Y-H, Kuo S-Y, Lo Y-T (2017) Portfolio optimization based on funds standardization and genetic algorithm. IEEE Access 5
8. Chen Y, Zhou A (2019) MOEA/D with an improved multi-dimensional mapping coding scheme for constrained multi-objective portfolio optimization. In: 2019 IEEE congress on evolutionary computation (CEC). IEEE
9. Xu S, Zou XB, Liu WL, Wang XX, Zhu HL, Zhao T (2010) Research of Particle Swarm Optimization Algorithm Based on Nelder-Mead Simplex and its application on partial discharge parameter recognition. In: 2010 IEEE international power modulator and high voltage conference. IEEE
10. Chang K-H, Lu H-K (2014) Solving Quantile-based Stochastic optimization problems with modified stochastic Nelder–Mead Simplex method. In: 2014 IEEE 18th international conference on computer supported cooperative work in design. IEEE

Smart Farming Techniques for New Farmers Using Machine Learning

Ramena Venkata Satya Rohit, Dhrati Chandrawat, and D. Rajeswari

Abstract Smart farming is a modern way to bring ease in the field of agriculture with less manpower and more modern equipment. To know the features and characteristics of various crop types, machine learning techniques can be used. In these past few years, it is developed and used effectively in various fields. Machine learning is a booming and challenging research field in agricultural data analysis. This paper uses sensing parameters like PH, contents of soil, temperature, rainfall and humidity. The project emphasis, machine learning-based real-time analytics are performed to suggest the most suitable crop, pesticides and technique. To get a better classification of crops, logistic regression and support vector machines (SVM) are used. The outcome demonstrates that the proposed technique provides 20% more accuracy than the existing algorithm. The complete project comes up with the idea to help farmers to start with zero farming knowledge and helps them to yield the maximum profit.

Keywords Smart farming · Potential of hydrogen · Support vector machine · Logistic regression

1 Introduction

Smart farming involves the mixing of leading technologies into existing farming practices to extend production efficiency and highlight the quality of agricultural products [1, 2]. This project is a Web application which allows the user to enter the parameter like PH, rainfall, temperature, humidity and soil moisture. These values

R. V. S. Rohit (✉) · D. Chandrawat (✉) · D. Rajeswari (✉)
Computer Science Engineering, SRM Institute of Science and Technology, Chennai, Tamil Nadu, India
e-mail: rohitkumarrohit817@gmail.com

D. Chandrawat
e-mail: dhratichandrawat_shivdan@srmuniv.edu.in

D. Rajeswari
e-mail: rajeswad@srmuniv.edu.in

R. P. Mahapatra et al. (eds.), *Proceedings of 6th International Conference on Recent Trends in Computing*, Lecture Notes in Networks and Systems 177,
https://doi.org/10.1007/978-981-33-4501-0_20

are taken as an input in support vector machine and logistic regression algorithm. This helps the individual algorithm to predict which crop is suitable for land and which pesticides are used to get a better yield, based on accuracy given by these algorithms. The user has to decide which crop should be picked between the crop suggested by the support vector machine and the logistic regression algorithm.

From the decade's farmers in India as well as in other countries are worried about how effective their crop yield is [3]. The discovery of problems associated with crop yield can be resolved using crop yield indicators to boost the crop yield rate. The crop selector may be applied to minimize losses and might be used to maximize crop selection rates. The two main factors that affect the crops yield rate are the seed quality that can be improved using the techniques called hybridization, and another reason is crop-based selection on favorable condition. A technique was introduced to maximize the crop yield rate [4] called a crop selection method (CSM). The main objective of the project is to predict pesticides and suitable crops for a specific land from previous data and also attract the people to pursue a career in agriculture. It enhances the knowledge to grow good crops and gives them a good percentage of yield.

Support vector machine (SVM) may be a set of related supervised learning methods utilized in diagnosis for classification and regression. SVM reduces the classification error and increases the geometric margin simultaneously. SVM is named as maximum margin classifiers, and it often efficiently performs nonlinear classification using kernel trick. An SVM model may be a representation of the examples as points in space, mapped so that the samples of the separate categories are divided by an outsized margin gap that is as wide as possible. Given labeled training data are given as data points to the SVM algorithm as inputs. The SVM classifier first maps the input vectors into a choice value, then performs the classification using an appropriate threshold value.

The productive analysis is done by logistic regression. It is also used for data description and explains the relationship in binary variables and more nominal, ordinal, interval and ratio level independent variables. A predicted variable is called as criteria variable and referred to as the base variable for prediction, which is called as X. When it has only one predictor variable, the method is called as simple regression. In simple regression, a straight line is formed on the basis of X and Y value.

In previous years SVM, logistic regression and other machine learning algorithms were used to analyze and monitor in case of a single crop type. Dataset was modeled and framed for the single crop type. But the aim of the project is to utilize this algorithm to analyze various crops at a time and suggest which crop is suitable for the land and suggest pesticides and techniques to increase crop productivity.

2 Literature Survey

Raghav Kumar et al. [5] discussed a smart agriculture system that can guide farmer in crop management. The development of intelligent smart farming focuses on machine learning algorithms, which are changing the face of agricultural production day by day. This not only increases the management skills but also making it cost-effective and reduces the wastage. The purpose of this paper is to introduce a machine learning-based farming system that can help farmer's crop management skill by obtaining live data. This paper also concentrates on efficient environmental monitoring which increases the overall yield and quality of the products.

Rohit Kumar Rajak et al. [6] discussed common difficulty exists among Indian farmers; they do not make the right choice of crop, based on their soil needs. Productivity has been affected because of this. The precision agriculture method is used to solve this. The database collected from the farm is characterized by agricultural experts (e.g., soil through the soil testing laboratory dataset). Datasets used in soil testing tell about the majority voting technique recommended for the crop. It helps to figure which type of parameter should be included in datasets.

The paper [7] has the main objective to improve farming practice. Precision farming is to emphasize farming practices in such a way as to improve accuracy in predicting crop growth. Agricultural soil quality includes soil properties like organic materials such as PH, Mg, N, C, and major nutrients for plants are Mg^{2+}, K^{+}, NH^{4+}, Ca^{2+}. The usage of elements like H^{+}, Na^{+} and Al^{3+} in crops is very harmful. The quality of the crop in fertilization is recommended, and it needs micronutrients such as P, K, MN, ZN for plant growth. The crops productive are dependent on properties mainly chemical, physical and forecasts. Therefore, the main objective is to analyze, adjust and establish the factors of growth and soil properties of crops.

Drummond et al. [8] discussed another method of testing the yield reaction involves an analysis of spatial, huge datasets. This paper collected various types of datasets in precision agriculture. The data are linearly analyzed that have been reported quite often. This paper used the backpropagation technique. The properties of two soils vary within and between the fields and may also exhibit strong temporal variability. The authors have not found the correlation in all the method which has been proposed.

3 Proposed System

This work concentrates on suggesting more accurate crop for various land types. To find more accurate content for soil temperature, humidity and soil conditions are taken into account. It also provides knowledge of specialized techniques to manage that particular crop. SVM and logistic regression are used to find the crop which is best for a particular area based on the datasets gathered in real time. The standard datasets contain all minimum necessities needed for a particular crop are maintained and used for the crop prediction. The dataset is also plotted in real time to let the user know about the current conditions.

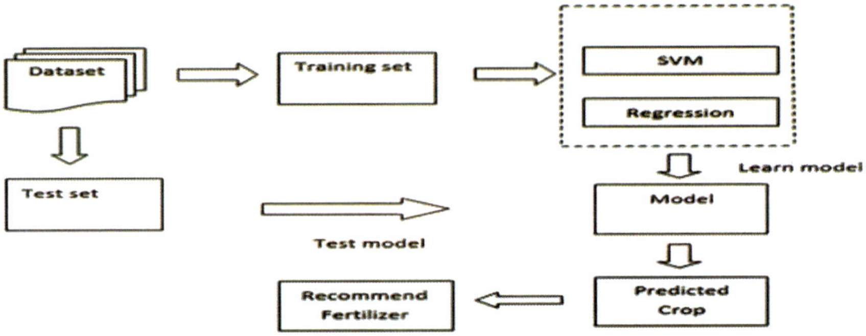

Fig. 1 System architecture

4 System Architecture

This process uses the dataset includes the pH, temperature, soil and humidity, rainfall from which the two algorithms classify the best crop for the particular land. This dataset comes under test sets, and every different crop is considered as a different model, then this test set helps the algorithm to predict the result accordingly and recommend the crop. In this project, 80% of the datasets used in training the model and 20% as our test datasets which make the model to work on more the complex situation, so that it can give a much accurate result, which is represented in Fig. 1.

5 Flowchart Diagram

5.1 Use Case Diagram

It is an interaction between user and system, which shows the user relationship among them and also showcases the use cases where the user is involved. It is used to identify the users individually involved to accomplish many other diagrams as well from the use case in Fig. 2.

Here, the farmer gives the dataset, and the processing will start. The first step is data cleaning which leads to split, train and test datasets. Then, the training of the model predicts the crop and recommends the fertilizer.

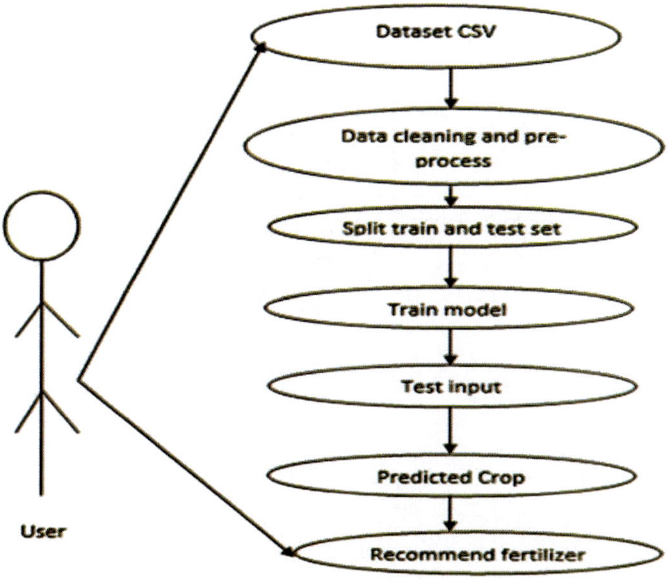

Fig. 2 Use case diagram

5.2 Deployement Diagram

This is a UML diagram that shows the architecture execution of the system. This also includes nodes such as software and hardware execution environment and the intermediately relating them. It helps us to visualize the physical hardware and software of the system (Fig. 3).

Here, farmer gives the dataset as the input on the Web application. Then, preprocessing is done using the training dataset. Then, SVM and logistic regression are applied, and it will predict the exact crop to give maximum output.

6 Performance Measures

6.1 Mean Square Error (MSE)

It is [9] a prediction that is obtained by using n-number of dataset values (e.g., as from a least-squares fit). The sample MSE predictor is computed by using Eq. 1.Where Y_i represents observed values, $\hat{Y}_i$ represents the predicted value.

$$\mathrm{MSE} = \frac{1}{n}\sum_{i=1}^{n}\left(Y_i - \hat{Y}_i\right)^2 \tag{1}$$

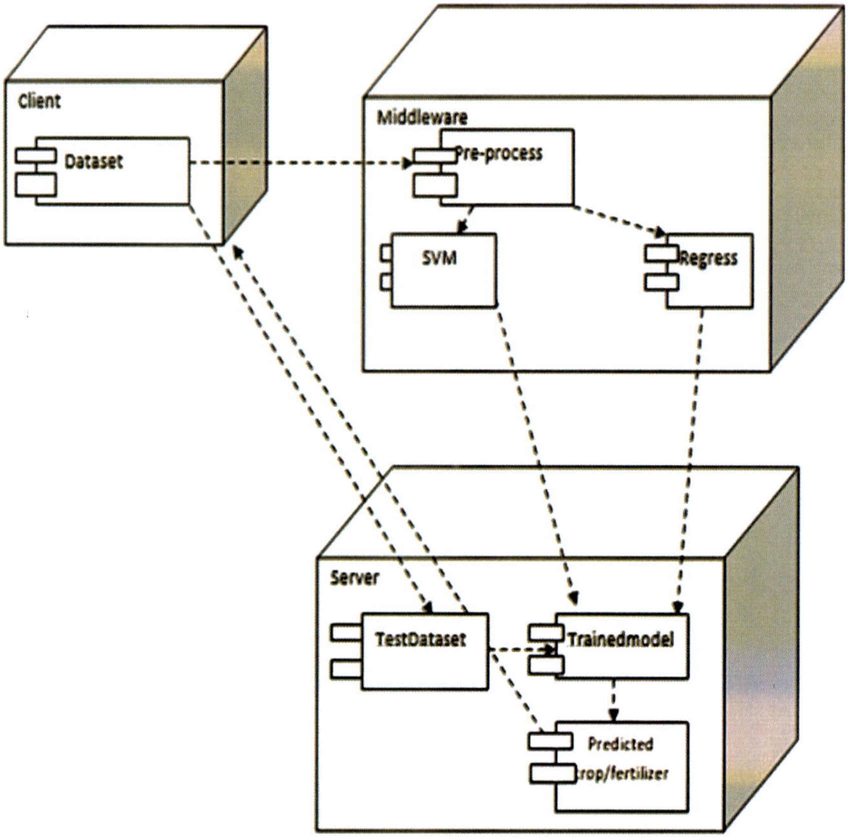

Fig. 3 Deployement diagram

6.2 Mean Absolute Error (MAE)

It [9] is obtained by the difference between two continuous variables. Let us consider x and y are variables of coupled observations that showcase the same circumstance. Let us contemplate a scatter plot, where the point I have coordinates (x_i, y_i). MAE is the average vertical distance between each point. MAE is also the mean distance between each point. The formula for MAE is presented in Eq. 2.

$$\text{MAE} = \frac{\sum_{i=1}^{n} |y_i - x_i|}{n} = \frac{\sum_{i=1}^{n} |e_i|}{n} \tag{2}$$

6.3 Root Mean Square Error (RMSE)

It is [10] a standard way to measure the error in predicting quantitative data presented in Eq. 3 is the actual value, is predicted value. N is the number of data.

$$\text{RMSE} = \sqrt{\sum_{i=1}^{N} \left|\widehat{y_i} - y_i\right|^2} \tag{3}$$

6.4 Root Mean Square (RMS)

The RMS is a collection of values, which is calculated by the square root of the average of squares of the values. The RMS is represented in Eq. 4. x_i is the number of observation occurred.

$$X_{\text{rms}} = \sqrt{\frac{x_1^2 + x_2^2 + x_3^3 \ldots x_n^2}{n}} \tag{4}$$

7 Results and Discussion

7.1 Dataset Collection

This gives us a detailed view of the dataset, that has attributes such as temperature, humidity, soil moisture, rain condition pH value. This dataset is analyzed to give better crop suggestions based on the attribute condition. Train_test_split function is used to split the dataset into two categories, one training data and second testing data. This dataset represents real-time observation required to segregate different crops that fall under these conditions. The given Fig. 4 shows the sample dataset considered without pre-processing (Table 1).

7.2 Code Snippet

Figure 5 represents the sample code to predict crops and pesticides. All the constraints are also presented.

In Fig. 4, Python language (variable) is used for setting the constraints. Python programming language facilitates to perform various functions and easy access to

```
csv_row = line.split(",")
print(csv_row[0],csv_row[1],csv_row[2],csv_row[3])
a=int(csv_row[0])
b=int(csv_row[1])
c=int(csv_row[2])
d=int(csv_row[3])
e=int(csv_row[4])
f="NO"
g=0
if (a >= 20 and a <= 35 and b >= 50 and b <= 90 and c >= 60 a
    f = "Wheat"
    g=1
elif (a >= 30 and a <= 35 and b >= 50 and b <= 60 and c >= 40
    f = "Oats"
    g=2
elif (a >= 20 and a <= 40 and b >= 60 and b <= 80 and c >= 50
    f = "Gram"
    g=3
elif (a >= 10 and a <= 25 and b >= 70 and b <= 90 and c >= 50
    f = "Pea"
    g=4
elif (a >= 15 and a <= 25 and b >= 60 and b <= 90 and c >= 60
    f = "Tea"
    g=5
elif (a >= 24 and a <= 30 and b >= 60 and b <= 80 and c >= 50
```

Fig. 4 Condition for crop suggestion

values present in variables and tried to set the constraint to the point which will give an accurate set of the dataset. It will help the farmer to predict better crops.

8 Software Used

8.1 Python

Python 3 is a programming language used in the project. This allows importing various packages like NumPy, pandas, matplotlib and sklearn which make reading and writing operation easily into the file and helps to train the machine learning model efficiently. This also helps to plot graphs based on the datasets.

Table 1 Sample dataset

A	B	C	D	E
Temp	Humidity	Soil moisture	Rain	pH
29	85	78	1002	7
30	85	81	1001	7
29	85	83	1002	7
29	85	81	1002	7
29	84	82	1002	7
29	84	79	1001	7
29	84	80	1002	7
30	84	79	1001	7
30	84	79	1002	7
29	84	81	1003	7
29	84	80	1003	7
29	85	83	1002	7
30	84	79	1001	7
30	84	79	1002	7
30	84	83	1002	7
29	84	83	1001	7
30	84	85	1002	7
30	84	84	1001	7
30	85	83	1002	7
30	84	82	1002	7
30	85	79	1002	7

8.2 Pycharm and Anaconda Prompt

These are the software used for implementing the project and made easy to write programming. It also allows for installing all the packages easily.

8.3 Python Tkinter

Python Tkinter is a window application that acts as an interface. It helps to link to the web page that the user will be interacting with.

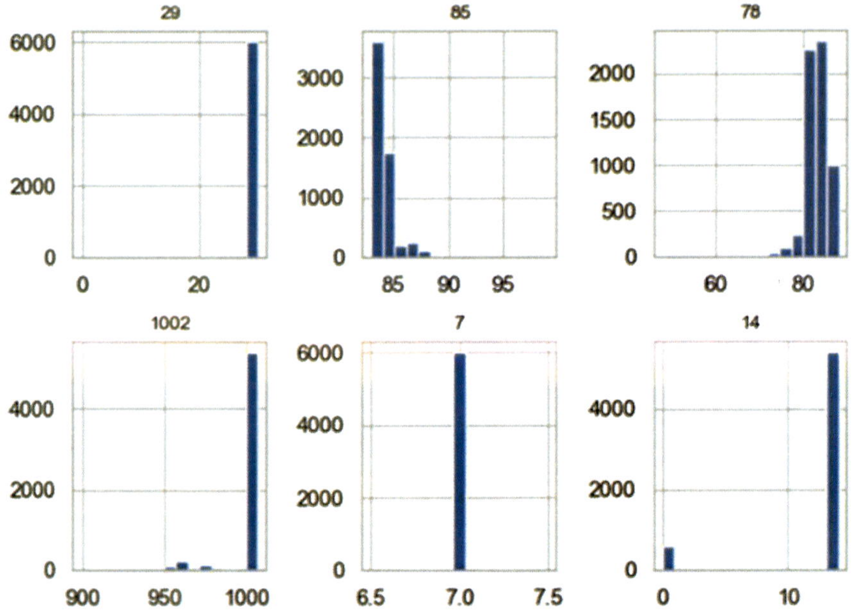

Fig. 5 Data histogram

9 Analysis of Project

9.1 Data Histogram

This gives the graphs based on the analysis of datasets and the constraints mentioned above in Fig. 4.

Data Histogram gives a detailed graphical representation on each parameter such as temperature, soil moisture, humidity, rainfall and pH. In Fig. 5, the y-axis represents the number of records formed, and the x-axis represents the data points.

9.2 Co-relation Matrix Diagram

As datasets have many columns, the co-relation matrix facilitates the best way to determine the correlation among various columns. Figure 6 represents the correlation matrix diagram.

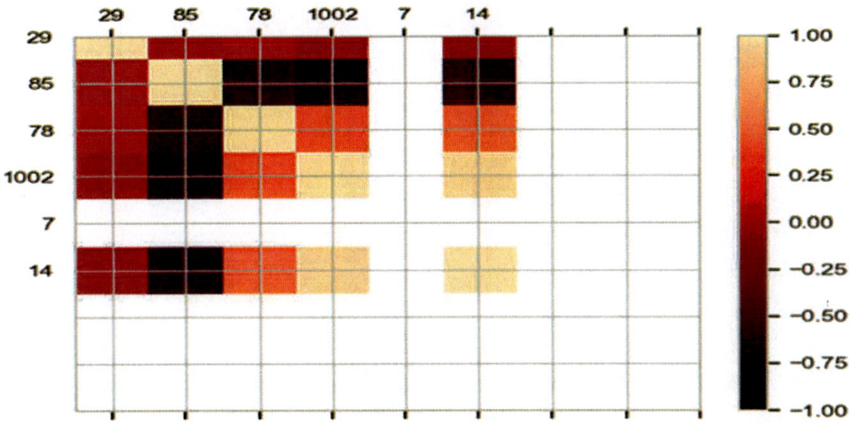

Fig. 6 Co-relation matrix diagram

10 Comparison Between Machine Learning Algorithm

SVM and logistic regression algorithms are trained with datasets. The dataset consists of value examined on real-time conditions such as soil moisture, pH, rainfall and temperature. As both, the algorithm follows different procedures to predict and prioritize constraints based on the result. The accuracy of the individual algorithm is determined by the MAE, MSE, RMSE and accuracy value. Individual algorithm produces their own accuracy value on the basis of test dataset. This accuracy value helps us to differentiate and decide which algorithm is more efficient.

11 Sample Output

Figure 7 shows the sample output of the SVM algorithm which is observed by processing the test dataset. It also shows the accuracy value obtained by RMSE, MSE and MAE.

Figure 8 shows the sample output of the logistic regression algorithm which is observed by processing the test dataset. It also shows the accuracy value obtained by RMSE, MSE and MAE. Figure 9 shows the sample screenshot of an interface where the user will enter the values in the GUI.

12 Conclusion

This project has been implemented to help the new farmer to proceed with their wish to farm with zero knowledge without leading it to failure. This project illustrates

```
[[29, 85, 78, 1002, 7, 'NC'], [30, 85, 81, 1001, 7, 'NC'],
[[29, 85, 78, 1002, 7, 14], [30, 85, 81, 1001, 7, 14], [29,
   29  85  78  1002  7  14
0  30  85  81  1001  7  14
1  29  85  83  1002  7  14
2  29  85  81  1002  7  14
3  29  84  82  1002  7  14
4  29  84  79  1001  7  14
C:\Users\Rohith\PycharmProjects\majorproject\venv\lib\site-
  ax.set_xlim(boundaries_list[j])
C:\Users\Rohith\PycharmProjects\majorproject\venv\lib\site-
  ax.set_ylim(boundaries_list[i])
C:\Users\Rohith\PycharmProjects\majorproject\venv\lib\site-
  ax.set_xlim(boundaries_list[i])
[[  30    85    81 1001]
 [  29    85    83 1002]
 [  29    85    81 1002]
 ...
 [  30    86    79  963]
 [  30    86    80  963]
 [  30    87    78  963]]
[14 14 14 ...  0  0  0]
MSE VALUE FOR SVM IS 2.203333
C:\Users\Rohith\PycharmProjects\majorproject\venv\lib\site-
MAE VALUE FOR SVM IS 0.158000
R-SQUARED VALUE FOR SVM IS 0.871142
  "the number of iterations.", ConvergenceWarning)
```

Fig. 7 Sample output of SVM

how efficiently the pesticides and related techniques, crops can be predicted to give a better yield. This will bring a new revolution in the agriculture sector. It improves the result of overall farming by 20%, which will help them to farm more and get better yield. This also helps to build the economy. It will also increase the interest of the present generation into farming and also help them to utilize land economically and feasibly. Future work could be extended by implementing this technique by including more datasets. Once the dataset has been increased, data mining technique can also be added.

```
C:\Users\Rohith\PycharmProjects\logistic\venv\:
[[  30   85   81 1001]
 [  29   85   83 1002]
 [  29   85   81 1002]
 ...
 [  30   86   79  963]
 [  30   86   80  963]
 [  30   87   78  963]]
[14 14 14 ...  7  0  7]
-----------------------------------------------
C:\Users\Rohith\PycharmProjects\logistic\venv\:
MSE VALUE FOR Logistic Regression IS 9.986000
MAE VALUE FOR Logistic Regression IS 1.698000
STOP: TOTAL NO. of ITERATIONS REACHED LIMIT.

R-SQUARED VALUE FOR Logistic Regression IS 0.1:
Increase the number of iterations (max_iter) o:
RMSE VALUE FOR Logistic Regression IS 3.160063
    https://scikit-learn.org/stable/modules/pr(
ACCURACY VALUE Logistic Regression IS 0.636667
```

Fig. 8 Sample output of logistic regression

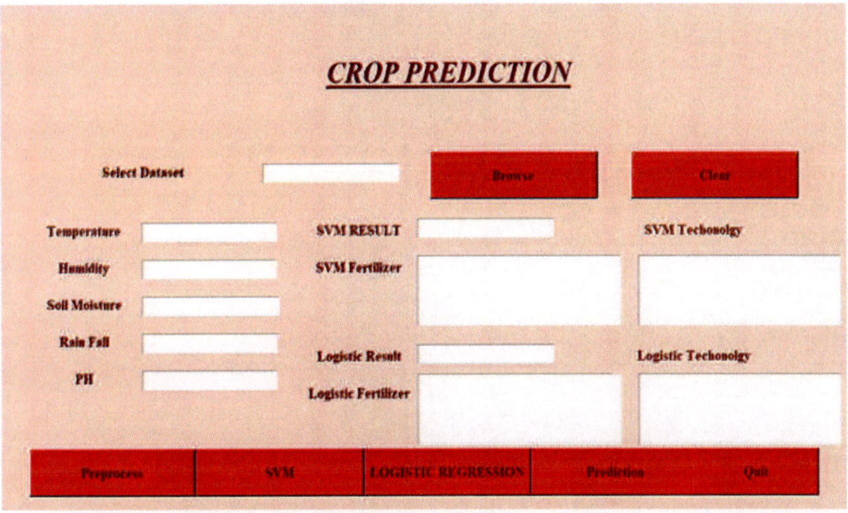

Fig. 9 Sample output of GUI

References

1. Bodake K, Ghate R, Doshi H, Jadhav P, Tarle B (2018) Soil based fertilizer recommendation system using internet of things. MVP J Eng Sci 01(01)
2. Kumar R, Singh MP, Kumar P, Singh JP (2015) Crop selection method to maximize crop yield rate using machine learning technique. In: International conference on smart technologies and management for computing, communication, controls, energy and materials (ICSTM)
3. Raghav Kumar T, Aiswarya B, Suresh A, Jain D, Balaji N, Sankaran V (2018) Smart management of crop cultivation using IOT and machine learning. Int Res J Eng Technol (IRJET) 05(11)
4. Rajak RK, Pawar A, Pendke M, Shinde P, Rathod S, Devare A (2017) Crop recommendation system to maximize crop yield using machine learning technique. Int Res J Eng Technol (IRJET) 04(12)
5. Ghosh S, Koley S (2014) Machine Learning for soil fertility and plant nutrient management using back propagation, Neural network. Int J Recent Innov Trends Comput Commun 02(02)
6. Drummond ST, Sudduth KA, Joshi A, Birrel SJ, Kitchen NR (2013) Statistical and neural methods for site-specific yield prediction. Trans ASAE 46(01)
7. Botchkarev A (2018) Performance metrics (error measures) in machine learning regression, forecasting and prognostics: properties and typology. Available online: https://arxiv.org/abs/1809.03006. Accessed on 9 Sept 2018
8. Jha K, Doshi A, Patel P, Shah M (2019) A comprehensive review on automation in agriculture using artificial intelligence. Artif Intell Agric 2
9. Majumdar J, Naraseeyappa S, Ankalaki S (2017) Analysis of agriculture data using data mining techniques: application of big data. J Big Data
10. Sujjaviriyasup T, Pitiruek K (2013) Agricultural product forecasting using machine learning approach. Int J Math Anal 07(38)

A Deep Learning Approach to Distinguish 2019-nCoV and SARS-CoV Sequences

Mayank Sharma

Abstract This paper presents a classification of protein sequences obtained from the 2019 Novel Coronavirus (2019n-CoV) and the 2003 SARS Coronavirus (SARS-CoV) using natural language processing. Very recent researches have indicated that the 2019-nCoV bears almost 79% sequence identity to the SARS-CoV but is sufficiently unique to be considered as the 7th entry to the list of human-infecting coronaviruses. 181 protein sequences of the 2019-nCoV (the number available right now due to limited number of cases) and 843 of the SARS-CoV were extracted from NCBI's GenBank. These were split into dimers, encoded into numerical features and trained on two deep learning models—the CNN and RNN. Results indicate an accuracy of 97.65% from both models, providing useful insights on sequence analysis of the 2019-nCoV and its differentiation from other coronaviruses.

Keywords Coronavirus · 2019-nCoV · Bioinformatics · Deep learning · Natural language processing · Protein sequence classification · Sequence prediction

1 Introduction

The 2019 Novel Coronavirus (2019-nCoV) outbreak that originated in Wuhan, China, has caused an emergency worldwide. Its first case was reported on December 31, 2019. At the time of writing the paper, more than 30,000 patients have been tested positive and more than 600 fatalities have occurred from the 2019-nCoV. Positive cases were first identified in China, but there has been subsequent emergence of cases tested positive in several countries including Thailand, Singapore, Italy, France, USA, Japan, and India. Researchers have estimated that the reproductive number for 2019-nCoV is 2.68 and that the epidemic doubling time is nearly 6.4 days [1].

Through the usage of next-generation sample sequencing of bronchoalveolar lavage fluid and cultured isolates from nine patients, recent researches have indicated that the 2019-nCoV is closely related to the SARS-CoV (a virus identified in

M. Sharma (✉)
Delhi Technological University, New Delhi, India
e-mail: mayanksharma8993@gmail.com

R. P. Mahapatra et al. (eds.), *Proceedings of 6th International Conference on Recent Trends in Computing*, Lecture Notes in Networks and Systems 177,
https://doi.org/10.1007/978-981-33-4501-0_21

2003) with sequence identity of about 79% [2]. It has also been suggested that the 2019-nCoV has a receptor-binding structure similar to that of the SARS-CoV. For a little background, the 2003 epidemic of SARS (Severe Acute Respiratory Syndrome) started in the Guangdong province of China and then spread to 26 countries, affecting more than 8000 people.

Linked to the Hunan seafood market in Wuhan, China, the 2019-nCoV, although having sequence identity related to the SARS-CoV, has been called sufficiently different to be treated as the seventh addition to the family of human-infecting coronaviruses. The most common symptoms at the onset of this disease include fever, cough, and fatigue. The laboratory diagnosis of the disease is confirmed with abnormalities in chest CT scans [3, 4].

In this paper, protein sequences obtained from both 2019-nCoV and SARS-CoV are extracted from NCBI's GenBank. Since the outbreak is relatively new, only 181 sequences could be obtained for the 2019-nCoV. For the SARS-CoV, 843 sequences were taken, bringing the dataset to a total of 1024 sequences. These are then broken down into dimers with the help of a simple user defined function, converted into numerical encodings and fed as inputs to two deep learning models, the CNN and RNN. Their performance metrics are compared to guide future researches pertaining to sequence analysis of the 2019-nCoV and its differentiation from other related coronaviruses.

2 Literature Review

Prediction and analysis of DNA/protein sequences using biological methods have attracted the eye of researchers since many years. More recently, the usage of machine and deep learning techniques has become popular in the field of sequence prediction. The authors in [5] built an SVM model to detect various types of cancers from tumor DNA signatures. A prediction of prostate cancer cells using principal component analysis was shown in [6]. DNA sequences of the Hepatitis B virus were classified into their genotypes (B, C1, C2, or C3) by researchers in [7]. In this research, feature extraction was done using the information gain criterion and three machine learning models were run to obtain an accuracy of about 70%. This number was deemed to have significant usefulness in performing an initial diagnosis of liver cancer.

A natural language processing approach to classification of G-Protein coupled receptors (GPCRs) from chi-square features and N-grams using decision tree and Naive Bayes' algorithm was discussed in [8]. The usage of K-Mers in prediction and analysis of sequences has further accelerated studies in the field of bioinformatics as shown in [9] and [10], where the authors incorporated features extracted from K-Mers along with other categorical features to build hybrid machine learning models for prediction. The usage of deep learning models for classification of mRNA and lncRNA sequences using K-Mers was shown in [11]. Similarly, a study on the usage of deep learning techniques to classify DNA sequences was presented in [12]. This research evaluated the performance of three models which were built

Table 1 Mean and standard deviations of the protein sequence lengths (in amino acids)

Virus	Mean	Standard deviation
SARS-CoV	919.80	1985.66
2019-nCoV	712.57	1543.86

using a combination of LSTM, convolutional layers, and fully-connected multi-layer perceptrons.

3 Dataset Used

The National Center for Biotechnology Information (NCBI)'s GenBank [13] is regularly updating extracted genome and protein sequences from 2019-nCoV cases in various countries of the world such as China, Thailand, Finland, Italy, and USA in it's database. Up until the time of submission of this paper, the data available for 2019-nCoV was little due to the recent emergence of the disease and limited number of cases. A total of 181 protein sequences of the 2019-nCoV was extracted. These included various types of proteins, such as non-structural, membrane, and envelope proteins. In conjunction to these, 843 protein sequences obtained from the SARS Coronavirus (SARS-CoV) were extracted from NCBI's GenBank too. This number was chosen so as to build a dataset of sufficient size (1024 sequences) to train the deep learning models. Descriptive statistics on sequence lengths of both the viruses have been shown in Table 1.

4 K-Mer Splitting and Numerical Encoding

K-Mer splitting is the process of breaking down protein/DNA sequences into smaller 'words' (with each word containing K successive alphabets in the sequence) that can be used as features for natural language prediction models. For example, the protein sequence 'GLFDIVK' (where the alphabets G, L, F, D, I, V and K refer to the one-letter codes of it's amino acid constituents), when split into K-mers with $K = 4$, will produce the following set of 'words': GLFD, LFDI, FDIV, and DIVK. In this research, sequences corresponding to the 2019-nCoV and SARS-CoV are split into dimers ($K = 2$) with the help of a user defined function. These are then translated into numerical feature inputs with the help of the *text-to-sequences* method available in Keras, a popular Python library used for creating deep learning models. To establish uniformity, all feature inputs are padded to attain a length of 100.

5 Deep Learning Models Used

The convolutional neural network (CNN) architecture used in this research is shown in Fig. 1. In the figure, the dimensions indicated at the top of each layer denote its output shape. The input layer is a Keras *embedding layer* of the size 100 × 100. It is followed by a one-dimensional *convolution layer* that has a kernel of size 100 × 10 with ReLU activation. This is followed by a *global max-pooling layer*, which is used to reduce the dimensionality of the embeddings. After this, a *dropout layer* is added to prevent overfitting. Finally, two *dense layers* are added, the first one with ReLU activation that reduces the dimensionality to 10 and the second one with softmax activation that is essentially the output layer and reduces the dimensionality to 2 (the number of output labels). ReLU and softmax activation functions are used because they can handle issues such as sparsity and vanishing gradient as compared to conventional activation functions like sigmoid and tan-sigmoid.

The ReLU activation function is given by (1), where $H(x)$ refers to the output of the ReLU function for an input variable x.

$$H(x) = \max(0, x) \tag{1}$$

The softmax activation function is given by (2) where $G(x_i)$ refers to the output of the softmax function for a given input feature vector x_i and $\sum_{j=1}^{n} e^{x_j}$ refers to the sum of exponents of all input feature vectors x_j where $j \in \{1, 2, \ldots n\}$.

$$G(x_i) = \frac{e^{x_i}}{\sum_{j=1}^{n} e^{x_j}} \tag{2}$$

The recurrent neural network (RNN) architecture used is shown in Fig. 2. The basic principle that differentiates RNNs from CNNs is that RNNs contain feedback connections and thus learn from outputs generated in previous iterations. The same *embedding layer* (as in CNN) is used in this model too. It is followed by an *LSTM* (Long short-term memory) layer, which is added because vanilla RNNs suffer from the problem of vanishing gradient during back-propagation. This is followed by two *dense layers* with ReLU and softmax activation, respectively. In between these two layers, a *dropout layer* is added to prevent overfitting.

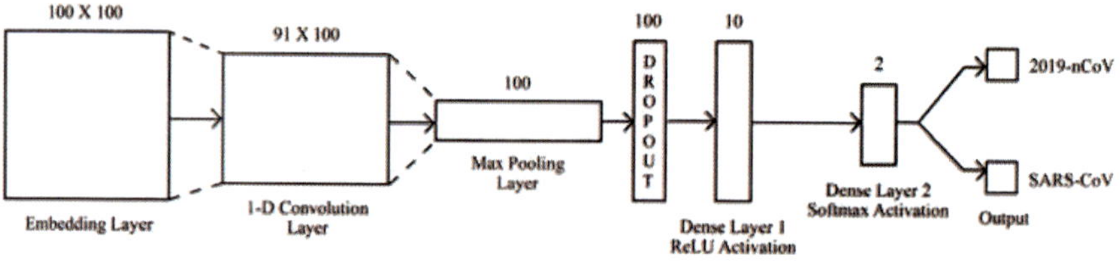

Fig. 1 Architecture of the CNN model used

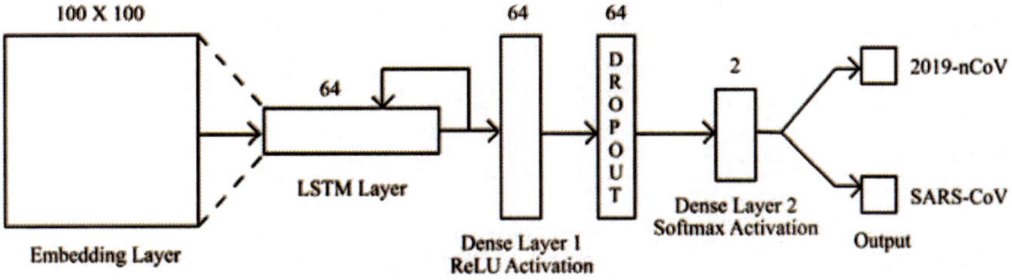

Fig. 2 Architecture of the RNN model used

6 Results and Discussions

Performance curves for the CNN and RNN models have been shown in Figs. 3 and 4, respectively. These indicate the training accuracy, validation accuracy, training loss, and validation loss with the number of epochs for both the models. The basic purpose of building these graphs is to identify when the neural network models overfit and thus they can be stopped in advance.

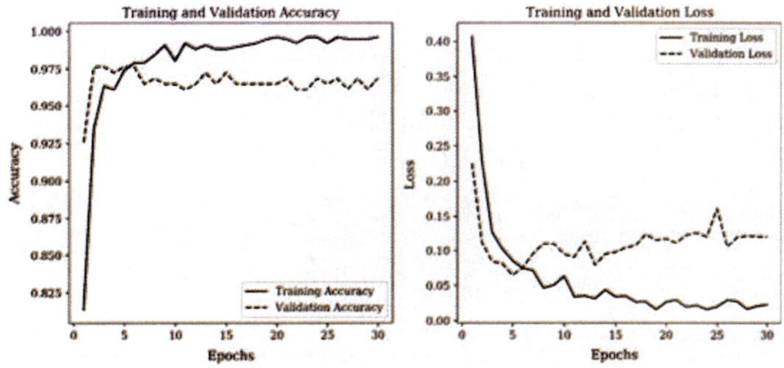

Fig. 3 Performance curves for the CNN model built on dimers

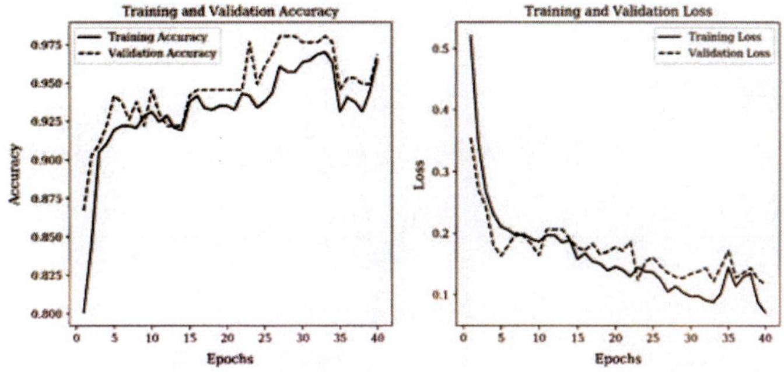

Fig. 4 Performance curves for the RNN model built on dimers

Table 2 Training and validation accuracies for the CNN and RNN model

Model	Training accuracy (%)	Validation accuracy (%)
CNN	97.91	97.65
RNN	94.14	97.65

A neural network model is said to overfit when its validation accuracy remains almost constant but its validation loss starts to increase with the number of epochs. As is clear from both the figures, the validation losses of the CNN and RNN models reach their minima at 6 and 23 epochs, respectively, after which they start increasing. Thus, the model trainings are stopped at these epochs in order to prevent overfitting.

The training and validation accuracies for both the models are indicated in Table 2. It can be seen that both models are able to classify the protein sequences with same validation accuracy, which is 97.65%. The difference between training and validation accuracies also indicates that the model does not overfit. Although both the models classify sequences with the same accuracy, CNN serves as a better classifier as compared to RNN because of the following three reasons—it takes less number of epochs for model training (6 as compared to 23), gives a steady performance curve (as seen in Fig. 3) and gives a very small difference between the training and validation accuracies, which indicates perfect fitting.

Table 3 and 4 indicate the various performance metrics (Precision, Recall, and F1-Score) for the two deep learning models. It can be seen that both the models are able to distinguish between the labels with good F1-Scores. This is particularly impressive because the dataset was imbalanced. On the whole, the CNN model provides higher F1-Scores as compared to the RNN model with 0.99 and 0.91 for the SARS-CoV and 2019-nCoV, respectively.

Table 3 Performance metrics for the CNN model

Virus	Precision	Recall	F1-Score
SARS-CoV	0.99	0.98	0.99
2019-nCoV	0.89	0.94	0.91

Table 4 Performance metrics for the RNN model

Virus	Precision	Recall	F1-Score
SARS-CoV	1.00	0.97	0.98
2019-nCoV	0.82	0.97	0.89

7 Conclusion

The 2019 Novel Coronavirus (2019-nCoV) outbreak that originated in Wuhan, China, just over a month ago has sparked a worldwide emergency. Researchers across the world have isolated genomic sequences of the virus from the patients affected and have found out that the sequence identity of the 2019-nCoV is almost 79% similar to that of the 2003 Severe Acute Respiratory Syndrome Coronavirus (SARS-CoV). However, with limited data and information available about this new disease, there is an immediate need of researches on the virus and it's similarity to other coronaviruses. In this paper, a natural language processing-deep learning approach to classification of protein sequences from the 2019-nCoV and the SARS-CoV was discussed. Two deep learning models (CNN and RNN) were built using 1024 sequences extracted from NCBI's GenBank. Although both models were able to classify the sequences with the same accuracy (97.65%), CNN serves as a better classifier because it takes less number of epochs to train, gives a steady performance curve and provides better F1-Scores. Through this paper, the author aimed to contribute to the research on sequence classification of the 2019-nCoV and it's distinction from other coronaviruses, a field that is relatively very new considering the recent outbreak of the disease.

References

1. Wu JT, Leung K, Leung GM (2020) Nowcasting and forecasting the potential domestic and international spread of the 2019-nCoV outbreak originating in Wuhan, China: a modelling study. The Lancet
2. Lu R, Zhao X, Li J, Niu P, Yang B, Wu H, Wang W, Song H, Huang B, Zhu N, Bi Y, Ma X, Zhan F, Wang L, Hu T, Zhou H, Hu Z, Zhou W, Zhao L, Chen J, Meng Y, Wang J, Lin Y, Yuan J, Xie Z, Ma J, Liu WJ, Wang D, Xu W, Holmes EC, Gao GF, Wu G, Chen W, Shi W, Tan W (2020) Genomic characterisation and epidemiology of 2019 novel coronavirus: implications for virus origins and receptor binding. The Lancet
3. Zhu N, Zhang D, Wang W, Li X, Yang B, Song J, Zhao X, Huang B, Shi W, Lu R, Niu P, Zhan F, Ma X, Wang D, Xu W, Wu G, Gao GF, Tan W (2020) A novel coronavirus from patients with pneumonia in China, 2019. New England J Med
4. Huang C, Wang Y, Li X, Ren L, Zhao J, Hu Y, Zhang L, Fan G, Xu J, Gu X, Cheng Z, Yu T, Xia J, Wei Y, Wu W, Xie X, Yin W, Li H, Liu M, Xiao Y, Gao H, Guo L, Xie J, Wang G, Jiang R, Gao Z, Jin Q, Wang J, Cao B (2020) Clinical features of patients infected with 2019 novel coronavirus in Wuhan, China. The Lancet
5. Soh KP, Szczurek E, Sakoparnig T, Beerenwinkel N (2017) Predicting cancer type from tumour DNA signatures. Genome Med 9 (2017)
6. Ghosh A, Barman S (2013) Prediction of prostate cancer cells based on principal component analysis technique. Procedia Technol 10:37–44
7. Leung K-S, Lee KH, Wang J-F, Ng EYT, Chan HLY, Tsui SKW, Mok TSK, Tse PC-H, Sung JJ-Y (2011) Data mining on DNA sequences of Hepatitis B Virus. IEEE/ACM Trans Comput Biol Bioinf 8:428–440
8. Cheng BYM, Carbonell JG, Klein-Seetharaman J (2005) Protein classification based on text document classification techniques. Proteins Struct Funct Bioinf 58:955–970

9. Phan D, Nguyen NG, Lumbanraja FR, Faisal MR, Abapihi B, Purnama B, Delimayanti MK, Kubo M, Satou K (2017) Combined use of k-Mer numerical features and position-specific categorical features in fixed-length DNA sequence classification. J Biomed Sci Eng 10:390–401
10. Kishk A, Elzizy A, Galal D, Razek EA, Fawzy E, Ahmed G, Gawish M, Hamad S, El-Hadidi M (2018) A hybrid machine learning approach for the phenotypic classification of metagenomic colon cancer reads based on Kmer frequency and biomarker profiling. In: 2018 9th Cairo international biomedical engineering conference (CIBEC). IEEE
11. Wen J, Liu Y, Shi Y, Huang H, Deng B, Xiao X (2018) A classification model for lncRNA and mRNA based on k-mers and a convolutional neural network. BMC Bioinf 20
12. Lo Bosco G, Di Gangi MA (2017) Deep learning architectures for DNA sequence classification. In: Fuzzy logic and soft computing applications. Springer International Publishing, pp 162–171
13. National Center for Biotechnology Information, https://www.ncbi.nlm.nih.gov

Image Dehazing by Image Enhancement and Multi-scale Laplacian Pyramid Fusion

Subhash Chand Agrawal and Rajesh Kumar Tripathi

Abstract When atmospheric particles, such as dust, smoke or water droplets, accumulate in the air, a haze phenomenon occurs. The inclement weather conditions diminish the visibility, reduce the contrast and change the color of an image. The poor visibility may cause a serious problem to various computer vision applications such as object recognition in surveillance systems, transportation systems, etc. Therefore, image dehazing is an essential task, used to remove the influence of weather factors from the image. In this paper, a novel single image dehazing method is introduced which improves the visibility and contrast of the hazy image. To achieve a haze-free image, the original hazy image is processed through basic image enhancement-based operations such as log transformation, inverse log transformation and guided filtering. The resulting set of multi-exposure images is then merged into a single high-quality haze-free image through a multi-scale Laplacian fusion. The experimental evaluation is performed on a large set of challenging hazy images in terms of both qualitative and quantitative parameters. The results obtained through the proposed fusion-based method not only remove the haze effectively but also able to highlight the micro details of the objects in the image.

Keywords Over enhancement · Physical model · Image dehazing · Laplacian pyramid · Guided filter

---

S. C. Agrawal (✉) · R. K. Tripathi
Department of Computer Engineering and Applications, GLA University, Mathura, UP 281406, India
e-mail: subhash.agrawal@gla.ac.in

R. K. Tripathi
e-mail: rajesh.tripathi@gla.ac.in

R. P. Mahapatra et al. (eds.), *Proceedings of 6th International Conference on Recent Trends in Computing*, Lecture Notes in Networks and Systems 177,
https://doi.org/10.1007/978-981-33-4501-0_22

1 Introduction and Related Work

The adverse weather conditions such as haze, fog, rain, etc. diminish the visibility of the image and distort the color of the objects in the scene. Thus, these degraded images are not suitable for various computer vision applications such as video surveillance, weather forecasting, remote sensing, intelligent transportation systems, etc. These applications demand high-quality images for post-processing [1]. Therefore, dehazing process is essential to enhance the visual quality of the image and improve the contrast. The hazy image formation model is given by [2]:

$$H(x) = HF(x)\,t(x) + A\,(1 - t(x)) \tag{1}$$

where $H(x)$ is the captured hazy image and $HF(x)$ is the hazy-free image. A is the atmospheric light. $t(x)$ is the transmission map, portion of the light, reaching to the camera without scattering, and it can be calculated as:

$$t(x) = e^{-\beta\, d(x)} \tag{2}$$

where, β is the atmospheric scattering coefficient and $d(x)$ is the distance between camera and the image. The value of $t(x)$ lies in the range of [0–1]. The Eq. 1 is the sum of two terms: attenuated reflected light and airlight. When light is reflected from the object, by the suspension of these particles in the air, it is attenuated. Second term introduces in the image due to scattering of atmospheric light. The process of hazy image formation is shown in Fig. 1.

Image dehazing is a challenging problem, which had drawn the attention of several researchers in the past. The methods for dehazing can be divided in to three categories: (1) image restoration-based method, (2) image enhancement-based method and (3) image fusion-based method. Some popular methods of each category are

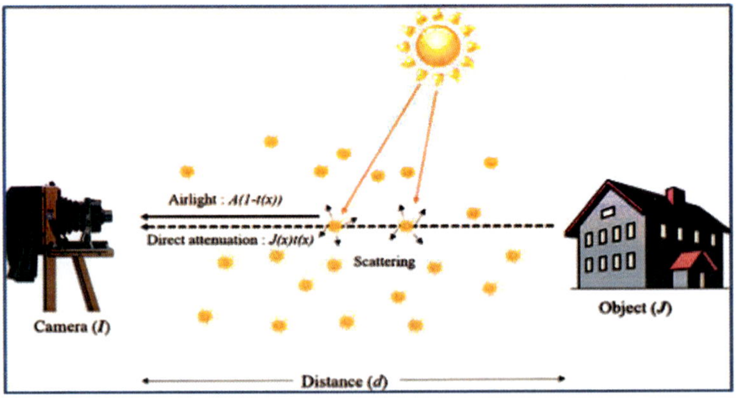

Fig. 1 Physical model of haze formation [3]

discussed in this section. Image restoration-based method uses optical model of image formation. These methods estimate the transmission map or depth map based on some assumptions or priors taken from haze-free image. He et al. [4] proposed a method based on dark channel prior based on clear image. It takes minimum intensity value of R, G and B channels in a haze-free image, centered at a local patch, and it is close to zero. Based on this assumption, author calculated the transmission map. This method produces good result; however, it suffers from the problem of over enhancement. Berman et al. [5] presented a method using a non-local prior. According to this paper, every haze-free image can be clustered using few hundred colors in RGB space. However, this method fails when atmospheric light is brighter than the scene because in this situation, it is very difficult to find haze lines. Zhu et al. [6] suggested a linear model based on the relationship of colors with the depth of the image using a supervised learning. However, training procedure is complex and parameter is highly dependent on training data and also fails in presence of dense haze.

In image enhancement-based methods for dehazing, Jia and Yue [7] also used wavelet decomposition to enhance the visibility. However, this approach can remove the effect of haze from image captured from a short distance. Liu et al. [8] proposed a method using multi-scale wavelet-based approach in frequency domain. The limitation of this method is that it produces halo artifacts in the haze-free image.

In fusion-based method, Ancuti et al. [9] generated two images from the original hazy image. First input image is obtained by white balancing. Second image increases contrast in faraway regions. Furthermore, three weight maps luminance map, weight map and chromatic map are used. Finally, these weight maps are fused in multi-scale fashion using a Gaussians and Laplacian pyramids. Galdran [10] proposed fusion-based method which utilized gamma correction operations on input hazy image and local histogram equalization to improve the contrast. The transmission of unknown hazy image can also be estimated by machine learning model from simple linear model [6] to convolution neural networks [11, 12].

The methods based on the physical model rely on some assumptions or priors taken from haze-free images. Therefore, these methods fail, if prior information or assumptions do not hold true. The main contribution of this paper is to solve single image dehazing problem without inversion of the physical model. This paper proposes a new method for single image dehazing using spatially-varying image enhancement techniques and fusion procedure. Thus, the proposed method avoids the need for estimating the transmission map and its costly refinement process. It directly enhances the visual quality of the image and is able to highlight the micro-details of the objects in the image.

The remaining part of this paper is structured as follows: Sect. 2 describes the proposed methodology in detail. Section 3 presents an extensive study of the proposed method on various challenging hazy images. Finally, Sect. 4 discusses the conclusions and future work.

2 Proposed Methodology

The intensity values of a hazy image are increased due to the global atmospheric light in the captured image. Thus, any solution to $HF(x)$ must contain intensity values lower than $H(x)$. This can be shown by simply rearranging the Eq. 1 in following form:

$$t(x) = \frac{A - H(x)}{A - HF(x)} \tag{3}$$

Since $t(x)$ is in the range of [0–1], it follows from Eq. 3 that $A - H(x) <= A - HF(x)$ and this inequality will satisfy only when $HF(x) \leq H(x)\ \forall x$.

Based on this observation, this paper proposes a method to utilize this information available in a set of multi-exposed images $R = \{I_1, I_2, I_3, \ldots I_n\}$. The proposed methodology consists of two modules: (1) resulting image generation and (2) fusion of resulting images. In the first module, three resulting images are generated using simple image enhancement-based operations like log transformation, inverse log transformation and guided filtering. In the second module, these images are fused into a single high-quality haze-free image using multi-exposure fusion scheme. The block diagram of the proposed method is shown in Fig. 2.

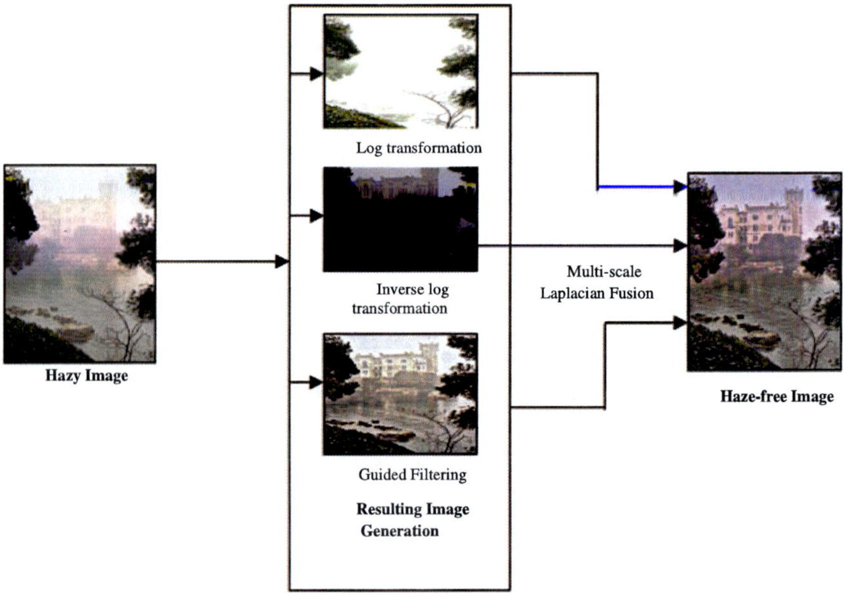

Fig. 2 Block diagram of proposed methodology

2.1 Resulting Image Generation

Before resulting image generation, first we normalize the image in the range of [0–1] then we apply the image enhancement technique. The images are generated in the following ways.

2.1.1 Log Transformation

The first image is generated using log transformation. It increases the details or contrast of lower intensity values in a hazy image by brightening the intensity values of an image, as shown in Fig. 3. This transformation is used in the proposed work to get details of nearby regions or objects in the images which are close to the camera, as visually shown in Fig. 4. It is defined as follows:

$$I_1 = c \log(1 + x) \tag{4}$$

When we take $c = 1$, lower intensities are not changed, they remain the approximately same, but higher intensities are decreased. In $c = 3$, details of lower intensities are increased very much and most of higher intensities are transformed to 1. Therefore, we select $c = 2$ to get better details of lower-intensity pixels close to the camera

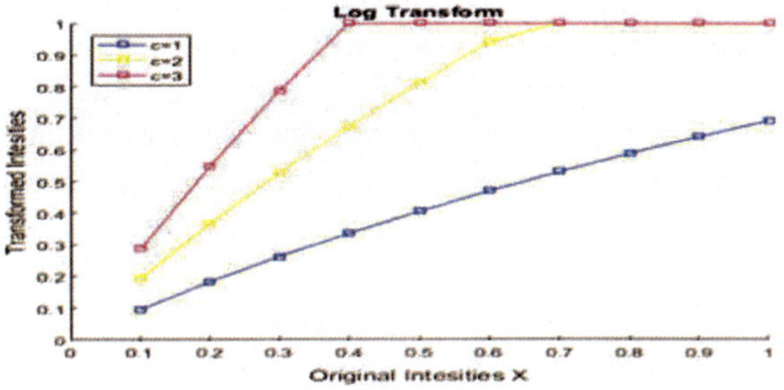

Fig. 3 Example of log transformation on pixel values [0.1 0.2 0.3...0.1]

(a) (b) (c) (d)

Fig. 4 Visual analysis on a hazy image for different values of c in log transformation: **a** Hazy image. **b, c, d** Images generated using $c = 1$, $c = 2$ and $c = 3$ respectively at a moderate pace. However, in this process, some higher-intensity values will be transformed to 1, but the fusion process will select that lost information from other two images.

2.1.2 Inverse Log Transformation

Inverse log transformation just does the opposite of log transformation. To get the details of higher-intensity pixel values in faraway regions of the hazy image, we apply inverse log transformation. We take the inverse of log transformation (Eq. 4). In the first step, replace $f(x)$ by y and then interchange x and y and in the second step, log expression is converted into exponential expression and y is replaced by an inverse notation $f^{-1}(x)$ as follows:

$$\begin{aligned} \exp(x/c) - 1 &= y \\ \exp(x/c) - 1 &= f^{-1}(x) \end{aligned} \tag{5}$$

Therefore,

$$I_2 = \exp(x/c) - 1 \tag{6}$$

The process of intensity transformation by the inverse log is shown in Fig. 5. We can observe in the figure that lower intensities are reduced very much; therefore, we lost the details of nearby regions. However, these details can be found in the first image obtained by log transformation. But, we can improve the visibility details of higher pixels in the faraway regions as intensities of these regions are reduced, which is the key idea of this paper for removing haze from the image. The same analysis is visually shown in Fig. 6.

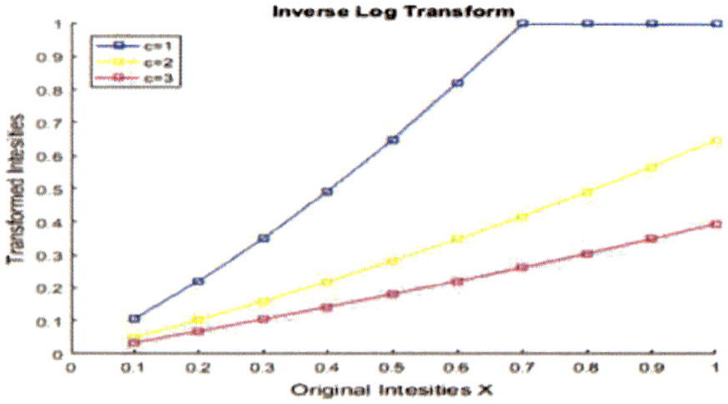

Fig. 5 Example of inverse log transformation on pixel values [0.1 0.2 0.3…0.1]

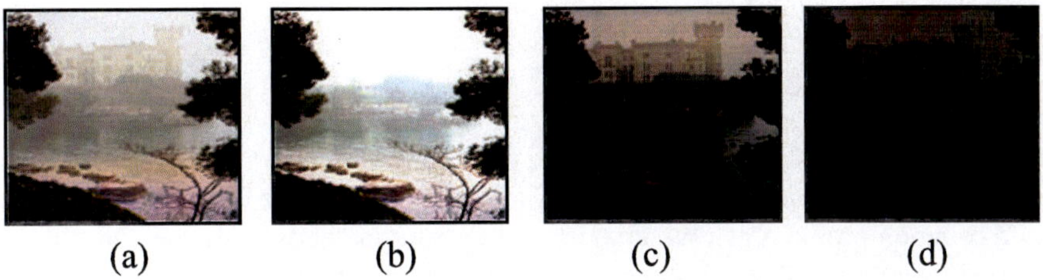

Fig. 6 Visual analysis on a hazy image for different values of c in inverse log transformation: **a** Hazy image. **b, c, d** Images generated using $c = 1$, $c = 2$ and $c = 3$ respectively

2.1.3 Guided Filtering

Due to haze, edges of the image are degraded. Therefore, we apply the edge-preserving smoothing operator. In the proposed work, guided image filtering [13] is utilized to preserve the edges. It has various applications in image enhancement, image matting, image dehazing and so on. It generates the filtering output based on the guidance image. The guidance image is the hazy image itself and output f a pixel P_i is computed as follows:

$$q_i = \sum_j Wij(I)\, Pj \tag{7}$$

where P is the input image and I is the guidance image. i and j are the pixel indexes of output and input images, respectively. Wij is the filter kernel of guidance image I and assumes q is a linear derivation of I in a window l_k, centered at pixel k:

$$q_i = x_k * I_i + y_k, \quad \forall i \in l_k \tag{8}$$

where (x_k, y_k) are the linear coefficients that remain fixed in l_k. These values are calculated using the cost function given by a linear ridge regression model, a square window of radius r. I_3 image is generated after applying guided filtering on the hazy image as shown in Fig. 7d. The drawback of guided filtering is that it distorts the color of the image.

Fig. 7 Resulting images of module I and II: **a** original hazy image. **b** Image I_1-log transformation. **c** Image I_2-Inverse log transformation. **d** Image I_3-Guided image filtering. **e** Fused haze-free image

2.2 *Multi-exposure Fusion*

The resulting images generated in the first module are fused into a single haze-free image using multi-scale Laplacian pyramid decomposition [14]. The proposed work employs this technique because the fused haze-free image is free from artifacts, which is a common problem of several existing dehazing methods. This method utilizes three features: contrast, saturation and well-exposedness to find correct exposed regions of the image from several images before performing multi-scale fusion. The weight-map function W is defined as a product of these three measures, guided by a power function as given:

$$W_{k,ij} = (\mathrm{con}_{k,ij})^a * (\mathrm{Sat}_{k,ij})^b * (\mathrm{Exp}_{k,ij})^c \tag{9}$$

where Sat, con and Exp are saturation, contrast and well-exposedness, respectively, and their corresponding power factors are a, b and c. The subscript k, ij refers to pixel (i, j) in kth image. If any of the power factors is 0, then its corresponding measure is not considered in the fusion process. The resulting image is a weighted sum of all images in the stack. The process of these two modules is shown in Fig. 7.

3 Experiment Results and Discussion

To test the effectiveness of the proposed method, it is applied on various challenging hazy images. The performance is recorded against the recent state-of-the-art methods. These methods are AMEF [10], CAP [6], NLID [5] and DCP [4]. Furthermore, analysis of the proposed methods is done in both ways: qualitative and quantitative.

3.1 *Qualitative Evaluation*

Comparative results of the proposed method in terms of visual quality with existing methods are shown in Fig. 8. Image 1 is an example of long-short hazy picture, i.e., objects are located at the far distance. The result obtained by various methods suffers from the problem of over-saturation around the leaves and tree region. The color of the building is also changed in NLID. The dehazed image produced by the proposed method is free from over enhancement, and the color of the building is also retained.

Dehazed result of Image 2 by NLID and DCP suffers from the problem of over enhancement as the face of dog can be noticed in dehazed image. Image 3 is an example of mild haze. The effectiveness of the proposed method can be confirmed by face of women and details (leaves, flowers, etc.) on the right side in the dehazed image with other methods. Dehazing performance for image 4, NLID and the proposed method give a quite satisfactory result for the objects located in nearby regions but

Fig. 8 Results of various methods: **a** Hazy image, **b** AMEF [10], **c** CAP [6], **d** NLID [5], **e** DCP [4], **f** The proposed method

at the same time, color cast problem is being seen in method of the NLID. Methods of AMEF and CAP are not able to remove haze properly in the image for the object located at the far distance. DCP has better image contrast as compared to the proposed method, but visibility of the proposed method is more.

3.2 Quantitative Analysis

Haze often degrades the edges and texture information of the image. To measure this information, the most famous approach is visible edge-based method proposed by Hautiere et al. [15]. It compares the haze-free image with hazy image with three indexes: e (visible edges ratio), saturated pixel ratio (σ) and gradient ratio (r). The formulas for these indexes are given below:

$$e = \frac{n_r - n_0}{n_0} \tag{10}$$

$$r = e^{-\left(\frac{1}{n_r} \sum_{P_i \in \phi_r} \log r_i\right)} \tag{11}$$

$$\sigma = \frac{n_s}{\dim_x \times \dim_y} \tag{12}$$

where n_r and n_0 are the number of visible edges in haze-free and hazy image respectively. Φ_r is the set of visible edges in haze-free image, P_i are the pixels of the visible edges, r_i is the Sobel gradient ratio of P_i and corresponding pixels of the hazy image, n_s is the number of saturated pixels (black or white) and $\dim_x$ and $\dim_y$ are the height and width. For good dehazing performance, there must be the larger value of e and r while smaller value of σ.

The comparative value of e, σ and r with other methods are shown in Table 1, Table 2 and Table 3, respectively. From these tables, it is clear that methods which are competing proposed method are DCP and NLID. Although these methods have a large value of e for some images but resulting images are over-saturated, i.e., some pixels are truncated to black or white. For example, for image 1, NLID has the highest e value but at the same time, σ value is quite high.

Table 1 Comparison of value of e for images shown in Fig. 8

Image No./methods	AMEF [10]	CAP [6]	NLID [5]	DCP [4]	The proposed method
Image 1	0.455	0.399	0.813	0.538	0.483
Image 2	2.062	1.522	2.896	4.457	3.09
Image 3	0.865	0.229	1.000	0.926	1.37
Image 4	2.390	1.511	3.296	4.663	3.997
Average	1.443	0.915	2.001	2.646	2.235

Table 2 Comparison of value of σ for images shown in Fig. 8

Image No./methods	AMEF [12]	CAP [7]	NLID [6]	DCP [5]	The proposed method
Image 1	0.425	0.021	0.285	0.024	0.003
Image 2	0	0.000	0.060	0.015	0.003
Image 3	0.001	0.000	0.017	0	0
Image 4	0.000	0.000	0.014	0.005	0
Average	0.107	0.005	0.094	0.011	0.002

Table 3 Comparison of value of r for images shown in Fig. 8

Image No./methods	AMEF [10]	CAP [6]	NLID [5]	DCP [4]	The proposed method
Image 1	1.844	1.159	1.868	1.818	1.973
Image 2	2.580	1.400	3.884	3.083	5.62
Image 3	1.642	1.082	1.976	1.349	3.12
Image 4	2.601	1.696	3.649	3.663	4.562
Average	2.167	1.334	2.844	2.478	3.819

Table 4 FADE score on a corpus of 500 foggy images

Methods	DCP [4]	NLID [5]	AMEF [10]	CAP [6]	The proposed method
FADE score	0.388	0.314	0.559	1.048	0.375

3.2.1 Evaluation Using Fog-Aware Density Estimator (FADE)

We have also compared the proposed method on Fog-Aware Density Estimator (FADE) score on 500 hazy images. The blind assessment quality metrics are failed to measure the subjective opinion about image appearance [16]. Therefore, we apply FADE metric, specially designed for evaluation of dehazing algorithms, proposed by Choi et al. [17]. The FADE score is machine learning-based metric which computes the fog-aware features across the image. The comparative result of FADE score with other techniques is illustrated in Table 4. The lower FADE score is an indicator of good dehazing performance and improved visibility.

Among all the compared methods, NLID has the lowest FADE score. The proposed method is in the second place, and DCP is at the third place. However, NLID and DCP suffer from the problem of over enhancement, i.e., color becomes darker and some time generate the over-saturated result. Lower FADE score again proves that the proposed method has good dehazing capability.

4 Conclusion and Future Work

In this paper, we have proposed a new image dehazing method using basic image enhancement operations like log transformation, inverse log transformation and guided filtering. The proposed method can remove the effect of haze without over-enhancement and color distortion. The experimental results demonstrate that the proposed method competes for the recent state-of-the-art methods. Finally, multi-scale Laplacian fusion scheme helps to fuse images in artifacts free manner. Future work will focus on investigating more interesting image-processing operations that can be utilized in single-image dehazing.

References

1. Wang W, Yuan X (2017) Recent advances in image dehazing. IEEE/CAA J Automatica Sinica 4:410–436
2. Cozman F, Krotkov E (1997) Depth from scattering. In: Proceedings of the IEEE computer society conference on computer vision and pattern recognition, pp 801–806
3. Singh D, Kumar V (2017) Comprehensive survey on haze removal techniques. Multimedia Tools Appl, 1–26
4. He K, Sun J, Tang X (2011) Single image haze removal using dark channel prior. IEEE Trans Pattern Anal Mach Intell 33:2341–2353

5. Berman D, Avidan S (2016) Non-local image dehazing. In: Proceedings of the IEEE conference on computer vision and pattern recognition, pp 1674–1682.
6. Zhu QS, Mai JM, Shao L (2015) A fast single image haze removal algorithm using color attenuation prior. IEEE Trans Image Process 24:3522–3533
7. Jia J, Yue H (2014) A wavelet-based approach to improve foggy image clarity. In: Proceedings of IFAC World Congress, pp 930–935
8. Liu X, Zhang H, Cheung YM, You X, Tang YY (2017) Efficient single image dehazing and denoising: an efficient multiscale correlated wavelet approach. Comput Vis Image Understanding 162:23–33
9. Ancuti CO, Ancuti C (2013) Single image dehazing by multi-scale fusion. IEEE Trans Image Process 22:3271–3282
10. Galdran (2018) Artificial multiple exposure image dehazing. Signal Process 149:135–147
11. Ren W, Liu S, Zhang H, Pan J, Cao X, Yang MH (2016) Single image dehazing via multi-scale convolutional neural networks. In: Proceedings of computer vision—ECCV 2016, Lecture Notes in Computer Science, Springer, pp 154–169
12. Cai B, Xu X, Jia K, Qing C, Tao D (2016) DehazeNet, "An end-to-end system for single image haze removal". IEEE Trans Image Process 25:5187–5198
13. He K, Sun J, Tang X (2013) Guided image filtering. IEEE Trans Pattern Anal Mach Intell 35:1397–1409
14. Mertens T, Kautz J, Reeth FV (2007) Exposure fusion. In: Proceedings of 15th Pacific conference on computer graphics and applications, pp 382–390
15. Hautiere N, Tarel JP, Aubert D, Dumont E (2008) Blind contrast enhancement assessment by gradient ratioing at visible edges. Image Anal Stereol J 27:87–95
16. Ma K, Liu W, Wang Z (2015) Perceptual evaluation of single image dehazing algorithms. IEEE international conference on image processing
17. Choi LK, You J, Bovik AC (2015) Referenceless prediction of perceptual fog density and perceptual image defogging. IEEE Trans Image Process 24:3888–3901

Eigenvector Component Calculation Speedup Over *NumPy* for High-Performance Computing

Shrey Dabhi and Manojkumar Parmar

Abstract Applications related to artificial intelligence, machine learning, and system identification simulations essentially use eigenvectors. Calculating eigenvectors for very large matrices using conventional methods is compute-intensive and renders the applications slow. Recently, the *Eigenvector-Eigenvalue Identity* formula promising significant speedup was identified. We study the algorithmic implementation of the formula against the existing state-of-the-art algorithms and their implementations to evaluate the performance gains. We provide a first of its kind systematic study of the implementation of the formula. We demonstrate further improvements using high-performance computing concepts over native *NumPy* eigenvector implementation which uses LAPACK and BLAS.

Keywords Eigenvalues · Eigenvectors · Eigenspace · Symmetric matrix · Hermitian matrix · NumPy · LAPACK · BLAS · High-performance computation · Vectorization · Parallelization

S. Dabhi (✉) · M. Parmar (✉)
Technology and Strategy Department, Innovation & Incubation Business Unit, Robert Bosch Engineering and Business Solutions Private Limited, Bengaluru 560095, Karnataka, India
e-mail: 16bit039@nirmauni.ac.in

M. Parmar
e-mail: manojkumar.parmar@bosch.com; manojkumar.parmar@hec.edu

S. Dabhi
Department of Computer Science and Engineering, Institute of Technology, Nirma University, Ahmedabad 382481, Gujarat, India

M. Parmar
HEC Paris, 1 Rue de la Libération, Jouy-en-Josas 78350, France

R. P. Mahapatra et al. (eds.), *Proceedings of 6th International Conference on Recent Trends in Computing*, Lecture Notes in Networks and Systems 177,
https://doi.org/10.1007/978-981-33-4501-0_23

1 Introduction

In mathematics, eigenvectors are fundamental to many matrix operations. They form the base of all the dimensionality reduction operations in artificial intelligence and machine learning. Moreover, eigenvectors are of extreme importance in system simulation and calculation of many real-world phenomena. They are the key to analyze the physics of rotating bodies, the stability of physical structures, oscillations of vibrating bodies, computational biology, economics, etc.

In linear algebra, an eigenvector is the characteristic vector of a linear transformation. It is a nonzero vector that changes at most by a scalar factor when that linear transformation is applied to it. The corresponding eigenvalue is the factor by which the eigenvector is scaled. Geometrically, an eigenvector, corresponding to a real nonzero eigenvalue, points in a direction in which it is stretched by the transformation and the eigenvalue is the factor by which it is stretched. If the eigenvalue is negative, the direction is reversed. This complexity causes the computations and thereby the applications to slow down.

In the following sections, we discuss our approach and the processes we applied to conclude. In Sect. 2, we explain the concept of eigenvalues and eigenvectors in brief. We also discuss the drawbacks of conventional methods and the current state-of-the-art algorithms. We then review the existing implementations of the *Eigenvector-Eigenvalue Identity* formula to formulate our strategy for the systematic improvement and evaluation. In Sect. 3, we shed light on the actual experimental setup and present algorithms for the baseline (given by Denton et al. [1, 2]) and the most performant variants of the implementations out of all the variants that we tested. We also present our thoughts on why exhibit Algorithm 1 gives the highest amount of speedup over *NumPy's* implementation. In Sect. 4, we discuss the results of our experiments. In Sect. 5, we give the direction for future researchers to improve over our work.

2 Related Works

The eigenvalues of an $n \times n$ matrix A are the roots of the characteristic equation:

$$\det(A - \lambda I) = 0 \tag{1}$$

where I is an $n \times n$ identity matrix. Eq. (1) has n solutions. The polynomial left-hand side of the characteristic equation is known as the characteristic polynomial. Hermitian matrices have real eigenvalues, i.e. the roots of Eq. (1) are real. A real and symmetric matrix is simply a special case of a Hermitian matrix. Traditionally, eigenvectors are calculated substituting the value of λ in the characteristic polynomial.

These polynomials are nth-order polynomials and solving them directly to calculate the eigenvectors can be very compute-intensive.

Currently, the state-of-the-art algorithm for the computation of eigenvectors of Hermitian matrices is the divide and conquer algorithm proposed by Cuppen et al. [3]. There are only two known implementations of this algorithm: one written in FORTRAN [4] and another one written in C language [5]. This algorithm first converts the Hermitian or real symmetric matrix into a tridiagonal matrix, and then recursively divides the matrix into two tridiagonal matrices until the lowest possible size is reached. Then, it calculates the eigenvectors for the smallest matrices and adds rank 1 corrections while accumulating the results to compensate for the loss of information incurred during the division of the matrix. The original recursive algorithm was later improved upon by various researchers for increasing the speed [6–8].

A very simple formula has been recently brought to light again after it was first discovered by Jacobi in 1834 and then subsequently rediscovered by many independent research teams throughout history. The new formula is proposed by Denton et al. [1] and formally named as the "*Eigenvector-Eigenvalue Identity*". They have traced the appearance of identity Eq. (2) throughout history and to prevent further disappearance and rediscovery of the identity, decided to formalize it finally in the form mentioned above. Various mathematical proofs of the identity have been covered in [1], however, computational gains have not been investigated systematically. The formal definition of identity has been given below.

If A is an $n \times n$ Hermitian matrix with eigenvalues $\lambda_i(A), \ldots, \lambda_n(A)$ and $i, j = 1, \ldots, n$, then the jth component $v_{i,j}$ of a unit eigenvector v_i associated to the eigenvalue $\lambda_i(A)$ is related to the eigenvalues $\lambda_1(M_j), \ldots, \lambda_{n-1}(M_j)$ of the minor M_j of A formed by removing the jth row and column by the formula

$$\left|v_{i,j}\right|^2 = \frac{\prod_{k=1}^{n-1}\left(\lambda_i(A) - \lambda_k\left(M_j\right)\right)}{\prod_{k=1,k\neq i}^{n}\left(\lambda_i(A) - \lambda_k(A)\right)} \tag{2}$$

In its current form, identity does not provide the direction of the components of the eigenvectors, but just the magnitude. This makes it unsuitable for applications requiring directions. But as mentioned in [1], it is possible to infer the directions of the components through various methods. It has been suggested by Ashok et al. [9] that for small matrices, the directions of eigenvectors can often be inferred by direct inspection of the eigenvector equation $Av_i = \lambda_i(A)v_i$. In general, we can apply Eq. (2) on multiple bases to retrieve the direction.

The basic implementation for identity formula (Eq. 2) was provided by Denton et al. [1, 2]. We improved upon this implementation by vectorising, parallelizing and adding batched processing capabilities. Despite the observations of Amdahl's law, we did not achieve tangible speedup in execution time in certain cases, due to the possible overhead of thread management.[1]

In this paper, we focus on implementing identity Formula (2) as efficiently as possible and attempt to match the accuracy, speed, and robustness of the state-of-the-art implementations. The majority of the popular machine learning and linear algebra libraries uses the low-level FORTRAN implementations from BLAS packages and LAPACK under the hood, for Eigen decomposition. All the machine learning libraries just provide high-level wrappers which abstract out a lot of complexities and provide sensible defaults for the required hyper-parameters to achieve the desired speed and accuracy from the Eigen decomposition algorithms from LAPACK.

3 Experiments

The baseline implementation of identity formula Eq. (2) as provided by Denton et al. [2] is quite slow as it repeatedly computes eigenvalues of the same set of matrices implying that it does not provide any advantage over other existing algorithms. The pseudo-code for the baseline implementation is provided in exhibit Algorithm 1. We used concepts of high-performance computing to improve on it. We conceived, designed and implemented five variations of the algorithm and the baseline version. The comparison of the performance of each version for three different tasks is discussed in the results section. The pseudo-code for the best performing variant is given in exhibit Algorithm 2.

Charis et al. [10] provide an alternative implementation for the Formula (2). They calculate only a particular eigenvector for the given matrix. A similar implementation in Python does not provide improvements over *NumPy* [11, 12].[2] We then decided to test the speedup in the calculation of one of the components of a particular eigenvector of the given matrix.

[1] Amdahl's Law gives the theoretical speedup in the execution time of a routine at a fixed workload that can be expected out of a system whose resources are improved. In our case, we are increasing the computational power of the system by allowing our routine to utilize multiple CPU cores in parallel.

[2] One of the reasons can be the lack of native parallel for loops in Python.

```
Function EigenComponentBaseline(matrix, i, j):
    n ← matrix.shape
    minor ← del(matrix, i)
    matrixEV ← EigenValues(matrix)
    minorEV ← EigenValues(minor)
    numerator ← 1.0
    for k = 0; k < n − 1; k ← k + 1 do
        numerator ← numerator * (matrixEV[j] − matrixEV[k])
    end
    denominator ← 1.0
    for k = 0; k < n; k ← k + 1 do
        if k ≠ j then
            denominator ← denominator * (matrixEV[j] − minorEV[k])
        end
    end
    component ← numerator ÷ denominator
return component
```

Algorithm 1. Baseline pseudo-algorithm

The experiments to measure the speedup were carried out on a high-performance workstation running Windows 10 operating system. It is equipped with 32 GB of random access memory and Intel® Xeon® CPU E3-1270 v6 with four physical cores and eight logical cores.

We use Python 3.7.6 compiled for MSC v.1916 for a 64-bit processor. The computation was run 10 times for each matrix size using each variation of the implementation. The execution time was measured using cProfile utility provided in Python's standard library. The mean values for the 10 runs have been reported.

For calculating all the components of all the eigenvectors, the baseline implementation is the slowest. For an $n \times n$ matrix, the algorithm calls the function $2n^3$ times as shown in exhibit Algorithm 1. By vectorising the algorithm, we can reduce the number of calls, therefore, providing speedup.

We expected that parallel computation would reduce the execution time, but the possible overhead of thread creation and management leads to an increase in the time complexity.

The baseline and vectorized versions have limited performance for very large matrices (of the order of 150×150 and greater). One of the reasons could be the intermediate calculations leading to a datatype underflow or overflow. We batched all the calculations to combat this issue and also checked if computing all the batches in parallel would lead to any significant performance gains as well.

```
Function EigenComponentOptimized(matrix, i, j, batchSize):
    n ← matrix.shape
    minor ← del(matrix, i)
    matrixEV ← EigenValues(matrix)
    eigenValue ← matrixEV[j]
    matrixEV ← del(matrixEV, j)
    minorEV ← EigenValues(minor)
    batches, nBatch ← PrepareBatches(matrixEV, minorEV, batchSize)
    for k = 1; k < nBatch; k ← k + 1 do
        tNumer[k], tDenom[k] ← dispatch(BatchProcessor(batches[k]))
    end
    component ← 1.0
    for k = 0; k < nBatch; k ← k + 1 do
        component ← component * (join(tNumer[k]) ÷ join(tDenom[k]))
    end
return component
```

Algorithm 2. Optimized pseudo-algorithm

LAPACK's implementation of Cuppen's algorithm [3] used by *NumPy* [11, 12] only returns the complete set of all the eigenvectors. Hence, we further explored if we could benefit from the calculation of single vectors or a single component of any one vector.

4 Results

In Fig. 1a, b, identity refers to the batched vectorized implementation of Eq. (2) and identity parallelized refers to exhibit Algorithm 2. In all the figures, X-axis represents the size of the matrix for which we are calculating the Eigen decomposition and the Y-axis represents the time taken in seconds.

Figure 1a shows the speedup over *NumPy's* implementation for the calculation of a single component of a particular eigenvector for the given matrix. Figure 1b shows the performance of our implementation against the current state-of-the-art, i.e., *NumPy*, for calculating a complete eigenvector for the given matrix.

Figure 1c, d shows how we step-by-step implemented the concepts of high-performance computing to improve the performance of the formula. We achieved speedup with each iteration of optimization and ended up with exhibit Algorithm 2 as the most optimized and performant implementation of Formula (2).

From Table 1, we can infer that by using Eq. (2) and exhibit Algorithm 2 with an increase in the size of the matrix, we can achieve up to 4.5× speedup over *NumPy*.

The formula is more efficient than the state-of-the-art algorithms only when we need a few components of a particular eigenvector. Currently, only the applications such as Web indexing, Web search, signal pre-processing, etc., which only require partial eigenvectors, can benefit from the implementation of this identity.

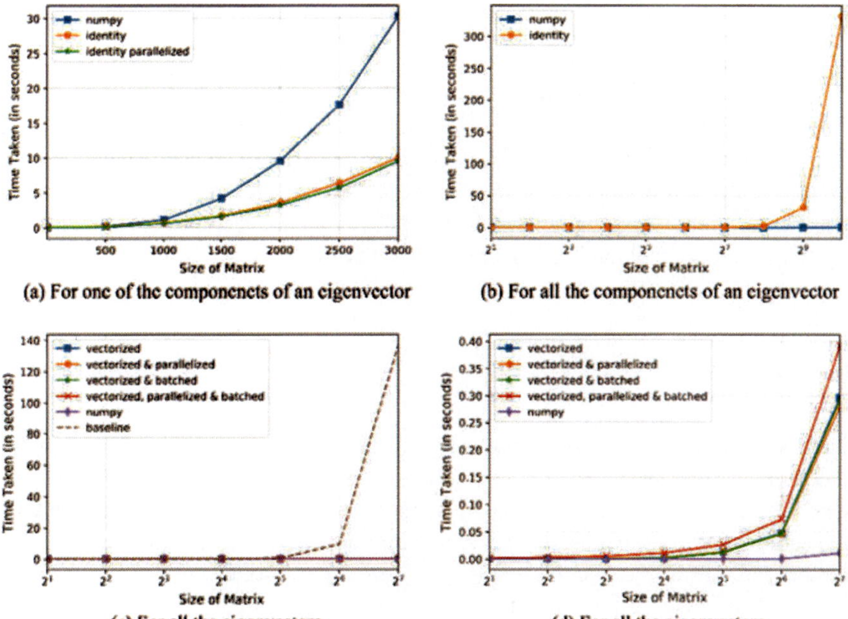

(a) For one of the componenets of an eigenvector

(b) For all the componenets of an eigenvector

(c) For all the eigenvectors

(d) For all the eigenvectors

Fig. 1 Comparison of the execution time of all the variants of the algorithm and LAPACK's implementation of Cuppen's algorithm

Table 1 Time in seconds for calculating 1 eigenvector component

Size of matrix	NumPy[a]	Exhibit Algorithm 2
2	0.000057	0.000233
502	0.170962	0.119722
1002	1.100800	0.595935
1502	4.165680	1.603579
2002	9.522480	3.212001
2502	17.632952	5.707374
3002	30.342656	9.522745
3502	47.404600	11.886900
4002	69.955400	16.775200
4502	98.680800	22.943500
5002	134.324200	30.547200
5502	177.629000	39.741700
6002	229.338600	50.682400

[a] Always computes the entire set

5 Future Work

It might be possible to further improve the performance if the formula is implemented in a lower-level language like C or FORTRAN to reduce overheads. LAPACK has also been written in FORTRAN, and, therefore, it can achieve such high performance. LAPACK algorithms also benefit from clever machine level multi-threading, which is extremely limited in a higher level language like Python, despite that we are getting a glimpse of the potential improvements as seen in Fig. 1a.

6 Summary

The *Eigenvector-Eigenvalue Identity* formula theoretically holds a promise to achieve speedup over the current state-of-the-art algorithms for the calculation of eigenvectors. We tried to realize this promise in an optimized algorithm. Our implementation can achieve up to $4.5\times$ speedup over *NumPy* for applications requiring partial eigenvectors. We are of the opinion that this algorithm can further provide a speedup if implemented using lower-level languages.

Acknowledgements We want to thank Sri Krishnan V, Mohan B V, Palak Pareshkumar Sukharamwala and Tanya Motwani from Robert Bosch Engineering and Business Solutions Private Limited, India, for their valuable comments, contributions and continued support to the project. We are grateful to all experts for providing us with their valuable insights and informed opinions ensuring completeness of our study.

References

1. Denton PB, Parke SJ, Tao T, Zhang X (2019) Eigenvectors from eigenvalues: a survey of a basic identity in linear algebra. arXiv preprint arXiv:1908.03795
2. Peterdenton/eigenvector-eigenvalue-identity, https://github.com/PeterDenton/Eigenvector-Eigenvalue-Identity. Last accessed 2020/01/30
3. Cuppen JJM (1980) A divide and conquer method for the symmetric tridiagonal eigenproblem. Numer Math 36(2):177–195
4. Anderson E, Bai Z, Bischof C, Blackford S, Demmel J, Dongarra J, Du Croz J, Greenbaum A, Hammarling S, McKenney A, Sorensen D (1999) LAPACK users' guide, 3rd edn. Society for Industrial and Applied Mathematics, Philadelphia, PA
5. masyagin1998/divide-and-conquer-eigenvalues, https://github.com/masyagin1998/divide-and-conquer-eigenvalues. Last accessed 2020/01/30
6. Rutter JD (1994) A serial implementation of Cuppen's divide and conquer algorithm for the symmetric eigenvalue problem. EECS Department, University of California, Berkeley
7. Gu M, Eisenstat SC (1995) A divide-and-conquer algorithm for the symmetric tridiagonal eigenproblem. SIAM J Matrix Anal Appl 16(1):172–191
8. Tisseur F, Dongarra J (1999) A parallel divide and conquer algorithm for the symmetric eigenvalue problem on distributed memory architectures. SIAM J Sci Comput 20(6):2223–2236

9. Mukherjee AK, Datta KK (1989) Two new graph-theoretical methods for generation of eigenvectors of chemical graphs. Proc Indian Acad Sci—Chem Sci 101(6):499–517
10. cgyurgyik/eigenvectors-from-eigenvalues, https://github.com/cgyurgyik/eigenvectors-from-eigenvalues/. Last accessed 2020/01/30
11. Oliphant TE (2006) A guide to NumPy, vol 1. Trelgol Publishing, USA
12. van der Walt S, Colbert SC, Varoquaux G (2011) The NumPy array: a structure for efficient numerical computation. Comput Sci Eng 13(2):22–30

Realtime Handwritten Digit Recognition Using Keras Sequential Model and Pygame

K. Senthil Kumar, Suman Kumar, and Aabhash Tiwari

Abstract Handwritten recognition has received the greater attention in deep learning research community due to its vast applications and ambiguity in learning methods. CNN in deep learning is now becoming one of the most appealing approaches and has been a crucial factor in the variety of recent success and challenging machine learning applications such as object detection. This research paper is about the extended application of handwritten digit recognition, i.e., realtime detection of the handwritten digits. CNN is used as the model for the classification of the image and more specifically Keras sequential model is used as a classifier. The interface is created by Pygame. The layout of the interface is kept simple and is divided into two frames one for input and other for output. The image pre-processing is the most important step which has done with the help of OpenCV and Scipy. MNIST is the dataset used for training and testing. Handwritten digit recognition has various real-life time uses. It is used in the detection of vehicle number, banks for reading cheques, post offices for arranging letter, and many other tasks.

Keywords Sequential model · Convolutional neural network (CNN) · MNIST

1 Introduction

Handwritten digit recognition is a technology that is very much needed in this world as of today. It is used in the detection of vehicle number, banks for reading cheques, and post office for arranging letters. In this paper, the extended application of handwritten recognition, i.e., real-time recognition of handwritten digits is implemented with the

K. Senthil Kumar · S. Kumar · A. Tiwari (✉)
Computer Science and Engineering, SRM Institute of Science and Technology, Chennai, India
e-mail: abhashtiwari.97@gmail.com

K. Senthil Kumar
e-mail: senthilkumar.k@ktr.srmuniv.ac.in; gowrisenthil2k4@gmail.com

S. Kumar
e-mail: sk2257682@hotmail.com

R. P. Mahapatra et al. (eds.), *Proceedings of 6th International Conference on Recent Trends in Computing*, Lecture Notes in Networks and Systems 177,
https://doi.org/10.1007/978-981-33-4501-0_24

help of CNN and Pygame. The classifier used in this project is a sequential model with a 4-layer CNN. The dataset used in this paper is MNIST dataset. The MNIST dataset has 70,000 handwritten digits. Among these, 60,000 are training examples and 10,000 are testing examples. Each image in this data is set to represented as an array of 28 * 28. The aim of this proposed endeavor was to make path toward digitization comprehensible by giving high accuracy and speedy computation for recognizing digits. That is why this problem was approached using CNN as they provide better accuracy over such tasks. To make real-time recognition, an interface is needed to draw, so to build that interface Pygame was used as a tool. The input will be drawn on the left side of window and final output will be appeared on the right side.

2 Literature Survey

Numerous researchers blatant their contribution in the field of digit recognition. Based upon their outcome and features, weights were assigned and it was executed on character recognition system by Hanmandlu and Murthy [1]. Graves and Schmidhuber [2] used a hidden Markov model that used a recurrent neural network. This helped in determining the sequence of characters in scripts that are handwritten. This was implemented for classifying the handwritten Arabic words. This gave an accuracy of about 91%. Multilayer perceptron was used by Pal and Singh [3] for recognizing English characters that are handwritten and an accuracy of 94% was successfully achieved. It was also able to improve the computation time for training the dataset. Neves [4] implemented the support vector machine algorithm and compared this with the MLP algorithm. This recognized the offline handwritten characters with a much better accuracy. The tests and training were done on the standard dataset NIST SD19. Although MLP is a better classifier for all nonlinear classes segmentation, it unfortunately gets trapped in a local minima. This lets the support vector machine get a better accuracy. Younis and Alkhateeb used the MNIST Dataset and extracted features without any preprocessing. It addressed the handwritten OCR problem by implementing a deep neural network. The accuracy achieved was 98.46% [5]. Dutt [6] also used the MNIST Dataset. But they also used the Keras and Theano libraries and used multilayer CNN and got an accuracy of 98.7%. Ghosh and Maghari [7] received 98.08% and demonstrated that DNN is the best algorithm, after completing a comparative study of three neural networks. But there may be similarity in the shape of digits, and for this reason, some amount of error rate will be present. CNN should be the best classifier in this case as compared to the other algorithms such as support vector machine, KNN, and the random forest classifier for HDR. Reference [8] implemented a DeepLearning4j framework which was incorporated with a rectified linear units activation which was never used before. The major aim of the paper was to recognize and determine digits that are handwritten with a higher accuracy and low computation time as well. CNN is used for fault detection and also for classification purposes. Recognition of handwritten digits is a big issue of interest in the community of researchers. Many researches have already been done on this topic and the flow of papers is still on. This topic is a big interest topic in the field of machine learning. Highest accuracy has been achieved using deep learning algorithms and also be using different libraries such as theano and keras. Tensorflow has also been used. This enables the accuracy and performance to be better than other algorithms. Convolutional neural network is being used in various areas such as Natural Language Processing, Video analysis and many other research areas as well. Sentiment analysis has also gained traction among researchers and people are trying to do more and more research in this field and move deeper into the research area. The research will go deeper in the coming days as well.

3 Dataset

A huge set of vision dataset which is used for training and testing is [Modified National Institute of Standards and Technology (MNIST)] (Fig. 1).

The dataset contains 60,000 images for training and 10,000 images used for testing. All images are grayscale and arranged in a explicit style of 28 * 28 pixels. Then, it can be compressed into a 28 * 28 = 784-dimensional vector. Each element of the vector has a binary value which specifies pixel intensity.

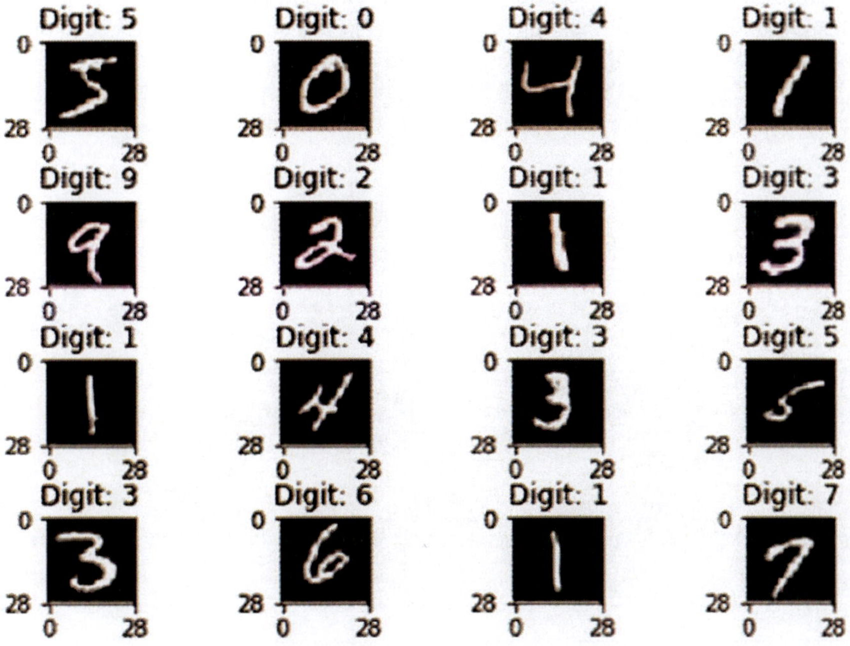

Fig. 1 Sample images of MNIST dataset

4 Implementation

4.1 Convolutional Neural Networks for Digit Recognition

In pattern recognition and image processing problems, neural network are often used. The most promising tool to carry out this is CNN. It is a deep learning technique which is inspired by the neuron connectivity pattern in animal visual cortex. CNN is mainly used in image recognition, object recognition, etc. The CNN model for recognizing the digits is constructed using the KERAS library of Python and TensorFlow as a backend. Sequential model is used as classifier which consists of linear stack of layers.

In CNN, the neurons have learnable weights and biases. Compared to others, CNN requires minimum preprocessing. The input is always a vector in neural network but in CNN the input will be a multi-channelled image.

The CNN comprises of an input layer, hidden layers, and an output layer. The hidden layer constitutes the convolutional layer, rectified layer unit (ReLU), i.e., activation function, pooling layers, normalized layers, and fully connected layers (Fig. 2).

The input layer consists of 28 by 28-pixel images which means it contains the network of 784 neurons as input data. The layer next to input layer is convolutional layer which receives the output from the input layer. A convolution operation is performed on the input in convolutional layer. The operation performed in this layer helps in reducing the number of free parameters. And design each filter such that it slides above the input data to have pixels with the utmost intensity.

The rectified linear unit (ReLU) is a activation function which is a linear function, easy to use and achieves better performance. Also, it does not saturate.

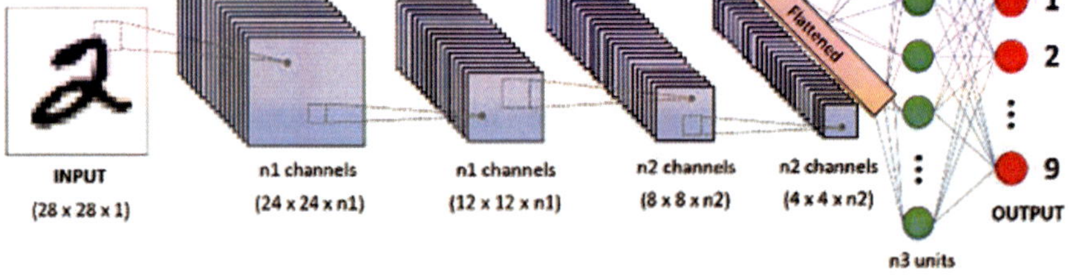

Fig. 2 A seven-layered CNN architecture for digit recognition

The concept of pooling layer is to merge the product of neuron at one layer into single neuron at the next level. The fully connected layer reckons the score of input digits. At the end, a SoftMax classifier is applied at the end which returns the probabilities of all output classes. From all these values, the class with the largest value is selected as a final classification. To work on Keras API, it is converted into four-dimensional NumPy arrays. The normalization is done by dividing RGB code by 255. Adam optimizers is used to update the neuron weights and it require little memory.

4.2 Image Preprocessing

After training the model, it will be saved as a classifier. Now the image preprocessing is done on the input image obtained by the interface. The image processing is done with the help of Open Cv2 and Scipy.

The steps taken are:

1. Read image:—In this, the path of our image is stored into a variable and then a function is created that converts images into array.
2. Resize image:—Some images captured by camera vary in size. The base size needs to be established for all images fed into our system.
3. Remove Noise:—By using Gaussian Blur () function, image noise may be used.
4. Segmentation:—More techniques to smoothen and reduce further noise is done.
5. Scaling and Padding:—The image is then scaled and padded to make it suitable for prediction (Fig. 3).

4.3 Application Interface

The application interface is created with the help of Pygame. Pygame is a group of modules in Python which is highly portable and can be used for building video games.

For opening the Pygame window, run the interface. On the left side the digits will be drawn by the user and the corresponding output will be displayed on the right side. To draw the digit left click your mouse and once finished release the left click. To reset the screen, click right button.

The output will be displayed using a bounding box and the recognized digit will be displayed in small letters at the top. Multiple numbers can be recognized as well in real time (Fig. 4).

The image obtained after drawing the input is now fed into the pre-processing stage and classifier predicts the digit and classifies it into one of the 10 classes.

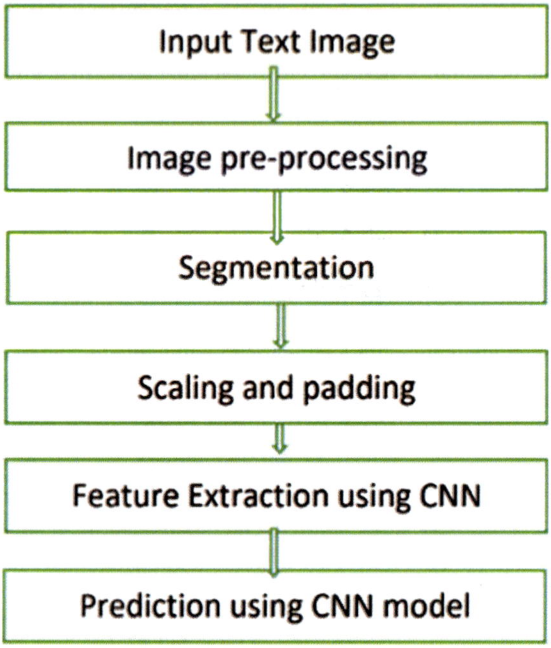

Fig. 3 Flow diagram of proposed method

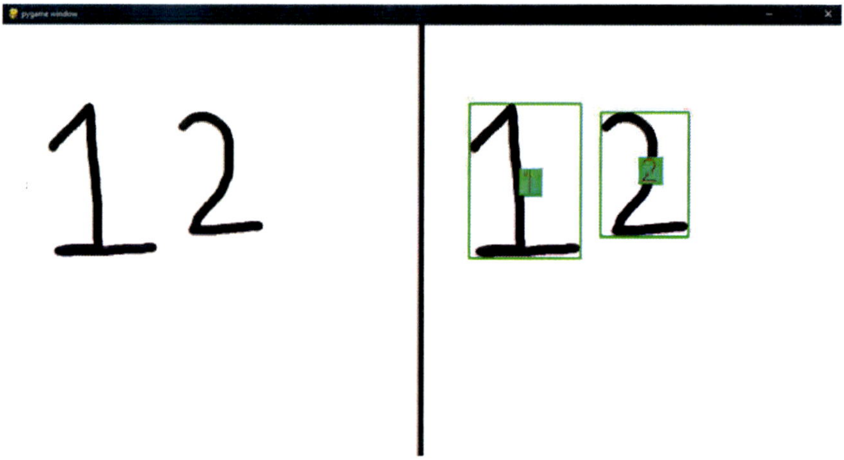

Fig. 4 Pygame application interface

5 Result

See Table 1 and Fig. 5.

Table 1 Test accuracy and loss for 15 epochs

CNN layers	Accuracy (%)	Loss
2	98.68	0.068
3	98.97	0.032
4	99.25	0.025

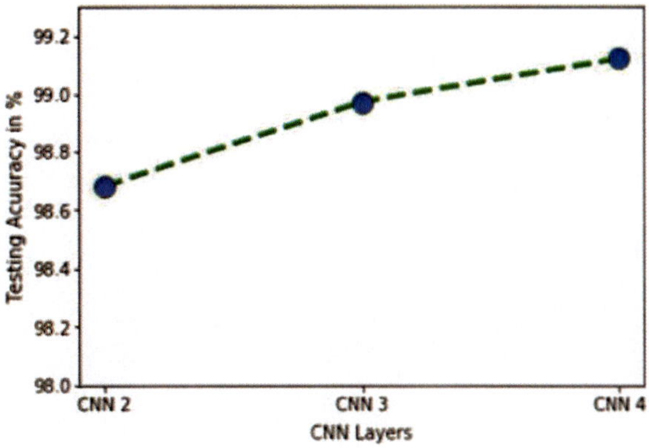

Fig. 5 Testing accuracy for 15 epochs

The above graph shows the accuracy of the test data when trained with 15 epochs. The four-layered CNN (Fig. 6).

The above graph shows us the losses when the number of epochs is 15 (Fig. 7).

The above figure shows us the accuracy of the test data when the number of epochs is 5.

In the below three figures, the graph has been made for testing accuracy against the training accuracy for 15 epochs with different layered CNNs (Figs. 8, 9, 10 and 11).

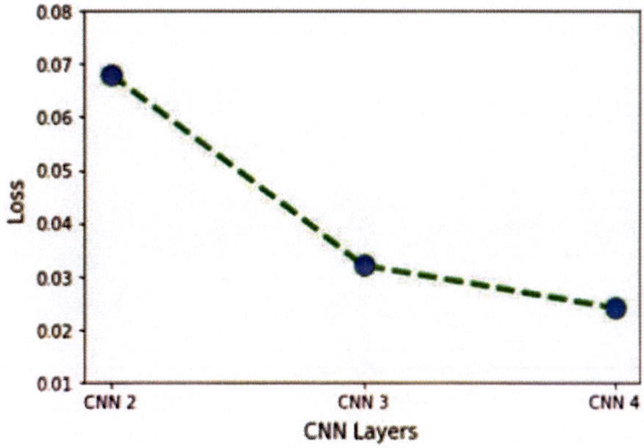

Fig. 6 Loss for 15 epochs

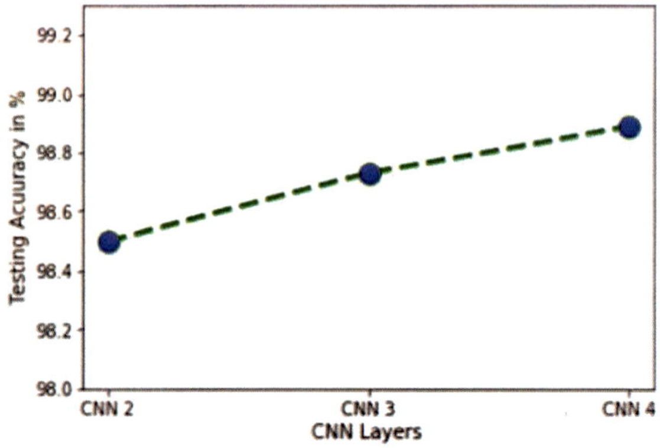

Fig. 7 Testing accuracy for 5 epochs

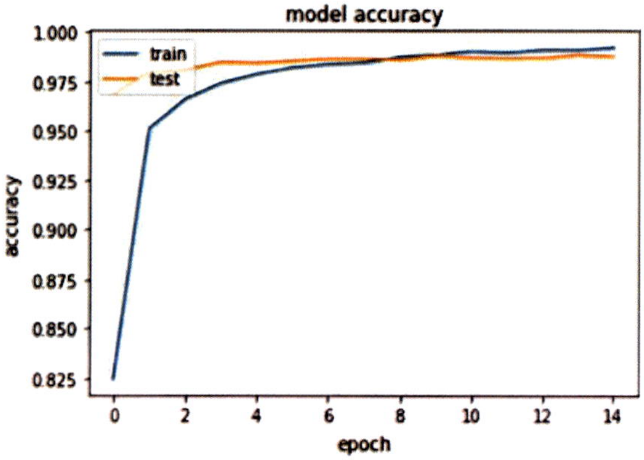

Fig. 8 Testing accuracy versus training accuracy for 15 epochs in two-layer CNN

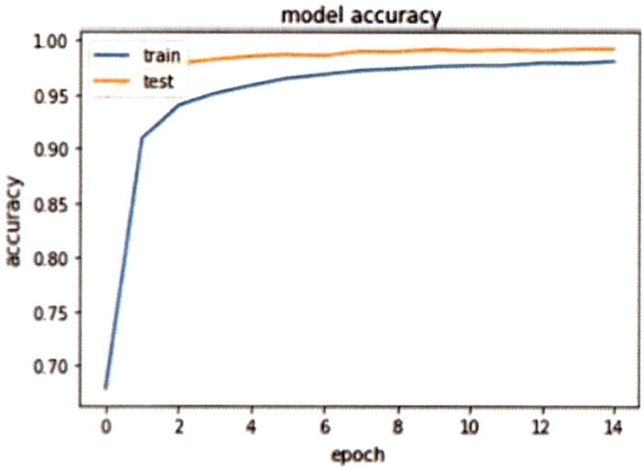

Fig. 9 Testing accuracy versus training accuracy for 15 epochs in three-layer CNN

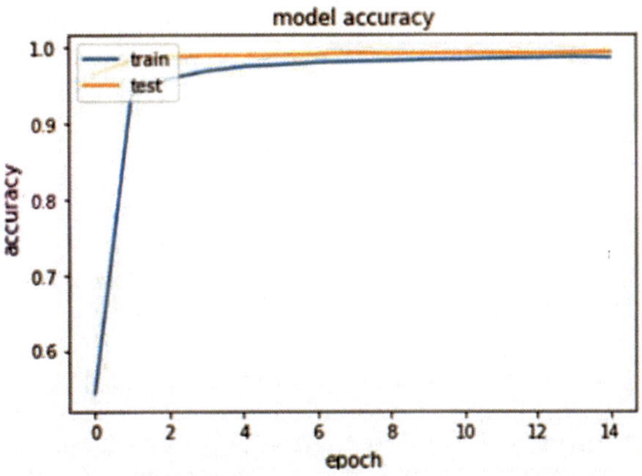

Fig. 10 Testing accuracy versus training accuracy for 15 epochs in four-layer CNN

Fig. 11 Final output

6 Conclusion

The accuracy for the test data was checked for different CNN Layers, namely two-layered, three-layered, and four-layered CNN. The four-layered CNN gave the most accuracy that was 99.25% with the least loss among all the layers compared. This was in case of 15 epochs.

A comparative study was also done with 5 epochs. The accuracy was highest with the four-layered CNN which is 98.89%, but when compared with the 15 epochs, the accuracy has been less. This is mainly because when 15 epochs are being used, the algorithm is able to train itself more efficiently and thus a greater accuracy and a lesser loss. Besides this, the API built is also able to display the output perfectly for every digit that is written by the user.

This will enable in the accurate recognition of the digits by online mechanism with a good accuracy. Users may simply type in a digit in free hand in the console window, and the recognized digit would be seen in the next window. The whole process can be improved by using different algorithms such as a multilayer perceptron. This would help in optimization of the code. There are other algorithms as well that may be used.

References

1. Hanmandlu M, Murthy O (2007) Fuzzy model-based recognition of handwritten numerals. Pattern Recogn 40:1840–1854
2. Graves A, Schmidhuber J (2009) Offline handwriting recognition with multidimensional recurrent neural networks. In: Koller D, Schuurmans D, Bengio Y, Bottou L (eds) Advances in neural information processing systems 21. Curran Associates, Inc., pp 545–552
3. Singh APD (2010) Handwritten english character recognition using neural network. Int J Comput Sci Commun 1:141–144
4. Neves RFP, Lopes Filho ANG, Mello CAB, Zanchettin C (2011) A svm based off-line handwritten digit recognizer. In: 2011 IEEE international conference on systems, man, and cybernetics, pp 510–515
5. Younis KS, Alkhateeb AA (2017) A new implementation of deep neural networks for optical character recognition and face recognition. In: Proceedings of the new trends in information technology, Jordan, pp 157–162
6. Anuj Dutt AD (2017) Handwritten digit recognition using deep learning. Int J Adv Res Comput Eng Technol (IJARCET) 6
7. Abu Ghosh MM, Maghari AY (2017) A comparative study on handwriting digit recognition using neural networks. Int Conf Promis Electron Technol (ICPET) 2017:77–81
8. Ali S, Shaukat Z, Azeem M, Sakhawat Z, Mahmood T, Rehman K (2019) An efficient and improved scheme for handwritten digit recognition based on convolutional neural network. SN Appl Sci 1

Discrete Cuckoo Search Algorithm for Tour Recommendation System

Tejaswini Bhosale and Shashank Pushkar

Abstract This paper presents route planning for a tourist as tourist trip design problem (TTDP) using discrete cuckoo search algorithm (DCS). The DCS algorithm is based on brood parasitic nature of cuckoo bird which is characterized by levy flight. The experiment of proposed DCS is conducted on orienteering problem (OP) benchmark instances and their performance compared with existence data of iterative local search (ILS) and greedy randomized adaptive search procedure (GRASP). Result showed that proposed DCS is required less computational time as compared to ILS and GRASP. It concluded that DCS is efficient method for exploring and exploiting search space in reasonable amount of time.

Keywords Cuckoo search algorithm · Tourist trip design problem orienteering problem · Nature-inspired algorithm

1 Introduction

Nowadays, Internet era come major revolutionary changes in tourism system. Tourist route planning becomes more customer-oriented and information-intensive. The route planning problem is falling in the category of tourist trip design problem (TTDP) as a part of orienteering problem (OP). The effective route gets planned for tourist to visit multiple points of interest (POIs) which match tourist satisfaction and preferences subject to budgeting constraints like traveling distance, visiting, and traveling cost, etc. daily [1].

Many times, tourists have limited time and budget, thus not able to visit all POIs. Planning an effective tour with the personal interest of tourists is a crucial job. Various methods are used to solve this problem; some exact methods provide a good solution

T. Bhosale (✉) · S. Pushkar
Department of Computer Science, Birla Institute of Technology, Mesra, Ranchi, India
e-mail: bhosale.tejaswini@gmail.com

S. Pushkar
e-mail: shashank@bitmesra.ac.in

R. P. Mahapatra et al. (eds.), *Proceedings of 6th International Conference on Recent Trends in Computing*, Lecture Notes in Networks and Systems 177,
https://doi.org/10.1007/978-981-33-4501-0_25

for small no. of POIs, but it gets failed as vertex increases. Heuristics provide a solution for large number of vertex but do not guaranty to be an optimal solution. Metaheuristics group of natured-inspired algorithms helps to solve a combinatorial optimization problem like TTDP in a reasonable amount of time with less parameters.

Many metaheuristics provide a different solution approach to TTDP like evolutionary algorithms, bee colony, particle swarm optimization, ant colony, etc. Our goal in this paper is to propose an effective discrete cuckoo search which mimic the behavior of cuckoo bird. Cuckoo search employs a good solution methodology for TTDP.

This paper is arranged in following manner; Sect. 2 contains related work and literature of TTDP. In Sect. 3, we provide a problem formulation for TTDP and metaheuristics CS followed by the proposed discrete CS approach. In Sect. 4, we present the results and analysis, then we end with our conclusion.

2 Related Work

The orienteering problem is an outdoor sport played in a mountain area where the player must collect maximum profit by visiting control points under the given time budget [2]. OP is a combination of two complementary objectives as profit maximization and traveling cost minimization, so it is traveling salesman problem with profit along with other specified routing problem like vehicle routing problem. This problem is an NP-hard problem [3] and gives birth to too many routing and scheduling applications like tourist trip design (TTDP), mobile crowd sourcing, etc. TTDP as an orienteering problem belongs to the class of NP-hard combinatorial optimization problems as it is computationally expensive to calculate all perturbations for large instances in bounded time limits [4]. OP provides a feasible solution with the exact method branch and bound [5] and branch and cut [6] for a small number of POIs. Computational cost increases exponentially as the value of *N increases.* Heuristic approach provide a near-optimal solution like tabu search proposed [7] form an iterative cluster and remove chain so it very less probability get trap in local optima. The first heuristic for OP proposed by [2] which is based on the center of gravity approach [3] proposed heuristics used a greedy approach to find path construction and he performed path improvement heuristics like 2-opt. Metaheuristics are mainly inspired by nature, utilizes the various behavioral aspect of animals, birds, insects, etc. Metaheuristics are robust to problem size so it easy to get a global optimal solution for large no. of POIs. Memetic algorithm [8] improved a local search technique which helps to directly improve route quality. Figure 1 represents the taxonomy for various solution methodology for TTDP.

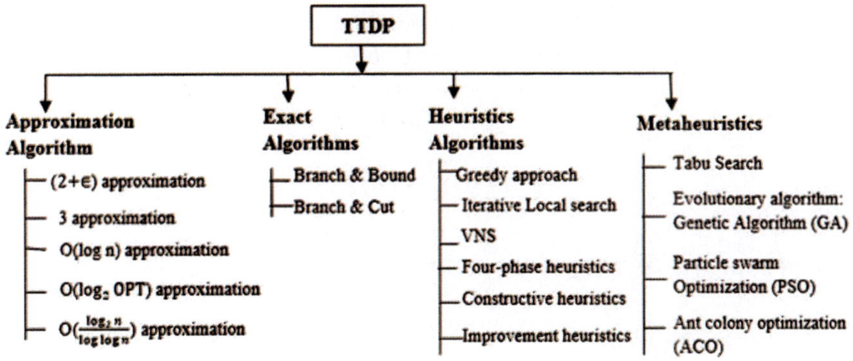

Fig. 1 Taxonomy of TTDP methods

3 Problem Definition and Formulation

TTDP is an NP-hard and cannot be solved efficiently by any known algorithm in a practically acceptable time scale. As problem size increases, the number of combinations increases exponentially so it becomes difficult to solve in optimal time. The TTDP is to find an itinerary for a tourist, visiting each POI at most once, requiring a total time (distance) do not exceeding a given bound $C_{\max}$, and collecting a maximum total profit. POIs in city represented as a vertex of complete undirected graph $G = (P, R)$ such that $P = \{p_1, p_2, \ldots, p_n\}$ and connecting roads represent arc of graph $R = \{(p_i, p_j): p_i, p_j \in R\}$.

The TTDP is formulated as a single-day tour with profit minimization subject to budgeting constraint of travel cost as follows [4]:

$$\max \sum_{i=1}^{n-1} \sum_{j=2}^{n} B_i . P_{ij} \tag{1}$$

Subject to:

$$\sum_{j=2}^{n} P1_j = \sum_{i=1}^{n-1} P\text{in} = 1 \tag{2}$$

$$\sum_{i=1}^{n-1} Pik = \sum_{j=2}^{n} Pkj \leq 1, \text{ for all } k = 2, \ldots, n-1, \tag{3}$$

$$\sum_{i=1}^{n-1} \sum_{j=2}^{n} C_{ij} P_{ij} \leq C\text{max} \tag{4}$$

$$P_{ij} \in \{0, 1\}, \text{for all } i, j = 1, \ldots, N. \tag{5}$$

Equation 1 represents the objective function in which maximization profit collected from visited POIs. Equations 2–5 represent the various constraints on route formation. The route starts from vertex 1 and end to vertex N indicated in Eq. 2. Equation 3 makes sure that each vertex visited at most once. Finally, Eq. 5 is an integrity constraint on route formation.

3.1 Proposed Methodology

3.1.1 Cuckoo Search

Levy the cuckoo bird obeys the brood parasitism behavior. Cuckoo finds a new nest via levy flights so it can cover the larger territorial space. Levy flight models a random walk characterized by step length which obeys power-law distributions.

Yang and Deb in 2009 [9] developed a metaheuristics search algorithm named as cuckoo search (CS). CS was designed with some basic assumption about cuckoo behaviors as follows:

- Cuckoo lay single egg each time and nest selection done randomly.
- High quality (Egg) nest generates a new generation cuckoo.
- Total no. of the nest is fixed and Pa probability of remove the non-fitted nest and replaced with a new one.

If the abandon parameter Pa is set as 0.25 means, it removes 25% of the not fittest nest and will be replaced by a new randomly generated nest. Cuckoo tries to reduce Pa with nature gifted ability like imitate color and patterns of the host egg. It matches the timing of laying an egg with a host bird. Cuckoo chicks get more feeding chance as it produces same and strong hungry call as host chicks, etc. CS is characterized by levy flight to form a new solution.

3.1.2 Levy Flight

Levy flight is widely used in many scientific applications and fields of optimization. Levy flight uses both local and global search to give a well stabilization between intensification and diversification. It is a random walk with no. of instantaneous moves chooses from probability density function which characterizes by the heavy tail. A cuckoo generates a new solution $\mathrm{Sol}_i^{(t+1)}$ via levy flights, according to Eq. (6)

$$\mathrm{Sol}_i^{t+1} = \mathrm{Sol}_i^{(t)} + \alpha \oplus \mathrm{levy}(S, \lambda) \tag{6}$$

where α is ($\alpha > 0$) step length. $\oplus$ is a pair-wise multiplication. Step length followed by levy distribution.

$$\text{Levy}(S, \lambda) \sim S^{-\lambda} \quad (1 < \lambda \leq 3) \tag{7}$$

Which as infinite variance with infinite means. S is step size drawn from levy distributions. Listed alphabetically according to their surnames in the author index.

3.2 Discrete Cuckoo Search

The original CS was developed with the intention of application in continuous optimization problems. This algorithm can therefore not be used directly to solve a problem in a discrete domain (TTDP). So, in this paper, we extend the original CS to the discrete domain through the change in algorithm.

The solution for TTDP is a permutation representation of POIs. Each POI associated with some benefit value. To adapt the CS for TTDP, we reinterpreted the key element and operators. In CS, number of the nest are fixed and it is the size of the population. Each cuckoo lay one egg in one nest which represents a single egg and equivalent to a feasible permutation of POIs. So, by the projection of TTDP, we do not distinct wish the egg, nest, cuckoo. BDCS is a population-based search. Following are the step of global optimal solution generation.

As our discrete cuckoo search algorithm is a stochastic search algorithm, we can summarize it as the following key steps:

(1) Initial population generate randomly. Initial nest population is n, $n = (p_{01}, \ldots, p_{0n}, p_{11}, \ldots, p_{1n}, \ldots, p_{n1}, \ldots, p_{nn})$ each with fitness value $A = \{\boldsymbol{a}_{01}, \boldsymbol{a}_{02}, \ldots, \boldsymbol{a}_{0n}\}$, calculate global best z_0^t **(Using Algorithm 1)**.
(2) Update the new solutions/positions by Eq. 6.
(3) Apply abonnement parameter p_{a} Update $_{\boldsymbol{a}}^{i(t+1)}$. Then, find the new nest which replace abundant nest, evaluate fitness value to get new global best z^{t*} **(Step 2 and 3 using Algorithm 2)**.
(4) Continue till stopping criterion encounter, then declare z_t^* as the global optimal solution. Otherwise, return to Step 2.

```
Algorithm 1: Initial Population
For i=1 to n do
      Pi node is selected randomly from subset of nodes;
      For j= 2 to n-1 do
          Pj node selected with Nearest Insert Method
      End For
Construct a route by ordering selected nodes;
End For
```

```
Algorithm 2: Discrete Cuckoo Search
Initialize CS Parameters
Produce the First generation of N nest using Algorithm.1
While t < Max Generation or Stopping Criteria
         Obtain a New Solution Soljt+1 with levy flight
distribution.
         Compute Fittness function fj of newly generated
solution and compare with old fitness fi;
        If fj > fi then
          Replace the old Soli with new Solj
      End if
          A fraction of abundance Pa apply to remove worst
fitted Nest & replace with new nest
          Find the global optimal solution.
End while
Post process result & visualization
```

4 Result and Discussions

For validation of proposed DCS, experiments were carried out on Python3.7 with Intel core i5-3470 3.2 GHz processor and 4.0 GB memory to calculate the average computational time and it was compared with existing algorithms ILS and GRASP for TTDP.

4.1 Benchmark Instances

OP instances were selected two database which consist of three datasets namely OP21, OP32, OP33 consists 49 instances from Tsiligrides-1984 [2] and two datasets namely OP64, OP66 with 40 instances from Chao-1996 [3]. All benchmark instance was downloaded from http://www.mech.kuleuven.be/en/cib/op. For the performance evaluation of proposed DCS, additional benefit values were assigned to test instances.

The proposed algorithm and already implemented ILS, GRASP algorithms were applied to the two selected OP standard database to evaluate average computational

Table 1 Comparison of the mean and standard deviation of the computational time

Instances	ILS		GRASP		DCS	
	Avg. Time	Std. Dav.	Avg. Time	Std. Dav.	Avg. Time	Std. Dav.
OP21	2.24	1.63	2.18	1.45	2.12	1.22
OP32	2.85	1.83	2.69	1.89	2.56	1.76
OP33	3.42	1.94	3.48	1.92	3.42	1.92
OP64	177.9	175.2	156.3	148.3	149.6	138.5
OP66	240.5	228.1	233.5	215.8	229.3	214.4

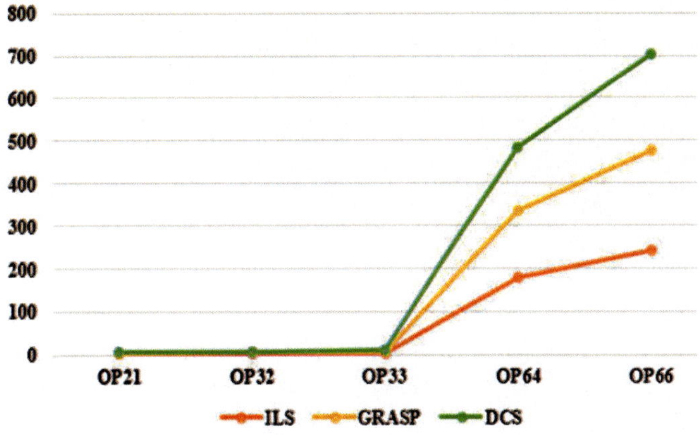

Fig. 2 Comparative analysis of avg. computational time of DCS, ILS, GRASP

time and compared with each other. Table 1 presents comparison of mean average time and standard deviation (Std. Dav.) of the computational times for ILS, GRASP, and DCS using benchmark instances. It shows that DCS is required less computational time as compared to ILS and GRASP. Figure 2 shows the graph plotted for comparision. It indicates that the DCS algorithm is capable to find optimal solution for TTDP in minimal time.

5 Conclusion

This paper presents a metaheuristics solution technique for TTDP using DCS algorithm to a plan tourist route with a visit number of POIs in cities. It helps the tourist to more satisfying visit. It showed that proposed DCS is required less computational time as compared to ILS and GRASP. It proves that DCS is efficient method for exploring and exploiting search space in reasonable amount of time using levy flight strategies.

References

1. Vansteenwegen P, Souffriau W, Van Oudheusden D (2011) The orienteering problem: a survey. Eur J Oper Res 209(1):1–10
2. Tsiligirides T (1984) Heuristic methods applied to orienteering. J Oper Res Soc 35(9):797–809
3. Chao IM, Golden BL, Wasil EL (1996) A fast and effective heuristic for the orienteering problem. Eur J Oper Res 88(3):475–489
4. Gavalas D, Konstantopoulos C, Mastakas K, Pantziou G (2014) A survey on algorithmic approaches for solving tourist trip design problems. J Heur 20(3):291–328
5. Ramesh R, Yoon YS, Karwan MH (1992) An optimal algorithm for the orienteering tour problem. ORSA J. Comp 4(2):155–165
6. Fischetti M, Gonzffalez JJS, Toth P (1998) Solving the orienteering problem through branch-and-cut. INFORMS J. Comp 10(2):133–148
7. Sylejmani K, Dorn J, Musliu N (2012) A tabu search approach for multi constrained team orienteering problem and its application in touristic trip planning. In Pro.12th Inte. Conf. Hybrid Intell. Sys. (HIS'2012) 1, pp 6–14
8. Bouly H, Dang, DC, Moukrim A (2010) A memetic algorithm for the team orienteering problem. 4OR: Quart J Oper Res 8:49–70
9. Yang XS, Deb S (2010) Engineering optimisation by cuckoo search. Int J Math Modell Numer Optim 1(4):330–343

Generate Signature for Polymorphic Worm: A Real-Time Honeypot Approach

Avijit Mondal, Sayan Nath, Purbasha Bose, and Radha Tamal Goswami

Abstract In Internet infrastructure, information security is the most major concern for any individual or any organization. The general regular signature-based approach fails to detect unknown attacks promptly. Our research approach construes a unique honeypot system which will be able to detect the suspected traffic from other ordinary traffic and captures relevant particulars about different worm operation without attacker's information. Our structural approach is automated and generates a high-quality signature for mainly unknown polymorphic worms. We are using a probabilistic approach to generate an attack pattern which deals with low-false alarm. Our observation with artificially produced polymorphic worms illustrates the proposed setup can be qualified to improve the efficiency of the intrusion detection system signature database and also able to increase the possibility of encounter different kinds of polymorphic worms with high expertise.

Keywords Honeynet · Honeypot · Intrusion detection system · Polygraph · Polymorphic worm · Signature · Token · Worm

---

A. Mondal (✉) · S. Nath · P. Bose
Techno International Batanagar, Kolkata, India
e-mail: avijitmondal88@yahoo.com

S. Nath
e-mail: sayan2016.nath@gmail.com

P. Bose
e-mail: boseriya1991@gmail.com

R. T. Goswami
Techno International New Town, Kolkata, India
e-mail: rtgoswami@tict.edu.in

R. P. Mahapatra et al. (eds.), *Proceedings of 6th International Conference on Recent Trends in Computing*, Lecture Notes in Networks and Systems 177,
https://doi.org/10.1007/978-981-33-4501-0_26

1 Introduction

In a short period, worms spread through the Internet. It replicates from one system to another. The traditional intrusion detection system can detect worms after a significant loss of assets as they do not have any previous signature pattern. As intranet consists of more than one LAN, and this distinct LAN has different vulnerabilities that is why one security mechanism for detecting worm is very difficult. Also, in the initial phase, it is difficult to differentiate normal traffic and malicious traffic as they both look quite the same. So the defense system must be very strong enough to detect the malicious signature of worms. Therefore, we propose a honeypot signature generation approach which is effective within the local autonomous system.

Honeypot is a resource of information whose value repose in unauthorized or illicit uses of that resource [1]. Anything that will come through a honeypot will be taken as a probe. It is generally two types: high interaction and low interaction honeypot. High interaction honeypot gives the user access to the real-time operating system in which nothing is restricted and low interaction honeypot does not provide any security, only collects information of the attack. Detection of traffic in honeypot is signature and anomaly-based.

Anomaly-based detection captures traffic based on previous statistic value sometimes called heuristic technique. This approach requires an understanding of normal traffic behavior to detect malicious traffic, which requires proper training in real-time scenarios. This approach is effective but produces huge false alarm. A signature-based mechanism does not look into host behavior that requires a sequence of byte string in every data packet. Any similarity found with the stored signature will capture it as malicious activity. The signature-based approach is much easier to implement online. But it identifies only that pattern which is already generated and manually stored to the signature generator by the researcher. As the manually signature generation process is quite slow, now researchers much interested in automatically generation process for known as well as unknown worms.

In this research work, we used a honeypot system to segregate both vicious and routine traffic for the collection of material to detain traffic and also forward the information for the generation of the automated signature.

This paper is organized as follows. Section 2 describes the way of a different polymorphic worm. In Sect. 3, we discussed honeypot architecture, data control, and also data capture technique. Section 4 describes the signature generation methodology of polymorphic worms. Section 5 discusses the experimental result and at last Sect. 6 derives the conclusion.

2 Form of Polymorphic Worm

A polymorphic worm differently appears as each time replicates itself. Many possible paths are there to form polymorphic worms [2]. A mechanism is self-encryption by using a variable key that encrypts the whole structure of a worm and erases the signature and statistical characteristics of that worm string. A copy of that worm with decryption pattern and also key are sent to a victim device, where the encrypted body is converted into a regular worm program by the decrypting pattern. Another technique to make polymorphic worm is by inserting a garbage code. That is why we are more concerned to detect polymorphic worm instances instead of the normal worm byte sequence.

3 System Structure

3.1 Architecture

Figure 1 interprets the system architecture of the honeypot system. It contains two different virtual honeypot—honeynet1, honeynet2. This honeypot emulates services like File Transfer Protocol, Hypertext Transfer Protocol, Simple Mail Transfer Protocol, Post Office Protocol, Dynamic Host Configuration Protocol, etc. Simulation is carried out at layer 7 (application layer) of the open systems interconnection model. The system where the honeypots is operating can be addressed as a server

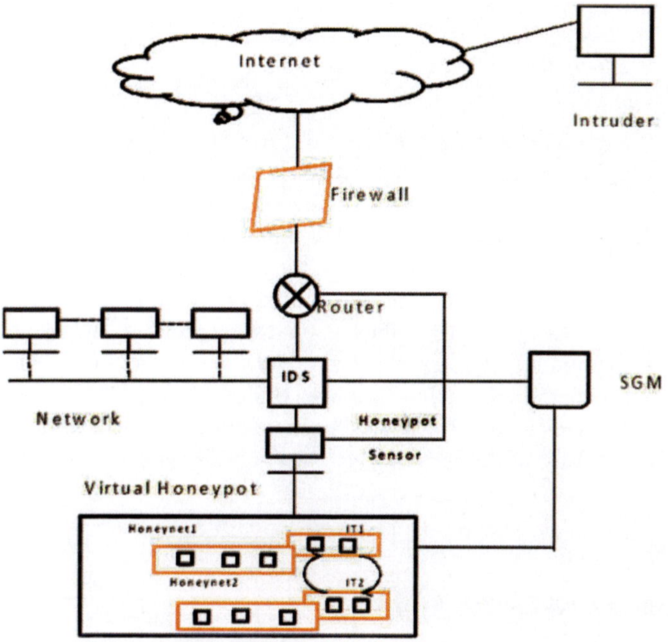

Fig. 1 System architecture of the system

in a network. The intention is to perform their normal activity is record all activities of the intruder without their prior knowledge. The information captured by the honeypot can be forwarded to the local firewall to generate a high-quality signature for unknown worms.

3.2 Data Control

Honeypot works at layer two bridges. That is why it is very difficult for the intruder to trace physical address, logical address, or routing path. These two bridges permit the intruder to come into the honeynet, but when tries to get off, its activity will be recorded and trapped [3].

Internal translators 1 and 2 are implemented in the router. Internal translator 1 intercepts all outbound connections and redirects them to the honeynet1. When honeynet1 is jeopardized, it will strive for an outbound connection because of the self-replication property of worm. And the honeynet2 also tries to make an outbound connection through internal translator 2. As per our threshold value, we set 5 connections. More than 5 connections will be automatically blocked. The advantages are, enough instances of malicious activity can be collected and the chance of DOS attack will also be minimized.

3.3 Data Capture

To generate signature only, single layer defense is not sufficient. For that, our proposed system contains a multilayer defense mechanism. The first layer we set firewall which records information about any unauthorized activity through source address, a destination address, port number, timestamp, etc. The second layer is a layer two bridges which is set in between the router and honeypot sensor. It can identify suspicious activities through its anomaly-based detection technique. The last layer is our honeypot system which captures the relevant data required for the signature generation of polymorphic worms. Honeypot unlocks the containing host port and waits for the connection to make. Honeypot listens to the port of both connection-oriented control protocol and connectionless protocol which also able to interact with a visitor of the network. For each activity, an event log is also created which can be used to capture malicious traffic.

4 Signature Generate Approach

The signature generating process depends on the conviction that several invariant consecutive strings exist in each possible variant of polymorphic worm [4, 5]. These strings are called tokens. The form of the signature is provided in Eq. (1).

$$F = \{(t_1, n_1), (t_2, n_2), (t_3, n_3) \ldots (t_k, n_k),\} \quad (1)$$

where

$$\left.\begin{aligned} t_i &= \text{token} \\ n_i &= \#\text{Occurence} \end{aligned}\right\} \forall i \in [1, k]$$

Any signature should equate with a flow (F) (as long as it consists of the minimum n_k copies of t_k as a substring).

Equation (2) presents the polymorphic worm token.

$$\begin{aligned} &[(\backslash\text{'XKK}\backslash\text{XBK'}, 1.0), (\text{'POST'}, 1.0), (\text{'}\backslash S\backslash N\text{'}, 5.0), \\ &(\text{'HTTP/1.2}\backslash S\backslash N\text{'}, 1.0), (\text{'}\backslash S\backslash N\text{ Host'}, 2.0)] \end{aligned} \quad (2)$$

The form of the signature is taken as: (token, number of occurrences).

Our proposed signature generation algorithm reproduces in Figs. 2 and 3, respectively.

Our approach extracts different tokens that account for λ fraction in fishy traffic. To avert all stings to be included as tokens, we set the minimum length as 2.

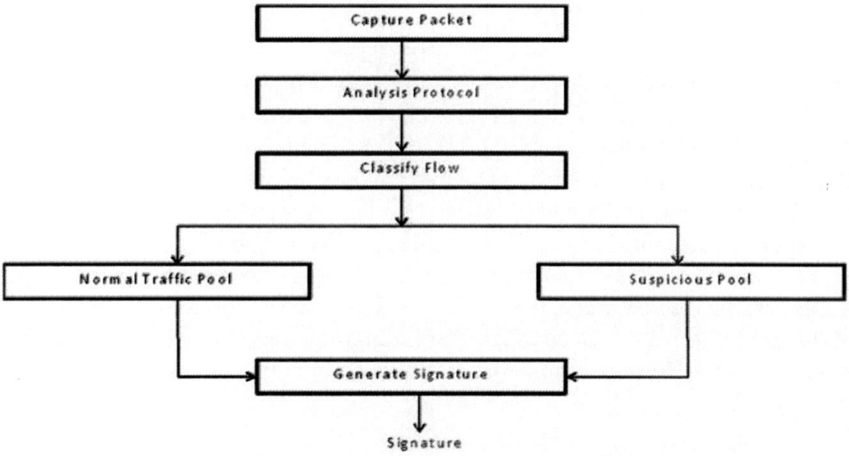

Fig. 2 Architecture of signature generation

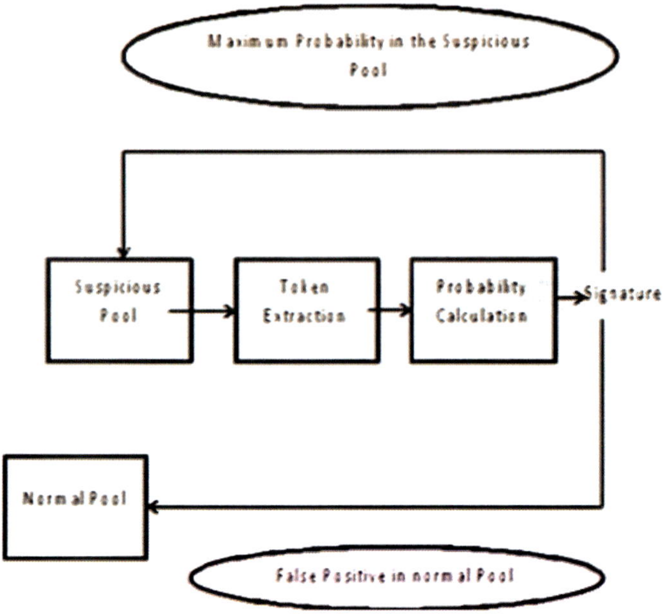

Fig. 3 Generation of signature

4.1 *Token Extraction Process*

Hamsa [6] utilized deep shallow sort [7] for the extraction of tokens. In comparison with polygraph [4], the suffix array-based algorithm extracts token at a much faster speed. We have used the suffix array based technique for our token extraction, as it performs a much faster way and enhances memory efficiency [4]. Here, we stored the combination in the form of: {(token), (number of occurrences)}.

Here, t_1 is a sub-token of t_2, if $t_1 \neq t_2$ and t_1 are also substring of t_2.

We transfer every sub-token into signature, if this satisfies the minimal length constraint and also coverage $> \lambda$.

We have shorted the token as per descending order according to the length. We nominate every {(token), (number of occurrences)} pair as a unique identifier.

4.2 *Probability Calculation for Signature*

Our system architecture is based on double honeypot system and we expect that major suspicious flow will be worm attack. Some inoffensive flow will take part in the honeypot system by default.

We assume the multi-set invariant in this following Eq. (3):

$$T = \left\{t_1^{\wedge}, t_2^{\wedge}, t_3^{\wedge}, \ldots, t_k^{\wedge}\right\} \tag{3}$$

Now, we determine the chances of token to appear in this sample, which gives the proper classification of worm, i.e., for every token t_i, we calculate probability and choose t like the first token with the highest probability in this following Eq. (4).

Hence,

$$\begin{aligned} &P_r\{t_i\} \geq P_r\{t_i\}\backslash - /i \\ &P_r\{t_1, t_2\} \geq P_r\{t_1, t_i\}\backslash - /i > 1 \\ &P_r\left\{t_1, t_2, \ldots, t_j\right\} \geq P_r\left\{t_1, t_2, \ldots, t_j - 1, t_i\right\}\backslash - /j\backslash - /i > j - 1 \end{aligned} \tag{4}$$

Herewith, the token t_1 with maximum probability present in this sample with all other tokens which is available in this list. t_1 associated with t_2 with maximum probability includes every pair of mentioned list. The proposed greedy approach will generate a signature (S), resulting in the probability where each token of this set S crossing a threshold probabilistic value in this following Eq. (5).

$$P_r(\mathrm{S}) \geq P_{\mathrm{rthreshold}} \tag{5}$$

Type 2 error (false positive value) is computed in a regular traffic signature pool is obligated by the mentioned condition. If the condition does not satisfy, then a new threshold value can be determined in Eq. (6).

$$\mathrm{EP}_s \leq \mathrm{FPmax} \tag{6}$$

5 Experiment and Result

We launch this experiment in our college laboratory local area network. We have installed *KFSensor* to build up a honeynet system is depicted in Sect. 3.1. *KFSensor* is fundamentally an emulated server. *KFSensor* servers listen to transmission control protocol and user datagram protocol ports and communicate with visitors (normal or malware traffic) and produce a signature. It is a graphical user interface which acts as security software and we also used antivirus and firewall. Our goal is to originate a signature for unknown polymorphic worm through the facts which are captured by the virtual honeypot.

We capture a port view and even view of *KFSensor* server system. Port view shows the current working port and event view displays all events and visitor list corresponding to the port (Figs. 4, 5, 6, 7, 8, and 9).

We used MySQL for recording activity log for analysis. Synthetically generated polymorphic worm we used in our experiment. BIND-TSIG [8] uses DNS protocol

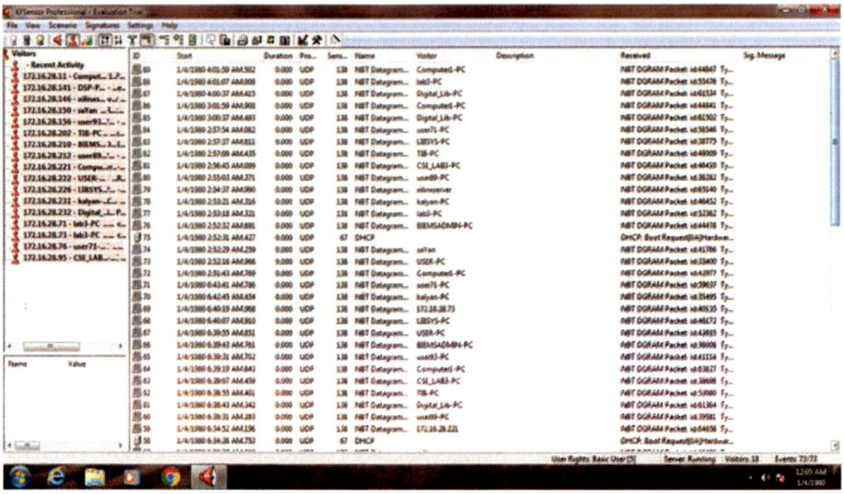

Fig. 4 Port view 1

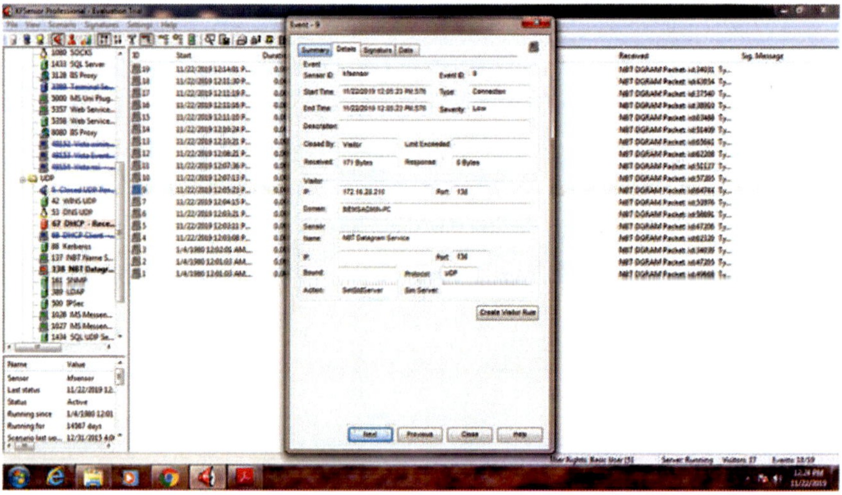

Fig. 5 Port view 2

and Apache-Knacker [9] to practice text-based HTTP protocol. In the interest of comparison here same polymorphic worm we are using, which is also used by polygraph [4]. In our experiment, we used two whole days of network traffic traces for DNS and HTTP protocol. We run this experiment in our college laboratory network which contains around 300 hosts and the speed of the network is moderate. Among all other traffic, we only count the polymorphic worm as suspicious traffic [2]. As worm has self-replication property, it tries to make an outbound connection and we

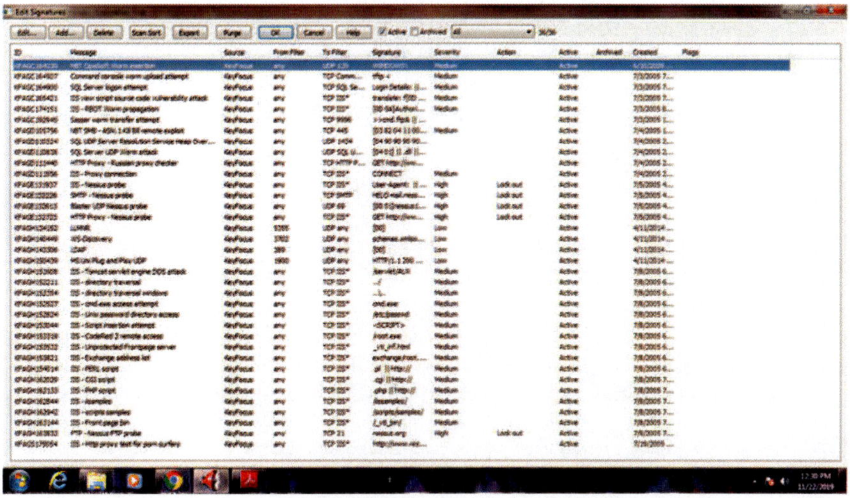

Fig. 6 Port view 3

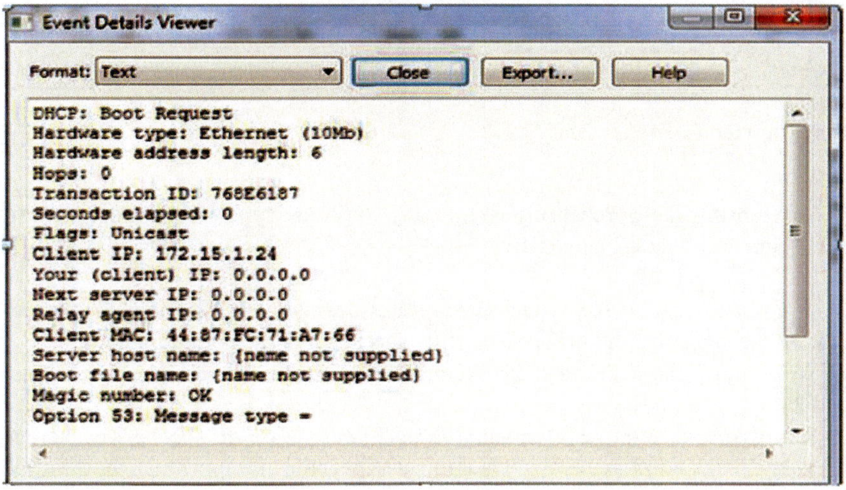

Fig. 7 Analysis of event

run a signature generation algorithm to the worm sample of a polymorphic worm. Our sample suspicious pool size was in the range between 3 and 40. A signature that is generated by using two different suspicious pools considered unreliable where type 2 error is around 100%.

We chose the minimal length for the token is $l_{\min} = 2$, each token is required to cover 15% of the suspicious pool, so the value of $\lambda = 0.15$.

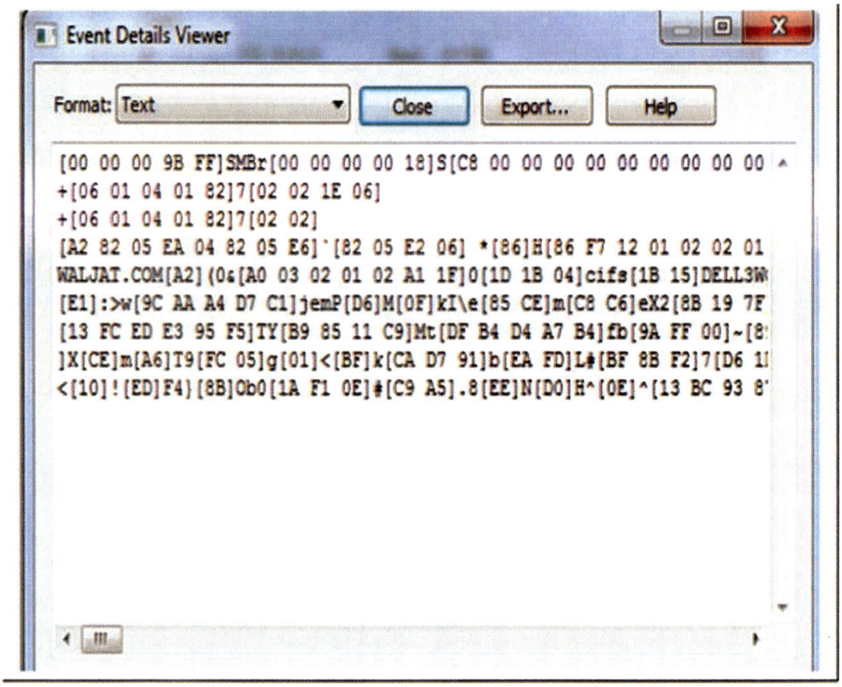

Fig. 8 Text format event

We grant threshold probability as $P_{r\text{threshold}} = 0.75$ and **type 2 error** (false positive) maximum adequate value = 0.01.

i.e., $\text{FP}_{\max} = 0.01$;

In normal traffic, if the **type 2 error** (false positive) is more than 1%, then we discard the signature and probability of occurrence in a suspicious pool <0.75.

We used Apache-Knacker and BIND-TSIG for an experiment is shown in Tables 1 and 2.

Our experimental probability value = 0.830 in the case of Apache-Knacker and the case of BIND-TSIG = 0.880. The value we are getting is higher than the threshold one.

The false alarm = 0.00201% for Apache-Knacker and false alarm = 0% for BIND-TSIG. We found the probabilistic signature, we are getting is as good as a polygraph signature [10]. The memory efficiency and the speed of the signature generation of this strategy are much higher due to the SA-IS-based array algorithm [4] like token extraction. Though the speed along with memory efficiency is methodically validated in [4], the experimental establishment is our future scope.

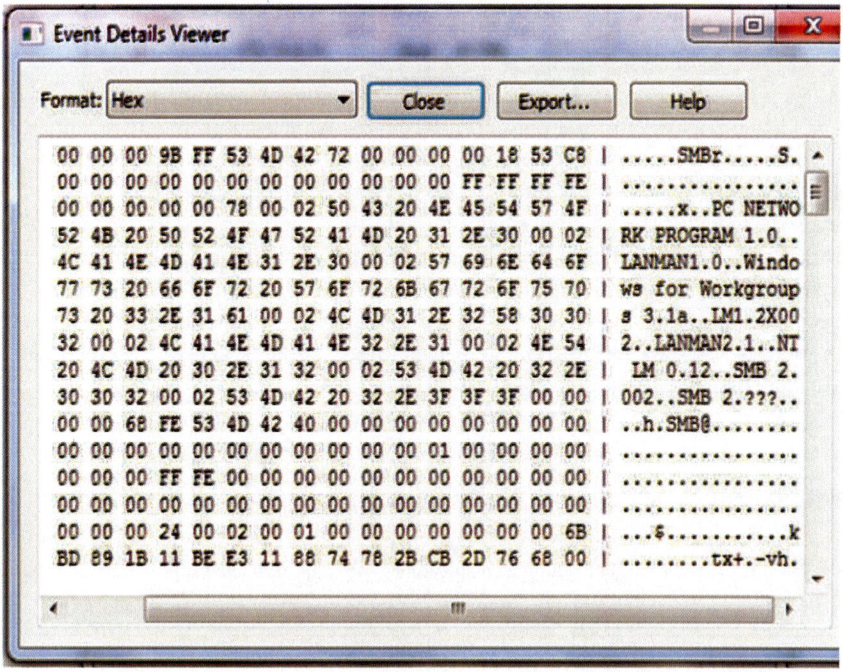

Fig. 9 Hex format event

Table 1 Apache-Knacker worm details result analysis

Category	False alarm (%)	Type 2 error (%)
Signature generate for polygraph	000.00241	0
Subsequent-token signature generate	0.00083	0
Probability wise signature generate	0.00201	0
[(\'XKK\XBK', 1.0), ('POST', 1.0), ('\S\N', 5.0), ('HTTP/1.2\S\N', 1.0), ('\S\N Host', 2.0)] ***Calculated probability = 0.830***		

Table 2 BIND-TSIG worm details result analysis

Category	False alarm (%)	Type 2 error (%)
Signature generate for polygraph	0	0
Subsequent-token signature generate	0	0
Probability wise signature generate	0	0
[(\'XKK\XBK'/'1.2\S\N'\XKK' ***Calculated probability = 0.880***		

6 Conclusion

In this research work, we have discussed the signature generation approach of the novel honeypot system for a polymorphic worm. We used a probabilistic approach to generate a signature [11]. Our proposed model captures suspicious traffic and analyzes them and forwards them to the signature generator. We generate a multi-sets byte sequence by using a greedy probabilistic approach and extract the token from sample attack and used them for detecting polymorphic worm. Here, we select **type 2 error** (false positive) value lower than the permitted value for signature. We recorded all the data through our own college network. We also did an event analysis for the attack and event log generated. This log also used to generate a high-quality efficient signature for polymorphic worm.

References

1. Tang Y, Chen S (2007) An automated signature-based approach against polymorphic Internet worms. In: IEEE transaction on parallel and distributed systems, pp 879–892
2. Tang Y, Chen S (2010) Defending against Internet worms: a signature-based approach. Department of Computer & Information Science & Engineering, University of Florida, Gainesville, FL, USA., pp 32611–6120
3. Spitzner L (2003) The Honeynet project: trapping the hackers. IEEE S&P, pp 15–23
4. Newsome J, Karp B, Song D (2005) Polygraph: automatically generating signatures for polymorphic worms. In: IEEE security and privacy symposium, 2005
5. Li Z, Sanghi M, Chen Y, Kao M, Chavez B (2006) Hamsa: fast signature generation for zero-day polymorphic worms with provable attack resilience. IEEE S&P, pp 33–47
6. Li Z, Sanghi M, Chen Y, Kao M, Chavez B (2006) Hamsa: fast signature generation for zero-day polymorphic worms with provable attack resilience. In: Proceedings IEEE S&P, pp 33–47
7. Manzini G, Ferragina P (2004) Engineering a lightweight suffix array construction algorithm. Algorithmica 40(1)
8. SANS Institute, Lion worm. Available: http://www.sans.org/y2k/lion.htm
9. C.CAN-2003-0245, Apache apr-psprintf memory corruption vulnerability. Available: http://www.securityfocus.com/bid/7723/discussion
10. Zhou J, Heckman M, Reynolds B, Carlson A, Bishop M (2007) Modelling network intrusion detection alerts for correlation. ACM Trans Inf Syst Secur (TISSEC) 10(1):1–31
11. Bio intrusion detection system. Int J Inf Secur Res (IJISR) 1(1/2). Available: http://www.bro-ids.org/ (2011)

Exploration of IoT-Based Real-Time Healthcare Monitoring System

Yajie Liu and R. Annie Uthra

Abstract In recent years, the development of Internet of things (IoT) has developed so fast, it has already applied in different areas such as industry, smart home, smart city, and so on. IoT-based healthcare is developed a lot among all these fields at present. By using Internet of things (IoT) technology, the health field has already updated and taken revolutions for the whole monitoring system in order to improve the efficiency and save cost. In this paper, we explore the different IOT-based technologies used in monitoring health systems, discuss the various technologies and benefits of these technologies. A simple IoT framework is developed in this paper to analyze the human health parameters. The health abnormality of the human is identified using classification techniques KNN and SVM. The results of KNN and SVM are compared in terms of accuracy, in which SVM outperforms KNN.

Keywords Internet of things (IoT) · Health monitoring · Machine learning

1 Introduction

Internet of things (IoT) is a technology that improves the quality of life by allowing all physical objects to communicate to each other. Apart from that, it allows to share some information that are useful without human intervention or less. Besides, IoT is defined as the Internet working of hardware devices that uses the standard communication protocol to provide services to the end-users [1]. In recent years, technologies developed so fast and applied in different areas, especially in healthcare field. Nowadays, human beings are much take care of the quality of life and their health conditions. Compared with traditional healthcare system, IoT-based healthcare

Y. Liu (✉) · R. Annie Uthra
Department of Computer Science and Engineering, SRM Institute of Science and Technology, Chennai, Tamil Nadu, India
e-mail: 2932028014@qq.com

R. Annie Uthra
e-mail: annieuthra@gmail.com

R. P. Mahapatra et al. (eds.), *Proceedings of 6th International Conference on Recent Trends in Computing*, Lecture Notes in Networks and Systems 177,
https://doi.org/10.1007/978-981-33-4501-0_27

system increased a lot in present days which improved the traditional healthcare system [2] in terms of monitoring, analysis, identification, storage, etc.

There is a rapid growth in the number of physical objects connected to the Internet that causes a lot of challenges in IOT field. As most of the devices relay on renewable energy and have small-sized batteries, the devices should be implemented by using low-power design. In present, IoT is widely used in different areas, such as industry, smart city, smart home, and so on. For example, one of the applications of IOT is to monitor the temperature and humidity in a special space, such as a mall, workshop, industry, etc., to maintain the suitable surrounding environment. It could give high accurate statics to determine ventilation in time. In recent years, there are numbers of wearable devices which are used for patients in order to monitor their health signal condition. Another application of IoT is to monitor the quality of air in the environment as air pollution is severe in these years and provide reliable suggestions to prevent the negative affect on the health conditions of human beings. The collected data from sensors and other IoT devices could be stored in cloud which is reliable for the concern authorities to assess the stored data. If some data values overpass the set threshold values, then the end-users will receive an alert message through SMS [3].

2 IoT for Healthcare

The IoT-based real-time health monitoring system mainly monitors the heartbeat, because cardiovascular disease (CVD) is the principal reason of death in the world. The common IoT-based monitoring system implemented a three-tier novel health monitoring architecture to detect arrhythmias [4] and implement a Web-based GUI to view the health status of patients. The health data which are collected from the patient's body will be transmitted to the IOT cloud through Wi-Fi which could be accessed from the IoT cloud, a Web-based GUI or phone app is developed to achieve that. Real-time healthcare monitoring system is required in order to carrying out immediate actions in case of unhealthy situations. Therefore, the system could use any cloud service like Alibaba cloud and ThingSpeak cloud as the IoT platform. It also could use some portable API monitoring devices. Arduino microcontroller is preferred to be used with its open-source flexibility and integrated with Wi-Fi module as its network communication medium. The IOT concept is leveraged using serial communication technology to transfer the sensor data to monitor the parameters remotely. Whenever the measured values exceed the limits, an SMS will be sent to the end user to alert the possible environmental health effect [5].

LoRa wireless network and Mysignals hardware [6] are used along with sensors like body temperature sensor, pulse rate sensor, ECG sensor, and oxygen saturation sensor. In this framework, MySignals platform is used to gather the sensor data and transmitted the sensor data to the personal laptop through LoRa wireless network. The results of this proposed system displayed correctly and the MySignals platform can be successfully interacted with other sensors.

The authors proposed a health monitoring system [4] with three-tier Internet of things (IoT) for heartbeat monitoring to determine the cardiovascular disease. The method used AD 8232 ECG sensor and STM32F429 microcontroller as a gateway for monitoring body signals such as: temperature, ECG, and blood pressure. Besides the data could be stored in cloud for future processing.

The authors presented [7] a DIY (do it yourself) solution for a adaptable and sustainable patient-oriented infrastructure development. A body sensor network (BSN) is developed to monitor the patients' physical and physiological situations and transfer sensed-data to the centralized location through wireless or wired network. It used Raspberry Pi (RPi) as the microcontroller platform which is low-power, portable, and affordable device for the patients. Besides, the authors addressed the security of healthcare information and listed several methods to ensure the quality and availability.

In [8], authors proposed a system of healthcare based on Internet of things (IoT), and this system could monitor the patients' body condition and display the data through Internet. If any data exceeded the initialized threshold, the system could also automatically send these critical conditions to not only doctors but also family member in order to detect any diseases in the early stage and take medical treatment or other actions. In [9], authors presented an IoT security risk model (ISRM) to alleviate the risk of IoT used in healthcare. In this paper, it not only discussed the several IoT technologies used in healthcare field, but also gave the current challenges especially in security area for IoT applied in healthcare aspect. The ISRM contains five different layers, from inner to outside is authentication, encryption, secure boot and IPS, firewall, and education and policies. By using this proposed risk management model, it can help to achieve the expected standard in IoT security risk. The proposed ISRM model [8] is displayed in Fig. 1.

In [10], authors proposed a system in which wireless implantable sensor can be imbedded in people's body. This kind of sensor has subcutaneous solar energy harvesting module designed together with the body sensor. Considering the life of the battery, the main aim is to test this sensor to identify which part of our body can harvest the solar energy effectively to keep the long-time running in people's body. The result of this paper found that the best position of our body to implanted the sensor is between shoulder and neck. In [11], authors introduced the positive aspect of backscatter communications (BackCom). In this paper, it not only introduced the basic principles techniques but also the application of next-generation BackCom. The main advantage of BackCom is that the users do not worry about the battery life of the sensor, so such sensor could be deployed in any places like hazardous environment so that no sensors need to recharge or replace. It could be used in spread fields.

In [12], authors design and implement a system by using the International standard (ISO/IEEE 11073 PHD and CoAP) to reduce the data loss-rate and improve the interoperability among different devices. In [13], authors investigated different classification algorithms' performance based on liver disease. The tested classification methods contain logistic regression, decision tree, k-nearest neighbor algorithm (KNN), and support vector machine (SVM) algorithm. The result showed that various machine leaning techniques could help to detect and find the liver disease in

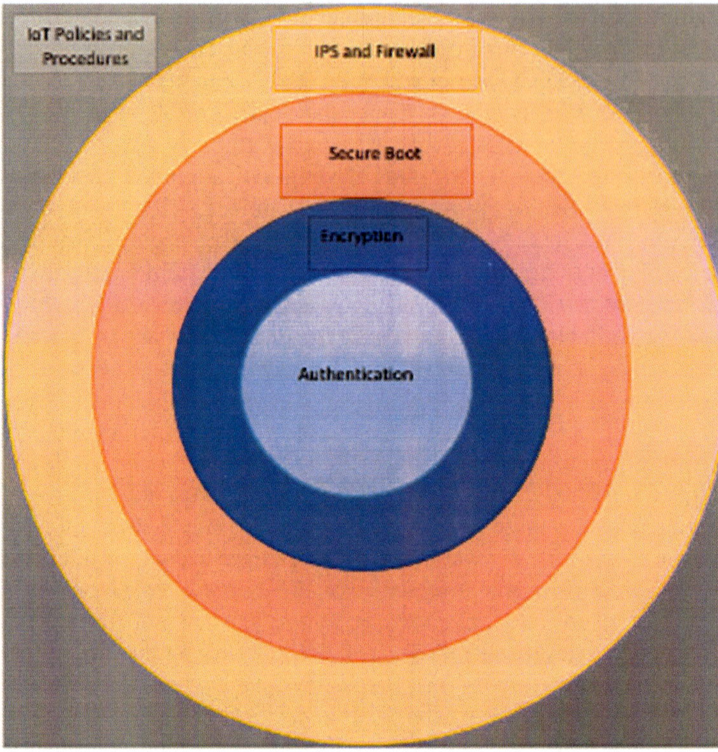

Fig. 1 IoT in healthcare security risks management propose model references

the early stage. In [14], authors proposed a system to detect diabetic foot ulcer based on Internet of things. In this system, it used the open-source hardware, like Node MCU, the flexi-force sensor, SD58C pulse sensor and the body temperature sensor. This system could monitor the patient's body signal and transmit the sensed-data to the cloud. If any abnormality data is detected, the system will give immediate alert to the patient to pay attention his or her body.

In [15], authors introduced the role of big data analysis in healthcare field. Besides, it presented different techniques of big data used in healthcare. It discussed the importance of applicability in healthcare. Furthermore, machine learning technique is applied in diabetic patient data analysis based on Internet of things (IoT). The result of this paper showed that using big data analysis for dealing with a large amount of data to get useful results to help people to do decision-making process. In [16], authors proposed a system of monitoring the sleep status by detecting Obtrusive Sleep Apnea (OSA) disorders. In this system, it not only directly monitors the patients' body condition by checking several parameters: sleep status, sleep environment, physiological, and physical activities but also gave the corresponding treatment by using big data techniques to process the monitored data on cloud computing based on the open data available in smart cities. The result of this paper showed that using big data techniques advance to guide the OSA treatment in detecting abnormity in the early stage.

3 IoT-Based Healthcare Data Analysis

In this paper, we presented the system framework that couples IoT technology with machine learning (ML) technology after researching lots of recently published articles and considering the reality of it. Machine learning is used to make the system much smarter and help to make decisions. Figure 2 depicts the proposed architecture diagram of the IoT-based healthcare monitoring system couples with machine learning techniques.

3.1 Hardware Setup of the Proposed System

The system aims to gather the important body signal like temperature, electrocardiogram (ECG), and blood pressure and assess the human health state. The proposed system uses LM35 body temperature sensor, ECG sensor, and blood pressure sensor. LM35 sensor is used to collect the temperature of human body which is easy to attach to body and has high accuracy. MKB0805 sensor is a integrated and stabilized module which could be used for the human body heart rate and blood pressure signal acquisition. Blood pressure includes the systolic pressure and diastolic pressure. Apart from that, MKB0805 sensor provides automatic optimization function. The proposed system uses the IoT data which is given in Table 1.

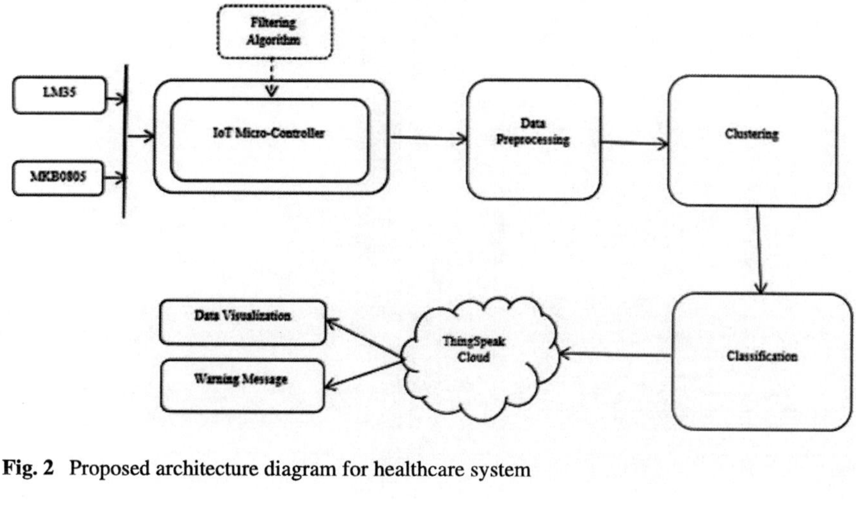

Fig. 2 Proposed architecture diagram for healthcare system

Table 1 IoT data

No.		Parameters
1		Human body temperature
2		Heart rate
3		Systolic pressure
4		Diastolic pressure

In the proposed system, the microcontroller used is Arduino UNO and ESP8266 Node MCU Wi-Fi (Wireless Fidelity) controller. The sensor data are generated through Arduino UNO microcontroller and it will continue to send the sensor data to Node MCU Wi-Fi microcontroller after eliminating the record of invalid data generated, including incomplete data and outlier data through filtering algorithm. The sensor data are sent from Arduino to Node MCU WiFi module by generating JSON (JavaScript Object Notation) document with ArduinoJson version 6. JSON document encapsulates the sensor data and writes it to the queue in JSON format. There are two main functions used in this procedure: serializeJson() and deserializeJson(). Node MCU is used as a gateway to transfer the data to the ThingSpeak cloud. Before storing the data on to the cloud, the serial data are pre-processed by data cleaning to fill the missing values and data transformation to select the used features. The real-time hardware developed to fetch the parameters like human body temperature, heartbeat, systolic pressure, and diastolic pressure. The real hardware developed is shown in Fig. 3.

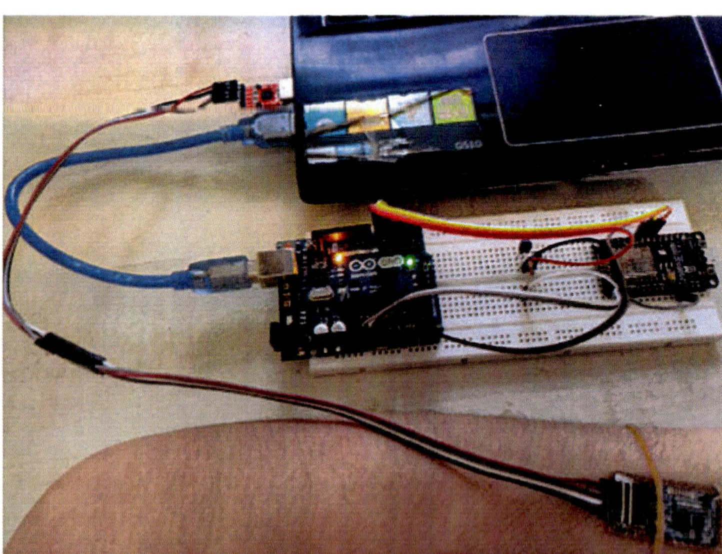

Fig. 3 Hardware setup of proposed system

3.2 Healthcare Data Analysis

Besides, machine learning algorithms are applied to detect the human health state. The proposed framework used k-means algorithm to perform clustering, and k-nearest neighbors (KNN) algorithm for classification, classification is a typical algorithm in machine learning field, it segments the big volumes of data into several groups and each group represents the same class. Using K-means the data set is classified into two clusters namely normal and abnormal health state and the data set is trained using KNN for health state prediction. KNN method is widely used in classification to classify the new data through voting by the given nearest data. Then the set of data are pushed to the cloud, the proposed system uses ThingSpeak cloud. When the data is classified as abnormal, warning message is sent to the phone as soon as the data is pushing to the cloud which could be used for future analysis.

In order to assess the collected sensor data set is normal or not, the proposed system uses some machine learning algorithms such as clustering techniques and classification techniques. Initially, k-means algorithm is used for the clustering as the data set is not labeled. After clustering, we would get four different clusters, i.e., the abnormal data is further classified into abnormal heart rate(abnormal-HR), abnormal blood pressure (Abnormal-BP), abnormal temperature (Abnormal-T). The result of this method is that each row of data is classified as Normal, Abnormal-HR, Abnormal-BP, and Abnormal-T which is represented with labels: 1, 2, 3, and 4, respectively, as shown in Table 2.

K-nearest neighbors (KNN) algorithm is used on the labeled data set for the classification. KNN is a widely used basic machine learning algorithm in the world. It is an efficient for a small dimensional dataset. The result of classification is around 95 percent accuracy rate in this experiment. The once result of classification compared with class label of the clustered dataset as shown in Fig. 4.

Table 2 Types of class label

Class label		Represent No.
Normal		1
Abnormal-HR		2
Abnormal-BP		3
Abnormal-T		4

Fig. 4 Confusion matrix of classification

```
               model
data_test_label  1  2  3  4
              1 58  0  3  0
              2  0 62  2  0
              3  2  0 66  0
              4  0  0  3 44
```

The proposed system also used support vector machine (SVM) to classify the data set. As the data set is linearly separable SVM out performs KNN. The results of these two methods are also compared.

4 Results and Discussion

The performance of the proposed system is working successfully. In this paper, we analyzed the collected information from human body like heartbeat, temperature, and blood pressure with the help of sensors and other hardware devices. The sensors we used are easier to attach to human body and has higher stability & accuracy rate compared with previous experiments. This system makes a meaningful contribution compared with current health monitoring system in case of its low-cost of construction, installation, and easy access to monitoring sensor data in real time through the ThingSpeak Web application. We stored all the monitored results in the cloud through network. Besides, we applied machine learning algorithms like clustering and classification techniques on the data which are stored in the cloud. By using these algorithms, we are able to find abnormal behaviors in heartbeat, blood pressure. With the help of this, we are able to detect abnormal condition in the early stage which helps patients to be more cautious about their health status. In future work, security concerns are to be taken care about the data that is stored in cloud. The results of one week monitored data of a patient is showed as Fig. 5.

	Morning	Evening	Morning	Evening	Morning	Evening	Morning	Evening	Morning	Evening	Morning	Evening	Morning	Evening
systolic	128	132	141	142	136	140	142	138	136	136	138	138	136	142
diastolic	78	79	88	86	83	83	85	83	80	82	84	86	82	86
heart rate	71	75	72	70	72	72	80	70	76	71	70	76	70	73
temperature	28	33	35	36	42	43	36	36	33	36	42	40	37	42

(a)

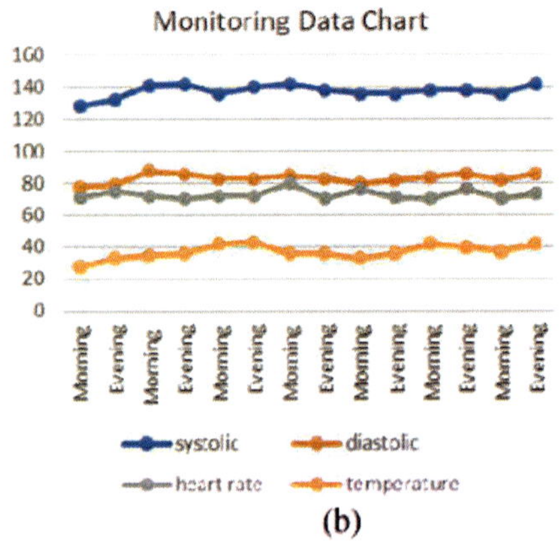

(b)

Fig. 5 One week patient data: **a** data set, **b** data chart

In the machine leaning part, we set different k values in the k-nearest neighbors (KNN) algorithm, the accuracy rate for this method is also different, the accuracy rate corresponding to each k value is depicted in Fig. 6. From this diagram, we could see that when k equals to 5, it has the highest accuracy rate.

In addition, for the classification part, we used another algorithm called support vector machine (SVM). We compared the accuracy rate between KNN and SVM methods. The result shows that SVM algorithm has higher accuracy rate compared to KNN algorithm. The result of comparison between SVM and KNN is depicted as Fig. 7.

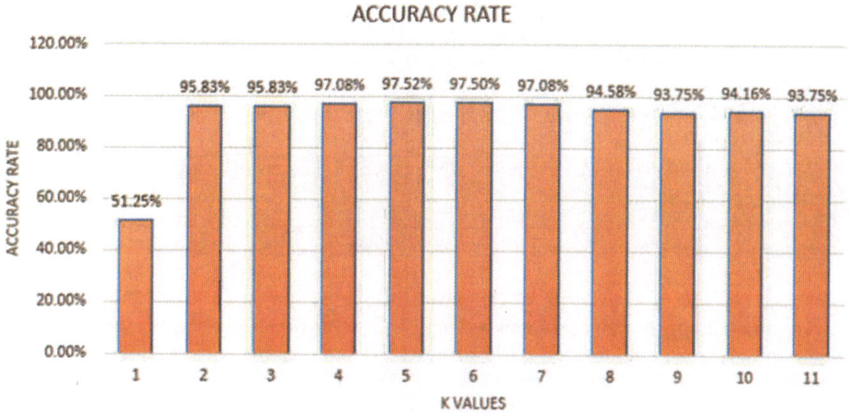

Fig. 6 Accuracy rate corresponding to each *k* value

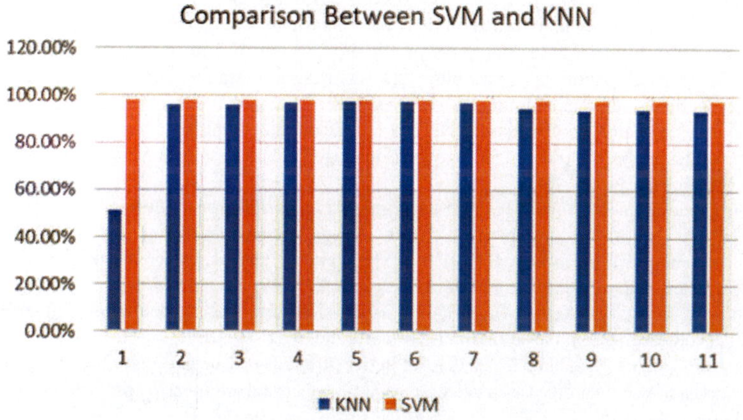

Fig. 7 Comparison between SVM and KNN algorithms

5 Conclusion

In this paper, we proposed a system of healthcare monitoring system couples with machine learning algorithms like clustering and classification techniques based on Internet of things (IoT). This proposed system is affordable and portable for the patients at home, Arduino Uno as the microcontroller could collect the sensor data and send the monitored data to the ThingSpeak cloud through Node MCU WiFi module. The machine learning algorithms such as clustering technique and classification technique are applied on the data stored on cloud. When the system detects abnormal data, it will send alert message to the patients or family members.

References

1. Chaudhuri A (2018) Internet of Things and its potential. Internet of Things, for Things, and by Things, pp 3–16
2. Agana MA, Ofem OA, Ele BI (2019) A framework for a fuzzy smart home IoT e-Health support system. Lecture Notes in Networks and Systems Advances in Information and Communication, pp 432–447
3. Wan J, Al-Awlaqi MAAH, Li M, O'Grady M, Gu X, Wang J, Cao N (2018) Wearable IoT enabled real-time health monitoring system. EURASIP J Wirel Commun Network 1:2018
4. Deepika N, Anand M, Jerald F (2019) A novel three tier internet of things health monitoring system. Indones J Electr Eng Comput Sci 15, 631–637
5. "Health Monitoring System Using Internet Of Things," Int J Eng Res Adv Technol, 2018
6. Islam MS, Islam MT, Almutairi AF, Beng GK, Misran N, Amin N (2019) Monitoring of the human body signal through the Internet of Things (IoT) based LoRa wireless network system. Appl Sci 9(9):1884
7. Maksimovic M, Vujovic V, Perisic B (2015) A custom Internet of Things healthcare system. In: 2015 10th Iberian conference on information systems and technologies (CISTI)
8. Ahmed ZU, Mortuza MG, Uddin MJ, Kabir MH, Mahiuddin M, Hoque MJ (2018) Internet of Things based patient health monitoring system using wearable biomedical device. In: 2018 international conference on innovation in engineering and technology (ICIET)
9. Bakar NAA, Ramli WMW, Hassan NH (2019) The internet of things in healthcare: an overview, challenges and model plan for security risks management process. Indones J Electr Eng Comput Sci 15(1):414
10. Wu T, Redoute J-M, Yuce MR (2018) A wireless implantable sensor design with subcutaneous energy harvesting for long-term IoT healthcare applications. IEEE Access 6:35801–35808
11. Liu W, Huang K, Zhou X et al (2019) Next generation backscatter communication: systems, techniques, and applications. J Wirel Com Netw 2019:69
12. Ge S-Y, Chun S-M, Kim H-S, Park J-T (2016) Design and implementation of interoperable IoT healthcare system based on international standards
13. Singh P, Chahal D, Kharb L (2019) Predictive strength of selected classification algorithms for diagnosis of liver disease. In: Lecture notes in electrical engineering proceedings of ICRIC 2019, pp 239–255
14. Gupta P, Pandey A, Akshita P, Sharma A (2019) IoT based healthcare kit for diabetic foot ulcer. In: Lecture notes in electrical engineering proceedings of ICRIC 2019, pp 15–22
15. Rghioui A, Lloret J, Oumnad A (2020) Big Data classification and Internet of Things in healthcare. Int J E-Health Med Commun 11(2):20–37
16. Yacchirema DC, Sarabia-Jacome D, Palau CE, Esteve M (2018) A smart system for sleep monitoring by integrating IoT with big data analytics. IEEE Access 6:35988–36001

REA-RP: Residual Energy Aware Routing Protocol for Wireless Sensor Network

Arunita Kundaliya and D. K. Lobiyal

Abstract In a wireless sensor network (WSN), energy constraint on sensor nodes is primarily due to limited supply of power from the battery. Major share of battery power is utilized in communication within the network. Thus, introducing a method that supports the efficient usage of the available energy in each node is the fundamental design issue. An on demand residual energy aware routing protocol (REA-RP) is proposed to work at the network layer. The protocol targets improvement in the network's lifetime by maximizing lifetime of the sensor nodes in the network. To enhance lifetime of the network, it is important to balance out the energy consumption of the nodes along the routing path. Usage of nodes with lower residual energy should be minimized. In the protocol, the routing decisions are facilitated by taking into account a node's residual energy and the hop distance. The REA-RP protocol's performance is compared with another on demand routing protocol, AODV in NS3 simulator. The analysis of simulation results in terms of metrics such as throughput, network lifetime, routing overhead indicate that REA-RP has performed better than AODV.

Keywords Networks lifetime · REA-RP · Residual energy · WSN

1 Introduction

Wireless sensor network (WSN) consisting of tiny sensor nodes is equipped with a limited energy source in the form of battery. The nodes are capable of sensing, processing and communicating data. Largely, the nodes in WSN are deployed in control or monitoring application where replacing or recharging of batteries is neither warranted nor always possible. Once nodes are deployed for operation, it is expected

A. Kundaliya (✉) · D. K. Lobiyal
School of Computer and System Sciences, Jawaharlal Nehru University, New Delhi 110067, India
e-mail: arunita@yahoo.com

D. K. Lobiyal
e-mail: lobiyal@gmail.com

R. P. Mahapatra et al. (eds.), *Proceedings of 6th International Conference on Recent Trends in Computing*, Lecture Notes in Networks and Systems 177,
https://doi.org/10.1007/978-981-33-4501-0_28

that they last longer or remain operational for a considerable amount of time interval. The information gathered by the sensor nodes is disseminated between the communicating nodes, that is, source and the sink. The sink may be many hops away from the source. Two main factors that affect consumption of energy in the node are:

- The transceiver state- Energy consumption is high during transmission of data, reception of data, or when the node is idle (or overhearing), and it is low when the node is sleeping.
- Events- Events associated with successful transmission, forwarding, collision or retransmission of data packets and control packets transmission, results in consumption of energy in a node.

In case of multi-hop networks, the node's sensor spent their energy largely in communication of data/control packets. Only those nodes having enough energy can participate in data communication, i.e., either for forwarding, transmission or reception of packets from other nodes. This make energy-efficient methods at network layer, an imperative component for energy saving in the wireless sensor network. The primary design objective in a WSN is reduction in the consumption of energy in every node to the minimum and also to improvise certain network performance parameters such as network lifetime, delay, protocol overhead, throughput, etc. Uneven energy consumption, load balancing, and maintaining or improving network parameters are challenges in WSN. Uneven usage of energy in the network may cause fast depletion of energy level in some nodes in the network, resulting in a problem called energy hole problem. This problem may cause network partitioning, which hinders performance of the network. Many energy-based strategies such as radio communication and duty cycling at physical layer, energy aware MAC and routing protocols [2, 4, 5, 11, 13, 14] were suggested in the past to minimize the power consumption in sensor nodes thus, enhancing lifetime of the network. Here, in the proposed routing protocol, a node uses information about the residual or remaining energy of its neighboring nodes as decision criteria for routing so that, traffic can be rerouted through nodes having higher residual energy. Nodes with lower residual energy level can thus preserve their energy for usage in future. In this way, the proposed scheme tries to carry out balance in the consumption of the energy in the network.

2 Related Work

Energy is a limited and crucial resource in a WSN as everything, be it sensing or processing or communication, can happen only when node has energy. An important question arises- how a multi-hop path between the source and the sink node, which is energy-efficient too can be found for communication? Different metrics for routing are often used by the routing protocols in improvement of energy efficiency. Guojun et al. [1] method enhances the lifetime of wireless sensor networks by balancing the energy usage of the nodes through proposed a transmission power control method. The energy-harvesting state and the extent of remaining energy in a sensor node are utilized by the method to select the next hop. The energy-harvesting speed and the remaining energy level both are used to classify transmission power into various levels. Priority is given to the nodes with high energy level and are fast charging over the uncharged nodes. Liping Lv et al. [2] developed a multi-path method for routing which is aware of the residual energy of nodes. For main path selection, algorithm uses hop count and the residual energy of the next hop nodes. The algorithm avoids nodes with low energy level to balance the consumption of energy in a network, and enhance the network life time. In case of any break in the main path, a new data transmission path can be preferred by the source node which is unrelated to the main path. Here, in [3], selection of next cluster head in the network meant for an IoT application, is based on the initial energy and residual energy of the node along with an optimum value of cluster heads. In [6], the authors proposed a method which pairs a transmitter with lowest energy consumption to a receiver with highest remaining energy to decide the relay node. In [7], the routing protocol uses performance metric Link Quality Indicator (LQI) and the residual energy of node to select next hop node. In [8], remaining energy (RE) of nodes have been used for deciding next-hop in routing protocol developed for "Low Power and Lossy Network". The RE is estimated through calculation of current energy consumption in each state of node and duration for which that state is maintained. Energy-efficient routing protocols proposed in [9, 12] localized decisions are taken by sensor nodes for packet forwarding, which are based on the two communicating nodes' distance, and their residual energy. In [15], demonstrated routing method uses a combination of Fuzzy approach and A-star algorithm to enhance network's lifetime. The optimum routing path between source and sink is the path which has the maximum remaining energy, least number of hops and lower traffic loads. In [16], proposed energy aware routing protocol uses a tree for sleep scheduling of sensor nodes. The sink node is used as root node of the tree. While the leaf nodes of the tree are made to sleep, the non-leaf nodes remain awake so that a continued path between any nodes in the tree to the sink node can be provided. Sink node is selected as the root node. The remaining energy of the nodes in tree is used to periodically reconstruct the tree keeping in mind the balancing of energy usage of nodes in the network. The reconstruction of tree deletes the non-functional nodes from the tree. In [17], authors proposes an energy aware routing protocol based on ant colony optimization method. The algorithm selects the next hop by taking into account the residual energy of the next hop, sink node's distance from the current node and the average energy of the path. In [18], the algorithm used for determining next-hop is based on energy density and weight of link. The link's weight is calculated on the basis of nodes' energy, length of link and traffic load on link. In [19], the proposed routing protocol selects optimum energy routes based on the current energy level of nodes along the routing path and the cost of total transmission energy. In multi-hop transmission, path selection is done by selecting nodes with higher remaining energy than a given threshold. It uses the minimum transmission power in the network. Idea is balancing the consumption of transmission energy in the network by skipping nodes with depleted energy along the shortest routing path.

3 REA-RP Protocol for Routing

In the proposed protocol, the multi-hop routing path between sink and source node is laid only when the source node generates a packet for transmission. This is also the unique feature of AODV protocol, which has been used in the proposed protocol. The proposed protocol aims at enhancing the throughput and the network's lifetime by selecting the most suitable node, as next hop at each forwarding stage. To enhance the lifetime of sensor nodes in the wireless network, sensor nodes with lower residual energy were used as less as possible. The protocol balances consumption of energy in sensor nodes by distributing data traffic on alternative routes, depending on nodes' residual energy, provided an alternative route exists. When routes are being established, if more than one node has an equal amount of residual energy, then the node which has the shortest path to the sink is selected as next hop. The relationship between the residual energy, initial energy and energy spent by sensor node in monitoring and communication is given by the following equation:

$$\mathrm{RE}_i = \mathrm{IE}_i - \mathrm{ECM}_i$$

where
RE_i represents the *i*th node's residual energy.
$\mathrm{IE}i_l$ represents *i*th node's initial energy.
ECM_i represents energy spent on communication and monitoring by *i*th node.

3.1 Working of Protocol

- To allow flow of data, control packets (ROUTE REQUEST and ROUTE REPLY) are exchanged between the nodes. Before transmission of data packet, a control packet ROUTE REQUEST is sent to the receiver and an acknowledgement to route request in the form of control packet ROUTE REPLY is sent from receiver to the sender. The structure of ROUTE REQUEST control packet is same as that of AODV ROUTE REQUEST control packet. The structure of ROUTE REPLY control packet used in the proposed protocol is described in Fig. 1.
- The request for route is broadcasted by a node in the network, whenever it has a packet ready to transmit.
- Upon receipt of a route request, it is again rebroadcasted by the node, if the receiving node is not the sink node.
- The sink node, on reception of the route request, generates a route reply meant for the packet's originator node, i.e., source. The route reply is relayed back to the originator, where each intermediate node on the path sends its residual energy too in the route reply.

(Size in bytes) 1	1	1	1
Type	Flags	Size of prefix	Hop count
IP Address of Sink Node			
Sequence Number of Sink Node			
IP address of packet Source node			
Residual energy of node			
Packet lifetime			

Fig. 1 Structure of ROUTE REPLY control packet

- On reception of the route reply by the intermediate nodes, routing path is established to the sink node by choosing as next hop, a sensor node having maximum remaining energy.

4 Simulation Results

NS3 (ver 3.27), a free and open source, discrete event simulator, is selected for studying and comparing the performance of REA-RP protocol with AODV protocol. The comparative analysis of the two routing protocols is done on the following performance metrics:

- *Throughput*: It is total number of successfully received packets per sec at the sink.
- *Routing overhead*: It is total number of control packets transmitted in the network per second.
- *End-to-end delay*: It is time taken by a packet to reach from source to destination, across a network.
- *Network lifetime*: It is calculated as maximum time period within which the sensor node are functional to monitor an event.

Table 1 lists the parameters along with their corresponding values, which are used for simulation of the protocols.

To study the proposed protocol, we simulated the network consisting of variable number of sensor nodes such as 25, 50, 75, and 100. Distribution of the sensor nodes in the area selected for simulation is random. Node's position coordinates are randomly generated by a uniform random variable generator and the node's position does not change throughout the simulation period. In the network, there are ten source–sink pair. Each source node generates at the rate of 2048 bps, a packet of size 64 bytes. The packets both control and data, are transmitted at 4.5 dBm. The proposed protocol is an energy aware protocol, i.e., operation of the protocol is dependent on the current energy level of the node. This requirement necessitates that each node must have a source of energy, which represents node's power supply. The "BasicEnergySource" and Wifi Radio Energy Model of NS3 [20] are used as the energy source and device energy model respectively in the simulation. More

Table 1 Simulation parameters

Operating system	Ubuntu 16.04
Total simulation time (s)	250
Simulation area (m)	1025 × 500
Number of nodes	25, 50, 75, 100
Number of source-sink pair	10
Mobility model	Constant position
Data packet size (bytes)	64
Data rate (bps)	2048
MAC protocol	802.11b
Radio propagation loss	Two ray ground model
Topology	Random
Power levels (dBm)	4.5
Initial node power (J)	100
Transport protocol	UDP

than one energy source and more than one device energy models could be installed on a node. To draw power from the source, the device energy model is connected to the energy source in the node. The "BasicEnergySource" model class provides functions which keeps track of remaining or residual energy, decrease in the residual energy, and notifies the device energy model when energy is completely depleted in the node. The Wifi radio devices have energy model defined in Wifi Radio Energy Model class. The Wifi radio devices can be in any one of the states such as idle, busy, transmitting, receiving, and switching all of whom are described in the WifiPhy class of NS3. The WiFi PHY class (YansWifiPhy) of NS3 is fully supported by WifiRadioEnergyModel class. To compare energy efficiency of the two protocols, each node in the network is initialized with energy of 100 J. In the simulation, to monitor and report back the packet flows, the "FlowMonitor" model [20, 21] of NS3 is used. The FlowMonitor class gathers end-to-end flow statistics and coordinates efforts of the flow probes. The flow stats are stored in data structures kept in memory. These stats can be accessed by user via "getter" functions defined in the class named FlowMonitor. The flow stats between source–sink pair are collected in two separate data structures in NS3, namely FlowMonitor::FlowStats and FlowProbe::FlowStats. The FlowProbe::FlowStats collects flow stats for probes that are specifically created in the network nodes, so that a record of the packets which are exchanged between the nodes can be maintained. Packets are categorized depending on their association with the flow. Each flow in the network can be described by the probe's characteristics for example, an IP flow is described by those packets who use same protocol, IP address of source, source port address, IP of sink, and sink port address. For each flow, the collected statistics are exported in a XML format. The probes can be accessed directly by the user in order to access specific stats about each flow. Each flow probe classifies the packets at four points- that is, when a packet is sent by the source, forwarded by the intermediate nodes and received or dropped by the destination node. From the class FlowMonitor::FlowStats, following attributes has been used to calculate performance metrics:

- "timeFirstRxPacket": It is the time when the first packet is received by the sink node, in a flow [21].
- "timeLastRxPacket": It is the time when the last packet is received by the sink node, in a flow [21].
- "delaySum": It is sum of end-to-end delays, for all received packets which belong to the same flow [21].
- "rxPackets": It is the sum of all the received packets of the sink node in a flow [21].

$$\text{Average End to end delay} = \text{delaySum}/\text{rxPackets}$$

$$\begin{aligned}&\text{Average Throughput(kbps)}\\&\quad = 8 * \text{rxBytes}/(\text{timeLastRxPacket} - \text{timeFirstRxPacket})/1024\end{aligned}$$

The simulation is run for a total of 250 s. For each performance metric, the simulation is repeated with five different seed values of the uniform random variable generator. An average of the values obtained for each seed in the simulation is used to arrive at the final value of the metric. A comparison in the performance of the two protocolg REA-RP with AODV is shown in Fig. 2.

5 Conclusion

In this paper routing protocol, REA-RP, is introduced which primarily depends on the residual energy of nodes in a network. The residual energy of sensor nodes is used to establish the active routes in the network. The results of simulation show that the average throughput of REA-RP protocol is significantly higher than the AODV protocol. It is also observed that, in case of AODV, as the node density varies from 25 to 100 the average end-to-end delay (measured in seconds) varies largely from a higher delay value to a lower delay value or from a lower delay value to a higher delay value. In case of REA-RP, as the node density varies from 25 to 100, the average end-to-end delay first increases steadily to a point. This value is still lower than the highest delay value in AODV and then decreases. The routing overhead is lower in REA-RP and the network lifetime has increased in REA-RP when compared with AODV. The simulation results indicate that network traffic distribution based on residual energy, enhances network's lifetime.

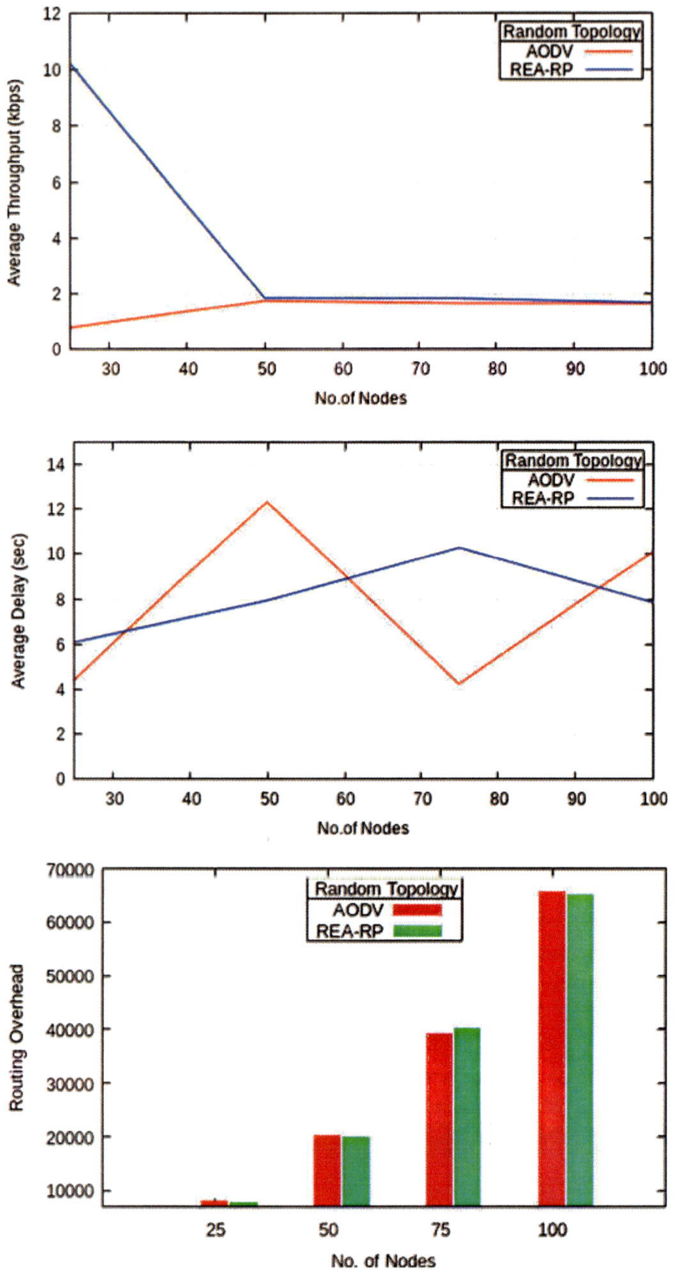

Fig. 2 Average throughput, end-to-end delay, routing overhead, network lifetime

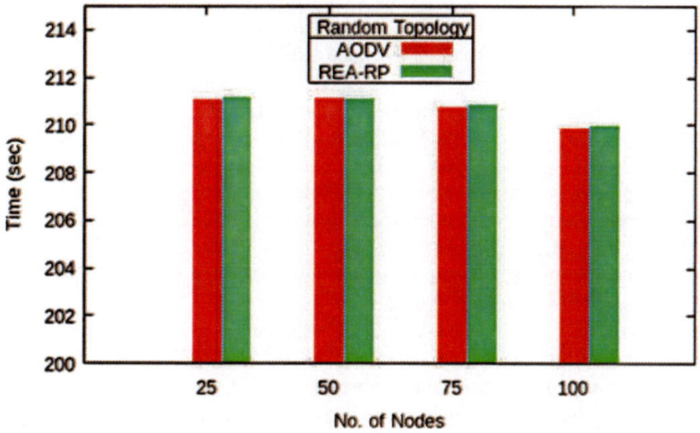

Fig. 2 (continued)

References

1. Lv L (2017)An energy aware multipath routing algorithm for wireless sensor networks. Int J Online Eng 13(4)
2. Yahya B, Ben Othman J (2009) Towards a classification of energy aware MAC protocols for wireless sensor networks. Wirel Commun Mob Comput 9(12):1572–1607
3. Soua R, Minet P (2011) A survey on energy efficient techniques in wireless sensor networks. In: 2011 4th joint IFIP wireless and mobile networking conference (WMNC 2011). IEEE, pp 1–9
4. Yoo H, Shim M, Kim D, Kim KH (2010) GLOBAL: a gradient-based routing protocol for load-balancing in large-scale wireless sensor networks with multiple sinks. In: The IEEE symposium on computers and communications, Riccione, pp 556–562. https://doi.org/10.1109/ISCC.2010.5546784
5. Hu T, Fei Y (2010) QELAR: a machine-learning-based adaptive routing protocol for energy-efficient and lifetime-extended underwater sensor networks. IEEE Trans Mob Comput 9(6):796–809. https://doi.org/10.1109/TMC.2010.28
6. Han Z, Wu J, Zhang J, Liu L, Tian K (2014) A general self-organized tree-based energy-balance routing protocol for wireless sensor network. IEEE Trans Nucl Sci 61(2):732–740. https://doi.org/10.1109/TNS.2014.2309351
7. Jaradat T, Benhaddou D, Balakrishnan M, Al-Fuqaha A (2013) Energy efficient cross-layer routing protocol in wireless sensor networks based on fuzzy logic. In: 2013 9th international wireless communications and mobile computing conference (IWCMC), Sardinia, pp 177–182. https://doi.org/10.1109/IWCMC.2013.6583555
8. Dai G, Qiu J, Liu P, Lin B, Zhang S (2012) Remaining energy-level-based transmission power control for energy-harvesting WSNs. Int J Distrib Sens Netw. https://doi.org/10.1155/2012/934240
9. Behera TM, Mohapatra SK, Samal UC, Khan MS, Daneshmand M, Gandomi AH (2019) Residual energy based cluster-head selection in WSNs for IoT application. IEEE Internet of Things J
10. Panigrahi B, De S, Panda BS, Lan Sun Luk J-D (2010) Energy-efficient greedy forwarding protocol for wireless sensor networks. In: 2010 IEEE 71st Vehicular Technology Conference. IEEE, pp 1–5.

11. Yan C, Hu J, Shen L, Song T (2009) RPLRE: a routing protocol based on LQI and residual energy for wireless sensor networks. In: 2009 first international conference on information science and engineering, Nanjing, pp 2714–2717. https://doi.org/10.1109/ICISE.2009.1041
12. Wahid A, Lee S, Kim D (2011) An energy-efficient routing protocol for UWSNs using physical distance and residual energy. In: OCEANS 2011 IEEE—Spain, Santander, pp 1–6. https://doi.org/10.1109/Oceans-Spain.2011.6003615
13. Yong Z, Pei Q (2012) A energy-efficient clustering routing algorithm based on distance and residual energy for wireless sensor networks. Procedia Eng 29:1882–1888. https://doi.org/10.1016/j.proeng.2012.01.231. ISSN 1877-7058
14. AlShawi IS, Yan L, Pan W, Luo B (2012) Lifetime enhancement in wireless sensor networks using fuzzy approach and A-star algorithm, p 82
15. Swain AR, Hansdah RC, Chouhan VK (2010) An energy aware routing protocol with sleep scheduling for wireless sensor networks. In: 2010 24th IEEE international conference on advanced information networking and applications. IEEE, pp 933–940
16. Cheng D, Xun Y, Zhou T, Li W (2011) An energy aware ant colony algorithm for the routing of wireless sensor networks. In: International conference on intelligent computing and information science. Springer, Berlin, Heidelberg, pp 395–401
17. Zytoune O, Aboutajdine D (2010) A uniform balancing energy routing protocol for wireless sensor networks. Wirel Pers Commun 55(2):147–161
18. https://www.nsnam.org/
19. Carneiro G, Fortuna P, Ricardo M (2009) FlowMonitor—a network monitoring framework for the Network Simulator 3 (NS-3). https://doi.org/10.4108/ICST.VALUETOOLS2009.7493.
20. Kamgueu PO, Nataf E, Ndié TD, Festor O (2013) Energy-based routing metric for RPL. [Research Report] RR-8208, INRIA, pp 14 ⟨hal-00779519⟩
21. Zhang D, Li G, Zheng Ke, Ming X, Pan Z-H (2013) An energy-balanced routing method based on forward-aware factor for wireless sensor networks. IEEE Trans Industr Inf 10(1):766–773

A KNN-Based Approach for Myoelectric Arm

Parul Rani, Vikash Shaw, Yogesh Kumar, Ranjan Kumar Jha, and Sanjeev Kumar

Abstract Passive and body-powered prosthetic hand widely used by upper limb amputation. But these techniques cannot fulfil the basic requirements of the amputee. Direct control myoelectric arm works on the on–off, proportional level control and threshold level and capable of performing only a few gestures of the real human hand. This paper presents a pattern recognition-based approach for myoelectric arm to work near natural hand. A seven electrodes band for the forearm has been designed to acquire electromyogram (EMG) signal and classify four different gestures which mostly uses by human being in their daily life. A KNN algorithm was explored for classifying these gestures. It was observed that with the proposed methodology classifier achieved 89.5% average accuracy and 0.93 average F1-score.

Keywords EMG signal · EMG band · Active electrode · SEMG · KNN · Myoelectric arm · Machine learning

P. Rani · V. Shaw · Y. Kumar · R. K. Jha · S. Kumar (✉)
CSIR-Central Scientific Instruments Organisation, Chandigarh 160030, India
e-mail: virdi205@gmail.com; virdi205@csio.res.in

Academy of Scientific & Innovative Research (AcSIR), Ghaziabad 201002, India

P. Rani
e-mail: rishalparul93@gmail.com

V. Shaw
e-mail: vikashshaw.aec@gmail.com

Y. Kumar
e-mail: yogeshh305@gmail.com

R. K. Jha
e-mail: ranjan.drmgr@gmail.com

R. P. Mahapatra et al. (eds.), *Proceedings of 6th International Conference on Recent Trends in Computing*, Lecture Notes in Networks and Systems 177,
https://doi.org/10.1007/978-981-33-4501-0_29

1 Introduction

In a developing country like India, the life of limb amputees is very challenging as they have to face emotional, physical as well as social sympathy and embarrassment. In a world population of approximately 6.7 billion, 10 million are amputees in which 3 million (30%) are armed amputees [1]. Moreover, in the 3 million people, 58% are transradial (blow elbow) arm amputees. Although passive and body-powered prosthetic hand widely used by upper limb amputation, these techniques cannot fulfil the basic requirements of the amputee [2], 3]. To overcome the limitations of these techniques, myoelectric prosthesis is used. Electric property of muscles is known as myoelectric. Myoelectric-controlled-based prosthetic provides an optimum solution for the natural appearance and functionality of the human like hand [4]. In addition, myoelectric prosthetic device is controlled by two techniques such as non-pattern recognition and pattern recognition (PR). In the non-PR (direct control), EMG signal obtained by flexor and extensor muscles, which control opening, closing and gripping of the hand with the proportional force exerted by EMG [5]. On the other hand, non-PR myoelectric arm works on the on–off, proportional level control and threshold level which is capable of performing only a few gestures of the hand [6]. However, EMG-PR-based techniques have increased facility of myoelectric arm and overcome limitations of conventional prosthetic arm (non-PR). EMG-PR works on extracting multi-features from the signal rather than entirely depends on the amplitude of the signal. Also, PR-based myoelectric arm can perform numerous gesture of hand, but success rate of PR depends upon selection of features and classifier [7, 8]. It works on the assumption that EMG signal pattern is repeatable for the same gesture and distinct for different gesture [9]. Till now PR-based system gives high functionality and high accuracy on offline and limited up to laboratory demonstrations [10, 11], but when examining in the real time, it becomes dysfunctional [7]. Hence, an algorithm which can perform efficiently for offline as well as real time is the need of the hour. Moreover, the performance of the algorithm is affected by several factors such as position of the electrode, EMG crosstalk, arm position, fatigue of the muscles [12]. In addition, seven muscles of forearm namely Flexor carpi radialis, Palmaris longus, Flexor carpi ulnaris, Extensor Carpi Radialis Longus, Extensor Carpi Ulnaris and Extensor Digitorum are mainly responsible for a maximum gesture of the hand [13].

This paper proposed a seven electrodes band to acquire EMG signals from the seven anterior and posterior superficial muscles of forearm with the KNN machine learning algorithm for the myoelectric arm in order to work near natural hand.

2 Proposed Methodology

In general, the EMG signal ranges between 10 μV and 150 mV. To detect such small order of signal, an active electrode has developed. The process of the proposed system depicts in Fig. 1. While, for the data acquisition, a seven active electrodes

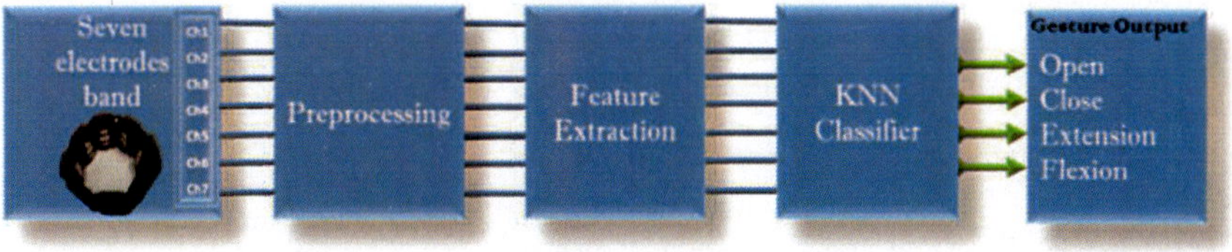

Fig. 1 Pattern recognition-based control strategy for upper extremity prosthesis

band in which electrodes were placed at a distance of 25 mm has been designed to acquire EMG signal for different gestures.

To avoid positional effect of electrodes, random placement of band was proposed, and the data was acquired for 10 s for each gesture with ten random placements of band. Furthermore, one second time resolution was taken for feature extraction. Different time domain features were extracted to train and test the classifier. KNN machine learning algorithm with Euclidean distance calculation technique was used as classifier because it can be developed in real-time scenario.

3 Data Acquisition

Seven channels EMG data were recorded in reference single ended mode with NI-6003 USB DAQ at 1 kHz sampling rate. To operate myoelectric arm, EMG signal has been acquired from the forearm of the subject. This happens because EMG signal generated through the connection between muscle and central nervous system is unique to an individual [14] which makes the myoelectric arm user-specific. With this assumption, the signals were acquired for a single subject. In order to acquire the data, subject was asked to seat in a chair at normal position and perform different gestures such as open, close, extension and flexion. Moreover, each gesture was performed for ten seconds with the time gap of 1 min. The sequence of gestures which were asked to perform to the subject is shown in Fig. 2. The green block shows the relax portion of one minute after each gesture performed.

Fig. 2 Experimental procedure

Table 1 Features extracted from the preprocessed data

Features	Formula	Parameters
Root mean square (RMS)	$\sqrt{\frac{1}{N}\sum_{i=1}^{N} x_i^2}$	$N = 1000 =$ length of sEMG signal $X_i =$ ith sample of the sEMG signal $\mu =$ mean $s =$ standard deviation
Variance (VAR)	$\frac{1}{N-1}\sum_{i=1}^{N}\left(x - \mu_i^2\right)$	
Average amplitude change (AAC)	$\frac{1}{N}\sum_{i=1}^{N-1}\lvert x_{i+1} - x_i\rvert$	
Difference absolute standard deviation value (DASDV)	$\sqrt{\frac{1}{N-1}\sum_{i=1}^{N-1}(x_{i+1} - x_i)^2}$	
Kurtosis	$\frac{\sum_{i-1}^{N}(X_i-\mu)^4/N}{s^4}$	
Skewness	$\frac{\sum_{i-1}^{N}(X_i-\mu)^3/N}{s^3}$	

3.1 Preprocessing

A 6th order Butterworth bandpass filter having lower cut-off frequency 10 Hz and higher cutoff frequency 450 Hz is applied because EMG signal ranges between this frequency band. The line frequency (50 Hz) lies in the band, which may affect the low-level EMG signal, thereby, in order to avoid this effect, stop band (49-51 Hz) filter is applied to each channel. The features were extracted from the preprocessed signal.

3.2 Feature Extraction

Six-time domain features were extracted from the preprocessed data. In sharp contrast, frequency domain features required domain transformation (time to frequency) which may increase the computational cost. Features were extracted for all seven-channel preprocessed EMG signal, and dataset was created. The given Table 1 shows the six-time domain features extracted to train and test the classifier.

3.3 Dataset

One second data was considered for feature extraction and classification. Single run in the dataset comprised of 1×42 features, because there are seven electrodes and each electrode contributes six features. Ten seconds data was recorded for each gesture, so the dataset size was 100×42 for one gesture. Thus, total dataset size was 400×42 for four gestures.

Table 2 Confusion matrix [15]

		Actual	
		Positive	Negative
Predicted	Positive	True Positive (TP)	False Positive (FP)
	Negative	False Negative (FN)	True Negative (TN)

4 Evaluation Parameters

Performance of KNN classifier was explored with the created dataset. Accuracy of the classifier was tested for different value of *K*. In order to avoid biasing in the validation, 5-cross validation was done for the value of K which gave highest accuracy. Along with the accuracy, other parameters viz. PPV, TPR, FPR, F1-score of classifiers were studied to evaluate the performance of the classifier (Table 2).

$$\text{Accuracy} = \frac{\text{TP} + \text{TN}}{\text{TP} + \text{FN} + \text{FP} + \text{TN}}$$

$$\text{TPR} = \frac{\text{TP}}{\text{TP} + \text{FN}}, \quad \text{PPV} = \frac{\text{TP}}{\text{TP} + \text{FP}}, \quad \text{FPR} = \frac{\text{FP}}{\text{FP} + \text{TN}},$$

$$\text{F Score} = 2 * \text{PPV} * \frac{\text{TPR}}{\text{PPV} + \text{TPR}}$$

5 Results

The following graph shows the accuracy of four gestures for different values of *K*. From the graph (Fig. 3a), it is clear that for $K = 3$ system gave good accuracy for all the gestures. Additionally, for $k = 1$ system gave highest accuracy for wrist flexion but not for hand open and close as compared to the $k = 3$. The chance of confusion for KNN classifier may get increased as the value of *K* increases which decreases the classifier accuracy. Same can be seen in Fig. 3b as the value of K increases the accuracy gets decreased.

The overall accuracy for the classifier for different values of *K* shown in Fig. 3b. For $k = 1$, the overall accuracy was 92.5, while for $K = 3$, the accuracy was 93.75. Therefore, $K = 3$ was taken for further calculation and parameters study.

At $K = 3$, five-cross validation was done to avoid biasing in the selection of training and testing dataset. Table 3 shows the fivefold result. As consequence, the minimum, maximum and average accuracy achieved by the classifier was 83.7, 93.75 and 89.5, respectively.

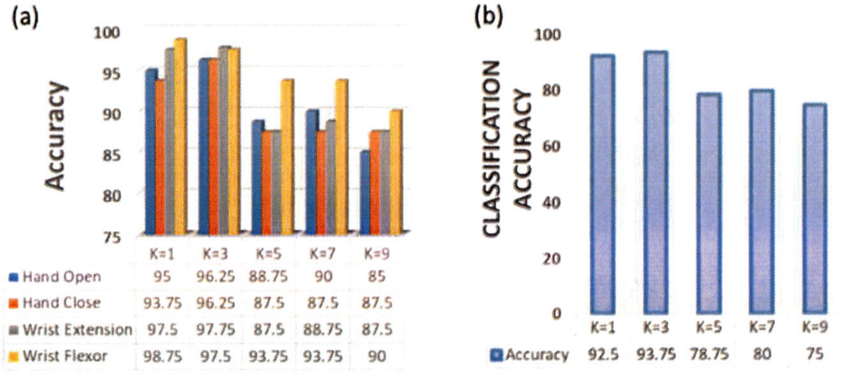

Fig. 3 **a** Accuracy for different gesture for different value of *K*, **b** classifier accuracy for different value of *K*

Table 3 Average classification accuracy of four classes using fivefold cross validation for $K = 3$

Fold number	Test accuracy
Fold-1	91.25
Fold-2	83.75
Fold-3	90.00
Fold-4	88.75
Fold-5	93.75
Average	89.5

To train and test the classifier, 80% (320 trials) and 20% (80 trials) data were selected randomly. The given table shows the confusion matrix for the test data. From the confusion matrix shown in Fig. 4a, different parameters like TPR, FPR, PPV and F1-score were calculated with the formula mentioned above and the plot shown in Fig. 4b. Below graph shows the evaluation parameters for the classifier at $K = 3$. The classifier F1-score for all four gestures was approximately 0.93 which signifies that classifier has effective precision as well as recall ability. The classifier has maximum 0.087 FPR which shows that the classifier has low chance of misclassification.

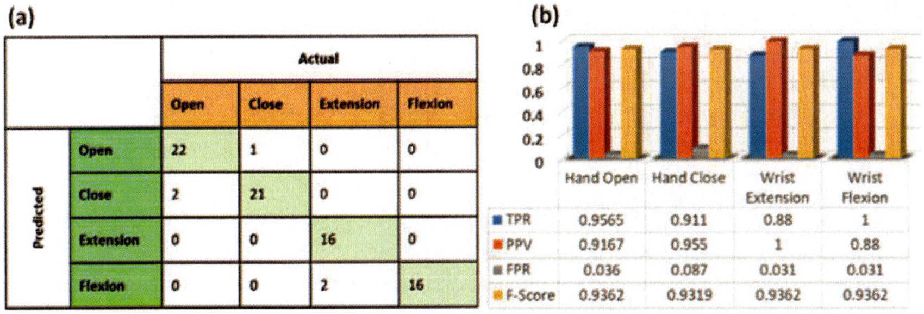

Fig. 4 **a** Confusion matrix, **b** evaluation parameters (TPR, PPV, FPR, F1-score) for the classifier

6 Conclusion and Discussion

This paper presents an approach for a myoelectric arm to work near natural arm. A seven electrodes band along with KNN machine learning algorithm is proposed based on the assumption that mainly seven forearm's anterior and posterior muscles are involved in the different hand gesture. The effect of different value of K was also studied, and the result shows that for discussed six times domain features, classifier outperforms for $K = 3$. The classifier average F1-score is approximately 0.93 which signifies high precision and recall property.

7 Future Scope

Same classifier will be tested online with able people and with amputees. The accuracy can be improved further by increasing the features. Also, different classifier, which can also work for the real-time process, will also be explored.

References

1 Leblanc M (2015) 'Give Hope—Give a Hand'—The LN-4 Prosthetic Hand, pp 2–4
2. Smith LH, Hargrove LJ, Lock BA, Kuiken TA, Member S (2011) Determining the optimal window length for pattern recognition-based myoelectric control: balancing the competing effects of classification error and controller delay 19(2):186–192
3. Maat B, Smit G, Plettenburg D, Breedveld P (2018) Passive prosthetic hands and tools: a literature review
4. Balciunas A, Murphy P, Zdobinski Z (2019) Viability and function of myoelectric prosthetics as compared to passive, body-powered, and electrically powered prosthetics
5. Hindu B, Rourkela T (2012) Design of below elbow electronic robotic arm with proportional grip force (June 2015)
6. Naik GR, Member S, Al-timemy AH, Nguyen HT, Member S (2016) Transradial amputee gesture classification using an optimal number of sEMG sensors: an approach using ICA clustering. IEEE Trans Neural Syst Rehabil Eng 24(8):837–846

7. Parajuli N et al Real-time EMG based pattern recognition control challenges and future implementation
8. Resnik L, Huang HH, Winslow A, Crouch DL, Zhang F, Wolk N (2018) Evaluation of EMG pattern recognition for upper limb prosthesis control: a case study in comparison with direct myoelectric control, pp 1–13
9. Hudgins B, Parker P, Scott RN (1993) A new strategy for multifunction myoelectric control. IEEE Trans Biomed Eng 40(1):82–94
10. Kuiken TA et al (2011) NIH public access 301(6):619–628
11. Hargrove LJ, Miller LA, Turner K, Kuiken TA (2017) Myoelectric pattern recognition outperforms direct control for transhumeral amputees with targeted muscle reinnervation: a randomized clinical trial. Sci Rep (October):1–9.
12. Samuel OW (2019) Intelligent EMG pattern recognition control method for upper-limb multifunctional prostheses: advances, current challenges, and future prospects. IEEE Access 7:10150–10165
13. Zonnino A, Sergi F (2020) Model-based estimation of individual muscle force based on measurements of muscle activity in forearm muscles during. IEEE Trans Biomed Eng 67(1):134–145
14. sureshm S, Holi MS (2011) Electromyography analysis for Person identification. Int J Biometrics Bioinforma (5):172
15. "Confusion matrix—Wikipedia. [Online]. Available: https://en.wikipedia.org/wiki/Confusion_matrix. [Accessed: 24-Feb-2020].

Implementation of Smart Diet Assistance Application

Myron Carvalho, Prajwal Kotian, Hima George, Dipti Pawade, Ashwini Dalvi, and Irfan Siddavatam

Abstract The number of health and fitness platforms has increased drastically over the years, because of the people's desire to live more healthy lifestyles. Keeping this need in mind, we have proposed a mobile application that would be seen as a pocket assistant to track meals, get suggestions and get a check on one's personal goals and health. The application aims provide user friendly solution in order to give the best guidance and advice on nutrition plans and tracking them. Along with that meal recommendation feature is provided so that user can get the healthy food suggestion based on his eating pattern. In some cases, user may want to explore the nearby hotels where he can get suggested dishes, to cater this requirement we have incorporated the location-based hotel recommendation feature. The main objective of our study is to provide a diet monitoring application which is not only easy to use but also capable enough to track the nutrition record and provide solution accordingly.

Keywords Diet monitoring · Food recommendation system · Location-based hotel recommendation system · Voice input processing · Barcode scanning · Pearson correlation

M. Carvalho (✉) · P. Kotian · H. George · D. Pawade · A. Dalvi · I. Siddavatam
Department of Information Technology, K. J. Somaiya College of Engineering, Vidyavihar, Mumbai, India
e-mail: myron.c@somaiya.edu

P. Kotian
e-mail: prajwal.kotian@somaiya.edu

H. George
e-mail: hima.george@somaiya.edu

D. Pawade
e-mail: diptipawade@somaiya.edu

A. Dalvi
e-mail: ashwinidalvi@somaiya.edu

I. Siddavatam
e-mail: irfansiddavatam@somaiya.edu

R. P. Mahapatra et al. (eds.), *Proceedings of 6th International Conference on Recent Trends in Computing*, Lecture Notes in Networks and Systems 177,
https://doi.org/10.1007/978-981-33-4501-0_30

1 Introduction

Healthy eating plays a critical role in our daily well-being and is indispensable in preventing and managing conditions such as diabetes, high blood pressure, cancer, mental illnesses, asthma, etc. As time has progressed people have started realizing the fact that to solve all this, they just need to tune one routine of their daily life, that's what they eat throughout the day. But understanding this need was just the first step of the whole process, and unfortunately, the most crucial one where most give up. They lack the knowledge of how to go about from there. As most of us lack the right knowledge of what, when, how much, and in what situations is to be eaten or avoided; which is the whole game from here, we all find the need for someone to assist us all time. Ever since then year after year developers and companies have come up with different solutions to cater this problem. To get an overview, we have compared the existing apps which are primarily trying to offer a solution to the emerging problem. For comparison we have considered the following parameters:

- Set Goal: allow user to set their health goals such as weight gain/loss or remain fit
- Diary: keep a track of activities/exercises done throughout the day
- Control food craving: warn user if they are consuming unhealthy food
- Add photos of yourself: upload pictures to notice changes in the body
- Challenges: Exercises/Activities recommended by app to challenge the user
- Download Progress: download the report in pdf form
- Point/Score System: Assign scores to show the progress to the user
- Full week diet plan: Recommend a healthy diet plan according to users' goal.

Table 1 Comparison of existing mobile application

	Applications					
	My-FitnessPal [1]	Healthy food [2]	HealthifyMe [3]	Fabulous [4]	My Diet Coach [5]	Diet Assistant [6]
Set goal	Y	Y	Y	Y	Y	Y
Diary	Y	N	N	Y	Y	N
Control food craving	N	N	N	Y	Y	N
Add photos	N	N	N	Y	Y	Y
Challenges	N	N	N	Y	Y	N
Download progress	Y	N	N	Y	N	N
Point/score system	N	N	N	Y	Y	N
Full week diet plans	Y	N	Y	N	N	Y

After comparing all the above apps, we could draw a conclusion, that few apps acted like a static catalog of data decorated with great visual designs. But they lacked something - intelligence. The main constraint of these applications is inconvenience caused, because the user has to manually enter all the meal details on a daily basis which leads to the user slowly drawing back from using the application. Thus, the user in the end finds it cumbersome and painstaking to use it daily. Hence, we are providing multiple input options for the user to select from. These input methods would be voice, barcode or the usual way that is through textual input. In addition to this we have a unique feature that separates our application from others, a recommendation system which would be dynamic in suggesting a meal based on the user's previous meals. This is further enhanced by providing nearby restaurants options for the recommended meal.

2 Literature Survey

A considerable amount of work has been done in the field of fitness and nutrition when it comes to mobile or web applications. Some of the notable works in this field are as follows:

Engin et al. [7] proposed a system that acquires acceleration data via a sensor worn by the user on the wrist. This data is sent to the mobile device via Bluetooth. Acceleration data is then smoothened using a Gaussian smoothing filter to remove noise effects. Based on this, real-time feedback such as the total number of bites, bites-taken rate, and eating speed is presented to the user during the meal. The user is warned by the system if he is eating too fast. Krisana et al. [8] put forth a health monitoring application for obese Children. Personal information such as height, weight, gender, age, etc. along with diet goals is given as input to the system. The system then tracks daily food consumption and helps the user to reduce weight and control obesity. It summarizes the result of application on a weekly basis and recommends proper food consumption to the users. The authors of [9] have proposed a platform wherein the clients can manually upload their health data through the mobile health tracking application. Then, with client's consent, RDNs (Registered Dietitian Nutritionists) can have access to the physical activity records of their clients. Based on all the data, RDNs will prepare diet plans for their clients. This system enables direct communication with RDNs and hence provides motivation for the client to achieve his/her fitness goal with the aid of feedback or progress updates provided by the RDNs. Another system called Lucy [10] which is a nutritionist assistant conversational agent, acquires users' personal information, log of meals and available food information through a text-based input mechanism and tracks the user behavior. The input of food item received is broken down into its nutritional values and is compared with users' goal, daily requirements, and food alternatives. After processing, it provides weekly/monthly reports of food intake, calories consumed, time of food consumption and gives warning messages. The authors in [11] propose a system which can recognize the contents of your meal from a single image and then predict its nutritional content. Food detection is done by using Semantic image segmentation CNN-based multi-label classifier with CRF smoothing. Volume and quantity of food is evaluated using preliminary results on estimating the depth of each pixel from a single RGB image, using a CNN. Training of dataset is done by GoogLeNet CNN. The authors of [12] conducted a survey to understand the effects of crowdsourcing on obesity management and prevention of people. For this purpose they included 12 experts in the field of nutrition and 12 non-expert people. This study has clearly shown that crowdsourcing has increased potential in health, including obesity research and practice. The authors of [13] have proposed a system where the user gives a voice input of food which is first converted to text. Then the sentence is tokenized to get individual words. Each word is given a tag based on some criteria. Based on the tags the system would identify the food, portion size and meal type. After that either exact or approximate string matching from the USDA nutrition database is carried out. Based on the food retrieved from database, calorie breakdown of food is given. The authors of [14] propose a system which includes actual cooking and serving utensils. The components used in this system are Smart Dietary Monitoring System, Smart Plate, Smart Cooking Vessel, Smart Drinking Glass, and Smart Hub. The system provides a health chart which act as a basis for providing timely notifications to the user. The Cloud-based Health Analytics also aim to provide calculated health notifications for the users regarding the selection of food items after studying the present and past health records. After studying the aforementioned research papers and analyzing the features of existing applications we can conclude that every proposed solution lacked some vital features like dynamic meal recommendation and suggesting a nearby restaurant for the same through location tracking which we have incorporated in our app.

3 Methodology

The application has three main modules:

1. **Accepting the input and calculating the nutrition value**

Initially, on launching the app, the user would be asked to sign in using his Google account. When the user selects an account, it would be validated in the backend. The new user is asked to enter his age, height, weight, gender and goal plan. Here we have provided three goal plans viz. weight loss, mass gain, and maintain weight. From the above information, the BMI of the user is calculated. Based on the BMI and the goal plan chosen, user's daily calorie consumption is monitored. The next step is to accept information regarding food consumed by the user. There are three ways through which the user can provide the information about the food consumption viz. through barcode scanning, as voice input and the common one is through text input.

- Voice input:—The user needs to speak out the food consumed by him which is fed as an input to Google speech input dialog. Then the individual words spoken

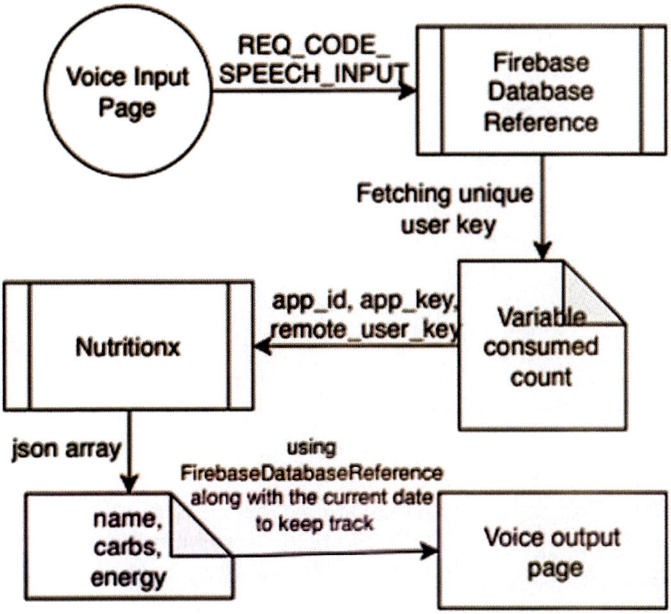

Fig. 1 Voice input processing

by the user are figured out at the server side. The words that have similar pronunciation but different meanings are figured out once, other words are recognized and their context is understood. Then the appropriate words are finalized to form the sentence spoken by the user into a string form. This meal string can consist of multiple food items and quantities. The nutrients breakdown for that meal will be calculated using Nutritionx which is a natural language processing API. Figure 1 depict the steps involved in processing the voice input.

- Barcode:—Through this option the user can scan the barcodes of the packaged foods, and then their nutrient values would be stored. This uses the "open food facts" API to process the barcodes. When a user clicks on the barcode input button, a barcode scanning page will open which detects the barcode of the product. This barcode is then used to fetch product details from "open food facts" API in JSON format, which will be displayed to the user in an appropriate manner. Figure 2 gives brief overview of steps involved in scanning the barcode and extracting the nutrient values.
- Manual/Text: When the user clicks on the manual input button, he will be redirected to a page where he has to enter the meal food items and quantities. Rest of the processing is same as the voice input.

2. **Meal recommendation**

Another feature under consideration is meal recommendation. Here, based on the user's previous meal, food recommendations are provided. Figure 3 depicts the flow

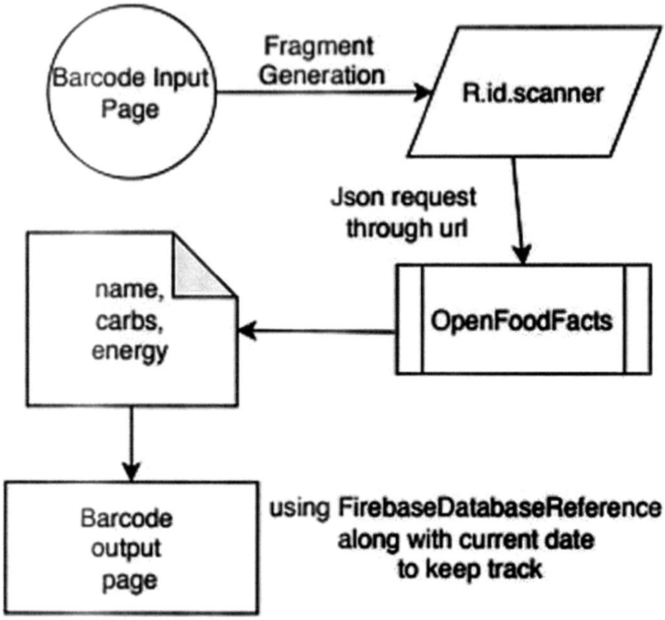

Fig. 2 Barcode input processing

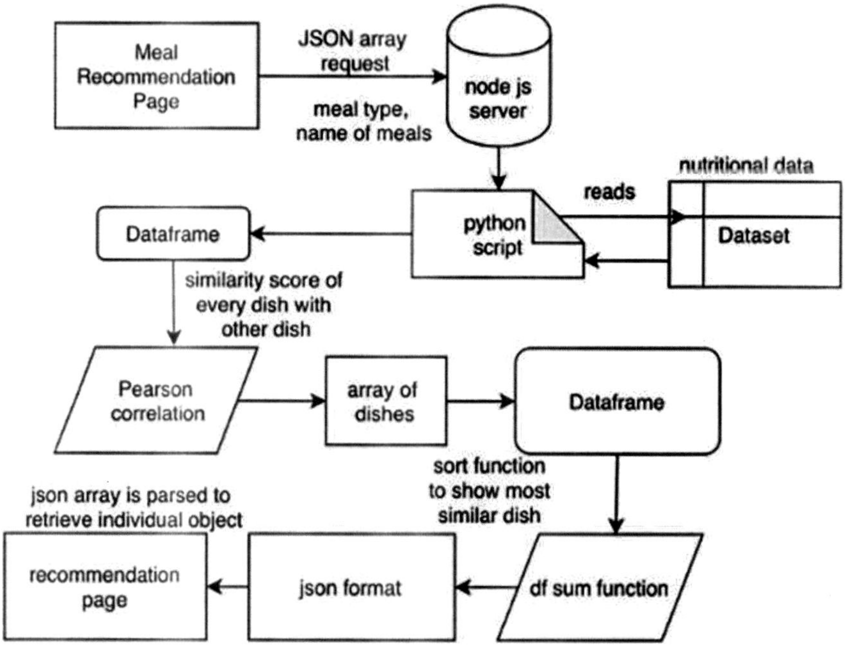

Fig. 3 Meal recommendation system

of activities in a meal recommendation system. When the user clicks on the recommendation button, a request would be sent through the android device to a server which will run a python script and return the result in JSON format. This request consists of the meal type, i.e., breakfast, lunch or dinner, along with the names of the meals that were consumed by the user for the past 5 days in that category. This would be sent as a JSONArrayRequest as we are expecting an array of JSON as the output. This request is received by the node.js server which reads the input parameters such as the meal type and the names of the meals. The server then calls the python script by spawning a child process and passes on the inputs to that script. In this script, firstly, the appropriate dataset is read based on the meal type received as input. The dataset consists of the names of the dishes as well as the calories, carbohydrates, fat, and protein content of each of these dishes. This is saved as a dataframe and then the Pearson correlation (measure of the linear correlation between two variables X and Y.) is applied to this data which then calculates the similarity score for each dish with every other dish and stores it as a matrix. Pearson correlation is given in Eq. 1.

$$\rho_{X,Y} = \text{cov}(X, Y)/(\sigma_X.\sigma_Y) \quad (1)$$

Since the dataset contains name of the dish and their calories, carbohydrates, fats and protein content, hence, the Pearson correlation can be found out using the formula given in Eq. 1. Then, all the dishes that we received are stored into an array. Now, for every element in this array we retrieve, the row of similarity score from the precalculated matrix and store it in a dataframe. The similarity scores for each dish is aggregated and then sorted in descending order to get the most similar to the last consumed 5 dishes. Then, the associated calories, fats, carbohydrates, and protein values are appended to the dataframe, and thereafter, it is converted into JSON format which is then returned to the android class. The JSON array received is then parsed to retrieve the individual JSON objects. The data from these JSON objects is then used to display the meal recommendations to the user.

3. **Hotel recommendation**

This is an optional facility provided to the user in case he wants to go outside and enjoy the food in nearby hotels. Here, user can get the hotel recommendation for selected food in nearby vicinity. Initially, the location permissions of user's mobile device are checked. If permissions are not given, then dialog pop up asking for the permissions. Once the permissions are given, then last known latitude and longitude of the user device is retrieved. After that dish name selected by the user from the recommendation list is fetched. Since the string of dish names would have spaces in between, it is replaced with "%20" so that there would be no errors in passing in the URL. Then, a request URL string is created that will have latitude, longitude, and dish name. Create a new Volley request that will provide classes to make JSON requests. Then, we create a model which will be used in fetching restaurant results. Create a JsonObjectRequest with parameters request method, request URL, and the response listener. In response listener, we process the output that we received from

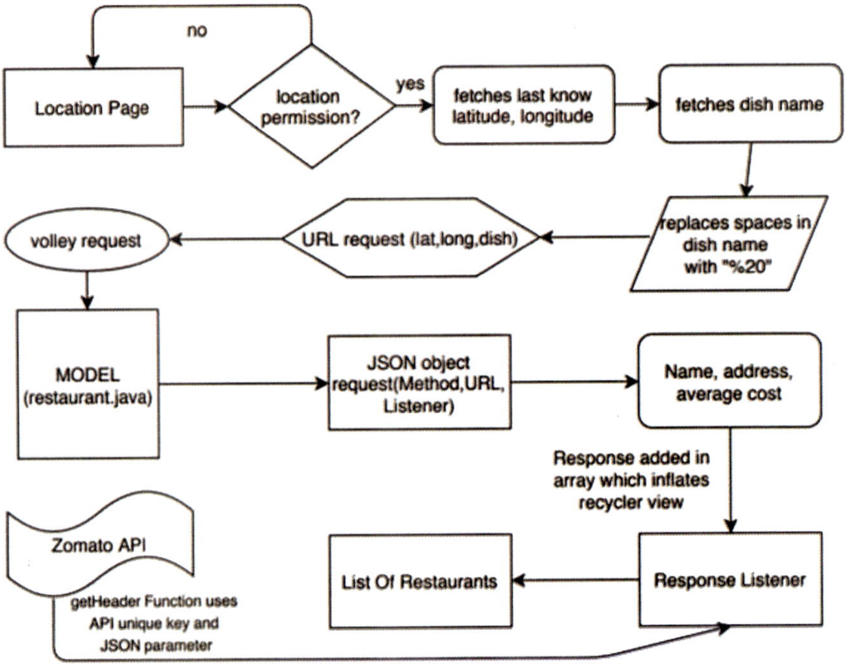

Fig. 4 Hotel recommendation system

the request (i.e., we fetch the JSON Array and its JSON Objects). From each JSON object, we fetch the restaurant name, address, average cost for two, restaurant images (if available), popularity rating, distance. Using the distance, popularity, and cost, we can apply filters and show the results accordingly (optional). In response listener, we have specified the user-key(our unique Zomato API key) and the response type (JSON). Finally, the JSON request is added to the Volley request queue to process it. The response would be added into an array which would be passed to the recycler view to inflate the list of restaurants. The complete process is shown in Fig. 4.

4 Result and Discussion

For the experimental purpose, we have accepted different food consumed by different people via voice input, barcode or textual input and calculated the nutritional breakdown of each meal. Figure 5 shows the sample output for voice input. Along with this report, application also gives remark about the calories consumed in a day and health alters according to the goal plan. For the recommendation system, consider that a user had the following 5 dishes for breakfast on each of the previous 5 days viz. Dhokla, Malpoha, Upma, Aloo Paratha, Masala Dosa. These 5 dishes are passed as

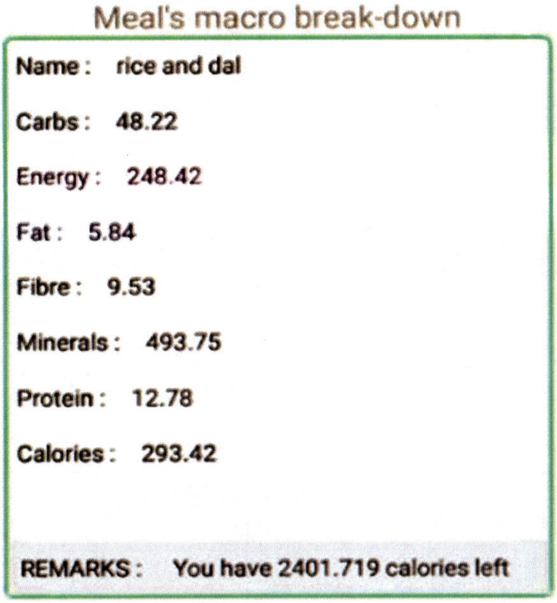

Fig. 5 Meal's nutrition breakdown

input during the call to the server that hosts the python script. For breakfast recommendation, firstly, we have calculated the similarity matrix for all the dishes in the database using Pearson's correlation. In our database, a total of 200 dishes are there, so the matrix of size 200 × 200 is created. But for explanation purpose, we have considered sample 10 dishes. We calculate the similarity of each dish with every other dish using Pearson's correlation. This is stored in a matrix form as shown in Table 2.

Once this is done, we then reduce the above matrix so that it contains only the last five consumed dishes by the user, comparing it to the dishes in dataset(consider 10 dishes for now). We then take a sum of similarity scores of dishes in datasets shown in Table 3. So, here we will recommend Masala dosa to our user. Along with this data, we append other information such as carbohydrates, calories, fats and protein content and convert it into a JSON format and send it to the android device. Similarly, it works for recommending lunch and dinner to the users. The screenshot is shown in Fig. 6.

After this, if the user wishes to order the food from hotel, then he can click on the meal name to get the nearby restaurants that serve that meal. The meal name would be stored in a string when the user clicks on it: dish = "Masala dosa." Then, we use Android location services to get the longitude and latitude of the user. These services use the last known location of device as latitude, longitude.

Let us assume, longitude = 72.8887593 and latitude = 19.0802481. Application will parse these values along with dish name to Zomato API and from all the restaurant

Table 2 Similarity matrix for dishes in database

	Kanda Poha	Idli Sambhar	Upma	Masala dosa	Malpoha	Dhokla	Aloo Paratha	Puri Bhaji	Bread butter	Bread jam
Kanda Poha	1.0	0.996	0.999	0.993	0.33	0.67	0.78	0.79	0.23	0.36
Idli Sambhar	0.996	1.0	0.998	0.997	0.58	0.7	0.9	0.84	0.64	0.6
Upma	0.999	0.998	1.0	0.997	0.87	0.92	0.92	0.9	0.76	0.78
Masala dosa	0.993	0.997	0.997	1.0	0.77	0.9	0.88	0.85	0.61	0.57
Malpoha	0.33	0.58	0.87	0.77	1.0	0.83	0.78	0.73	0.79	0.83
Dhokla	0.67	0.7	0.92	0.9	0.83	1.0	0.85	0.82	0.52	0.59
Alloo Paratha	0.78	0.9	0.92	0.88	0.78	0.85	1.0	0.93	0.43	0.55
Puri Bhaji	0.79	0.84	0.9	0.85	0.73	0.82	0.93	1.0	0.7	0.73
Bread butter	0.23	0.64	0.76	0.61	0.79	0.52	0.43	0.7	1.0	0.97
Bread jam	0.36	0.6	0.78	0.57	0.83	0.59	0.55	0.73	0.97	1.0

Table 3 Similarity matrix for previous meal with dishes in database

		Dishes from database									
		Masala dosa	Upma	Dhokla	Aloo Paratha	Puri Bhaji	Idli Sambhar	Kanda Poha	Malpoha	Bread jam	Bread butter
Previous 5 dishes by user(breakfast)	Dhokla	0.997	0.998	1.0	0.83	0.83	0.7	0.67	0.83	0.59	.0.52
	Malpoha	0.996	0.995	0.998	0.78	0.73	0.58	0.33	1.0	0.79	0.83.
	Upma	0.999	0.993	0.997	0.92	0.9	0.998	0.999	0.87	0.78	0.76.
	Masala dosa	1.0	0.993	0.997	0.88	0.85	0.997	0.993	0.77	0.57	0.61.
	Aloo Paratha	0.999	0.990	0.995	1.0	0.93	0.9	0.78	0.78	0.55	0.43.
	Total similarity score	4.991	4.969	4.987	4.41	4.24	4.175	3.772	4.25	3.28	2.63

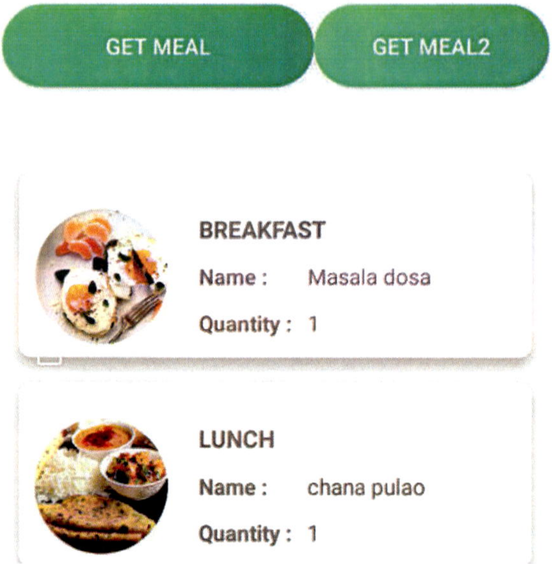

Fig. 6 Meal recommendation

related data available, the restaurants nearby vicinity which serves have the required dish (Masala dosa) are filtered out, giving the sample list of restaurants as shown in Table 4. Now, the results are sorted on the basis of real distance from user using the latitude and longitude of user and the restaurants giving and the appropriate details are then displayed.

To evaluate the performance, the application was beta tested by group of 10 users and then their feedback was taken. Through the feedback evaluation, success rate, satisfaction rate, error rate, and net promotion score are calculated whose values are 86.89%, 81.04%, 13.11%, and 7.43, respectively. This applications not only provide the ease to maintain the diet information but also come up as an effective way to get the food recommendations and places from where we can get that food.

Table 4 List of restaurants serving "Masala dosa" fetched from Zomato API

Sr. No.	Restaurant name	Latitude	Longitude
1	Sadguru Kitchen	19.08	72.8899
2	Cream & Kranch	19.08	72.8888
3	Hotel Popular	19.08	72.8896
4	Karnika Mehak Kitchen	19.079	72.8897
5	Foodtopia	19.0793	72.88539
6	Kohinoor Elite Restaurant	19.081	72.8853

5 Conclusion

In this paper, we presented a mobile application that can act as an assistant to track and suggest meals based on our current health and personal goals. We compared various existing applications and noticed the need for dynamic and interactive nature in the applications that is why we have these features in our application. One of the key features of this application is accepting the input in various modes and the consistency in results. This help us to achieve user convenience and interactiveness. Additional features like food recommendation and restaurant suggestions is what makes this application useful and different than others. All of the information to users is provided in real-time thus showcasing the dynamic nature of results. The proposed solution could be used for helping people with obesity, overweight, and other eating disorders to promote healthy lives without having to invest a large chunk of their time.

References

1. https://play.google.com/store/apps/details?id=com.myfitnesspal.android&hl=en_IN. Retrieved 19 April 2019
2. https://play.google.com/store/apps/details?id=com.oajoo.healthyfood. Retrieved 19 April 2019
3. https://play.google.com/store/apps/details?id=com.healthifyme.basic. Retrieved 19 April 2019
4. https://play.google.com/store/apps/details?id=co.thefabulous.app. Retrieved 19 April 2019
5. https://play.google.com/store/apps/details?id=com.dietcoacher.sos. Retrieved 19 April 2019
6. https://play.google.com/store/apps/details?id=com.aportela.diets.view. Retrieved 19 April 2019
7. Mendi E, Ozyavuz O, Pekesen E, Bayrak (2017) Health monitoring web platform for real-time expert-user interaction. In: 8th International conference on information, intelligence, systems & applications (IISA), Larnaca, Cyprus, 27–30 Aug 2017
8. Boonphan K, Soontornpipit P, Viwatwongkasam C, Sillabutra J, Sativipawee P (2018) Health monitoring application of obese children in perspective for guardian in Thailand. In: International electrical engineering congress (iEECON), Krabi, Thailand, Thailand, 7–9 March 2018
9. Achilleos A, Yeratziotis A, Mettouris C, Evripidou K, Hadjinikolaou P, Papadopoulos GA (2017) Health monitoring web platform for real-time expert-user interaction. In: 8th International conference on information, intelligence, systems applications (IISA 2017), August 2017
10. Parra MO, Favela J, Castro LA, Morales A (2018) Monitoring eating behaviors for a nutritionist E-assistant using crowdsourcing. Computer 51(3):43–51. https://doi.org/10.1109/MC.2018.1731078
11. Myers A et al (2015) Im2Calories: towards an automated mobile vision food diary. In: 2015 IEEE international conference on computer vision (ICCV), Santiago, 2015, pp 1233–1241. https://doi.org/10.1109/iccv.2015.146
12. Moorhead A, Bond R, Zheng H (2015) Smart food: crowdsourcing of experts in nutrition and non-experts in identifying calories of meals using smartphone as a potential tool contributing to obesity prevention and management. In: IEEE international conference on bioinformatics and biomedicine (BIBM), Nov 2015

13. Hezarjaribi N, Mazrouee S, Ghasemzadeh H (2018) Speech2Health: a mobile framework for monitoring dietary composition from spoken data. IEEE J Biomed Health Inf 22(1):252–264
14. Nabi J, Doddamadaiah AR, Lakhotia R (2015) Smart dietary monitoring system. In: 2015 IEEE international symposium on nanoelectronic and information systems, Indore, pp 207–212. https://doi.org/10.1109/iNIS.2015.20

Urban Sound Classification Using Machine Learning and Neural Networks

Isha Agarwal, Parul Yadav, Neha Gupta, and Sarita Yadav

Abstract Noise pollution is rapidly becoming a reason to worry about in the growing fast world. If we look around ourselves, we find many types of sounds, some of them are soothing, but some are not. All these sounds which occur around us are classified as Urban Sounds. These sounds are a prominent part of the life of city dwellers who listen to them every day—these sounds range from a dog barking to the honking of cars on the road. The population is a major reason which adds up to the increasing decibels levels in the city. Although research in this field is still in early stages, but with the newly generated Urban Dataset, things have accelerated. We can now classify sounds into a variety of classes. A combination of audio informatics, feature extraction and deep learning techniques are used to process a stream of audio and then label it into one of the ten classes defined. The process of classification is done by changing the sound wave into digital format. It uses bit depth, frequency and amplitude to convert the stream into an array of numbers depicting the sound. This can be done using mel-frequency cepstral coefficients (MFCC), chromagram, spectral contrast and tonal centroid features. This feature set is used in various models like CNN, RNN, DNN, random forest and SVM. Unlike images, sounds offer a variety of challenges like background music, no level of organization and distorted pitch and amplitude to name a few. This makes it difficult to detect the type of sound and classify them. The results show that CNN shows a comparatively high result than all the other machine learning models.

Keywords Audio informatics · Decibels · MFCC · Random forest

I. Agarwal (✉) · P. Yadav · N. Gupta · S. Yadav
Bharati Vidyapeeth's College of Engineering, New Delhi, India
e-mail: Ishagar98@gmail.com

P. Yadav
e-mail: Parul.yadav@bharatividyapeeth.edu

N. Gupta
e-mail: Neha.gupta@bharatividyapeeth.edu

S. Yadav
e-mail: sarita.yadav@bharatividyapeeth.edu

R. P. Mahapatra et al. (eds.), *Proceedings of 6th International Conference on Recent Trends in Computing*, Lecture Notes in Networks and Systems 177,
https://doi.org/10.1007/978-981-33-4501-0_31

1 Introduction

Machine learning has evolved as the most used technology. It can be used to make many projects that solve common day problems. It is one of the subfields of artificial intelligence and plays an important role in many applications such as data mining, natural language processing, image recognition and expert systems [1–3]. Machine learning works in a unique way, it fits data into models that can be analyzed and used by people. It lets computer train data inputs and show them figuratively and statistically [4]. Finally, we dash out predictions by giving out outputs that fall in a specific range. Machine learning is divided into many types with deep learning is known as deep neural network [1]. The best part about deep learning is that it has the capability to produce new features and results from a limited set of features and datasets [1, 5]. It is used in speech recognition, image recognition, etc. [1]. Classification of sound is being done for a long time now but sorting out Urban Sounds is challenging. The dataset not only contains human voice but also environmental sounds like car, construction sites and animals. This dataset is cleaned of background music and other noises. For sound classification, the first step is the extraction of features after that various models are fit to take out on various feature sets. These feature sets are the crucial factors which determine the correctness of the model. Thus, in the past few years, we have tried to use various feature extraction techniques to uplift the accuracy. Models such as random forest, deep neural network and convolutional neural network are used, and accuracy is calculated. The one having highest is considered.

2 Literature Review

Many models and ways have been used for classification. There are many works published on this. Some of them are discussed here.

Chih-Wei Chang et al. [1] designed a paper for extracting the features of sound. The main problem with sound was it cannot be represented in numeric form, and just images can be divided into pixels. Thus, to convert voice into vector format, we use feature extraction. Therefore, he divided two methods for feature extraction. In the first method, the time–frequency is kept constant, and in the second, we expand the signal characteristics [1, 3, 6]. These characters are extracted frame by frame and then sorted into various classes according to the features. Urban Sound 8 K is used. It has approximately 8723 real-time recordings of sounds [1, 7]. In the first type, all the classes have the same feature and duration as the time and frequency are kept the same. The technique which is used to separate these features is mel-frequencies cepstral coefficients (MFCC). It is widely used as a competitive and reliable technique. The second technique retains the time series. The dataset is tested on recurrent neural network (RNN), deep neural network (DNN) and a convolutional neural network (CNN) [1]. The results showed that CNN showed higher accuracy as compared to others. It also has a higher probability as seen by others. CNN is used for fewer data.

Karol J Pczak [8] designed a paper which evaluates the CNN further and uses it to classify environment sound, on small data. This has two convolutional layers with max-pooling and two fully connected layers [5]. This all uses three datasets, namely the Urban Sound, ESC-50 and ESC-10 [9, 10]. The implementation uses MFCC features. Again, CNN outnumbered the other methods, and it can be used in environmental sounds even though we have very fewer data.

Ian McLoughlin et al. [11] designed a paper which compares the waveform from spectrogram and MFCC. These are then analyzed using deep neural network classifiers. Different levels of noise and sound are used to detect amplitude, frequency and other features. In this, many different features were used in different models, and accordingly, it was seen the best model and the best type of feature. SVM was concluded to be the appropriate model with DNN back-end [12].

Huan Zhou et al. [13] used ConvNet to classify the sound. This is used as a 2D graph as its input layer. In this, Urban Sound 8 K dataset along with multiple network inputs was used with at different time. This turned out to be a good practice as the variation with time showed a different performance every time [14]. Obtained results confirmed the benefaction of time resolutions on performance of classification is obtained for input with moderate time resolution [1].

Bibek Luitel et al. [15] designed a paper to eradicate the traffic noise. It helps blind people by recognizing the type of sound, for example, bus horn, car horn or a motorbike. This is done by taking recordings by self-annotating the dataset. It focused on trying to measure traffic levels and tone them down as it leads to many health issues [1]. Many different classifiers were used and feature extractions. It turned out that the traffic noise was harmful for many people and a way should be seen to lessen it.

Khine Zar Thwe et al. [16] used time frequency to depict the various stages of the process. Spectrogram is used to create a visual representation of the wave. His model had mainly three parts. The first was converting the wave into digital format and a visual representation of frequencies [1, 17]. Second part shows the features which are useful for us and on which we can build our model. Last part is the model which is applied on the features. The classifier used is support vector machines (SVM) [1, 18].

3 Dataset

We have evaluated our model on Urban Sound 8 K dataset which has 8732 typical urban noises [19, 20]. This dataset contains audio snippets of ten classes: Air Conditioner, Jack Hammer, Children Playing, Street Music, Dog Bark, Siren, Engine Idling, Drilling and Gun Shot. These sounds are free of background music and are specially noise free to help in classification of the urban sounds [13] (Table 1).

Table 1 Dataset configuration

Classes	Sample	Duration (s)
Children playing	1000	6
Engine idling	1000	9
Jack hammer	1000	7.5
Dog bark	1000	6.5
Street music	1000	6
Drilling	1000	9
Air conditioner	1000	6
Siren	929	3
Car horn	429	1.5
Gun shot	374	3.5

The dataset is divided into test (25%) and train (75%) data. The training data is fed into the machine learning model which is further tested using samples from test data.

The length of the sound varies from 1 to 2 s (gun shots) to 30 s (jackhammers) [19].

4 System Design and Implementation

Architecture diagram helps us to analyze the workflow in a better way and shows the user and system requirements [21]. It is a flowchart which helps us to get a pictorial view of the workflow, and it can be understood by everyone easily.

Figure 1 is a flowchart depicting the process of urban sound classification using supervised learning [1]. In the training phase, audio inputs of ten different classes are used as dataset. These audio files are given to the feature extractor which uses mel-frequency cepstral coefficients (MFCC). This produces thirteen coefficients which resemble the human auditory system [15, 22, 23]. As a result, MFCC is extensively used in speech and audio. All the thirteen coefficients are used for identification and

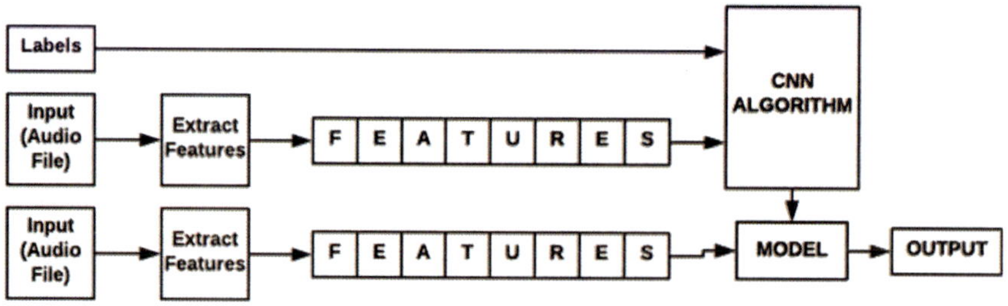

Fig. 1 Diagram showing flowchart of process

are enough to identify the sound of each class. Thus, it reduces the complexity and feature dimension.

These obtained features are extracted from individual frames to create a feature vector [24, 25]. Further, the feature vector is divided into sets for training and testing. Finally, three different supervised classifiers such as convolutional neural network (CNN), multilayer network (MLP) and support vector machines (SVM) are used to compare the performance of sound detection [15, 26]. Out of them, better performance is observed with CNN classifier.

5 Result Analysis

From the past few papers, we have seen that CNN gives the best accuracy among other models. This is because CNN although being an extension of multilayer perception model shows a higher accuracy due to the presence of convolution layer and pooling operations [8]. The convolution layer contains hidden units, which instead of processing on the whole input process on a small part of the whole input. This drastically reduces the parameters to estimate but also improves in terms of robustness. Further, the pooling layer merges the features and improves the accuracy [8] (Fig. 2).

A comparison between three important models is shown. The performance is calculated as per a value out of 10. The graphical representation shows that convolutional neural networks give us the best performance till now [27]. It gives us an outlook and helps us to see where we need to improve in case of machine learning models (Fig. 3).

Fig. 2 Graph showing the various performance rates till now

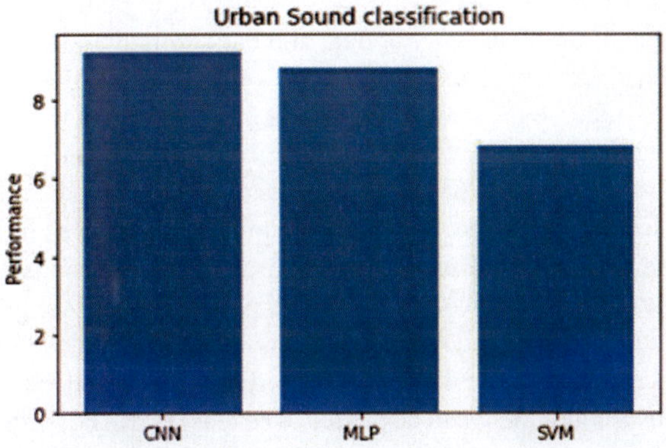

[[150	5	2	2	1	1	0	0	1	0]
[0	67	0	1	0	0	1	0	0	1]
[1	2	140	8	2	1	4	0	2	6]
[0	0	3	135	0	0	2	0	2	1]
[1	0	0	1	167	0	1	2	2	1]
[5	1	1	1	0	26	0	0	0	0]
[8	0	0	1	1	0	120	1	5	0]
[1	0	0	0	1	0	1	160	4	0]
[1	1	1	1	5	1	0	0	133	0]
[0	0	3	8	2	2	1	6	0	129]]

Fig. 3 Confusion matrix for a given set of labels and input data

6 Conclusion

The project 'Urban Sound Classification' using machine learning techniques and neural networks' helps us to classify noise around it into labels so that we can then further minimize it, predict it and analyze it. Different models were used, and CNN gave the highest accuracy (95%) till now with MLP (92%) and SVM 62%). The major problem is fewer data and background noise in the data [1]. One of the issues that occurred is when the sound of one class is the same as others and this happens to increase the complexity much more. We must think of ways so that more features can be used and what features will lead to accurate classification of the sounds. Also, many sounds do not have volumes of different pitches and thus sometimes mix up. So, features are important criteria which help, and this must be further improved which will eventually improve the accuracy [28]. We can use new models and increase the layers in the convolutional neural network so that we get better accuracy [29, 30]. The data can be cleaned, and all the noise can be removed from the background. We can also increase the dataset by recording sounds and try to increase the classes so that a better judgment can be passed. The classes can be further divided into categories, and we can train on various types of labels. This will ensure better usage, and outreach will be more.

7 Future Work

Future work concerns with developing services which can compare the performance in noisy and noiseless conditions using parameters which have not been explored yet. This will highlight its use in realistic conditions. Removal of noise is one of the main aspects which will work toward detection of numerous events and estimates traffic levels. All this will in turn reduce human health issues [15].

References

1. Pooja RK, Shetty S, Suhani M, Janardhana DR (2018) Sound classification using machine learning and neural networks. Int J Innov Res Technol 4(12):783–786
2. Breiman L (1999) Pasting small votes for classification in large databases and on-line. Mach Learn 36(1):85–103
3. Blaha M, Rumbaugh J (2005) Object oriented modelling and design with UML, 2nd edn, Pearson Education, pp 21–157
4. Kons Z, Toledo-Ronen O (2013) Audio event classification using deep neural networks. In: Proceedings of INTERSPEECH, pp 1482–1486
5. Fatahillah NR, Suryati P, Haryawan C (2017) Implementation of Naive Bayes classifier algorithm on social media (Twitter) to the teaching of Indonesian hate speech. In: 2017 international conference on sustainable information
6. Harte C, Sandler M, Gasser M Detecting harmonic change in musical audio in AMCMM
7. Albarakati N, Kecman V (2013) Fast neural network algorithm for solving classification tasks: batch error back-propagation algorithm. In: 2013 Proceedings of IEEE Southeastcon, Jacksonville, FL, pp 1–8
8. Piczak KJ (2015) 2015 IEEE international workshop on machine learning for signal processing environmental sound classification with convolutional neural networks. IEEE Int Work Mach Learn. signal Process Boston, USA, 2015, https://doi.org/10.5281/zenodo.12714
9. Salamon J, Bello JP (2015) Unsupervised feature learning for urban sound classification. In: 2015 IEEE international conference on acoustics, speech and signal processing (ICASSP). IEEE, pp. 171–175
10. Justin S, Pablo BJ (2015) Feature learning with deep scattering for urban sound analysis. In: Signal processing conference (EUSIPCO), 2015 23rd European. IEEE, pp 724–728
11. McLoughlin I, Zhang H, Xie Z, Song Y, Xiao W (2015) Robust sound event classification using deep neural networks. IEEE Trans Audio, Speech Lang Process 23(3):540–552. https://doi.org/10.1109/taslp.2015.2389618
12. Defreville B, Pachet F, Rosin C, Roy P (2006) Automatic recognition of urban sound sources. In: Audio engineering society convention 120. Audio Engineering Society
13. Zhou H, Song Y, Shu H (2017) Using deep convolutional neural network to classify urban sounds. In: IEEE Region 10 annual international conference proceedings/TENCON, vol 2017 Dec, pp 3089–3092. https://doi.org/10.1109/tencon.2017.8228392
14. Grangier D, Bengio S (2008) A discriminative kernel-based approach to rank images from text queries. In: IEEE transactions on pattern analysis and machine intelligence, 30(8):1371–1384
15. Luitel B, Murthy YVS, Koolagudi SG (2016) Sound event detection in urban soundscape using two-level classification. In: 2016 IEEE International Conference on Distributed Computing, VLSI, Electrical Circuits and Robotic Discover 2016—Proceedings, pp 259–263. https://doi.org/10.1109/discover.2016.7806268
16. Thwe KZ, War N (2017) Environmental sound classification based on time-frequency representation. In: Proc.—18th IEEE/ACIS international conference on software engineering, artificial

intelligence, networking, and parallel/distributed computing SNPD 2017, pp 251–255. https://doi.org/10.1109/snpd.2017.8022729
17. Hirano T, Maekawa T (2013) A hybrid unsupervised/supervised model for group activity recognition. In: Proceedings of the 2013 international symposium on wearable computers. ACM, pp 21–24
18. Aucouturier J-J, Defreville B, Pachet F (2007) The bag-of-frames approach to audio pattern recognition: a sufficient model for urban soundscapes but not for polyphonic music. J Acoust Soc Am 122(2):881–891
19. Salamon J, Jacoby C, Bello JP (2014) Urban sound datasets. In: MM'14 Proceedings 22nd ACM International Conference Multimedia, vol 3, pp 1041–1044. https://doi.org/10.1145/2647868.2655045
20. Zhou H, Song Y, Shu H (2017) Smart energy and environment cluster institute for Infocomm research. In: IEEE region conference (TENCON), Malaysia, 5–8 Nov 2017
21. Bleeck S, Ives T, Patterson RD (2004) Aim-mat: the auditory image model in matlab. Acta Acust United Acust 90(4):781–787
22. Walters TC (2014) Aim-c, a C++ implementation of the auditory image model. [Online]
23. Hinton GE, Osindero S, Teh Y-W (2006) A fast learning algorithm for deep belief nets. Neural Comput 18(7):1527–1554
24. Gupta R, Audhkhasi K, Lee S, Narayanan S (2013) Paralinguistic event detection from speech using probabilistic time-series smoothing and masking. In: Proceedings of interspeech, Lyon, France, pp 173–177
25. Huang Z, Weng C, Li K, Cheng Y-C, Lee C-H (2014) Deep learning vector quantization for acoustic information retreival. In: Acoustics, speech and signal processing. In: 2014 IEEE international conference on ICASSP 2014 proceedings IEEE, pp 1364–1368. [31] Glasberg BR, Moore BC (2002) A model of loudness applicable to time-varying sounds. J Audio Eng Soc 50(5):331–342
26. Suuny S, David Peter S, Poulose Jacob K (2013) Performance of different classifiers in speech recognition. IJRET
27. Morgan N (2012) Deep and wide: multiple layers in automatic speech recognition. IEEE Trans Audio Speech Lang Process 20(1):7–13
28. Moore BCJ (1992) An introduction to the psychology of hearing. Academic Press
29. Moore BC, Patterson RD (1986) Auditory frequency selectivity. Plenum Press
30. Bergeaud F, Mallat SG (1995) Processing images and sounds with matching pursuits. In: SPIE's 1995 international symposium on optical science, engineering, and instrumentation. International Society for Optics and Photonics, pp 2–13

Android-Based Driver Alert System for Accident Avoidance

Nidhi Sinha

Abstract Android-based driver alert system is a real-time application which is used for detecting driver drowsiness and alerts the driver to reduce the possible accidents. In this paper, the detection is based on the results of face and eye region or the pupil of the eyeball to determine whether the driver is in the drowsy condition or not. A new approach is introduced for detecting driver drowsiness by using Google API and OpenCV library. My system uses a modified implementation of the Viola-Jones algorithm for faces and eye detection of the driver. Detection is performed on the basis of the symptoms related to eye and face region of driver. In this method, face detection module, motion detection module, on road motion detection module, GPS module are used for extracting the symptoms of driver drowsiness. Android-based smart phones are equipped with several sensors that can help to assist in safety development for drivers on the road. Comparison between OpenCV library and Google API of both face and eye detection as shown in this paper using graph. The average correction detection ratio of face detection is 100% and eye detection is 85%, which shows my proposed system is feasible, cost effective, and reliable.

Keywords Drowsiness detection · Eyes detection · Face detection

1 Introduction

Most of the studies and researches showed that the vehicle drivers involved in accidents are highly related to driver's drowsiness, for example, non-stop driving behaviors. According to the recent survey, approximately 1.2 million people die on the world's road accident each year [1]. Driver inattentiveness is one of the main factor for causing road accidents. Driver inattentiveness means concentrating on secondary task like talking with someone while driving, yawning frequently, mentally depressed, having difficulty for keeping eyes open, lack of sleep [2]. The enhancement in the

N. Sinha (✉)
HKBK College of Engineering, Bangalore, India
e-mail: nidhi10nitp@gmail.com

R. P. Mahapatra et al. (eds.), *Proceedings of 6th International Conference on Recent Trends in Computing*, Lecture Notes in Networks and Systems 177,
https://doi.org/10.1007/978-981-33-4501-0_32

vehicles manufacturing over the last decades have made our vehicles more controlling, easier to drive and manage safer more energy efficient and more environmentally gracious. Nowadays, most of the accidents caused by cars are mostly due to the driver drowsiness. In the year of 2013, approximately 10,076 people were died in alcohol-injured-driving crashes, an average of one alcohol-injured-driving fatality occurred every 52 min. These alcohol-injured-driving fatalities accounted for 31% of the total motor vehicle traffic fatalities in the U.S. [3]. In this paper, I propose utilization of android-based mobile phones as the platform for abnormal driver detection system. As a self-sufficient apparatus, mobile phone are the combinations of hardware and software environment for the development of active abnormal driver monitoring system. The minimum requirement for such a mobile phone platform is the presence of simple sensors, e.g., accelerometer sensor. At present, many android-based smart phones fulfill this requirement. They include various types of sensors, including accelerometers sensors, and also GPS (global positioning system) facilities. Their communication component and speakers are unsurprisingly good enough for alerting the driver while driving. Approximately, 53 million smart phones were sold in the year 2014 in India [4], and their popularity is predictable to constantly increases in the upcoming year due to decreasing price.

1.1 Related Work

Improvement of public safety and to reduce the possible accidents is the main goal of android-based driver alert systems. Cameras of android-based mobile phones are used to detect the driver drowsiness. Several studies related to driver drowsiness are described as follows. Authors Sharma et al. [5] and Horng et al. described the utilization of numbers of pixels in the image of eye which is used to determine the eye is open or close. Author Horng et al. [6] described to locate eyes location by establishing an edge map. Later on many researches has been done in this area. Author [7–9] described about feature-based eye detection. This method includes various advance approaches like image projection and image binarization, which assume that the eye is darker than the face. More and more difficulties in processing is needed to detect the location of eyes because this method gives high error rates but this method is simple in detection. Later on, more and more researches have been done in eye detection. In [10], author described the eye detection method by using a geometrical face model having some features which was used for detecting eyes. In [11], authors described hybrid method for eye detection. In [12], authors described one of the accurate and fast methods for eye detection. Infrared spectrum is used for the detection. In this method, optical properties and physiological properties of the pupil are used in infrared spectrum.

1.2 Problem Definition

Driver drowsiness is the major cause of traffic crashes and financial losses. In the recent years, number of road accidents has increased manifold due to stress/mobile phone uses while driving, etc., resulting in lack of concentration of drivers. At present, there is no effective low cost mechanisms to prevent this. The vehicles are becoming smarter and robust but it is not resulting in lowering in accidents causing due to human failures. In India, numbers of road accident due to abnormal driving is as high as 6% of the world road traffic accidents.

The rate of world road traffic accident in India is one of the highest in the world. In Table 1, it shows the drunk driving statistics.

A. *Abnormal State*
   Abnormal state is a state where the below mentioned symptoms will apply:

   (1) *Sleepy/Drowsy or Drunk Driving State*:
   (2) *Tending to children*:
   (3) *Drunk above limit*:
   (4) *Yawning frequently*:
   (5) *Using a cellular phone.*

B. *Normal and Attentive State*
   Normal State is a state where the below mentioned symptoms will apply:

   (1) *Attentive State*:
   (2) *Relaxed + Attentive*:
   (3) *Drunk within limit + Attentive*:
   (4) *Tired + Attentive*:
   (5) *Talking with someone + Attentive*:
   (6) *Listening to song + Attentive.*

Table 1 .

Year	Total number of fatalities	Total number of alcohol related fatalities
1990	44,599	22,587
1991	41,508	20,159
1992	39,250	18,290
1993	40,150	17,908
1994	40,716	17,308
1995	41,817	17,732
1996	42,065	17,749
1997	42,013	16,711
1998	41,501	16,673
1999	41,717	16,572
2000	41,945	17,380

1.3 Proposed System

In the proposed system, the driver drowsiness is detected only by the processing of face and eye region as shown in Fig. 1. The main reason of driver drowsiness and distraction happened only due to driver's sleeping or not attentive while driving, so our main aim is to make driver attentive. In this paper, android-based mobile phones are used for detecting eyes and face of driver. We put the mobile phone at the top of the car with the help of stand. A new approach is introduced for detecting driver drowsiness by using Google API and OpenCV library as shown in Fig. 2. Our system uses a modified implementation of the Viola-Jones algorithm for faces and eye detection of the driver. Detection is performed on the basis of the symptoms related to eye and face region of driver. In this method, face detection module, motion detection module, on road motion detection module, global positioning system (GPS) module is used for extracting the symptoms of driver drowsiness. Android-based smart phones are equipped with several sensors that can help to assist in safety development for drivers on the road. Comparison between OpenCV library and Google API of both face and eye detection as shown in this paper using graph. The average correction detection ratio of face detection is 100% and eye detection is 85%, which shows our proposed system is feasible, cost effective, and reliable.

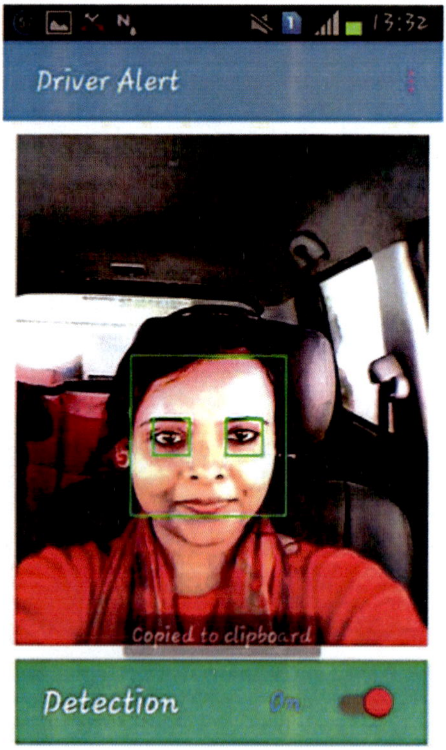

Fig. 1 Face detection and eye detection

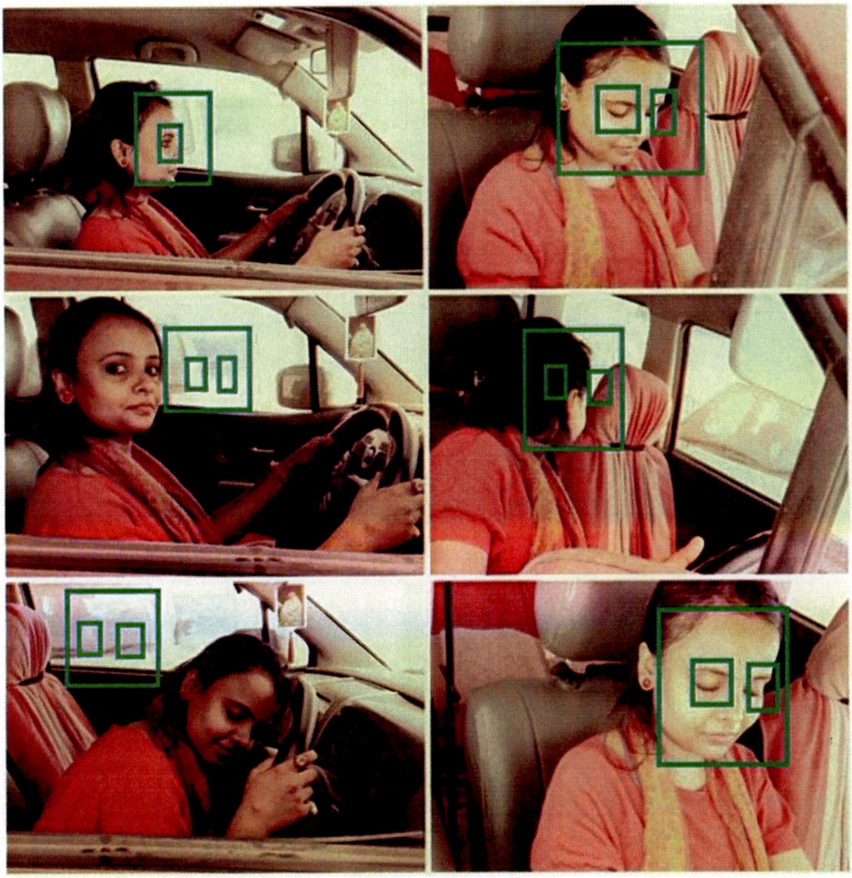

Fig. 2 Detection of driver's face and eye

A. *Process of Proposed System*

In the android-based driver alert system, it detects the driver's drowsiness and fatigueness using android-based smart phones as shown in Figs. 3 and 4. In this proposed system, we detect face and eye region of driver and monitor the eyelid closure using a threshold value as shown in Fig. 2. For eye closure duration, a common method is used for this called as "PERCLOS"—"Percentage of eye closure." In this method, percentage of time eyes are 75–99% closed over in interval of time. If the eyes are not detected and the "PERCLOS" exceeded the threshold, then the driver is considered as a abnormal or in sleepy state. The driver alert system is mainly focus on the recognition of face and eye as shown in Fig. 3. It consists of four major sections: (1) Image acquisition; (2) pupil and face recognition; (3) Blink Detection; and (4) activate alarm.

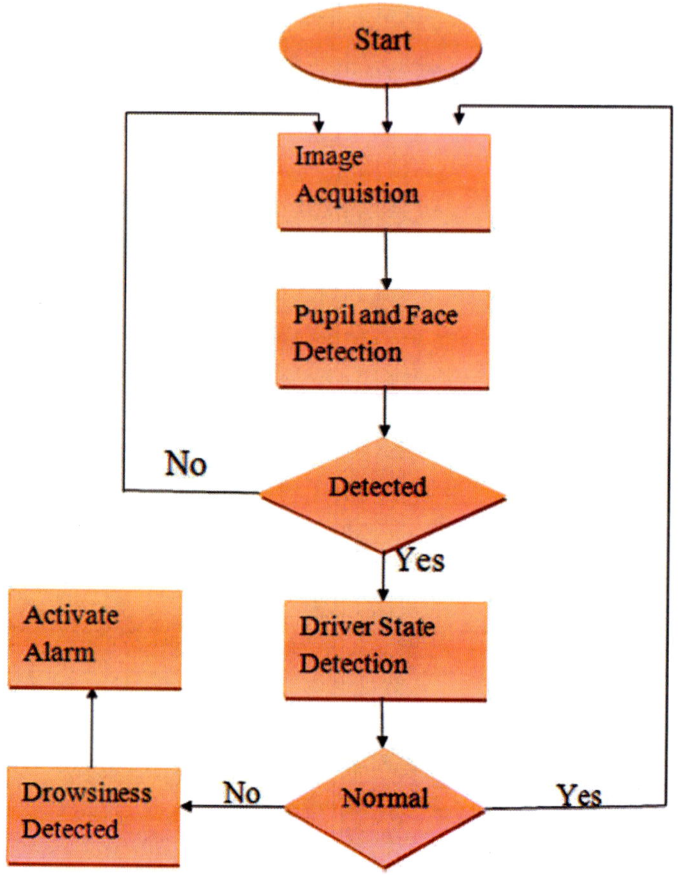

Fig. 3 System architecture for driver alert system

(1) Image Acquisition System: The purpose of this stage is to obtain high quality video images of the driver's face and eye. It is a major issue to acquire a high class image of eye and face with adequate details. There are many constraints for implementing such a system. However, such a system with the desired uniqueness up to an extent can be implemented. The images used in this study are taken from the record. The image is converted from RGB to gray level for color processing.

(2) Pupil and Face Recognition: In this part, the pupil and face detection method are the first level of drowsiness detection technique. The OpenCV library is used in this level for face and eye detection. We initiate a very fast and memory efficient face detection algorithm based on Viola-Jones method. This detection method proposed by Viola and Jones [8]. It is a real-time method which is very popular for face detection and implemented in our driver drowsiness detection system.

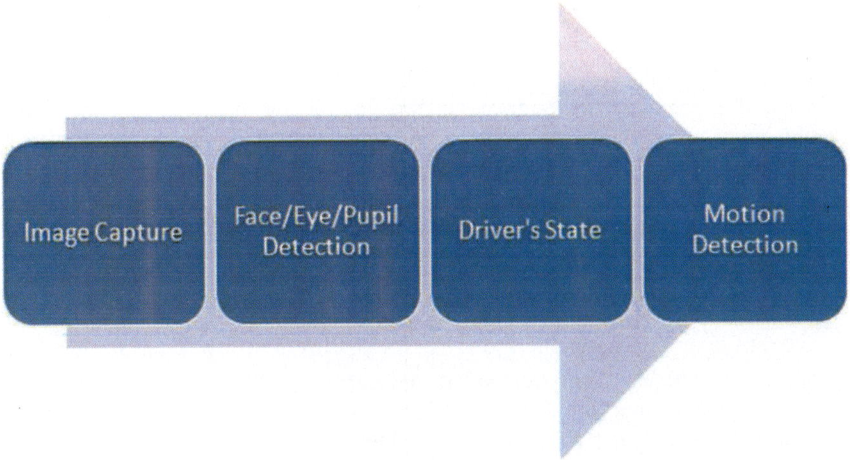

Fig. 4 Driver drowsiness system architecture

The Haar-like facial appearance can be proficiently calculated by the demonstration of the image with integral image. The cascaded AdaBoost classifier ensures high face and eye detection rate and low false alarm rate. The cascade classifiers reduce the time for searching the face area by using AdaBoost technique which is an optimal way. Several weak classifier become a strong classifier by the help of boosting classifier using this cascade algorithm.

(3) Blink Detection: Blinking of an eye is detected while the change of black pixels in the eye area. An open eyeball will have bigger visible pupil than a blinking eye. The pupil is often darker than the color of the rest of the eye. Thus, an open eye should have more black pixels. In the detected eye region, the eye image is converted to a binary image by using image thresholding. Then, the ratio of black pixels in the binary image of the eye is calculated. The ratio of black pixels is used as the criterion for eye blinks.

(4) Activate alarm: When the driver feels drowsy or not attentive towards driving, an alarm is activated to wake up the driver. The speakers of android-based mobile phones are good enough for alerting the driver while driving.

1.4 Symptoms Correlated to Tiredness and Disruption

In android-based driver alert system, useful symptoms for tiredness and disruption detection can be divided into four general categories:

A. **Indication Correlated to Eye Area**

The most important and dynamic part where every types of visualization of drowsiness and fatigueness occurs. Some indications like sleepy and drowsy are the indication of driver which shows the driver is drowsy. It is used only by sensing the driver eyes and face monitoring. As we all knows that eyes are the important part and the most important area. Most of the indication like drowsiness of driver is extracted by the eyes and the pupil.

1. Eye Closure: Eye closure is the simple indication which is used for conveying a message that the driver is drowsy. Eye Closure provides very useful information about driver anesthesia and drowsiness. Eye closure of two types for detecting drowsiness of driver, (1) continuously eye closure (2) For a limited period of time calculating the percentage of eye closure. In first part, it describes a continuously eye closure, if eyes are continuously closed for limited time then drowsiness is detected and an alarm is activated.
2. Eye Blinking Speed: Eye blinking speed is the rate of opening and closing period of eyelids movement during one frame. Drowsiness and distraction can be point out due to the reduced rate of eye blinking speed. This eye blinking speed is the time which spends in the period of an opening and closing of the eyelids happens during a blink event and for detecting the status of drowsiness of the driver, this indication is used. If the blinking speed crosses the threshold value which has range of about 0.5–0.8 s, then the status of the drowsiness of the driver will results as true.
3. Eye Saccadic Action: From the Gaze route or direction discussed before, one of the indications extracted is the eye saccadic rate and its unequal movements. The driver often has a quick review of environments while the driver has normal consciousness. Some asymmetrical and rapid saccadic movements are responsible for the occurrence of this rapid view. Quantity of such eye saccadic movements should be in a specified field of view.
4. Eye Saccadic Action: From the Gaze route or direction discussed before, one of the indications extracted is the eye saccadic rate and its unequal movements. The driver often has a quick review of environments while the driver has normal consciousness. Some asymmetrical and rapid saccadic movements are responsible for the occurrence of this rapid view. Quantity of such eye saccadic movements should be in a specified field of view.

B. **Symptoms Correlated to Mouth Region**

(1) Yawning: Another indication of drowsiness is yawning, which is occurred due to mouth gap. Driver talking, i.e., the continuous movement of its lips can be restrained as a symptom of drowsiness. That's why, the detection of driver's drowsiness and distraction can be measured with the symptoms related to the mouth region. The opening of mouth was sensed based on the proportion of width to height of mouth. It will be less if the mouth is closed as compared to when the mouth is open.

C. **Symptoms Correlated to Head**
An effective symptom for the recognition of the drivers drowsiness is the determination of head orientation and the estimation of driver intent to traffic lane deviations. In common, a 3-dimensional representation of the head is prerequisite intended for resolution of head orientation.

(1) Fixed Head: Over the extended epoch of time, the fixed position of the head is the symptom of drivers distraction. While driving, if driver focuses on something else except his/her driving task in his/her mind, then its head position will be stable for that period of time. And, this will be indication of the symptoms of driver drowsiness.

D. **Symptoms Correlated to Face Region** Sensations, effective behavior, conversations like verbal and non-verbal are resultant of facial appearance. For a period of time, the facial muscles changes results as a feature of changes in face image. Face features are extracted and handled via a spatial-temporal algorithms and data structures. Facial expressions are basically segregated into a number of uncomplicated expressions such as ordinary, joyful, depressed, surprised and annoyed, and so on.

2 Implementation

These are the main modules present besides the helper modules, which is responsible to make our system "Driver Alert" to work.

A. Image Capture

This module is responsible to provide input to other modules for processing. Mobile devices may have more than one camera device present, namely the front and the rear. When the mobile device is setup in the car to work, the front camera is used to detect drivers face, eyes and the pupil for further processing. The rear camera is then selected to be used by the motion detection module. Selecting the proper camera for the modules is also handled by this module. After the selection of the proper camera, this module provides the images from the camera feed to different modules for further processing as shown in Fig. 5.

B. Face/Eye Detection Module

The next module, one of the most important module. This module process the received (from the previous module) image to determine the drivers face, eyes, and pupil. This module leverages the functionality provided by the OpenCV library to do so. The detection of the said features is required to ascertain the state of the driver and triggering of the alarm. The OpenCV library provides an implementation of the Viola- Jones algorithm to detect face and eyes. It uses the Haar-Cascade classifier to detect both the features, i.e., face and eyes. Face detection not only helps in speeding

Fig. 5 Abnormal driving

up the process of detecting eyes but also aids in determining the attentive state of the driver. This module, further, detects the state of the pupil to improve the accuracy of our project. Pupil detection is done by a simple image processing technique as OpenCV library does not provide any functionality to detect pupil. The technique involves converting the image to a single channel image, preferably gray scale, and then a thresholding technique is applied to convert the image to binary and then from this binary image the pupil is detected. Pupil detection also helps in determining the blinking status of the driver whether the blinking rate/speed is normal or abnormal. Thus, OpenCV for android, i.e., OpenCV SDK (v2.4.10) for android is used to process the above-mentioned features.

2.1 Drivers State Detection

This module process the information passed to it from the previous modules and reduces the state of the driver based on his/her face movement, eye detection, and blinking rate. The state can be represented as:

- Abnormal State: -For drunk (above limit) driving
- Sleepy/Drowsy State
- Attentive State
- Unknown State.

This method uses a simple algorithm devised by us to determine the state. Testing were carried out to collect sample, containing face/eye movements. These samples were used to find a pattern representing these states and also to find some threshold values to decide the state in real time. The above-mentioned pattern conveys some information like if the driver is in attentive state his head movement will be negligible most of the time, thus it can be used to change the alarm trigger time accordingly. Similarly, if the person is feeling drowsy his head movement will be downwards either in left, right or straight direction or may be straight up, and if the driver is drunk, then his head movement will be haphazard most of the time, and thus, these can be used to modify the alarm trigger duration accordingly.

2.2 Motion Detection

Since the driver can stop the car and move away from the mobile phone camera before turning off the application, this may lead to false activation of the alarm. Thus, for such cases as well as making it convenient for the user, our motion detection module detects motion before triggering the alarm. There are three sub-modules which try to detect motion independently and then each of those sub-modules return a result according to their processing. Then, a majority function returns the motion detection result according to the majority of the result obtained from the sub-modules; i.e., majority function will return true if two of three sub-modules returns true as well. The three modules are:

- Accelerometer sub-module.
- On road motion detection sub-module.
- GPS sub-module.

3 Results

A. Battery Consumption Graph

Our system is currently implemented in java to run on Android OS, thus for any feasible use in real-life scenario, our project must efficiently run on the resource-constrained mobile devices without draining too much of the battery life. We tested two approach namely the Google API and the OpenCV library, both have certain subtle differences in their implementation. A comparative graph is shown above which depicts the battery consumed by each of the approach. The test was run on the Google Nexus 7 (NVidia Tegra 3) device running Android 4.4.4 Kit Kat. Our sample application was kept running for 30 min, the device battery level was noted before running the application and also after the testing was complete. The difference between noted levels was the drop in the battery level (percentage drop). The above graph shows that the OpenCV library (22%) consumes more battery than the Google API (15%). This may be contributed to the fact that the OpenCV implementation does a lot of image format conversion before and after processing while those conversion are not required in the case of Google API as shown in Fig. 6.

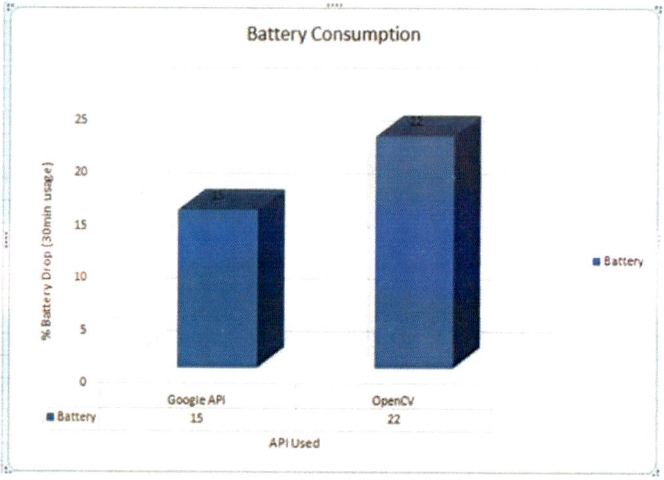

Fig. 6 Battery consumption graph

3.1 Performance Graph

Mobile devices are constrained in hardware resources, so one of the main concerns about any application is to utilize the resources efficiently. Thus, we did a comparative testing of both our approaches, namely the Google API and the OpenCV library. The testing was carried out on Google Nexus 7 (NVidia Tegra 3) device running Android 4.4.4 Kit Kat. This test looks out for the performance differences between the two approaches, the sample application shows the resultant image feed after processing, and the fps is recorded as the performance metric. The faster the implementation, the higher is the fps. The above graph shows that the Google API (21 fps) has better performance than the OpenCV library (15 fps), this may be due to the image format conversion taking place in OpenCV library while no such conversion is required in case of the Google API as shown in Fig. 7.

4 Conclusion

This paper employs motion capture technology to build a distracted driving drowsiness that detects and alerts drivers. Many modern technologies exist to detect the driver fatigue. Although the products made by FORD, TOYOTA, etc., are quite satisfactory in performance but they are not cost effective, since India has a majority of middle class population thus, this system is not popular among the people. To resolve such kind of issues, we proposed this model "Android-Based Driver Alert System For Accident Avoidance" which is cost effective, reliable, and efficient.

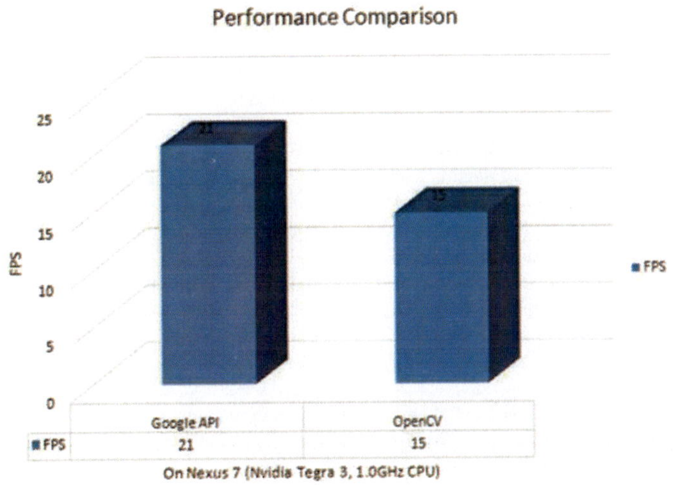

Fig. 7 Performance graph

5 Future Scope

In future many applications are likely to be implemented using brake support. By using sensors in brake and seat belt, we can manifold the accuracy and the rate of detection in future. We can use an embedded system for enhancing the accuracy and time of detection. Much work needs to be implement in order to make it cost effective.

References

1. Ji Q, Zhu Z, Lan P (2004) Real-time nonintrusive monitoring and pre-diction of driver fatigue. IEEE Trans Veh Technol 53:1052–1068
2. Fasel B, Luettin J (2003) Automatic facial expression analysis: a survey. Pattern Recogn 36:259–275
3. Huang D (2011) Local binary patterns and its application to facial image analysis a survey. Bin Patterns 41:765–781. http://www.scholarpedia.org/article/Local
4. http://timesofindia.indiatimes.com/tech/tech-news/Smartphone-sales-in-India-to-cross-53-million-in-2014/articleshow/45284438.cms
5. Xiong (Bill) Yu PE (2012) Non-contact driver drowsiness detection system. Safety IDEA
6. http://www.daimler.com/dccom/0-5-1210218-1-1210332-1-0-0-1210228-0-0-135-0-0-0-0-0-0-0-0.html
7. Hankey JM, Dingus TA () The development of design evaluation tool and model of attention demand. Virginia Tech Transportation Institute Monk C.A (Science Applications International Corporation)
8. Viola P, Jones M (2001) Rapid object detection using a boosted cascade of simple features. In: Conference on computer vision and pattern recognition, 2001, pp 1–9

9. Dai J, Teng J, Bai X, Shen Z (2010) Mobile phones based drunk driving detection. In: Proceedings 4th International Conference on PervasiveHealth no permissions, 2010, pp 1–8
10. Lee J, Li J, Liu L (2006) A novel driving pattern recognition and status monitoring system. In: First Pacific Rim symposium, PSIVT, 2006, pp 504–512
11. Sahayadhas A, Sundaraj K (2012) Detecting driver drowsiness based on sensors a review. Malaysia pp 16937–16953. ISSN 1424-8220
12. Chan CY (2002) On the detection of vehicular crashes-system characteristics and architecture. IEEE Trans Veh Technol 51(1):180–193
13. Bergasa LM, Sotelo MA, Barea R (2006) Real-time system for monitoring driver vigilance. IEEE Trans Intell Transp Syst 7
14. Torkkola K, Massey N, Wood C (2004) Driver inattention detection through intelligent analysis of readily available sensors. In: IEEE intelligent transportation systems conference U.S.A, Oct 2004, pp 326–331
15. Boer ER (2000) Behavioral entropy as an index of workload. In: Proceeding of the IEA/HFES,2000, pp 125–128
16. Albu AB, Widsten B, Wang T, Lan J (2008) A computer vision-based system for real time detection of sleep onset in fatigued drivers. In: IEEE intelligent vehicles symposium, June 2008, pp 25–30
17. Park I, Ahn J-H, Byun H (2006) Efficient measurement of eye blinking under various illumination conditions for drowsiness detection systems. In: Proceedings of the 18th international conference on pattern recognition
18. Desai AV, Haque MA (2006) Vigilance monitoring for operator safety: a simulation study on highway driving. J Saf Res 37:139–147
19. Krajewski J, Sommer D, Trutschel U (2006) Steering wheel behavior based estimation of fatigue. In: Proceeding of the 5th international driving symposium on human factors in driver assessment, training and vehicle design, 2006, pp 118–124
20. Lecce VD, Calabrese M (2008) Experimental system to support real time driving pattern recognition. In: Advanced intelligent computing theories and applications with aspects of artificial intelligents, Ann Emerg Med 37:1192–1199
21. http://www.toyota-global.com
22. Lecce VD,Calabrese M, Yawning detection for determining driver drowsiness. IEEE international workshop VLSI design and video tech. Suzhou, China, May 28–30
23. Shivangi R, Somani MSB (2012) EEG signal processing and classification of sensorimoter rhythm-based BCI. Int J Eng Res Technol (IJERT) 1(4)
24. Li L, Xie M, Dong H (2011) A method of driving fatigue detection based on eye location. Proc IEEE 3rd ICCSN, China, 2011, pp 480–484

Assessment and Analysis of Action Degeneracy Due to Blackhole Attacks in Wireless Sensor Networks

T. Saravanan and P. Sasikumar

Abstract Wireless sensor systems (WSN) have turned into a developing zone of innovative work because of the huge number of uses that can significantly profit by such frameworks and has lead to the advancement of minor, modest, expendable and independent battery controlled sensor nodes. One of the real challenges of WSN is security. The issue of security is because of the remote idea of the sensor systems and the natural power and battery constraints of sensor nodes make common security arrangements which are unfeasible. In this manner, the system is vulnerable to attacks by cruel routes and the information is dropped in attacks like blackhole attack. In this paper, it is proposed to assess the presentation of a system under the effect of hierarchical cluster-based data dissemination (HCDD) convention with Markov model of inductive routes.

Keywords Wireless sensor network (WSN) · Blackhole attack · Attacks · Performance

1 Introduction

A wireless sensor network (WSN) comprises of many sensor nodes. Each sensor node has a sensing capability as well as limited energy supply, compute power, memory and communication ability. Besides various applications, WSN may be used to monitor weather monitoring and wildlife habitats, the structural integrity of bridges and buildings, building security, location of valuable assets, traffic, and so on. However, realizing the huge potential of WSNs posses many research challenges

---

T. Saravanan (✉)
Department of Computer Science and Engineering, Galgotias College of Engineering and Technology, Greater Noida, Uttar Pradesh, India
e-mail: tsaravcse@gmail.com

P. Sasikumar
School of Computing Science and Engineering, Galgotias University, Greater Noida, Uttar Pradesh, India
e-mail: sasi.mca@gmail.com

R. P. Mahapatra et al. (eds.), *Proceedings of 6th International Conference on Recent Trends in Computing*, Lecture Notes in Networks and Systems 177,
https://doi.org/10.1007/978-981-33-4501-0_33

ranging from hardware and architectural issues to programming languages and operating systems for sensor networks, to security concerns, to algorithms for sensor network deployment, operation, and management [1–3].

The WSN do communicate a few shared behavior to general unplanned arranges. In this way, convention structure for sensor systems must record for the properties of specially appointed systems, including the accompanying like lifetime limitations forced by the constrained vitality supplies of the routes in the system, unreliable correspondence because of the remote medium and need for self-design, requiring practically zero human mediation.

Consolidating these exceptional highlights of sensor systems into convention configuration is significant so as to productively use the restricted assets of the system. Simultaneously, to keep the conventions as lightweight as could reasonably be expected, numerous plans center on specific subsets of these criteria for various sorts of uses [4]. This has prompted a significant number of various conventions from the information connection layer up to the vehicle layer, each with the objective of enabling the system to work independently for whatever length of time that conceivable while keeping up information channels and system handling to give the application's required nature of administration.

WSN falls in three important classifications called intermittent detecting, inquiry detecting and occasion detecting. In intermittent detecting observing is constantly done by the sensor which screens physical condition and persistently revealing estimation to the sink. In occasion detecting sensor work in a quiet checking state and are modified to tell about the occasion, for example, the nearness of the item in interruption recognition [5]. At the point when an occasion happens in the sensor field source route created information and make the declaration to the sink that buying in the information and this entire procedure is known as information spread. Numerous information spread methods have been proposed for sensor systems to investigate in-organize question preparing, circulated information stockpiling and estimate strategies [6].

Customary (or level) directing conventions for WSN may not be ideal regarding vitality utilization. Grouping can be utilized as a vitality proficient correspondence convention. The goals of bunching are to limit the all-out transmission power totalled over the routes in the chose way, and to adjust the heap among the routes for drawing out the system lifetime. Bunching is an example of layered conventions where a system is made out of a few bunches (or groups) of sensors. Each bunch is overseen by a unique route or pioneer, called group head (CH), which is in charge of planning the information transmission exercises of all sensors in its cluster. All sensors in a group speak with a bunch head that goes about as a nearby facilitator or sinks for performing intra- transmission course of action and information accumulation. Group heads in tern transmit the detected information to the worldwide sink. The transmission separation over which the sensors send their information to their bunch head is littler contrasted with their particular separations to the worldwide sink. Since a system is portrayed by its constrained remote channel transfer speed, it would be valuable if the measure of information transmitted to the sink can be diminished [7, 8].

To accomplish this objective, nearby cooperation between the sensors in a group is required so as to decrease data transfer capacity requests.

Various level approaches break the system into grouped layers. Routes are gathered into bunches with a group head that has the responsibility of directing from the group to the next group heads or base stations. Information passes through from a lower group layer to a higher group. In spite of the fact that, it jumps from starting with one route then onto the next route, yet as it bounces starting with one layer then onto the next layer, then it covers bigger separations. This moves the information quicker to the basestation. Hypothetically, the idleness in such a model is considerably less than in the multihop model [9]. Grouping gives intrinsic reorganization abilities at the group heads. In the group-based progressive model, information is first calculated in the group at that point sent to a higher-level group head. As it moves from a lower level to a higher one, it voyages more important separations, consequently decreasing the movement time and idleness. This model is superior to the one bounce or multi-jump model [10].

A group-based chain of command moves the information quicker to the base station in this manner lessening inertness than in the multi-jump model. Further, in bunch-based model just group heads performs information conglomeration while in the multi-jump model of the road route performs information collection. Therefore, the group-based model is more appropriate for time-basic applications than the multi-jump model. In any case, it has one downside, to be specific, as the separation between bunching level expands; the vitality spent is relative to the square of the separation. This expands vitality use. Regardless of this disadvantage, the advantages of this model far exceed its downside. A group-based various leveled model offers a superior way to deal with directing for WSNs.

In organizing sensor application, sensors are to be put (or sent) in order to screen a district or a lot of focuses. It is attractive that the sent accumulation of sensors has the option to speak with each other, either straightforwardly or in a roundabout way by means of multi bounce correspondence. WSN are defenseless against security attacks because of the communication idea of the transmission medium. Fundamentally attacks are delegated dynamic attacks and latent attacks. The checking and tuning in of the correspondence channel by unapproved aggressors are known as remote attack. The attacks against security is inactive in nature. The unapproved assailants screens, tunes in to and adjusts the information stream in the correspondence channel are known as dynamic attack.

Numerous sensor system steering conventions are very basic, and therefore are now and then considerably increasingly vulnerable to attacks against general specially appointed directing conventions. Most system layer attacks against sensor systems can be categorized as one of the accompanying classifications:

- Spoofed, adjusted or replayed directing data
- Sinkhole attacks
- Selective sending
- Sybil attacks
- HELLO flood attacks

- Wormholes
- Acknowledgement mocking.

Basic system attacks incorporate blackhole, dim opening and wormhole attacks and IP ridiculing. Blackhole attacks are those when noxious routes will not advance traffic. Outer attacks can be anticipated by utilizing security systems like firewalls and encryption. Inner attacks are progressively extreme as pernicious insider routes of the system which are approved and subsequently ensured by the system's security administrations. Such vindictive insiders who can work in gatherings may utilize security intends to ensure their attacks. Such noxious gatherings are called traded off routes, as their activities bargain arrange security.

In blackhole opening attacks, a malevolent route uses directing conventions to promote itself as having the most limited way to a goal route or to a parcel it needs to capture. Such antagonistic routes promote accessibility of crisp courses without checking the steering table. In this manner, an aggressor route consistently has the accessibility to answer to the course demand along these lines blocking and holding the information parcel. In flooding, a pernicious route's answer will be gotten by the mentioning route before receipt of answer from the genuine route prompting a malignant and produced course creation. At the point when this course is set up, it is currently up to the route to guarantee whether to drop all bundles or forward it to the obscure location [11].

In this paper, it is proposed to assess the presentation of a WSN under the effect of blackhole opening attack. Recreations are led to assess the exhibition debasement of WSN because of malevolent route movement. Segment 1 managed the essentials of the WSN and attacks in WSN, segment 2 audits some related works accessible in the writing. Area 3 subtleties the strategies utilized for assessment; segment 4 gives the re-enactment result and talks about the equivalent. Segment 5 finishes up the paper.

This paper presents a design for a 3D model of a TFG with a connecting diamond-shaped frame with the simple structure to create and guarantee the anti-phase mode in the first mode of the proposed model. In addition, the dynamic characteristic parameters of the proposed TFG are determined. Moreover, the mechanical response of the proof-mass when applying the angular velocity is considered in this paper. The results are obtained by using ANSYS software and used to be the input parameter for the dynamic analysis of the TFG in the continuous researches.

2 Related Works

In hello flood attack, a foe, which was not a lawful route in the system, can flood hi solicitation to any authentic route and break the security of WSN. The present answers for these sorts of attacks are primarily cryptographic, which experience the ill effects of substantial computational unpredictability. Thus, they were less reasonable for remote sensor systems. Singh et al. proposed a technique dependent on sign quality to identify and forestall hi flood attack. Routes had been named companion and outsider dependent on the sign quality. Short customer confuses that require less computational power and battery power had been utilized to check the legitimacy of suspicious routes.

Sinkhole attack was among the most dangerous directing attacks for WSNs. It might make the interloper bait all or the vast majority of the information stream that must be caught at the base station. Steering convention Mint Route was utilized to actualize this attack. When sinkhole attack has been actualized and the foe route has begun to fill in as system part in the information directing, it can apply some more dangers, for example, sending just the particular parcels. At last this drop of some significant information parcels can disturb the sensor organizes totally. Tiwari et al. attempted to archive all the realized security issues in remote sensor arranges and talks about a wide variety of attacks and their characterization instruments in WSN and various protections accessible to deal with them including the difficulties confronted, and responded to the call and have proposed a coordinated far reaching security that will give security administrations to all administrations of sensor organize. The remote correspondence innovation likewise gets different sorts of security dangers [12].

Wazid et al. proposed a similar exhibition investigation of two WSN's topologies, for example, tree and mesh under blackhole attack was finished. On the off chance that there was a WSN inclined to blackhole attack and requires time proficient system administration for data trade at that point tree topology was to be picked. In the event that it requires throughput effective and steady administration in the system, and at that point, mesh topology was prescribed. A calculation named as Topology Based Efficient Service Prediction (TBESP) calculation has been proposed relying on the examination done which will help in picking the most appropriate topology according to the system administration prerequisite under blackhole attack [13].

Tripathi et al. give a review of LEACH, the most prevalent bunched steering convention of WSN and how LEACH can be undermined by blackhole and gray hole assailant. "High vitality edge" idea was utilized to mimic these attacks on NS-2. The presentation of WSN enduring an onslaught was completely examined, by applying it on different system parameters with different route densities. It was seen that the impact of the blackhole attack was more on the system execution when contrasted with the gray hole attack [14]. Sheela et al. propose a lightweight, quick, proficient and versatile operator innovation-based security arrangement against blackhole attack for WSNs. WSN has a unique topology, irregular network, and asset obliged gadget routes. A parcel drop attack or blackhole attack was a sort of forswearing of-administration attack achieved by dropping bundles [15].

Gondwal et al. proposed a strategy to distinguish the blackhole attack utilizing different base-stations and a check operator-based innovation. This strategy was energy proficient, fast, lightweight and reduces message unpredictability. A compelling arrangement was suggested that uses numerous base stations to improve the conveyance of the parcels from the sensor routes coming to at any rate one base station in the system, and in this way, guaranteeing high bundle conveyance achievement. The proposed strategy was more productive than the past systems and gives better outcomes. Steering through various base stations calculation was possibly initiated when there was an opportunity of blackhole attack on the system [16].

Yadav et al. exhibited a fluffy-based choice to check a route was tainted by blackhole attack or route. The proposed framework will distinguish the attack over the route just as give the answer for decrease the information misfortune over the system. The proposed strategy will initially distinguish the blackhole opening route utilizing fluffy principle. The proposed calculation gives better answer for lessening the information misfortune over the system [17]. Zhang et al. proposed a hierarchical role-based data dissemination approach, named HRDD, for enormous scale WSNs with different portable sinks. In HRDD, a progressive bunch-based structure to find and keep up the steering ways for appropriating information to the portable sink, and relegate two jobs, named Indexing Agent and Gateway Agent, while Gateway Agents add to diminish vitality utilization and the telecom messages. The reproduction results legitimize that HRDD has the capacity to lessen the vitality utilization in the remote sensor systems and to draw out system lifetime [18, 19].

Lin et al. proposed a hierarchical cluster-based data dissemination scheme, named HCDD, to scatter information to the adaptable sink with less control overhead. In HCDD, sensor routes were self-sorted out to discover the course without the learning of route's area data. The re-enactment results demonstrate that HCDD plans can incredibly mitigate the control overhead, and, simultaneously, accomplish longer organize lifetime and tantamount number of got information with past works, for example, the TTDD-like strategies [20]. Shin et al. proposed a steady spine tree development calculation utilizing multi-jump bunches for WSNs. Spine routes, which were in charge of performing and overseeing multi-jump correspondence, could lessen the correspondence overhead, for example, control traffic and limit the quantity of dynamic routes. Past spine development calculations, for example, HCDD and multi-cluster, mobile, multimedia radio system (MMM), devour vitality rapidly [21] (Table 1).

A large-scale information spread application was described by countless data streams and data customers. Shoppers were keen on various, yet covering, subsets of the streams. Multicast was utilized to convey subsets of the streams to subsets of the buyers. Since multicast gatherings were a constrained asset, every buyer must sift through an enormous number of unneeded streams. Tock et al. proposed mitigate the end-route sifting load by utilizing progressive bunching of streams to move layer sessions, and grouping of sessions to organize layer multicast gatherings. This plan takes into account progressive separating of streams at the beneficiaries. Detailing a cost capacity that models and underscores the separating procedure, and propose calculations for the arrangement of the various leveled mapping issue. Execution assessment demonstrates a critical decrease of end route sifting cost contrasted with a non-hierarchic methodology.

Table 1 Comparison table of various protocols and its cons

Approach (Year)	Objective	Attacks (Cons)
Hanane Kalkhaa, Hassan Satoria and Khalid Satori (2018)	Identify malicious nodes in the networks through prevention of blackhole attack	Blackhole (Shortest path results selected by a starting node to communicate with a ending node)
Khan and Jamil (2017)	Detects malicious nodes and prevent and overcome blackhole attacks	Blackhole, Wormhole, Grayhole and Sybil (Number of mechanisms reviewed are very few)
Gurang and Chauhan (2017)	Review of various techniques	Blackhole (Limited only to typical blackhole attack)
Dubey and Saxena (2016)	A TSDRP Protocol is proposed to secure from attacks	Blackhole (Categorization is not presented, covers only typical blackhole attack, no trust-based scheme in discussed)
Mitchell and Chen (2014)	A survey of intrusion detection(IDS) in wireless network applications	Blackhole, D-DoS (All the schemes are categorized as IDS, main topics of Routing, logical inference, learning schemes were not discussed)
Mandela et. al. (2013)	Survey of intrusion detection (IDS) in wireless network applications	Blackhole, Wormhole (Only a few categories of mitigation techniques were discussed)

3 Proposed Methodology

The hierarchical cluster-based data dissemination (HCDD) with Markov model convention characterizes various leveled group engineering to keep the area of versatile sinks and discover ways for the information spread from the sensors to the sink. Each group is formed by a bunch head, a few portals and common sensors as found in Fig. 1. At the point when a portable sink crosses the system, it registers itself to the closest bunch head. A warning message is then proliferated to all group heads. During this system, each bunch head records the sink ID and its sender with the end goal that future information reports transmission can be effectively performed from sources to sink. The principle disadvantage of the spine-based methodology is the need to keep up the structure. Moreover, the problem area issue can happen as the traffic is focused over a gathering of group heads or pioneers.

In blackhole opening attack, a malicious route utilizes its directing standard so as to publicize itself for having the brief way to the final route or to the information it needs to capture. This opposed route promotes its accessibility of new courses regardless of checking its steering table. Along these lines, assailant route will consistently have the accessibility in answering to the course solicitation and subsequently block the information parcel and hold it. In convention dependent on flooding, the malignant

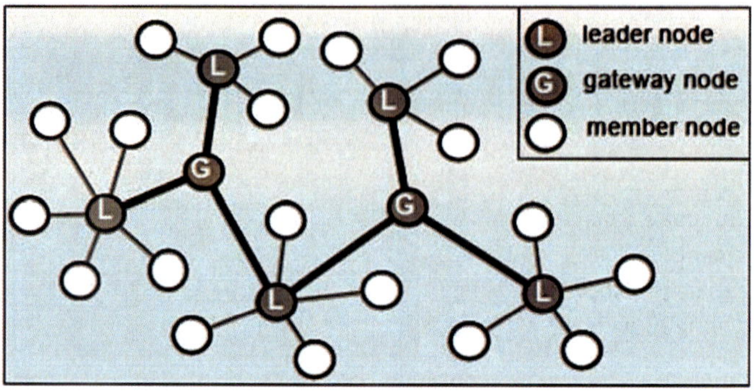

Fig. 1 Hierarchical cluster-based data dissemination

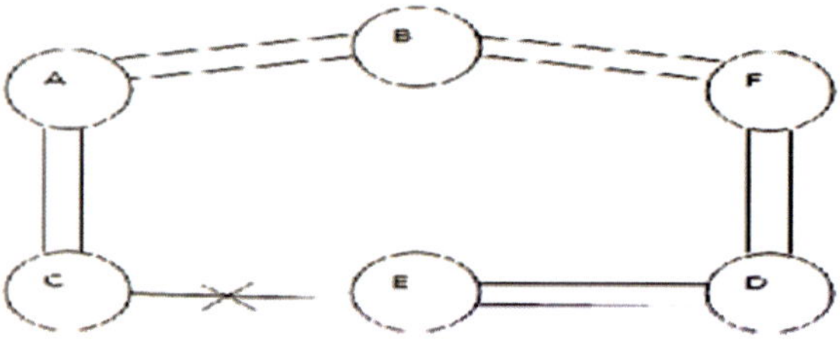

Fig. 2 Blackhole attack

route answer will be gotten by the mentioning route before the gathering of answer from genuine route; subsequently a noxious and fashioned course is made. At the point when this course is set up, presently it is up to the route whether to drop every one of the bundles or forward it to the obscure location.

The technique how malevolent route fits in the information courses changes. Figure 2 shows how blackhole opening issue emerges, here route “A” need to send information parcels to route “D” and start the course disclosure process. So the off chance that route “C” is a malevolent route then it will guarantee that it has dynamic course to the predefined goal when it gets RREQ parcels. It will at that point send the reaction to route “A” preceding some other route. Along these lines route “A” will imagine this is the dynamic course and accordingly dynamic course disclosure is finished. Route “A” will disregard every single other answer and will begin seeding information parcels to route “C”. Thusly, every one of the information parcel will be lost expended or lost.

4 Result and Discussion

In this work, the effect of blackhole opening attack is seen in WSN. The trial arrangement comprises of 40 routes appropriated over 1.5 km^2. Various leveled group-based information scattering is utilized. Two analyses are led the first without pernicious routes and the second with 5% of the routes being malicious. The vindictive routes are intended to haphazardly drop bundles independent of the source or goal address. The attack reproduced is black opening attack. Figures 3, 4 and 5 demonstrate the system execution regarding throughput, start to finish postponement and information dropped separately. Every one of the yields plotted are in time normal organization.

Figure 3 demonstrates the consequences of the throughput for WSN with and without pernicious nodes. The throughput diminishes by a normal of 34.27% within the sight of 5% vindictive routes. It is seen from Fig. 5 that the start to finish postpones increments by a normal of 50.39% because of the blackhole attack in WSN.

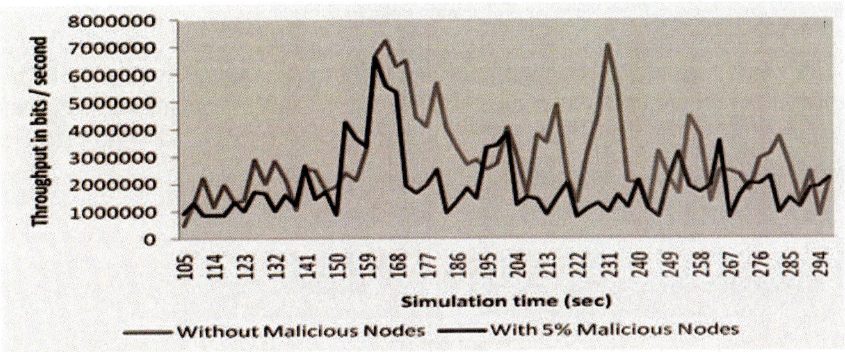

Fig. 3 Throughput

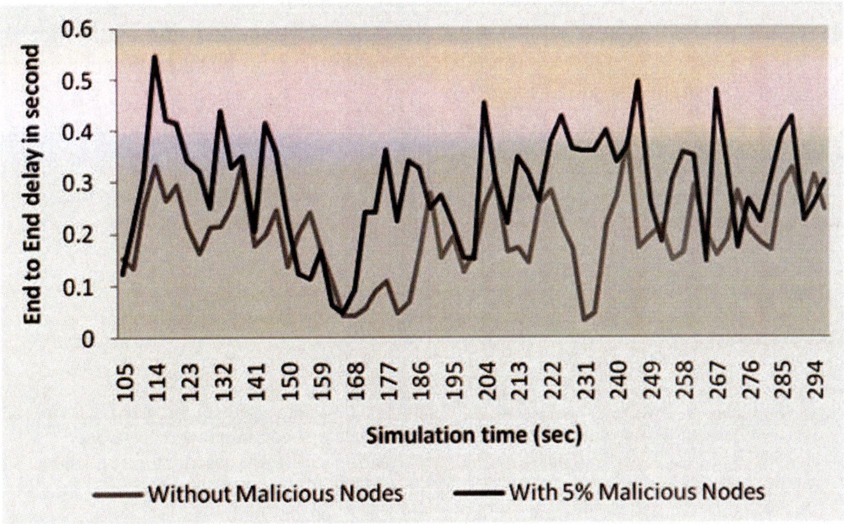

Fig. 4 End-to-end delay

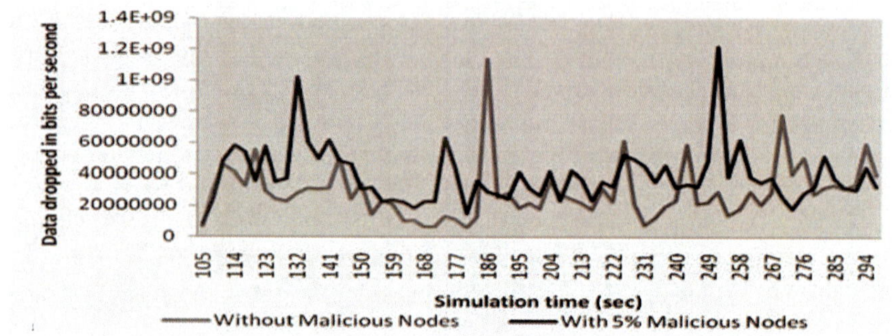

Fig. 5 Dropping of data

In a dark opening attack, the aggressor swallows (for example gets yet does not advance) every one of the messages he gets. Subsequently, any data that enters the dark hole district was caught. It is found in Fig. 5 that a normal of 35.3% information parcels are dropped or caught by the malevolent routes.

5 Conclusion

The most advances of equipment producing innovation and the improvement of effective programming calculations make in fact and financially attainable a system made out of various, little, ease sensors utilizing remote correspondences. In this paper, it is proposed to assess the exhibition of a system under the effect of destructive routes by hierarchical cluster-based data dissemination (HCDD) convention with Markov model. The spiteful routes are intended to randomly drop bundles independent of the source or goal address. The attack re-enacted is blackhole attack. Reproduction results demonstrate that the throughput diminishes by a normal of 34.27% and information dropped increments by a normal of 35.3% within the sight of 5% malignant routes. Further examinations to distinguish malevolent route and approaches to moderate them are basic.

References

1. Tripathi A, Mohapatra AK (2016) Mitigation of blackhole attack in MANET. In: 8th international conference on computational intelligence and communication networks
2. Patel HB, Jinwala DC (2019) Blackhole detection in 6LoWPAN based internet of things: an anomaly based approach. In: TENCON 2019—2019 IEEE Region 10 conference
3. Saravanan T (2014) An efficient multi channel query scheduling in wireless sensor networks in IJCSNS. Int J Comput Sci Network Secur 14(2):71–77
4. El-Semary AM, Diab H (2019) BP-AODV: blackhole protected AODV routing protocol for MANETs based on chaotic map. In: IEEE Access 2019

5. Sharma DK, Ahmad MS, Kshitij Sharma GH (2018) Performance analysis of opportunistic network protocols in blackhole environment. In: Second international conference on inventive communication and computational technologies
6. Elmahdi E, Yoo S-M, Sharshembiev K (2018) Securing data forwarding against blackhole attacks in mobile ad hoc networks. In: IEEE 8th annual computing and communication workshop and conference
7. Kshirsagar D, Patil A (2013) Blackhole attack detection and prevention by real time monitoring. In: Fourth international conference on computing. Communications and Networking Technologies
8. Bala A, Bansal M, Singh J (2009) Performance analysis of MANET under blackhole attack. In: First international conference on networks and communications
9. Saravanan T, Nithya NS (2018) Integrating wireless sensor networks with mobile application. Int J Eng Res Technol (IJERT) 6(8):1–3
10. Mohite VG, Ragha L (2015) Security agents for detecting and avoiding cooperative blackhole attacks in MANET. In: International conference on applied and theoretical computing and communication technology
11. Patidar K, Dubey V (2014) Modification in routing mechanism of AODV for defending blackhole and wormhole attacks. In: Conference on IT in business, industry and government
12. Mukesh Tiwari KV, Arya R, Choudhari KSC (2009) Designing intrusion detection to detect black hole and selective forwarding attack in WSN based on local information. In: Fourth international conference on computer sciences and convergence information technology
13. Wazid M, Katal A, Sachan RS, Goudar RH, Singh DP (2013) Detection and prevention mechanism for blackhole attack in wireless sensor network. In: International conference on communication and signal processing (2013)
14. Meenakshi Tripathi MS, Laxmi GV (2013) Comparing the impact of black hole and gray hole attack on LEACH in WSN. In: Procedia computer science 19:1101–1107 (2013)
15. Sheela D, Mahadevan G (2012) Mollifying the effect of cloning, sink hole and black hole attacks in wireless sensor networks using mobile agents with several base stations. Int J Comput Appl
16. Gondwal N, Diwaker C (2013) Detecting Blackhole attack in WSN by check agent using multiple base stations. Am Int J Res Sci Technol Eng Math (2013)
17. Yadav P, Gill RK, Kumar N (2012) A fuzzy based approach to detect black hole attack. Int J Soft Comput Eng
18. Dighe PG, Vaidya MB (2013) Counter effects of black hole attack on data transmission in wireless sensor network with multiple base stations. Int J Eng Innovat Technol (2013)
19. Zhang J (2008) Black hole quantum tunnelling and black hole entropy correction. Phys Lett B(2008)
20. Lin K, Yang SZ (2009) Fermions tunneling of higher-dimensional Kerr–Anti-de Sitter black hole with one rotational parameter. Phys Lett B
21. Morris S, Shin HS (2004) Liquidity black holes. Rev Finance 8(1):1–18

A Skellam Distribution Inspired Trust Factor-Based Selfish Node Detection Technique in MANETs

M. Deva Priya, A. Christy Jeba Malar, J. Sengathir, and T. Akash

Abstract The trust of nodes in Mobile Ad hoc NETwork (MANET) is the potential factor that needs to be investigated for facilitating predominant network performance. The network performance influencing trust is degraded due to the presence of intentionally behaving selfish nodes that do not forward data packets to their adjoining cooperative mobile nodes. The selfish nodes lessen packet forwarding based on the energy level for conserving their energy and remain active in the network. In this paper, Skellam Distribution Inspired Trust Factor-based Selfish Node Detection Technique (SDITF-SNDT) is propounded for enforcing effective and efficient detection, and isolation of selfishly behaving mobile nodes from the network. The proposed SDITF-SNDT initiates the process of selfish node detection by determining the mean packet deviation through which the Standard Deviation (SD) and Variance are estimated for computing the Skellam Distribution Inspired Trust Factor (SDITF). This computation aids in reliability estimation of nodes in order to categorize them into selfish and cooperative nodes. From the investigations conducted for the proposed SDITF-SNDT approach, excellent improvement in Packet Delivery Ratio (PDR) and phenomenal minimization in energy consumption are confirmed for increased number of mobile nodes in the network.

M. Deva Priya (✉) · T. Akash
Department of Computer Science and Engineering, Sri Krishna College of Technology, Coimbatore, Tamil Nadu, India
e-mail: m.devapriya@skct.edu.in

T. Akash
e-mail: 18tucs010@skct.edu.in

A. Christy Jeba Malar
Department of Information Technology, Sri Krishna College of Technology, Coimbatore, Tamil Nadu, India
e-mail: a.christyjebamalar@skct.edu.in

J. Sengathir
Department of Information Technology, CVR College of Engineering, Mangalpally, Vastunagar, Hyderabad, Telangana, India
e-mail: j.sengathir@gmail.com

R. P. Mahapatra et al. (eds.), *Proceedings of 6th International Conference on Recent Trends in Computing*, Lecture Notes in Networks and Systems 177,
https://doi.org/10.1007/978-981-33-4501-0_34

Keywords MANET · Selfish nodes · Trust · Energy · Skellam distribution

1 Introduction

In Mobile Ad hoc NETwork (MANET), the reliability of mobile nodes towards packet forwarding is determined to be indispensable for ensuring superior packet forwarding rate [1]. This reliability of mobile nodes is influenced by their selfish behaviour towards data forwarding in order to remain active in the network [2]. The existence of selfish nodes in the network gradually increases the routing overhead since the packets are dropped by the nodes [3–11]. The selfish nodes attribute towards poor performance in the network with increased energy consumption and delay [12]. Thus, the selfish nodes are to be significantly detected and isolated with an aim of focusing on facilitating excellent performance of the network [13]. Further, most of the existing selfish node detection schemes focus on the incentive, token and acknowledgement methods for achieving selfish node detection and isolation [14]. However, the detection rate of selfish nodes is determined to be comparatively lower than the reputation schemes considered for detecting and isolating selfish nodes in the network [15]. The reputation-based selfish node detection schemes depend on the extraction of packet forwarding information that are related to each mobile node in the network. Moreover, reputation mechanisms that use reliability factors for selfish node detection and isolation are determined to be superior to other evidence theory-based mitigation schemes [16]. Skellam distribution is found to be efficient in detecting and investigating selfish nodes' modelling parameters that are used as discrete random variables [13]. Skellam distribution is also considered for performing exact quantification of the negative impact imposed by the selfish nodes under routing. Thus, a selfish node detection and isolation approach that depends on Skellam distribution is essential for effective improvement in the performance of the network in terms of Packet Delivery Ratio (PDR) and throughput.

In this paper, Skellam Distribution Inspired Trust Factor-based Selfish Node Detection Technique (SDITF-SNDT) is proposed for enforcing effective and efficient detection and isolation of selfishly behaving mobile nodes from the network. SDITF-SNDT uses the key benefits of Skellam distribution for ensuring potential selfish detection. The proposed technique is determined to be significant in detection as it has the maximum probability of discovering the key factors of selfish behaviour in a multi-perspective manner. Investigations are done for quantifying the predominance of the proposed SDITF-SNDT approach in terms of PDR, throughput, total overhead and energy consumption for varying number of mobile and selfish nodes in the network.

The remaining sections of this paper are organized as follows. Section 2 deals with the survey of few predominant selfish node detection approaches propounded in the literature in the latest decades. Section 3 depicts the complete view on the proposed SDITF-SNDT approach and the related algorithm. Section 4 shows the results and discussions of the proposed SDITF-SNDT approach analyzed based on

PDR, throughput, total overhead and energy consumption for varying number of mobile and selfish nodes in the network. Section 5 presents the major contributions of the propounded SDITF-SNDT approach with its merits.

2 Related Work

A MAC selfishness detecting mechanism is proposed based on Time-To-Live (TTL) for facilitating higher degree of identification accuracy [17]. This TTL-based selfish detection method is considered to be the best in discovering the possible dimensions that are accountable for confirming a nodes' selfishness. Packet latency and energy consumptions of the TTL-based selfish detection are determined to be better when compared to ADB-SNDM and FT-FTS approaches taken for analysis. A Flexible Trust based Fault Tolerance Scheme (FT-FTS) is proposed for selfish node detection and data recovery [18]. FT-FTS approach is estimated to be potent in eliminating the number of packet drops by analyzing the buffer overflow features of the mobile node under consideration. This FT-FTS approach uses the adaptive characteristics in analyzing the deviations between the certification process timestamps. The FT-FTS approach is determined to minimize the false detection rate and identification delay even under heavy load conditions. Further, an Adaptive Decision Boundary-based Selfish Node Detection Model (ADB-SNDM) is proposed for accurate classification of mobile nodes into cooperative and selfish nodes [19]. This ADB-SNDM approach is capable of detecting selfish nodes by the process of grading that determines the reputation of mobile nodes involved in routing. ADB-SNDM is identified to be predominant in quantifying the exact degree of non-cooperation determined at the current point of time independent of the past period of routing. This ADB-SNDM scheme uses the potential of a leader node in the network for appropriate detection of selfish nodes with reduced energy consumption used for detection.

Furthermore, Split Half Reputation Cooperation-based Detection Mechanism (SHRCDM) is proposed for facilitating significant cooperation between the mobile nodes [20]. SHRCDM aids in better discrimination of cooperative and selfish nodes for performance improvement in the network. Throughput, control overhead and total overhead of SHRCDM are determined to be better when compared to LST-SNDA, ADB-SNDM and FT-FTS approaches. Laplace Stleltjes Transform-based Selfish Node Detection Approach (LST-SNDA) is proposed for systematic detection of selfish nodes in the network [21]. LST-SNDA utilizes the conditional probability of successive events for deciding selfishness possessed by the nodes in the network. LST-SNDA is determined to be useful in formulating the upper and lower thresholds of selfish node detection such that exact impact can be resolved with utmost care and enforcement. Energy consumption, packet delay and total overhead are determined to be superior in contrast to the ADB-SNDM and FT-FTS approaches taken for comparison.

A Semi-Markov Process Mechanism using Selfish Node Detection Technique (SMPM-SNDT) is proposed for effective forecasting of malicious activity based on the current status of packet forwarding rate [22]. SMPM-SNDT is estimated for enforcing superior performance in terms of enhanced throughput and reduced total overhead and control overhead. A selfish node detection scheme using energy level and ratio of node communication is propounded for identifying the malicious selfish activity of the mobile nodes in the network [23]. The energy level and ratio of node communication-based selfish node detection scheme are estimated to improve the rate of detection such that the selfish intent of the mobile nodes is reduced to a maximum degree. The energy level and ratio of node communication-based selfish node detection scheme are determined to be excellent in terms of PDR, throughput, total overhead and energy consumption. A Hierarchical Location Aware Hash Table-based Selfish Node Detection Mechanism (HLAHT-SNDM) is designed based on local and global reputations found by continuous monitoring [24]. HLAHT-SNDM is considered for integrating the benefits of discrete and continuous monitoring for better discrimination of nodes. The performance of HLAHT-SNDM is determined to be excellent in terms of throughput, total overhead and control overhead. A local trust factor-based selfish node detection approach is proposed for monitoring the forwarding potential of mobile nodes accurately [25]. In the local trust factor-based selfish node detection approach, the designated nodes are used for analyzing the characteristics of mobile nodes so as to determine the degree of intentional misbehaviour attributed by them. MANET is prone to Distributed Denial of Service (DDoS) which is distributed, and numerous malicious nodes flood the network with huge amount of traffic. Karthik et al. [26] have proposed a mitigation scheme that uses node isolation method, wherein the malicious node is detected and removed from the cluster, thereby reducing the packet drop due to selfish node attack. The performance is evaluated for Ad hoc Distance Vector (AODV), Destination Sequenced Distance Vector (DSDV) and Ad hoc On-demand Multipath Distance Vector (AOMDV) routing protocols.

In MANETs, nodes should cooperate to ensure security. Some prevalent schemes assume that nodes follow protocols without deviation. Nevertheless, these networks are exposed to diverse malicious attacks. As there are numerous models of attacks, trust routing methods can assure security in the network. Beghriche and Bilami [27] have proposed a novel trusted routing model for dealing with attacks in MANETs. Elleithy and Loud [28] deal with Quality of Service (QoS) issues like bandwidth utilization and delay. The authors have dealt with improving the average delays between nodes while considering the arrival, collision avoidance and back-off schemes. The main aim is to specifically transfer packets while selfish nodes are employed or disregarded during packet delivery. Selfish nodes and malevolent attackers are detected using watchdogs, where the selfish nodes are categorized into three different degrees. The thresholds of nodes are determined to classify nodes into normal and selfish nodes, wherein selfish nodes are not allowed to transmit. An algorithm is proposed to alleviate selfish nodes. Selfish nodes do not share resources with their neighbours and conserve their resources. Hasani and Babaie [29] have proposed a fuzzy-based detection method for ad hoc networks that is based on social network concepts. The positions of the nodes based on three variables namely, hop count, residual energy and cooperation history are determined using fuzzy interface to isolate probable selfish nodes from the network and preserve more number of active nodes in the network.

3 Proposed Skellam Distribution Inspired Trust Factor Based Selfish Node Detection Technique (SDITF-SNDT)

The proposed Skellam Distribution Inspired Trust Factor-based Selfish Node Detection Technique (SDITF-SNDT) is a statistical reliability factor-based mitigation mechanism proposed for mitigating selfish nodes. In SDITF-SNDT, selfish nodes are detected and isolated through three steps viz., (a) Estimation of mean packet deviation, (b) Calculation of variance and Standard Deviation (SD) for computing SDITF and (c) Identification and segregation of selfish nodes using computed SDITF.

3.1 *Estimation of Mean Packet Deviation*

In the proposed SDITF-SNDT, an ad hoc network with mobile nodes are monitored by the neighbouring nodes for ‘s’ sessions. Let the number of packets forwarded and received by each node ‘i’ be $\mathrm{PF}_{(1)}, \mathrm{PF}_{(2)}, \mathrm{PF}_{(3)}, \ldots, \mathrm{PF}_{(S)}$ and $\mathrm{PR}_{(1)}, \mathrm{PR}_{(2)}, \mathrm{PR}_{(3)}, \ldots, \mathrm{PR}_{(S)}$, respectively, as monitored by each of its neighbours in ‘*s*’ sessions. The deviation in the number of packets received to the number of packet forwarded by each mobile node to its neighbours as recommended through neighbour-based interaction in each session ‘c’ is given by,

$$\mathrm{DEV}_{C}^{\mathrm{PACKET}} = \mathrm{PR}_{C} - \mathrm{PF}_{C} \tag{1}$$

Further, the Packet Forwarding Capability (PFC) of mobile nodes in each session ‘c’ as identified by the neighbour is,

$$\mathrm{P}_{C}^{\mathrm{PFC}} = \frac{\mathrm{PF}_{C}}{\mathrm{PR}_{C}} \tag{2}$$

The mean packet deviation attributed by each mobile node as identified by ‘k’ neighbours is given by,

$$\mathrm{MDEV}_{C}^{\mathrm{PACKET}} = \sum_{c=1}^{s} \frac{\mathrm{DEV}_{C}^{\mathrm{PACKET}}}{s} \tag{3}$$

3.2 *Calculation of Variance and Standard Deviation (SD) for Computing SDITF*

When the value of 'P_C^{PFC}' and '$MDEV_C^{PACKET}$' are computed, the Variance and SD attributed by each mobile node towards data packet dissemination are calculated using,

$$SD_{DETECT} = P_S^{PFC} * (1 - P_S^{PFC}) \tag{4}$$

$$VAR_{DETECT} = \sum_{c=1}^{s} (MDEV_C^{PACKET} - DEV_C^{PACKET})^2 \tag{5}$$

SDITF computed based on (4) and (5) is,

$$(CARF)\alpha_{DETECT} = \frac{S}{S-1}\left(1 - \frac{\sum_{C=1}^{S} SD_{DETECT}}{VAR_{DETECT}}\right) \tag{6}$$

3.3 *Detection and Isolation of Selfish Node Misbehaviour using Computed SDITF*

When SDITF is found for each node, the decision pertaining to the detection and isolation of selfish nodes' misbehaviour is initiated. The mobile nodes identified with SDITF value less than 0.35 (obtained from simulation) are identified as compromised selfish nodes and are isolated from the multicasting process.

Algorithm

The following algorithm shows the steps involved in identifying selfish nodes using SDITF and isolating them from routing.

Algorithm 1: Skellam Distribution Inspired Trust Factor based Selfish Node Identification and Isolation
N - Number of mobile nodes in the network
GN - Group of nodes on the routing path
SN - Source Node
DN - Destination Node

Establish routing path by sending `RREQ' messages from the SN to all the nodes
Nodes respond to the SN by sending `RREP' messages
for every node (u) belonging to the list GN that uses 't' sessions for transmission
    Estimate the deviance between the number of packets received and forwarded by each mobile node through neighbour based communication
    $\mathbf{DEV_C^{PACKET} = PR_C - PF_C}$
    Determine packet forwarding capability of mobile node in each session 'c'
    $\mathbf{P_C^{PFC} = \frac{PF_C}{PR_C}}$
    Compute mean packet deviation attributed by each mobile node
    $\mathbf{MDEV_C^{PACKET} = \sum_{c=1}^{s} \frac{DEV_C^{PACKET}}{s}}$
    Calculate SD and Variance in mobile node behaviour using values of 'P_C^{PFC}' and '$MDEV_C^{PACKET}$'
    $\mathbf{SD_{DETECT} = P_S^{PFC} * (1 - P_S^{PFC})}$
    $\mathbf{VAR_{DETECT} = \sum_{c=1}^{s} (MDEV_C^{PACKET} - DEV_C^{PACKET})^2}$
    Calculate SDITF using
    $\mathbf{(CARF)\alpha_{DETECT} = \frac{S}{S-1}\left(1 - \frac{\sum_{C=1}^{S} SD_{DETECT}}{VAR_{DETECT}}\right)}$
    if (SDITF (u) < 0.35) then
        Node 'u' is selfish
        Invoke Selfish_Node_Mitigation (u)
    else
        Node 'u' is reliable
    End if
End for
End

The threshold of the proposed SDITF-SNDT scheme is found to be 0.35 based on investigations in which maximum number of selfish nodes is identified with optimum performance enhancement in the network.

4 Simulation Results and Experimental Analysis

The predominance of the proposed SDITF-SNDT is investigated based on the simulation experiments conducted using ns-2.33. The potential of the proposed SDITF-SNDT approach is compared with the existing SMPM-SNDT, HLAHT-SDNM and SHRCDM schemes in terms of PDR, throughput, total overhead and energy consumption for varying number of mobile and selfish nodes. The simulation area used for implementing the proposed SDITF-SNDT approach is 1500×1500 square metres. A total number of 100 mobile nodes are deployed randomly. Data packets of 512 bytes are used for data transmission. Constant Bit Rate (CBR) data traffic is taken as the data traffic pattern. A simulation time of 300 sec with pause time of 20 sec is taken.

Figure 1 exemplars the potential of the proposed SDITF-SNDT approach in terms of PDR and throughput examined for varying number of mobile nodes. The PDR of the proposed SDITF-SNDT is established to be 8%, 12% and 15% better when compared to the existing SMPM-SNDT, HLAHT-SDNM, and SHRCDM approaches. Similarly, the throughput of the proposed SDITF-SNDT approach is seen to be 11%, 13% and 16% excellent when compared to the existing SMPM-SNDT, HLAHT-SDNM and SHRCDM approaches. This predominant enhancement is mainly due to the capability of Skellam distribution in exploring the parameters that attribute towards the detection of selfish nodes.

Figure 2 unveils the predominance of the proposed SDITF-SNDT approach in terms of total overhead and energy consumption examined for varying number of mobile nodes.

The total overhead of the proposed SDITF-SNDT approach is determined to be significantly minimized by 11%, 14% and 17% in contrast to the existing SMPM-SNDT, HLAHT-SDNM and SHRCDM approaches. Similarly, the energy consumption of the proposed SDITF-SNDT approach is found to be excellently reduced by

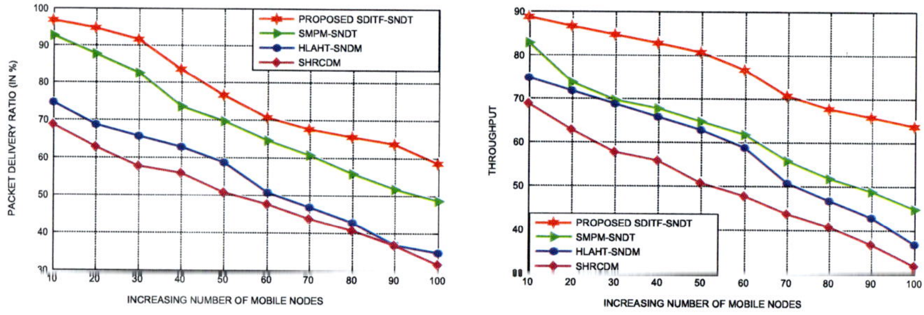

Fig. 1 Packet Delivery Ratio and Throughput of the Proposed SDITF-SNDT for Increasing Number of Mobile Nodes

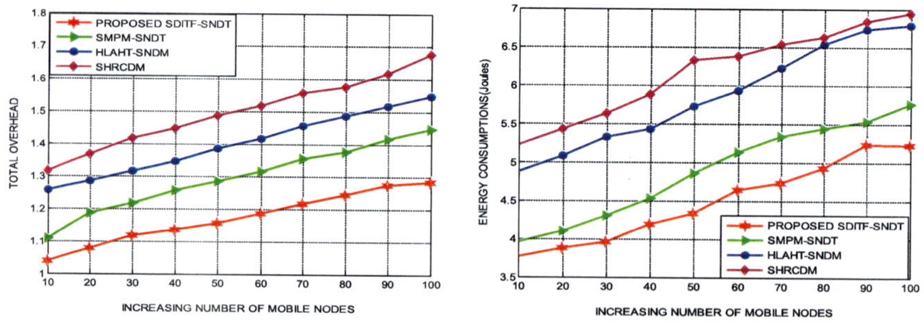

Fig. 2 Total Overhead and Energy Consumption of the Proposed SDITF-SNDT for Increasing number of Mobile Nodes

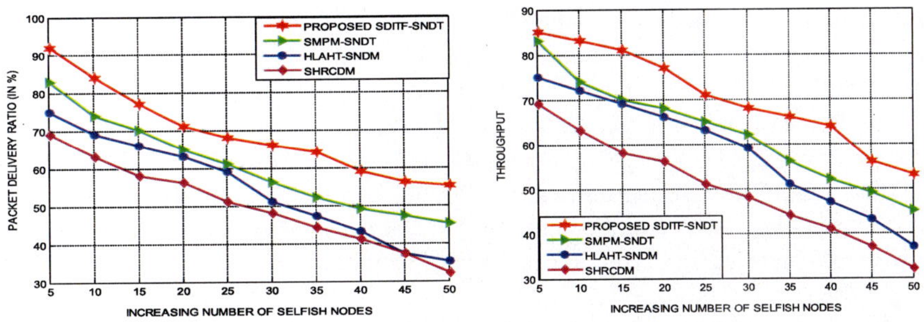

Fig. 3 Packet Delivery Ratio and Throughput of the Proposed SDITF-SNDT for Increasing Number of Selfish Nodes

9%, 12% and 15% in contrast to the existing SMPM-SNDT, HLAHT-SDNM and SHRCDM approaches. This predominant reduction in the total overhead and energy consumption of the network is due to the significance of Skellam distribution in reducing the number of packet retransmissions in the network.

Figure 3 shows the potential of the proposed SDITF-SNDT approach using PDR and throughput examined for varying number of selfish nodes. The PDR of the proposed SDITF-SNDT approach for increased number of selfish nodes is seen to be 9%, 13% and 16% better in contrast to the benchmarked SMPM-SNDT, HLAHT-SDNM and SHRCDM approaches. Similarly, the throughput of the proposed SDITF-SNDT approach for varying number of selfish nodes is found to be 10%, 12% and 15% excellent when compared to the existing SMPM-SNDT, HLAHT-SDNM and SHRCDM approaches. This improvement in the PDR and throughput for increasing number of selfish nodes is mainly due to the classification capability of the incorporated Skellam distribution in examining the decision parameters that attribute towards the identification of selfish nodes.

Figure 4 highlights the significance of the proposed SDITF-SNDT approach in terms of the total overhead and energy consumption examined for varying number of

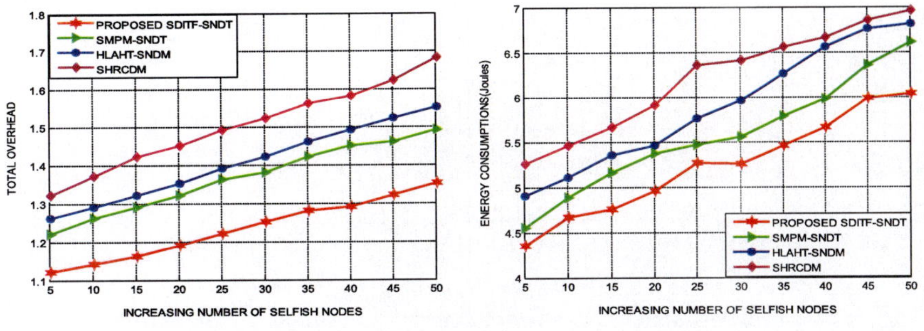

Fig. 4 Total Overhead and Energy Consumption of the Proposed SDITF-SNDT for Increasing Number of Mobile Nodes

selfish nodes. The total overhead of the proposed SDITF-SNDT approach is determined to be significantly minimized by 12%, 15% and 18% in contrast to the existing SMPM-SNDT, HLAHT-SDNM and SHRCDM approaches. Similarly, the energy consumption of the proposed SDITF-SNDT approach is found to be exceptionally reduced by 8%, 11% and 14% in contrast to the existing SMPM-SNDT, HLAHT-SDNM and SHRCDM approaches. This significant decrease in the total overhead and energy consumption of the network is due to the significance of Skellam distribution in balancing the detection rate with increasing rate of selfish nodes in the network.

In addition, Figs. 5 and 6 highlight the predominance of the proposed SDITF-SNDT explored using percentage increase in the detection rate explored for varying number of mobile and selfish nodes. The percentage increase in the PDR of the proposed SDITF-SNDT approach for varying number of mobile nodes is proved to be 7%, 9% and 11% excellent in contrast to the benchmarked SMPM-SNDT, HLAHT-SDNM and SHRCDM approaches. Similarly, the percentage increase in the PDR of the proposed SDITF-SNDT approach for varying number of selfish

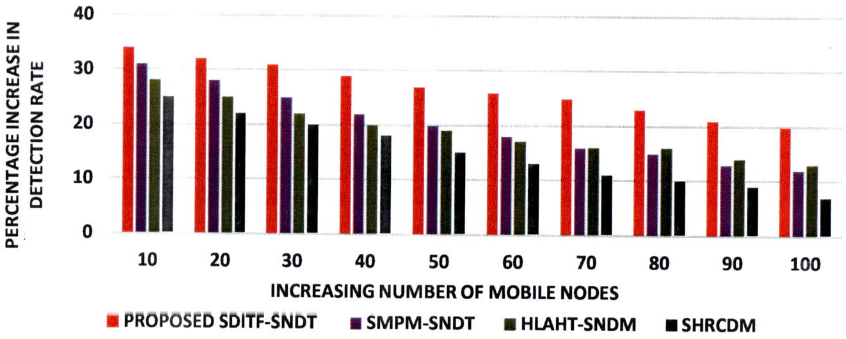

Fig. 5 Detection Rate of the Proposed SDITF-SNDT for Increasing Number of Mobile Nodes

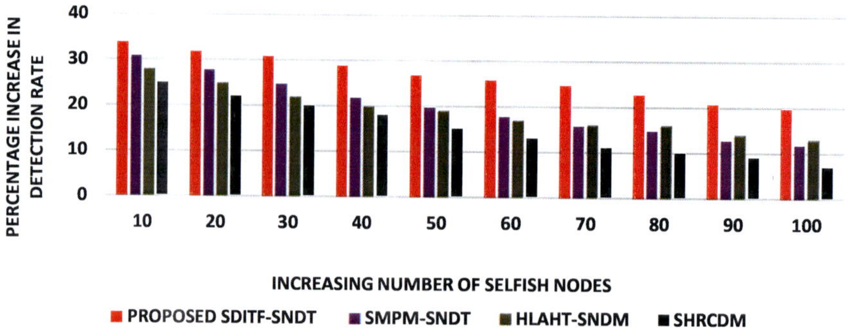

Fig. 6 Detection Rate of the Proposed SDITF-SNDT for Increasing Number of Selfish Nodes

nodes is determined to be excellent by 8%, 11% and 14% when compared to the existing SMPM-SNDT, HLAHT-SDNM and SHRCDM approaches.

5 Conclusion

The proposed Skellam Distribution Inspired Trust Factor Based Selfish Node Detection Technique (SDITF-SNDT) is mainly contributed for utilizing the significant merits of Skellam distribution towards trustworthy detection and segregation of selfish nodes in MANETs aiming at increasing the network performance. The proposed SDITF-SNDT is based on Skellam distribution-based variance and mean packet forwarding rate for estimating the degree of impact induced by the selfish characteristics of mobile nodes in the network such that the accurate detection and segregation of selfish nodes can be imposed for enhancing network performance. The simulation experiments of the proposed SDITF-SNDT confirm a mean improvement rate of PDR and throughput by 21% and 18%, and minimized energy consumption and routing overhead rate by 23% and 15% in contrast to the selfish node isolation schemes contributed in the literature. The mean rate in detection of the proposed SDITF-SNDT is also determined to be enhanced by 20% when compared to the selfish node isolation schemes.

References

1. Rukhande S, Shete P (2015) Optimized routing by excluding selfish nodes for MANET. Commun Appl Electron 3(5):43–49
2. Kurkure AM, Chaudhari B (2014) Analysing credit based ARAN to detect selfish nodes in MANET. In: Proceedings of IEEE international conference on advances in engineering & technology research, pp 1–5
3. Punithavathani DS, Sankaranarayanan K (2009) IPv4/IPv6 transition mechanisms. Eur J Sci Res 34(1):110–124
4. Balamurugan A, Purusothaman T (2012) IPSD: new coverage preserving and connectivity maintenance scheme for improving lifetime of wireless sensor networks. WSEAS Trans Commun 11(1):26–36
5. Kumar VS, Thansekhar MR, Saravanan R, Amali SMJ (2014) Solving multi-objective vehicle routing problem with time windows by FAGA. Procedia Eng 97:2176–2185
6. Kowsigan M, Rubasri M, Sujithra R, Banu HS (2017) Data security and data dissemination of distributed data in wireless sensor networks. Int J Eng Res Appl 7(3 part 4):26–31
7. Jayarajan P, Maheswar R, Sivasankaran V, Vigneswaran D, Udaiyakumar R (2018) Performance analysis of contention based priority queuing model using N-policy model for cluster based sensor networks. In: Proceedings of the IEEE international conference on communication and signal processing, pp 0229–0233
8. Madhavan P, Malathi P (2016) Intelligent framework for qos optimization in manet using soft computing models. Int J Adv Engg Tech 7(2):890–893
9. Priya MD, Aishwarya R, Anushya S, Keerthana S (2017) Detection and avoidance of malicious nodes on MANETs. Int J Sci Res Comput Sci Eng Inf Technol 2(2):434–441

10. Sengathir J, Manoharan R (2017) Co-operation enforcing reputation-based detection techniques and frameworks for handling selfish node behaviour in MANETs: a review. Wireless Pers Commun 97(3):3427–3447
11. Priya MD, Janakiraman S, Sandhya G, Aishwaryalakshmi G (2019) Efficient pre-authentication scheme for inter-ASN handover in high mobility MANET. Wirel Networks 1–15
12. Maharaja S, Jeyalakshmi R, Sabarish Kanna AV, Deva Priya M (2019) "Secured Routing in Mobile Ad hoc Networks (MANETs)", Int J Sci Res Comput Sci Engi Inf Technol (IJSRCSEIT), ISSN : 2456–3307, 5(2), 277–289
13. Kampitaki DG, Economides AA (2016) Novel routing protocol for mobile ad hoc networks with selfish and altruistic nodes. In: Proceedings of IEEE international conference on telecommunications and multimedia (TEMU), pp 1–5
14. Kampitaki DG, Karapistoli ED, Economides AA (2014) Evaluating selfishness impact on MANETs. In: Proceedings of IEEE international conference on telecommunications and multimedia (TEMU), pp 64–68
15. Lamba GK, Chaurasia A, Lala A (2016) Varying number of selfish nodes based simulation of AODV routing protocol in MANET using reputation based scheme. Int J Eng Comput Sci 5(6):56–65
16. Janakiraman S, Rajendiran M (2016) An Erlang factor-based conditional reliability mechanism for enforcing co-operation in MANETs. Serb J Electr Eng 13(2):265–284
17. Tarannum R, Pandey Y (2012) Detection and deletion of selfish MANET nodes-a distributed approach. In: Proceedings of the 1st IEEE international conference on recent advances in information technology (RAIT), pp 152–156
18. Verma A, Khare A (2013) TFT technique with adaptive thresholding for selfish attack prevention in MANET. Int J Comput Appl 63(16)
19. Akhtar AK, Sahoo G (2013) Classification of selfish and regular nodes based on reputation values in MANET using adaptive decision boundary
20. Sengathir J, Manoharan R (2013) A split half reliability coefficient based mathematical model for mitigating selfish nodes in MANETs. In: Proceedings of the 3rd IEEE international advance computing conference (IACC), pp 267–272
21. Sengathir J, Manoharan R (2014) Laplace stleltjes transform based conditional survivability coefficient model for mitigating selfish nodes in MANETs. Egypt Inf J 15(3):149–157
22. Sengathir J, Manoharan R (2015) A futuristic trust coefficient-based semi-Markov prediction model for mitigating selfish nodes in MANETs. EURASIP J Wirel Commun Networking 2015(1):158
23. Kumari DS, Sikamani KT (2015) Communication based clustering to detect selfish nodes in MANET. Indian J Sci Technol 8(20):1
24. Ramya K, Kavitha T (2016) Deterring selfish nodes using hierarchical account-aided reputation system in MANET. In: Proceedings of the IEEE international conference on computing technologies and intelligent data engineering (ICCTIDE'16), pp 1–7
25. Priyanka R, Ramkumar P (2016) Trust based detection algorithm to mitigate the attacker nodes in MANET. In: Proceedings of the IEEE international conference on computing technologies and intelligent data engineering (ICCTIDE'16), pp 1–6
26. Karthik PB, Nagesh HR, Chiplunkar NN (2017) Mitigation and performance evaluation mechanism for selfish node attack in MANETs. In: Proceedings of the IEEE international conference on computing, communication, control and automation (ICCUBEA), pp 1–6
27. Beghriche A, Bilami A (2018) A fuzzy trust-based routing model for mitigating the misbehaving nodes in mobile ad hoc networks. Int J Int Comput Cybern
28. Elleithy A, Loud V (2019) Selfish nodes mitigation in mobile Ad-Hoc networks. In: Proceedings of the IEEE long Island systems, applications and technology conference (LISAT), pp 1–5
29. Hasani H, Babaie S (2019) Selfish node detection in ad hoc networks based on fuzzy logic. Neural Comput Appl 31(10):6079–6090

Prediction of Breast Cancer Using Different Machine Learning Algorithms

Ritu Agarwal and Shivam Latiyan

Abstract Breast cancer is one of the most critical medical problems in our society. It is caused by numerous factors which are ignored in our daily life. The diagnosis process is tedious and lengthy. It mainly classifies cancer into malignant or benign. Malignant cancer spreads in our body at an exponential rate and its early detection can save the life of the victim person. When several tests are conducted on the body, the mean area, clump thickness, cell shape, mean radius, all measures are evaluated for the detection of cancer. Artificial intelligence is used as the diagnostic tool for the detection of cancer by evaluating the features extracted from the MRI scan of the breast area. This paper mainly focuses on comparing the accuracy, precision, recall, and other measures of different classification algorithms in the prediction of breast cancer. The UCI repository data set is used in comparing all the algorithms. Thus, machine learning comes into the picture to solve this problem. In machine learning, previous reports and data sets are used to predict. With the help of data sets, the system is trained and then it tells whether there is any tumor growth or not in the patient. With the help of various algorithms, we are able to predict whether someone is affected or not. In this paper, we are using dynamic prediction. Dynamic prediction helps future proofs the algorithm, since if in the future, the data sets change and new features are introduced, the algorithm would be able to adjust itself with the changes. The algorithms used in this paper are decision tree, k-nearest neighbor, Naives bayes, and support vector machine. After applying the above algorithm in dynamic prediction, we found that support vector machine is the best algorithm for our data set to predict the correct result. On applying support vector machine, it predicted roundly correct result.

Keywords Malignant · Machine learning · SVM · Precision · Accuracy

R. Agarwal (✉) · S. Latiyan
Delhi Technological University, Shahbad Daulatpur, Bawana Road, New Delhi, India
e-mail: ritu.jeea@gmail.com

S. Latiyan
e-mail: Shivamlatiyan98@gmail.com

R. P. Mahapatra et al. (eds.), *Proceedings of 6th International Conference on Recent Trends in Computing*, Lecture Notes in Networks and Systems 177,
https://doi.org/10.1007/978-981-33-4501-0_35

1 Introduction

Cancer is one of the deadliest diseases all over the world. It is characterized by tumorous growth in cells of the organ. Breast cancer begins in the breast tissue and may spread in mammary glands, lobes, ducts, and other regions of the breast. The symptoms of breast cancer vary from female to female but it includes painless lump, nipple discharge, Paget disease of nipples, redness rash, and dripping of the tissues in the targeted region.

Hence, detection of cancer is very time consuming and costly affair. The problem can be solved by machine learning.

Cancer detection can be performed in two ways:

1. Machine learning.
2. Image processing based.

Breast detection using the second way requires breast area image processing. The image is first cleared with use of filters like adaptive filters and then it is segmented so that tumorous growth could be easily detected. The curvature of the breast area is the essential component in such type of diagnosis. But in this, we are using machine learning-based approach. In machine learning-based algorithm, previous reports are used. A report generally consists of data sets. Data sets are used, and with the help of data sets, the system is trained and then it tells whether there is any diabetes or not.

Machine learning works in the same way as the mind works. Machine learning can be understood by understanding how it is consisted. It is consisted of:

1. Features.
2. Labels.

In machine learning, features and labels are passed. And on the basis of this, machine is trained. So, whenever next time a feature is passed to machine, it would predict labels based on the learning from previous data sets given to it. This research paper focuses mainly on the diagnosis of old age person and pregnant females. The paper mainly extracts the essential components for the prediction of the breast cancer.

2 Related Work

Cruze [1] gathered information about the efficiencies of different machine learning algorithms and found that SVM and ANN are the best algorithms for the prediction of survivability and recurrence of breast cancer. Mr. Sourabhe researched that sequential optimization [2] is best classification technique for the breast cancer detection. Hiba asari and Thomas noel compare the various classification algorithm for the prediction of breast cancer. They evaluated all the classification algorithm with the help of the kappa statistics and found that support vector machine shows the lowest possible error of all the algorithms [3]. Adel aloraniie also carried [4] out the study on the accuracy of the breast cancer prediction by using j48, and other decision tree algorithms gave better results than naïve Bayesian classifiers. Micalisee. v carried out that heterogenous data [5] can be essential component for better prediction of the breast cancer in short duration of time.

3 Methodology Used

3.1 Model Diagram

The proposed process is summarized in Fig. 1 as a diagram. The figure shows the flow of research done in the construction of the model.

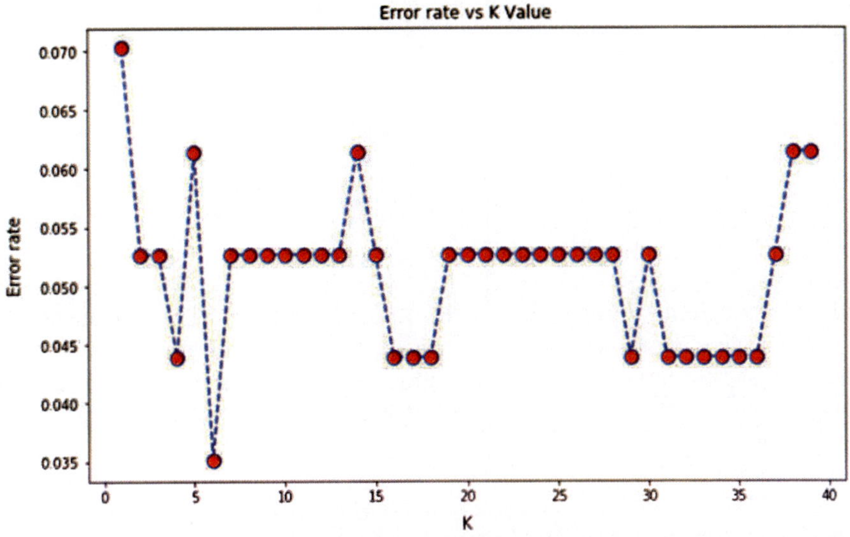

Fig. 1 Error rate dependency with *K* value

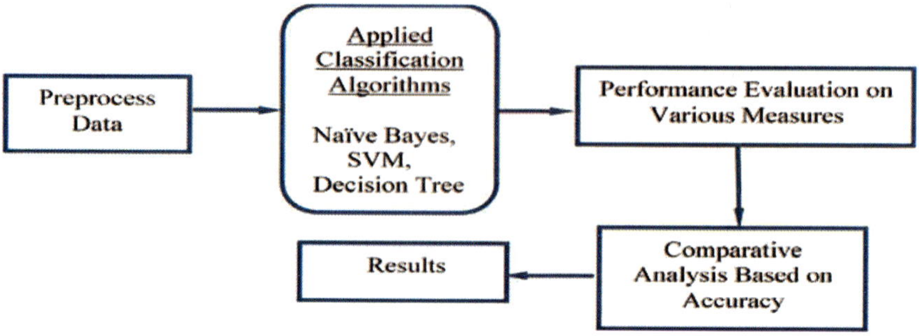

3.2 Brief Description of Algorithms

For all of our methods, we created sets of training, validation, and test data. For training data, we used unmodified user medical histories and diabetic outcome. For validation and test data, we took the medical history and split it into two halves: one with the outcome and other with different features present in medical history. The first half was used to test our algorithms, while the second half was used to estimate our results.

3.2.1 K-Nearest Neighbors (KNN)

We use x to describe a feature and y to describe the target which has to be predicted.

KNN comes under the supervised learning. The goal is to learn a function such that given a new observation x, function can strongly predicts the respective output y. KNN classifier has properties of a non-parametric and example-based learning algorithm.

Non-parametric: It creates no default assumptions regarding the current form of function. This prevents the problems of mismodeling the distribution of the data. In a condition, when our data is non-Gaussian but the learning models, we choose have a requirement of a Gaussian form. In such cases, the algorithm would make extremely poor predictions for the problem.

·Instance-based learning: The algorithm do not learn a model by itself. Actually, it decides to learn the training instances which are source of information in the prediction phase. Substantially, this means that only when a query to the database system is done (when we ask it to recommend a label provided an input). It is important to note that the minimal training phase of KNN is achieved at a memory cost, because we have to store a potentially big data set, plus a computational cost during test time because classifying a provided observation needs a run down for the full data set. So actually, this is unwanted because we usually want fast responses.

This figure denotes that error rate is minimum with lower value in breast cancer detection and its shows the $k = 6$, the error rate is minimum, and this value is chosen for the classification of the data. The k estimate is independent of the data set but it varies on the variation of the attribute values and correlation between the attributes. This figure shows that as k value gest the error rate also increases sufficiently. In this paper, all the k values are tested and $k = 6$ is selected for the classification of the breast cancer data set. Error rate show the maximum peak in k values in multiples of 15. In KNN classifier, the k parameter is the key component for the classification of the Wisconsin breast cancer data set. K values [6] varies for the heterogeneous data and spatial data.

3.2.2 Gaussian Naive Bayes

Gaussian is one of the three types of Naive Bayes variant. If our data is of discrete nature then gaussian is not used. It is used on the data sets which show a continuous nature in the features provided. Gaussian is a perfect match for continuous nature type of data. For the features of our data which are of continuous nature, we use Gaussian distribution to predict our result. It is also called normal distribution. Mean, standard deviation, and variance are the most important tool in finding Gaussian distribution (Fig. 2).

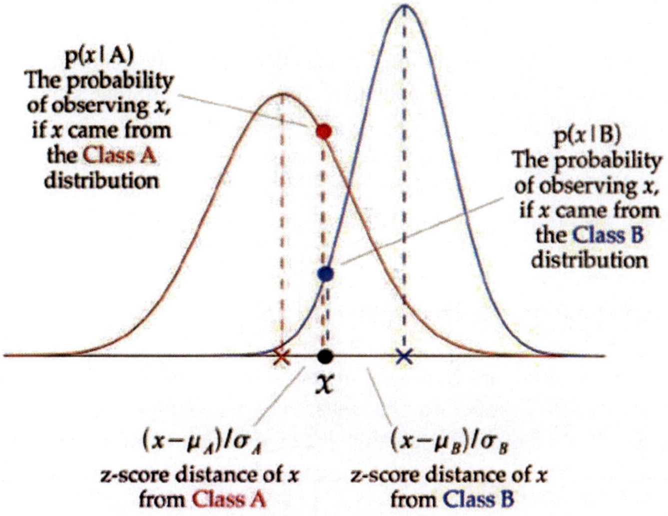

Fig. 2 Gaussian distribution curve

3.2.3 Decision Tree

Decision tree in machine learning is utilized for directed learning [classification and regression]. Decision tree misuses relationship among features and nonlinearity in the features. The basic method to construct a decision tree is to utilize a greedy approach. Think of you as are greedy on the number of decision nodes. The number of decision nodes ought to be negligible. By testing component esteem, the data set is broken into sub-data sets, with a condition that the split gives most extreme advantage to the order, i.e., the element esteem considered (among all the conceivable element esteem mixes) is the best accessible to sort the given informational collection into two subsets. In each sub-data set, another element esteem mix is picked, as in the previous split, to separate it into littler sub-data sets, with a similar condition that the split gives most extreme advantage to the grouping. The procedure is rehashed until a decision node is not required to add part the sub-data set, and practically the majority of the examples in that sub-data set have a place with a single class.

3.2.4 Support Vector Machines (SVMs)

SVM is a discriminant classifier. It processes different types of data and then classifies accordingly. Its working is based on the input of more than or equals to 2 classes. Firstly, it is using a separating line which acts as a border between two classes. The distance between support vector and separating line should be maximum and because of this, the system accuracy must increase. If a data set is not able to perform in 2D, then we convert it into a three-dimensional plane and then map the line with the help of kernel (with the help of sklearn library). For multidimensional, we would use a hyper plane for mapping. SVM should be described as a type of discriminative classifier officially characterized by an isolating hyper plane. As such, provided named preparing information (directed learning), the computation provides an ideal hyper plane that classifies new precedents.

3.2.5 SVM Hyperparameters Optimization

Support vector machine has hyperparameters which can be processed and get the high accuracy of the classifier. In this research paper, GridSearchCv is used, which is a dictionary, and it has parameters that can be used in the training of the classifier. In this paper, we used mainly two parameters for improving the accuracy of the classifier.

Gamma Parameter

It is the parameter that tells how the single training affects the processing of the classifier. In large gamma, closer data points will have extra weight. But in case of small gamma value more generalized solution is given more priority. Gamma parameter optimization is done in this research paper with GridSearchCv.

C Parameter

They are the parameter which are required in the classifying training points. *C* parameters is directly linked with a penalty of misclassification [7] of the classifier. The small *c* indicates that there will be a low penalty of misclassification. The bigger c indicates that the cost of misclassification will be higher. These parameters are also optimized in the research with the GridSearchCv library in python (Fig. 3).

This figure denotes the margin of the c parameters associated with the classifier's cost of misclassification. The support vector machine follows the principal of the maximum margin plane which is taken into account for the optimization of the classifier. The unscaled data is correctly classified by using optimal *C* parameters for the training and testing of the data. The c value is directly proportional to the penalty associated with misclassification.

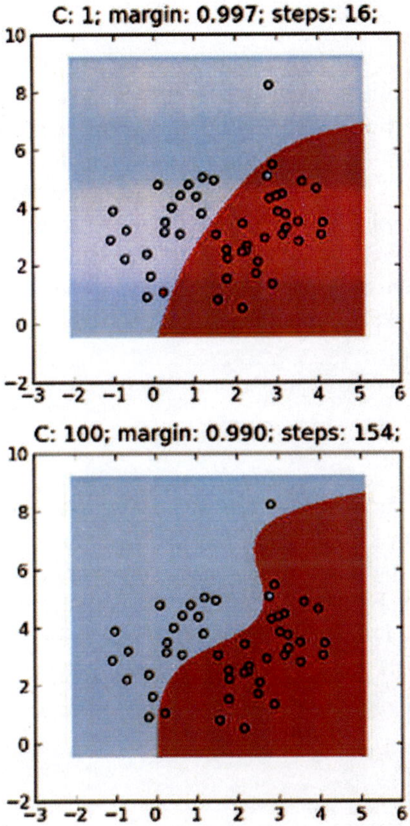

Fig. 3 *C* parameter with margins

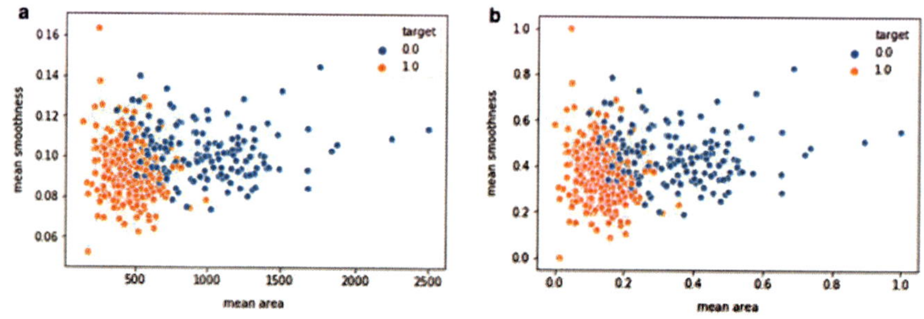

Fig. 4 **a** Unnormalized data **b** Normalized data

3.3 *Data Processing Methods*

3.3.1 Data Normalization

Normalization is a technique which is used as the critical steps in data preprocessing it allows the proper scaling of the data and alter the values of numeric columns in the data set to a middle scale without affecting the divergence in values. Normalization is essential only when the data has different ranges in the features extracted from the data set. Data normalization is done in this prediction to improve the accuracy measures of the classifiers like SVM, KNN etc. The spatial data is eliminated by the data normalization. It allows the uniformity [8] in the data. Mean area and mean smoothness are the two attributes which shows the importance of normalized data in prediction of the breast cancer (Fig. 4).

3.3.2 Data Standardization

Standardization is the technique useful for the gaussian data. The data should have attributes with the gaussian distribution. They should have differing means and standard deviations to a standard gaussian distributions with the mean values of 0. The another component is standard deviation which must be equal to 1. The breast cancer data has some of the essential attributes like mean area, mean radius which follows the gaussian distribution so all these attributes must be standardized to the standard gaussian distribution.

4 Results

Table 1 indicates the accuracy of the SVM classifier to be better than other algorithms. SVM gives the maximum efficiency by setting the parameters gamma and kernel values. The accuracy achieved in this research 96.00% is at par compared

Table 1 Comparative performance of classification algorithms on various measures

Classification algorithms	Precision	Recall	F-measure	Accuracy (%)
Naive Bayes	0.960	0.900	0.920	94.00
SVM	0.940	1.00	0.970	96.00
Decision tree	0.770	0.410	0.530	59.00
KNN	0.890	1.00	0.940	93.00

to algorithms of machine learning. The lowest accuracy achieved is of the decision tree classifier of 59.00%. Both KNN and naive Bayesian classifier give accuracy slightly lower than support vector machine. The support vector machine gives a recall of 1.0 which is ideal for the classification of the data set. The naïve Bayes classifier is the classifier which gives the highest precision of 0. 960. The SVM shows the significant increase in accuracy by hyperparameters optimization of the c and gamma parameters.

4.1 Result Parameters

1. **Accuracy**: Accuracy of the classifier is the percentage of correct prediction over the total number of the prediction made by the classifier. Each classifier is evaluated on the basis of this criteria. The figure below illustrates the accuracy of all the classifier used in breast cancer prediction. We can conclude the support vector machine has highest accuracy of 97.00% (Fig. 5).
2. **Precision**: The precision is a metric measure that tells the exact percentage of people having cancer of total cancer predicted patients by the classifier. In this

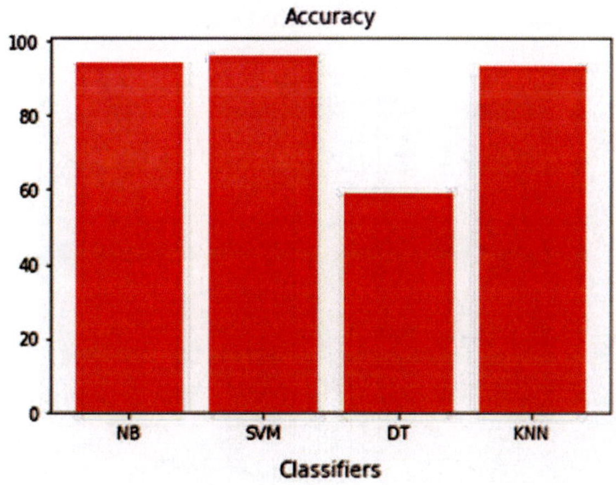

Fig. 5 Accuracy of different classifiers

paper, the naive Bayes algorithms have the highest precision of all the classifiers. The plot of the precision of all the classifiers is shown below in the figure (Fig. 6).

3. **Recall**: This metric tells us the percentage of correctly predicted breast cancer patients out all the actual breast cancer patients. The figure shown below plots the recall of the algorithms used in the paper. SVM classifier is the algorithm with the highest recall out of all (Fig. 7).
4. **F1 Score**: It is a measure that shows the relationship between recall and precision in a classifier. It is a weighted mean of precision and recall. The plot is shown for all the classification algorithms in the figure given below (Fig. 8).

Table 2 mainly gives the details about the number of correctly identified cancer of all the target victims. The SVM classifier is one of the best of all the classifier

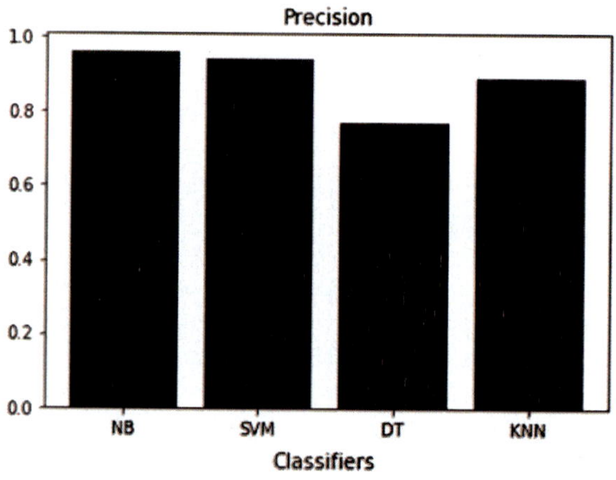

Fig. 6 Precision of different classifiers

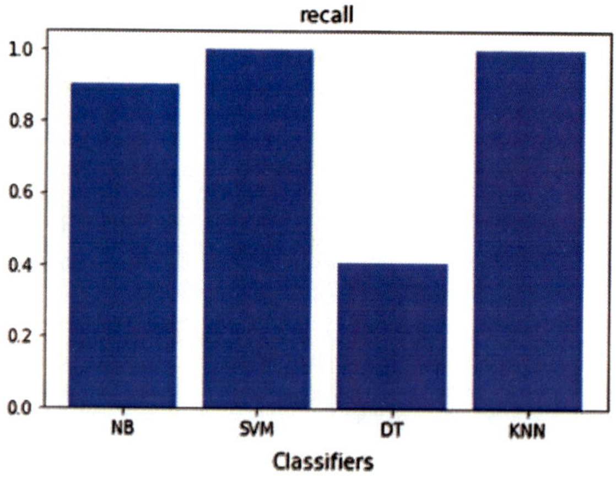

Fig. 7 Recall of all the algorithms

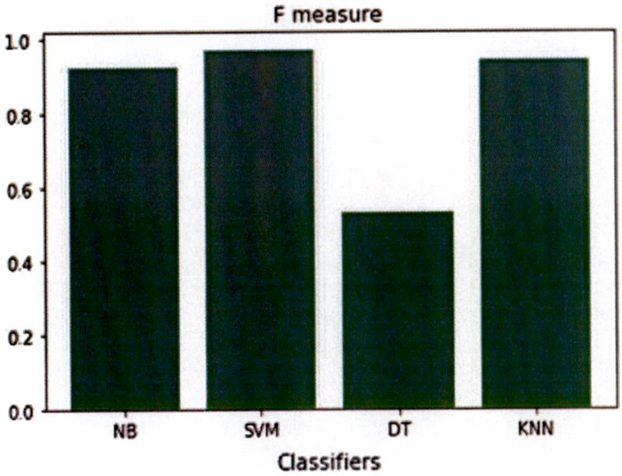

Fig. 8 F-measure of all the algorithms

Table 2 Classifier's performance on the basis of classified instances

Total number of instances	Classification algorithms	Correctly classified instances	Incorrectly classified instances
114	Naive Bayes	107	007
	SVM	110	004
	Decision tree	67	47
	KNN	106	008

algorithm its shows the high accuracy and high precision of all malignant cancer data. The test size sample is taken as 20% for the higher accuracy of all the classifiers. The training sample is taken as 80% from the 569 instances of the UCI data set; it mainly consists of 32 attributes out of which mean area and mean smoothness are the essential attributes which are used in the training process of the classifier (Figs. 9, 10, 11 and 12).

Confusion Matrix on Wisconsin data set

The confusion matrix is plotted for all the classifiers. In all the plots shown above, the support vector machine classifies the highest correct instances of malignant cancer patients. The support vector machine classifies 110 test samples out of all 114 samples taken for the creation of the confusion matrix. These plots depict that there is a high influence of the accuracy of the decision tree classifier with data normalization of the Wisconsin data set used in this paper. The decision tree classifier shows the worst performance of all the classier. It mainly classifies 47 samples wrong out of a total of 140 samples. The plots are made with 20% as sampling size for testing from the 569 instances of all the records. It can be inferred from the tables, plots that SVM is one of the best classifiers of all the machine learning algorithms and it is independent of the uniformity in the data set. It outperforms all the other machine learning algorithms

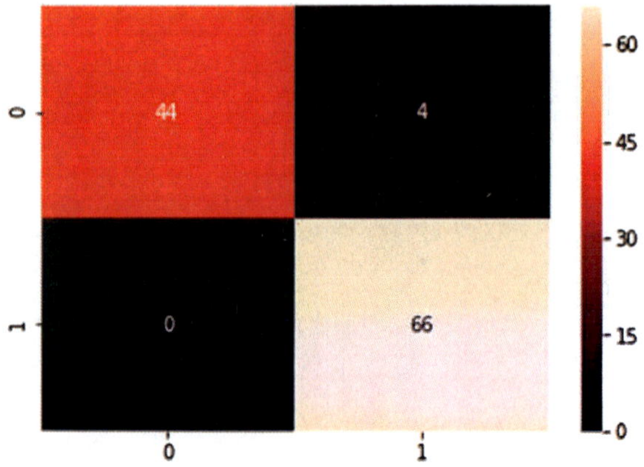

Fig. 9 SVM

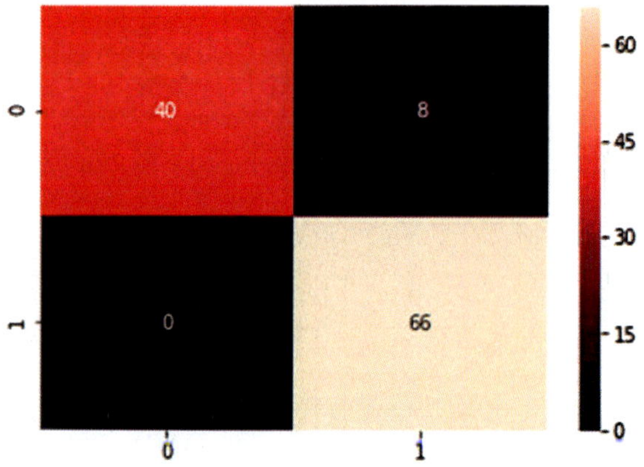

Fig. 10 KNN

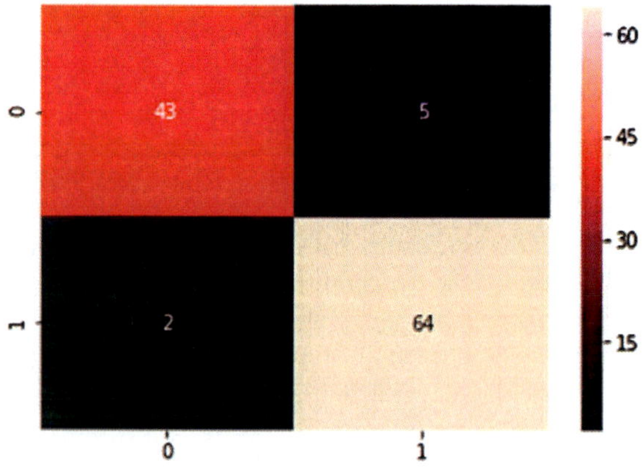

Fig. 11 Naïve Bayes classifier

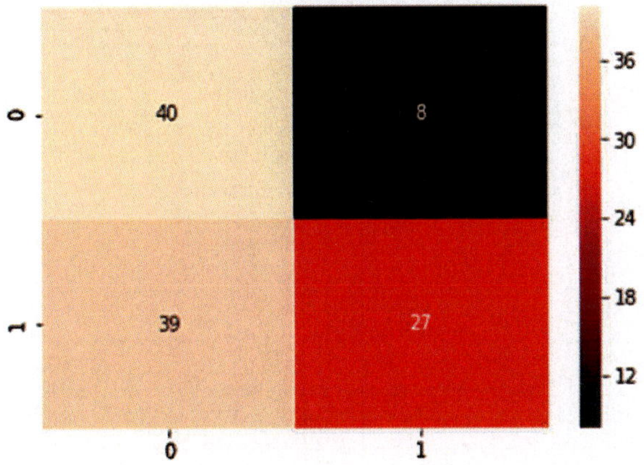

Fig. 12 Decision tree classifier

in all the accuracy parameters like precision, recall F1 score, etc. The KNN classifier also shows a significant increase in the accuracy with $k = 6$ it gives 93.00% accuracy. In the paper, it is also concluded that there is a decrease in accuracy for the KNN classifier with increasing K values. The Roc plot shown below for the support vector machine which shows the high sensitivity of the classifier. Roc plot is a measure relies on two factors the sensitivity and specificity of the classification algorithms (Figs. 13 and 14).

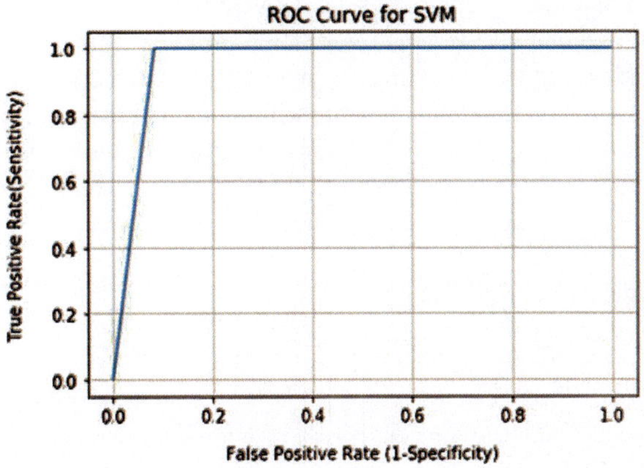

Fig. 13 Roc plot of SVM

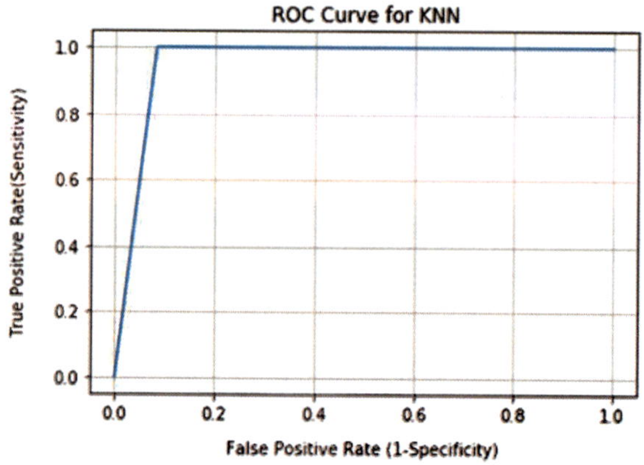

Fig. 14 Roc plot of KNN

5 Conclusion

The analyses of the medical data should be accurate and efficient for proper clinical treatment. In this paper, we used mainly four different algorithms: NB, SVM, DT, KNN, for the classification of the Wisconsin breast cancer data sets. Data was normalized before the classification and then all the algorithms were evaluated on all the accuracy measures. The support vector machine on average gave 97.50% accuracy which is the highest of all the algorithms. This paper can be further extended to genes engineering it will allow us new facts related to cancer. The lump growth factors can be easily monitored and all the gene inheritance could be studied in depth for better prediction of breast cancer in a woman at an early stage. In conclusion, the support vector machine and KNN are the best algorithms for the Wisconsin Breast cancer data set. We can also infer that hyperparameters optimization shows the significant increase in accuracy of the support vector machine. In the future, we would like to work on more efficient algorithms which will give better accuracy and precision on cancer data sets.

References

1. Cruz JA, Wishart DS (2006) Applications of machine learning in cancer prediction and prognosis. Cancer Inform 2:59–77. https://doi.org/10.1177/117693510600200030
2. Chaurasia V, Pal S (2014) A novel approach for breast cancer detection using data mining techniques. Int J Innov Res Comput Commun Eng 2(1):2456–2465
3. Asri H, Mousannif H, Al Moatassime H, Noel T (2016) Using machine learning algorithms for breast cancer risk prediction and diagnosis. Proced Comput Sci 83(Fams):1064–1069. https://doi.org/10.1016/j.procs.2016.04.224
4. Iagnosis D (2012) Different machine learning algorithms for breast cancer 3(6)

5. Kourou K, Exarchos TP, Exarchos KP, Karamouzis MV, Fotiadis DI (2015) Machine learning applications in cancer prognosis and prediction. Comput Struct Biotechnol J 13:8–17. https://doi.org/10.1016/j.csbj.2014.11.005
6. Cheng D, Zhang S, Deng Z, Zhu Y, Zong M (2014) k NN algorithm with data-driven k value k NN algorithm with data-driven k value. no. December, 2014. https://doi.org/10.1007/978-3-319-14717-8
7. Lessmann S, Stahlbock R, Crone SF (2005) Optimizing hyperparameters of support vector machines by genetic algorithms. In: Proceeding 2005 international conference artificial intelligence ICAI'05, vol 1, no May 2014, pp 74–80
8. Sola J, Sevilla J (1997) Importance of input data normalization for the application of neural networks to complex industrial problems. IEEE Trans Nucl Sci PART 3 44(3):1464–1468. https://doi.org/10.1109/23.589532

Classification of Cardiovascular Disease Using Wearable Device and Machine Learning Over Cloud

Naman Chauhan, Shiv Kumar, and Medhavi Malik

Abstract With the change in the lifestyle and food habits of the people nowadays, the heart-related diseases or the cardiovascular diseases (CVDs) are increasing among the population not just in India but around the globe. There is always a need of accurate and reliable predicting system for these life-threatening diseases in time for proper treatment. The healthcare industry is producing a large amount of data but it is not always made use to the full extent and is often underutilized. Various machine learning (ML) algorithms and techniques have been applied to various medical datasets to automate the analysis of large and complex datasets which in turn can detect, predict and even cure such diseases. The medical data such as heart rate, blood pressure, cholesterol and so on can be used as input and these features are modelled for classification. The data like blood pressure and heart rate are being taken from the user through the sensor on a wearable device that sends the data in real time to the cloud database. ML Platform as a Service (PaaS) hosts custom made ML model that is trained using several algorithms to pick the best accurate one. The classification results are showed to the user on his/her wearable device which is very ergonomic. This system can be used by normal people of all ages or patients who are concerned about health.

Keywords Cardiovascular disease · Heart rate sensor · Internet of things · Machine learning · Prediction model · SVM · ESP8266 · K-nearest neighbour · Logistic regression

N. Chauhan (✉) · S. Kumar · M. Malik
Department of CSE, SRMIST, Chennai, India
e-mail: chauhannaman98@gmail.com

S. Kumar
e-mail: kumar.shiv25298@gmail.com

M. Malik
e-mail: medhavimalik28@gmail.com

R. P. Mahapatra et al. (eds.), *Proceedings of 6th International Conference on Recent Trends in Computing*, Lecture Notes in Networks and Systems 177,
https://doi.org/10.1007/978-981-33-4501-0_36

1 Introduction

1.1 Cardiovascular Diseases (CVDs), Reasons, Risks and Symptoms

In the past half century, the changing lifestyle with the ease of doing the physically exhausting daily routines and food habits of the people led their bodies to become incubator for various health-related issues. According to the World Health Organization (WHO), CVDs are the number one cause of death globally [1].

Heart is predominantly, one of the most important organs in human body which is responsible for transporting blood to various organs. If it fails to work, all other organs will also start dying due to lack of supply of oxygen and nutrients.

Reasons and Symptoms of CVDs. Because of our day-to-day lifestyle which comprises of hectic routine and busy schedule, people are becoming prey to stress and anxiety. Also, bad food habits of modern world are leading to cholesterol and obesity in people. Factors like age, gender, diabetes, BMI, blood pressure, heart rate and habits of drinking alcohol and smoking lead to various CVDs.

Symptoms of cardiovascular diseases are [2]:

- Chest tightness.
- Shortness of breath.
- Nausea, indigestion, heartburn, or stomach pain.
- Sweating and fatigue.
- Pressure in the upper back pain that spreads to the arm.

Types of CVDs and its effects. Coronary heart disease or coronary artery disease is the narrowing of the coronary arteries. The coronary arteries supply oxygen and blood to the heart. It is one of the most infamous of all the heart diseases which leads to death of a large number of people annually.

Who is at risk for CVDs? There are risk factors that increase the potential to develop plaque within coronary arteries and cause them to narrow. Factors that put people at increased risk for heart diseases are [3]:

- Smoking
- High blood pressure (hypertension)
- High cholesterol
- Diabetes
- Obesity
- Family history of heart problems, especially heart attacks and strokes.

1.2 Diagnosis of CVDs

The symptoms of the CVDs play key role in the diagnosis which have been discussed above in this paper. Electrocardiogram (ECG), stress test, echocardiography, nuclear imaging, cardiac computerized tomography and cardiac catherization are few tests which are done for the diagnosis usually [3]. It is important to know about the associated symptoms including shortness of breath, sweating, nausea, vomiting, and indigestion, as well as malaise or fatigue.

So, this paper is based on the factors mentioned above. We try to predict the risk of the CVDs using several parameters or attributes using the machine learning techniques and algorithms. The aim of this paper is to propose a device and its system which shall be more ergonomic and user friendly but as well as it is accurate and efficient too.

2 The Proposed System

2.1 Overview

The whole system consists of two main components—wearable device and cloud. The first component is the interface which is directly accessible by the user. Device contains sensor and microcontroller. The device uses the IEEE 802.11 Wireless Fidelity connectivity at 2.4 GHz frequency to establish a connection to the internet through an access point or Wi-Fi router. Second component of the system is the cloud. Cloud has a real-time database based on Not-only SQL (NoSQL) to store the data is key-value pair. The other part is the cloud computing component. The model has been trained for classification predictions using the machine learning algorithms like support vector machine, logistic regression using cross validation and k-nearest neighbour. The prediction result of the model is updated in the database and then the wearable device takes the update and displays it to the user.

2.2 Components of System

Wearable device. The wearable device has sensor to record the heart rate of the user continuously. This sensor is a pulse sensor which is developed based on PPG (Photoplethysmography) techniques. This is a simple and low-cost optical technique that can be used to detect blood volume changes in the microvascular bed of tissues. It is relatively easy to detect the pulsatile component of the cardiac cycle according to this theory [4]. It is controlled by ESP8266 which works as the brain and it handles the data processing and data transmission. It connects with any Wi-Fi router or access point to make a connection with the cloud so that it can send and receive data.

Cloud. The cloud has a real-time database which is based on Not-only SQL or NoSQL. It actually saves data in a key-value pair fashion rather than a tabular one [5]. In this way, the values can be easily accessed by tracking the child paths of each entity. This makes it easier for device to directly GET or PUSH data to a specific child path directly. The seconds part of the cloud has machine learning model which is trained using various classification algorithms. The prediction results of the model are saved in the database to specific key-value pair using its corresponding child path.

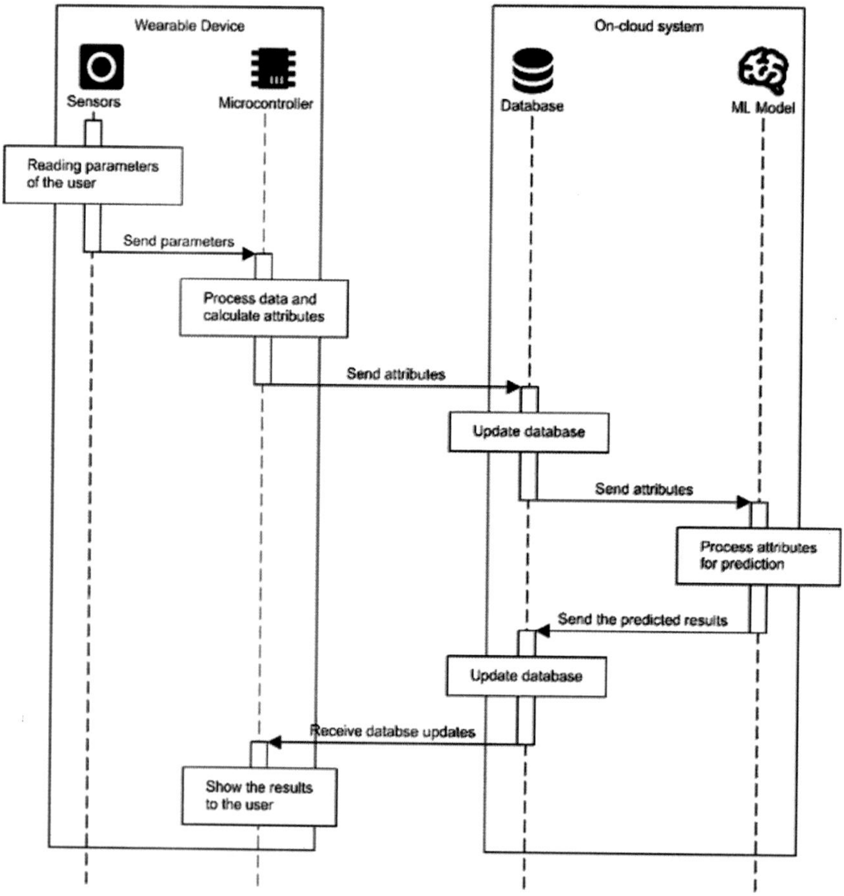

Fig. 1 Sequence diagram of the proposed system

2.3 Components of System

Wearable device and attribute calculation. Figure 1. explains the working of the system through a sequence diagram easily. The sensor in the wearable device works on the method of photoplethysmography which uses optical technique to detect the blood flow in veins and translates the data into the pulses. This analogue data is sent to the microcontroller which takes the input signals and calculates the heart rate and blood pressure of the user. Blood pressure is calculated by the microcontroller itself using a set of formula and algorithm mentioned below [6].

$$\text{Stroke volume} = \pi * (\text{LVOT diameter})^2/2 * \text{LVOT VTI} \quad (1)$$

$$\text{Cardiac output} = \text{Stroke volume} * \text{heart rate} \quad (2)$$

$$\text{Resistance} = \text{MAP} * 80/\text{Cardiac output} \quad (3)$$

$$\text{Blood pressure} = \text{Cardiac output} * \text{resistance} \quad (4)$$

Sensor has already sent the heart rate to the controller. Thus, we now easily use the values to put them into equations and calculate it. Using Eq. (4), we have the final value of blood pressure. This calculation goes using the following algorithm:

1. Initially, the blood flow will be sensed using the sensor.
2. After this, the heart rate will be calculated.
3. When the heart rate is calculated, the calculation of stroke volume, cardiac output and resistance.
4. Finally, the blood pressure is calculated.

The final values are then sent to the database. The values are updated periodically to the kay-value pairs using the specific child paths and values.

Cloud Database and Computing. On cloud, the NoSQL real-time database stores all the values received from the controller in a key-value pair fashion. This database is highly reliable and scalable. Each one of the values can be easily accessed by its child path for inserting, updating or deleting any value. The online cloud-based machine learning engine uses the values saved in the database to predict the outcome based on the data being sent from the user's end. When cognitive APIs fall short of requirements, you can leverage ML Platform as a Service (ML PaaS) to build highly customized machine learning models. For example, while a cognitive API may be able to identify the vehicle as a car, it may not be able to classify the car based on the make and model. Assuming you have a large dataset of cars labelled with the make and model, your data science team can rely on ML PaaS to train and deploy a custom model that's tailormade for the business scenario [7]. The custom model sends the predicted results to the database and saves it to its corresponding child path. The updated results are read by the controller from the child path and it shows the result to the user about the condition of his/her heart in realtime.

Algorithms used in the model training and dataset handling. The model has been trained using algorithms including support vector machine, logistic regression and k-nearest neighbour for training. The attributes used for model training are age, sex, chest pain type(cp), resting blood pressure(trestbps), maximum heart rate achieved(thalach), cholesterol serum(chol) and fasting blood sugar(fbs).

Dataset has been cleaned and unnecessary attributes were removed from the dataset. Also, no null values were found in the dataset. So, no substitution of values was done here. To find the correlation, a subplot graph was plotted but none of this was found among the attributes of dataset.

K-nearest neighbour algorithm. When KNN algorithm was used to train the model for the classification purpose, the highest accuracy of KNN came out to be 73.77% at $K = 11$. Jacquard score and F1 score of the model were found to be 0.6086 and 0.6941 respectively. The dataset was split into 80% and 20% dataset for training and testing respectively.

Support vector machine (SVM). The kernel used for SVM was Radius Basis Function (RBF). Gamma value was taken as 3. The confusion matrix was used to find the highest accuracy. 66.67% dataset was then split for training purpose and 33.33% was taken for testing purpose. The value of F1 score, i.e. ratio of correctly predicted to the all predicted values was 0.7070. The jacquard score was 0.645. The accuracy of the SVM is 70.70%.

Logistic regression. This supervised algorithm works on the probability generated for the target class. The class is assigned to the data point to the one whose probability is closer to 1. Confusion matrix was used to determine precision, recall and F1 score. F1 score is 0.8608, and jacquard index is 0.7741. The log loss was minimal to 44%. The accuracy of the model was achieved to 85.833%. Dataset was split into 66.67% and 33.33% for training and testing purposes, respectively (Fig. 2).

The highest accuracy of the model comes when logistic regression is used which is 85.833%. Thus, we use it here as our primary algorithm for prediction.

Fig. 2 Results of the algorithms used for training and testing

Algorithm	Jaccard	F1-score	LogLoss
KNN	0.6086	0.6941	NA
SVM	0.6455	0.7070	NA
LogisticRegression	0.7741	0.8608	0.44

3 Conclusion

3.1 Results

In the above paper, we have discussed various classification algorithms that can be used for the purpose of classification of heart disease datasets. Also, we have trained several classification models based of different algorithms and the accuracy was obtained by them. The highest accuracy was given by the logistic regression model which was accounted to be around 80%. The investigation tells us that several other technologies were used in other papers with some other count of attributes which yielded different accuracies depending upon the tools designed for execution. The accuracy of the structure can further be upgraded by creating various combinations of data mining and machine learning techniques. sugar(fbs).

The wearable device uses ESP8266 Wi-Fi microcontroller primarily which is much power efficient and cost effective as compared to other microcontroller with similar specifications. Wearable device can be powered by powered by batteries as small as button cells as ESP8266 only requires 3.3 V power supply to work.

3.2 Future Prospects

The wearable device can be improved with more ergonomic design which is easily worn around the wrist of the user. We have discussed above that there is a single sensor which is used to record the heart rate of the user and that heart rate is also used to calculate the blood pressure. If a sensor which can be used for reading blood pressure in the device, it can be utilized and device can be improved in the future. Also, there are several techniques which are useful for putting the ESP8266 to sleep mode when not in use and when the reading is required, an external interrupt using a button can be called to wake up the microcontroller and again go to sleep mode when the work is done.

The model has been trained using seven attributes but several other datasets can be used which have better training of the classification model. As of now, this system only predicts the CVDs but further datasets can be used to create and train ML models to detect any other disease like influenza, malaria, dengue, etc. Also, we can create a smartphone application which can connect with cloud database and improve the user interface and experience.

References

1. Cardiovascular diseases (CVDs). https://www.who.int/news-room/fact/details/cardiovascular-diseases-(cvds). Last Accessed 02 Feb 2020
2. Marimuthu M, Deivarani S, Gayathri R (2019) Analysis of heart disease prediction using various machine learning techniques. In: International conference on artificial intelligence, smart grid and smart city applications, (AISGSC 2019). PSG College of Technology, Coimbatore, p 1
3. Types of heart diseases, symptoms, Risk Factor, Causes and Stats. https://www.medicinenet.com/heart_disease_coronary_artery_disease/article.htm-who_is_at_risk_for_heart_cardiovascular_disease. Last accessed 06 Feb 2020
4. A review on wearable photoplethysmography sensors and their potential future applications in health care, https://www.ncbi.nlm.nih.gov/pmc/articles/PMC6426305/. Last accessed 09 Feb 2020
5. NoSQL Databases Explained| MongoDB. https://www.mongodb.com/nosql-explained. Last accessed 23 Feb 2020
6. Poornachandra S, Nivedha V, Arun S, Shankar M (2018) Blood pressure measurement using ARDUINO. Int J Res Appl Sci Eng Technol (IJRASET 2018), p 2470
7. An Executive's Guide To Understanding Cloud-based Machine Learning Services. https://www.forbes.com/sites/janakirammsv/2019/01/01/an-executives-guide-to-understanding-cloud-based-machine-learning-services/#1aefefda3e3e. Last accessed 28 Jan 2020

Indoor Scene Recognition with Depth Estimation over Monocular RGB-D Sequences

Geetanjali Bhola, Mohit Gupta, Saif Haque, and Rahul Singh Rawat

Abstract Object detection and recognition has been a field of extensive research over the past years and still continues to be explored. In our work, we have implemented indoor scene recognition with depth estimation of the nearest object. Our proposed work has been experimented over the input images and videos taken from our college lab. SSD Inception V2 model has been extensively trained to work for different objects from lab setup and then depth estimation is done on monocular RGB images using encoder-decoder architecture. Our proposed method uses DeepLabV3 for semantic segmentation. Our proposed work shows that simple depth estimation architecture outperforms current CNN methods with relative error of 12.6%, rms error of 46.4% and average log error of 5.4%. The mean average precision of object detection module is 37%.

Keywords Depth estimation · Semantic segmentation · CNN · Object detection

1 Introduction

We humans have been gifted by god to understand any image just by looking at it. But this task is very difficult for machines. A lot of research and endeavour have been done to make machines good at perception and with the arrival of computer vision and deep convolutional networks, success has been hit. This paper is the result of our research where we implemented object detection, wall and floor detection

G. Bhola (✉) · M. Gupta · S. Haque · R. S. Rawat
Department of Information Technology, Delhi Technological University, Delhi, India
e-mail: geetanjali.getz@gmail.com

M. Gupta
e-mail: mohitguptamt123@gmail.com

S. Haque
e-mail: haque.saif1995@gmail.com

R. S. Rawat
e-mail: rahulrawat1210@gmail.com

R. P. Mahapatra et al. (eds.), *Proceedings of 6th International Conference on Recent Trends in Computing*, Lecture Notes in Networks and Systems 177,
https://doi.org/10.1007/978-981-33-4501-0_37

using semantic segmentation and depth estimation of monocular RGB images in a restricted environment using real-time video processing and a single camera. The proposed work has been extensively tested to operate inside college campus buildings which include floors, walls, corridors, and objects present in labs.

The first contribution of this paper deals with floor and wall detection using semantic segmentation. Semantic segmentation is the computer vision technique to assign every pixel of image to a class label. It tries to model the relationship between pixels. For Example, there is a high probability for nearby pixels to have the same label. One may wonder why not use object detection for floors and walls. The reason behind this is that the object detection model may learn insignificant features like color, texture which is a computationally infeasible task. On the other hand, semantic segmentation discards unnecessary features like color, texture, etc and retains significant features for floor and wall detection like edges thereby enhancing the model performance.

We have implemented semantic segmentation using Google's DeepLab ResNet [1] model. One main motivation for using Deeplab [1] is that image segmentation is done along with controlling signal decimation-decreasing amount of data and number of samples that the system may take as input. Another motivation is that Deeplab [1] enables multi-scale contextual learning of features-collecting features from images of different scales. An Imagenet [2] pretrained residual neural network (ResNet [3]) is used for feature extraction. Deeplab [1] uses atrous convolutions with varying dilation rates to help capture multi scale contextual information.

The second contribution of this paper is in the area of object detection. Object detection is the area of image processing responsible for localization and classification of objects in an image. Localization is done by creating bounding boxes around objects. Tensorflow Object Detection API has been used to train pretrained models using transfer learning.

A pretrained model is a model which has been trained earlier on a large dataset. The model is retrained on different dataset using transfer learning. A pre-trained model works best in a generic environment. It can hinder performance when applied to a specific environment. So, we have used transfer learning to tailor this model to enhance performance in our restricted environment.

The third contribution of paper is to estimate depth of monocular images using transfer learning. Depth estimation is the task of estimating the depths in a two dimensional image. The task requires an input RGB image and outputs a depth image. The depth image includes information about the distance of the objects in the image from the viewpoint, which is usually the camera taking the image.

Transfer learning is used to retrain the image classification model which can be used as feature encoder for depth estimation model. The upper hand of this approach is that it provides a more modular structure where upcoming tasks in one area are efficiently moved to the problem of depth estimation.

Accurate depth estimation of objects in images is a basic step in scene understanding and reconstruction applications. We have used a CNN [4] based model for generating a high quality depth color map by using transfer learning. A typical encoder decoder structure is followed. The encoder is initialized using high performing pre-trained networks to extract features. Augmentation and various training strategies are used to produce more accurate results. Although our decoder is simple, it can produce detailed high-resolution depth maps. The NYU dataset [5] used in this method has an upper bound of 10 meters. The depths estimated by this method are relative to the deepest point assumed.

The rest of the paper is organized as follows: Sect. 2 discusses Related Work. Section 3 explains the proposed method. Section 4 discusses the experimental results. Finally, Sect. 5 gives the conclusion.

2 Related Work

Our problem consists of 3 subproblems namely object detection, semantic segmentation, and depth estimation. There has been comprehensive research in these fields and many algorithms have been proposed like fuzzy logic-based object recognition [6] and automatic mean-shift clustering-based image segmentation [7], but with some drawbacks.

Shaoqing Ren et al. proposed a method (Faster-RCNN) [8] for object detection which consists of two networks: region proposals are generated using region proposal network (RPN) and the objects are detected by another network making use of these proposals. The region boxes are ranked by the RPN and the ones most likely having the objects are proposed. However, Faster-RCNN is slower compared to the SSD [9] model. For semantic segmentation, there is a region-based method (R-CNN [10]) that follows the "segmentation using recognition" pipeline. Freeform regions from the image are extracted and described, followed by region-based classification. But it faces a couple of drawbacks such as the feature extracted is not compatible with the segmentation task and it lacks spatial information for accurate boundary generation.

Many CNN [11–15] methods have tried to solve the problem of Monocular depth estimation where they treat it as a regression problem to find the depth color map of a single image (RGB). General problems in these methods are poor quality and resolution of generated depth maps. The aim of our method is to produce good quality maps with accurate borderlines with the help of standard neural network architectures.

3 Proposed Method

3.1 Floor and Wall Detection Using Deeplab Resnet Architecture

We used semantic segmentation for building the floor and wall detection model. Unlike Image Classification, in semantic segmentation decisions are made for every

Table 1 mAP for different semantic segmentation models in various dataset competitions [16]

Model	2012 PASCAL VOC (*mIoU*)	PASCAL-Context (*mIoU*)	2016 COCO (*AP*)	2016 COCO (*AR*)	2017 COCO (*AP*)	Cityscapes (*mIoU*)
FCN	62.2	X	X	X	X	X
ParseNet	69.8	40.4	X	X	X	X
Conv and Deconv	72.5	X	X	X	X	X
FPN	X	X	X	48.1	X	X
PSPNet	85.4	X	X	X	X	80.2
Mask-R-CNN	X	X	37.1	X	41.8	X
DeepLab	79.7	45.7	X	X	X	70.4
DeepLab v3	86.9	X	X	X	X	81.3
DeepLab v3+	89.0	X	X	X	X	82.1
PANet	X	X	42.0	X	46.7	X
EncNet	85.9	52.6	X	X	X	X

pixel in an image. Each pixel of a digital image is assigned an individual label. It is important to note that semantic segmentation does not assign different labels to different instances of the same object. Classification DCNNs are usually used for semantic segmentation. They consist of four main operations: convolutions, activation function, pooling, and fully connected layers.

Regular DCNNs such as the AlexNet [17] and VGG [18] are not suitable for dense prediction tasks. Presence of many layers in these models, designed to reduce spatial dimensions of the input features, lead to highly decimated feature vectors lacking sharp details. Also, fully connected layers having fixed sizes loose spatial information during computation. DeepLab [1], on the other hand, provides an architecture that is capable of controlling signal decimation and learning multi-scale contextual features.

Apparently from Table 1, Google's DeepLab [1] V3 model has better performance as compared to others.

3.1.1 Architecture

As shown in Fig. 1, ImageNet [2] pre-trained Resnet [3] is used by DeepLab as its main feature extractor network. However, a new Residual block is proposed for multi scale feature learning. The last ResNet block doesn't use regular convolutions but it makes use of atrous convolution. Also, different rates of dilation are used by every convolution in this new block to capture the multi-scale context. Atrous Spatial Pyramid Pooling [1] (ASPP) is used on top of this new block.

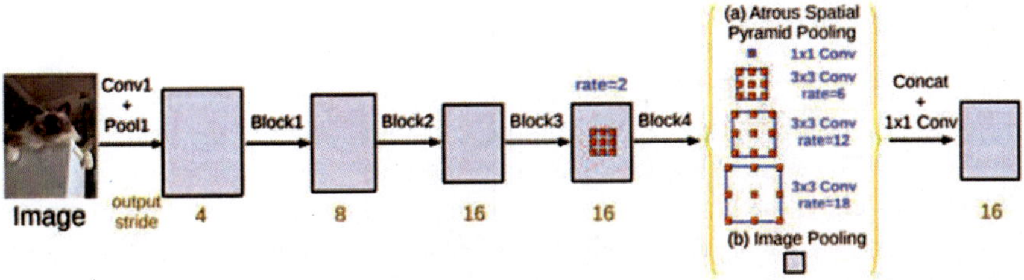

Fig. 1 Google's deeplab resnet architecture [19]

Atrous Convolutions are just regular convolutions possessing the capability to expand the filter's field of view. The working of atrous convolutions is like that first, the convolution filter is dilated according to the dilation rate. Secondly, the empty spaces are filled with zeros–creating a sparse filter.

The model is provided multi scale contextual information by **Atrous Spatial Pyramid Pooling** [1] **(ASPP).** For this purpose, a series of atrous convolutions are added with different dilation rates. These rates capture long-range context.

3.1.2 Learning and Inference

Weighted cross-entropy (WCE) is a variant of CE where all positive examples get weighted by some coefficient. It is used in the case of class imbalance.

$$\mathrm{WCE}\left(p,\hat{p}\right) = -\left(p\log\left(\hat{p}\right) + (1-p)\log\left(1-\hat{p}\right)\right) \tag{1}$$

where $p = P(Y = 0)$ and $\hat{p} = \mathbf{P}(\hat{Y} = 0) = \frac{1}{1+e^{-x}}$.

Set $\beta > 1$ for decreasing the number of false negatives. For decreasing the number of false positives, set $\beta < 1$

Dice Loss:

$$\mathrm{DL}\left(p,\hat{p}\right) = 1 - \frac{1+2p\hat{p}}{\hat{p}+1+p} \tag{2}$$

where $p = P(Y = 0)$ and $\hat{p} = \mathbf{P}(\hat{Y} = 0) = \frac{1}{1+e^{-x}}$.

3.2 Object Detection Using SSD Inception V2

We did custom training of the Single Shot MultiBox Detector (SSD [9]) Inception Model for our restricted environment via transfer learning. The Dataset was prepared using Google's Open Images Repository [20] for the training.

3.2.1 Architecture

VGG-16 [18] model which is pre-trained on ImageNet [2] Dataset is used as a base model by SSD [9] for extracting useful image features. Figure 2 shows that SSD [9] puts several Convolutional Feature layers of decreasing sizes on top of the VGG-16 [18] model. These feature layers are seen as a pyramidal depiction of images at various scales. The large smooth feature maps are good at capturing tiny objects whereas small unsmooth feature maps are good for detecting large objects. Detection is done at every tapered layer aiming objects of different sizes in SSD [9].

3.2.2 Learning and Inference

The training speed and the overall model performance can be hugely affected by considering different loss functions. We defined a loss function that balances both the above-mentioned factors.

The **Localization Loss** is a flat L1 loss between the ground truth values and the predicted bounding box correction [16]:

$$
\begin{aligned}
\mathcal{L}\mathrm{loc} &= \sum i,j \sum_{m \in \{x,y,w,h\}} \mathbb{1}_{ij}^{\mathrm{match}} L_1^{\mathrm{smooth}} (d_m^i - t_m^j)^2 \\
L_1^{\mathrm{smooth}}(x) &= \begin{cases} |x| - 0.5 & \text{otherwise} \\ 0.5x^2 & \text{if } |x| < 1 \end{cases} \\
t_x^j &= (g_x^j - p_x^i)/p_w^i \\
t_y^j &= (g_y^j - p_y^i)/p_h^i \\
t_w^j &= \log(g_w^j/p_w^i) \\
t_h^j &= \log(g_h^j/p_h^i)
\end{aligned}
\tag{3}
$$

where $\mathbb{1}_{ij}^{\mathrm{match}}$ indicates whether the ith bounding box with coordinates $(p_x^i, p_y^i, p_w^i, p_h^i)$ is compared to the coordinates $(g_x^j, g_y^j, g_w^j, g_h^j)$ of jth ground truth box for any object.

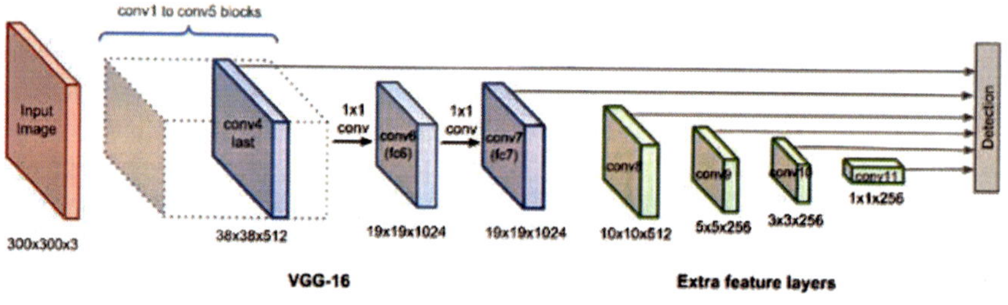

Fig. 2 Single shot multibox detector architecture [21]

d_m^i, $m \in \{x, y, w, h\}$ are the predicted correction terms. Figure 5 shows Localization Loss of the model during training.

The **Classification Loss** is a multi-class softmax loss [16]:

$$\mathcal{L}cls = -\sum i \in \text{pos} \mathbb{1} ij^k \log(\hat{c}_i^k) - \sum i \in \text{neg} \log(\hat{c}_i^0), \text{ where } \hat{c}_i^k = \text{softmax}(c_i^k) \quad (4)$$

where $\mathbb{1}_{ij}^k$ shows if the ith bounding box and the jth ground truth box are compared for an object in class k, pos is the collection of compared bounding boxes (NN items in total) and neg is the collection of negative samples. Figure 5 shows Classification Loss of the model during training.

The **Loss Function** is the sum of these two losses:

$$\mathcal{L} = \frac{1}{N}(\mathcal{L}_{\text{cls}} + \alpha \mathcal{L}_{\text{loc}}) \quad (5)$$

Figure 5 shows Total Loss of the model during training.

3.2.3 Speed Versus mAP Tradeoff

As apparent from the Fig. 3 SSD [9] model is faster than many regional-based models like Faster-RCNN [8] and is sufficiently accurate at detecting large objects. Since our scenario requires fast and accurate detection of large objects like the staircase, door, etc, we selected the SSD [9] model as the pre-trained model for custom training.

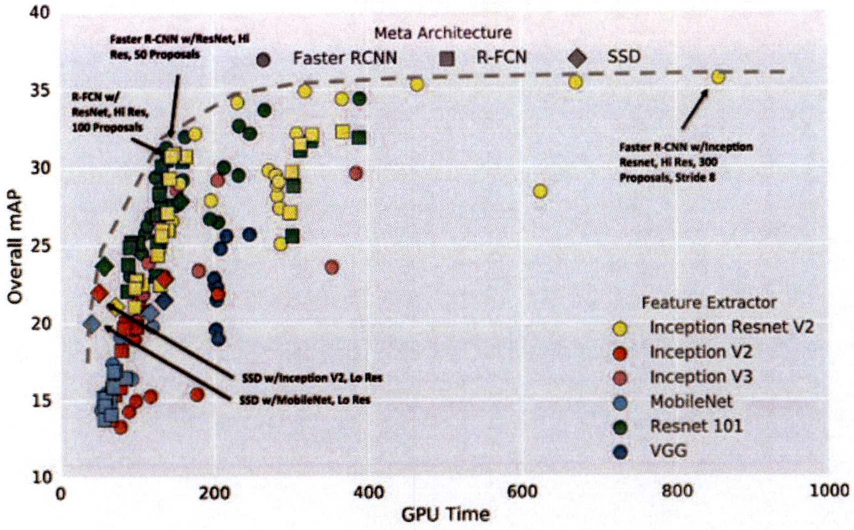

Fig. 3 Speed versus mAP tradeoff for various object detection architectures [22]

3.3 *Depth Estimation Using DenseNet 169*

Many depth estimation papers use an encoder-decoder type neural network. The input for the task of depth estimation is an RGB image and the output is a depth map which is a scaled-down version of the input image. We decided to work with the following deep learning architecture because of its simplicity, higher accuracy and less training time.

3.3.1 Architecture

We have used a pre-trained DenseNet 169 [23] as the encoder in the paper. The encoder is present before the fully connected layer and it comprises of 4 dense blocks. Unlike other depth models, it uses a very simple decoder. A simple bi-linear upsampling layer followed by 2 convolutional layers is present in each decoder block. This model concatenates the upsampling layers with the corresponding layers in the encoder.

3.3.2 Learning and Inference

For training our network, loss is defined as the weighted sum of 3 loss functions:

$$L(z, \hat{z}) = \lambda L_{\text{depth}}(z, \hat{z}) + L_{\text{grad}}(z, \hat{z}) + L_{\text{SSIM}}(z, \hat{z}) \tag{6}$$

L_{depth} is Least Absolute Deviations (LAD) of every point on the depth values:

$$L_{\text{depth}}(z, \hat{z}) = \frac{1}{n} \sum_{p}^{n} \left| z_p - \hat{z}_p \right| \tag{7}$$

L_{grad} is Least Absolute Deviations (LAD) defined over gradient of depth image:

$$L_{\text{grad}}(z, \hat{z}) = \frac{1}{n} \sum_{p}^{n} \left| g_x \left(z_p, \hat{z}p \right) \right| + \left| g_z \left(z_p, \hat{z}_p \right) \right| \tag{8}$$

L_{SSIM} uses the Structural Similarity (SSIM) term metric which is generally used for image reconstruction tasks. We define it as a loss L_{SSIM} [24] as follows:

$$L_{\text{SSIM}}(z, \hat{z}) = \frac{1 - \text{SSIM}(z, \hat{z})}{2} \tag{9}$$

4 Results

4.1 Floor and Wall Detection Using Deeplab Resnet Architecture

4.1.1 Dataset

ADE20K [25, 26] is a dataset of 20 k scene centric images that are annotated with objects and object parts. It consists of a total of 150 semantic categories including stuff like the sky, road, grass, person, car, etc. We trained our DeepLab [1] ResNet Model using transfer learning in TensorFlow. Our dataset consists of 20210 images in the training set and 2000 images in the testing set. We re-labeled 11 classes into 2 classes in order to easily classify the wall and floor. We combined window, door, fence, pillar, signboard, bulletin board classes into a single class WALL. We also combined road, ground, field, path, runway classes into a single class FLOOR. The input image size was 640×480 in size (Fig. 4).

4.1.2 Implementation Details

We implemented our floor and wall detection module using Google's DeepLab ResNet [1] Model and trained on NVIDIA-SMI 418.67 GPU with 12.8Gb RAM. We used semantic segmentation Technique to detect the wall and floor. ResNet [3] is a very popular DCNN. ResNets provide a framework for the training of deeper models. The batch size was 12 and RELU [27] was used as the activation function.

Fig. 4 Floor and Wall in the above sample images by our model (Green color represents Floor and Pink represents Wall)

4.1.3 Performance Metrics

Means IoU: The IoU measure quantifies the common number of pixels between the prediction and target kernel divided by the total number of pixels in both kernels.

$$\text{mean IoU} = \frac{\text{prediction} \cap \text{target}}{\text{prediction} \cup \text{target}} \tag{10}$$

After custom training of DeepLabV3 [1] Model on ADE20K dataset [25, 26], the Mean IoU comes out to be 0.7394.

4.2 *Object Detection Using SSD Inception V2*

4.2.1 Dataset

Google's Open Images Dataset V4 [20] consists of 9 M images annotated with image-level labels, object bounding boxes, and visual relationships. It contains 14.6 M bounding boxes for 600 object classes on 1.74 M images in the training set. We trained our SSD [9] inception V2 model using transfer learning in TensorFlow. Our dataset 20 K images in the training set and 2K images in the testing set. We incorporated 10 classes of objects in our dataset. Google OID [20] consists of images of random sizes but our preprocessing task normalizes the size to 300 × 300 (Fig. 5).

4.2.2 Implementation Details

We implemented our object detection model using TensorFlow and trained on NVIDIA-SMI 418.67 GPU with 12.8 Gb RAM. We have used Single Shot Multibox Detector (SSD [9]) for object localization and Inception V2 as our feature extractor. For the training, we used the ADAM [28] optimizer with a learning rate of 0.004. The batch size was set to 12. The training was performed for 20 M iterations (Table 2).

4.2.3 Performance Metrics

Mean Average Precision (mAP) measure represents the model's power. mAP is vastly used in multi class classification and Information retrieval settings. Average precision is calculated for each object class based on our model predictions. Area under the precision recall curve is defined as average precision. mAP is calculated by dividing the sum of individual class precision by the total number of classes. The output for object detection is shown in Fig. 6.

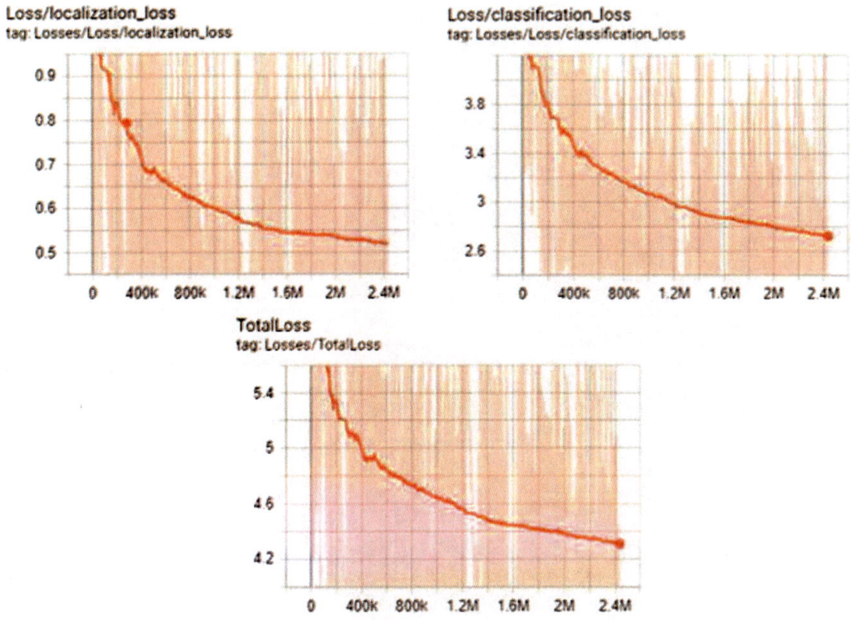

Fig. 5 Localization loss, classification loss and total loss during training

Table 2 Mean average precision scores for object detection module

Class	mAP@IUC = 0.5
Chair	0.142291
Computer monitor	0.474028
Desk	0.172609
Dog	0.841964
Laptop	0.698190
Luggage and bags	0.420981
Person	0.093841
Stairs	0.435304
Table	0.132276
Door	0.306319
Average	0.371780

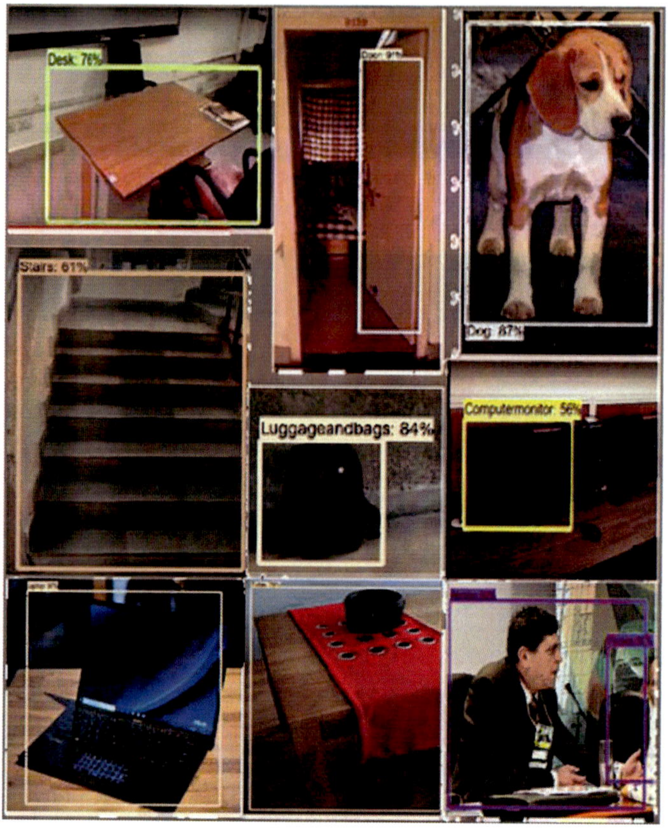

Fig. 6 Classes detected and localized by our trained model (L to R): Desk, Door, Dog, Stairs, Luggageandbags, Computermonitor

4.3 Depth Estimation Using DenseNet 169

4.3.1 Dataset

NYU Depth V2 [5] Dataset contains depth color maps and images for various indoor scenarios of resolution of 640 × 480. The dataset 120 K images in the raining set and 654 images in the testing set. We trained our model 50 K images. 10 m of upper bound was used in-depth maps. Our network produced output images of half the resolution as of the input image.

4.3.2 Implementation Details

We used DenseNet 169 [23] classification model as feature encoder for our depth estimation model and trained on NVIDIA-SMI 418.67 GPU with 12.8 Gb RAM. We used the ADAM [28] optimizer with a learning rate of 0.0001 for the training. The batch size was set to 5. The training was performed for 20 epochs each containing 10 K iterations.

4.3.3 Performance Metrics

We used the following measures to evaluate the performance of our model:

1. Average Relative Error (rel):

$$\frac{1}{n}\sum_{p}^{n}\frac{|y_p - \hat{y}_p|}{y} \tag{11}$$

Apparently our method has better performance than DenseNet121 and ResNet50

2. Root Mean Squarred Error (rms):

$$\sqrt{\frac{1}{n}\sum_{p}^{n}(y_p - \hat{y}_p)^2} \tag{12}$$

3. Average ($\log_{10}$) Error:

$$\frac{1}{n}\sum_{p}^{n}\left|\log_{10}(y_p) - \log_{10}(\hat{y}_p)\right| \tag{13}$$

The output for depth estimation is shown in Fig. 7 (Table 3).

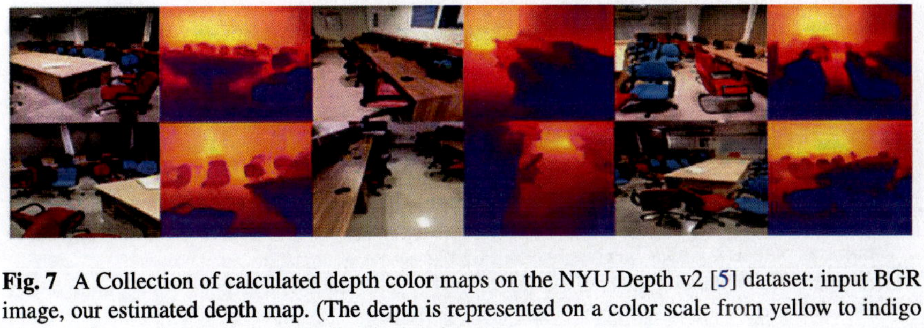

Fig. 7 A Collection of calculated depth color maps on the NYU Depth v2 [5] dataset: input BGR image, our estimated depth map. (The depth is represented on a color scale from yellow to indigo where yellow is the deepest and indigo is the closest)

Table 3 Comparison of our model with other models

Model	rel	rms	$\log_{10}$
Our method	0.126	0.464	0.054
DenseNet 121 [23]	0.131	0.497	0.057
Resnet50 [3]	0.157	0.559	0.067
Laina [13]	0.127	0.573	0.055

5 Conclusion

In this paper, we proposed a machine vision-based system that consisted of 3 modules namely floor and wall detection, object detection and depth estimation, to operate inside college campus buildings. Considering the trade-off between speed and accuracy, DeepLab ResNet architecture and SSD Inception V2 architecture proved better than region-based methods like R-CNN for semantic segmentation and object recognition respectively. The experimental results show that our simple encoder decoder structure to estimate depth outperforms many of the current CNN methods. Future scope of the paper is to explore the system's use in applications like robotic navigation and blind people mobility.

References

1. Chen L-C, Zhu Y, Papandreou G, Schroff F, Adam H (2018) Encoder-decoder with atrous separable convolution for semantic image segmentation. CoRR. arxiv: abs/1802.02611
2. Deng J, Dong W, Socher R, Li L-J, Li K, Fei-Fei L (2009) ImageNet: a large-scale hierarchical image database. In: CVPR09
3. He K, Zhang X, Ren S, Sun J (2015) Deep residual learning for image recognition. CoRR. arxiv:abs/1512.03385
4. O'Shea K, Nash R (2015) An introduction to convolutional neural networks. CoRR. arxiv: abs/1511.08458
5. Silberman N, Hoiem D, Kohli P, Fergus R (2012) Indoor segmentation and support inference from rgbd images, In ECCV
6. Susan S, Chandna S (2013) Object recognition from color images by fuzzy classification of gabor wavelet features. pp 301–305
7. Susan S, Kumar A (2016) Auto-segmentation using mean-shift and entropy analysis. In: 2016 3rd international conference on computing for sustainable global development (INDIACom), pp. 292–296
8. Ren S, He K, Girshick RB, Sun J (2015) Faster R-CNN: towards real-time object detection with region proposal networks. CoRR. arxiv: abs/1506.01497
9. Liu W, Anguelov D, Erhan D, Szegedy C, Reed SE, Fu C-Y, Berg AC (2015) SSD: single shot multibox detector. CoRR. arxiv: abs/1512.02325
10. Girshick RB, Donahue J, Darrell T, Malik J (2013) Rich feature hierarchies for accurate object detection and semantic segmentation. CoRR. arxiv:abs/1311.2524
11. Fu H, Gong M, Wang C, Batmanghelich K, Tao D (2018) Deep ordinal regression network for monocular depth estimation. CoRR. arxiv:abs/1806.02446

12. Hao Z, Li Y, You S, Lu F (2018) Detail preserving depth estimation from a single image using attention guided networks. CoRR. arxiv:abs/1809.00646
13. Laina I, Rupprecht C, Belagiannis V, Tombari F, Navab N (2016) Deeper depth prediction with fully convolutional residual networks. CoRR. arxiv:abs/1606.00373
14. Xu D, Ricci E, Ouyang W, Wang X, Sebe N (2017) Multi-scale continuous crfs as sequential deep networks for monocular depth estimation. CoRR. arxiv: abs/1704.02157
15. Xu D, Wang W, Tang H, Liu H, Sebe N, Ricci E (2018) Structured attention guided convolutional neural fields for monocular depth estimation. CoRR. arxiv: abs/1803.11029
16. https://medium.com/@arthur_ouaknine/review-of-deep-learning-algorithms-for-image-semantic-segmentation-509a600f7b57
17. Krizhevsky A, Sutskever I, Hinton GE (2012) Imagenet classification with deep convolutional neural networks. In: Pereira F, Burges CJC, Bottou L, Weinberger KQ, (eds) Advances in neural information processing systems, vol 25. Curran Associates, Inc., pp 1097–1105
18. Simonyan K, Zisserman A (2014) Very deep convolutional networks for large-scale image recognition. cite arxiv: 1409.1556
19. Chen, L-C, Papandreou G, Schroff F, Adam H (2017) Rethinking atrous convolution for semantic image segmentation. CoRR. arxiv: abs/1706.05587
20. Kuznetsova A, Rom H, Alldrin N, Uijlings JRR, Krasin I, Pont-Tuset J, amali S, Popov S, Malloci M, Duerig T, Ferrari V (2018) The open images dataset V4: unified image classification, object detection, and visual relationship detection at scale. CoRR. arxiv:abs/1811.00982
21. https://lilianweng.github.io/lil-log/2018/12/27/object-detection-part-4.html#ssd-single-shot-multibox-detector
22. Huang J, Rathod V, Sun C, Zhu M, Korattikara A, Fathi A, Fischer I, Wojna Z, Song Y, Guadarrama S, Murphy K (2016) Speed/accuracy trade-offs for modern convolutional object detectors. CoRR. arxiv: abs/1611.10012
23. Huang G, Liu Z, Weinberger KQ (2016) Densely connected convolutional networks. CoRR. arxiv:abs/1608.06993
24. Wang Z, Bovik AC, Sheikh Hamid R, Simoncelli Eero P (2004) Image quality assessment: from error visibility to structural similarity. IEEE Trans. Image Proc 13(4):600–612
25. Zhou B, Zhao H, Puig X, Fidler S, Barriuso A, Torralba A (2016) Semantic understanding of scenes through the ade20k dataset. arXiv preprint arXiv: 1608.05442
26. Zhou B, Zhao H, Puig X, Fidler S, Barriuso A, Torralba A (2017) Scene parsing through ade20k dataset. In: Proceedings of the IEEE conference on Computer vision and pattern recognition
27. Agarap AF (2018) Deep learning using rectified linear units (relu). CoRR. arxiv: abs/1803.08375
28. Kingma DP, Ba J (2015) Adam: a method for stochastic optimization, 2014. cite arxiv:1412.6980. Comment Published as a conference paper at the 3rd International Conference for Learning Representations, San Diego

Cognitive Outline Study of Electroencephalogram Signal Patterns Towards Monitoring and Analyzing Sleep Behavioural Disorders

Shivam Tiwari, Deepak Arora, and Puneet Sharma

Abstract A better consciousness depends on how well your sleep is. Study shows that, sleep disorder become a crucial problem day by day due to work pressure and other different causes. This continuous lack of sleep or uneven sleep pattern can cause to develop different major health issues in humans. These sleep patterns are difficult to detect, analyze, and experience as they exhibit at the time when people are in their deep sleep phase. This research work is focused on studying, methods for analyzing and identifying different sleep patterns based on EEG patterns. Every stage of sleep shows a different electroencephalogram pattern. These EEG patterns show that sleep can be observed as a combination of different cyclic sleep stages. This work shows different sleep stage patterns analysis found in different sleep disorders. Based on these patterns, monitoring of quality sleep can be performed. In this paper, authors have discussed 10–20 electrodes systems to capture brain signals, electroencephalogram signals, and its categorization. This study also represent the detailed description of several methods identified for monitoring sleep behavioural patterns in order to understand the experimental basis and forming a strong theoretical background towards analyzing these EEG signals.

Keywords RAM · NRAM · Electroencephalogram · Sleep stages · Quality sleep

S. Tiwari (✉) · D. Arora · P. Sharma
Department of Computer Science and Engineering, Amity School of Engineering and Technology, Amity University, Lucknow Campus, India
e-mail: shivam.tiwari.info@gmail.com

D. Arora
e-mail: deepakarorainbox@gmail.com

P. Sharma
e-mail: puneetgrandmaster@gmail.com

R. P. Mahapatra et al. (eds.), *Proceedings of 6th International Conference on Recent Trends in Computing*, Lecture Notes in Networks and Systems 177,
https://doi.org/10.1007/978-981-33-4501-0_38

1 Introduction

In recent years, there has been lots of research performed on the quality of sleep. Research related to sleep depends on how the data is collected. It is unlikely that the data collected from a wearable device or non-wearable device would give completely accurate results [1]. Collecting data from anyone method there is some raw factors added to the data in form of subject's uncomfortableness with respect to wearable device or environmental noise factor in terms of non-wearable device. For this type of research, data collection is often a challenging task.

The quality of sleep depends on the pattern of its stages. To understand these stages better, EEG signals are classified based on their frequency. 8–13, 13–30, 4–8, and 0.5–4 Hz are considered as Alpha, Beta, Theta, Delta, respectively. The EEG signals is like a recorder that records electrical impulses that are originating in the brain while two or more brain cells are communicating to each other. These impulses are detected with the help of the 10–20 system of electrodes that are attached to the scalp. While designing the wearable electrode scalp, it considers a ratio of 10–20 that means every 20% of half scalp must find an electrode to measure electrical impulses. To deal with the sleep disorder, it is necessary to understand its EEG pattern. Sleep is divided in two parts (i) Rapid eye movement (RAM) (ii) Non-rapid eye movement (NRAM). Non-rapid eye movement is further categorized in sleep stages like S0, S1, S2, and S3 where S0 for basic light sleep, S1 and S2 for light sleep, and S3 is considered as deep sleep. Every stage of sleep shows a different EEG (electroencephalogram) pattern. To measure sleep quality EEG pattern is considered as input parameter. In this paper, a cognitive outline of electroencephalogram in monitoring of sleep behavior is drawn.

2 Literature Review

According to Kagawa et al. [1], a correct result can be measured by measuring the changes that occur in the sleep stages by the non-contact method. Nine, a 24 GHz microwave radar system, is placed under the mattress to measure the movement of the sleeping person on the bed. For better estimation of sleep stages, a body-movement index is determined. To classify wake and sleep position, LDF (Linear discriminant function) is used and obtained 79.3% accuracy. Radar impulses are the base of calculating the sleep stages (rapid eye movement, non-rapid eye movement). In view of the increasing sleep problem, the author has suggested to the medical department to evaluate and analyse real-time sleep data. Sleep is a cyclic process in which the stages often repeat at an interval of every 90 min. Sleep-related study comes under Polysomnography (PSG). Under PSG, various sensors are installed on the body to detect the impulses reading. Still, it is more difficult to sleep to the person who wears the sensory device on the head and other parts of the body, so these readings cannot obtain an accurate result. Because of this author, introduce the unwearable system to measure the precise results. In this paper, polysomnography and radar system accuracy are examined by taking 11 male data whose age was around 22 years. According to Barsocchi et al. [2], explaining the process of sleep with a wearable device is quite complicated. Often people are uncomfortable due to these devices, which make it difficult to get good results. In this paper, an unwearable system is introduced, which has 48 force sensing resistor (FSR) grid. This grid is installed under the mattress, and it is connected to a computer through a data cable. For practical approach author installed a video recorder at night mode to check the ground reality and took almost 2 m height bed and 70 kg subject. Here, multiclass logistic regression is used on WEKA to obtain the result. In this experiment, a large amount of data is collected, so authors have calculated the result by devising the data in three parts. In result, a better performance is shown with CCI 95.8%, 88.5%, 90.2% on 4.2%, 11.5%, 9.8% error rate, respectively. This whole research is based on FSR that measures the pressure points of subjects. According to Adami et al. [3], measuring the movement of the leg in bed is an important step in experiments conducted to assess the quality of sleep. The author has used the linear classifier to separate the body movement and the leg movement. To measure the leg or body movement, a load cell sensors grid is installed under the mattress. Here, author supports the unobtrusive process to measure sleep quality. The base of periodic leg movement measurement is conducted from the World Association of Sleep Medicine. EMG data of 17 people (29–74 years old, 10 male, and 7 female) is considered for polysomnography. In result, 97% accuracy is obtained while classifying the leg and body movement. The base part of this experiment is to calculate the sleep quality based on PLM (periodic leg movement). According to Tseng et al. [4], sleep quality has been tested with a bluetooth-based EEG device. EEG device calculates the brain signals that are produced in the subject's brain and shares to the cloud storage for further analysis. Because of brain, signals have very low voltage; the author used the DAQ100 module to capture the accurate frequency of signals. In this research, the wearable device is used to measure sleep quality. This study explains the sleep disorder ratio of Taiwan society. Based on the Chinese sleep survey, 19.3%, 21.8% of insomnia patients were there in 2013 and 2009, respectively. According to the author, sleep is divided in two parts (1) RAM—it takes 25% of total sleep time. (2) NRAM—it is further classified in four parts, each part takes 5%, 45%, 45%, 13% respectively. EEG reading for each stage is different, so to identify the sleep quality, the author compares the EEG readings. According to Cheng et al. [5], Sleep plays a very important role in human life. Humans spend 1/3 of their entire life in sleep. Realizing the importance of sleep, it becomes necessary that the quality of sleep be tested on the right parameters. According to the Medical Sleep Organization of Taiwan report, insomnia patients increased from 11.5% to 21.8% in the short period of 2006 to 2009.

Collecting sleep data has always been a challenging task. Most of the devices are very expensive, and the discomfort of the people during the test spoils the sleep data. Here, author collected two types of data (1) subjective—this is the bundle of subject's information like medical history etc. (2) objective—this is recorded information during sleep. The author believes that subjective information is as important as the objective one. Here, more than one types of sensors are used to collect data like acceleration sensors are used to measure the body movement. To understand the sleep pattern, it is important to calculate a quality sleep index value also. Author modified the sleep quality index value (defined by Zeo Inc., ZQ = { (TOTAL SLEEP DURATION * 1 + DEEP SLEEP DURATION * 1.5 + REM SLEEP TIME * 0.5)— (TOTAL AWAKEN TIME * 0.5 + (AWAKENING TIMES / 15))} * 8.5) with some challenges. Some challenges are: subject may be a short sleeper; sleep time varies age to age, so the values of sleep quality index should not be constant. The author modified the formula with some weight value. This new formula has weight value as per requirement. In the end, the author verified his formula with better accuracy than the traditional formula. According to Surantha et al. [6], in the era of growing technology, the Internet of things (IOT) played a miraculous role. In recent years, Internet of things has changed the way devices work. This is an important change, the way in which machine is interacting with another machine. All the experiments were done in the quality of sleep collect sleep data manually with the help of sensors. In this type of data collection technique, sometimes, some noise signals are added, which somehow spoil the data. Such problems can be dealt with by using IOT technology. The author explained the workflow through IOT technology in sleep assessment. According to Zaffaroni et al. [7]. There are many types of research to understand the quality of sleep. Among them, the experiment from non-wearable devices maintains a distinct identity. In this paper, ultrasonic technology is used to measure sleep quality. Here, a special application has been designed that can measure the sonar generated by the movement of a person who is asleep in bed. Such experiments often claim less expensive and good results. It has been tested in the laboratory on 38 subjects to test the veracity of the application. Realizing the importance of sleep in human life, it becomes important that its use should very cheap and of good accuracy. Sonar measurement application provides this opportunity to measure sleep data in a very efficient cost or zero cost. The quality of sleep can be tested in two ways (1) through the wearable device (2) through the non-wearable device. The wearable device is capable of producing accurate results, but the discomfort of the tested person often worsens its results. In this experiment, one S + device, a mobile application that is installed in Samsung S7 is used to measure the sleep data set. There are 62 subjects to collect sleep data. This research is done in Berlin under an advanced sleep research organization. To test the quality of sleep, six parameters have been considered, which are as follows -

(1) Total sleep time (TST), (2) Total rapid eye movement time (TREMT), (3) Total deep sleep time (TDS), (4) Total light sleep time (TLST), (5) Wake time, and (6) Sleep latency. In result, accuracy is calculated for each that are 87%, 89%, 84% for wake time, REM, and deep sleep time, respectively. Realizing the importance of quality of sleep, this experiment is the father of revolutionary results. Its cost is the lowest and the best result is witnessed. Such experiments are often able to make human life comfortable. According to Yan et al. [8], realizing the importance of sleep, it is important that every phase of sleep be closely examined. This research examines the movement of the body to test each phase of sleep differently. Different sleep stages are very close to each other; testing these stages separately is useful for sleep-related research. Here, some sensors are placed under the bed, which generates

frequency rays based on the movement of the subject. Brain signals that are measured by the EEG that supports to define sleep stages are separated through support vector machine. The formula that is used to define sleep stage is:

Deep sleep index (D) = 0.55 * number of small movements + 1.05 * medium movements + 1.02 * large movements.

Here, D should be less than 0 1 for deep sleep stage. 0.02936, 0.2385, and 0.04826 are the threshold value of small, medium, and large movements. In results, 95.7%, 90.6%, and 66.3% accuracy is obtained for deep, light, and wake stage of sleep, respectively. According to Tataraidze et al. [9], there can be many parameters to test the quality of sleep; among them, testing the breathing pattern can be an important component. In this research, each stage of sleep has been understood based on breathing patterns. Quality of sleep focuses on breathing patterns. In this research, a data set has been created by measuring the breathing pattern from sensors with non-wearable devices. Polysomnography has been an important process initially for accurate measurement of sleep quality. However, the discomfort of subjects hinders accuracy. Here, a sleep disorder-breathing algorithm was written under which the results were generated. Algorithm obtained result 93%. According to Dhamchatsoontree et al. [10], a grid of 48 pressure sensors has been installed in the bed to assess the quality of sleep accurately. The author gave a unique name to the force sensor is I-sleep sensor. Sleep quality index is calculated based on body movement. The machine-learning algorithm obtained 86.7% accuracy to detect the sleep posture. The author explained two technology, (1) sensor-based (2) camera-based. In order to result 95.2%, 94.9%, 94.7%, 95.8% accuracy is obtained through random forest, K-nearest neighbor, logistic regression, and support vector machine, respectively. The data set contained 30 subjects (15 males and 15 females). According to Liao et al. [11], the quality of sleep is based on different readings for different populations, such as location, gender, and age. In this research, EEG readings on different genders have been examined. EEG patterns have been introduced in the frequency form, and each frequency range is given a name such as 0.5–4, 4–8, 8–12, and 12–32 Hz to delta, theta, alpha, and beta, respectively. These frequency ranges give different readings for different sleep stages and show some bigger changes in their fluctuation while considering different gender, age, and location. Frequencies of beta range and delta range were found to be increased during the non-rapid eye movement. During the experiment, it was found that the delta fluency range was higher in females than males. In the result, it was found that sex differences should be considered as an important parameter for testing the quality of sleep. The author also explained the limitations of the experiment and gave the future aspect of this experiment. In this experiment, author did not consider high-frequency range Gamma so that results that are obtained can vary while considering Gamma range frequencies. This experiment was based on wearable 10–20 system to detect the brain signals and applied weighted phase lag index machine learning algorithm to classify the frequency range. According to Ravan et al. [12], realizing the complexity of the process of sleep quality research, many experiments are associated with low cost, one of them being vagus nerve therapy. This experiment raises a silver lining for epilepsy patients. Through Vagus therapy, it was found that the sleep quality of epilepsy patients could be increased. Here the data of 23 people based on the EEG readings taken from Physio bank has been used. The system is trained with 22 epilepsy patients' data that are recorded in a month. Support vector machine algorithm is used to classify the sleep stages. In result support, vector machine gives 90% accuracy. This research proved that vagus nerve therapy could play an important role in quality sleep. According to Cheng et al. [13], an intelligent system is designed to test the quality of sleep showing the stages of sleep through smart phone application. Oximeter has been used in this experiment, based on which the quality of sleep is measured through a smart phone application, which is specially designed for classifying the sleep stages based on heart rate and other parameters. Author explained the architecture of the system which has four steps (1) data collection through oximeter or other devices, (2) Pre-analysis by algorithm, (3) smart phone application role, and (4) obtain result. This research is based on a wearable device but in a less expensive mode. According to Terzano et al. [14], to understand the process of sleep, it is necessary to examine its pattern and its duplication after a time interval. To understand the whole sleep process, it is necessary to examine the changes in EEG readings with respect to change of physical patterns of subject occurring in a time interval. To measure the quality sleep index author emphasizes to examine the EEG pattern during periodic *K*-alpha (cyclic alternative pattern). Authors have established a relationship among electrooculogram, electromyogram, electrocardiogram and EEG. Every alternative changes in EEG pattern is examined and compared with electrooculogram, electromyogram and electrocardiogram readings. In result, author proved EEG pattern is able to calculate the sleep quality index. According to Poryazova et al. [15], cyclic alternative pattern (CAP) is an important parameter that helps to understand sleep quality. This research examines the CAP pattern of narcolepsy patients. Narcolepsy patients were founded with less movement compared to a healthy human when the CAP patterns were tested. A summary of systematic review is shown in Table 1.

3 Background

Sleep is a natural process that is a part of important process like refreshing memory, testing immunity system, recovery of cell health. Sleep plays an important role in person's mental stability and physical strength. Sleep at the right time and for a right duration is essential for healthy living. Sleep plays a different role in every age stage, as physical growth and building their immunity system in children while in old sleep is beneficial in recovery of cell health [11]. Lack of sleep or lack of a quality sleep has been a common problem in the last few years. Lack of a quality sleep leads to fatal consequences such as physical and mental debility. In view of the increasing sleep patients, many sleep study centers have been established in many countries. According to a Taiwan report, there has been an increase in the number of sleep patients from 11.5% to 21.6% in a gap of 3 years from 2007 to 2009 [5]. Such reports are worrisome. Sleep is classified into two stages based on quality of sleep such as (1) Rapid eye movement (RAM), (2) Non-rapid eye movement (NRAM).

Table 1 Summary of literature review

Sensor/technology used	Assessment results	References
Grid of 24-GHz microwave radar, linear discriminant function	79.3% accuracy to classify sleep stages	[1]
Force sensing resister, logistic regression	>90% accuracy in classification	[2]
Linear classifier, load cell sensor grid	97% accuracy to classify leg and body movement	[1, 3]
Bluetooth based EEG device, DAQ100	RAM takes 25% of total sleep, NRAM (5%, 45%, 45%, 13% of NRAM takes N1, N2, N3, N4 respectively)	[4]
Mathematical calculation of sleep index	Modified sleep index formula	[5]
Theoretical framework of Internet of things	A model is proposed to measure sleep quality	[6]
Sonar measurement	>84% accuracy to classify sleep stages	[7]
EEG device, force sensor	Deep sleep index formula is calculated	[3, 8, 14, 16, 17]
Sleep disorder breathing algorithm is written and implemented	93% accuracy to classify breath disorder	[9, 16]
A grid of 48 pressure sensor, KNN, logistic regression, SVM, random forest	86.7% accuracy to detect sleep posture, 95.2%, 94.9%, 94.7%, 95.8% accuracy obtained by KNN, logistic regression, SVM, and random forest	[2, 10]
10–20 system	Different gender, age shows different EEG pattern	[11, 17]
Vagus nerve therapy, SVM	90% accuracy to classify epilepsy patient	[12]
Oximeter	System design concept	[13]
EEG device	CAP pattern is important parameter for sleep study	[14, 16, 17]
EEG device	Cyclic alternative pattern is found lesser in narcolepsy patient	[14, 15, 17]

Rapid eye movement is that state of sleep in which a person's brain functions almost like a waking state. The movement of eyes in this state is exactly like the waking state. A person often dreams in this state of sleep. Moreover, the movement of eyes keeps on according to the film running in the dream. While Non-rapid eye movement is considered a state of complete rest. In this state of sleep, signals of low frequency are found in the brain. NREM is further classified in N1, N2, N3 where N1 and N2 for light sleep and N3 for deep sleep [16, 17].

3.1 System and Sensors

Electroencephalogram is a process by which every electrical impulses generated during communication between neurons in the brain can be examined. A human brain is of 1.5 kg and have almost 10 ^ 20 neurons. Every Neuron is connected to the rest of the neurons. In the human brain, each nerve communicates to each other in the form of electrical signals or impulses. These impulses are generated even though in sleeping mode. The voltage of these impulses are very low up to 10–100 μv [17] so to examine the impulses it should be amplify. For better understanding of these EEG impulses, firstly, these are passed through an amplifier then readings are examined for further study. To better understand of EEG impulses it is important to understand human brain sections shown in Fig. 1 and working areas as shown in Table 2. As per Fig. 1, human brain is divided in 10 sections and each section is having its

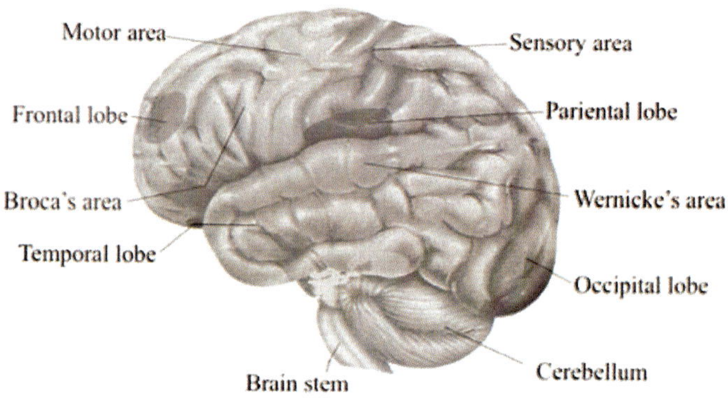

Fig. 1 Human brain internal structure and their working area

Table 2 Human brain section name, symbolic representation, and working area

Brain section name	Sensor's name	Working area
Frontal lobe	Fp1, Fp2, F3, Fz, F4, F8, F7	Movement, problem solving, concentrating, thinking, behavior, personality, mood
Motar area	–	Control of voluntary muscles
Sensory area	–	Skin sensations
Broca's area	–	Speech control
Temporal lobe	T3, T4, T5, T6	Hearing, language, memory
Brain stem	–	Consciousness, breathing, heart rate
Parietal lobe	P3, Pz, P4	Sensations, perception, body awareness
Occipital lobe	O1, O2	Vision
Wernicke's area	–	Language comprehension
Cerebellum	C3, Cz, C4	Posture, balance, coordination of movement

Fig. 2 10–20 system

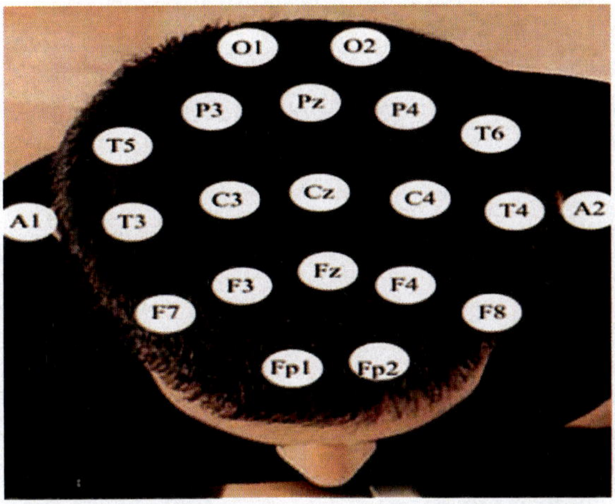

responsibility. At the particular action, the particular section activates and the nerves in that section communicate through electrical signals. These electrical signals are detected by EEG device. There is a need of a generalized system to measure electrical impulses that are generated in human brain.

3.2 10–20 System

10–20 system is a generalized system in which every 20% of every half-scalp has an electrode as shown in Fig. 2 that measures every brain impulse [17]. Different size brains are found in different age groups, genders, and locations because of that designing of a fixed system is too difficult. To overcome from a problem, a generalized system is designed which do not have any specific electrode for specific section of brain but it have a fixed parameter that is every 20% of every half scalp has an electrode. This system is internationally recognized. With the help of this 10–20 scalp, system EEG measures the brain impulses. Every animal have two types of electrical signal (1) skin impulse, (2) internal nerve signal [17]. Here, sensor A1, A2 measure the skin impulse. So when EEG detects the brain impulse then additional skin impulse is founded in actual brain impulse due to externally wearable 10–20 system. So to measure the accurate result, skin impulse reading is subtracted from the total reading. On behalf of brain impulse, pattern sleep quality is measured [16, 17]. To better understanding of brain impulses generally frequency form of signal is used [14, 17]. On the basis of frequency range, some categories are defined such as: 8–13, 13–30, 4–8, and 0.5–4 Hz is considered as Alpha, Beta, Theta, and Delta, respectively [17] shown in Fig. 3.

A generalized result is frequencies of beta range and delta range were found to be increased during the non-rapid eye movement [11].

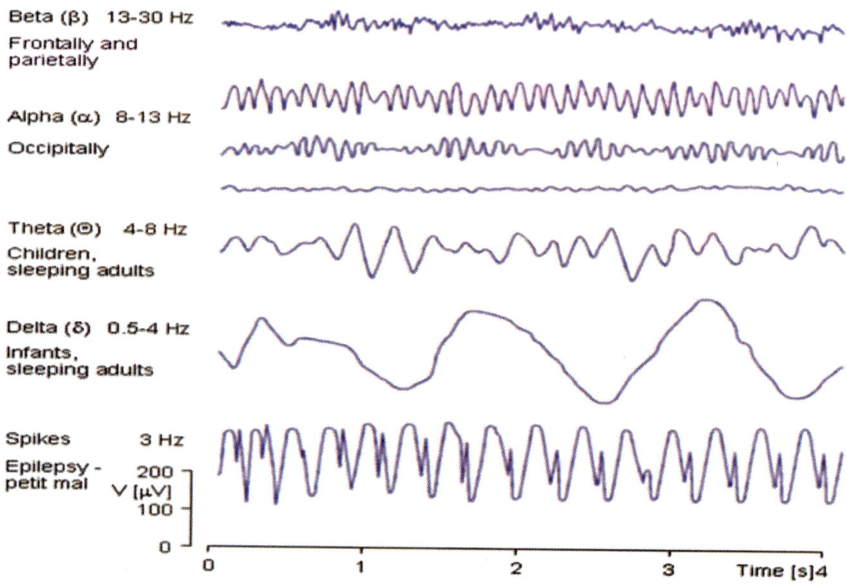

Fig. 3 Electrical impulses categorization based on frequency range

4 Discussion and Findings

This paper includes both types of researches (1) through wearable device (2) through non-wearable device. Both methods have their own advantages. Some researches support non-wearable device [2, 3, 6, 7 and 9] because Subject feels somewhat uncomfortable with wearable device when testing sleep quality so that there may be chances of some raw factors are added to the data. However, we cannot ignore the accuracy of collecting data through wearable device [4, 11, 12 and 13]. Because of subject's uncomfortableness with wearable device researchers, collect data from same subject in 3–4 days by which the raw factors in data can easily be measured and removed it [13]. However, there are always some chances to add raw factors in data. Now, talk about non-wearable device, data collection through non-wearable device is a critical job. There is more chances to add raw factors as noise in data from the environment. There are some process to remove noise from data but it depends on the accuracy rate of the method.

5 Conclusion

Sleep is an important aspect in the living world. Literature shows the direct connectivity between many of diseases with the quality of sleep. Early diagnosis can be one of the major remedial step towards preventing disease caused be different sleep disorders. There has been many researches have been performed by different researchers to understand sleep disorder patterns. These research works have been performed by majorly with two types of devices wearable and non-wearable. Each have its own advantages and disadvantages. This research, paper includes deep study about both wearable and non-wearable devices and sensors in order to understand their applicability and scope of all possible reduction for getting higher degree of accuracy. Authors have studied and analyzed the available literature work to identify the gaps in the area of sleep pattern identification for detecting sleep disorders. Authors also discussed all sensors and systems used and signal analysis regarding identifying hidden sleep patterns in detail, which can further be helpful for other researchers in the this field also.

References

1. Kagawa M, Sasaki N, Suzumura K, Matsui T (2015) Sleep stage classification by body movement index and respiratory interval indices using multiple radar sensors. In: Proceeding of 2015 37th annual international conference of the IEEE engineering in medicine and biology society, INSPEC accession number: 15584678. https://doi.org/10.1109/EMBC.2015.7320153
2. Barsocchi P, Bianchini M, Crivello A, La Rosa D, Palumbo F, Scarselli F (2016) An unobtrusive sleep monitoring system for the human sleep behaviour understanding. In: Proceeding of 2016 7th IEEE international conference on cognitive infocommunications INSPEC accession number: 16564236. https://doi.org/10.1109/CogInfoCom.2016.7804531
3. Adami AM, Adami AG, Hayes TL, Beattie ZT (2013) A system for assessment of limb movements in sleep. In: Proceeding of 2013 IEEE 15th international conference on e-health networking, applications and services INSPEC accession number: 14060943. https://doi.org/10.1109/HealthCom.20136720712
4. Tseng HW, Huang C-D, Yen L-Y, Liao Y-D, Lee Y-W (2016) A method of measurement sleep quality by using EEG. In: Proceeding of 2016 IEEE international conference on consumer electronics-Taiwan, INSPEC accession number: 16177900. https://doi.org/10.1109/ICCE-TW.2016.7521000
5. Cheng S-P, Mei H (2012) A personalised sleep quality assessment mechanism based on sleep pattern analysis. In: Proceeding of IEEE. https://doi.org/10.1109/IBICA.2012.55
6. Surantha N, Kusuma GP, Isa SM (2016) Internet of things for sleep quality monitoring system: a survey. In: Proceeding of 2016 IEEE 11th international conference on knowledge, information and creativity support systems, INSPEC accession number: 16969170. https://doi.org/10.1109/KICSS.2016.7951426
7. Zaffaroni A, Coffey S, Dodd S, Kilroy H, Lyon G, O'Rourke D, Lederer K, Fiete I, Prenzel T (2019) Sleep staging monitoring based on sonar smartphone technology. In: Proceeding of 2019 41st Annual international conference the IEEE engineering in medicine and biology society. INSPEC accession number: 19127036. https://doi.org/10.1109/EMBC.2019.8857033
8. Yan Q, Xu C (2018) A method of sleeping state recognition based on pressure-body movement-sleeping model. In: Proceeding of 2018 IEEE 3rd international conference on cloud computing and big data analysis, INSPEC accession number: 17844178. https://doi.org/10.1109/ICCCBDA.2018.8386539
9. Tataraidze AB, Anishchenko LN, Korostovtseva LS, Bochkarev MV, Sviryaev YV (2018) Non-contact respiratory monitoring of subjects with sleep-disordered breathing. In: Proceeding of 2018 IEEE international conference. quality management, transport and information security,

information technologies. INSPEC accession number: 18262535. https://doi.org/10.1109/ITMQIS.2018.8525001

10. Dhamchatsoontree S, Sirisin C, Proncharoensukkul M, Wongpatikaseree K (2019) i-Sleep: intelligent sleep detection system for analyzing sleep behaviour. In: Proceeding of 2019 IEEE 4th international conference on information technology. INSPEC accession number: 19211584. https://doi.org/10.1109/INCIT.2019.8912047
11. Liao Y, Zhou G, Liang J, Zhang X, Guo X, Luo Y (2019) Overall population generalities, sex differences, and individual differences in sleep electroencephalography functional connectivity. In: Proceeding of IEEE, 7. https://doi.org/10.1109/ACCESS.2019.2950066
12. Rvan M, Begnaud J (2019) Investigating the effect of short term responsive VNS therapy on sleep quality using automatic sleep staging. Proc IEEE Trans Biomed Eng 66(12). Accession number: 19189150. https://doi.org/10.1109/TBME.2019.2903987.INSPEC
13. Cheng S-H, Huang J-C (2017) An intelligent sleep quality detection system based on wearable device. In: Proceeding of 2017 IEEE international conference on machine learning and cybernetics, INSPEC accession number: 17372997. https://doi.org/10.1109/ICMLC.2017.8108966
14. Terzano MG, Parrino L, Smerieri A, Chervin R, Chokroverty S, Guilleminault C, Hirshkowitz M, Mahowald M, Moldofsky H, Rosa A, Thomas R, Walters A (2002) Atlas, rules, and recording techniques for the scoring of cyclic alternating pattern (CAP) in human sleep. In: Proceeding of Elsevier https://doi.org/10.1016/s1389-9457(01)00149-6
15. Poryazova R, Werth E, Parrino L, Terzano MG, Bassetti CL (2011) Cyclic alternating pattern in narcolepsy patients and healthy controls after partial and total sleep deprivation. Proc Clinic Neurophysiol 122:1788–1793. https://doi.org/10.1016/j.clinph.2011.02.028
16. Marcus CL, Carroll JL, Donnelly DF, Loughlin GM (2008) Sleep in children—development changes in sleep pattern. In: Proceeding of lung biology in health and disease, Second edition, Publisher: Informa healthcare USA, Inc. (2008) ISBN-10:1-42006080-5 and ISBN-13:978-1-4200-6080-5
17. Pollak CP, Thorpy MJ, Yager J (2010) The encyclopaedia of sleep and sleep disorder 3rd edn. Publisher: Facts On File, An imprint of infobase publishing, New York NY 10001, ISBN-13:978-08160-6833-3, ISBN-10:0-8160-6833-X and ISBN: 978-1-4381-2577-0 (2010)

Position Estimation of Small UAV from Monocular Camera by Using Matched Features

Hariom Dhungana, Francesco Bellotti, Riccardo Berta, and Alessandro De Gloria

Abstract This paper focuses on position estimation of a small unmanned aerial vehicle (UAV) using a monocular camera. Features from accelerated segment test (FAST) descriptors are used as a matched pattern to estimate differential change in position of the UAV. Visual simultaneous localization and mapping (V-SLAM) is a probabilistic filter-based method and a prominent real-time positioning method in robotics. V-SLAM performs drift-free tracking of the pose of a UAV on long run but the prediction states have limited certainty because of using sparse number of features used in real-time position estimation. Visual odometry (VO) is a deterministic positioning method and is more accurate to estimate the relative UAV position from adjacent frames without a persistent map. VO gives high drift error on the long run because of the accumulation of drift error at each frame transformation. Bundle adjustment (BA) and loop closure are two error optimization techniques to reduce drift error in VO. Due to limited computation resources available in the small-scale UAV, the optimization techniques are not appropriate in the small UAV positioning. In this work, VO with fractional V-SLAM is proposed to reduce the drift error on position estimation from matched features. The obtained results show the positioning estimation from proposed method works in an outdoor environment and that overperforms than VO and V-SLAM.

Keywords Positioning · UAV · Visual odometry · Visual SLAM

H. Dhungana (✉) · F. Bellotti · R. Berta · A. De Gloria
DITEN, University of Genova, 16145 Genova, Italy
e-mail: hariom.dhungana@elios.unige.it

F. Bellotti
e-mail: franz@elios.unige.it

R. Berta
e-mail: berta@elios.unige.it

A. De Gloria
e-mail: adg@elios.unige.it

R. P. Mahapatra et al. (eds.), *Proceedings of 6th International Conference on Recent Trends in Computing*, Lecture Notes in Networks and Systems 177,
https://doi.org/10.1007/978-981-33-4501-0_39

1 Introduction

Small UAVs are used in various application scenarios such as search and rescue, monitoring and security, tracking and observation. Due to the applicability of such UAVs in civil applications [1], their importance grows day by day. Traffic monitoring, fire detection, pipeline/power line inspection, industrial inspections are the common civil applications that are performed by UAVs. The performance of application scenarios depends on the accuracy of the UAV position. Coordinated search, virtual views, and image stitching are usual activities of the small UAV in an outdoor environment. In order to cover the interested area in a short duration by multiple UAVs, there is a need for coordination. The coordinated search needs sharing of position information to neighboring UAVs [2]. Quick generated virtual views and 3D reconstruction help in the analysis of damages from an earthquake. Minimum number of images captured by the UAVs to cover interest region makes the swiftness in 3D reconstruction [3]. Appropriate position selection of the camera gives the full coverage with minimum number of images, which helps quick formation of 3D map. Mosaicking of images gives a big picture of visual information in a single frame. Image mosaicking suffers from metadata error and camera orientation error [4]. Accurate location and orientation information of camera helps to reduce metadata error in image stitching. These are the motivation to get accurate positioning of the UAVs.

Defining an UAVs location and orientation with some reference in 3D space is called positioning and requires six parametric values to represent the complete information. Positioning can be defined in two ways: relative and absolute positioning. The relative positioning deals with comparing to nearby objects or periphery as reference and is commonly used in human perception. The absolute positioning is defined with respect to some global reference. Remarkable achievement of the computational power in small circuits makes it easy to implement visual-based positioning in small UAVs. Since, there is already a camera in a small UAV; it can be used as a positioning sensor. A camera is a passive device in terms of radio interference so useful in passive observation. Images and videos are the output data from the camera. Visual positioning is the process of calculating differential change in position of the UAV by analyzing repeated patterns that appeared on the consecutive image frames. Our assumptions for visual position estimation are availability of sufficient illumination in the environment; sufficient scene overlaps between successive frames and dominance of static scene over moving objects. Position sensing is a fairly mature field; however, there is a need for research for cost minimization, infrastructure reduction, scalability improvement, and flexibility on implementation as stated in [5].

Davisons monocular SLAM [6] and Nisters VO [7] are perhaps the parent of our work and were far ahead of their time. Our contribution in this paper is to adapt the related work of Extended Kalman Filter (EKF)-based monocular SLAM [6], VO [7], inverse depth representation of features [8], and camera centric coordinate system [9] to estimate the position of the small UAV. Furthermore, we have incorporated the combination of random sample consensus (RANSAC) with EKF as in [10] to speed up the capability to detect and reject a high number of spurious features.

2 Related Works

The real-time VO with low delay was proposed in [7] using the Harris corners. The disparity between the matched features depends on the speed of the camera and is tested in different ranges. Minimum five matched points in consecutive image frames are needed to estimate the six parameters of position from a calibrated camera. In practice, more than five points are used for the motion estimation to get robust and accurate. More than eight points matching leads to an overdetermined system to solve in the least squares sense and provides a degree of robustness to noise [11]. In [12], Scaramuzza and Siegwart demonstrated real-time algorithms for calculation ego-motion of a ground vehicle relative to the road using an omnidirectional camera. VO on the ground plane using only one feature correspondence was tested in [13]. Computation time per frame was less than one millisecond and the number of iterations for outlier removal was seven. Circular motion model of two degrees of freedom (DoF) was used for camera motion. Although the method was fast and efficient, it was insufficient to represent all position parameters of the UAV from single-feature correspondence. VO system based on two-frame estimates of instantaneous relative motion can work in constant time, but will inevitably exhibit drift because of accumulation of small errors in the inter frame motion estimates. Bundle adjustment [14] and loop closure are two error optimization techniques to reduce drift error in VO. BA reduces the drift compared to two view VO because it incorporates constraints between several frames. The computational complexity of BA increases cubically $O((qN+lm)^3)$ with number of parameter involve in BA, where q represents number of parameters to describe feature points, N represents number of feature points, l represents number of camera poses and m represents number of parameters to describe camera poses. For this reason, BAis not feasible in the small UAV positioning. The loop closer is not always available in outdoor UAV missions.

EKF was first successfully applied to the SLAM problem in [15]. In SLAM, the update time and computational complexity of maintaining the coupled pose and scene covariance of the EKF algorithm scales quadratically $O(N^2)$ with increasing the number of features (N). This in turn limits the number of feature matches available at any instant to that map so position accuracy is limited. SLAM with Compressed EKF (CEKF) was implemented in [16] by using a laser scanner in an outdoor environment to make computation cost constant until needed to update the whole map. The comparative performance evaluations of SLAM algorithms on the basis of position accuracy and processing time requirements were done in [17]. Among different versions of KF, the CEKF was more efficient for an outdoor environment where a lot of features are present. Inverse depth parametrization for features was proposed in [8] that can handle both close and distant features within the standard EKF framework. It remains well behaved for features at both stages (initialization and tracking) of

SLAM processing, but has the drawback in computational terms that each point is represented by a 6D state vector as opposed to the 3D of a Euclidean XYZ representation. Dense scene flow combined with stereo V-SLAM was implemented in [18] and considerable positioning error was improved in moving objects. Nearby features give less linearization error on position estimation from V-SLAM so a camera centric coordinate system was proposed to reduce linearization error in [9].

In [19], Alcantarilla et al. proposed VO priors for robust EKF-SLAM. Constant linear and angular velocity model of the camera was replaced by VO prior and that performed quick pose estimation using the two stages RANSAC: a two-point algorithm for rotation followed by a one-point algorithm for translation. The result was better than constant velocity model in stereo vision model but very good prior information in the camera rotation was needed. Hwang and Song [20] used multiple features for stable monocular SLAM. Three types of features such as corners, line, and illumination were used to make the landmark state vectors. That test was verified inside the room. The stable localization from multiple groups of features can be obtained but the arbitrary detection and association of different features on outdoor scenes is difficult to realize. Williams and Reid [21] proposed the V-SLAM system with persistent map by incorporating the additional information available from VO style measurements into the filter. A summary of their work was to produce a VO system using the monocular SLAM on each frame. Our objective, in different ways, is to run V-SLAM and VO separately and fusion of position information. V-SLAM runs in lower frequency than the VO so the computational complexity of maintaining the coupled pose and scene covariance become less than the related work. Hernandez et al. [22] used V-SLAM with partial odometry information from IMU by using inverse depth representation of features and that work was performed in a structured environment with uniform landmarks and the range within a meter.

3 Proposed Approach

The block diagram of the proposed method is shown in Fig. 1. The camera works as the positioning sensor and the image sequences $\boldsymbol{I_{0:n}} = \{\boldsymbol{I_0}, \boldsymbol{I_1}, \boldsymbol{I_2}, \ldots, \boldsymbol{I_n}\}$ are the input of the system. The VO and V-SLAM are two subsystems of the proposed method and run respective algorithms to deduce position information of UAV from image sequences. VO uses every consecutive image frame to extract relative change in position from 2D-2D transformation. The estimated position from VO is notated as $\boldsymbol{O_f}$. V-SLAM uses interleaved image frames for the position estimation to reduce the computational burden and the estimated position is notated as $\boldsymbol{S_f}$. As the interleaved distance increases, the computational cost decreases but there is a need for a minimum number of matched features to estimate position. For this reason, we cannot go beyond the threshold in normal movement of the UAV. The estimated position from two subsystems is fused after every interleaved frame by simply averaged and notated as $\boldsymbol{P_f}$. This new value is used as reference to estimate the next position in both subsystems. Two statements on proposed approaches can be claimed based on

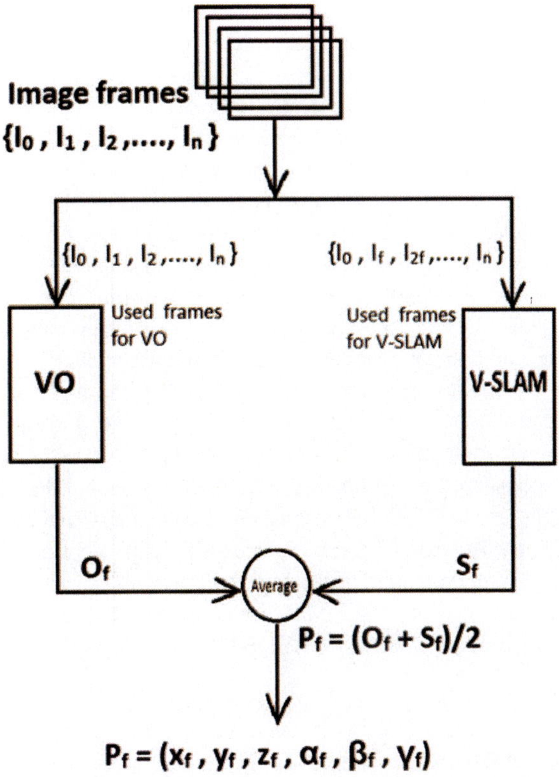

Fig. 1 Block diagram of visual odometry combined with fractional V-SLAM

hypotheses stated on related works. First one is uncertainty on position estimation in our work will be less than V-SLAM because of we have used the more number of features than in [6]. Second, the computational cost will be less in our approach than in [21] because of SLAM runs in interleaved input image sequences.

4 Experimental Setup

The experiment is performed on data available from Library for Visual Odometry 2 (LIBVISO2) [1], Karlsruhe Institute of Technology. Reason of selecting this data is higher feature density, high feature matching speedup of factor, and it supports monocular ego-motion estimation. Beside dataset, those libraries contain camera calibration parameters of respective sequences so it makes it easy to test our approach in a long outdoor environment. 759 image frames of resolution 1344×391 pixels with portable gray map (.pgm) are used for position estimation. A notebook computer with core i5 2.6-GHz processor, 8 GB memory, and 64 bit operating system is used to perform position estimation from visual data. The proposed approach is processed in MATLAB R2011b. Corner features are detected by the FAST descriptor because of its short detection and matching time. Minimum number of landmarks used to perform visual position in this work is 25, which is more than sparse V-SLAM so this leads to more accurate position estimation. Interleaved distance 5 is used and weights of fusion are tuned fifty-fifty for VO and V-SLAM by empirical basis. Qualitative analysis of various weights on VO and V-SLAM is tested, and among them, average value gives better results.

Accuracy of estimated position is the main evaluating parameter of this work. The accuracy of the estimated position depends on the utility of the computation resource such as memory, processor and time. Therefore, computation resources are another evaluating parameter, which signifies how much computing resource that has been spent to achieve a certain accuracy level. Trajectory plot evaluates accuracy of estimated position in this work. The estimated position consists of six parameters of the UAV location information. The starting position of the UAV is origin with orientation aligned with respective axes. The accuracy of estimated position is observed by comparing with the approximated ground truth. Performing the test on the same data by varying one parameter and keeping all other fixes give relative comparisons. Relative comparison by adjusting the number of features is performed. We have compared trajectory evaluation and computing time evaluation of proposed approach with V-SLAM and VO without bundle adjustment. The comparison and the discussion are discussed in the following section.

5 Obtained Results

Figure 2 shows the comparison of trajectory by three different methods. The experiment is performed on the Karlsruhe sequences. The top left window shows the dataset, the matched inliers features, matched outlier features, and unmatched features are pointed by three different color red, magenta, and blue, respectively. The position estimation is performed on the basis of matched inliers points. Other three windows show the trajectory of the estimated position by three techniques. One unit of axes is scaled to a hundred meters. The triangle represents the estimated position of UAV. The black straight line shows the approximate ground truth. The camera moves hundreds of meters in the forward direction. The top right windows show VO trajectory, the bottom left shows V-SLAM trajectory, and the bottom right shows our approach. Due to deficiency of optimization, the estimated position starts to deviate after traveling around 75 m far from origin. The orientation error of the camera greatly increases so the heading direction deviates in the wrong direction. Trajectory of the camera from V-SLAM is comparatively far better than VO. Due to the consideration of a limited number of features, counted as 25, the trajectory is rough. In our approach, the trajectory is more smooth and straight compared to the rest of others. So from this result, we can say that the accuracy of position estimation from our approach is superior to VO without pose optimization and V-SLAM with the same number of features.

Table 1 Computation time with respect to number of features

	Number of features	15	17	19	21	23	25
Computation time (s)	VO	14	15	15	16	16	17
	V-SLAM	362	437	522	615	712	830
	Our approach	155	184	218	255	296	341

Table 1 shows the variation of computation time with respect to the number of features that are used to perform position estimation by visual odometry, V-SLAM, and our approach. The comparison is done by using data set from Karlsruhe sequences of 1351 × 374 pixels and 200 frames. Based on the number of features used to perform position estimation, the computing time is compared. The computation time to perform position estimation in VO does not increase significantly with varying the number of features from 15 to 25. But in V-SLAM, the computation time varies from 362 to 830 s if the number of features increases from 15 to 25. The rate of increasing shows nonlinearity relation between the computation time and the number of features. In these two experiments, both VO and V-SLAM run in every sequence. In our approach, the V-SLAM runs in a fractional manner so the computation time is less compared to V-SLAM. By minimizing the number of features, the computation time by our method is considerably reduced. Our approach gives better computing time saving than V-SLAM with the same rate of varying the number of features. In summary, based on the results from Fig. 2 and Table 1, it can be said that our approach overperforms than the individual VO and V-SLAM.

6 Conclusion

The key information from the above experiments is VO combined with V-SLAM works on outdoor environments for position estimation via single camera. The proposed approach on this work over performs than the individual VO without optimization and V-SLAM. The computing cost of the proposed approach is also less than V-SLAM, but there must be sufficient number of matched features in interleaved frames. Outdoor environments with less sharp corner features need different image features to handle visual positioning. Blur in the image frame and motion of the image feature creates a high number of unmatched features so decreasing correlation threshold can be a better alternative. There is a need for a different approach in visual position estimation if the large portion of scene is moving.

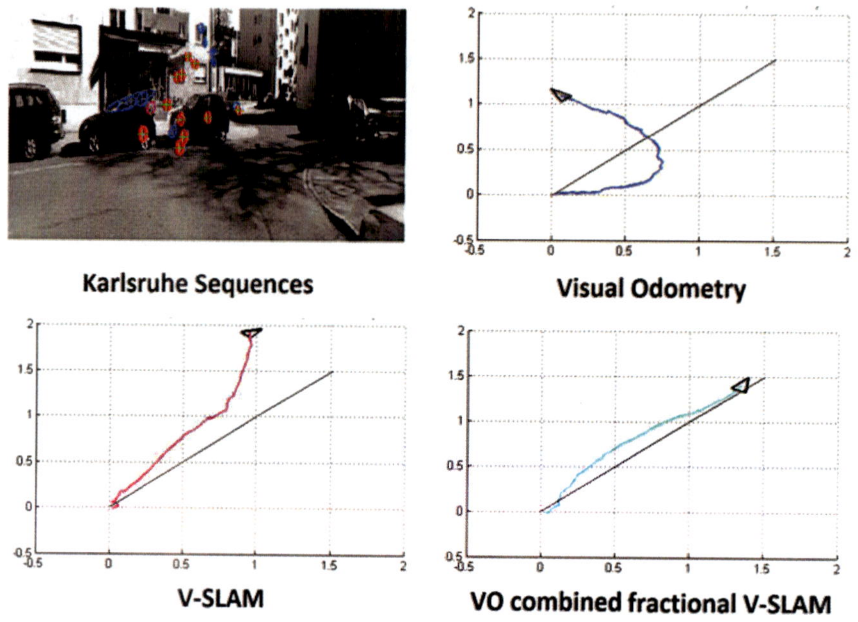

Fig. 2 Comparison of trajectory by three methods VO, V-SLAM and our approach using the same 750 image frames. The top left window shows datasets and the remaining three windows show trajectories by three different methods, both axes represent the distance and scaled to 100 m on unit scale

References

1. Shakhatreh H, Sawalmeh AH, Al-Fuqaha A, Dou Z, Almaita E, Khalil I, Othman NS, Khreishah A, Guizani M (2019) Unmanned aerial vehicles (UAVs): A survey on civil applications and key research challenges. IEEE Access 7:48572–48634
2. Shen C, Chang TH, Gong J, Zeng Y, Zhang R (2020) Multi-UAV interference coordination via joint trajectory and power control. IEEE Trans Sig Proc
3. Mostegel C, Wendel A, Bischof H (2014) Active monocular localization: towards autonomous monocular exploration for multirotor mavs. In: 2014 IEEE international conference on robotics and automation (ICRA). IEEE, pp 3848–3855
4. Ghosh D, Kaabouch N (2016) A survey on image mosaicing techniques. J Vis Commun Image Represent 34:1–11
5. Aqel MO, Marhaban MH, Saripan MI, Ismail NB (2016) Review of visual odometry: types, approaches, challenges, and applications. SpringerPlus 5(1):1897
6. Davison AJ, Reid ID, Molton ND, Stasse O (2007) Monoslam: real-time single camera slam. IEEE Trans Pattern Anal nd Mach Intell 29(6):1052–1067
7. Nisteŕ D, Naroditsky O, Bergen J (2004) Visual odometry. In: Proceedings of the 2004 IEEE computer society conference on computer vision and pattern recognition (CVPR), vol 1 IEEE, pp I–652
8. Civera J, Davison AJ, Montiel J (2008) Inverse depth parametrization for monocular slam. IEEE Trans Robot 24(5):932–945
9. Castellanos JA, Naira J, Tardós JD (2004) Limits to the consistency of ekf-based slam 1. In: 5th IFAC symposium on intelligent autonomous vehicles

10. Civera J, Grasa OG, Davison AJ, Montiel JMM (2010) 1-point ransac for extended kalman filtering: application to real-time structure from motion and visual odometry. J Field Robot 27(5):609–631
11. Scaramuzza D, Fraundorfer F (2011) Visual odometry [tutorial]. Robot Autom Magaz IEEE 18(4):80–92
12. Scaramuzza D, Siegwart R (2008) Appearance-guided monocular omnidirectional visual odometry for outdoor ground vehicles. IEEE Trans Rob 24(5):1015–1026
13. Scaramuzza D, Fraundorfer F, Siegwart R (2009) Real-time monocular visual odometry for on-road vehicles with 1-point ransac. In: IEEE international conference on robotics and automation (ICRA). IEEE, pp 4293–4299
14. Triggs B, McLauchlan PF, Hartley RI, Fitzgibbon AW (2000) Bundle adjustment a modern synthesis. In: Vision algorithms: theory and practice. Springer, pp 298–372
15. Smith R, Self M, Cheeseman P, Smith R, Self M (1987) A stochastic map for uncertain spatial relationships. In: 4th international symposium on robotic research, pp 467–474
16. Folkesson J, Christensen H (2003) Outdoor exploration and slam using a compressed filter. In: ICRA, pp 419–426
17. Tuna G, Gulez K, Cagri Gungor V, Veli Mumcu T (2012) Evaluations of different simultaneous localization and mapping (slam) algorithms. In: IECON 2012–38th annual conference on IEEE industrial electronics society. IEEE pp 2693–2698
18. Alcantarilla PF, Yebes JJ, Almazán J, Bergasa LM (2012) On combining visual slam and dense scene flow to increase the robustness of localization and mapping in dynamic environments. In: 2012 IEEE international conference on robotics and automation (ICRA). IEEE, pp 1290–1297
19. Alcantarilla PF, Bergasa LM, Dellaert F (2010) Visual odometry priors for robust ekf-slam. In: 2010 IEEE international conference on robotics and automation (ICRA). IEEE (2010), pp 3501–3506
20. Hwang S-Y, Song J-B (2011) Monocular vision-based slam in indoor environment using corner, lamp, and door features from upward-looking camera. Industr Electron, IEEE Trans 58(10):4804–4812
21. Williams B, Reid I (2010) On combining visual slam and visual odometry. In: 2010 IEEE international conference on robotics and automation (ICRA). IEEE, pp 3494–3500
22. Hernandez E, Ibarra JM, Neira J, Cisneros R, Lavin JE (2011) Visual slam with oriented landmarks and partial odometry. In: 2011 21st International conference on electrical communications and computers (CONIELECOMP). IEEE, pp 39–45

Convolutional Neural Network for Seizure Detection Using Scalp Electroencephalogram (EEG)

Sudesh Kumar, Rekh Ram Janghel, and Satya Prakash Sahu

Abstract Epilepsy is one of the most common neurological brain dysfunctions after stroke, Alzheimer and migraine in humans. Epilepsy is a disease that affects approximately 50 million people worldwide. Seizure is occurred due to abnormal electrical discharges of the neurons in the brain cell. EEG is commonly used to accommodate information about the electrical activity of the brain, an automated epilepsy seizure detection and classification using EEG signals in deep learning techniques. As the rapid innovation in the field of healthcare increases, various biomedical signals, namely electrocardiogram, electroencephalogram (EEG) and electromyogram, play a crucial role for the measurement of various diseases such as cardiovascular diseases, brain disorders, etc. We have proposed a novel deep convolutional neural network model, which classifies the EEG signals into two distinct classes namely normal and seizure. The experiment is performed on a publicly available benchmark CHB-MIT database. The performance of the model was evaluated in terms of sensitivity, specificity and accuracy. The experimental results that have been obtained are 87.40% accuracy, 88.10% sensitivity and 87.10% specificity, which are then compared to other existing works of literature and are better than the existing work.

Keywords Epilepsy · Machine learning · Electroencephalogram (EEG) · Deep learning · Convolutional neural network (CNN) · Classification

S. Kumar (✉) · R. R. Janghel · S. P. Sahu
National Institute of Technology Raipur, Raipur, India
e-mail: Sudeshk5548@gmail.com

R. R. Janghel
e-mail: rrjanghel.it@nitrr.ac.in

S. P. Sahu
e-mail: spsahu.it@nitrr.ac.in

R. P. Mahapatra et al. (eds.), *Proceedings of 6th International Conference on Recent Trends in Computing*, Lecture Notes in Networks and Systems 177,
https://doi.org/10.1007/978-981-33-4501-0_40

1 Introduction

The electroencephalogram (EEG) is a low abundancy sign produced in the mind, because of data stream during the correspondence of a few neurons. Subsequently, cautious examination of these sign could be valuable in understanding numerous human cerebrum issue diseases [1]. Human cerebrum contains a large number of neurons which are in charge of data stream. Because of this progression of data, a human body acts in like manner. We all know that in our brain, millions of neuron take place. So, when one neuron hits another neuron, an electric charge is produced which cannot be detected as it is of very low frequency, so different kinds of mechanism are present in taking care of this type of mechanism [2].

Epilepsy issue is a lot of regular neurological states of the mind recognized by intermittent unjustifiable seizures which are an outcome of abrupt blasts of unusual electrical releases in the cerebrum. As indicated by report of World Health Organization that 50 million individuals were impacted by this disease all over the world, nearly one in each 100 people are affected by a spasm in their life expectancy, by and large 2.4 million new occurrences of this disease are represented reliably worldwide [3]. As of not long ago, the sign of this disease is whimsical and its course of procedure has been less comprehended. The patients that suffer from this disease experience shifting indications relying upon the segment and the span of the mind that is influenced. The seizures that cause the disease of epilepsy can lead to physical, social and even mental problems that incorporate the sudden loss and damage to the brain. EEG signals have been classified into numerous rhythms dependent on their frequencies.

An EEG is especially useful once in a while when the cerebrum is in risk by offering a sensitive hint of cerebral working. Early examinations have shown confirmation of this abnormal development to be an invaluable guide in identifying evidence of epilepsy and diagnostic tumors, such between time is guideline speaking of long time periods, subsequently an extended EEG recording. Nowadays, EEG sign is used to get information critical to the assurance, representation and fix of these strange case. EEG is enlisted utilizing cathodes organized on the head and has modest amplitudes in the scope of 20 V [4]. The anodes are put by the 10–20 global framework which can be seen on the physionet.org.

Countless strategies have been created for programmed seizure recognition, utilizing EEG signals. Removing highlights that best portray the conduct of EEGs is vital for programmed seizure location framework's exhibition. A few element extraction and choice strategies have been accounted for in the writing. The vast majority of them use hand-fashioned highlights in the time–space recurrence area, time–recurrence space or in some cases in a blend of two areas. In any case, these area-based strategies experience three fundamental difficulties. To start with, they are delicate (not vigorous enough) to intense varieties in seizure designs. This is on the grounds that the EEG information is non-stationary and its factual highlights change crosswise over various subjects and after some time for a similar subject. Besides, EEG information securing frameworks are entirely powerless to a different scope of relics, for example, muscle exercises, eye-flickers and ecological background noise. Everyone of these wellsprings of clamor can adjust the real EEG highlights and thus truly influence the exhibition precision of seizure discovery frameworks. The creators of have contemplated the effect of high clamor levels on the acknowledgment execution of epileptic seizures. It merits featuring that identifying seizures from loud EEG information defiled with a medium-level commotion has brought about a drop of 10% in the seizure discovery exactness. At long last, most existing seizure discovery frameworks have been prepared on little scale EEG datasets gathered from couple of explicit patients, making them less down to earth in clinical applications.

2 Literature Review

EEG is a kind of technique that analysis or detect activity in your brain signals. It contains small metal disks that called electrodes to attach with wires and paste into scalp. This electrode catches tiny electrical signals that come from the brain cells. It records electrical patterns in the brain. It was very helpful for diagnosing a condition like headaches, epilepsy, dizziness, head injuries, sleeping problem and brain tumors [5]. It has been described below, first measurements of EEG were made by a German psychiatrist Hans Berger in the year 1929.

In 2017, Sachin Talathi [6] proposed an Epileptic Seizure Detection system based on deep recurrent neural networks which have achieved an accuracy of up to 99.6%. In view of this paper, author does firstly normalized the CHB-MIT database. Then it does layer deep convolutional neural network which contains main three layers which are convolutional layer, pooling layer and fully connected layer. One new model introduced for focal and non-focal EEG classification using deep neural network [7]. It is different the focal and non-focal EEG signals using softmax classifier. This classification is helpful for surgery decision of epilepsy patient. CHB-MIT Children's Hospital Boston database use for feature extraction and classification using DNN algorithm. They differ on aligned heartbeat. Performance evaluated from the specificity and sensitivity [8].

Hybridization concept comes when novel deep learning algorithm stuck on handling complexity. One of the paper does 3D CNN for human action reorganization. This model firstly does extracts features from both type spatial and temporal dimension. One paper proposed a hybrid CNN with restricted Boltzmann machine (RBM) model to recognize face in wild condition [9]. The top layer is used by RBM for function extraction from various convolutionary classes. DL is the concept that comes from the ANN. In past few decades, ANN becomes more popular in research area. ANN work on the concept of neural network and generating real values by time can take more time in many computation steps. Backpropagation algorithm used for effective gradient descent algorithm. When Backpropagation is applied on any dataset, the testing accuracy does not give satisfactory results as compared to training accuracy. BPA is stuck on the local optima when it applies on random initialize node. In very large dataset, BPA suffers with overfitting problem [10]. So that further machine learning algorithms like SVM, logistic regression, KNN and decision tree are used to reach global optima.

In 2018, Troung et al. [11] extracted the features using short-term Fourier transform (STFT) and four different features are extracted, and for classification, CNN classifier is used for the seizure detection and his proposed methodology achieved a sensitivity of 81.2% while in 2017, Khan et al. [12] extracted the statistical features and used CNN classifier for epilepsy detection and achieved a sensitivity of upto 88%. In 2019, Ozcan et al. [13] extracted the statistical features and used CNN classifier for epilepsy detection and achieved a sensitivity of 87.01%. Many researchers have been focusing on CNN classifier due to the more number of layers involved in using this classifier, and in 2018, Zhou et al. [14] have extracted the features using time domain signals and used this classifier and achieved an accuracy of 97.5%. Other than CNN, researchers have also used recurrent neural network, deep belief network, autoencoder in detecting the epilepsy.

In 2018, Tsiouris et al. [15] used recurrent neural network by extraction of features using wavelet coefficients and achieved a sensitivity and specificity of 99.84% and 99.86%, respectively, while in 2018, Xinghua et al. [16] have achieved a sensitivity of 88.80% and specificity of 88.60% by applying Bi-LSTM which is a type of recurrent neural network. Many algorithms have been proposed by many researchers but only a few algorithms have achieved the desired result for detection of epileptic seizure.

3 Materials and Method

3.1 Dataset Description

The CHB-MIT dataset which we used in our research paper is publically available on the physionet.org [17]. The dataset contains 686 scalp signal from 23 different patients which had been treated at Children's Hospital in Boston. The subjects had been put on seizure medication and the records were taken days after the medication for proper estimation of the seizure. There were 5 male and 17 female subjects, and for one male, 2 records were taken. They all are between 1.5 and 22 years of age. Some of the dataset records are one hour long while some of the dataset records are 4 to 6 h long depending on the nature of the seizure records. There were a total of 686 records as a whole in which only 198 records contain seizure [18].

3.2 Dataset Preparation

While preparing the EEG dataset, we used a window which contains 2048 samples of 8 s long of ictal, inter-ictal and pre-ictal classes. All the subjects' records were

Table 1 Ictal, inter-ictal and pre-ictal subdivision

Class	Time (s)	Patients	Seizure records	Total segments
Ictal (Seizure)	8	14	198	1200
Inter-ictal (normal)	8	6	240	1200
Pre-ictal (partial seizure)	8	3	248	1200

used while preparing these samples of 8 s long. Total of 1200 samples of each class were prepared.

Since there are a total of 198 seizure records available in the original dataset, we used the best 87 seizure periods while preparing our dataset. For each class, we used different time periods. Different time periods are as follows:

1. For ictal class, the experts have given the time periods in which the seizure occurs, so while keeping those time periods in mind, we have collected the best 8 s segments.
2. For pre-ictal, class experts have given that 15–20 min prior to the ictal time period is the pre-ictal class period. So, we have collected 8 s samples from that time period.
3. For inter-ictal class, we have taken the segments which are not in the pre-ictal and ictal class. So, a total of 1200 segments of each class is prepared in our dataset for detection and prediction of epileptic seizure.

Alternatively, we can prepare dataset while using the subset of three classes as we have prepared dataset using ictal and inter-ictal class; also we can prepare dataset using ictal and pre-ictal class as shown in Table 1.

3.3 *Proposed Methodology*

3.4 *Normalization*

In this paper, our dataset is having some variables of large variance and some small. So, when data is considered as a random variable, normalizing means converting to normal distribution, essentially normalization usage. We apply z-score normalization as preprocessing the dataset. Normalization is applied for data preparation for learning [19]. The main objective of normalization is to diversity the value of numeric columns in the dataset to use a common scale, without misinterpreting difference in the ranges of values. Used or needed usually only when features have different ranges. Normalization is a more radical transformation.

$$Z = (x - \mu)/\sigma \tag{1}$$

where z is z-score, x is random variable, μ is mean and σ is standard deviation.

3.5 *Convolution Neural Network (CNN)*

DL is a machine learning technique that takes feature directly from input data. The deep learning method is now more accurate than other classification for large dataset. That means you have the network that provides input from the weight of the output and update when it does not get its desired output hat needed [20]. Deep learning is often referred to as end to end learning. The basic concept behind CNN is very same as LeNet-5. That has three layers which are convolutional layer [21], pooling layer [22], and fully connected layer. Figure 1 shows the architecture of the proposed CNN model.

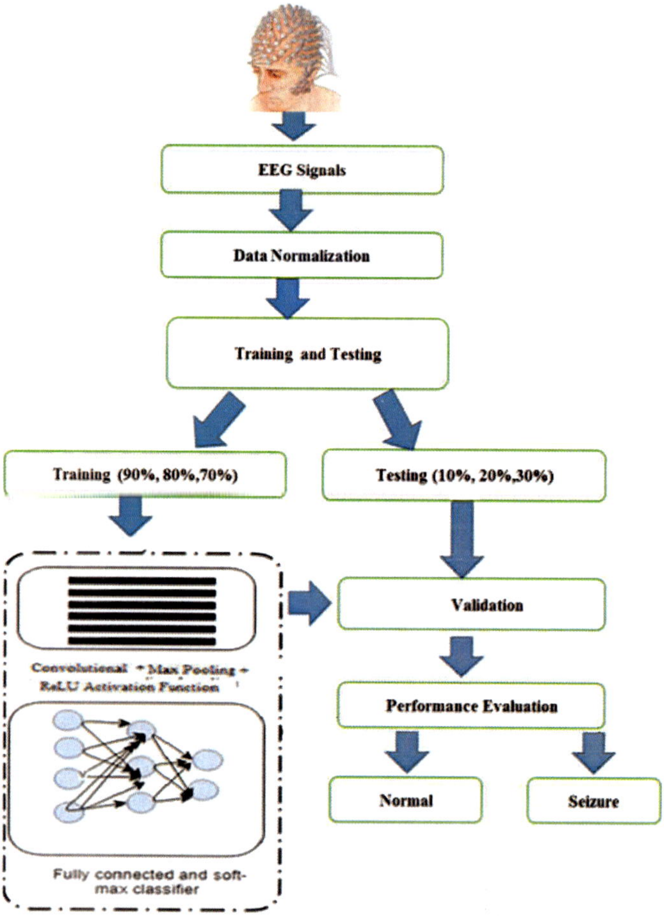

Fig. 1 Proposed methodology

3.5.1 Pooling Layer

Pooling operation is used for reducing a dimensionality of output for the next convolution layer and also called as down sampling layers. There are three types of pooling like average pooling, max-pooling and sum pooling [23]. In sum pooling, we sum all the elements in this feature map. In average pooling, we do an average of all sum elements and map that feature. Mostly, we apply max-pooling which takes maximum elements for feature mapping.

3.5.2 Fully Connected Layer

In a fully connected layer, it is similar to a neural network given output from max-pooling layer. Now we flattened our matrix into a vector and feed it into a fully connected layer like a neural network [24].

3.5.3 Softmax

It is the final output layer of CNN Model. Softmax function provides the classification output. Mathematical formula of softmax is

$$\sigma(z)_i = \frac{e^{(z)_i}}{\sum_{j=1}^{k} e^{(z)_i}} \tag{2}$$

3.6 Proposed CNN Architecture

Given Table 2 shows the summary of details of the proposed CNN model Structure. In this CNN model having 12 layers. We give 3200 inputs and it includes four convolutional layer, four max-pooling layers and three fully connected layers. Every layer we give filter (kernel) size required. CNN model starts with convolution layer that takes input 3200 data with 40 kernel size. Than max-pooling layer takes half of the values 1600 from that convolutional layer and takes 1600 inputs with 30 kernel size. Next convolutional layer takes max-pooling inputs 1000 with 10 kernel size and max-pooling layer takes inputs 991. Similarly, last convolutional layer takes an input of 991 and max-pooling as 495. Flatten layer use with relu activation feature for connecting with fully connected layer. After this last three fully connected layers and last fully connected with a softmax layer.

Table 2 Layer-wise details of CNN model

Layers	Type	Number of neurons	Kernel size for each output
1 Layer	Convolution	3200 × 10	40
2 Layer	Max-pooling	1600 × 20	2
3 Layer	Convolution	1500 × 10	30
4 Layer	Max-pooling	1000 × 10	2
5 Layer	Convolution	991 × 10	10
6 Layer	Max-pooling	495 × 10	2
7 Layer	Convolution	492 × 10	4
8 Layer	Max-pooling	246 × 10	2
9 Layer	Flatten	984	–
10 Layer	FC	150	–
11 Layer	FC	50	–
12 Layer	FC	3	–

4 Experimental Setup

In this proposed methodology, we trained 90, 80, 70% of data and rest for testing and validation. Also we do tenfold cross-validation approach while taking the learning rate 0.001. Firstly, it randomly divided signals into ten equal pats. Then 9 out of 10 signals parts are used to training. Remaining one-tenth part of signals perform testing. This strategy repeated ten times by shifting the test and training dataset. It works as a convolutional backpropagation with a 34 number of batch size. Calculation of the loss function gradient with respect to the weight in the backpropagation process. It gives feedback for error signals backwards in the network. During training, it does weight update to a network. Batch size is used for getting a number of signals for each training update.

5 Experimental Results and Discussion

Sensitivity, specificity, F1-Score and accuracy are the classification performance evaluation indices used in the present work as shown in Table 3. The above mentioned performance evaluation indices are calculated as:

$$\text{Sensitivity} = \frac{T_{\text{P}}}{T_{\text{P}} + F_{\text{N}}} \times 100 \tag{3}$$

$$\text{Specificity} = \frac{T_{N}}{T_{\text{N}} + F_{\text{P}}} \times 100 \tag{4}$$

Table 3 Performance measures with learning rate—0.001

Random state	Size ratio	Sensitivity	F1 score	Accuracy	Specificity	Precision
2	90–10	84.23	84.92	84.92	84.23	84.98
4	90–10	84.98	84.01	84.01	84.78	84.45
6	90–10	85.12	85.45	85.45	84.98	84.76
8	90–10	86.14	86.78	86.78	85.02	85.89
10	90–10	**88.10**	**87.40**	**87.40**	**87.10**	**86.98**
11	90–10	83.23	83.92	83.92	83.23	83.98
4	80–20	81.98	80.01	80.01	80.78	81.45
6	80–20	82.12	80.45	80.45	80.98	81.76
8	80–20	82.14	80.78	80.78	81.02	81.89
10	80–20	82.20	80.90	80.90	81.10	81.98
11	80–20	81.98	80.01	80.01	80.78	81.45
2	70–30	77.98	76.08	76.08	77.12	77.23
4	70–30	78.12	76.26	76.26	77.18	77.30
6	70–30	78.01	76.22	76.22	77.12	77.24
8	70–30	78.15	76.27	76.27	77.19	77.30
10	70–30	78.98	76.54	76.54	77.21	77.45
11	70–30	78.15	76.27	76.27	77.19	77.30
12	70–30	78.12	76.26	76.26	77.18	77.30

Bold indicates highest calculated performance metrics

$$\text{Accuracy} = \frac{T_P + T_N}{T_P + T_N + F_P + F_N} \tag{5}$$

Here, T_P, T_N, F_P, and F_N, respectively, are true positive events, true negative events, false positive events and false negative events. The present work has two motivations, (i) to classify seizure and non-seizure EEG signal, (ii) to classify ictal and inter-ictal EEG signal. The EEG dataset used in the present work is comprised of 23 pediatric patients, respectively. The duration of each channel is 8 s with a sampling frequency of 173.61 Hz. The datasets that have the seizure are known to be an ictal class while the dataset which is prior to the ictal class segments is known to a pre-ictal class, and before pre-ictal, all the dataset is known to be as an inter-ictal class.

The software which is used while preparing our algorithm is Anaconda 2019.10, in which we have used Python 3.6 as our programming language on Windows 8.1. Libraries such as Keras, TensorFlow have been installed for achieving the better results. The hardware which we have used is Lenovo ThinkStation P330 Tower-2xIntel Xeon SILVER 4114-512 GB SSD PCIe + 1 TB 7200 rpm-128.0 GB RAM. In order to reduce the time, we have used this workstation which takes 2 min for an epoch to run while comparing it with the normal system which takes 8 min for an epoch to run. The whole algorithm has been executed for 200 epochs in order to

Table 4 Comparative analysis with state-of-the-art methods

Authors	Features	Classifier	Sensitivity (%)
Truong et al. [11]	STFT	CNN	81.2
Khan et al. [12]	Wavelet transform coefficient	CNN	87.8
Ozcan et al. [13]	Spectral power, statistical moments	CNN	85.7
Rajamanickam, et al. [18]	Wavelet transform coefficients	CNN	86.29
Proposed work	**Time domain signals**	**CNN**	**88.10**

Bold indicates highest calculated performance metrics

achieve high accuracy, high sensitivity and high specificity while comparing with the existing methods. Various tuning parameters have been set up in order to get the best results as shown in Table 4. The best result is achieved by tuning the parameters as random state 10, size ratio of 90–10 and learning rate of 0.001 which gives us the high performance measures as compared to other tuning parameters.

Performance comparison of the present work with the existing EEG classification techniques is discussed in Table 4. Table 4 presents the classification performance of seizure and non-seizure EEG data with the state-of-the-art methods. From Table 4, it is observed that the proposed technique achieved sensitivity of 88.10% which is better when compared to state-of-the-art methods. Table 4 clearly indicates that the present study produces a better performance by achieving sensitivity, specificity and accuracy of 88.10%, 87.10% and 87.40%, respectively, in comparison to existing techniques discussed in the literature.

6 Conclusion

In this study, we have worked on the convolutional neural network model for automatic seizure detection using EEG signals. Considering the state-of-the-art method, this model has a capability that learns better and done effectively discriminate between seizure and normal EEG. We have performed all ratios of training and testing dataset and saw that when our model is getting the training dataset of lower proportions then it gives us lower accuracy when compared with the training dataset of higher proportions. The main reason for lower accuracy is that lesser proportion of training dataset. While learning time, it automatically does take feature extraction and selection, so it wants more number for data for learning. We have done binary class classification in our proposed methodology and it obtained an average accuracy of 87.40%, sensitivity of 88.10% and specificity of 87.10%. The proposed model has been examined on publically available CHB-MIT EEG dataset and compared with several baseline methods. The result determines the effectiveness and superiority of the proposed model in detecting epileptic seizures. One more advantage of this model is that it does not require any feature extraction and selection methods. The performance of this technique can be improved by providing more number of dataset.

References

1. Satapathy SK, Dehuri S, Jagadev AK (2017) EEG signal lassification using PSO trained RBF neural network for epilepsy identification. Inform Med Unlocked 6(June 2016):1–11
2. Satapathy SK, Dehuri S, Jagadev AK (2017) ABC optimized RBF network for classification of EEG signal for epileptic seizure identification. Egypt Inform J 18(1):55–66
3. Kerr MP (2012) The impact of epilepsy on patients' lives. Acta Neurol Scand 126(S194):1–9
4. Zhou W, Gotman J (2004) Removal of EMG and ECG artifacts from EEC based on wavelet transform and ICA. In: Proceedings of annual international conference of the ieee engineering in medicine and biology society, vol 26 I, pp 392–395
5. Sadati N, Mohseni HR, Maghsoudi A (2006) Epileptic seizure detection using neural fuzzy networks. In: IEEE international conference on fuzzy systems, pp 596–600
6. Talathi SS (2017) Deep recurrent neural networks for seizure detection and early seizure detection systems. arXiv Prepr. arXiv: 1706.03283
7. Kim L (2018) DeepX : deep learning accelerator for restricted boltzmann machine artificial neural networks, vol 29, no 5, pp 1441–1453
8. Xu SS, Member S, Mak M, Member S, Cheung C, Member S (2018) Towards end-to-end ecg classification with raw signal extraction and deep neural networks. IEEE J Biomed Heal Inform 8:1
9. Chen M et al (2017) Deep feature learning for medical image analysis with convolutional autoencoder neural network, vol 7790, no c, pp 1–10
10. Kumar S, Dehuri S, Kumar A (2017) ABC optimized RBF network for classification of EEG signal for epileptic seizure identification. Egypt Inform J 18(1):55–66
11. Truong ND et al (2018) Convolutional neural networks for seizure prediction using intracranial and scalp electroencephalogram. Neural Netw 105:104–111
12. Khan H et al (2018) Focal onset seizure prediction using convolutional networks. IEEE Trans Biomed Eng 65(9):2109–2118
13. Ozcan AR, Erturk S (2019) Seizure prediction in scalp EEG using 3D convolutional neural networks with an image-based approach. IEEE Trans Neural Syst Rehabil Eng
14. Zhou M et al (2018) Epileptic seizure detection based on EEG signals and CNN. Frontiers in Neuroinformatics 12:95
15. Tsiouris KM et al (2018) A long short-term memory deep learning network for the prediction of epileptic seizures using EEG signals. Comput Biol Med 99:24–37
16. Yao X, Cheng Q, Zhang G-Q (2019) Automated classification of seizures against nonseizures: a deep learning approach. arXiv preprint arXiv:1906.02745
17. Alickovic E, Kevric J, Subasi A (2018) Performance evaluation of empirical mode decomposition, discrete wavelet transform, and wavelet packed decomposition for automated epileptic seizure detection and prediction. Biomed Signal Process Control 39:94–102
18. Yuvaraj R et al (2018) A deep learning scheme for automatic seizure detection from long-term scalp EEG. In: 2018 52nd asilomar conference on signals, systems, and computers. IEEE
19. Meng L, Tan A, Member S, Wunsch DC (2016) Adaptive scaling of cluster boundaries for large-scale social media data clustering 27(12):2656–2669
20. Jahankhani P, Kodogiannis V, Revett K (2006) EEG signal classification using wavelet feature extraction and neural networks. In: IEEE John Vincent Atanasoff 2006 international symposium on modern computing (JVA'06), pp 120–124
21. Li Y, Wang X, Luo M, Li K, Yang X, Guo Q (2018) Epileptic seizure classification of EEGs using time—frequency analysis based multiscale 22(2):386–397

22. Zhang L, Zhang D (2017) Evolutionary cost-sensitive extreme learning machine 28(12):3045–3060
23. Vidyaratne L, Glandon A, Alam M, Iftekharuddin KM (2016) Detection , pp 1202–1207
24. Acharya UR, Oh SL, Hagiwara Y, Tan JH, Adeli H (2018) Deep convolutional neural network for the automated detection and diagnosis of seizure using EEG signals. Comput Biol Med 100:270–278

Short-Term Load Forecasting Using Artificial Neural Network and Time Series Model to Predict the Load Demand for Delhi and Greater Noida Cities

Nikita Singh, Pratyush Sharma, Nikhil Kumar, and Mini Sreejeth

Abstract Forecasting of load refers to prediction of power demanded by the targeted geographical area based on the trends and patterns of previous load demands. To forecast the load accurately, is one of the biggest challenges of all electrical utilities and load dispatch centres. In this paper, artificial neural network (ANN) and time series (TS) models are used to study short-term load forecasting (STLF). It aims to predict the day-ahead peak load as well as average 24-h load demand for the city of New Delhi and Greater Noida. Both types of forecasting techniques are compared based on mean-squared logarithmic error (MSLE), mean-squared error (MSE) and mean absolute error (MAE). With the advancement of deep learning techniques, huge amounts of load demand data can be used to forecast the demand from minutes ahead to years ahead.

Keywords Artificial neural network · Mean-squared logarithmic error (MSLE) · Regression · Sequential pattern · Short-term load forecasting · Time series

1 Introduction

To forecast load for the future minutes, hours, days, years or even further, is one of the main priorities of load dispatch centres across the globe. One of the main reasons for this is the constant struggle for them to manage between over-drawing and under-drawing of power because of varying demand for the loads [1]. Both these situations

N. Singh (✉) · P. Sharma · N. Kumar · M. Sreejeth
Delhi Technological University, New Delhi, India
e-mail: writetonikita7@gmail.com

P. Sharma
e-mail: prats9199@gmail.com

N. Kumar
e-mail: nk64860@gmail.com

M. Sreejeth
e-mail: minisreejeth@dce.ac.in

R. P. Mahapatra et al. (eds.), *Proceedings of 6th International Conference on Recent Trends in Computing*, Lecture Notes in Networks and Systems 177,
https://doi.org/10.1007/978-981-33-4501-0_41

lead to losses to power distribution centres and they thrive to handle such situations in order to maintain the frequency of load dispatched and reliability of power. Electrical utility needs forecasting for unit commitment, to schedule device maintenance plan and reduce spinning reserve capacity [2]. As the conventional forecasting techniques, based on statistical modeling [1], are slowly being replaced by deep learning models, the prediction of future load is becoming better in terms of accuracy, which will help all electrical utilities in general and also in reducing operational costs and improving security [3]. Many upcoming artificial intelligence models and some classical models previously used for STLF include linear regression, fuzzy-neural models, expert systems, fuzzy inference, tree-bagged technique, exponential smoothing method, kalman filter method, use of support vector machine and artificial neural networks [3, 4]. Forecasting can be based on different techniques like time series forecasting, i.e. learning from the nonlinear trend varying with time, intelligent forecasting, i.e. adapting to nonlinear relationship between input features and output or combination forecasting, i.e. grouping of different approaches [5]. Various time series models like auto-regressive integrated moving average (ARIMA), ARMA, Box-Jenkins models and interactive time series modeling (ISTM) have been used by many researchers for forecasting non-deterministic data series that have random component [6, 7].

This paper is focused on artificial neural network(ANN) and time series modeling to carry out short-term load forecasting for two cities—New Delhi and Greater Noida. The real-time load demand data, at every fifteen-minute interval, for a week in September 2019 is used for training the models, to predict the peak load data as well 24-h average load data for the day following that week. For the ANN model, several factors including time of the week, time of the day and weather dependencies like humidity, precipitation, temperature, windspeed, etc., have been considered, whereas for the time series analysis, a univariate system is considered. Univariate system considers a single variable that varies sequentially or in a particular pattern over a given time frame. Time series model detects the pattern between successive time series elements to detect the future sequence in that variable [5]. Keras, a deep learning library over TensorFlow framework is used for designing and training the deep learning models. Keras is an easy-to use open-source deep learning framework that supports multiple backend engines and can be easily deployed across a range of platforms; hence, it is preferred.

This paper has five sections. Section 2 describes the forecasting techniques used and their principle of working. Section 3 explains about the data preprocessing, modeling and analysis. Section 4 represents the results and comparison between the two forecasting techniques used here. Section 5 is the conclusion.

2 Forecasting Techniques

In this paper, two different forecasting techniques have been used to forecast the day-ahead 24-h average load and peak load demand. These techniques are artificial neural network based on deep learning and time series modelling.

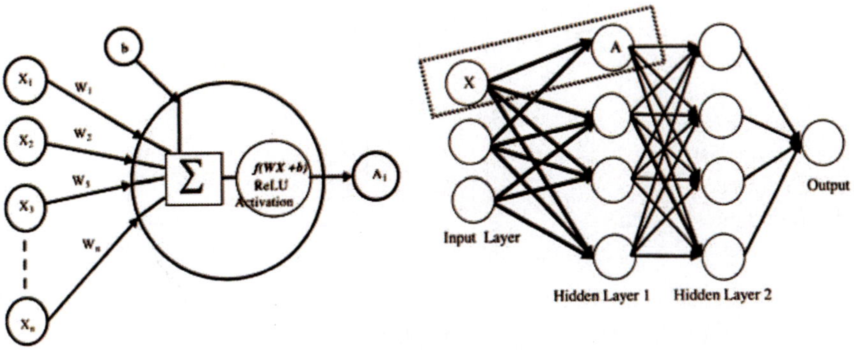

Fig. 1 Working of a single node in a neural network and general architecture of ANN

2.1 *Artificial Neural Network*

An artificial neural network is a data processing system consisting of a large number of simple highly interconnected processing elements called artificial neurons. This massive parallel distributed processor has neurons and weights for each of these neurons, as shown in Fig. 1, to depict the effect each neuron has on the final result [3]. In supervised learning, for a given set of input and output pairs as training examples, the ANN is able to detect nonlinear functions relating to the input and output vectors. This is due to the presence of hidden layers, each having an activation functions to account for the non-linearity between the connecting nodes of adjacent layers. This information processing characteristic makes them capable of learning from training examples and generalizing it for the test examples. Each feature in a given input vector is assigned weights which are updated after each iteration or epoch using the backpropagation algorithm of ANN. These weights describe the amount of effect each feature has to get a desired change in the output. Details of the model architecture used here are given in Sect. 3.2.

2.2 *Time Series Sequence Model*

Time series is typically defined as an ordered sequence of values that are usually equally spaced over time. Time series analysis extracts meaningful statistics and characteristics of serially correlated data. This analysis helps to forecast based on past values of data, in imputation, to detect anomalies and to spot patterns in the discrete time data that determine the series itself. Using machine learning techniques, neural networks can be trained based on the time series to predict future values of sequential data having seasonality, trend and noise [8]. Components of a time series can be expressed from the following:

$$Y_t = T_t + C_t + S_t + R_t \quad (1)$$

Where

T_t Secular trend
C_t Cyclic variation
S_t Seasonal variation
R_t Residual variation
Y_t Value of variable at time instant 't'.

Given the dataset used in this paper, for each day, both the peak load and the 24-h average load in the load forecast are single values at each time step, and thus they are taken as the single variable that varies sequentially over a certain time period. Hence, a univariate time series model is used to forecast both of them for a day-ahead. The nonlinear trend in the power load data can be modelled with a polynomial or an exponential or logarithmic equation.

$$Y_t = a + b_1 t + b_2 t^2 + \cdots + b_n t^n + e_t \textbf{ or } Y_t = \exp(a + b_1 t + e_t) \quad (2)$$

where a, b_k, e_t are constants of the polynomial or exponential equation

An alternate method to model the trends in time series is by a weighted moving average model [1]. It is used with data that has a definite rhythmic pattern of fluctuations and helps to smoothen it. An exponential weighted moving average is a model which allocates positive weights to current and past values of the time series in such a way that value of the weights exponentially decrease as the series goes from current value to farther back in time [1].

$$S_1 = Y_1; S_t = wY_t + (1 - w) S_{t-1} \quad (3)$$

Therefore $\hat{y}_{t+1} = S_t$
where

S_t exponentially weighted average for value at time step 't'
W weight parameter
$\hat{y}_{t+1}$ forecasted value of time series at time step '$t + 1$'
Y_1 actual value of the time series at first time step.

The time series model used for load forecasting here, takes one training set example by taking load inputs of three days as three input features to predict the next fourth day load. The next training set examples are formed by continuing the above in a cyclic manner for all load values of consecutive days. A visual description of the formation of training and test sets from a week's load data is shown in Fig. 2. This training set is given to the neural network, which in this case will have load data of three consecutive days as three input features to predict the same sequential variable for the day-ahead of the week taken for training. This time series modelling helps the neural network to detect the trend and sequential pattern, which is the cause for variation of load over consecutive time steps.

Training set input matrix				Training set output matrix	
$x^{\{1\}}$	$day\ 1$	$day\ 2$	$day\ 3$	$y^{\{1\}}$	$day\ 4$
$x^{\{2\}}$	$day\ 2$	$day\ 3$	$day\ 4$	$y^{\{2\}}$	$day\ 5$
$x^{\{3\}}$	$day\ 3$	$day\ 4$	$day\ 5$	$y^{\{3\}}$	$day\ 6$
	⋮	⋮	⋮		⋮
$x^{\{n\}}$	$day\ 5$	$day\ 6$	$day\ 7$	$y^{\{n\}}$	$day\ 7$

Test set input matrix				Test set output	
$x^{\{test\}}$	$day\ 6$	$day\ 7$	$day\ 8$	$y^{\{test\}}$	$day\ 9$

Fig. 2 Visualization of training and test sets with time series model as used in this paper

3 Load Forecasting Methodology

This paper aims to apply Short Term Load Forecasting to load data within India. Peak load data for Delhi city is collected from official website of Northern Regional Load Dispatch Centre (NRLDC) [9]. The load data at every fifteen-minute interval for Greater Noida city is also collected from the load dispatch centre of Noida Power Corporation Limited (NPCL). Forecasting load demand is a regression type, supervised learning problem. This is explained more in Sect. 3.3.

3.1 Data Preprocessing and Analysis

All the data was collected for the first week of September 2019. Daily load for one week, i.e. from Sunday to next Sunday was taken as training data. Load was forecasted for the following Monday, i.e. the ninth day immediately after the training set. Based on the data collected, three forecasts were made:

Model 1: ANN model to predict the 24-h average load of the day ahead for Greater Noida city. This model includes the effect of weather on load demand for that specific duration. Weather data was recorded at Safdarjung Airport station, New Delhi [10]. Hence, the following ten features were considered relevant to be impacting the input average power load:

• Average load data from 00:00–06:00	• Average humidity
• Average load data from 06:00–12:00	• Average temperature
• Average load data from 12:00–18:00	• Average windspeed
• Average load data from 18:00–24:00	• Average due point
• Weekday or weekend/holiday	• Precipitation

(continued)

(continued)

• Average load data from 00:00–06:00	• Average humidity

Model 2: Time series model to predict 24-h average load data of Greater Noida city. It is univariate in nature as it takes the daily average load as the single variable varying sequentially with each time step.

Model 3: Univariate time series model to predict the peak load data of Delhi city taking the single variable as the daily peak, which varies with each time step.

3.2 Model Architecture

All models are shallow neural networks with an input layer having ten neurons, one for each of the features considered (as shown in Sect. 3.1), a hidden layer having multiple nodes (five) and an output layer with a single node. To forecast load, activation functions used are rectified linear unit (ReLU) and Linear. ReLU activation function is used for the hidden layer as it overcomes the vanishing gradient problem and allows models to learn faster. As forecasting is a regression problem, i.e. to output a real-valued output, the linear activation function is used with the output layer. Weights are randomly initialized using the uniform Keras Initializer, also known as Glorot uniform or Xavier uniform initializer. The optimizer used for updating the weights (W_t) after back propagation is Adaptive Moment Estimation (Adam). The equation for updating parameters using Adam optimizer is expressed as:

$$W_t = W_{t-1} - \eta \frac{\widehat{m_t}}{\sqrt{\hat{v}_t} + \epsilon} \tag{4}$$

where

W_t weight updated at time step tW
W_{t-1} weight at previous time step $t - 1$
η learning rate
$\hat{m}_t$ bias corrected first moment (the mean) estimate of gradients
$\hat{v}_t$ bias corrected second moment (the uncentered variance) estimate of gradients.

Adam is essentially combination of root mean square propagation and momentum. It is a very effective optimizer for a wide variety of neural networks. Time series model used in this paper is explained in Sect. 2.2. The architecture used for time series consists of three input neurons, as visualized in Fig. 2, four hidden layer neurons and one output neuron.

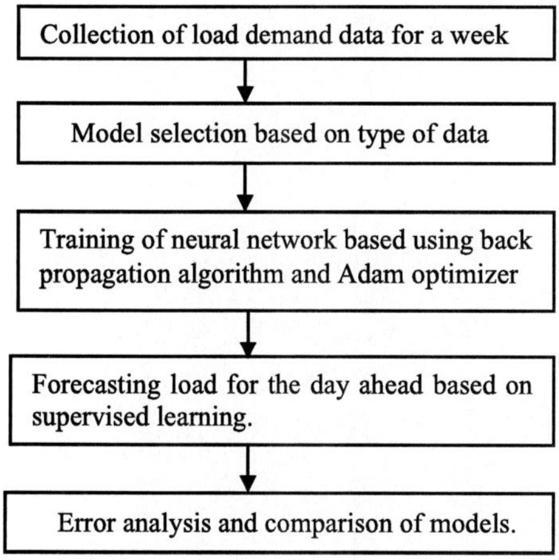

Fig. 3 Flowchart depicting the working methodology for both forecasting techniques

3.3 Working Methodology

Load data at an interval of fifteen minutes and statistics are collected and analysed. Appropriate features based on the pattern of data and factors affecting the domestic demand for power are noted [8]. To understand the *nonlinear* relationship between load and features affecting it, an artificial neural network is used. It uses backpropagation algorithm to train the weights given to each feature based on the output load required. This is called supervised learning. After training the model for thousand iterations, load is forecasted for the day ahead. This is a real-valued output and not a classification; hence, forecasting is a regression type of machine learning problem. The working methodology to train on the dataset and predict on the test set is explained in the flowchart in Fig. 3.

4 Results and Error Analysis

All the three models are trained using data for eight days, i.e. from Sunday to next Sunday. The load quantity for the ninth day is forecasted. A comparison of all the predicted load values and target outputs is calculated using three loss measures as given below:

Mean-squared logarithmic error (MSLE)

$$\text{MSLE} = \left| \frac{1}{n} \sum\nolimits_{i=1}^{n} (\log(y_i + 1) - \log(\hat{y} + 1))^2 \right| \quad (5)$$

Mean-squared error loss (MSE)

$$\text{MSE} = \left| \frac{1}{n} \sum\nolimits_{i=1}^{n} (y_i - \hat{y})^2 \right| \tag{6}$$

Mean absolute error loss (MAE)

$$\text{MAE} = \left| \frac{1}{n} \sum\nolimits_{i=1}^{n} |y_i - \hat{y}| \right| \tag{7}$$

where

- y_i Target output (Power load)
- $\hat{y}$ Predicted output
- n Total number of training samples.

Table 1 depicts the final values of the loss measures shown in (5), (6) and (7) after training the three models for 1000 iterations. Two types of errors are calculated:

1. Training set error—error on applying the learning model to the training set data, after 1000 epochs and
2. Test set error—error on applying the trained model to the test set (day whose load demand has to be forecasted) after 1000 epochs.

The outputs of each of the three models (as explained in Sect. 3.1) are obtained and graphical variations in the loss functions over 1000 generations are plotted. Figure 4 represents the graph of three loss measures after training model 1. Similarly, Figs. 5 and 6 represent the same for model 2 and model 3. The training set error reduces as the model eventually learns from the training set. Also, the performance on test set improves as shown by reduction of test set error after each epoch. An approximate difference between the two indicates if there may exist any unwanted bias or variance in the model. Analysis of the same is important to make the model safe from overfitting and underfitting the training data. Across all three models, mean-squared logarithmic error (MSLE) has proven to be a more accurate loss function to determine the performance of the model considered. Among the different models studied in this paper, a comparison is made by viewing the difference between training set errors and test set errors as shown in Table 1. It depicts that model 1 is slightly better than the time series model in terms of improvement in test set error over 1000 iterations. This is perhaps because of a larger neural network in model 1 to detect the non-linear relationship between the various features and load demand in Greater Noida city.

Table 1 Comparison of the three models based on the three error types considered for training and test sets

S. No.	Model type	Mean-squared logarithmic error		Mean-squared error		Mean absolute error	
		Training error	Test set error	Training error	Test set error	Training error	Test set error
1	ANN model to forecast 24-h average load of Greater Noida	$3.5 * 10^{-4}$	$3.72 * 10^{-1}$	$2.08 * 10^{-4}$	$1.9 * 10^{-3}$	$8.7 * 10^{-4}$	$2 * 10^{-1}$
2	Time series model to forecast 24-h average load of Greater Noida	$1.69 * 10^{-2}$	$1.02 * 10^{-2}$	$1.30 * 10^{3}$	$1.88 * 10^{3}$	$1.68 * 10^{1}$	$8.08 * 10^{1}$
3	Time series model to forecast peak load data of New Delhi	$2.3 * 10^{-3}$	$4.6 * 10^{-3}$	$6.33 * 10^{4}$	$1.33 * 10^{5}$	$1.66 * 10^{2}$	$4.09 * 10^{2}$

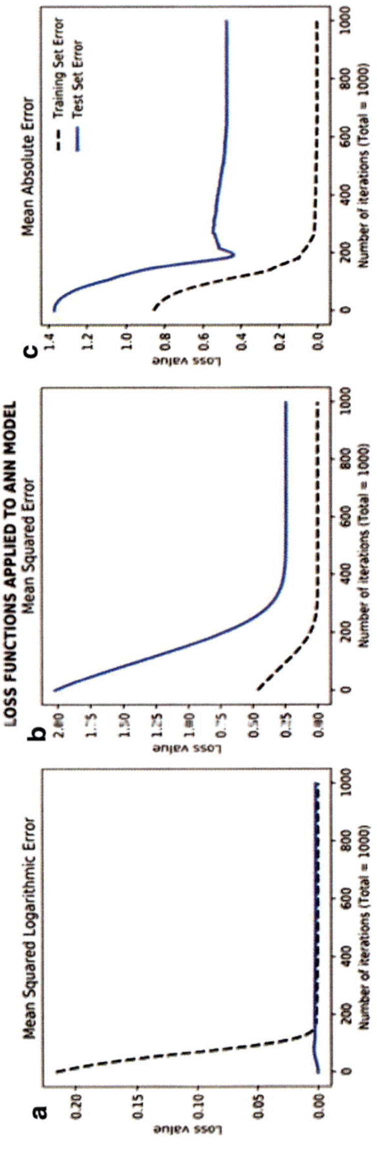

Fig. 4 Graph of all three errors **a** MSLE **b** MSE **c** MAE on training and test sets, varying over 1000 epochs of training on Model 1

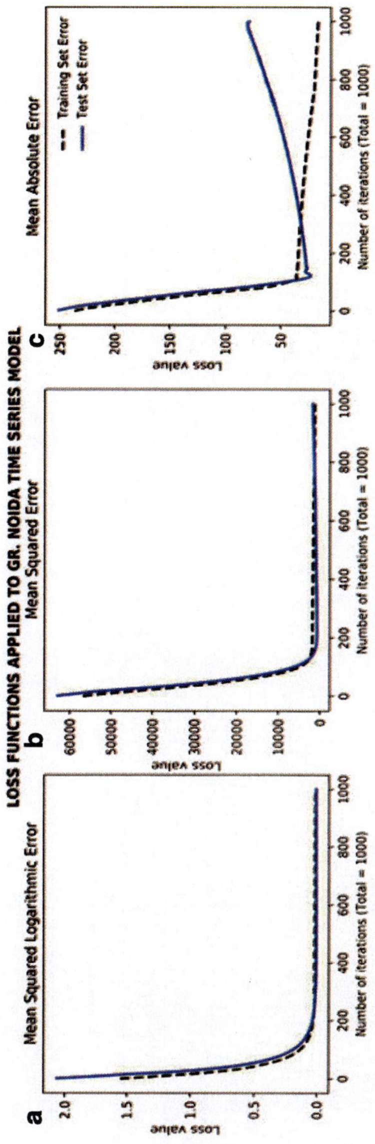

Fig. 5 Graph of all three errors **a** MSLE **b** MSE **c** MAE on training and test sets, varying over 1000 epochs of training on model 2

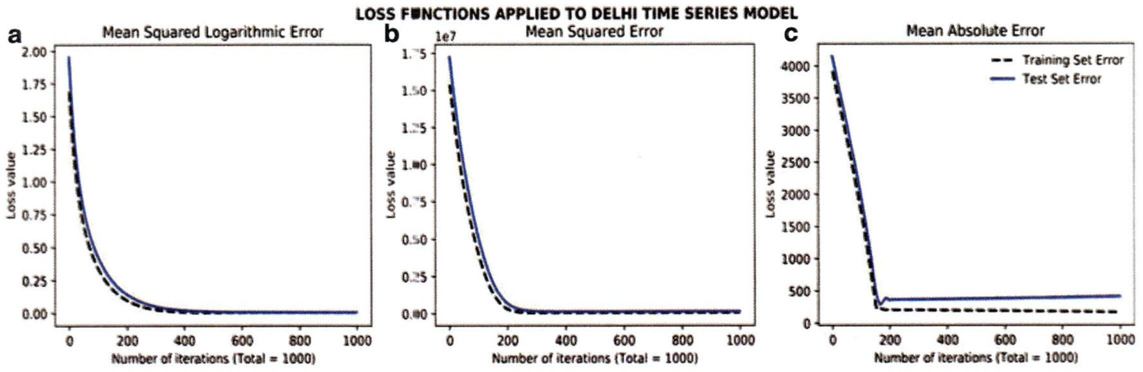

Fig. 6 Graph of all three errors **a** MSLE **b** MSE **c** MAE on training and test sets, varying over 1000 epochs of training on model 3

5 Conclusion

The forecast of the day-ahead is predicted using artificial neural network and time series modeling. Sequential data such as 24-h average load demand or peak load of a day are taken as stand-alone variables for a time series model. This is because these variables are single valued at each time step and show a sequential trend varying with time. Hence, a univariate time series model is used to forecast such type of load data. Another model shown in this paper is the ANN model which considers multiple features, like the weather recorded during the day, such as precipitation, humidity and windspeed, particular time of the week such as weekend or weekday and peak power demand timings for a given day, that affect the demand of power load in a region and are used to train the neural network. This paper focuses on implementing these forecasts in India as these forecasting models are trained using historical load data and weather-related dependencies of Delhi NCR. Both the forecasts from the ANN and the time series model on Greater Noida load data show that ANN is slightly more accurate than time series. The results conclude that mean-squared logarithmic error is better loss measure than MSE and MAE to compare models of different forecasting techniques.

References

1. Samuel I, Ojewola T, Awelewa A, Amaize P (2016) Short-term load forecasting using the time series and artificial neural network methods. IOSR J Electr Electron Eng (IOSR-JEEE). 11(1 Ver. III):72–81. e-ISSN: 2278–1676, p-ISSN: 2320-3331
2. Mahmoud Elgarhy S, Othman MM, Taha A, Hasanien HM (2017) Short term load forecasting using ANN Technique. In: 19th international middle east power systems conference (MEPCON), Menoufia University, Egypt, 19–21 December (2017)
3. Singh S, Hussain S, Abid Bazaz M (2017) Short term load forecasting using artificial neural network. In: Fourth international conference on image information processing (ICIIP)
4. Patel H, Pandya M, Aware M (2015) Short term load forecasting of indian system using linear regression and artificial neural network. In: 5th Nirma University international conference on engineering (NUiCONE)
5. Lv X, Cheng X, Shuang Y, Yan –mei T (2018) Short-term power load forecasting based on balanced KNN. IOP Conference series: materials science and engineering, vol 322, pp 072058
6. Gross G, Galiana FD (1987) Short-term load forecasting. The IEEE 75(12):1558–1573
7. Amjady N (2001) Short-term hourly load forecasting using time-series modeling with peak load estimation capability. IEEE Trans Power Syst 16(3):498–505
8. Razak FA, Shitan M, Hashim dan AH, Abidin IZ (2009) Load forecasting using time series models. In: Razak FA et al (ed) Jurnal Kejuruteraan, vol 21, pp 53–62
9. Website link. https://nrldc.in for Delhi city load data
10. Website link. https://www.wunderground.com for Weather data

Attainment of Green Computing in Cloudlet-Based Mobile Cloud Computing Model Using Squirrel Search Algorithm

Kushal Anjaria and Nirmal Patel

Abstract Utilization and exploration of mobile devices to enable mobile users to gain benefits of the cloud computing paradigm by an environment-friendly method is an efficient strategy for meeting current industrial demands. However, energy drain, performance, and limited connectivity are the challenges faced by mobile cloud computing (MCC). The present work aims to solve these challenges by using cloudlets that possess the capabilities of the Squirrel Search Algorithm (SSA). The present work examines SSA by a simulation of practical scenarios and provides reliable results for the evaluations.

Keywords Cloud computing · Green computing · Optimization · Squirrel search algorithm

1 Introduction

Nowadays, cloud computing has posed a serious question to the environment as data centers hosting cloud applications devour vast amounts of electrical energy, playing a part in high operational costs and carbon footprints to the environment. Therefore, we need green cloud computing solutions that not only reduce the environmental effect but also increase performance. Many approaches are ranging from software-hardware management, policy design, decision making, and legal issues to achieve green computing in the cloud paradigm [1, 2]. In the present study, the proposed

Note Both the authors have contributed equally towards this paper. Names of the authors are mentioned in alphabetical order.

K. Anjaria (✉)
IT and Systems Area, Institute of Rural Management Anand (IRMA), Anand, Gujarat, India
e-mail: kushal.anjaria@gmail.com

N. Patel
Computer Engineering Department, Gujarat Power Engineering and Research Institute (GPERI), Mehsana, Gujarat, India
e-mail: nicpatel963@gmail.com

R. P. Mahapatra et al. (eds.), *Proceedings of 6th International Conference on Recent Trends in Computing*, Lecture Notes in Networks and Systems 177,
https://doi.org/10.1007/978-981-33-4501-0_42

approach emphasis on the technical side that leverages communication and nature-inspired optimization algorithm.

In the cloud computing paradigm, mobile cloud computing (MCC) is an exposure of many internet-based technological developments, which enables mobile users to acquire benefits of cloud computing and attain green computing by using their mobile devices. At this time, MCC is a widely used technology in cloud computing paradigms [3]. As a result, the present work focuses on the MCC model and the achievement of green computing in MCC architecture. Based on the industrial and commercial needs, our advanced deployment offers a unique mechanism to avoid energy waste in MCC when users are suffering a complicated and unstable networking environment. In a nutshell, the present work aims to not only optimize energy consumption, but also improve the performance of the MCC paradigm.

Nowadays, it is a prevalent scenario when a large number of data centers and mobile devices available at a single geographical location. In such a situation, aims are to select such a data center which provides fast service to the mobile device, establish a connection between mobile devices and data centers that consume less energy and if a mobile device is continuously moving, then the transition of the mobile device from one network to the other should be smooth. Moreover, in real-time, mobile devices can be dynamically added in the networked region of any cloud as well as some mobile devices can immediately go beyond the wireless signal range of clouds. In such a constantly changing real-time scenario, it becomes complex to take care of the energy of each mobile device and its connection. Thus, the problem of finding a systematic wireless network link for a mobile device that does not degrade its performance becomes an NP-Hard problem [4]. The NP-Hard problem means as hard as the hardest problems in NP. Blum and Li [5] suggested that to solve the NP-Hard problem, swarm intelligence, and heuristic-based algorithms bring good bargains. So, in the present work, the Squirrel Search Algorithm (SSA) has been used to achieve energy-aware green computing in the MCC paradigm.

Many researchers and scholars have already attempted research in the field of energy-aware green MCC architecture. Gai et al. [6] have used a dynamic programming-based method to deploy green computing in MCC. However, time consistent nature of the dynamic programming makes the solution more timeconsuming [7]. Kushal et al. [8] have applied the ant colony optimization (ACO) algorithm to obtain energy-aware green MCC. ACO is a nature-inspired algorithm. The present work also aims to apply a nature-inspired algorithm, i.e., Squirrel Search Algorithm (SSA), to the same green cloud computing problem. The present paper intends to apply one more nature-inspired optimization problem to achieve green computing as the scientific community is still developing new optimization techniques for solving novel and more complex optimization problems under the ideology of continuous improvement in order to achieve a better outline. Moreover, as per the "no free lunch" (NFL) theorem, there is no single nature-inspired optimization technique, which can effectively solve all optimization problems [9]. So, along with the application of SSA to obtain green computing in the mobile cloud, the present work will also compare SSA and ACO. Furthermore, the current work also aims to discuss the simulator designed to apply SSA to MCC.

In a nutshell, the contributions of the current work are as follows:

- Application of SSA to obtain green computing in MCC.
- Comparison of SSA and ACO.
- Discussion on a newly developed simulator for SSA.

After discussing the basics and related literature in the introduction section, the paper has been organized as follows: The second section discusses the background and explains the research problem. The third section elucidates the concept of SSA and its implication to MCC architecture. Section 4 demonstrates the developed simulator to implement SSA in the MCC context. It also explains experimental results. Section 5 concludes the proposed work.

2 Background and Description of the Addressed Research Problem

The present work utilizes the concept of cloudlets in MCC. The term cloudlet refers to the surface connecting mobile devices and cloud servers in MCC, which plays a mediator role by paying attention to the business logic [10]. MCC uses a cloudlet-based technique to assign, manage, and optimize the cloud-based infrastructure usages and services for achieving green computing. The cloudlet-based MCC follows three-tier architecture. Figure 1 describes the cloudlet-based MCC architecture:

As described in Fig. 1, mobile devices send service requests to the closest cloudlets before the requests reach the cloud servers. The mobile device communicates with the cloudlet layer using one of the wireless network protocols. The cloudlets have to

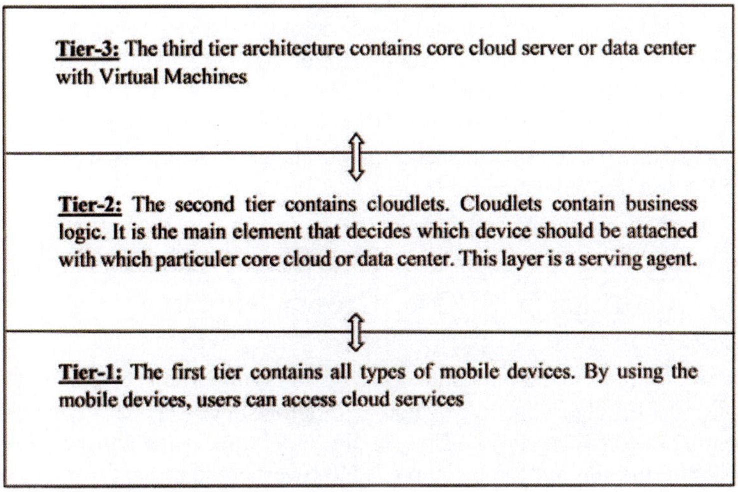

Fig. 1 Three-tier architecture of cloudlet-based MCC

allocate the cloud servers for better service performance determined by a group of constraints, such as nearby server locations or networking stability. The cloud server or data center is the final layer of the MCC architecture. It contains a collection of virtual machines (VMs) to complete any task.

SSA is applied in cloudlets for adapting to the continually changing context and assisting cloudlets in accessing cloud computing resources within a changing operational environment. So, in order to achieve green computing, two facets should be carefully handled: (1) Cloudlet should handle the transition of mobile devices, and (2) The cloudlet should allocate such a cloud server, which should give better service to mobile devices, and its connection with a mobile device should not be repeatedly broken. These two facets are crucial from green computing perspectives as continuous connection loss, transition of the mobile device, and lousy performance of cloud servers are the main three reasons for energy drain [11].

To achieve green computing, SSA has been applied to the three-tier architecture described in Fig. 1. The SSA is inspired by flying squirrels, which are a diversified group of arboreal and nocturnal types of rodents that are adapted for gliding locomotion [12]. Jain et al. [12] have pioneered the SSA for the first time as a nature-inspired algorithm for optimization. The squirrels have a parachute like membrane that helps squirrel glide from one tree to another. The algorithm is based on the nutritional requirements of squirrels. During autumn, squirrels eat acorn nuts, and during winters when nutritional demands are higher, then squirrels eat hickories. So, during autumns, squirrels perform two tasks: The first is to eat acorn nuts to meet regular nutritional requirements, and the second is to collect hickory nuts. Climatic conditions force flying squirrels to be less active in winters as compared to autumn.

In literature, SSA has been used for multi-region combined heat and power economic dispatch to incorporate renewable energy sources [13]. Further, it has been applied to the bin packing problem [14], transient stability constrained optimal power flow [15], and job shop scheduling [16]. Along with the application of SSA, researchers have also studied some of the variants of SSA. Zheng and Luo [17] proposed novel SSA for generalized optimization, Wang and Su [18] proposed SSA for global function optimization, Hu et al. [19] have combined SSA with invasive weed optimization and Wand, and Du [20] have presented SSA with external population and adaptive weight vectors adjustment. The next section discusses the details about SSA and its implications in MCC.

3 SSA Algorithm in MCC for Green Computing

SSA starts with the random initial location of flying squirrels similar to other population-based algorithms. A vector in n-dimensional search space represents the location of a flying squirrel. Thus, in algorithm also, using a matrix, the hyper dimension of the squirrel can be mapped. However, in the present work, to present simplicity, three-dimensional positioning of the squirrel has been utilized. The following matrix can represent the location of all flying squirrels:

$$SP = \begin{bmatrix} SP_{11} & SP_{12} & SP_{13} \\ SP_{21} & SP_{22} & SP_{23} \\ \ldots & \ldots & \ldots \\ \ldots & \ldots & \ldots \\ SP_{n1} & SP_{n2} & SP_{n3} \end{bmatrix} \tag{1}$$

where SP_{nm} describes the m_{th} dimension of the nth squirrel. The matrix *SP* in Eq. 1 can be useful in describing the location of the squirrel in general. However, it is crucial to allocate the initial location of each participating flying squirrel in the forest. The initial location of flying squirrels can be allocated using the uniform distribution described using Eq. 2:

$$SP_{\text{i}} = SP_{\text{L}} + U \times (SP_{\text{U}} - SP_{\text{L}}) \tag{2}$$

In Eq. 2, SP_{i} describes the initial position, SP_{U} describes upper bound and SP_{L} is the indicator of the lower bound of the *i*th flying squirrel in the *j*th dimension. Moreover, in Eq. 2, *U* is the uniformly distributed random number, and it can be considered in the range of [0,1]. In the normal distribution, the fitness of the location for each flying squirrel can be calculated by putting the values of decision variables into any user-defined fitness function. The fitness function may take care of each participating flying squirrel's location, quality of food source searched, i.e., an optimal food source (hickory tree), a normal food source (acorn tree), no food source, and finally, the probability of survival of flying squirrels. Based on the fitness function, the experimenter can declare squirrel with minimum fitness value on the hickory trees. The next three best flying squirrels can be considered on the acorn tree, and all remaining squirrels will be on the normal trees. The normal trees are the trees without any type of food.

In an experiment and implementation of SSA, the experimenter can decide the number of squirrels, a number of the trees, i.e., normal, hickory, acorn, and the number of the flying squirrels. For successful implementation of the SSA, it is assumed that more than 90% of the total population of the squirrel is on normal trees. The remaining squirrels are on the trees with the food source. In an experiment, the number of a food source can also be varied, as suggested by Jain et al. [12] and Zheng and Luo [13].

The SSA optimization algorithm is based on the movement of the flying squirrels. In the present work, we assume that a predator is not present in the forest, which can affect random movement of the squirrel. In the absence of a predator, random movement of the squirrels can be minimized. In such a scenario, the movement of the squirrels has three possibilities: (1) Flying squirrels, which are on acorn nut trees SP_{AT}, may move towards the hickory nut tree to save the nuts for winter. (2) Flying squirrels, which are on normal trees SP_{NT} may move towards acorn nut trees to fulfill their routine energy requirements (3) Some squirrels which are on normal trees and already consumed acorn nuts may move towards the hickory nut tree in order to store

Table 1 Symbols used in SSA

S. No	Symbol	Meaning
1	$SP_{AT}t+1$	The new location of the squirrels which were on acorn nut trees
2	SPAT t	The current location of the squirrels which are on acorn nut trees
3	SPHT t	The current location of the squirrels which are on hickory nut trees
4	SP_{NT} t + 1	The new location of the squirrels which were on normal trees
5	SPNT t	The current location of the squirrels which are on normal trees
6	d_g	It is a random gliding distance and pre-decided based on the nature of the experiment. Its value should be between 0 and 1
7	G_C	It is the gliding constant. Its value is very significant for the execution of the algorithm. In the present work, we have chosen the value of G_C as 1.901. The value is similar to the one calculated by Jain et al. [12] in their research

hickory nuts so that it can be consumed during the winter season. Formally, these possibilities can be described as follows:

- Movement of the squirrels from acorn nut trees SP_{AT} gliding towards hickory nut tree SP_{HT} can be described as:

$$SP_{\mathrm{ATt+1}} = SP_{\mathrm{ATt}} + [d_g \times G_c \times (SP_{\mathrm{HTt}} - SP_{\mathrm{ATt}})] \tag{3}$$

- Movement of the squirrels from normal trees SP_{NT} gliding towards acorn nut tree SP_{AT}, can be described as:

$$SP_{\mathrm{NTt+1}} = SP_{\mathrm{NTt}} + [d_g \times G_c \times (SP_{\mathrm{ATt}} - SP_{\mathrm{NTt}})] \tag{4}$$

- Movement of the squirrels from normal trees that have already consumed acorn nuts and moving towards the hickory nut tree can be described as:

$$SP_{\mathrm{NTt+1}} = SP_{\mathrm{NTt}} + [d_g \times G_c \times (SP_{\mathrm{HTt}} - SP_{\mathrm{NTt}})] \tag{5}$$

Table 1 describes the meaning of the symbols used in Eqs. 2, 3 and 4.

3.1 SSA for MCC with Cloudlet to Attain Green Computing

This sub-section aims to apply SSA in the MCC context. The merger of SSA and MCC will result in energy-aware green mobile cloud computing. In the MCC context, one of the assumptions is that the functional area of mobile devices will be considered as a forest. Further, all the cloudlets in the functional area will be considered as trees of the forest. The assumption is intuitive also as it is the responsibility of cloudlets to connect the mobile device with the data center, which provides better performance

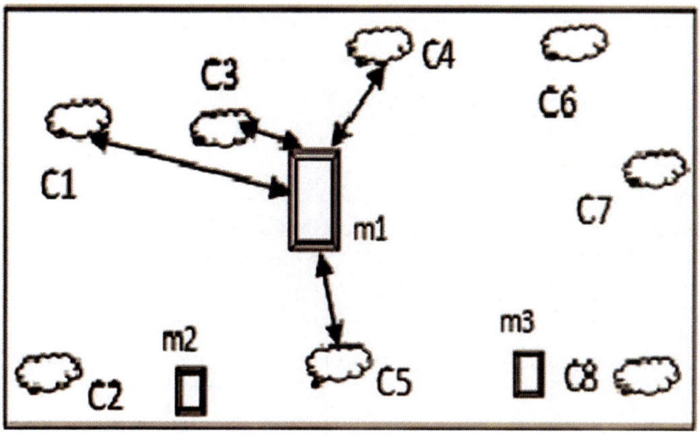

Fig. 2 Cloudlet-based MCC and its relevance with SSA

and energy efficiency. All the mobile devices which aim to obtain services from the cloud are considered as squirrels. The squirrels aim to find hickory trees. In the context of MCC, the nearest and efficient cloudlet will be the hickory tree. All the remaining cloudlets around the mobile devices which are not the closest and efficient but can serve the mobile devices if required are acorn nut trees. Finally, normal trees are all the cloudlets that are out of the range of mobile devices or which were serving mobile devices previously. The details have been described in Fig. 2: As described in Fig. 2, A is the forest. All the cloudlets are trees in the forest. Further, cloudlet C3 is the nearest cloudlet of mobile device m1. Figure 2 describes a mechanism for mobile device m1 out of available three mobile devices m1, m2, and m3. So, cloudlet C3 is a hickory tree. The mobile device m1 can also get connected to other cloudlets like C1, C4, and C5 if C3 fails. So, cloudlets C1, C4, C5 are acorn trees. Finally, cloudlets C2, C6, C7, and C8 are the cloudlets with which mobile device m1 was previously connected, or they are out of the reach of m1. As a result, cloudlets C2, C6, C7, and C8 are normal trees. In Fig. 2, mobile devices m1, m2, and m3 are considered squirrels. Figure 2 describes a limited number of mobile devices and cloudlets for better understanding. The current work considers a real-time scenario where a large number of mobile devices and cloudlets exist in any functional area of the MCC paradigm.

The present work considers the distance between mobile devices and cloudlets as a linear function of two entities. The first entity is the energy required for connecting and communicating mobile devices with the cloudlets. In literature, there are many measures and parameters available to calculate energy consumption [8, 21]. In the present work, the assumption is energy, and the distance between mobile devices and cloudlet has linear relation. The second entity considered is the number of free VMs in the data centers or the cloud servers. Thus, the distance described in Fig. 2 with a bi-directional arrow is f (LD, -VM), where f is the distance function, LD is a linear distance, and VM is the number of free VMs. In the above function, the negative sign denotes that we want to maximize the number of VMs and minimize the linear distance using SSA. The function suggests that the SSA aims to find such a cloudlet for a mobile device, which can reduce energy consumption and improve the performance and execution times. In other words, the cloudlet should allocate such a data center or cloud server to the mobile device, which has more number of free VMs so that cloud servers can serve mobile devices better.

From the above set-up of SSA, it is clear that if mobile devices increase in functional area A, then the serving capacity of the cloudlet may decrease gradually. The algorithm below describes how SSA can be implemented for MCC. Cloudlets will execute the algorithm after receiving the requests from the mobile devices. In the algorithm, it is assumed that networking protocols are working correctly, and there is no possibility of network loss. In the algorithm below, the symbol "//" denotes comments.

Algorithm 1 Pseudo code for SSA in MCC.

Require: - Generate initial locations of mobile devices that are equivalent to flying squirrels using equation-2.

- Based on the request received from the mobile device, sort the mobile devices based on the distance function f.
- Identify hickory nut trees, acorn trees, and normal trees. The assumption is that there is K number of squirrels.
- Identify seasonal constant s_n. The constant will help to identify the season, i.e., winter or summer. The behavior of squirrels changes during winter.

Pseudo-code: INITIALIZE $t = \emptyset$
while mobile device is in A **do**
for $t \leftarrow 1$ to K1 **do**
// K1 is the number of mobile devices, i.e., flying squirrels which are on or near acorn nut tree and moving towards hickory nut tree.

$SP_{AT}^{t+1} = SP_{AT}^{t} + [d_g \times G_c \times (SP_{HT}^{t} - SP_{AT}^{t})]$ ....equation(3) **end for for**
$t \leftarrow 1$ to K2 **do**
// K2 is the number of mobile devices, i.e., flying squirrels which are on or near normal trees and moving towards acorn nut tree.
$SP_{NT}^{t+1} = SP_{NT}^{t} + [d_g \times G_c \times (SP_{AT}^{t} - SP_{NT}^{t})]$ ....equation(4) **end for for**
$t = 1$ to $K3$ **do**
// K3 is the number of mobile devices, i.e., flying squirrels which are on or near normal trees and moving towards a hickory nut tree.
$SP_{NT}^{t+1} = SP_{NT}^{t} + [d_g \times G_c \times (SP_{HT}^{t} - SP_{NT}^{t})]$ ....equation(5) **end for**
Update the value of seasonal constant s_n after pre-decided time-period// after a specific time, the season may change.
The location of the required squirrel on a hickory nut tree is the final optimal solution. **end while**

The above algorithm shows that just like all nature-inspired algorithms, it also imitates two distinct features of nature, i.e., adaptability and the choice of the fittest. It is evident in the above algorithm that K1 + K2 + K3 = K. In the present work, to verify the above algorithm, we have designed a simulator for SSA in the MCC context. The next section discusses experimental results related to SSA in MCC.

4 Simulation, Experimental Results, and Comparison

In literature, it is discerned in many studies that so-called "novel" algorithms lack the rigorous theoretical background, and most of the nature-inspired algorithms are previous algorithms under new interpretations [22]. Hence, the criterion for releasing and implying a novel algorithm becomes much higher. There must be significant differences in the new algorithm from the others in order to release and imply a new algorithm [23]. Therefore, we have developed a simulator, performed experiments, and compared the results of SSA with ant colony optimization (ACO).

During the experiment, uniformity describing constant U of Eq. 2 has been considered as $U = 0.5$. Further, in the experiment, gliding distance d_g has been considered as 0.52, and gliding constant G_c has been considered as 1.9. Figure below describes the simulator of SSA for MCC.

As shown in Fig. 3, the users can drag and drop the required number of mobile devices and cloudlets from the panel at the right. With each mobile device, required processes to complete tasks of the mobile device are also described. Each cloudlet in Fig. 3 describes three parameters: (1) minimum distance from the connected mobile device (2) Total available VMs of the data center with which the cloudlet is associated and (3) The number of idle/free VMs. Here, mobile devices and cloudlets will only be associated if the mobile device requires optimal energy to establish a connection, and the data center which will complete the processes of a mobile device has a sufficient number of free VMs. The condition related to free VMs will ensure that mobile devices associated with such a data center, which has capabilities to be highly responsive. In the diagram, red lines indicate that a mobile device may associate with the cloudlet. However, after the execution of SSA in the background, a mobile device will associate with only one cloudlet. The red line in the diagram shows that after the execution of SSA, the mobile devices have been associated with which particular cloudlet. In the diagram, to calculate the distance among mobile devices and cloudlets, x and y-axis have been considered. Next, the SSA algorithm has been compared with the ACO algorithm for the MCC paradigm.

The experiments reveal that if there are less number of mobile devices and cloudlets, then the results obtained by ACO and SSA both are same. The similarity of the results of both the algorithms are described in Figs. 4 and 5:

Next, we reduced the number of cloudlets and increased the number of the mobile device. In such a scenario, it is observed that the result obtained by SSA and ACO shows a significant difference in selecting the cloudlets (Tables 2 and 3).

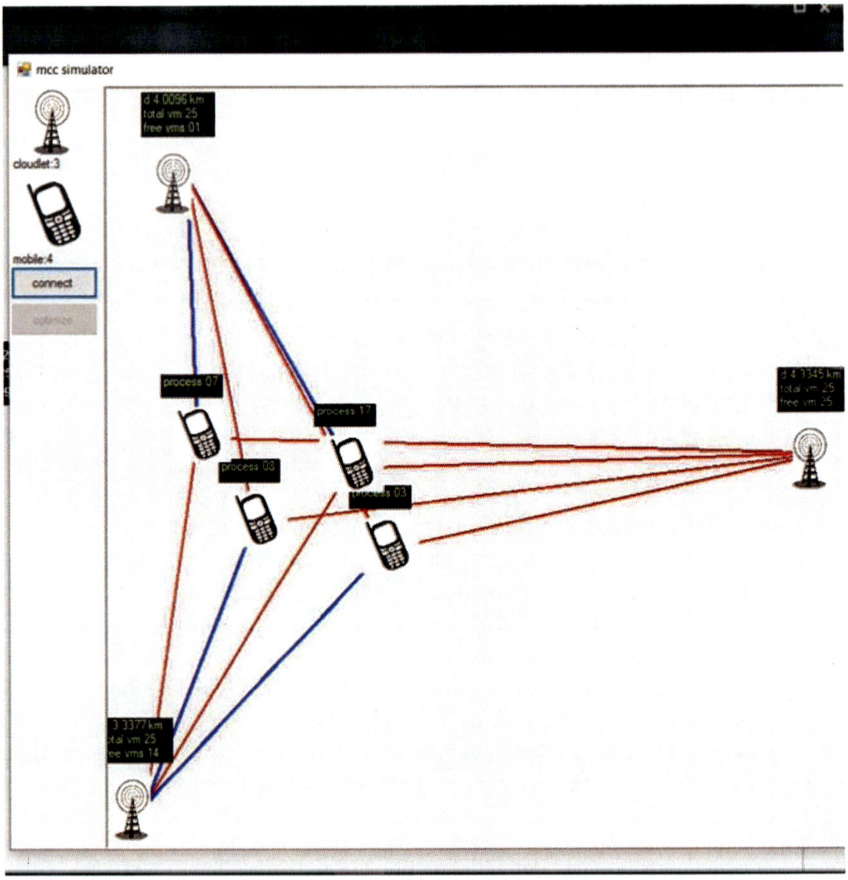

Fig. 3 Simulator for SSA in MCC paradigm

After repeated experiments and results described in Fig. 4, Tables 2, 3 and from literature survey, we found that the differences in results obtained using SSA and ACO exist due to the following reasons:

- ACO considers cumulative effects and choices of ants. On the other end, SSA uses sorted information and divides the pattern matrix initially into three regions, like a global optimum solution (SP_{HT}), near-optimal solutions (SPAT), and random solutions (SP_{NT}). Thus, these three solutions work together to find a solution. In other words, ACO finds the solution in a pattern matrix by a single strategy. On the other hand, SSA employs three strategies in different regions of the pattern matrix. As a result, SSA obtains a holistic view of MCC architecture and gets complete positioning of cloudlets.
- In ACO, there are several constants, like parameters ρ, α, and β [24]. On the other hand, in the SSA, only gliding sequence is the only constant which is considered

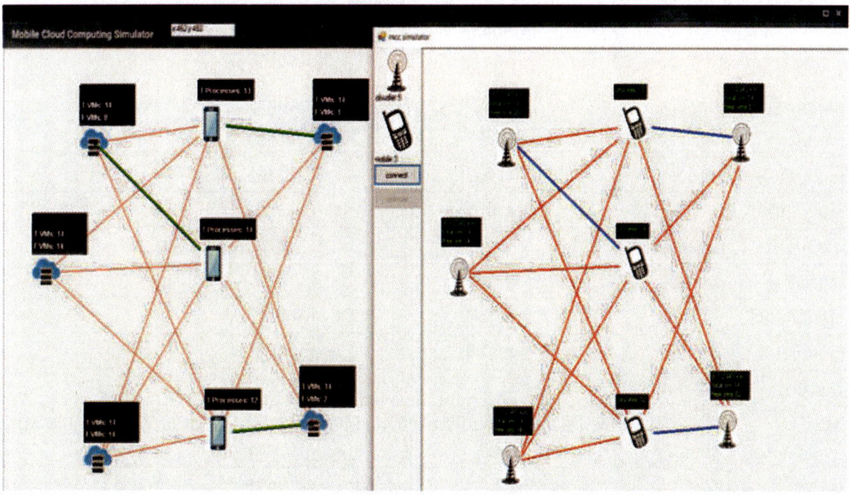

Fig. 4 Simulator for SSA in MCC paradigm

Table 2 Input and output to the simulator when the number of mobile devices and cloudlets are less.

Input:

mobile:01	process:13	x:353	y:81
mobile:02	process:12	x:357	y:533
mobile:03	process:14	x:357	y:284
cloudlet:01	vms:14	x:137	y:108
cloudlet:02	vms:14	x:560	y:100
cloudlet:03	vms:14	x:145	y:572
cloudlet:04	vms:14	x:050	y:296
cloudlet:05	vms:14	x:536	y:519

Output:

Mobile	Cloudlet selected by ACO	sCloudlet selected by SSA	sDis C1 (In km.)	Dis C2 (In km.)	Dis C3 (In km.)	Dis C4 (In km.)	Dis C5 (In km.)
m1	c2	c2	2.1868	2.0787	5.3324	3.7153	4.7469
m2	c5	c5	4.7857	4.7822	2.1556	3.8784	1.7955
m3	c1	c1	2.8174	3.5761	3.5761	3.0723	2.9541

while solving the problem using SSA. This fact helps SSA in achieving better results than ACO.

- The simulation of flying squirrel behavior provides an opportunity to introduce a seasonal condition in SSA. The seasonal conditions invoke the algorithm several

Table 3 Input and output of simulator when the result of ACO and SSA differs

Input:				
mobile:01	process:10	x:589	y:088	
mobile:02	process:11	x:753	y:169	
mobile:03	process:12	x:465	y:219	
mobile:04	process:13	x:665	y:290	
mobile:05	process:14	x:321	y:425	
mobile:06	process:15	x:269	y:233	
cloudlet:01	vms:50	x:137	y:138	
cloudlet:02	vms:50	x:489	y:404	
Output:				
Mobile	Cloudlets selected by ACO	Cloudlets selected by SSA	Dis C1 (In km)	Dis C2 (Ink m))
m1	c2	c2	4.5476	3.3145
m2	c1	c2	6.1678	3.5344
m3	c1	c2	3.3785	1.8655
m4	c2	c2	5.4944	2.0097
m5	c1	c1	3.4092	1.6931
m6	c2	c1	1.6263	2.7864

times to initiate a search in different directions, and thus, the algorithm does not get stuck in locally optimal solutions.

Next section concludes the present work.

5 Conclusion

This paper proposed a novel approach of using a dynamic cloudlet-based MCC model to gain the benefits of green computing. The experimental results proved that our SSA-based approach was an effective mechanism that could solve unconstrained optimization problems and enable mobile users to address green IT within a dynamic, complicated wireless environment. It is observed from the comparative statistical analysis that SSA achieves the global optimum solutions with remarkable accuracy in comparison to the ACO.

References

1. Airehrour D, Cherrington M, Madanian S, Singh J (2019) Reducing ICTcarbon footprints through adoption of green computing. In: Academy of economic studies in Bucharest. department of economic informatics and cybernetics https://doi.org/10.12948/ie2019.04.17
2. Masdari M, Zangakani M (2019) Green cloud computing using proactive virtual machine placement: challenges and issues. J Grid Comput 1–33
3. Fernando N, Loke SW, Rahayu W (2013) Mobile cloud computing: a survey. Future Gener Comput Syst 29:84–106
4. Průša D, Werner T (2019) Solving LP relaxations of some np-hard problems is as hard as solving any linear program. SIAM J Opt 29(3):1745–1771
5. Blum C, Li X (2008) Swarm intelligence in optimization. Berlin, Heidelberg, In Swarm intelligence Springer, pp 43–85
6. Gai K, Qiu M, Zhao H, Tao L, Zong Z (2016) Dynamic energy-awarecloudlet-based mobile cloud computing model for green computing. J Network Comput Appl 59:46–54
7. Bhowmik B (2010) Dynamic programming–its principles, applications, strengths, and limitations. Criterion, 4:7
8. Kushal A, Raj P, Naitik P, Tapan P, Anjali T (2019) Dynamic SourceACO algorithm for energy-aware cloudlet-based mobile cloud computing architecture. In: 2019 3rd international conference on computing methodologies and communication (ICCMC), IEEE (March 2019) pp 41–48
9. Wolpert DH, Macready WG (1997) No free lunch theorems for optimization. IEEE Trans Evol Comput 1(1):67–82
10. Kandavel N, Kumaravel A (2019) A novel royal seal cloudlet for security enhancement in mobile cloud computing. Int J Comput Sci Inf Secur (IJCSIS) 17(1)
11. Beloglazov A, Buyya R (2010) Energy efficient resource management in virtualized cloud data centers. In: 2010 10th IEEE/ACM international conference on cluster, cloud and grid computing IEEE 826–831
12. Jain M, Singh V, Rani A (2019) A novel nature-inspired algorithm for optimization: squirrel search algorithm. Swarm Evol Comput 44:148–175
13. Basu M (2019) Squirrel search algorithm for multi-region combined heat andpower economic dispatch incorporating renewable energy sources. Energ 182:296–305
14. El-Ashmawi WH, Elminaam DSA (2019) A modified squirrel search algorithm based on improved best fit heuristic and operator strategy for bin packing problem. Appl Soft Comput 82:105565
15. Saha A, Dey P, Bhattacharya A, Marungsri B (2019) SSA-A New metaheuristic algorithm for solving transient stability constrained optimal power flow
16. Martìnez CAL, Rianõ HEH, de la Vega MJS (2019) Un squirrel search algorithm discreto aplicado al problema Job Shop con operadores calificados. INGE CUC 15
17. Zheng T, Luo W (2019) An improved squirrel search algorithm for optimization. Complexity
18. Wang Y, Du T (2019) An improved squirrel search algorithm for global function optimization. Algorithms 12:80
19. Hu H, Zhang L, Bai Y, Wang P, Tan X (2019) A Hybrid Algorithm based on squirrel search algorithm and invasive weed optimization for optimization. IEEE Access 7:105652–105668
20. Wang Y, Du T (2020) A Multi-objective improved squirrel search algorithm based on decomposition with external population and adaptive weight vectors adjustment. Phys A 542:123526
21. Kumar K, Lu YH (2010) Cloud computing for mobile users: can offloading computation save energy? Computer 43:51–56
22. Fister Jr I, Mlakar U, Brest J, Fister I (2016) A new population-based nature-inspired algorithm every month: is the current era coming to the end. In: Proceedings of the 3rd student computer science research conference, University of Primorska Press 33–37

23. Sörensen K (2015) Metaheuristics—the metaphor exposed. Int Trans Oper Res 22:3–18
24. Anjaria K, Mishra A (2017) Thread scheduling using ant colony optimization: an intelligent scheduling approach towards minimal information leakage. Karbala Int J Mod Sci 3:241–258

Forged Multinational Currency Recognition System Using Convolutional Neural Network

Megha Jadhav, Yogeshkumar Sharma, and Gayatri Bhandari

Abstract Forged multinational currency recognition and identification system is developed to verify that the currency is genuine or fake currency. In current scenario of counterfeiting is affecting the banking sector and other fields as well. This paper introduces as advance convolutinal neural network technique for identification and detection of forged currency banknotes using image dataset. This technique is considered as relatively superior to earlier techniques. Proposed approach uses deep learning algorithm as nowadays deep learning is being measured as incredible achievement in image classification tasks. This approach can also be helpful for normal human being in recognizing a forged currency note through an image in real time of various countries. Currently, this approach is implemented on various denominations of Indian and Saudi Arabian Country. This systems success rate can be measured in terms of both precision and time. The key objective behind this system is to concentrate on those minute parameters that will be adequate to differentiate between forged and original banknotes.

Keywords Convolutional neural network · Identification · Recognition · Deep learning

M. Jadhav · Y. Sharma (✉) · G. Bhandari (✉)
Department of Computer Science and Engineering, Shri JJT University, Chudela, Jhunjhunu, Rajstan, India
e-mail: dr.sharmayogeshkumar@gmail.com

G. Bhandari
e-mail: gayatri.bhandari1980@gmail.com

M. Jadhav
e-mail: meghajadhav004@gmail.com

R. P. Mahapatra et al. (eds.), *Proceedings of 6th International Conference on Recent Trends in Computing*, Lecture Notes in Networks and Systems 177,
https://doi.org/10.1007/978-981-33-4501-0_43

1 Introduction

In today's world, in general, cash money is used in rare case due to the increased use of online mode of transaction, cash transactions are still considered to be major part in the world market. Banknotes are used to carry out this sort of financial activities. In such unavoidable conditions, almost all country faces a major issue of counterfeit identification of banknotes. Banknotes are used for fundraising activities. In such unavoidable conditions, almost each country from all over the world faces a major problem of counterfeit identification of banknotes. Due to their use of highly developed technology, the counterfeiters become more difficult to track down. In old days, generating forge banknote was possible for printing house only, but now any person can print forged banknotes simply at home only by making use of various advances in the field of scanners and a laser printer. Differentiating between genuine and fake banknotes is difficult for a normal person, particularly since they have a lot of similar features and are designed to look exactly the same as a genuine banknote with great precision. The entry into circulation of falsified banknotes should be stopped from carrying out smooth cash transactions. Entry into circulation of forged banknotes should prevent the smooth cash transactions from continuing. Consequently, the problem of effectively distinguishing forged banknotes from genuine or fake is becoming increasingly important as it causes great loss and damage to society and also affects the economy of almost all countries. That enforcement requires an effective algorithm that can accurately predict whether a banknote is genuine or not and is easily accessible and efficient.

In deep learning, the proposed system is developed to distinguish whether the currency is genuine or fake through the use of a neural network. Deep learning is a function of artificial intelligence which works similar to human brain for processing data and to create pattern for making use of decision [1].

It may be a step in the recent demonetization drive toward eliminating fraud and black money but the problem of counterfeit currency is not yet discussed. As shown in Fig. 1, in general, neural network is a computational model which operates similar to human brain neurons. Growing neuron takes an input, performs some operations on it, and moves the output to its hidden layer afterwards [2, 3].

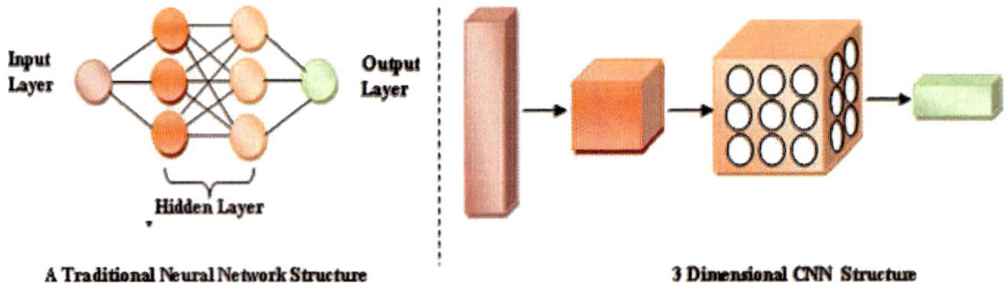

Fig. 1 Neural network structure

2 Literature Survey

Several techniques were proposed by various authors for Forged Banknote Detection. In this scenario, a brief evaluation of some important contributors to existing literature is presented shortly.

Gouri Sanjay et al. (2018) presented the paper 'Detection of Fake Indian Currency' using various minutiae parameters just like a deep learning algorithm. This will be achieved using image processing techniques such as CNN, pattern recognition, SURF and some algorithms, which will be adequate to recognize the difference between fake and original currency notes. The suggested solution extracts multiple features from Indian currency and makes use of them to identify fake currency [4].

The paper presented by Mohammad H Alshayeji, 'Detection System for Counterfeit Currency Based on Bit-Plane Slicing Technique,' has a technique to extract the most important data from fake banknote images using an edge detector algorithm, using the bit-plane slicing technique. This technique is to decompose 256 grayscale original images into their equivalent of eight binary images. Using canny edge detection algorithm upper order bit rate is calculated for grayscale banknote image [5].

Da-Costa presented the paper 'A Bank Note Recognition System for Recognizing Multiple Banknotes in Various Perspectives and Scale.' He introduced detection, description, and matching features that are exercised in recognizing output to increase confidence. Yet because of high processing power, the proposed model is not suitable for mobile phones [6].

Taking above survey into account analyzed that we should pick algorithms in terms of time complexity and accuracy which should be able to handle large amounts of dataset efficiently. CNN's advantage over image processing is it is a type of neural network for input data processing that has a grid like topology inherent in it. CNN is defined as a neural network with convolution in at least one of layers instead of general matrix multiplication.

A convolutinal neural network helps to search important patterns which help to get more appropriate results. While the image processing disadvantages include primary cost may be high depending on the device used for different purpose like battery consumption, high cost, etc. [7].

3 Proposed System

To build a model for identification of forged banknote, we simulated operation using CNN.

3.1 Suggested Solution

We created a convolutional neural network which is trained on fake and original currency dataset based on the proposed algorithm and can be tested whether the currency is genuine or fake. By training an artificial neural network on currency image dataset, testing can be done by making use of neural network to determine which class the image belongs to when it sees an image with genuine or fake note next time. Convolutional neural network is widely used in pattern recognition and image recognition problems today [8, 9]. In Convolution, neural networks generally make use of min 5 to max 25 layers of pattern recognition distinct. As in CNN, the feature extraction and characteristic classification by default in one step, raw data is taken as input for first time independent preprocessing.

3.2 CNN Architecture

In Internet from lot of pre-trained available models, Alex Net is chosen to incorporate and train the deep learning in proposed approach. For this purpose, we need to install the webcam and Alex Net add-ons using Add-On Explorer. After installing Deep Learning Toolbox Model for Alex Net, we can use it to classify images. Alex Net is a pre-trained convolutional neural network that can train on one million images and categorize images into thousands of categories [10]. For Alex net, it is superior, as it has 12 layers with parameters like weight and bias. In such extremely deep networks, the total number of parameters with phase in all layers of convolution is set to 1. The Alex Net uses multiple 3×3 convolution layers to represent the properties of the complements (Fig. 2).

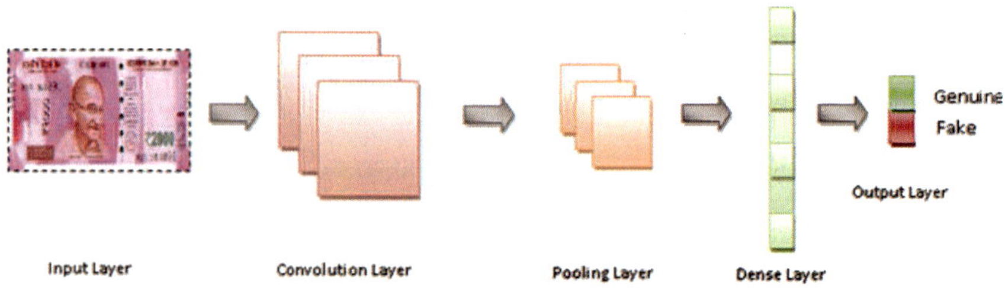

Fig. 2 Convolutional neural network architecture

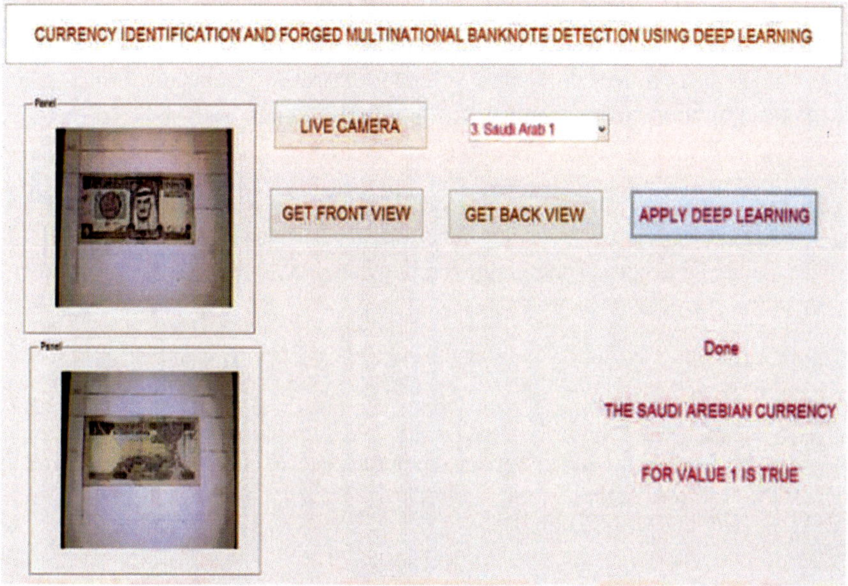

Fig. 3 Saudi currency original note 1 Saudi Riyal

3.3 Data-Set

We have collected dataset of notes of various countries' denominations. Dataset was built for front side view and backside view of note. The photos were generated using different sources like Internet or 'Children Bank of India' and 'Churan Lable'. Later on this dataset, genuine notes and forged notes are separated into two parts (Fig. 3).

In this, we exercised 70% of the images in the dataset for training purposes and the remaining 30% for the benefit of the analysis. After these, training images are supplied to the model as input, and the model is trained accordingly.

As discussed above, we trained our model. Next step following to training and testing performed is set epoch, that raise precision by producing value of epoch.

3.4 Preprocessing

Preprocess dataset is the easiest way to get that data without downloading above and below fitting. Key benefit of pre-processing dataset is to add value to base value generated by the dataset and get better training set. The Alex Net input is an RGB 256 × 256 image. Then apply the following kinds of image preprocessing to the initial datasets [10–12].

3.4.1 Re-scaling

Input image need to rescale picture to render dataset in a standard format, which helps to do training faster and reliable.

3.4.2 Shearing

Shave the image to better and reliable training dataset. Share range is in degree with shear angle in clockwise direction.

3.4.3 Transformations

A perspective transformation is applied on training data to zoom in image to correctly obtain the accurate results.

3.5 Training Phase

After choosing Alex Net which is pertained network, we fine-tuned it which speed up the process of preparation. In order to train CNN, generally it is good to begin with a poorly network and get network with improved accuracy. At the end of training, we want loss level very lower to achieve improved accuracy. This means that our neural network has a high accuracy learning efficiency. Even if there are several algorithms that perform activation and optimization, the use of ReLu Nonlinearity is an important feature of Alex Net. Activation functions Tanh or sigmoid used to be the standard way to train a model of a neural network. Alex Net showed that deep CNNs could be trained much faster using nonlinearity of ReLU than using the saturating activation functions such as tanh or sigmoid. 'Adam' is taken as optimization method, because by speeding up training it enhances the neural networks. And it is also easy to compute a ReLU move. In order to conclude input image belongs to genuine or fake, Alex Net model's accuracy fine-tuned with the created dataset is observed. On the corresponding test set, it was approximately 55%, while in initial phase by considering less, i.e. up to 250 images; results are good. When image dataset is increased by means of real-world samples, the model can be trained with high accurately that helps to get improved result. Notwithstanding the perfect score, the outcome could be due to dataset over fitting which enforce the need of examining accuracy and failure pattern after preprocessing by adjusting parameters [10, 12].

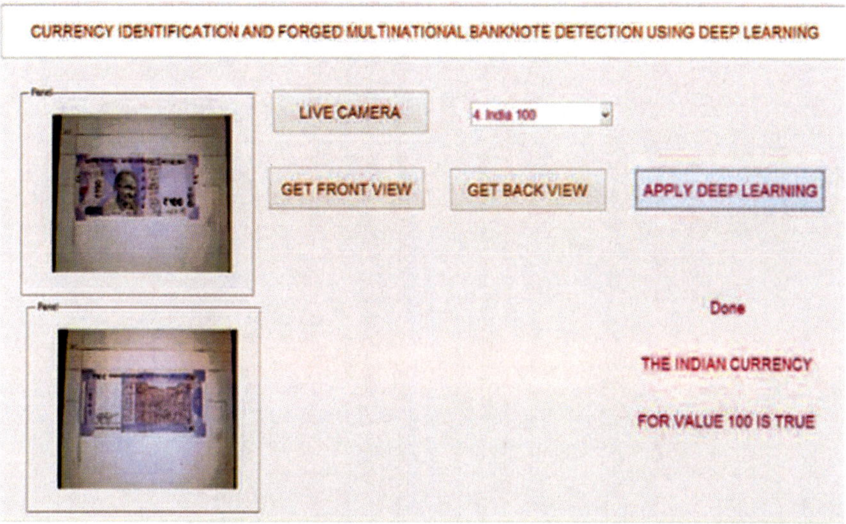

Fig. 4 Indian currency value of 100 rupee

4 Results and Analysis

We may examine the performance patterns in terms of loss trends as training and validation, respectively, and accuracy trends also as training and validation, respectively. Our model is said to be under-fitted if training loss is greater than validation loss, and training accuracy is lower than validation accuracy and over-fitted if training loss is lower than validation loss, and training accuracy is greater than the validation accuracy [10, 13]. So with respect to above analysis, our model is said to be over-fitted. After understanding the fitted model from the above study, we recognized the currency shown in the following figures of currency Saudi Arabian Saudi riyal note 1,10 and Indian rupee note 100, 500 is as captured from our experimental model, capable of clearly identifying currency with denominations and distinguish between fake and genuine (Figs. 4, 5 and 6).

5 Conclusion

In this paper, the model suggested, shows the probability of using CNN with an Alex Net. Even though considered dataset is of less images and not compare to the forged currency data-set's real world situation, it is considered very positive throughout the experiment. In trained model, the method of differentiating between genuine and fake note is quick and simple. In addition, through these experiments, we can ensure that the proposed model can be properly trained under the real and huge dataset and

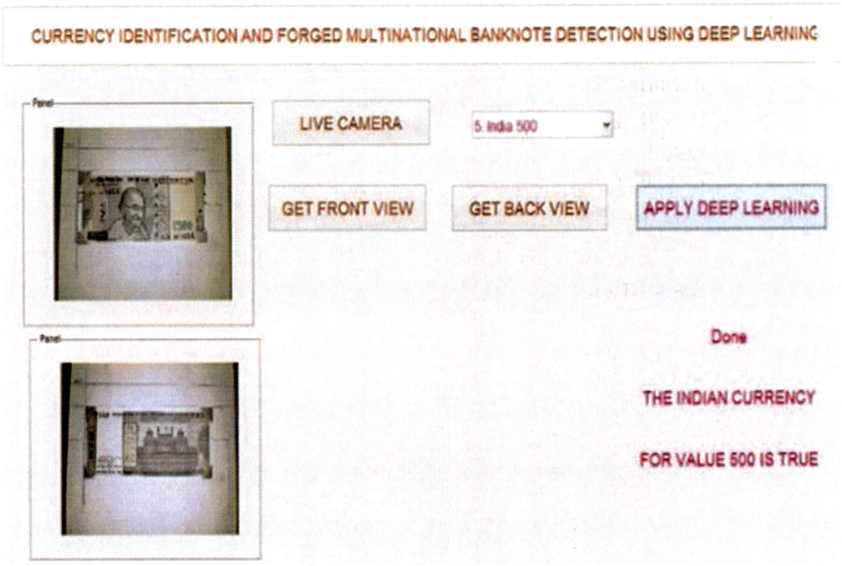

Fig. 5 Indian currency value of 500 rupee

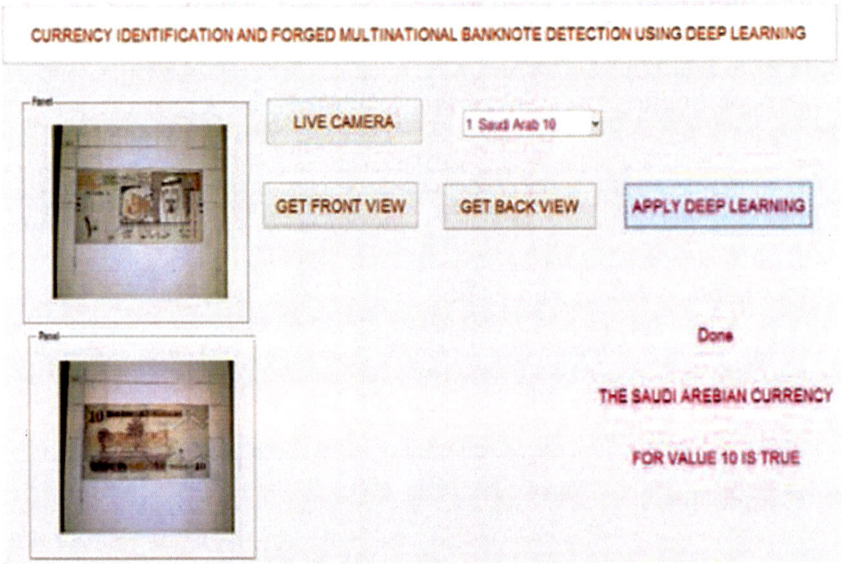

Fig. 6 Saudi currency original note 10 Saudi Riyal

also provides perfect results that can allow people to recognize the currency notes as counterfeit or genuine. Future research may include implementing the model as an application in the smart phone and making it easier for society and people to recognize the counterfeit currency. The findings of proposed model can also be tested for specific CNN which will have higher error rate as compare to this and improved using other pre-processing technique edge detection to cut image which will helps to yield superior result.

References

1. Mirza R, Nanda V (2012) Design and implementation of Indian paper currency authentication system based on feature extraction by edge based segmentation using Sobel, operator. Int J Eng Res Dev 3:41–46
2. Nagpure PR, Ghotkar T, Shetty S (2016) Recognition and fake note detection. Int J Innov Res Comput Commun Eng 4
3. Nishikage T, Takeda F (2016) Multiple kinds of paper currency recognition using neural network and application for Euro currency. 27–47
4. Tele GS, Mahakalkar S, Kathalkar AP, Sahoo B, Dhamane V (2018) Detection of fake Indian currency. Int J Adv Res Ideas Innov Technol 4(2):170–176
5. Arshad SM, Sudagar DS, Nausheeda B (2017) Fake Indian currency detection using image processing. Int J Latest Trends Eng Technol 598–602
6. Da-costa (2014) Multiview banknote recognition with component and shape analysis
7. Akamatsu N, Fukumi M (1996) A method to design a neural pattern recognition system by using a genetic algorithm with partial fitness and a deterministic mutation. In: IEEE conference
8. Gunaratna DAKS, Kodikara ND, Premaratne HL (2008) ANN based currency recognition system using compressed gray scale and application for Sri Lankan currency notes-SLCRec. World Acad Sci Eng Technol Int J Comput Inf Eng 2:2957–2962
9. Chowdhury A, Jahangir N (2007) Bangladeshi banknote recognition by neural network with axis symmetrical masks in masks. In: 10th international conference on computer and information technology
10. Navya Krishna G, Sai Pooja G, Naga Sri Ram B, Yamini Radha V, Rajarajeswari P (2019) Recognition of fake currency note using convolutional neural networks. Int J Innov Technol Explor Eng 8:58–61. ISSN: 2278-3075
11. Gidveer P, Darade SR (2016) Automatic recognition of fake Indian currency note. In: International conference on electrical power and energy systems (ICEPES), Bhopal, India, pp 290–294
12. Sharma YK (2018) Security-in-digital-images-using-visualcryptography-scheme. J Comput Inf Syst 14(6):49–57. ISSN: 1553-9105
13. Sharma YK (2018) An empirical research framework for encryption authentication with image and video. J Adv Res Dyn Control Syst 10(8):59–71. ISSN: 1943-023X

A State of the Art Bibliometric Analysis of Predictive Maintenance: A Quantitative Analysis

Amitkumar Patil and Gunjan Soni

Abstract A bibliometric analysis of publications associated with Predictive Maintenance (PM) is shown in this article. It is a quantitative analysis of publications associated with PM throughout the history. For encompassing all the important and relevant literature, Web of Science, one of the largest databases of peer-reviewed research publications, is used as source of quantitative data. Quantitative insights based on this analysis are discussed in this article.

Keywords Predictive maintenance · Prognostics · Diagnosis · Bibliometric analysis

1 Introduction

Predictive Maintenance (PM) also known as condition-based monitoring is a popular maintenance technique which helps in predicting the failure of a system even before it occurs. Condition-based monitoring is a maintenance strategy that utilizes the data obtained while monitoring the condition of a system or system components to propose maintenance for it [1]. It is used to ensure certain level of reliability during system life. Predictive maintenance follows the following fundamental steps: Collect data from the system or its components without affecting the running operations, diagnose and analyze trends regarding the considered physical asset and identify potential failures through previous analysis of similar system. Jardine et al. [2] have categorized condition-based monitoring in three steps: (1) Data acquisition, (2) Data processing, and (3) Maintenance decision-making. In this process, diagnostics and prognostics are also crucial parts of a predictive maintenance policy. Diagnostics is concerned with fault detection, fault isolation, and fault identification when it occurs while

A. Patil (✉) · G. Soni
Malaviya National Institute of Technology Jaipur, Jaipur, Rajasthan, India
e-mail: 2019rme9109@mnit.ac.in

G. Soni
e-mail: gsoni.mech@mnit.ac.in

R. P. Mahapatra et al. (eds.), *Proceedings of 6th International Conference on Recent Trends in Computing*, Lecture Notes in Networks and Systems 177,
https://doi.org/10.1007/978-981-33-4501-0_44

prognostics is predicting the time when an impending failure will occur. Ahmad and Kamaruddin [3] present comparison of time-based maintenance and condition-based maintenance. It is helpful in understanding the necessity and applications of both maintenance strategies. Heng et al. [4], Alaswad and Xiang [5], Garg and Deshmukh [6], Goyal and Pabla [7], Lee et al. [8], Si et al. [9], Taylor et al. [10] give the overview of PM or condition-based maintenance techniques, applications and potential aspects of future research. Jardine et al. [2] and Heng et al. [11] show the various techniques of combining different sensors data to get optimum results. Prognostics and diagnostics of faults in machine components or any system is generally done to find Remaining Useful Life (RUL). Si et al. [9], Liao [12], Sikorska et al. [13] be referred to gain insights of various approaches to find RUL of a system. For the last few years, majority of the research related to PM or condition-based monitoring is done for rotary components such as gears, bearings, etc., only. To guide future research in a progressive direction, Sikorska et al. [13] have presented a detailed review of past and recent researches in prognostics.

Most of the times review articles are focused on a specific theme of research area but for a widely spread research areas such as PM, depicting an overall picture of research area becomes necessary to guide new researchers and find new research gaps. This article fulfills this objective in a very concise manner.

In this paper, a quantitative analysis of development of PM throughout the history is performed. The following sections are as follows: First source of quantitative data is mentioned followed by quantitative analysis of the publications associated with PM categorized based on year, research area, document type, source title, author, and country is presented. In the end, discussions and conclusion are presented.

2 Source of Quantitative Data

Data for this quantitative analysis has been collected till October 25 2019 from Web of Science, which is one of the largest databases of peer-reviewed scholarly journal, proceedings and books in sciences, humanities, social sciences, and arts, etc., (Web of Science Core Collection, 2019). On the basis of multiple aspects of this data, quantitative analysis has been presented in following sections.

3 Quantitative Analysis of Publications Related to Predictive Maintenance

3.1 Statistics: Year-Wise

With "Predictive Maintenance" or "Condition-Based Monitoring" or "Diagnostics and Prognostics" keywords, 4941 publications related to PM were extracted from

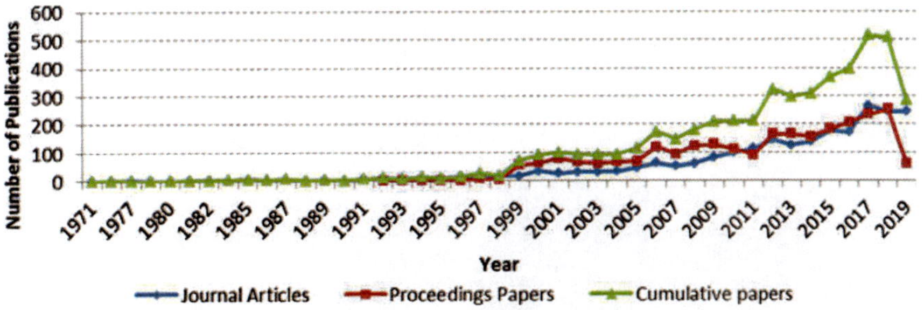

Fig. 1 Annual trend of quantity of documents related to PM (1970–2019)

Web of Science database (till October 25 2019). It was since year 1998 when a significant amount of increase in quantity of publications was noticed. 2017 and 2018 were the crucial years as 514 and 506 publications, were published each year, respectively, cumulating to 20% of all the publications associated with PM. From year 2009 to 2018, on an average 334 publication related to PM have been published. Year 2017 and 2012 recorded highest increments of 119 and 111 publications, respectively, (Fig. 1).

3.2 Statistics: Research Area-Wise

Of 4941 publications related to predictive maintenance, more than 71%, 3534 documents are related to the research field of engineering (Fig. 2). Top three fields in which large number of publications has been published are engineering, computer science, and operations research management science with 3534, 1033, and 553 publications, respectively. Majority of publications are interdisciplinary; hence, summation of number of publications does not add up to 4941. This is why top three fields represent lesser share in Fig. 2.

3.3 Statistics: Document Type-Wise

Table 1 and Fig. 3 show quantity of publications related to PM in different types of documents such as proceeding papers, journal articles, review papers, editorial materials, book chapters, and others. Table 1 shows that highest number of documents associated with PM is proceedings papers and journal articles with 2549 and 2314 counts, respectively, cumulates to 95% of all 4941 publications while review papers, editorial material, book chapters, meeting abstracts, news items, letters, etc., constitute rest 5%. Figure 1 shows annual trends of journal articles and proceedings papers from year 1970 to 2019. First proceedings paper was published in 1992 while

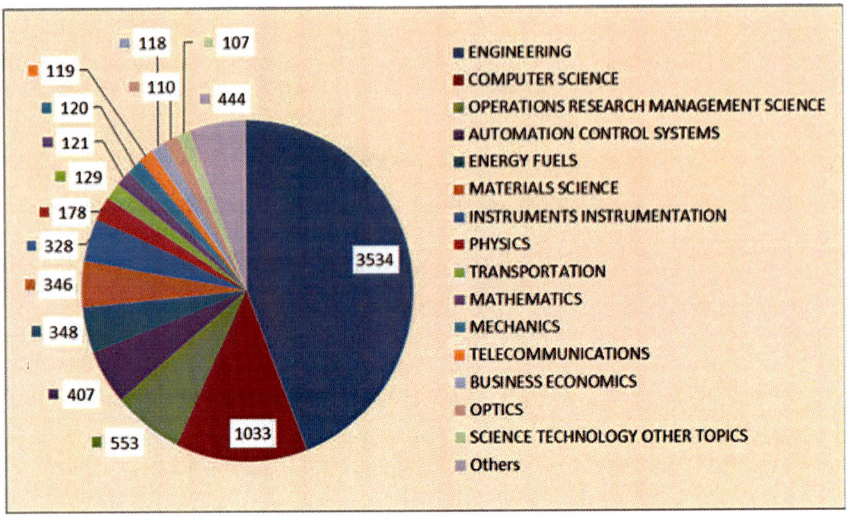

Fig. 2 Publications statistics: research area-wise

Table 1 Publication statistics: document type-wise

S. No.	Document type	Qty. of publications
1.	Proceedings papers	2549
2.	Journal articles	2314
3.	Review papers	129
4.	Editorial material	55
5.	Book chapter	46
6.	Meeting abstract	15
7.	News item	14
8.	Letter	9
9.	Early access	7
10.	Book	2
11.	Correction	1
12.	Correction addition	1
13.	Data paper	1
14.	Note	1

journal articles were begun to be published in year 1970. Proceedings papers have shown higher increment since 1999–2018 compared to journal articles which shows steady growth over the years.

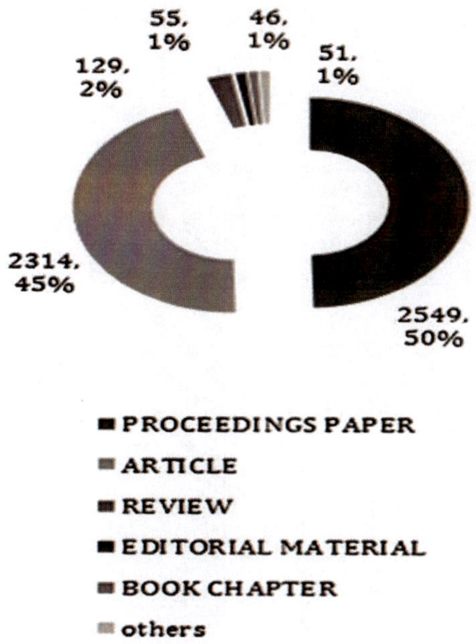

Fig. 3 Publication statistics: document type-wise

3.4 *Statistics: Source Title-Wise*

Top 15 source titles which published publications related to PM in Web of Science database are shown in Fig. 4. Out of top 15, source titles six are standard journals, i.e., reliability engineering system safety, mechanical systems and signal processing, journal of quality in maintenance engineering, international journal of advanced manufacturing technology, computers industrial engineering and international journal of production research while others are conference proceedings. Figure 5 shows trend of top 5 journal articles over the year.

It can be noticed from Fig. 5 that computers industrial engineering, mechanical systems, and signal processing and reliability engineering system safety are one the oldest journals in the field of PM. Reliability engineering system safety and mechanical systems and signal processing source titles have published on an average around six and four documents per year, respectively. Computers industrial engineering has shown steady growth till year 2011, after which it has shown significant increase in quantity of documents published associated with PM.

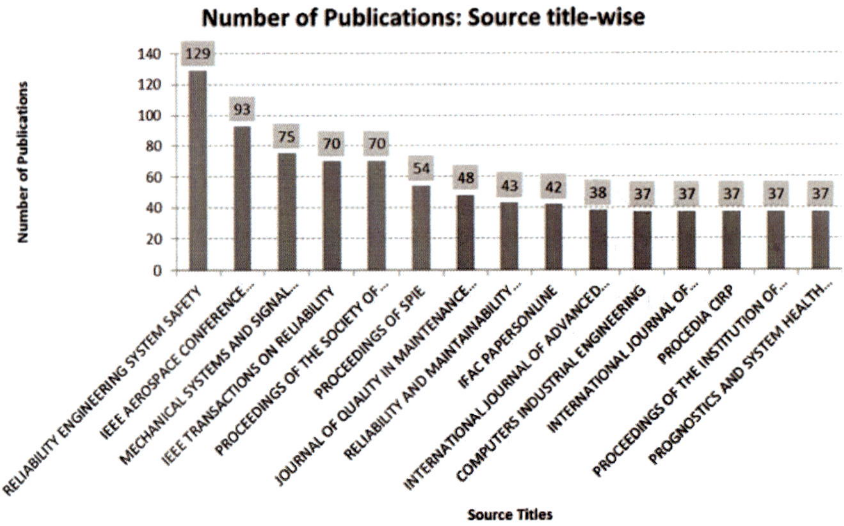

Fig. 4 Publications statistics: source title-wise

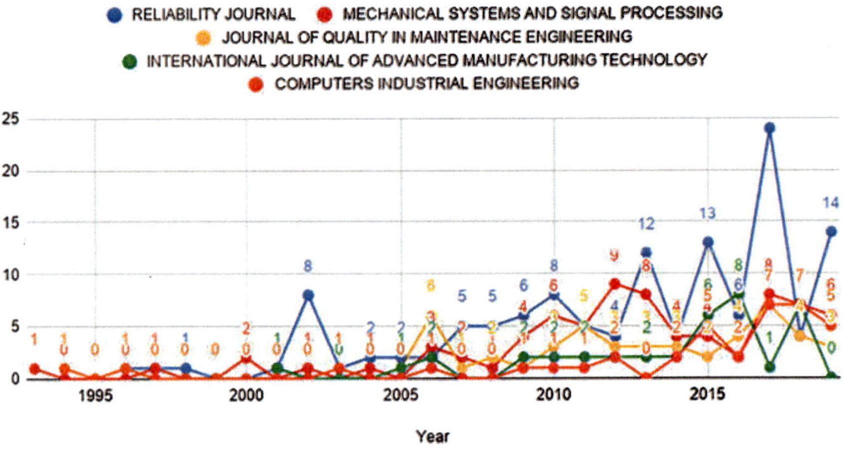

Fig. 5 Annual trends of publications associated with PM: top 5 source titles (1993–2019)

3.5 *Statistics: Author-Wise*

Figure 6 represents top 15 authors who published highest number of publications related to PM.

From Fig. 6, it can be noticed that top five authors, namely Berenguer C (from Univ. de Technologie de Troyes, France), Antoine Grall (from Univ. de Technologie de Troyes, France), Wang WB (from University of Science and Technology

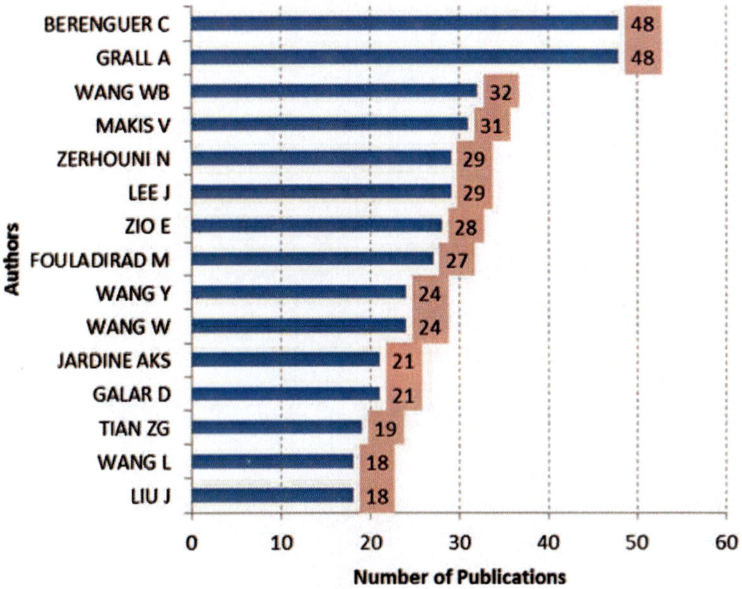

Fig. 6 Publications statistics: author-wise

Beijing, Beijing, PR China), Makis V (from University of Toronto, Toronto, Canada), and Zerhouni N (from National High School of Mechanics and Microtechniques, Besançon, France) have published 48, 48, 32, 31, and 29 publications, respectively. From the data collected for this analysis, 36 documents are anonymous.

4 Summary

Predictive maintenance or condition-based monitoring is four decades old technique widely used for failure diagnosis of systems. As it can be noticed in Sect. 3.1 that significant amount of publications (4941 publications) has been published since year 1970–2019. Major portion of documents published on PM are in the field of engineering, computer science, and operations research management science, as seen in (Fig. 2). Projections of quantity of publications related to PM from Fig. 1 indicate that in future, also number of publications is going to increase. Given that journal articles are more significant than conference proceedings papers, quantitative analysis of journal publications is discussed next. Among the top 15 source titles that published documents associated with PM, reliability engineering system safety has shown significantly higher growth and recorded highest number of publications, too (Figs. 4 and 5).

As seen from Fig. 7, Berenguer C and Wang WB started publishing their work

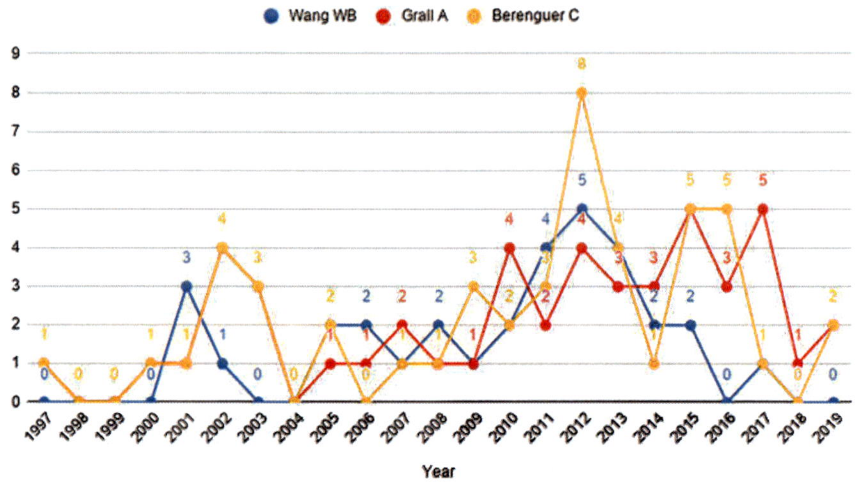

Fig. 7 Annual trend of publications associated with PM: top three authors (1997–2019)

in same year (1997) but the former shows consistency in his work compared to the other one. In contrast to that Grall A started publishing since year 2004, yet with much higher growth rate, he equaled total quantity of publications to Berenguer C (Fig. 7).

5 Conclusion

This article presents a bibliometric analysis of publications associated with Predictive Maintenance (PM). For beginners, it helps to identify prominent authors, journal titles, emerging research themes, and trends of publications throughout history of PM. For the past decade, number of proceeding papers associated with PM has increased which shows that researchers are trying to explore untapped research gaps through conferences.

References

1. Wakiru JM, Pintelon L, Muchiri PN, Chemweno PK (2019) A review on lubricant condition monitoring information analysis for maintenance decision support. Mech Syst Signal Process 118:108–132. https://doi.org/10.1016/j.ymssp.2018.08.039
2. Jardine AKS, Lin D, Banjevic D (2006) A review on machinery diagnostics and prognostics implementing condition-based maintenance. Mech Syst Signal Process 20:1483–1510. https://doi.org/10.1016/j.ymssp.2005.09.012
3. Ahmad R, Kamaruddin S (2012) Computers & Industrial Engineering An overview of time-based and condition-based maintenance in industrial application. Comput Ind Eng 63:135–149.

https://doi.org/10.1016/j.cie.2012.02.002
4. Heng A, Zhang S, Tan ACC, Mathew J (2009) Rotating machinery prognostics: state of the art, challenges and opportunities. Mech Syst Signal Process 23:724–739. https://doi.org/10.1016/j.ymssp.2008.06.009
5. Alaswad S, Xiang Y (2017) A review on condition-based maintenance optimization models for stochastically deteriorating system. Reliab Eng Syst Saf 157:54–63. https://doi.org/10.1016/j.ress.2016.08.009
6. Garg A, Deshmukh SG (2006) Maintenance management: literature review and directions. J Qual Maint Eng. https://doi.org/10.1108/13552510610685075
7. Goyal D, Pabla BS (2015) Condition based maintenance of machine tools—a review. CIRP J Manuf Sci Technol. https://doi.org/10.1016/j.cirpj.2015.05.004
8. Lee J, Kao H, Yang S (2014) Service innovation and smart analytics for Industry 4.0 and big data environment. Procedia CIRP 16:3–8. https://doi.org/10.1016/j.procir.2014.02.001
9. Si X, Wang W, Hu C, Zhou D (2011) Remaining useful life estimation—a review on the statistical data driven approaches. Eur J Oper Res 213:1–14. https://doi.org/10.1016/j.ejor.2010.11.018
10. Taylor P, Katipamula S, Brambley MR, Brambley MR (2011) Review article: methods for fault detection, diagnostics, and prognostics for building systems—a review. Part I Methods for fault detection, diagnostics, and prognostics for building systems—a review, Part I, pp 37–41
11. Corres M, Goicoechea J, Rodr Y (2010) Vibration detection using optical fiber sensors. J Sens 2010:10–17. https://doi.org/10.1155/2010/936487
12. Liao L (2014) Review of hybrid prognostics approaches for remaining useful life prediction of engineered systems, and an application to battery life prediction. IEEE Trans Reliab 63:191–207
13. Sikorska JZ, Hodkiewicz M, Ma L (2011) Prognostic modelling options for remaining useful life estimation by industry. IEEE Trans Reliab 25:1803–1836. https://doi.org/10.1016/j.ymssp.2010.11.018

An Approach for Integrating Enterprise-Level Applications

Ganesh Tanpure, Rakesh Jain, and Gunjan Soni

Abstract The manufacturing organizations focusing of sustainability of business have been adopting IT systems for achieving excellence of different business functions. However, multiple applications co-existing for achieving optimum results at specific function level are not desired while considering optimum deliverables of entire organization. The different applications hence need to be connected to each other in a systematic way to ensure that the end goal is achieved. The research focuses on identifying the integration needs and providing a methodology for achieving this. The authors have analyzed existing literature and available integration technologies for providing the approach. There integrations need to be categorized based on the requirement. These requirements need to be explored and validated through systematic research methodology. The authors have adopted a case-study-based approach in an automotive component manufacturing industry wherein the proposed methodology was validated by implementing CAD/CAM/CAPP system integration along with Requirement Management and Product lifecycle management (PLM) systems.

Keywords Integration · Manufacturing systems · ERP · PLM

1 Introduction

The competitive environment in today's world is witnessing challenges due to not only high product customization and scale of production but also due to types of information and sources, complexity of tasks, degree of their incompleteness or ambiguity inter-relations [1]. The continuous collaboration with different entities

G. Tanpure (✉) · R. Jain · G. Soni
Malviya National Institute of Technology Jaipur, Jaipur 302017, India
e-mail: 2015rme9037@mnit.ac.in

R. Jain
e-mail: rjain.mech@mnit.ac.in

G. Soni
e-mail: gsoni.mech@mnit.ac.in

R. P. Mahapatra et al. (eds.), *Proceedings of 6th International Conference on Recent Trends in Computing*, Lecture Notes in Networks and Systems 177,
https://doi.org/10.1007/978-981-33-4501-0_45

starting from early design stages of development is required and continues during mass production of the actual components. During this activity, the business perspective about the functional requirements is very important. Along with any such system concerns related the data security and authenticity of the data also arises. There are also multiple factors thus posing constraints on the systems. Thus, it needs carefully designed methodology to respond to current market environment.

With increased IT interference with manufacturing, there have been different systems which have facilitated the complex tasks of managing the product and process related information. The existing literature has provided insights into the increasing interest of manufacturing industries to adopt various IT applications. A study showed that Enterprise Resource Planning (ERP), Product Lifecycle Management (PLM), Supply chain Management (SCM), E-Business and many others are priority implementation industries [2]. These IT systems need to be aligned in order to have smooth collaborative setup [3]. Each of these applications is developed to focus and optimize on a specific function within the organization and has small overlap with other functions. To demonstrate this, we could analyze the capabilities offered by ERP systems in the domain of Manufacturing and Execution, such a system provides robust decision support, planning and control capabilities [4, 5] however is not equipped to manage the changes related to product data level. Thus, it becomes critical for an organization to synchronize the different process focused applications to derive optimum business focused results.

This research article is focused on following research questions,

1. Which are the important enterprise-level systems analyzed by extant literature?
2. Is there need for synchronizing different systems?
3. Could there be a framework for providing integrated systems approach?

The research questions identified above are motivated through the industry based studies. The enterprise-level applications are not being explored to benefit the organizational performance [6]. There is also correlation established between the industrial performance in terms of sales with the degree to which adoption of Information Technology practices has helped the organizations [7].

2 Literature Review

Figure 1 illustrates typical scenario faced by organization in globalized environment. Geographically scattered business and development pose challenges to collaborate across the different functions of organization (Internal Collaboration).

With the advent of many organizations adapting to outsourcing of component development complicates, these challenges much more as this needs smoother collaboration with agencies who are outside the organization. (External Collaboration). Thus, with multiple stakeholders and multiple business processes, there also exists multiple systems developed over the period of time. Thus, the authors adopted literature study to decode the current research work in this domain.

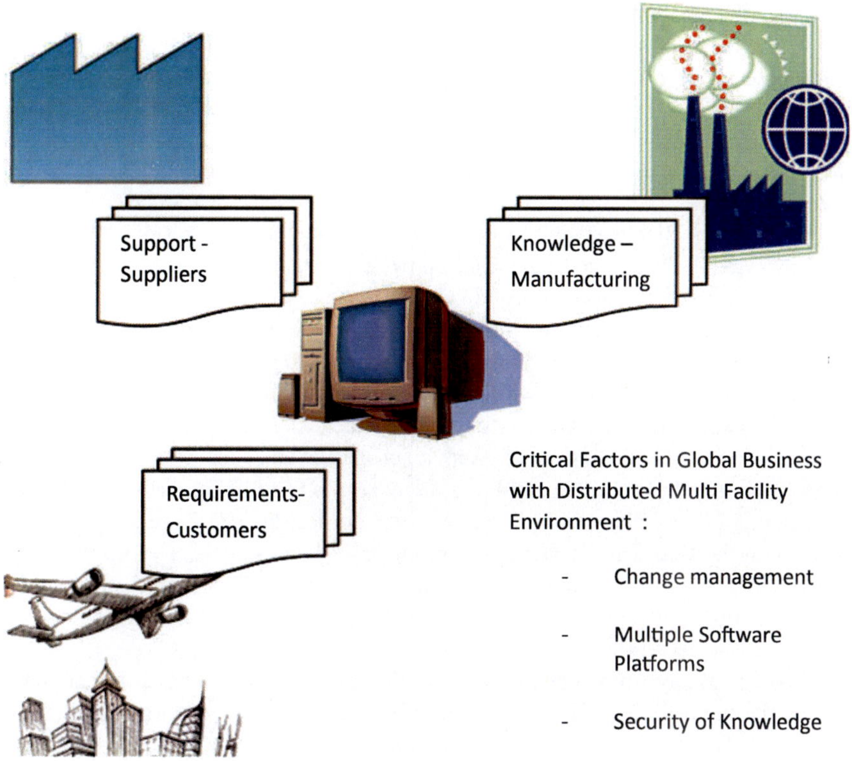

Fig. 1 Global business challenges

The extant literature related to enterprise-level applications adopted by manufacturing industries was studies by referring scientific databases such as Sage, Springer, Emerald, Science direct, Hindawi, IEEE, Inderscience and Scopus. The focused review revealed important applications used in the current industrial environment.

The basic area which needs to be studied prior to integration of different applications is related to data/information exchange. Since variety of applications deal with data in different formats, there exists a need for unified method of data exchange. The capturing of critical information related to product features during inception of the product design activities is an important aspect since it impacts the reusability, accessibility of such information [8]. With advances in Computer Aided Drafting (CAD) and Computer Aided Manufacturing (CAM) applications, there was need to integrate such application which are sub-applications of PLM systems. The use of neutral data formats such as STEP/IGES could provide the solution to integration of such technical issues. The research has reported integration of such system [9]. An important aspect which has generalized integration approach for multiple application is addressed by integration related research [10]. A methodology of adopting Extensible Mark-up Language (xml) for data exchange during integration could address the issue of addressing metadata exchange protocol between different applications. A specific mapping files could be created in order to maintain one to one relationship between information maintained in different systems. An aspect of complete integration by enabling a feedback mechanism is also discussed for certain application [11]. Another important aspect of integration is on the connectivity front, which needs identification of real-time integration or at discrete time intervals or on specific event trigger.

Along with the technology supporting the integration of applications, the authors also reviewed the functionalities expected from the business processes which could complement the integration. Information modeling, computer aided decision support, information system architecture and modeling framework are the areas which need focus. The literature in past [12] has presented an approach of simultaneous product and process design philosophy which is supported by Web-based systems. The bill of materials and process flowcharts are used for analysis of the product. Possibility of usage of internet based integrated system for co-design and concurrent activities have been investigate [13] with an emphasis on two types of tools for such an integrated approach, tools for visualization and tools for co-design. There are multiple tools which can be used for the visualization purpose; however, there is limitation on visualization tools as complexity of model is increasing. The requirement for co-design is fulfilled through different modules for design, manufacturability analysis, etc.

Existing research has focused on multiple activities to address different stages viz. Beginning of Life (BOL), Middle of Life (MOL), and End of Life (EOL). These stages are characterized by different sets activities undertaken by organizations for its products. In order to focus on the integration aspects, there needs to be good understanding related to the different activities occurring for a product during its life cycle. Figure 2 provides such activities already explored by extant research. However, it is important to note that there exists more than one enterprise-level application which plays a role in each of these states.

The analysis of the existing literature revealed an important requirement wherein distinct applications have been identified for integrating with PLM systems. PLM focuses on product data and hence it plays role in all the phases of product life unlike any other application focusing on certain aspects of product life. Also, in order to carve ways for the future technologies which are included in industry 4.0, it is necessary to rely upon a neatly designed product and process information source [14]. Table 1 presents the summarized results of the literature review, wherein the applications are identified for the need of integration for achieving optimum business results. It is evident that although the application listed in table are designed for achieving optimization of a specific function, it is not feasible to rely on local optimum's in business. It is important to integrate such different applications toward a common objective.

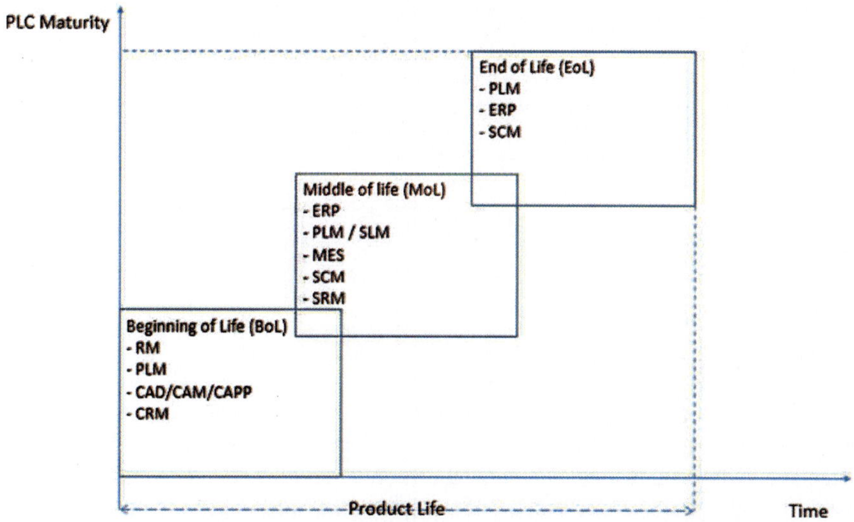

Fig. 2 Important activities in product life cycle though enterprise-level application

Table 1 Integration needs through existing literature

Application	Need of integration in literature	Remarks
Enterprise resource planning	[15–18]	These articles have identified the need to integrate with PLM systems in order to provide efficient/collaborative functioning
Manufacturing execution systems	[16, 18]	
Customer relationship management	[15–18]	
Supplier relationship management	[16]	
Supply chain management	[16, 17, 20]	
Service lifecycle management	[19]	

3 Methodology

The section provides focus on the integration methodology to be adopted for distinct software systems. The idea of single piece of software could not be adopted due to its restrictive nature [21] thus the authors focused on identifying critical areas for developing integration between each layer of application. There are multiple factors which can be considered for EAI approaches [22]; the authors have utilized the ones critical for the applications under considerations. The extant literature already identifies critical success factors for integration [23]. However, since the focus is developing and

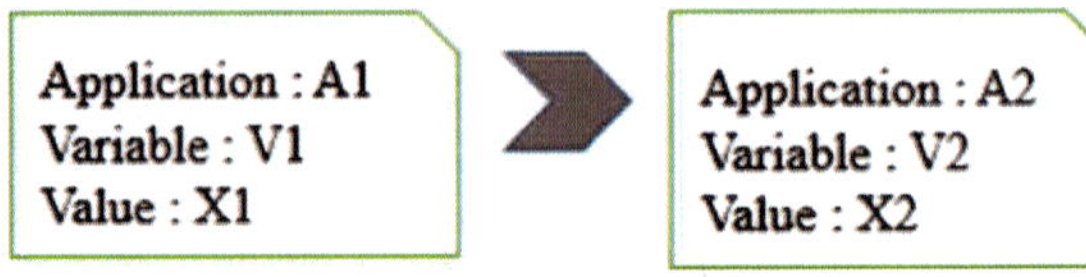

Fig. 3 Mapping of metadata

integration between existing pieces of systems relevant factors such as "IT-Business alignment" and "Technical infrastructure" were more relevant. The subject matter expert's opinion from industry was adopted to derive following parameters as basis of integration. It might be

1. **Integration technology**

These parameters are important to develop the interface between the integrating applications. The literature review has revealed the capability of utilizing formats such as STEP; IGES for such integration. However, there are also matured products available which could be readily configured for integration. There are also inbuilt adapters developed by application developers to integrate with other applications developed by same organization.

2. **Identify metadata/content to be transferred**

The information to be transferred via integrating systems needs to be identified and this activity is dependent on the business requirements. It is not mandatory to transfer all the available information in one system to be transferred to downstream application. However, it is important to identify the conversion rules required to be applied while making any integration, this process is termed as "mapping" in EAI approach.

Figure 3 describes the scenario while two integrating applications are to be considered. The transformation of information from one application to another is the area of concern which is addressed through this parameter.

3. **Feedback**

The authors also provided close attention to the frequency of transfer. The real-time transfer of data is possible in case the data is small; however, for large chunk of data approach needs to be planned based on requirement of availability. Techniques such as asynchronous or response based. This analysis helped in fulfilling resource requirement in terms of system hardware configuration.

4. **Mode of integration**

While analyzing the literature, the authors have understood the relationship between applications follows a predefined design. Based on the functional need, the applications to application integration are defined. The different systems identified for integration requirements need to be explored systematically for defining the integration level. A typical application and its relationship with its components and with the

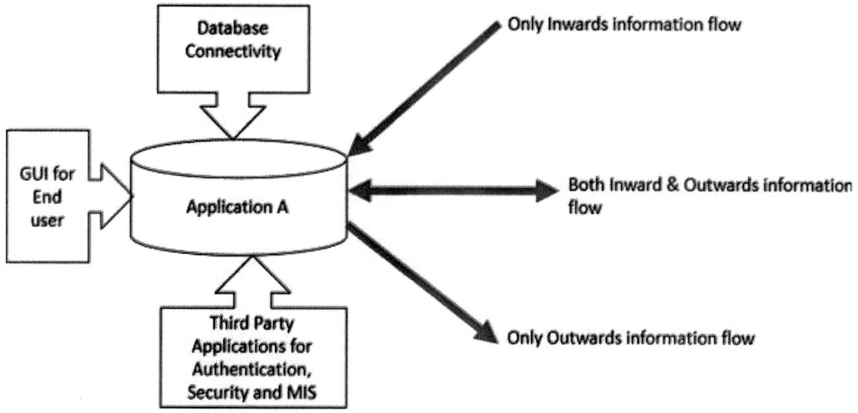

Fig. 4 Application integration design

external enterprise-level application are demonstrated by the relationship provided in Fig. 4.

Based on the integrating application, each of the integration approach could be explained as below. The master/slave communication relationship could be explored for integrating applications.

1. Only Inward Information Flow: This is a typical example of "Slave" application. The applications which are at "Slave" level would only be following the information received from master application. These applications are mostly focused on execution tasks based on the signal received.
2. Only Outward Information flow: This is a typical example of "Master" Application. The applications which are at "Master" level, would only be providing signals or information to slave applications for execution. These applications would be enriched with analytical abilities to synthesize information flowing in from single or multiple sources and break it down to simple instructions for slave applications.
3. Both Inward and Outward Information flow: These are truly collaborative applications. In a sense, these applications could still be characterized by certain aspects of "Master/Slave" communication model; however, they would still have a feedback mechanism which makes it different than typical "Master/Slave" configuration. Each of the integrating application is aware of the processes being employed at the other application.

Such a relationship needs to be defined between each of the integrating applications and then selection of suitable platform for integration needs to be adopted.

Since there would be multiple applications involved in an organization, it would be required to identify the level of integration between integrating applications. The case study approach could be helpful in revealing these relationships since case study could be explained as, "empirical inquiry that investigates a contemporary phenomenon within its real-life context, especially when the boundaries between phenomenon and context are not clearly evident" [24]. Thus, the authors would make an attempt to relate the academic research and real-life industry practice in order to establish a framework connecting these two. An interview with subject matter expert and case study-based study would be adopted to provide a framework of integration between important enterprise-level applications identified through the literature review.

4 Case Study

The automotive component manufacturer is one of the market leaders in Indian context and has operations in multiple locations throughout the world. The organizations had envisioned the importance of adopting IT systems to overcome the challenge of managing increasingly complex products. The interaction between different functions was characterized by communication gaps and thus it reduced the ability to respond to customer requirements in swift and flawless manner. The top management had willingness to deploy a fix for this solution and the middle management was playing active role during the implementation. The actual implementation of the entire system stretched over period of 6–8 months and included multiple stakeholders from distinct functions.

Thus, it experienced need for integrating the business through IT platform. The initial system comprised of CAD/CAM/CAPP system, requirement management system integration along with PLM. This approach enabled the organization to map the activity of translation of the customer requirements into product features through product design and process design. The actual realization of this was achieved via manufacturing activity supported by ERP.

Figure 5 describes the theoretical model for information flow from need identification to disposition of the product. It also marks the systems which are used during different phases. Figure 6 provides insights into activities taking place into the integrated system.

It should be noted that the integration was designed based on the parameters discussed in the methodology section and all the three systems are connected in the integrated systems via different protocols as shown in Table 2.

The existing literature has revealed instances of application integration; however, the focus is provided on integrating few applications only. However, the approach proposed by this article covers the entire business processes from inception to disposition of the product. The organization is further planning to increase the scope of integrated systems by integrating the current system with ERP and MES; however, only broad framework is available for this activity. The authors plan to take this activity up in future.

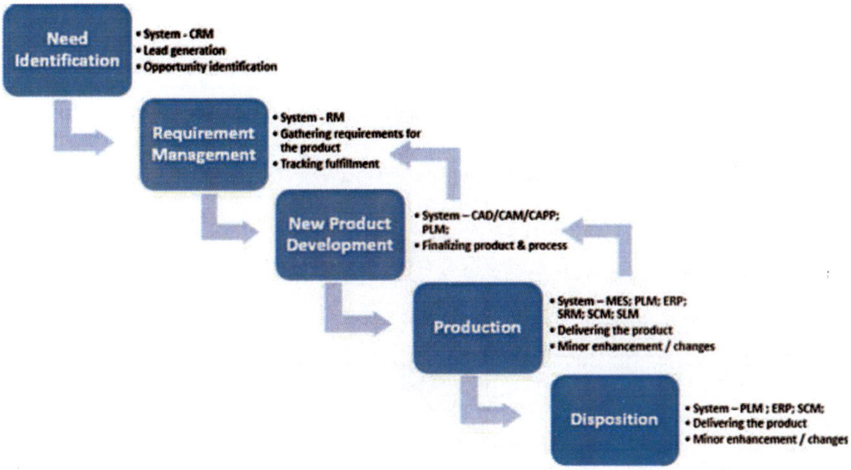

Fig. 5 Theoretical model for integration

5 Results and Discussion

The integrated approach has provided seamless connectivity between different applications. The integrated IT system-based approach has also enhanced first time right (FTR) percentage owing to elimination of manual errors. There are other benefits involved including faster product development cycle; improved traceability of customer requirements to product features. The approach could be extended further to include more application which are being used in the organization; however, it would include efforts from both the IT and business functions to achieve this.

The distinct systems which have been emphasized in literature were identified along with the need for integration. The authors have also provided a theoretical model and validated the results via case study. Thus, the research questions raised during the initial phase are resolved through the approach provided by the article.

6 Conclusion

It has been established through the literature review that there exists a need for integrating different applications and thus the next logical step in this direction would be to explore the relationship further. The extant literature has contributed for exploring the integrations with certain applications however an inception to disposition approach, covering entire business requirements is missing. Thus, the work presented by authors would enable industries to reap maximum benefits from the

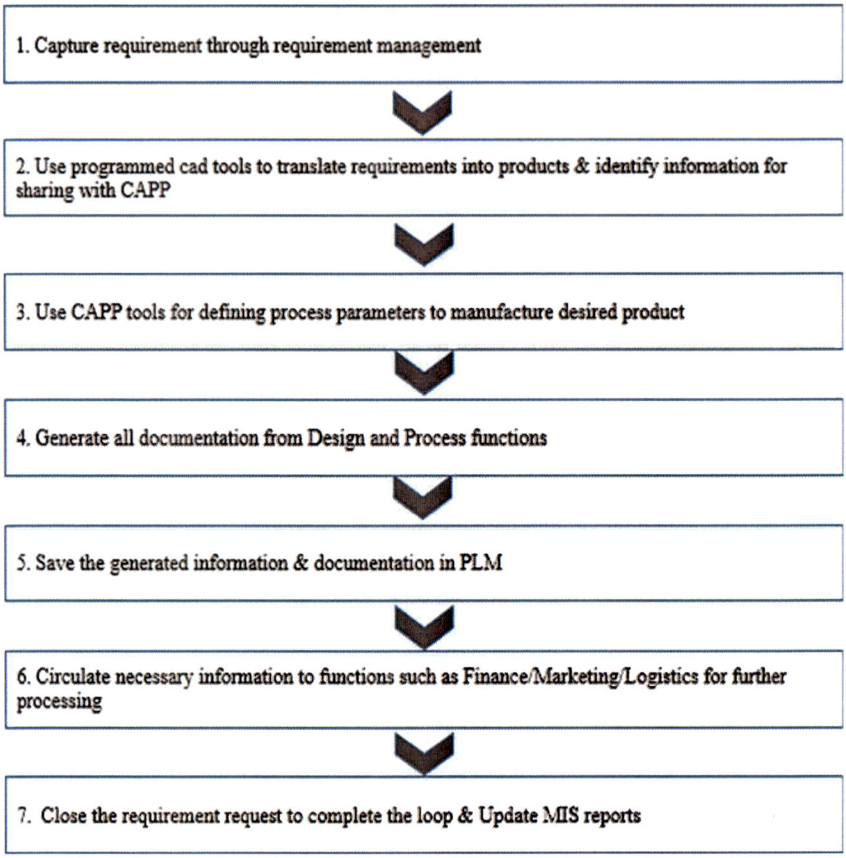

Fig. 6 Workflow in integrated systems

Table 2 Integration parameters for case study

Description	Integration technology	Metadata	Feedback	Mode
RM to CAD	.xml file	Product Specifications	Event triggered; feedback disabled	Outward
CAD to CAPP	Single DB	Product dimensions	Event triggered; feedback disabled	Outward
CAD/CAPP to PLM	Inbuilt adapters	Product dimensions and process parameters	Event triggered; feedback enabled	Both inward and outward

IT investments. A framework to establish such integration thus a framework established through SME interviews and its validation would helpful in taking this research further.

The authors recommend addressing following aspects for achieving a viable system integration approach,

Exploring functional relationship/data relationship between the integrating applications

1. Refining the existing methodology through SME interactions
2. Develop a generic framework of integrations
3. Validation with case study or pilot system.

The authors plan to undertake these activities of developing an integration further for enterprise systems in the future with help of the proposed approach.

References

1. Prasad B (1996) Concurrent engineering fundamentals, Preface, 19, New Jersey. Prentice Hall PTR, USA
2. Ameri F, Dutta D (2005) Product lifecycle management: closing the knowledge loops. Comput-Aided Des Appl 2(5):577–590
3. Garetti M, Terzi M, Bertacci N (2005) Organizational change and knowledge management in PLM implementation. Int J Prod Lifecycle Manage 1(1):43–51
4. Klaus H, Roseman M, Gable G (2000) What is ERP. Inf Syst Front 2:141–162
5. Markus ML, Tanis C (2000) The enterprise systems experience from adoption to success. In: Zmud R (ed) Framing the domains of IT research Glimpsing the future through the past. Pinnaflex Educational Resources, Inc., Cincinnati, pp 173–207
6. Kaminski PC, Oliveira AC, Lopes TM (2008) Knowledge transfer in product development processes: a case study in small and medium enterprises (SMEs) of the metal-mechanic sector from Sao Paulo, Brazil. Technovation, vol 28, pp 29–36
7. Bagchi-Sen S (2001) Product innovation and competitive advantage in an area of industrial decline: the Niagara region of Canada. Technovation 21:45–54
8. Sudarshan R, Fenves SJ, Sriram RD (2005) A product information modeling framework for product lifecycle management. Comput Aided Des 37(13):1399–1411
9. Nosenzo V, Tornincasa S, Bonisoli E et al (2014) Open questions on product lifecycle management (PLM) with CAD/CAE integration. Int J Interact Des Manuf 8(2):91–107
10. Balakrishna A, Suresh Babu R, Nageswara Rao D et al (2006) Integration of CAD/CAM/CAE in Product Development System Using STEP/XML. Concurr Eng Res Appl 14(2):121–128
11. Kiritsis D, Nguyen VK, Stark J (2008) How closed-loop PLM can improve knowledge management over the complete product lifecycle? Int J Prod Lifecycle Manage 3(1):54–77
12. Huang GQ, Mak KL (1999) Current practices of engineering change management in UK manufacturing industries. Int J Oper Prod Manage 19(1):21–37
13. Li WD, Fuh JYH, Wong YS (2004) An Internet-enabled integrated system for co-design and concurrent engineering. Comput Ind 55(1):87–103
14. Ghobakhloo M (2018) The future of manufacturing industry: a strategic roadmap toward Industry 4.0. J Manuf Technol Manage 29(6):910–936
15. Lv Y, Ni Y, Zhou H et al (2016) Multi-level ontology integration model for business collaboration. Int J Adv Manuf Technol 84(1–4):445–451
16. Ming XG, Yan JQ, Wang XH et al (2008) Collaborative process planning and manufacturing in product lifecycle management. Comput Ind 59(2–3):154–166

17. Chiang TA, Trappey AJC (2007) Development of value chain collaborative model for product lifecycle management and its LCD industry adoption. Int J Prod Econ 109(1–2):90–104
18. Jun H, Kirtisis D, Xirouchakis P (2007) Research issues on closed-loop PLM. Comput Ind 58:855–868
19. Wiesner S, Freitag M, Westphal I, Klaus-Dieter T (2015) Interactions between service and product lifecycle management. In: 7th industrial product-service systems conference—PSS, industry transformation for sustainability and business, vol 30, pp 36–41
20. Marchetta MG, Mayer F, Forradellas RQ (2011) A reference framework following a proactive approach for product lifecycle management. Comput Ind 62(7):672–683
21. Segonds F, Cohen G, Véron P, Josquin P (2016) PLM and early stages collaboration in interactive design, a case study in the glass industry. Int J Interact Des Manuf 10(2):95–104
22. Linthicum D (2003) Enterprise application integration. Addison-Weasley
23. Gericke A, Klesse M, Winter R, Wortmann F (2010) Success factors of application integration: an exploratory analysis. Commun Assoc Inf Syst 27(1):37
24. Yin RK (2009) Case study research: design and methods, 4th edn. Sage, Thousand Oaks

Application of Clustering Algorithms for Locating Distribution Centers of Logistics System

L. N. Pattanaik, Amit Kumar Gupta, Shifa Surin, and Aashish Pratap Singh

Abstract Locating distribution centers in logistics systems is a challenge as it can affect the delivery cost and time to a large extent. The present work addresses this problem of finding optimal location for distribution centers or warehouses for a typical supply network by applying data clustering algorithms. Two popular clustering algorithms, one each from hard clustering (k-means) and soft clustering (fuzzy c-means), are applied on sets of data representing latitude, longitude and population of ninety-five cities to identify required number of centroids. Ideally, each centroid can be a potential distribution center for the respective cluster. However, a city closest to the centroid is selected as a distribution center owing to geographical and practical constraints. Further, actual population of the cities are also considered for giving due weightage to the location during clustering.

Keywords Logistics system · Distribution center · Data clustering · K-means clustering · Fuzzy c-means clustering

L. N. Pattanaik · A. K. Gupta (✉) · S. Surin · A. P. Singh
Department of Production Engineering, Birla Institute of Technology, Mesra, Ranchi 835215, India
e-mail: a.k.gupta0813@gmail.com

L. N. Pattanaik
e-mail: lnpattanaik@bitmesra.ac.in

S. Surin
e-mail: shifajanet10@gmail.com

A. P. Singh
e-mail: aashishpratapsingh02@gmail.com

R. P. Mahapatra et al. (eds.), *Proceedings of 6th International Conference on Recent Trends in Computing*, Lecture Notes in Networks and Systems 177,
https://doi.org/10.1007/978-981-33-4501-0_46

1 Introduction

The present evolution in logistics is a result of changing market needs driven by globalization. The markets for both perishable goods and non-perishable goods are encouraging direct-to-consumer (D2C) deliveries. This enables e-commerce, especially online shopping by providing home delivery. Market share in the e-commerce space is mainly a function of speed, flexibility, security and the cost of delivery. Even though an increasing number of online shoppers are putting greater priority on the speed of delivery, they are also extremely price sensitive and generally not willing to pay more for faster delivery.

Thus, retailers will have to bear the extra cost associated with faster delivery in order to compete with other players in the market. Hence, all the retailers are trying to find novel means of product D2C deliveries. There have been various researches on the drone delivery of goods, which have proved to be a bane especially for the last mile delivery service. These drones go to places where a normal truck could not go because of the traffic and bad road condition. They have the required speed and are able to serve the purpose as well.

E-commerce retailers such as Amazon® and Alibaba® are planning for home delivery services partly or fully performed by drones. Amazon® company has plans for an ultra-rapid package delivery service titled Amazon Prime Air in which a customer within a ten-mile reach of performing Amazon order fulfillment center will be able to get their order within 30 min of order through an independently operated multi-rotor drone. Ever Google has made its plans for a drone-based package delivery system, and many other companies over the world are either devising such systems or finding related concepts. This sets a benchmark for other companies and markets as well to find a way to deliver the goods to customers efficiently in minimum time. This challenges the companies dealing with the logistics to find a feasible solution which will be beneficial for the customers as well as the suppliers.

The use of clustering algorithm for the locations of distribution center in a logistic system has not yet been explored completely. The two clustering algorithms used in the present paper identify clusters in a set of cities according to the criteria of minimum distance.

In Sect. 2, relevant literatures are reviewed to identify the status of the present research and gaps. In Sect. 3, the location problem is described along with the two clustering methodologies. In Sect. 4, the collected set of data representing coordinates of 95 cities and the applications of clustering algorithms are presented. Few computational steps of the clustering algorithm are also given in the same section. Result and discussion are provided in Sect. 5. Finally, conclusions and future scopes are given.

2 Literature Review

In this literature review, some relevant works of clustering in various fields have been presented. A credit card fraud detection problem was addressed by using a clustering approach [1]. The data which was generated randomly for credit card used the k-means clustering algorithm for detecting whether the transaction made was fraud or not. Clusters were formed to detect fraud in credit card transaction whether they are low, high, risky and highly risky. The k-means clustering algorithm was used which was found to be a simple and efficient algorithm for credit card fraud detection.

Four different clustering algorithms, namely k-means, density-based, filtered and farthest first clustering, were used for the clustering of retail sales data [2]. Data mining in the marketing business has played a vital role in finding a pattern in raw data so that some meaningful information could be derived from them.

In another work to improve the order picking so that the time and distance can be minimized to decrease material handling, a two-level approach where the first step clustered the parts and the second step found the location of parts was reported [3].

In order to find the gap between the amalgamation of supply chain management and industrial cluster, a clustering approach was applied [4].

Fuzzy clustering for multi-facility location problem was carried out using different approaches for making the capacity of each facility as unlimited [5]. Customers according to their geographical locations were grouped by spherical and elliptical fuzzy cluster, and after experimenting the different number of clusters, cluster centers were allotted to the facility centers.

Another work applied cluster analysis to determine location of distribution centers for a fast food restaurant chain [6]. The study was in three stages: compiling spatial data, cluster analysis and network analysis. Cluster analysis results were used to consider the location of the additional distribution center.

Another paper explored how and why firms with logistics-intensive operations such as fast-moving consumer good (FMCG) distributers benefit from residing in logistics clusters [7]. Recent work investigates the impact of supply chain biomass cluster on sustainable forest development [8].

From the literature review, successful applications of clustering algorithms to logistics design problem were found. The present paper also explores the suitability of two popular clustering algorithms to find optimal locations of distribution centers in a logistics system.

3 Location Problem and Methodology

Clustering is the work of segregating the population or data points into a number of groups such that data points in the same groups are more similar to other data points in the same group and dissimilar to the data points in other groups. It is essentially a compilation of objects on the basis of similarity and dissimilarity between them. The major types of clustering algorithms are partitional clustering, hierarchical clustering, fuzzy clustering, density-based clustering, model-based clustering and k-means clustering.

The present problem deals with finding optimum locations of distribution centers to cater the demands of many cities by providing a minimum distance of transportation, thereby minimizing the time and cost of delivery. The centroid of each cluster will be selected as a potential distribution center in order to minimize the product transportation cost. The two clustering algorithms take latitudes and longitudes of a given number of cities as inputs and identify locations for the distribution centers after clustering them into several groups. The two clustering algorithms implemented in the present work are briefly explained here.

3.1 Clustering Algorithms

K-means is a type of clustering algorithm in which the input number of clusters is decided randomly; then, the centroid is calculated for each and every cluster. In the next step, each and every data point is reassigned in different clusters based on the minimum distance. And finally, these steps get repeated until there is no cluster change.

Fuzzy algorithm works on the fuzzy logic by assigning membership to each and every data point on the basis of distance between the cluster centroid and the data points. The summation of membership of each data point should be equal to 1.

4 Data Set for Clustering

The data set (Table 1) used in this paper represents latitude, longitudes and population of 95 cities which are available in the public domain. For computational convenience, the latitudes and longitudes are recalculated with reference to an assumed origin. The population of each city which is an important parameter in deciding the location of distribution centers is also considered during the clustering.

The clustering of the above data is carried out by the two algorithms using Python 3 codes in Jupyter Notebook software.

5 Results

During implementation of k-means clustering, an optimum number of natural clusters can be obtained by using elbow method. In the present case, the output from the elbow method is presented in Fig. 1. It can be observed from the graph that the elbow occurs for k values in the range of 6–9. Hence, a k value of 8 is selected to find the

Table 1 Coordinates of 95 cities and their population [9]

*Latitude	*Longitude	Population	*Latitude	*Longitude	Population
6	4	2,684,487	7.7	22	110,608
1.3	30	2,914,253	10.7	36	102,456
1.5	28	2,693,477	9.8	11	101,520
2	30	2,062,330	9.1	9	99,931
2.2	26	1,492,073	6.4	16	109,000
2.9	21	947,334	7.9	12	87,357
3	24	716,259	11.6	14	106,000
3.2	25	1,042,886	7.8	18	2,811,569
3.2	19	2,445,474	8.1	17	2,562,012
3.7	21	949,443	7.2	32	75,736
3.9	30	900,422	6.9	27	241,000
4	11	599,578	4.3	21	62,201
4	6	531,885	7.2	34	202,000
4.1	35	949,413	9.2	10	65,437
4.5	7	959,124	3.9	26	14,538
4.9	35	1,042,886	5.7	34	51,849
5.1	11	48,141	4.6	18	52,849
5.3	11	1,321,442	10.6	13	50,475
5.9	25	180,748	8.3	28	44,507
6	5	45,508	9.1	31	49,864
6.8	12	899,200	6.3	14	54,482
7.1	26	599,813	7.1	7	62,927
7.9	19	31,035	8.5	34	52,892
8.2	17	32,869	6.8	10	55,336
8.7	28	32,899	10.2	19	48,411
9	21	461,790	11	28	78,679
10.2	31	2,046,652	7.5	12	34,056
9.1	17	470,839	6.8	26	34,044
12	23	410,210	7.6	13	41,587
11.3	27	393,724	8.3	32	37,776
8.3	32	267,000	6.7	34	40,665
4.6	14	261,000	9.1	21	38,815
7.7	18	251,000	6.3	15	30,170
8.6	15	202,000	8.4	16	40,068
4.2	9	241,000	9.9	24	31,155
12	16	213,000	4.2	21	27,302

(continued)

Table 1 (continued)

*Latitude	*Longitude	Population	*Latitude	*Longitude	Population
11	21	310,817	8.1	17	43,785
10.3	23	155,175	7.3	19	35,536
10.9	33	147,396	6.7	11	29,932
8.4	12	147,126	4.2	31	23,574
8.9	8	137,068	3.5	33	56,214
9.3	18	156,200	7.6	13	15,236
7.3	24	125,183	6.4	27	57,126
6.9	27	113,012	8.2	20	56,223
6.2	16	134,458	6.1	23	102,546
7.6	31	107,076	9.8	12	113,524
5.9	24	105,221	4.2	11	241,000
7.7	22	110,608	7.1	8	316,000

*Indicates modified coordinates with reference to the origin

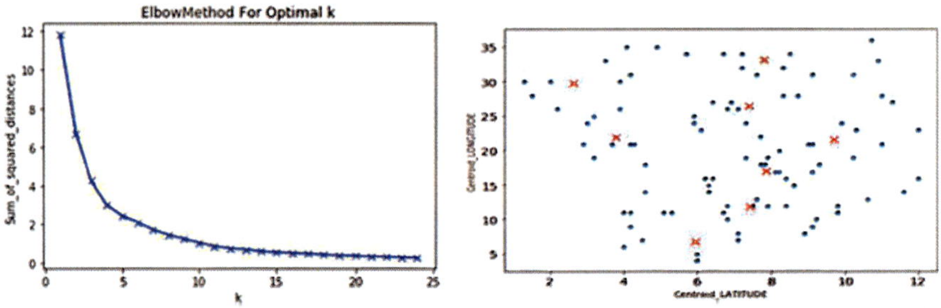

Fig. 1 Plot for optimal k in elbow method and 8 clusters identified through k-means algorithm

optimum clusters. After implementing the steps of k-means clustering algorithm, the 8 centroids representing the distribution centers are as obtained in Fig. 1 (marked as a cross). But in practice, the centroids may not be feasible as distribution centers owing to geographical constraints; therefore, the existing cities which are nearest in terms of Euclidean distance from the coordinates of centroid can be selected as distribution centers. In Fig. 2, these 8 distribution centers are marked as 1–8.

In the second method, fuzzy c-means clustering is applied on the same data to identify same number of clusters. The coordinates and centroids are presented in Fig. 3. Similar to previous case, the nearest cities are identified as distribution centers marked as 1–8.

	Centroid_LATITUDE	Centroid_LONGITUDE
0	9.711111	21.555556
1	7.226316	11.684211
2	7.526667	33.133333
3	7.708333	26.416667
4	4.000000	21.000000
5	8.312500	16.750000
6	5.975000	6.750000
7	2.775000	28.250000

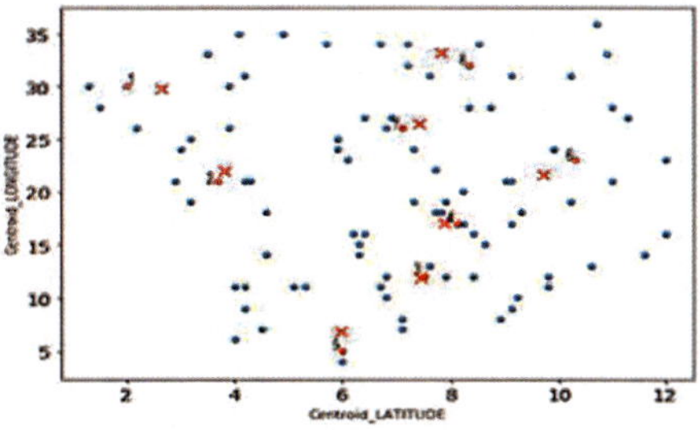

Fig. 2 Coordinates of 8 centroids found through k-means and identification of nearest cities as distribution centers

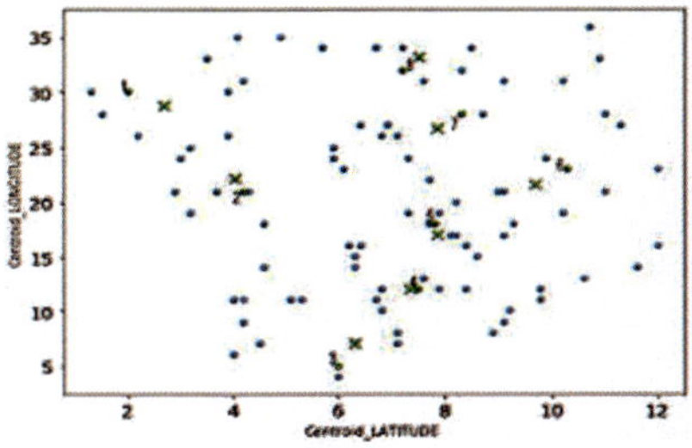

	Centroid_LATITUDE	Centroid_LONGITUDE
0	9.711111	21.555556
1	7.445000	11.800000
2	7.814286	33.142857
3	4.040000	22.500000
4	5.975000	6.750000
5	2.657143	29.714286
6	7.875000	17.000000
7	7.872727	26.636364

Fig. 3 Coordinates of 8 centroids found through fuzzy c-means and identification of nearest cities as distribution centers

6 Conclusion

In this paper, a location problem is addressed for a logistics system using 2 different clustering algorithms, namely k-means and fuzzy c-means. The coordinate data for 95 cities are used to identify 8 number of clusters whose centroids are selected as the potential distribution centers. But in real scenario, the centroid may not lie on any city; therefore in such case, the city with nearest centroid will be assigned as the distribution centers. The clustering algorithms are coded in Python 3 language which is very versatile and useful for such data analytics applications.

References

1. Esnaf S, Küçükdeniz T (2009) Credit Froud Detection. J Intell Manuf 259–265
2. Shrivastava, Arya PN (2012) Retail sales problem. Int J Comput Commun Network School Ind Eng. Purdue University, West Lafayette, IN, USA, Volume 1, No. 2
3. Tolossa NJ, Beshah B, Kitaw D, Mangano G, De Marco A (2013) The order picking in material handling. Int J Market Stud 5(6). Published by Canadian Center of Science and Education
4. Javadi M, Lehto MR (2016) The amalgamation of supply chain management and industrial cluster. Int J Prod Res
5. Vaishali (2016) Multi-facility location problem. Int J Comput Appl 0975–8887 98(3)

6. Widaningrum DL et al (2017) Location of fast food restaurant chain. Bina Nusantara University, Indonesia 11480, IOP Conf. Ser.: Earth Environ. Sci. 109 012021
7. Abushaikha I (2018) The influence of logistics clustering on distribution capabilities. Int J Retail Distrib Manage 46:577–594
8. Švažas M, Navickas V (2019) Impact of supply chain biomass cluster on sustainable forest development. J Int Stud 309–321
9. https://simplemaps.com/data/in-cities

Robustness Analysis of Behavioral Cloning-Based Deep Learning Models for Obstacle Mitigation in Autonomous Vehicles

Pranit Gopaldas Shah, Arvind R. Yadav, and Bhupendra Ramani

Abstract Maneuvering a steady on-road obstacle at high speed involves taking multiple decisions in split seconds. An inaccurate decision may result in a crash. One of the key decisions that need to be taken is can the on-road steady obstacle be surpassed. The model learns to clone the driver's behavior of maneuvering a non-surpass-able obstacle and pass through a surpass-able obstacle. No data with labels of "surpass-able" and "non-surpass-able" was provided during training. We have developed an array of test cases to verify the robustness of CNN models used in autonomous driving. Experimenting between activation functions and dropouts the model achieves an accuracy of 87.33% and run time of 4478 s with input of only 4881 images (training + testing). The model is trained for limited on-road steady obstacles. This paper provides a unique method to verify the robustness of CNN models for obstacle mitigation in autonomous vehicles.

Keywords End-to-end learning · Computer vision · Behavioral cloning · Autonomous vehicles · Activation function · Dropout · Obstacle mitigation

1 Introduction

Affordability of powerful computational hardware and advances in deep learning techniques has made vision-based autonomous driving an active research focus within the transport industry. Autonomous driving has drawn enormous attention in

P. G. Shah (✉) · B. Ramani
Department of Computer Science and Engineering, Parul Institute of Engineering & Technology, Vadodara, India
e-mail: pgopalshah@gmail.com

B. Ramani
e-mail: bhupendra.ramani2817@paruluniversity.ac.in

A. R. Yadav
Department of Electronics and Communication, Parul Institute of Engineering & Technology, Vadodara, India
e-mail: arvind.yadav@paruluniversity.ac.in

R. P. Mahapatra et al. (eds.), *Proceedings of 6th International Conference on Recent Trends in Computing*, Lecture Notes in Networks and Systems 177,
https://doi.org/10.1007/978-981-33-4501-0_47

vision-based applications apart from others like hand-gesture [1] recognition. There are considerable drawbacks of the existing techniques to overcome, even though the research worldwide has already taken a giant leap. Foremost downside is the inability to explicitly model each possible scenario. Driving needs responding to a large variety of complex environmental conditions and agent behaviors.

Maneuvering every steady on-road obstacle either surpass-able or non-surpass-able is cost-intensive. Again maneuvering a steady on-road obstacle at high speed involves taking multiple decisions in split seconds. An inaccurate decision may result in a crash. One of the key decisions that need to be taken is can the steady on-road obstacle be surpassed.

An autonomous driving vehicle successfully navigating through the streets should be able to follow the roadway as well as maneuver only in cases required. If the autonomous vehicle can surpass a steady on-road obstacle without being unstable, then it must do so. Therefore, authors herein propose an improved convolution neural network model. Overall, this research work makes the following contributions:

- Provides evaluation method for popular learning models by defining test cases to mitigate an on-road obstacle.
- Identify, evaluate, and validate the configuration producing optimum results by performing a combination of activation function and dropout.
- Improve prediction accuracy by validating statistical data with a visual saliency map.

The paper has been organized as follows: Sect. 2 gives an overview of related work done in the past and present. Section 3 describes the methodology used. Section 4 presents experimental setup, results, and discussion. Finally, Sect. 5 concludes this paper.

2 Related Work

The perception-driven method and end-to-end method are the two popular vision-based paradigms for self-driving cars. Both the perception and end-to-end methods have been reviewed extensively through a literature study and presented here. In the later part of this section, we have reviewed our study on object detection in the context of deep learning.

2.1 Perception-Driven Learning Methods

The traditional perception-based method has made remarkable achievements, in past decades, in the field of self-driving cars. Several methods of detections have been proposed to generate a description of the local environment. Depending on the technique used, current detection research can be broadly classified as shown in Fig. 1.

1. **Vision-based detection**

Object location through enclosing boxes has been adopted by researchers as a part of object detection task. This bounding box can be anything from a steady road signs, traffic signs, moving cars, moving bicycles, etc. The model is trained with labeled data of objects.

An et al. [2] used Deep Convolutional Neural Network (DCNN) for image classification into obstacle/pothole or non-pothole. The method compared Inception_v4, Inception_ResNet_v2, ResNet_v2_152, and MobileNet_v1 for comparison between color and grayscale images. This method is limited in its efficiency in processing a single frame of image. Canny edge detection/Hough transformation area is implemented to locate lane position through many lane detection methods. There are no specific geometric restrictions required to identify uneven lane boundaries in these methods.

2. **3D reconstruction**

There are three major approaches for 3D reconstruction of obstacle/pothole each with its drawback.

Li et al. [3] used infrared laser line projectors, a digital camera, and a multi-view coplanar scheme for calibrating the lasers. The method plotted more feature points in the camera's point of view and was much more cost-effective. Wang et al. [4] used a series of cameras. This method generated a 3D surface model through a series of 2D images captured. A high computation requirement was a key backdrop. Xbox Kinect sensor was used by Joubert et al. [5] and Moazzam et al. [6]. The method could not minimize the error and power for computing, even though equipment price was minimized.

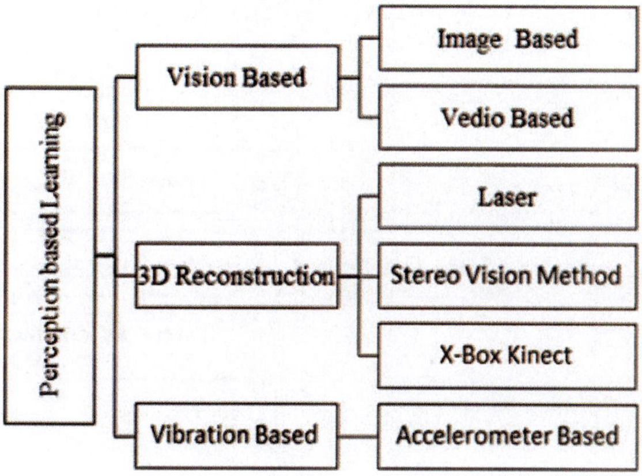

Fig. 1 Perception-driven learning

3. **Vibration-based detection**

Bhatt et al. [7] combined accelerometer, gyroscope, location, and speed data to classify road condition and detect potholes/obstacles. SVM with radial basic function (RBF) kernel was used for road condition classification. SVM and gradient boost were used for pothole/non-pothole classification. Failure attribute includes an inability to accurately classify between good and bad roads due to insufficient data for all road types. The key backdrop was the inability to distinguish between a bump, a manhole, or a pothole.

One key takeaway from all the above work done in perception-based learning provides proactive data to the driver regarding the obstacle. These methods do not provide any method for routing through the obstacle. Again, there is not classification if an obstacle can be surpassed or not.

2.2 *End-to-End Learning Methods*

Pomerleau [8] pioneered end-to-end training of a neural network to steer a car. In 1989, Pomerleau built the Autonomous Land Vehicle in a Neural Network (ALVINN) system. In the scenario of autonomous driving, one of the key requirement is an ability to identify salient objects. Refer Fig. 2 below for salient object based end-to-end learning methods.

1. **Sensitivity based**

After 25 years of advances in data and computational power DAVE-2 in [9] demonstrated the potential of end-to-end learning. A CNN-based model introducing PilotNet [10] network architecture designed by NVIDIA. The key differentiator for this architecture was its ability to identify salient objects without the need for hand-coded rules and instead learn by observing. However, Dave's PilotNet model admits that the convolutional layers were chosen empirically, and hence, the performance

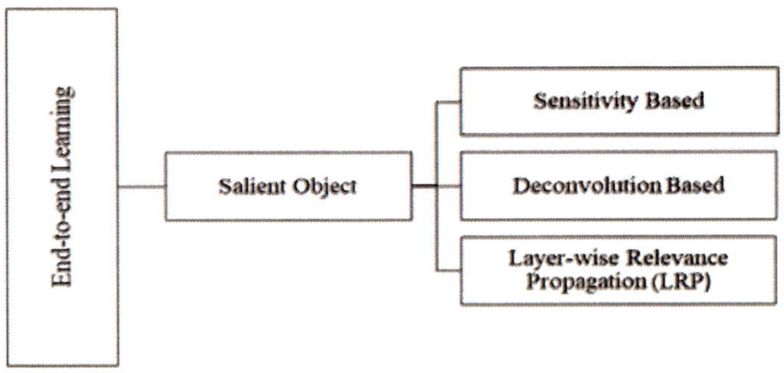

Fig. 2 End-to-end learning

was not sufficiently reliable to provide a full alternative to more modular approaches to off-road driving.

2. **Deconvolution based**

Zeiler [11] presented a method for mid and high-level feature learning like corners, junctions, and object parts. Zeiler [11] resolves the two fundamental problems found in image descriptors like SIFT, HOG, or edge gradient calculators followed by some histogram or pooling operation. First being invariance to orientation and scale. Secondly, a CNN models inefficiency in training each model for the input. Visualization available is for one activation per layer.

3. **Layer-wise relevance propagation**

Binder et al. [12] implemented layer-wise relevance propagation to compute scores for image pixels and image regions denoting the impact of the particular image region on the prediction of the classifier for one particular test image. Binder et al. [12] demonstrated controlling the noisiness of the heatmap; however, an optimal trade between numerical stability and sparsity/meaningfulness of the heatmap was kept as an item for future work. Salient object-based methods does identify the key features that impact the steering angles. However, these methods have not been explicitly developed/tested for identifying if an obstacle can be surpassed or not.

To overcome the above-mentioned limitations, we propose to perform extensive training and testing for a neural network to clone an obstacle mitigation policy. Even though there is a great deal of work and the literature on the task of steering angle prediction, our goal is not to propose yet another method for prediction, but rather provide a different perspective on on-road steady obstacles mitigation model. A model to not only detect an on-road steady obstacle but also to predict the obstacle can be surpassed to avoid unnecessary maneuvering.

3 Methodology

This section provides the details on CNN models used for validation and steps performed on data for accurate prediction.

3.1 Preprocessing

Network model continuously predicts the steering angle to clear all test cases with an input of raw pixels incorporating attention in an end-to-end manner. Our experiments and results must be independent of car geometry; hence, we represent steering command as inverse turning radius r_t^{-1} (r is turning radius at timestamp t). We use inverse to prevent numerical instability, singularity, and smooth transitions through

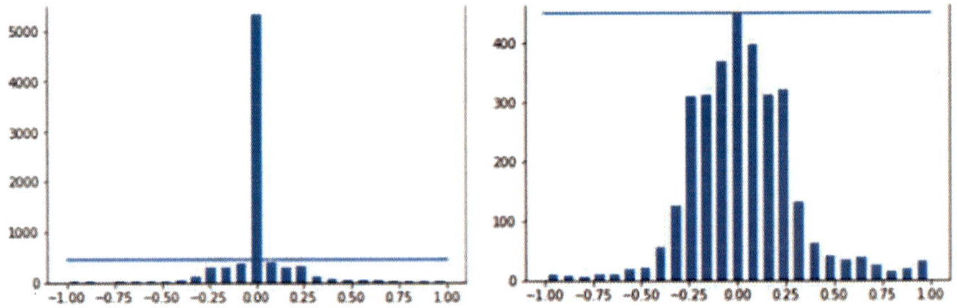

Fig. 3 Left: Data with 5000+ images having "0" steering. Right: Non-biased data with sample reduced to 400+ per bin to have a uniformly distributed steering

zero from left turns to right turns. The relation between steering angle θ_t and inverse turning radius can be given as

$$\theta_t = f_{\text{Steers}}(u_t) = u_t d_w K_s \left(1 + K_{\text{slip}} v_t^2\right). \tag{1}$$

where θ_t in degree, and v_t (m/s) is a steering angle and velocity at time t, respectively. K_s, K_{slip} and d_w are vehicle-specific parameters. K_s is steering ratio between the turn of the steering and the turn of the wheels. K_{slip} represents the relative motion between the front and rear wheels.

3.2 Data Bias Removal

The general tendency of every driver is to drive as steady as possible without a lot of maneuvering. However, in the case of behavioral cloning, if the driver drives steady during training, then all the model will learn which is to maintain a zero steering angle. Such training would generate results bias to zero output. To avoid output bias toward a −ve/+ve steering angle driving, the training is done for complete track in clockwise and then anticlockwise direction over the track. Additional recovery training tracks are also done, where the car is take off the center lane and then recovered to back. The data bias is removed by trimming samples per bin as shown in Fig. 3.

3.3 Encoder Model: Convolution to Clone Obstacle Mitigation Behavior

Convolutional Neural Networks (CNN) are adaptations of the multi-layer perceptron and are biologically inspired. In the context of self-driving cars, CNNs can learn the entire task of lane and road following without manual decomposition into road or lane marking detection, semantic abstraction, path planning, and control. The network learns to detect the outline of a road without the need for explicit labels during training. The following research implements variants of the CNN architecture established by NVIDIA for self-driving cars, the PilotNet [10], and DroNet [13].

4 Experimental Setup, Results, and Discussion

This section presents the basic description of an experimental setup for data collection, training, and testing. Further, authors elaborate on the configuration of hardware and software used. Later on authors enlist the training cases and test cases for model evaluation and the evaluation criteria, eventually, the results from our experiment.

Model training and testing are performed in a virtual environment Unity 2017 in the interest of research cost. Other software tools include visual studio for Unity, Atom, Jupyter Notebook, and Anaconda. GitHub for online code repository and Google Colab platform for online code execution. Programming is done in c# and python. Multiple packages including OpenCV, NumPy, matplotlib, Keras (model, optimizer, layers), pandas, and sklearn are used. Hardware includes laptop with Intel i5 processor@1.27 GHz, 8 GB RAM, and Intel HD Graphics family with 2 GB RAM. Virtual 3D models of non-surpass-able and surpass-able obstacles are created in unity as shown in Fig. 4.

Three cameras are mounted in front of the model car to capture the left, center, and right images as depicted in Fig. 5.

4.1 Training and Test Cases

Our network model is trained on only for 4 cases and is expected to pass 10 test cases based on training. These training cases are to train the network to clone driver's behavior while driving through both non-surpass-able and surpass-able obstacles. Training cases areas listed in Fig. 6.

Test cases are to verify the robustness of the network's ability to learning to maneuver through varied configurations of obstacles. Test cases are as listed in Fig. 7.

Fig. 4 Left: Car model alongside non-surpass-able obstacle. Right: Car model alongside surpass-able obstacle

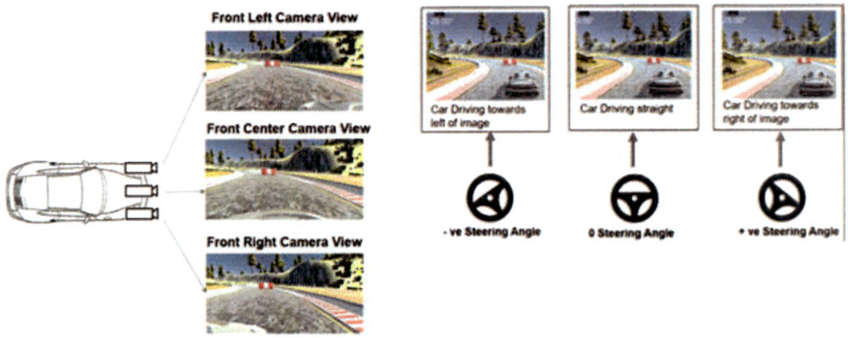

Fig. 5 Left: Front camera to capture images. Right: Steering angle and car driving direction

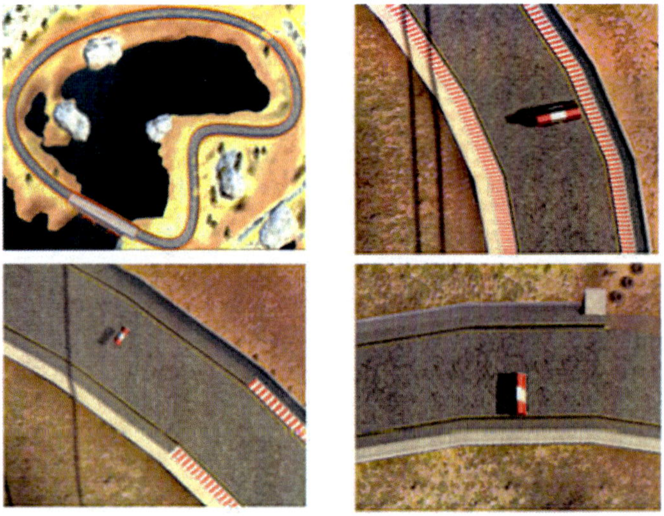

Fig. 6 Top Left: Training case-TRC01: The track is empty without any obstacles. The model is trained on this empty track. Top right: TRC02: The track has a non-surpass-able obstacle on the right of the simulated car. The model is trained to maneuver through the obstacle. Bottom left: TRC03: The track has a surpass-able obstacle in the center of the track. The model is trained to pass through the obstacle. Bottom right: TRC04: The track has a non-surpass-able obstacle on the left of the simulated car. The model is trained to maneuver through the obstacle

4.2 Model Accuracy Calculation

Each model is trained in 4 cases and tested for 10 cases. The model accuracy is defined as follows:

Number of test cases model passed:	*P*

(continued)

(continued)

Score for a test cases in which model self-recovered:	SR
Model accuracy:	MA
Total test cases:	T
Number of obstacles correctly maneuvered or surpassed:	SR_o
Total maneuverable and surpass-able obstacles:	T_o

$$\mathrm{MA} = ((P + \mathrm{SR})/T)/100 \tag{2}$$

$$\mathrm{SR} = \mathrm{SR}_o/T_o \tag{3}$$

4.3 Dataset Characteristics

The right size of the dataset is the key to an accurate predictable solution. Initial training was started with 30,000 images; however, due to resource constraints, we reached an optimal data size of 9970 images that produced reliable results. Zero steering bias images are removed from input images. Data augmentation is performed to increase the data size and accuracy. The final dataset is classified using test_train_split functionality in the sklearn library. All models are trained using 3904 samples and validated with 977 samples as shown in Table 1.

This paper represents two deep learning models tested with a combination of activation function and dropout on the same database. Table 2 shows the model used, code assigned for ease, configurations used, and val_loss achieved. Table 2 clearly indicated that model P1, PilotNet with elu and no dropout, achieved the best val_loss of 0.0250. Figure 8 left and right shows the prediction accuracy and model running time when using different configurations on two learning models using the same dataset. Model P1, PilotNet with elu and no dropout, achieves the highest prediction accuracy of 87.33%. Model P4 and P5 have the least accuracy of 10%. The highest accuracy achieved by both DroNet model D1 and D2 are 50% and 30%, respectively, with heavy processing time of 12,000 s and 8280 s, respectively. Model P1 achieves the highest accuracy consuming the processing time of 4478 s. Model P1, PilotNet with elu and no dropout, performed a self-recovery in test cases 6 and 7 as listed in Table 3.

Overall model accuracy for model P1 is calculated as below

$$\begin{aligned}\mathrm{MA}_{P1} &= ((P_{P1} + \mathrm{SR}_{P1})/T)/100 \\ &= ((8 + (0.33 + 0.4))/10)/100 \\ &= 87.33\%\end{aligned}$$

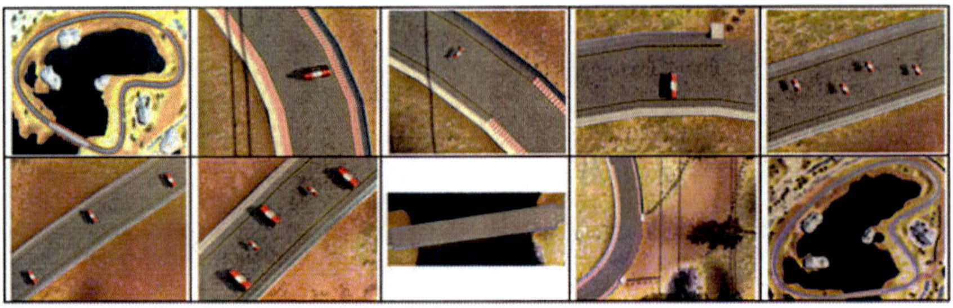

Fig. 7 Starting from top left first: Test Case-TC01: The track is empty without any obstacles. The model is supposed to drive through the entire track to call it has passed this test case. Second: TC02: The track has a non-surpass-able obstacle on the right of the simulated car. The model is supposed to maneuver through the obstacle and retain a steady drive to call it has passed this test case. Third: TC03: The track has a surpass-able obstacle. The model is supposed to pass through the obstacle with minimum vehicle instability and maintain a steady drive to call it has passed this test case. Fourth: TC04: The track has a non-surpass-able obstacle on the left of the simulated car. The model is supposed to maneuver through the obstacle and retain a steady drive to call it has passed this test case. Fifth: TC05: The track has an array of multiple surpass-able obstacles. The model is supposed to pass through the obstacles with minimum vehicle instability and maintain a steady drive to call it has passed this test case. Starting bottom left First: TC06: The model is supposed to maneuver through an array of left and right non-surpass-able obstacles and retain a steady drive to call it has passed this test case. Second: TC07: The track has an array of both multiple left and right non-surpass-able and surpass-able obstacles. The model is supposed to pass through surpass-able obstacles and maneuver through the obstacle to call it pass. Third: TC08: In this case, an unknown bridge, which has a different track, has to be crossed to call this case passes. Fourth: TC09: In this case, an unknown unpaved path has to be avoided by the model. The model has not been trained for this behavior. Fifth: TC10: In model is expected to clear all the obstacles, pass 7through unknown bridge and unpaved path all in a single stretch

Table 1 Data characteristics for different models

Model	Input data	Input data after removing bias		Trainable parameters
		Training samples	Validation samples	
DroNet	9930	3904	977	311,777
PilotNet	9930	3904	977	252,219

Table 2 Configuration and val_loss comparison for different models

Model	Code	Configuration	val_loss
DroNet	D1	Dropout + Relu	0.0295
	D2	No Dropout + elu	0.0731
PilotNet	P1	No Dropout + elu	0.0250
	P2	Dropout + elu	0.0454
	P3	No Dropout + Relu	0.0276
	P4	No Dropout + Sigmoid	0.0934
	P5	No Dropout + Softmax	0.0863

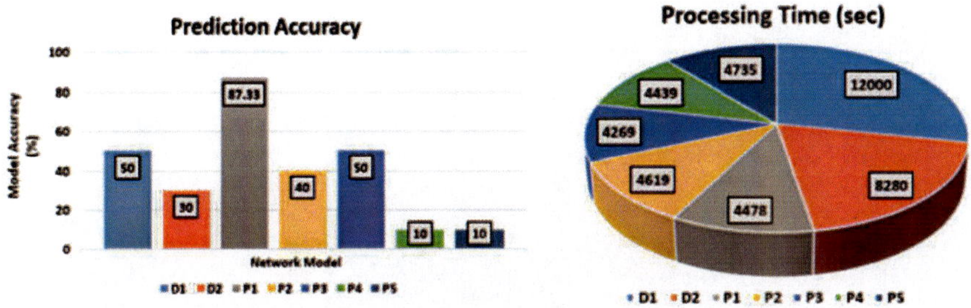

Fig. 8 Left: Prediction accuracy. Right: Processing time

Table 3 Path followed and individual test case accuracy for Model P1

	Test case 6	Test case 7
Path followed		
Description	The model maneuvers first skips second by going off the track and recovers back post skipping third obstacle	The model maneuvers first passes through second, skips third by going off the track and recovers back post skipping fourth and fifth obstacle
SR calculation	$SR_o = 1; T_o = 3; SR_6 = 0.33$	$SR_o = 2; T_o = 5; SR_7 = 0.4$

A saliency map is generated to co-related and validate the statistical results obtained. PilotNet models are taken for saliency mapping for identification of left non-surpass-able, surpass-able, and right non-surpass-able obstacles. As depicted in the table below, model P4 and P5 have failed completely which co-relates with only prediction accuracy of 10%. Model P2 has wrongly detected the saliency for surpass-able obstacles and has not detected the lane boundaries at all. Model P1 performs better than P3 with better saliency map both for obstacles and lanes as depicted in Table 4.

5 Conclusion

In this paper, we presented and compared the two most popular autonomous driving methods including DroNet and PilotNet. We experimented with combinations of different activation functions with-without dropout. The experiment has demonstrated the PilotNet model P1 can learn the entire task of non-surpass-able obstacle maneuvering and passing through a surpass-able obstacle. The experiment has provided us with a clear insight into the effect of each activation function and dropout

Table 4 Saliency map for PilotNet-based models

Model	Non-surpass-able left(alpha=0.004)	Surpass-able (alpha=0.003)	Non-surpass-able right(alpha=0.0045)
P2			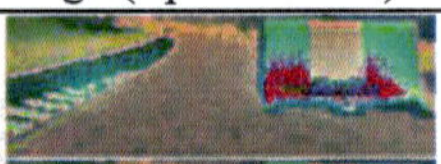
P3			
P1			
P4			
P5			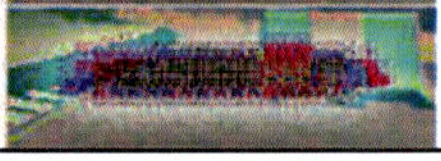

on steering angle prediction. PilotNet model P1 has the highest prediction accuracy, lowest val_loss, and reasonable processing time with the best visual saliency map for obstacles with the current dataset. The experiment has clearly concluded that PilotNet, with activation function elu without dropout, outperforms all other models and configurations. The system learned to mitigate through an obstacle without the need for explicit surpass-able and non-surpass-able obstacle labeling during training. In future work, we would like to optimize PilotNet to improve prediction accuracy. We would also like to introduce a custom-Net that would outperform all current autonomous driving methods.

References

1. Shah P, Pandya K, Shah H, Gandhi J (2019) Survey on vision-based hand gesture recognition. Int J Comput Sci Eng 7(5):281–288
2. An K, Lee S, Ryu S, Seo D (2018) Detecting a pothole using deep convolutional neural network models for an adaptive shock observing in a vehicle driving. 1–2. https://doi.org/10.1109/icce.2018.8326142
3. Li Q, Yao M, Yao X, Xu B (2009) A real-time 3D scanning system for pavement distortion inspection. Meas Sci Technol 15702–15709
4. Hou Z, Wang KCP, Gong W (2007) Experimentation of 3D pavement imaging through stereovision. In: International conference on transportation engineering, pp 376–381
5. Joubert D, Tyatyantsi A, Mphahlehle J, Manchidi V (2011) Pothole tagging system. In: 4th robotics and mechatronics conference of South Africa, pp 1–4
6. Moazzam KK, Mathavan S, Usman S, Rahman M (2013) Metrology and visualization of potholes using the Microsoft Kinect sensor. In: 16th international IEEE annual conference on intelligent transportation systems, pp 1284–1291
7. Bhatt U, Mani S, Xi E, Kolter J (2017) Intelligent pothole detection and road condition assessment
8. Pomerleau DA (1989) ALVINN, an autonomous land vehicle in a neural network. Technical report, Carnegie Mellon University

9. Bojarski M, Testa D, Dworakowski D, Firner B, Flepp B, Goyal P, Jackel L, Monfort M, Muller U, Zhang J, Zhang X, Zhao J, Zieba K (2016) End to end learning for self-driving cars
10. Bojarski M, Yeres P, Choromanska A, Choromanski K, Firner B, Jackel L, Muller U (2017) Explaining how a deep neural network trained with end-to-end learning steers a car. arXiv preprint arXiv:1704.07911
11. Zeiler MD, Taylor GW, Fergus R (2011) Adaptive deconvolutional networks for mid and high-level feature learning. In: 2011 international conference on computer vision, Barcelona, pp 2018–2025
12. Binder A, Lapuschkin S, Montavon G, Müller K-R, Samek W (2016) Layer-wise relevance propagation for deep neural network architectures, pp 913–922. https://doi.org/10.1007/978-981-10-0557-2_87
13. Loquercio A, Maqueda AI, del-Blanco CR, Scaramuzza D (2018) DroNet: learning to fly by driving. IEEE Robot Autom Lett 3(2):1088–1095

Retrieval of Trademark Images by Using Local Shape and SURF Feature Fusion

Rajesh Kumar Tripathi and Subhash Chand Agrawal

Abstract Trademark images are the copyright of any individual or commercial organizations. To protect the copyright, retrieval of trademark image is required which retrieves similar registered trademarks. This paper proposed a method which extract the shape features using Local Zernike Moments, orientation and description using SURF to retrieve the similar trademark images to protect the copyright. Then, a weighted similarity is computed by using the weighted average of Zernike moments and SURF to fuse. Euclidean distance is applied to retrieve the more similar images. Experiments have been carried out on proposed dataset and found that proposed approach works better and improves the accuracy.

Keywords Trademark image · Local Zernike moment · SURF · Image retrieval

1 Introduction

A trademark retrieval is a challenging area in the eld of image processing and image retrieval. Trademark is a unique identification of any commercial organization, individual or other authorized organizations to identify the services or products to the clients by a symbol or indication. The number of trademarks growing rapidly in different countries. A trademark can be a logo, symbol, image, phrase, design, text, or a combination of these. Trademark is mainly useful in making distinction of one commercial organization from other and their services or goods as well [13, 15].

In the recent scenario, trademarks are increasing day by day rapidly. Trademark image retrieval helps in unique trademark registration to any organization after confirming by matching proposed trademark with the retrieved trademark registered images. To protect the rights of registered trademarks, it is important to check the

R. K. Tripathi (✉) · S. C. Agrawal
Computer Engineering and Applications, IET, GLA University, Mathura, Uttar Pradesh, India
e-mail: frajesh.tripathi@gla.ac.in

S. C. Agrawal
e-mail: subhash.agrawalg@gla.ac.in

R. P. Mahapatra et al. (eds.), *Proceedings of 6th International Conference on Recent Trends in Computing*, Lecture Notes in Networks and Systems 177,
https://doi.org/10.1007/978-981-33-4501-0_48

uniqueness of the trademark by finding the similarity with the existing trademarks and trademarks that are to be registered.

Trademark image retrieval (TIR) can be performed by two methods: trademark image retrieval using description and image retrieval using content. Description based approach uses the text in which query is red in textual form [4].

In image retrieval based on content (CBIR) [11], a query is red through an image and its feature is generated to describe its content and retrieve similar trademark images. CBIR is very challenging because it requires a strong discriminating features such as texture feature, shape feature and color feature. To compute the similarity between query and gallery images of trademarks, first feature vectors are computed for query image and all the gallery trademark images stored in database.

We have proposed an approach to compute the local orientation, description and shape features of trademark images. We have applied Local Zernike moments [8, 12, 14] to extract shape feature and SURF (Speeded Up Robust Feature) for local texture orientation and description extraction based on key points to perform better.

Proposed approach is robust to trademark image retrieval with the following features:

1. Zernike Moment is a shape descriptor which is rotational invariant, robust to noise and minor changes in the shape and performs better to extract shape feature to discriminate the interclass similar images. The Zernike moments computation has been done on local regions of the image which makes retrieval of similar trademarks more better.
2. Local feature descriptor SURF is robust to illumination, rotation and scale invariant as well as better feature extractor. Proposed approach included Zernike moment as shape descriptor and SURF as a orientation and scale in-variant descriptor, which performs well in Trademark Image Retrieval (TIR).
3. The proposed approach and weighted similarity metric performed well to retrieve the similar images on proposed dataset and outperformed to Agrawal et al. [1].

Remaining part of the paper is structured as follows: Sect. 2 discusses the different image retrieval approaches, Sect. 3 has process of proposed work, Sect. 4 shows the result analysis of the proposed retrieval system of trademarks, and conclusion is presented in the Sect. 5.

2 Related Work

In this, we discuss the progress of the state-of-the-art in the retrieval of trademark images. Many researchers of this eld have developed many TIR systems to improve the performance.

Trademark [7], Star [17], and Artisan [5] are three important image retrieval systems of trademark images. There are different methods have been presented. The graphical feature vectors have utilized by a TIR system to interpret the trademark image automatically and computes the similarity score on the basis of human perception [7]. The Star system uses content-based TIR methods that include the moment invariants, grey level projection and Fourier descriptors. The Star system used spatial layout and shape components of an image [17]. However, it is di cult to recognize some significant components by an automated process [3]. Therefore, manual operation is required to segment some abstract trademark images to a certain extent. The Artisan system introduced a novel approach that adds principles derived from Gestalt psychology [5].

In 2001, Hsieh et al. [6] presented multiple classifier color image retrieval using region growing method. Wei et al. [16] proposed synthetic features to describe interior structure and global shape for the trademark image retrieval. Qi et al. [10] combined shape description and feature matching to retrieve trademark images. Anuar et al. [9] presented an integration of global-local descriptors to extract the features through Zernike's moments coefficients and edge gradiant co-occurrence matrix. Wu et al. [18] presented regional and boundary feature fusion for trademark image retrieval. Agrawal et al. [1] presented an architecture and combined texture and shape features to retrieve trademark images and performed well in case of rank-I identification while could not performed well in case of rank-5, rank-10 and rank-15. To improve the performance of TIR, we have used texture and local feature descriptor.

3 Proposed Approach

In this section, we described the step-sequences of proposed approach. Figure 1 shows the well defined sequence of our approach-local feature extraction through Zernike

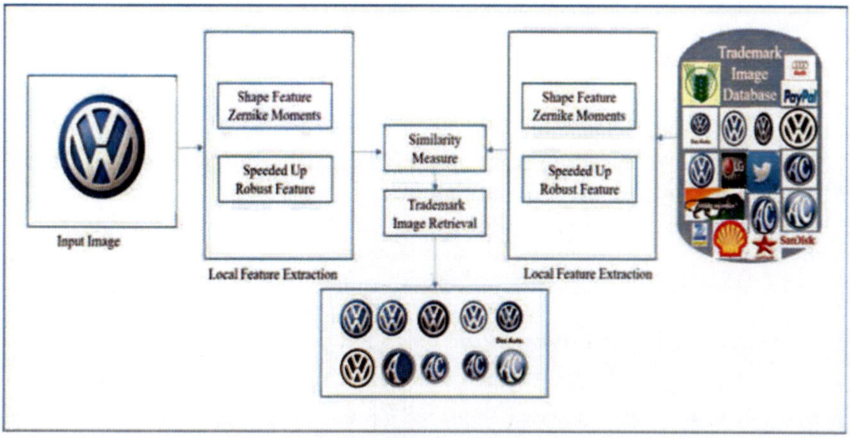

Fig. 1 Framework of the proposed approach presents feature extraction through Zernike Moment and SURF as a local feature descriptor, similarity measure using Euclidean distance and then retrieval of trademark images

Moment as shape feature of trademark image and SURF as a local feature based on scale and orientation invariant, similarity is measured using Euclidean distance and then rank-1, 5, 10, 15, 20 retrieval of trademark images.

3.1 Local Feature Extraction Through Zernike Moments (ZM)

In this subsection, we applied widely used Zernike Moment which is highly utilized in image retrieval (CBIR). We have following advantages of Zernike moments:

1. Magnitudes of Zernike moments are rotational invariant which is helpful to retrieve similar trademarks existing in different rotation.
2. Most important advantage of this shape descriptor is its robustness to noise as well as able to handle minor variations in shape.
3. It has minimum information redundancy due to its orthogonal property.

For the trademark images, if P_{xy} is the reference pixel, then local ZM is computed with given formula as below:

$$S_{u;v} = \frac{u+1}{X_x X_y} P_{xy} V_{uv}(x; y); \quad x^2 + y^2 \geq 1 \tag{1}$$

where $u = 0; 1; 2;$ ::::::::::::; 1 and v represents to angular dependence. $V_{uv}(r;) = R_{uv}(r)\exp^{jv}$ are the Zernike moments in polar coordinate system with some specified conditions, $x^2 + y^2 = 1$; $u \geq jvj$; $(uj\ vj)$ is even and $S_{uv} = S_{u;\,v}R_{uv}(b)$ is polynomial which is orthogonal, defined as:

$$R_{u;v}(b) = \frac{uvj}{1=0}(1)^{s1}\frac{s1(uv)^2}{s1!(u+jvj\,s1)!(uvj\,s1)}bu2s1 \tag{2}$$

$Z1 = jS_{uv}j$ yields the magnitude of ZM.

To estimate ZM of a trademark image, interested non-overlapping region is extracted and then mapping of that region is done in polar-coordinate system. To estimate, moments around every pixel is computed to form the feature vector.

3.2 SURF Feature Extraction

SURF feature descriptor is known as Speed Up Robust Features which is scale and orientation invariant point detector. There are two basic steps for feature extraction:

Orientation Computation. Firstly, we identify orientation of interest points to make feature extraction rotation invariant. First, responses of the Haar wavelet are obtained in x and y direction. Once the responses of wavelet are computed, weighted using a Gaussian which is centered at the point of interest, the responses are presented as vectors with the vertical strength along the ordinate and horizontal strength along the abscissa. The orientation which is dominant is computed by finding the addition of all the responses in a sliding window orientation which covers an angle of 3. The both vertical and horizontal responses are summed within the window. The sum of both responses provides a new feature vector.

Descriptor Components Computation. To extract the feature description, firstly a region of square dimension is constructed centered around the point of interest and along the selected orientation. The region is divided into smaller sub-regions of dimension 4 4 regularly which has important spatial information. We compute features at regularly spaced sample points for each sub-region. For simplification, we call hw_x as response of Haar wavelet in horizontal direction and $\mathrm{hw_y}$ as response of wavelet in vertical direction. Responses hw_x and hw_y are weighted using a Gaussian that is centered at the point of interest to improve the robustness towards localisation errors and geometric deformations.

After this, the hw_x and hw_y wavelet responses are added of each sub-region and used a first entries set to form the feature vector. To extract information about the polarity of the changes in intensity, we also compute the addition of the x and y direction responses, jhw_xj and jhw_yj. Hence, a 4-dimensional feature vector v for its intensity structure is calculated for each sub-region. This yields a descriptor vector of length 64 for all sub-regions of dimension 4 4. The descriptor is converted into unit vector to achieve invariance to contrast.

3.3 Similarity Measure

To generate the similarity score between query trademark image and stored trademark gallery images, Euclidean distance measure has been used. Euclidean distance measure yields better results in image retrieval scenario to nd similar images against the query. The Euclidean distance is shown as:

$$\mathrm{Dist}(q; D) = n_{(qi\ Di)^2} \tag{3}$$

In the above formula, Dist(q; D) is the distance computed between the feature descriptor vector of query q and trademark gallery images D.

In the proposed approach, we have computed the weighted similarity between Zernike moment's feature and SURF feature. Consider D_{zs} is the value of shape similarity and D_{ss} is the SURF similarity value. We computed weighted similarity D_{ws} as:

$$D_{ws} = \frac{w_{zs} D_{zs} + w_{ss} D_{ss}}{w_{zs} + w_{ss}} \tag{4}$$

where w_{zs} and w_{ss} are assigned weights of Zernike moments similarity and SURF similarity respectively.

4 Performance Analysis

To analyze the performance of the system, we have extended dataset [1] from 300 images to 750 trademark images. To validate the accuracy and effectiveness of the proposed TIR, we have tested on own dataset and existing dataset [1].

We have implemented TIR system on latest Intel system having processor i7 8th Gen CPU and 8 GB RAM, with operating system Microsoft Windows 10.

Matlab 2016a is used for the implementation and feature extraction and rank-1, 5, 10, 15, 20 retrieval. In the Graphical User Interface (shown in Fig. 2), pink button has two operation-first is to select the image that is to be select as input and second is to display the retrieved result.

The extended trademark database of [1] contains images with shape, color, illumination, size variations etc. The performance of the proposed approach is evaluated on the basis of precision. The precision [2] can be represented as:

$$\text{Precision} = \frac{N_A(q)}{N_R(q)} \tag{5}$$

Here, $N_A(q)$ is number of relevant trademark images retrieved and $N_R(q)$ is total retrieved trademark images as a result of query images.

Table 1 shows the rank-1, 5, 10, 15, 20 accuracy of the proposed approach and its components-Zernike moments and SURF descriptors.

Fig. 2 Eight correct retrieved results of trademark Cheetos from proposed dataset. First row is the query image and second and third row is the retrieved rank-10 matching results

Table 1 Performance of proposed approach on Own Trademark dataset

Approaches/rank	01	05	10	15	20
Shape descriptor	94.5	86.75	75.25	73.35	68.97
Speeded up robust feature (SURF)	91.5	82.67	73.13	72.73	66.89
Proposed approach	95.75	92.55	85.52	78.15	74.56

Fig. 3 Eight correctly retrieved results of Trademark of Audi from own dataset. First row is representing to query, second and third rows representing to the retrieved rank-10 matching results respectively

Figure 2 shows a GUI which has query image in first row and retrieved the results in second and third rows. Most of the images (8) are correctly retrieved while 2 images retrieved incorrectly in rank-10 retrieved images. This trademark image is text image and proposed approach works better with 80% accuracy.

Figure 3 shows the retrieved results of Trademark Audi. Retrieved results have 8 correct retrieved trademark images and 2 incorrect retrieved images while rings of Audi trademark are of different sizes with text. Proposed approach is robust to scale and orientation.

Figure 4 presents the retrieval results of Trademark Merritt College. Six images of this trademark have been successfully retrieved while four images retrieved incorrectly because there is only seven images of this trademark in the database. Proposed approach is robust to oriented images retrieval.

Figure 5 shows the retrieval results of Trademark ICAR. Six images of this trademark have been successfully retrieved while four images retrieved incorrectly because there is only six images of this trademark in the database. Finally, Proposed approach retrieves all the correct images related to query images existing in the dataset.

Fig. 4 Six correctly retrieved results of Trademark Merritt College. First row is representing to query, second and third rows representing to the retrieved rank-10 matching results respectively

Fig. 5 Six correctly retrieved results of Trademark ICAR. First row is representing to query, second and third rows representing to the retrieved rank-10 matching results respectively

Table 2 presents the performance of Shape descriptor-Zernike Moments, SURF, and proposed approach which fuse both and then compared with Agrawal et al. [1] on its trademark dataset.

Table 2 Performance analysis of different descriptors and proposed work on Existing Trademark dataset [1]

Approaches/rank	01	05	10	15	20
Shape feature	100	84.60	70.23	68.41	62.16
SURF	100	74.75	71.23	68.25	61.75
Agrawal et al. [1]	100	91.55	79.92	73.70	71.34
Proposed approach	100	92.15	83.17	76.15	73.35

5 Conclusion

This paper presents a robust approach to retrieve trademark images from a gallery images by ring a query in the form of an image. Shape descriptor and local descriptor performed well and provided a better result in comparison to existing approach. This approach handles orientation, scale, shape features to retrieve more similar trademark's images from the database. The proposed approach and its feature extractor performed well and yielded 100% accuracy in Rank-1 and more than 90% in Rank-5.

Rank-1 and Rank-5 accuracy is better in comparison to Rank-10, Rank-15 and Rank-20. In future, improvement is required in the approach to increase the recognition and retrieval accuracy of Rank 10, 15, 20 and more.

References

1. Agrawal D, Jalal AS, Tripathi R (2013) Trademark image retrieval by integrating shape with texture feature. In: International conference on information systems and computer networks, pp 30–33
2. Ahmed KT, Irtaza A, Iqbal MA (2017) Fusion of local and global features for effective image extraction. Appl Intell 47(2):526–543
3. Alwis S, Austin J (1998) A novel architecture for trademark image retrieval systems. In: Proceedings of the challenge of image retrieval, British Computer Society, pp 1–6
4. Chalechale A, Mertins A, Naghdy G (2004) Edge image description using angular radial partitioning. IEE Proc Vis Image Signal Process 151(2):93–101
5. Eakins JP, Graham ME, Boardman JM (1997) Evaluation of a trademark image retrieval system. In: Proceedings of the 19th annual BCS-IRSG conference on information retrieval research, pp 1–1. IRSG'97, BCS Learning & Development Ltd., Swindon, UK. http://dl.acm.org/citation.cfm?id=2227984.2227985
6. Hsieh IS, Fan KC (2001) Multiple classifiers for color ag and trademark image retrieval. Trans Image Proc 10(6):938–950. http://dx.doi.org/10.1109/83.923290
7. Kato T (1992) Database architecture for content-based image retrieval. In: Electronic imaging
8. Kim WY, Kim YS (2000) A region-based shape descriptor using Zernike moments. Sig Proc Image Commun 16:95–102
9. Mohd Anuar F, Setchi R, Lai YK (2013) Trademark image retrieval using an integrated shape descriptor. Expert Syst Appl 40(1):105–121. https://doi.org/10.1016/j.eswa.2012.07.031
10. Qi H, Li K, Shen Y, Qu W (2010) An effective solution for trademark image retrieval by combining shape description and feature matching. Pattern Recogn 43(6):2017–2027. http://dx.doi.org/10.1016/j.patcog.2010.01.007

11. Rajam IF, Valli S (2013) A survey on content based image retrieval. Life Sci J 10:2475–2487
12. Saryanidi E, Dal V, Tek SC, Tun B, Gkmen M (2012) Local Zernike moments: a new representation for face recognition. In: 2012 19th IEEE international conference on image processing, pp 585–588
13. Schietse J, Eakins JP, Veltkamp RC (2007) Practice and challenges in trademark image retrieval. In: Proceedings of the 6th ACM international conference on image and video retrieval. ACM, pp 518–524
14. Teague MR (1980) Image analysis via the general theory of moments. J Opt Soc Am 70(8):920–930. http://www.osapublishing.org/abstract.cfm?URI=josa-70-8-920
15. Tursun O, Denman S, Sivapalan S, Sridharan S, Fookes C, Mau S (2018) Component-based attention for large-scale trademark retrieval. CoRR. http://arxiv.org/abs/1811.02746
16. Wei CH, Li Y, Chau WY, Li CT (2009) Trademark image retrieval using synthetic features for describing global shape and interior structure. Pattern Recogn 42(3):386–394. https://doi.org/10.1016/j.patcog.2008.08.019
17. Wu J, Lam C, Mehtre B, Gao Y, Desai NA (1996) Content based retrieval for trademark registration. Multim Tools Appl 3
18. Wu M, Xiao W, Hong Z (2018) Similar image retrieval in large-scale trademark databases based on regional and boundary fusion feature. PLOS ONE 13:e0205002

Wind Power Forecasting Using Artificial Neural Networks (ANN) and Artificial Neuro-fuzzy Inference System (ANFIS)

Ujjwal, Shivam Kukrety, Likhitha Kakinada, and Shalini

Abstract Prediction as a matter of fact has always been a key area of interest for the researchers. Prediction of anything can be done in an approximate way, and it cannot be made precisely as anything that is yet to happen cannot be predicted precisely. Weather prediction has been amongst the recent developments in the field of predicting the future. Weather is one such natural phenomenon which till date is mostly predicted using highly sophisticated machines that measure the value of past and with the study of that data by humans gives us the future value. In this paper, we would like to make weather prediction a more autonomous thing requiring minimum interference by humans with the use of artificial neural networks (ANN) as well as fuzzy inference systems. Here the ANN is used basically to memorize the past data of the atmosphere while the fuzzy inference engine is used for decision making as to what is the outcome of a particular output given by the ANN. With the combination of ANN and fuzzy, we would be able to mimic the human brain to some extent and that way some of the tasks will be less time-consuming and a lot more economical.

Keywords Artificial neural network · Fuzzy logic · Soft computing · Forecasting · Fuzzification · Defuzzification

Ujjwal · Shivam Kukrety · Likhitha Kakinada · Shalini (✉)
Department of Electrical Engineering, Delhi Technological University, Rohini, Delhi, India
e-mail: shalini.chauhan41196@gmail.com

Ujjwal
e-mail: ujjwalthakur1507@gmail.com

Shivam Kukrety
e-mail: shivam.k@hotmail.com

Likhitha Kakinada
e-mail: vidyalikhitha.k@gmail.com

R. P. Mahapatra et al. (eds.), *Proceedings of 6th International Conference on Recent Trends in Computing*, Lecture Notes in Networks and Systems 177,
https://doi.org/10.1007/978-981-33-4501-0_49

1 Introduction

A wind power forecast corresponds to an estimate of the expected production of one or more wind turbines (referred to as a wind farm) in the near future [1, 2]. Wind forecasting is utterly needed w.r.t the fact that in the electricity grid, at any moment, the balance must be maintained between the electricity consumption and the generation; otherwise, disturbances in the power quality or supply may not occur. Wind generation is a direct function of wind speed and in contrast to conventional generation systems is not easily dispatchable [3, 4]. Fluctuations of the wind generation thus receive a great amount of attention, and the main factors include most prominently the constant changes in the weather and many other factors effecting it, thereby, respectively, including wind speed and hence the respective voltage, etc. So, instead of using the complex machinery and enormous human effort in order to forecast the respective wind data from the previously collected data, we have to use the ANN. Network so as to do so. The artificial neural network helps in the memorization of all the previously connected data and hence helps in the previous data recollection automatically via it [5, 6]. Hence, it can be stated that the ANN actually remembered each and every data from the respective previously collected data. So, actually helps in eliminating the works of the complex machineries completely. Whereas, the fuzzy inference by humans with the use of artificial networks as well as fuzzy inference systems [7, 8].

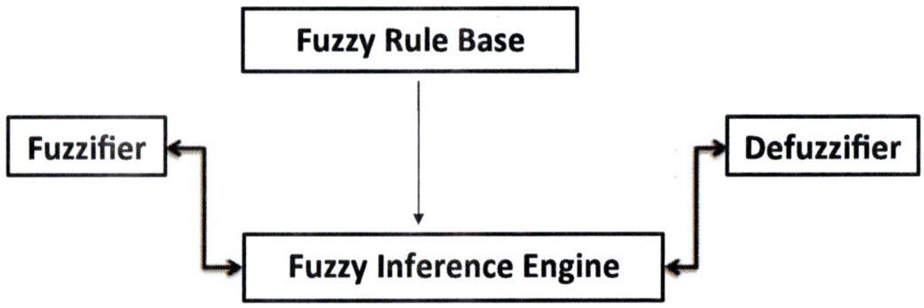

Fig. 1 Basic diagram of how a fuzzy inference engine works

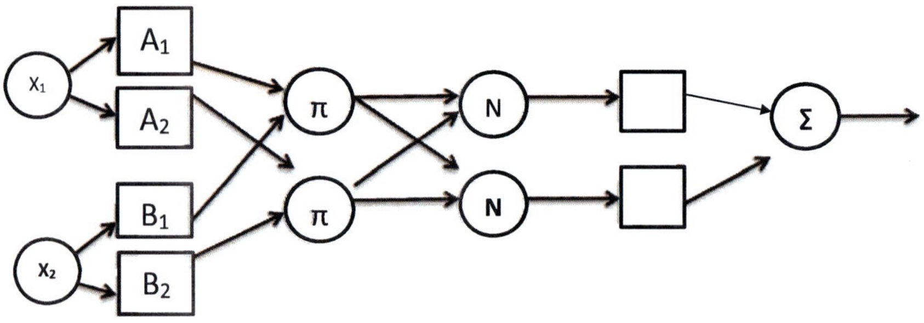

Fig. 2 A basic diagram of ANFIS

The proposed ANN evaluates the performance of the developed models by applying different transfer functions, hidden layers and neurons to predict temperature for 365 days of the year. The criteria used for appropriate model selection are mean square error (MSE) [9, 10]. Contrary to similar researches, the data model and workflow suggested in the paper generated lesser MSE (i.e. more accurate results) that too with reduced computational complexity (i.e. better performance). The advantage which ANN has over other weather forecasting method is that the ANN minimizes the error using various algorithms and gives us a predicted value which is nearly equal to the actual value. Such network is simulated over newer data to find out the weather trend in future course. The neural network fitting tool GUI nntool available in MATLAB 7.6.0 (R2008a) is used to carry out the analysis on the weather data using artificial feed-forward neural network with back-propagation principles. The neural network fitting tool GUI nntool available in MATLAB 7.6.0 (R2008a) is used to carry out the analysis on the weather data using artificial feed-forward neural network with back-propagation principles. The main aim is to create and train a network that can predict the individual weather components, e.g. maximum temperature, minimum temperature, wind speed, etc., for a particular station and a particular day given the weather for the previous day (target data) and the historical 3 year data of that particular day (input data) [11–13].

2 Algorithm

2.1 The Levenberg–Marquardt Algorithm

The Levenberg–Marquardt algorithm is an algorithm which is used to solve the nonlinear least-squares problems in computing [14–16]. This method is also called as damped least-squares (DLS) method.

The LMA is used in many software applications for solving generic curve-fitting problems. However, as with many fitting algorithms, the LMA finds only a *local minimum*, which is not necessarily the *global minimum*. The LMA interpolates between the *Gauss–Newton algorithm* (GNA) and the method of *gradient descent*. The LMA is more *robust* than the GNA, which means that in many cases, it finds a solution even if it starts very far off the final minimum. For well-behaved functions and reasonable starting parameters, the LMA tends to be a bit slower than the GNA. LMA can also be viewed as *Gauss–Newton* using a *trust region* approach.

The algorithm was first published in 1944 by *Kenneth Levenberg*, while working at the *Frankford Army Arsenal*. It was rediscovered in 1963 by *Donald Marquardt*, who worked as a *statistician* at *DuPont*, and independently by Girard, Wynne and Morrison, respectively.

Now, defining new parameter in Gauss–Newton algorithm, we can deal with convergence guarantee, convergence rate and matrix singularity.

$$\left(J(m^k)^T.J\left(m^k\right)+\lambda\mathbf{I}\right)\left(\boldsymbol{m}^{k+1}-\boldsymbol{m}^k\right)=-\boldsymbol{J}\left(\boldsymbol{m}^k\right)\boldsymbol{F}\left(\boldsymbol{m}^k\right)$$

On the choice of the value of λ:

$$m^{k+1}=m^k+\left(J^TJ+\lambda\mathbf{I}\right)^{-1}.J^T\Delta d$$

If $\boldsymbol{\lambda} \gg \geq \boldsymbol{\lambda}\mathbf{I} > J^T J$

$$m^{k+1}=m^k+(\lambda\mathbf{I})^{-1}.J^T\Delta d$$

If $\boldsymbol{\lambda} \ll\geq \boldsymbol{\lambda}\mathbf{I} < J^T J$

$$m^{k+1}=m^k+\left(J^TJ+\lambda\mathbf{I}\right)^{-1}.J^T\Delta d$$

Basically, the fuzzy system consists of three main parts described as fuzzification, inference and defuzzification and their interrelations can be hereby introduced via given below descriptive flowchart.

2.2 ANFIS

Talking about ANFIS, i.e. an adaptive neuro-fuzzy inference system or adaptive network-based fuzzy inference system (ANFIS), it basically integrates both neural networks and *fuzzy logic* principles, and it has potential to capture the benefits of both in a single *framework*. Its inference system corresponds to a set of fuzzy *IF-THEN rules* that have learning capability to approximate functions. Hence, ANFIS is considered to be an estimator. For using the ANFIS in a more efficient and optimal way, one can use the best parameters obtained by algorithm. It has uses in intelligent situational-aware energy management system [17, 18].

It is possible to identify two parts in the network structure, namely premise and consequence parts. In more details, the architecture is composed by five layers. The first layer takes the input values and determines the *membership functions* belonging to them. It is commonly called fuzzification layer. The membership degrees of each function are computed by using the premise parameter set, namely $\{a, b, c\}$. The second layer is responsible of generating the firing strengths for the rules. Due to its task, the second layer is denoted as "rule layer". The role of the third layer is to normalize the computed firing strengths, by diving each value for the total firing strength. The fourth layer takes as input the normalized values and the consequence parameter set $\{p, q, r\}$. The values returned by this layer are the defuzzificated ones, and those values are passed to the last layer to return the final output. The figure shows the ANFIS architecture [19–22].

3 Result

We have performed this experiment with the help of the following two approaches:

1. Basic artificial neural network
2. Artificial neuro-fuzzy inference system.

These two approaches have been explained in detail in the above introduction.

In the ANN model, we have given one year data for training of the neural network. To predict how close is the predicted output to the target value, we calculate the error between the two quantities. A graph of the error between the two has been plotted as shown in figure. From the figure, it is quite evident that the error is almost between 10 and 15%.

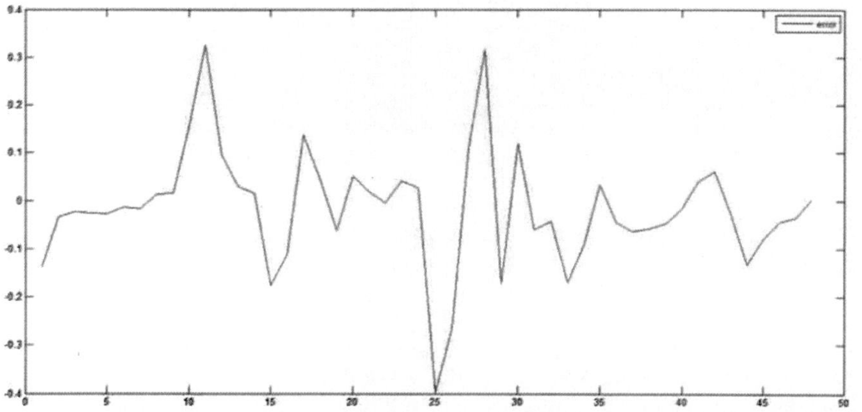

The second approach is by the use of ANFIS for wind prediction. In this, we give the data same as in neural networks and the processes are also quite similar to that of an ANN, but the output is scaled with the help of a fuzzy inference engine to give a better output with reduced error. The error in this case is significantly reduced to about 2–5%.

In both the above cases, i.e. the error is significantly lower because of optimization of data, if we use raw data there may be a significant amount of error jump example of which is that when ANN is used on raw data for prediction, the error is about 25–30%.

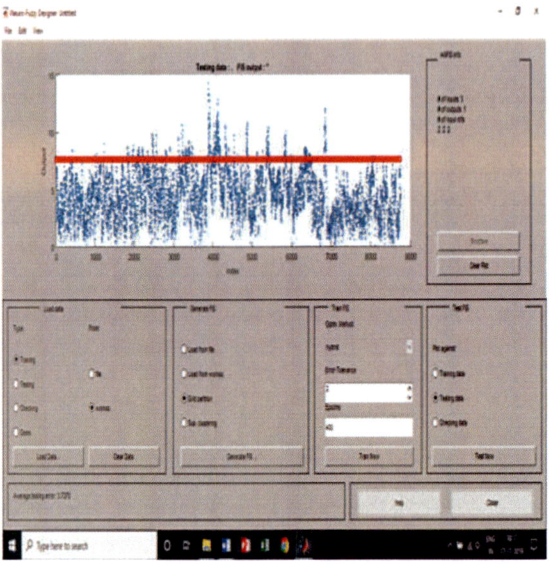

4 Discussion

As we all know wind speed varies time to time proportionally wind power also varies, so to predict value using ANN it requires lot of data. ANN can work efficiently on this much amount of data and have adaptability property; this makes ANN tool to predict optimum value. We have predicted value using ANN and ANFIS, and this gives us optimum results than previous research done. Previously work was done load prediction.

References

1. Barbounis TG, Theocharis JB (2006) Locally recurrent neural networks for long-term wind speed and power prediction. Neurocomputing 69(4–6):466–496
2. Filik UB, Gerek ON, Kurban M (2009) Hourly forecasting of long term electric energy demand using a novel modeling approach. In: Proceedings of 4th International Conference on Information Control (ICICIC). Innovation Computer, pp 115–118
3. Sideratos G, Hatziargyriou ND (2012) Probabilistic wind power forecasting using radial basis function neural networks. IEEE Trans Power Syst 27(4):1788–1796
4. Soman SS, Zareipour H, Malik O, Mandal P (2010) A review of wind power and wind speed forecasting methods with different time horizons. In: IEEE North American power symposium, pp 1–8
5. Li S, Wunsch DC, O'Hair EA, Giesselmann MG (2001) Using neural networks to estimate wind turbine power generation. IEEE Trans. Energy Convers 16(3):276–282

6. Mandal P, Senjyu T, Urasaki N, Funabashi T (2006) A neural network based several-hour ahead electric load forecasting using similar days approach. Electr Power Energy Syst 28:373; Mandal P, Senjyu T, Funabashi T (2006) Neural networks approach to forecast several hour ahead electricity prices and loads in a deregulated market. Energy Convers Manage 47:2128–2142. Dash PK, Liew AC, Rahman S, Fuzzy neural network and fuzzy expert system for loads
7. Ho KL, Hsu YY, Yang CC (1992) Short-term load forecasting using a multilayer neural network with an adaptive learning algorithm. IEEE Trans Power Syst 7(1):141–148; Peng TM, Hubele NF, Karady GG (1992) Advancement in the application of neural networks for short-term load forecasting. IEEE Trans Power Syst 7(1):250–256; Mohammed O, Park D, Merchant R, Dinh T, Tong C, Azeem A, Farah J, Drake C (1995) Practical experiences with an adaptive neural network short-term load forecasting system. IEEE Trans Power Syst 10(1):254–265
8. Soucek B, Soucek M (1988) Neural and massively parallel computers: the sixth generation. Wiley, New York; Guoqiang Z, Eddy Patuwo B, Hu MY (1998) Forecasting with artificial neural networks: the state of the art. Int J Forecasting 14:35–62; Widrow B, Stems D (1985) Adaptive signal processing. Prentice-Hall, New York
9. Gross G, Galiana FD (1987) Short term forecasting. Proc IEEE 75(12):1558–1573
10. Kandil N, Sood VK, Khorasani K, Patel RV (1992) Fault identification in an AC–DC transmission system using neural networks. IEEE Trans Power Syst 7(2):812–819
11. Kandil N, Wamkeue R, Saad M, Georges S (2006) An efficient approach for short-term load forecasting using artificial neural networks. J Electr Power Energy Syst 28:525–530. Ranaweera DK, Hubele NF, Papalexopoulos AD (1995) Application of radial basis function neural network model for short term load forecasting. IEE Proc Gener Trans Distrib 142(1)
12. Constantinopoulos C, Likas A (2006) An incremental training method for the probabilistic RBF network. IEEE Trans Neural Networks 17(4)
13. Focken U, Lange M, Mönnich K, Waldl H-P, Beyer HG, Luig A (2002) Short-term prediction of the aggregated power output of wind farms e a statistical analysis of the reduction of the prediction error by spatial smoothing effects. J Wind Eng Ind Aerodyn 90:231e46
14. Lange M, Focken U (2005) Physical approach to short-term wind power prediction. Springer, Berlin
15. Giebel G, Badger J, Landberg L, Nielsen HA, Nielsen T, Madsen H et al (2005) Wind power prediction ensembles. Report 1527. Risø National Laboratory, Denmark
16. Lang SJ, McKeogh EJ (2009) Forecasting wind generation, uncertainty and reserve requirement on the Irish power system using an ensemble prediction system. Wind Eng 33(5):433e48
17. Lang S, Möhrlen J, Jørgensen J, O'Gallachóir BP, McKeogh E (2001) Application of a multi-scheme ensemble prediction system for wind power forecasting in Ireland and comparison with validation results from Denmark and Germany. In: Proceedings of the European wind energy conference EWEC2001, Copenhagen, Denmark
18. Torres JL, García A, de Blas M, de Francisco A (2005) Forecast of hourly averages wind speed with ARMA models in Navarre. Solar Energy 79(1):65e77
19. El-Fouly THM, El-Saadany EF, Salama MMA (2006) Grey predictor for wind energy conversion systems output power prediction. IEEE Trans Power Syst 21:1450e2
20. Damousis IG, Dokopoulos P (2001) A fuzzy model expert system for the forecasting of wind speed and power generation in wind farms. In: Proceedings of the IEEE international conference on power industry computer applications PICA 01
21. Mandic DP, Javidi S, Goh SL, Kuh A, Aihara K (2009) Complex-valued prediction of wind profile using augmented complex statistics. Renew Energy 34(1):196e201
22. Negnevitsky M, Johnson P, Santoso S (2007) Short-term wind power forecasting using hybrid intelligent systems. In: Proceedings of the IEEE power engineering society general meeting, Tampa, Florida, USA

Implementation of Coordinate Rotation Digital Computer (CORDIC) Processing Unit by Using of VLSI Technology

Nitesh Kumar Sharma, Shanti Rathore, and M. R. Khan

Abstract In present, this era is based on 5G and advance computer vision; computer vision at present there is lots of mobile user are increase due to that there is need of fast processing unit which is able to represent video or audio data on any mobile usage in very less time. As we know in any processor, ALU unit is there which will perform the arithmetic operation but now days for all trigonometry-based calculation, there is need of CORDIC processor. Most of the computer vision algorithms need the computation consisting of two steps—namely arithmetic computation and trigonometric computation. The complexity is mostly by the trigonometric computations, which generate a need for separate algorithms which perform the entire trigonometric task. In this work, one well-known algorithm called Coordinate Rotation Digital Computer (CORDIC), alternatively known as digit by digit technique, and Volder's calculation has been addressed. In this, a novel processing strategy is utilized which is particularly reasonable for understanding the trigonometric connections associated with plane organize revolution and change from rectangular to polar directions. CORDIC calculation is additionally appropriate for square root, logarithmic, exponent capacity, and for computer-based computations.

Keywords Computer vision · DIP · DSP · CORDIC · Communication · Latency · VLSI · Hardware

N. K. Sharma (✉) · S. Rathore
Electronics and Communication Engineering, Dr. C.V. Raman University, Bilaspur, India
e-mail: sharma786.nitesh@gmail.com

S. Rathore
e-mail: rathoreshanti@gmail.com

M. R. Khan
Electronics and Telecommunication Engineering, Government Engineering College, Jagdalpur, India
e-mail: mrkhan@gecjdp.ac.in

R. P. Mahapatra et al. (eds.), *Proceedings of 6th International Conference on Recent Trends in Computing*, Lecture Notes in Networks and Systems 177,
https://doi.org/10.1007/978-981-33-4501-0_50

1 Introduction

We are living in the era of fast processing system, where every task will be finish in few second; currently, every single person is user of the mobile, laptop. Every single user of laptop and mobile technology need a system which will solve their problem in few second. In the present era of social media, where everyone want to upload pictures, video and multimedia data in good quality [1] generate the need of good quality image or video editors; based on computer vision technology, there is need of usage which help user to use these kind of filters which will improve the quality [2, 3]. Apart from consumer sector, if we are talking about the other sectors like medical [4], aerospace [5], defense [6] so there is need of real time computer vision algorithm which will improve the quality and analysis level. Now, if we are talking about that kind of computer vision algorithm like DCT, DWT, JPEG, etc. So there is one common thing which is calculation of angle and regarding that there is need of one separate algorithm which will perform the trigonometric analysis. The coordinate rotation digital computers (CORDIC) algorithm has the efficiency to perform the trigonometric analysis in quick time. The computer vision scene has for sometime been overwhelmed by the microprocessor with upgrades. While these processors are ease and offer extraordinary adaptability, they are regularly not quick enough for genuinely requesting CV or Digital Image Processing (DIP) undertakings. On existing CORDIC processing unit, here, we did the analysis by using of Verilog language, and all simulation is also perform on Verilog only, by using of Xilinx 14.2. Here, we did the analysis on sparten3, Spartan6, artix7 FPGA.

2 Literature Review

The first CORDIC calculation is made in 1959 right now urged answer for trigonometric capacity and turn computation, yet this strategy is looking with various issue like significant gear is crave, consistent term, and scale factor is also a critical issue [7]. Enlargement curious strategy is made [8], right now look at some curious new gatherings like log, exponential, and square root yet this access is furthermore defying a comparative difficult issues scale factor, colossal hardware, steady term. After the above methodologies, a huge amount of progress is done in CORDIC calculation, and many summarized strategy is urged for enlisting assorted limits like inverse of cosecant and secant [9, 10], changes [11, 12], models/inverse of anti-logarithms, square roots, Eigen regards [13], etc. Thing of the past [14], there has been noteworthy advances in the structure of the count to beat its genuine weaknesses. In [15–17], makers propose the usage of enthusiastic journey estimations for perceiving the scaled down scale turns.

The capability of these systems relies upon the probability of turn point which is the essential issue. Execution of it with respect to gear is inconvenient, and it is going up against another difficult issues. For reduction of scale factor issue, low multifaceted nature framework is used, and the system is Taylor plan advancement which in like manner has a few detriments. Extent of blend is a critical issue; for decrease of this issue, some new strategy is prescribed.

References [18, 19] are frameworks reliant on Taylor game plan approach. The past encounters low extent of association (RoC) which renders it unacceptable for rational usage, while the last grows the RoC yet them standing up to issue of reliable scale factor. The scaling free CORDIC and changed without scale CORDIC [20, 21] in [21] maker prescribed new technique for period of sine/cosine, right now discard a ROM and an enormous barrel shifter in the gear use of the CORDIC structure, yet this system encounters low extent of blending (RoC) which renders sit prohibited for useful usage.

In [22], maker urged a framework for abatement of scale factor and number of cycles. They focused on Radix-4 modified booth recording—Modification of CORDIC count is Radix-4 balanced corner recording. In past work, purpose of the without scaling CORDIC partition is between 0 to rad anyway right now degree is between $-\frac{\pi}{2}$ and $\frac{\pi}{2}$, [21, 22] in like manner facing steady scale factor issue. In [23], here, maker proposes the principle one piece distinguishing proof framework to perceive the little scope turns. The scale free arrangement of the urged count relies upon Taylor course of action expansion of the sine and cosine waves. In [24], maker use a comparative philosophy yet there they execute hyperbolic CORDIC work.

An equipment capable framework for creating inverse of cosecant and secant waves reliant on the Coordinate Rotation Digital Computer (CORDIC) calculation. Causo et al. [25] this urgeda technique right now utilizes driving one piece revelation methodology to perceive the littler scope turns. This strategy takes out complex request count. The scale-free arrangement of the estimation relies upon Taylor game plan breadth inverse of cosecant and secant. This urged figuring is up to fourth demand of Taylor game plan. This urged computation is moreover looking with various issues like augmentation in goof; number of accentuation increases; to diminish this issue, another procedure is prescribed [26]. Right now, using same strategy of [25] but they convert that building in to resemble structure, they urged CORDIC equal rotator use to be maximally overhauled for world class with the lower cost in an area eating up. Aggarwal et al. [21] this paper shows a zone time capable CORDIC count that absolutely discards the scale factor. And accomplishes the perfect extent of association. Other than we have urged an estimation to rename the simple plots for lessening the amount of CORDIC cycles. A summarized little scope turn assurance strategy subject to quick most basic 1-area hinders the mind boggling mission figurings for perceiving the scaled down scale upheavals. Causo et al. [22] exhort an ingenious idea type of the CORDIC figuring by familiarizing an equivalent rotator skilled with turn for more than one littler scope transformation point per time. Procedures for picking the progressive scaled down scale upheaval focuses consolidates similarly as the most appropriate plan for the urged approach are here reported. Aggarwal and Khare [25]: This paper shows a gear profitable plan for making inverse of cosecant and secant waves subject to the Coordinate Rotation Digital Computer (CORDIC) count. In this one of a kind structure, the CORDIC encounters noteworthy drawbacks like scale factor figuring, torpidity, and perfect assurance of little scope turns. The urged figuring beats all of these drawbacks. We use driving one piece revelation framework to recognize the little scope turns. The scale-free plan of the urged estimation relies upon Taylor game plan expansion of the sine and cosine waves. Aggarwal et al. [26] this paper exhibits a novel totally sans scaling CORDIC calculation in revolution mode for hyperbolic direction. We utilize most noteworthy 1 piece identification strategy for miniaturized scale turn succession age to decrease the quantity of cycles. By putting away the inverse of cosecant/and secant qualities at octant limits in a ROM, we can stretch out the scope of assembly to the whole arrange space. These are previous researches which are related to the CORDIC algorithm.

3 Research Issue and Future Scope

In this section, basically, we talk about research gap which need to be solved; as per detailed study, we found some of the common issue on most the previous existing approach, and those are followings:

1. **Issue of scale factor**: Many architecture are facing from scaling factor dilemma and which were focus on by many researchers in return the intermission which has been neglected in the improved approach.
2. **Time complexity**: According to previous existing approaches, most of the approaches are have the time complexity problem; they need large time to complete the task somehow [14] which is good in terms of time complexity aside from curtail intermission of hardware.
3. **Hardware complexity**: Numerous calculations crave multiplier factor some are crave steady multiplier factor some crave number of shifter rationale and big size adder and subs-tractor.
4. **Region of convergence**: It should be maximized with error resiliency.
5. **Lack in reliability**: Most of the approaches are not good in terms of reliability as generated related is not up to the mark.

Future Scope on CORDIC Algorithm

CORDIC calculation is significant in various usage like DSP, multimedia, and function generation yet sat the same time CORDIC processor faces various issues which I inspected in look into gap. So yonder there are succeeding goals which will extricate thenceforth:

1. **In VLSI, three parameters are most important which are:**

   (a) Power
   (b) Area
   (c) Speed
   There are no any approaches which are able to make justice with all these parameters.

2. **Improvement in time complexity**
3. **Reduction in scale factor**
4. **Need improvement in terms if reliability**
5. **Reduction in accuracy issues.**

4 Methodology and Implementation

In this section, we talk about the basic CORDIC algorithm and new CORDIC algorithm apart from that we also talk about the implementation of those approaches. Here, we also talk about the usage level analysis.

4.1 Conventional CORDIC Algorithm

CORDIC ALGORITHM the hidden guideline of the CORDIC calculation depends on two-measurement geometry. This calculation works revolution, succeeding straight, round, or hyperbolical directions. We center around revolution method of activity in roundabout direction.

1. Customary CORDIC algorithm let the vector *Va*[*Xa*, *Yb*] be derived by rotating the vector *Vb*[*Xa*, *Yb*] through an angle ⊖, then:

$$\begin{bmatrix} Xb \\ Yb \end{bmatrix} = Rp.\begin{bmatrix} Xa \\ Ya \end{bmatrix}, Rp = \begin{bmatrix} \mathrm{Cos}\Theta & -\mathrm{Sin}\Theta \\ \mathrm{Sin}\Theta & \mathrm{Cos}\Theta \end{bmatrix} \tag{1}$$

Here, Eq. (1) shapes the fundamental rule for iterative arrange calculation in CORDIC algorithm [1].

$$\Theta = \sum_{I=0}^{B} ui * ai \tag{2}$$

where ui $= -1, 1$; ai $= \tan^{-1} 2^{-i}$.

As we can see in Eq. (3), the scale factor *Ki* is free and not dependent of the direction of micro-rotation

$$Rp = Ki.\begin{bmatrix} 1 & -ui.2^{-i}\Theta \\ ui.2^{-i} & \mathrm{Cos}\Theta 1 \end{bmatrix} \tag{3}$$

Fig. 1 Angel classification

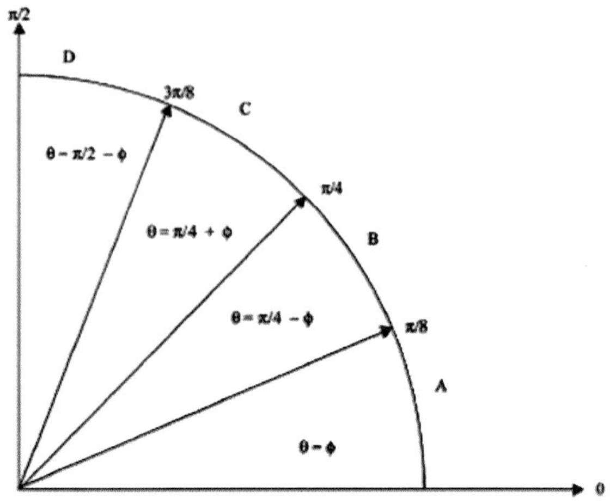

4.2 Scaling CORDIC

Scaling CORDIC was the principal endeavor to configuration scale-free coordinate. CORDIC conditions utilizing Taylor arrangement extension. Here, the smaller scale pivots are confined to against clockwise heading just, with the end goal that any point of turn is spoken to as the logarithmic whole of basic edges. The sine and cosine capacities are approximated to (Fig. 1):

$$\mathrm{Sin}ai = 2^{-i}, \mathrm{Cos}ai = 1 - 2^{2i+1} \tag{4}$$

$$\left[\frac{(b - 2.585)}{3}\right] \le i \le b - 1 \tag{5}$$

Region of Convergence = 22.5Θ.

4.3 Efficient CORDIC Algorithms

In this, author eliminates a ROM and a large barrel shifter in the hardware realization of the CORDIC system.

$$\lim_{i-\infty} \frac{\tan\left(2^{-i}\right)}{2^{-i}} = 1 \tag{6}$$

Here, K_0 is restricted to 0.707.

4.4 Taylor CORDIC Algorithms

The phase address θ is divided into the upper address ϕ and the lower address $\theta - \phi$. The second order Taylor series approximation computed around ϕ is expressed in

$$\sin\left(\frac{\pi}{2}\theta\right) = \sin\left(\frac{\pi}{2}\phi\right) + k1(\theta - \varphi)\cos\left(\frac{\pi}{2}\phi\right) + O(3) \tag{7}$$

Here, we use Verilog language, and tool is Xilinx 14.2 for designing and simulation. For analysis point of view, we use virtual FPGA Sparten 6.

5 Result and Analysis

In this section, we introduce the relative investigation of all with past existing methodology. For analysis point of view, we use these parameters:

1. Power
2. Speed
3. Hardware (LUT)

Time Analysis
See Fig. 2 and Table 1.

Power Analysis
See Fig. 3 and Table 2.

LUT Analysis
See Fig. 4 and Table 3.

As per the comparative analysis, we can see the efficient CORDIC is far better than other approaches but still there is lots of improvement is crave.

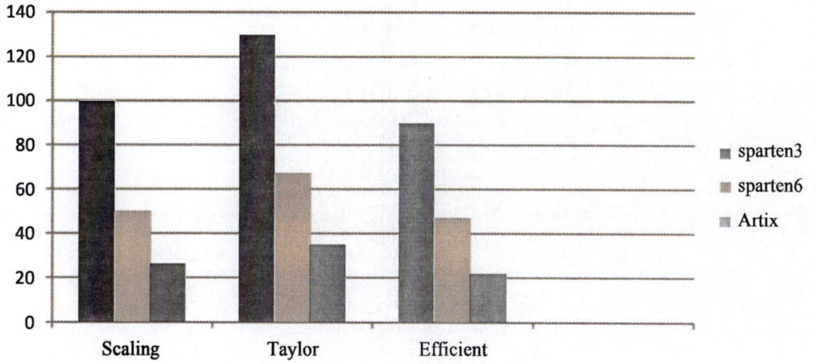

Fig. 2 Time analysis

Table 1 Comparative time analysis

FPGA type	Scaling free (μs)	Taylor based (μs)	Efficient CORDIC (μs)
Sparten 3	100	130	90.2
Sparten 6	50	67.23	47.2
Artix 7	26.36	35.23	22.12

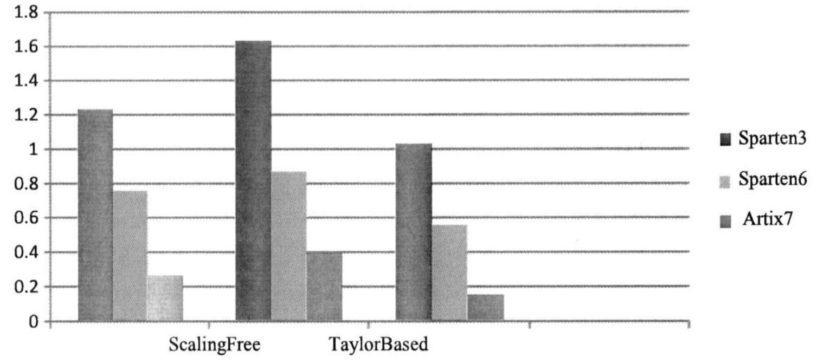

Fig. 3 Power analysis

Table 2 Comparative power analysis

FPGA type	Scaling free (mW)	Taylor based (mW)	Efficient CORDIC (mW)
Sparten 3	1.23	1.63	1.03
Sparten 6	0.76	0.87	0.56
Artix 7	0.26	0.39	0.156

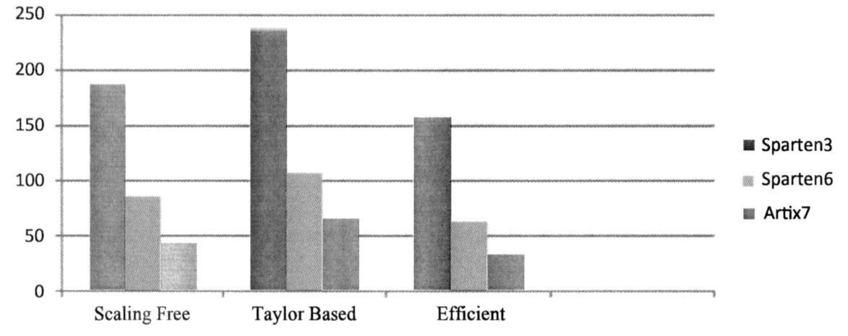

Fig. 4 LUT analysis

Table 3 Comparative LUT analysis

FPGA type	Scaling free	Taylor based	Efficient CORDIC
Sparten 3	187	237	157
Sparten 6	86	107	63
Artix 7	43	65	32

6 Conclusion

The greatness of CORDIC is its strength for united game plan for a colossal game plan of computational assignments including the appraisal of scientific and powerful limits, tally of duplication, division, square root and inverse of anti-logarithm, game plan of direct systems, QR-rot, and SVD, etc. Likewise, CORDIC is executed by a clear gear through reiterated move incorporate exercises. These attribute of CORDIC has chosen it an engaging option for avast assortment of uses. Over the latest fifty years, a couple of counts and models have been made to quicken the CORDIC by reducing its cycle checks and through its conduit execution. Additionally, its usage in a couple of various locales including signal planning, picture taking care of, correspondence, apply self-rule what's more, structures isolated from general sensible, and specific counts have been explored. Inaction of count, in any case, continues being the critical impediment of the CORDIC computation, since we don't have beneficial figuring's for its equivalent use. Right now did the similar investigation; thus, we discovered productive CORDIC [21] calculation which is best methodology.

References

1. Barozzi S et al (2019) Filtering images extracted from social media in the response phase of emergency events. In: 16th Conference on information systems for crisis response and management
2. Parmar Y, Sridharan K (2019) A resource-efficient multiplierless systolic array architecture for convolutions in deep networks. IEEE Trans Circ Syst II Express Briefs
3. Gurantz I, Goldenberg Y, Raghavan SA (1996) Demodulator for consumer uses. U.S. Patent No. 5,550,869, 27 Aug 1996
4. Wang S, Wen Z, Yu L (2006) High-performance fault-tolerant CORDIC processor for space usage. In: 2006 1st International symposium on systems and control in aerospace and astronautics. IEEE
5. Huang Y, Yang Z-Q, Guo L-J (2010) Usage of CORDIC algorithm in processing of tethered pose of aerostat radar. J Hefei Univ Technol (Nat Sci) 11
6. Hore A, Ziou D (2010) Image quality metrics: PSNR vs. SSIM. In: Twentieth international conference on pattern recognition. IEEE
7. Walther JS (1971) A unified algorithm for elementary functions. In: Spring joint computer conference, pp 379–385
8. Hsiao SF, Hu YH, Juang TB (2004) A memory efficient and high speed Sine/Cosine generator based on parallel CORDIC rotations. IEEE Signal Process Lett 11(2):152–155
9. Takagi N, Asada T, Yajima S (1991) Redundant CORDIC methods with a constant scale factor for sine and cosine computation. IEEE Trans Comput 40:989–995
10. Maharatna K, Banerjee S (2001) A VLSI array architecture for Hough transform. Pattern Recogn 34:1503–1512
11. Maharatna K, Dhar AS, Banerjee S (2001) A VLSI array architecture for realization of DFT, DHT, DCT and DST. Signal Process 81:1813–1822
12. Hu YH, Chern HM (1990) VLSI CORDIC array structure implementation of Toeplitz Eigen system solvers. In: Proceedings of IEEE international conference acoustics, speech, signal processing, NM, pp 1575–1578
13. Meher PK, Walls J, Juang T-B, Sridharan K, Maharatna K (2009) 50 years of CORDIC: algorithms, architectures and usage. IEEE Trans Circ Syst I Regul Pap 56(9):1893–1907

14. Wu CS, Wu AY (2001) Modified vector rotational CORDIC (MVR-CORDIC) algorithm and architecture. IEEE Trans Circ Syst II 48:548–561
15. Wu C-S, Wu A-Y, Lin C-H (2003) A high-performance/low-latency vector rotational CORDIC architecture based on extended elementary angle set and trellis-based searching schemes. IEEE Trans Circ Syst II Analog Digital Signal Process 50(9):589–601
16. Hu YH, Naganathan S (1993) An angle recoding method for CORDIC algorithm implementation. IEEE Trans Comput 42:99–102
17. Maharatna K, Banerjee S, Grass E, Krstic M, Troya A (2005) Modified virtually scaling-free adaptive CORDIC rotator algorithm and architecture. IEEE Trans Circ Syst Video Technol 15(11):1463–1474
18. Maharatna K, Troya A, Banerjee S, Grass E (2004) Virtually scaling free adaptive CORDIC rotator. IEE Proc Comput Digital Techn 151(6):448–456
19. Vachhani L, Sridharan K, Meher PK (2009) Efficient CORDIC algorithms and architectures for low area and high throughput implementation. IEEE Trans Circ Syst II 56(1):61–65
20. Jaime FJ, Sanchez MA, Hormigo J, Villalba J, Zapata EL (2010) Enhanced scaling-free CORDIC. IEEE Trans Circ Syst I Regul Pap 57(7):1654–1662
21. Aggarwal S, Meher PK, Khare K (2012) Area-time efficient scaling-free CORDIC using generalized micro-rotation selection. IEEE Trans Very Large Scale Integr (VLSI) Syst 20(8):1542–1546
22. Causo M, An T, de Barros Naviner A (2012) Parallel scaling-free and area-time efficient CORDIC algorithm. In: 19th IEEE International conference on electronics, circuits and systems, pp 149–152
23. Vankka J (2005) Digital synthesizers and transmitters for software radio. Springer, Dordrecht, The Netherlands
24. Jridi M, Alfalou A (2009) Direct digital frequency synthesizer with CORDIC algorithm and Taylor series approximation for digital receivers. Eur J Sci Res EuroJournals 30(4):542–553
25. Aggarwal S, Khare K (2012) Hardware efficient architecture for generating Sine/Cosine waves. In: 25th International conference on VLSI design
26. Aggarwal S, Meher PK, Khare K (2013) Scale-free hyperbolic CORDIC processor and its usage to waveform generation. IEEE Trans Circ Syst I 60(2):314–326

Dynamic Shadow Detection and Removal in Evening Time for Vehicle Counting System

Kalpesh R. Jadav and Arvind R. Yadav

Abstract Shadow detection and removal are an important process for the further progression in computer vision method, i.e., tracking, recognition, classification, etc. Vehicle counting system in traffic management is not able to count exact number of vehicles without shadow detection and removal especially during evening time. Due to strength of shadow in evening time, the shadow detection is quite difficult in traffic especially in evening time. Many researchers have published paper on shadow detection, but for daylight/noon time dataset. Hence, need arises for determination of shadow detection in evening time. In this paper, HSV color space-based non-deterministic model with background updating concept is proposed for shadow detection during evening traffic dataset. Investigational outcomes suggest that the proposed method has been able to detect dynamic shadows accurately.

Keywords Computer vision · Shadow detection · HSV color space · Classification · Background update

1 Introduction

In the evening time, especially after 5:30 pm the traffic volume is enormous on Vadodara-Mumbai highway. Also, in evening time, the strength of shadow is high than noon time and early morning, hence, the shadow detection becomes very complex for vehicle counting system. Over the years, many researchers have proposed motivating applications for traffic control on the road, detection and vehicle counting, using cameras, which are useful in many situations. Vehicle counting system consists of vehicle detection and under various conditions, i.e.,

K. R. Jadav (✉) · A. R. Yadav
Department of Electronics and Communication Engineering, Parul Institute of Engineering and Technology, Parul University, Vadodara, Gujarat, India
e-mail: kalpesh.jadav@paruluniversity.ac.in

A. R. Yadav
e-mail: arvind.yadav@paruluniversity.ac.in

R. P. Mahapatra et al. (eds.), *Proceedings of 6th International Conference on Recent Trends in Computing*, Lecture Notes in Networks and Systems 177,
https://doi.org/10.1007/978-981-33-4501-0_51

shadows, occlusion, illumination, reflection, and camera shaking. Among them, shadow is a prominent source of error in vehicles counting system and in detection.

Vehicle detection should be done by efficient method. Over the past many decades, background subtraction, frame difference, and optical flow are very useful methods of change detection in static background but change detection is also obtained from division method by proper thresholding and scalar value. The HSV color space method-based shadow detection and removal has been published by many researchers. The proposed algorithm employs HSV-based image division with background update, image enhancement, and morphological operation for shadow detection and removal.

The rest of the paper is organized as follows. Section 2 discusses the related work published by the researchers. Details of proposed method are given in Sect. 3. Experimental results of proposed method are discussed in Sect. 4. Section 5 discusses the quantitative metrics used for performance evaluation. At last, Sect. 6 provides a conclusion and future work of the work presented in this paper.

2 Literature Review

The region of shadow detection and removal has made excessive development in recent years by many researchers.

In 2015 Huang et al., presented dynamic threshold-based shadow detection method to accomplish maximum shadow detection and shadow discrimination rate by changing the radiance value [1]. They have used their own dataset which contains variable illumination value and found dynamic threshold value regarding variable illumination for shadow detection.

In 2009 Mosabbeb et al., presented deterministic non-model-based shadow detection algorithm (called it DNM1). Because HSV color space is very close to human perception of color [2], it can distinguish between object and shadow pixels more accurately. The DNM1 with k-mean clustering algorithm was determined by averaging shadow and foreground pixels determined by DNM1, using a new algorithm that returns valid results on non-convex datasets.

In 2012, Sabnis and Sharma presented four shadow detection approaches, i.e., SN, SNP and DNM1 and DNM2, umbra region is called as DNM, whereas the next approach which is more complete and deals with penumbra is called as DNM2. Statistical non-parametric (SNP), the statistics of shadow, and non-shadow are generated without considering any parameters, so it may depend on color consistency ability of human eye taking the light model for the relationship between irradiance and reflectance and exploiting distortion between chrominance and brightness statistical parametric (SP), the statistical information based on parameters is collected to form shadow and non-shadow. Further, they proposed two qualities of measures, Good Detection, and Good Discrimination. Good Detection specifies the low probability of miss classifying a shadow point. Whereas Good Discrimination specifies the low probability of classifying non shadow points as shadows. It is also called as low false alarm rate [3].

In 2013, Yang et al. presented hybrid method of RGB ratio-based and Blue-Red component ratio-based, moving foreground detection and post processing which achieved up to 80% shadow detection only on outdoor scene [4]. But the method is not appropriate for internal background scene where light are not from the sun and the sky.

3 Proposed Method: Shadow Detection

In the present effort, the ideas of image ratioing [5] with constant multiplier [6] have been used to implement a robust shadow detection and removal algorithm for vehicle counting system. The procedure of shadow detection is defined as follows.

1. In this implementation, the author considered self-built dataset of Vadodara–Mumbai express highway in evening time.
2. Performed background updates using given formula.

$$\text{BG update}(x, y) = \begin{cases} \text{BG, if Video Frame} = 1 \\ \text{otherwise, update} = (1 - \infty) \times \text{frame} + \infty \times \text{BG}(x, y) \end{cases} \tag{1}$$

3. Convert HSV color space video sequences from the dataset and take image ratioing of value video sequential frames.

$$\text{Value Frame}(:, :, 3) = \text{rgb2hsv}(\text{dataset Frames}) \tag{2}$$

$$\text{Value Frame Ratio}(x, y) = \frac{n + 1\text{th value Frame}(:, :, 3)}{n\text{th value Frame}(:, :, 3)} \tag{3}$$

4. For better output of shadow detection, the author shifted dark intensity of output from step 3 in suitable dynamic range with the help of image enhancement gamma correction method. The different dynamic range of shadow intensity in image histogram according to the value of gamma shown in Fig. 1.
5. Obtain shadow part by thresholding method
   If value ratio > Threshold 1 && value ratio < Threshold 2
   Value image contain shadow part
   Else value image does not contain shadow part.
6. Perform morphological dilation operation on binary output which increases the strength of change.

$$\text{shadow detection}(x, y) = \text{rgb2gray}(\text{equlized}(x, y) \oplus \text{SE}) \tag{4}$$

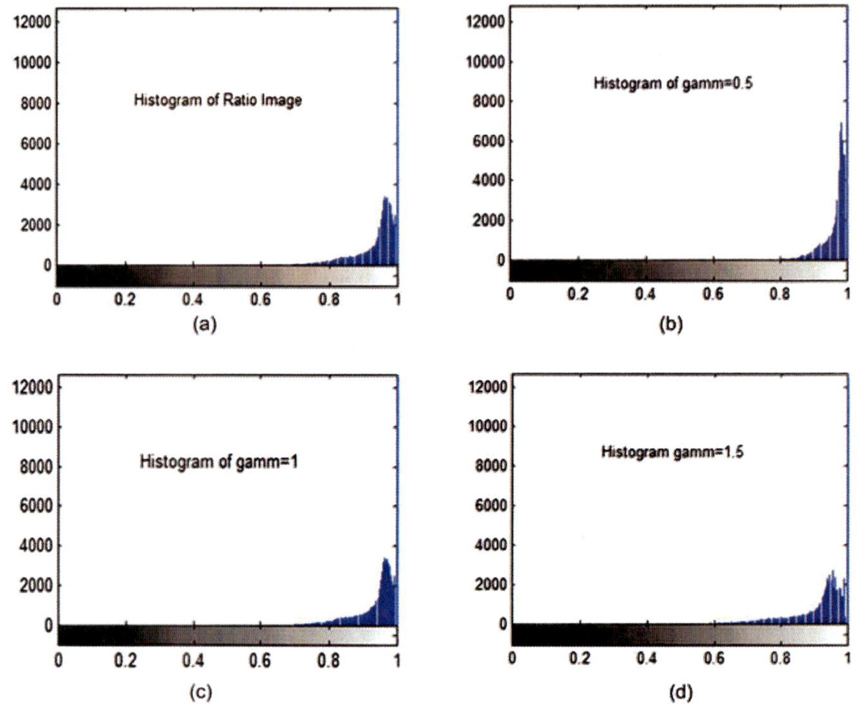

Fig. 1 **a** Original histogram of ratio, **b** histogram from gamma = 0.5, **c** histogram from gamma = 1, **d** histogram from gamma = 1.5

Where $SE = \begin{vmatrix} 1 & 1 & 1 \\ 1 & 1 & 1 \\ 1 & 1 & 1 \end{vmatrix}$, Symbol $\oplus$ denotes morphological dilation.

7. From connectivity assigned the label of neighborhood pixels and draw the bounding box from the region prop properties like bounding box, area, centroid, etc.

4 Experimental Results

The proposed system overview shown in Fig. 2 which contains image ratioing, background updating, image enhancement, and morphological operations. This part provides the outcomes of variant shadow detection that has been verified in obtained dataset by standard camera and used in experiments. The planned algorithms are comparing the existing standard methods. Statistical non-parametric (SNP) [3] approach and deterministic non-model-based (DNM) [4] approach are chosen.

The expected output flow wise shown in Fig. 3 of frame no. 18 and frame no. 60 by tested proposed method on a Dasktop-S1359LG device name PC operating Windows

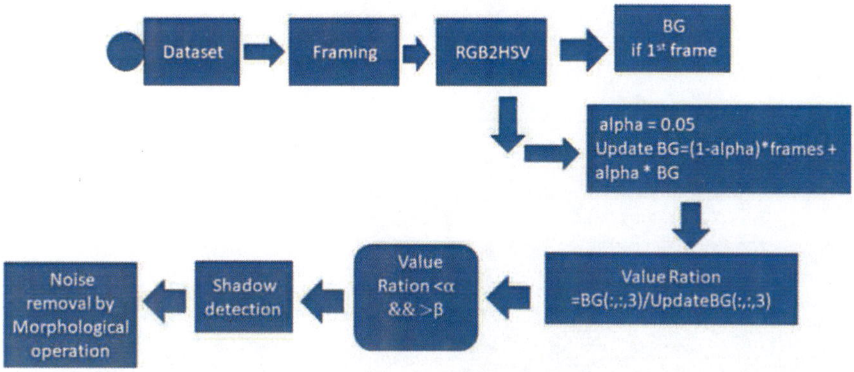

Fig. 2 Overview of proposed method

8.1 with a 2.40 GHz Intel® Core™ i3 CPU 4-GB RAM and 64-bit operating system. The author used MATLAB 2014 for the simulation. Our method can process around 3 frames per second.

Here, the author used the background update concept in implementation of proposed method it increase the strength of shadow region, the effect of background updating concept on proposed dataset are shown in Fig. 4.

The author tested other three dataset from the same concept and it gave good results in all condition with changing the threshold value. For future work, researcher need to design proper threshold for particular range of intensity of background. The experimental results of other dataset are shown in Fig. 5.

5 Performance Parameters

For the outcomes analysis, some standard quantitatively parameters have been used by many researchers [7] for shadow detection like shadow detection rate η, shadow discrimination rate ξ, precision, and specificity are used. They are defining as follow.

$$\eta = \frac{\mathrm{TP_s}}{\mathrm{TP_s} + \mathrm{FN_s}} \tag{5}$$

$$\xi = \frac{\mathrm{TP_f}}{\mathrm{TP_f} + \mathrm{FN_f}} \tag{6}$$

$$\text{precision} = \frac{\mathrm{TP}}{\mathrm{TP} + \mathrm{FN}} \tag{7}$$

$$\text{specificity} = \frac{\mathrm{TN}}{\mathrm{TN} + \mathrm{FP}} \tag{8}$$

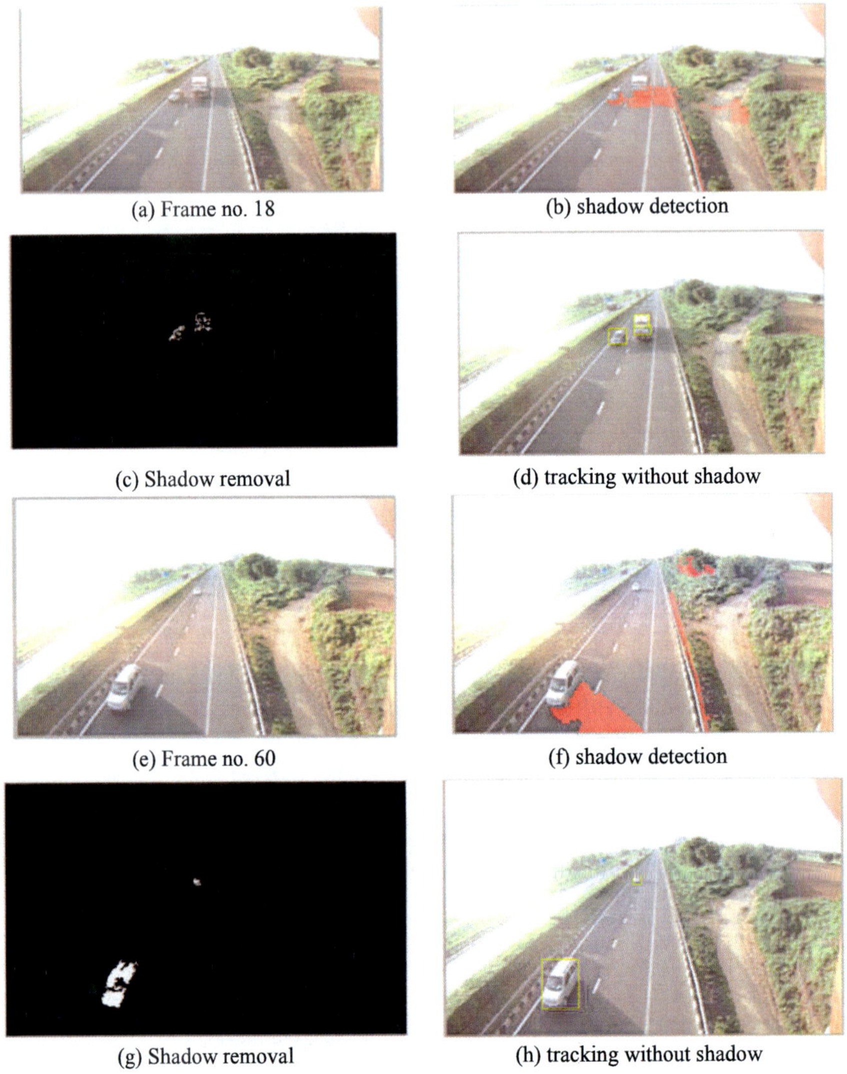

(a) Frame no. 18 (b) shadow detection (c) Shadow removal (d) tracking without shadow (e) Frame no. 60 (f) shadow detection (g) Shadow removal (h) tracking without shadow

Fig. 3 Result of frame 18 and 60 of the proposed method

In (5)–(8), the subscript symbol s stands for shadow and f for foreground, where TP_s and TP_f, respectively, denote the numbers of shadow elements and foreground elements correctly detected which are obtained by adding and subtracting shadow ground truth matrix and predicated matrix, FN_s and FN_f, respectively, denote the numbers of shadow elements and foreground elements incorrectly detected. The author tested and calculated processing time without background updating and without background updating in proposed algorithm the figure of processing time

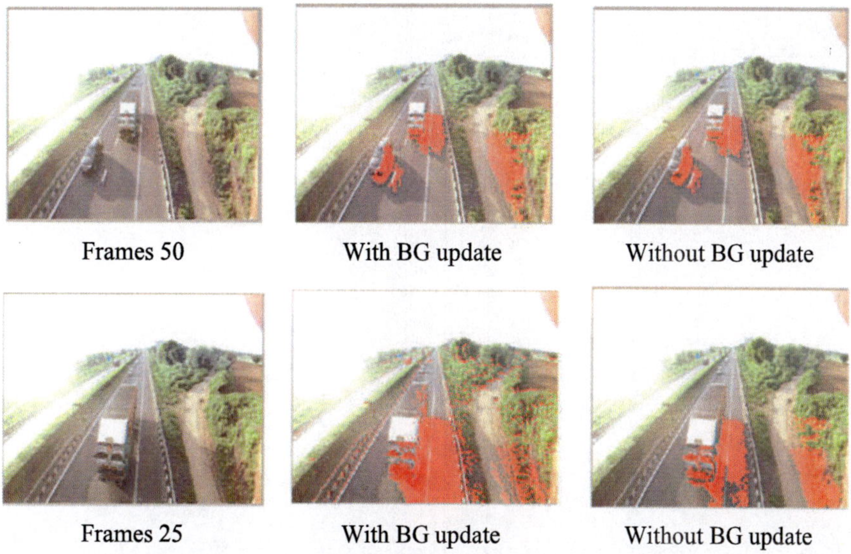

Fig. 4 Effect of background update with proposed method in obtained dataset

shown in Table 1. From the evolution, in less time proposed method give good result by background updating concept.

6 Conclusion and Future Work

In this paper, intensity (value) ratio from HSV color space-based method for dynamic shadow detection with background updating concept, image enhancement, and morphological operation on self-built dataset of Vadodara–Mumbai express highway is proposed. By examining the shadow evidence, the proposed method with background updating concept can detect shadow accurately in less time compare without background updating. By generating ground truths images of self-built dataset, the proposed method can be evaluate in terms of shadow detection rate η, shadow discrimination rate ξ, precision, and specificity.

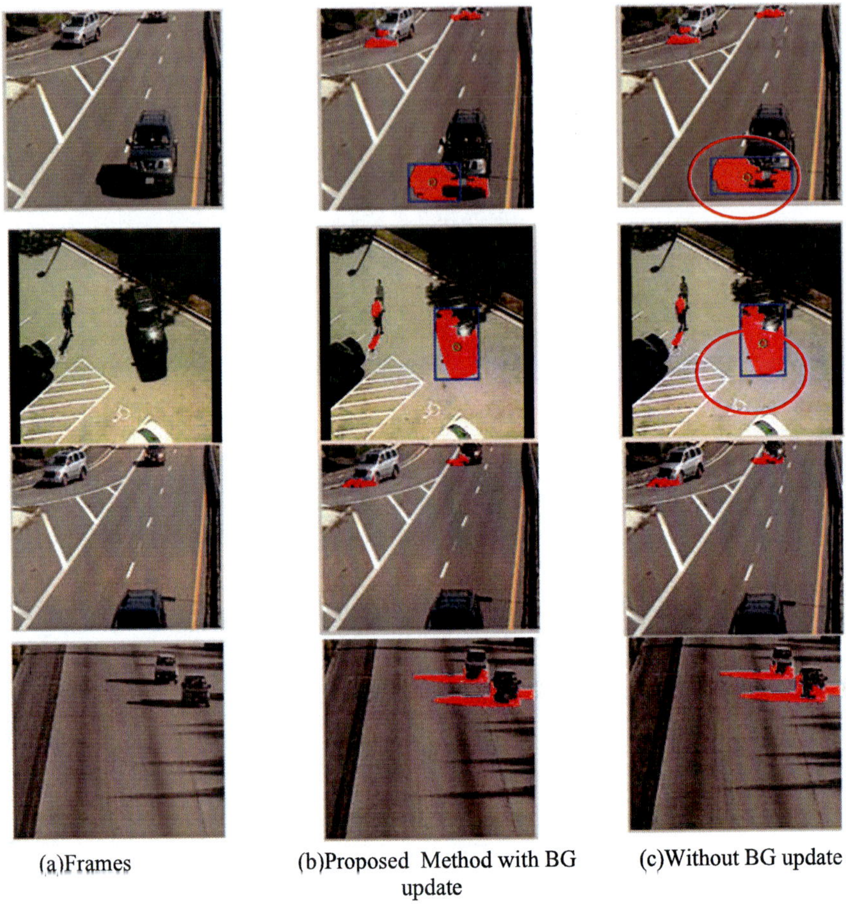

Fig. 5 Effect of background update with proposed method for other dataset

Table 1 Processing time of with background update and without background update

Proposed method	Processing time (s)
Without background update	10.666754
With background update	10.543264

References

1. Huang W, Kim KY, Yang Y, Kim YS (2015) Automatic shadow removal by illuminance in HSV color space. Comput Sci Inf Technol 3(3):70–75
2. Mosabbeb EA, Abbasian H, Fath M (2009) Clustering based non-parametric model for shadow detection in video sequences. In: International conference on image processing, computer vision, & pattern recognition, Las Vegas, Nevada, USA, vol 2, pp 440–445

3. Sabnis MK, Sharma A, Shadow detection, methods and its evaluation. Int J Electron Commun Comput Eng 3(6):2278–4209
4. Carmen D, Dale Joshua R, Cajote RD (2016) Moving shadow detection and removal for video-based traffic monitoring. In: Asia-Pacific signal and information processing association annual summit and conference (APSIPA). IEEE, pp 1–7
5. Lin HH, Pei JH, Liu DJ, Yang X (2008) A statistical parameter learning method for cast shadow model. In: International conference on machine learning and cybernetics, vol 4, 12 Jul 2008, IEEE, pp 2234–2239
6. Zigh E, Belbachir MF, Kadiri M, Djebbouri M, Kouninef B (2015) New shadow detection and removal approach to improve neural stereo correspondence of dense urban VHR remote sensing images. Eur J Remote Sens 48:447–463
7. Song R, Liu M, Wu M, Wang J, Liu C (2017) A shadow elimination algorithm based on HSV spatial feature and texture feature. In: International conference on emerging internetworking, data & web technologies, pp 585–591

Fractional Order Coefficient Optimization Using TLBO Method

Mrigank Singh, Nimish Agrawal, and Pragya Varshney

Abstract Fractional order differential equations can be utilized to model the physical world systems more efficiently than their integer order differential equation counterparts. Most of the problems in science and engineering are represented in the form of non-integer order differential equation. The computation of fractional order systems' step and impulse responses still pose an uphill task for the researchers. Thus, there is a need for an integer order approximation table that eases the process of fractional order simulation and realization of the real-world systems. In this paper, we have obtained approximated integer order coefficients for fractional order systems using teacher–learner-based optimization technique (TLBO). The TLBO technique performs better than the other optimization techniques as it is not bound by any algorithmic parameters. The obtained integer order approximation table using TLBO method is then analyzed and compared with step responses obtained by FOMCON toolbox and least square curve fitting method.

Keywords TLBO · Fractional order system · Huber loss · Optimization · FOMCON · Laplace transform

1 Introduction

The concept of fractional order system has prevailed since a long time, but recent years have seen tremendous work in the field due the contribution of fractional order calculus. Fractional order system has the capability to show long memory

M. Singh (✉) · N. Agrawal · P. Varshney
Division of Instrumentation and Control Engineering, Netaji Subhas University of Technology (Erstwhile NSIT), Sector 3, Dwarka, New Delhi 110078, India
e-mail: immriganksingh@gmail.com

N. Agrawal
e-mail: nimishagarwal12@gmail.com

P. Varshney
e-mail: pragya.varshney1@gmail.com

R. P. Mahapatra et al. (eds.), *Proceedings of 6th International Conference on Recent Trends in Computing*, Lecture Notes in Networks and Systems 177,
https://doi.org/10.1007/978-981-33-4501-0_52

or hereditary effect. It leads to development of new analysis and design method which practically suits real-time system better than the present techniques available. It provides a better solution to overshoot and resonance problems present in control systems. It also shows promising behavior in systems involving diffusion process and which demonstrate chaotic behavior. Fractional order controllers can be tuned more efficiently as they provide more degree of freedom. Fractional order system's hardware implementation is still not easily feasible, and thus, there is a need for the approximate integer order transfer function for research, simulation and hardware implementation. Some notable contributions in the field have yielded methods including continued fractional expansion (CFE) method, Oustaloup method [1], Carlson method [2], Matsuda method [3] and Charef method. Much work has been reported on the above-mentioned techniques but little has been mentioned on the least square curve fitting method [9].

In this paper, TLBO method is used to obtain approximate integer order transfer function for fractional order system. Initially, the inverse Laplace transform of $1/s^{\alpha}$ $(0 < \alpha < 1)$ is derived mathematically using Gamma table [10]. Then, TLBO method is used to fit the fifth degree time function on the obtained inverse Laplace function. Huber loss function has been used with TLBO to optimize the obtained coefficients of the time function by minimizing the error. The time function is then converted to Laplace domain using Laplace transform which eventually results into a fifth order approximate transfer function. The obtained transfer functions are compiled, and the performances of the time responses are thereby analyzed. The TLBO method has a lower degree of complexity owing to the less input parameters.

This paper is organized as follows: TLBO, Huber loss and FOMCON toolbox have been explained in Sects. 2, 3 and 4, respectively. The methodology of the proposed optimized solution in Sect. 5 and the analysis of the obtained results are shown in Sect. 6. Conclusion and further discussions have been presented in Sect. 7.

2 TLBO

The algorithm is a metaheuristic algorithm and imitates the process of learning and teaching skills of a classroom for the teacher and learner [4]. A group of students or learners in a class or section is called as a population in this process, and the development variables are the subjects or courses that are provided to the student (learners). The consequence of a student (learner) is similar to the numerical value of fitness, and the numerical value of objective function reflects a particular student's information. The best solution in this is similar to teacher, as the teacher is considered as the most skillful individual in the classroom. TLBO's process is split into two phases, teaching phase and learning phase.

The teaching phase involves the process of student enhancing their knowledge from the learning of the teacher, whereas the learning phase involves performance improvement from peer group study.

2.1 Teaching Phase

The mean vector μ^x is determined for each subject at generation x, which includes the average of the students or learners in the classroom. The mean vector μ^x is given by transpose of following matrix:

$$\begin{vmatrix} \text{mean}\left(\left[p^x_{(1,1)}, R, p^x_{(i,1)}, R, p^x_{(N,1)}\right]\right) \\ R \\ \text{mean}\left(\left[p^x_{(1,j)}, R, p^x_{(i,j)}, R, p^x_{(N,j)}\right]\right) \\ R \\ \text{mean}\left(\left[p^x_{(1,D)}, R, p^x_{(i,D)}, R, p^x_{(N,D)}\right]\right) \end{vmatrix}$$

This effectively gives us mean at iteration x for d number of courses or variables:

$$\mu^x = \left[\mu^x_1, \mu^x_2, R, \mu^x_j, R, \mu^x_d\right] \tag{1}$$

For the further iteration, the best vector with the least value of function is labeled as teacher (P^x_{Teacher}). The algorithm passes the learner's average to their instructor. From the current value of average and the ideal value of average vectors, a randomly weighted differential vector is generated and applied to the existing learner population for a new set of better learners.

$$P^x_{\text{new}(i)} = P^x_{(i)} + \text{rand.}\left(P^x_{\text{Teacher}} - T_F \mu^x\right) \tag{2}$$

where T_F is teaching factor which decides the value of average to be varied randomly at each iteration. Value of T_F has a range between two values: 1 or 2. T_F's value is calculated by the function:

$$T_F = \text{round}[1 + r_2] \tag{3}$$

where r_2 has a range between [0, 1] and is randomly selected. T_F is not a TLBO algorithm parameter. In the algorithm, T_F's value is not given as an input and is calculated using above-mentioned formula. The algorithm is found to work in the best form if the T_F value is either equal to 1 or 2, and the teaching variable is suggested to take either 1 or 2 depending on the rounding-up criterion provided by above formula to simplify the algorithm.

2.2 *Learning Phase*

This process consists of students learning by interaction with each other. The shared communication cycle tends to increase the learner's awareness. Each learner communicates with other learners at random and thus encourages the exchange of information. $P^x_{(i)}$ For a particular learner, is selected randomly picked and for another learner $P^x_{(r)}$ $(i \neq r)$. In the learner phase, the ith vector of $P^x_{\text{new}(i)}$ the matrix is given as

$$P^x_{\text{new}(i)} = \begin{cases} P^x_{(i)} + \text{rand.}\left(P^x_{(i)} - P^x_{(r)}\right), \text{ if } (P^x_{(r)} < P^x_{(i)}) \\ P^x_{(i)} + \text{rand.}\left(P^x_{(r)} - P^x_{(i)}\right), \text{ otherwise} \end{cases} \tag{4}$$

3 FOMCON Toolbox

FOMCON denotes the fractional order modeling and control toolbox, which operates on fractional order calculus [5–7]. The toolbox supports design and modeling of fractional order calculus and the analysis of SISO and MIMO fractional order complex systems. The tuning of fractional order PID controller can be efficiently achieved using FOMCON.

4 Huber Loss

Huber loss is a loss function given by Peter J. Huber in 1964 [8]. It is less sensitized than squared error loss to data anomaly. Its derivative also exists at 0. This is equal to an absolute error when there exists a significant error and is quadratic error when there is no or less error. How less this error should be in order to make it quadratic is decided by an adjustable parameter called hyperparameter that is denoted by δ (delta). Huber loss is defined piecewise by

$$\mathcal{L}_\delta(y, f(x)) = \begin{cases} \frac{1}{2}(y - f(x))^2, & \text{for} |y - f(x)| < \delta \\ \delta|y - f(x)| - \frac{1}{2}\delta^2, & \text{otherwise} \end{cases} \tag{5}$$

Delta choice is important as it decides what you are able to consider as an outlier.

4.1 Huber Loss's Merits

One major disadvantage of applying MAE to train neural networks is its constantly high differential slope, which can lead to missing of minimum levels at the end of the training using gradient differential. For MSE, as the cost reaches its peak, the gradient decreases, thus making it more accurate.

In such situations, Huber loss can be very beneficial as it bends around the minima that reduce the gradient. And outliers are more robust than MSE. It, therefore, blends strong MSE and MAE properties. The main disadvantage of Huber loss, however, is that we may need to train the hyperparameter, which is a recursive process.

5 Methodology

To find the integer order approximation of fractional order systems, the procedure followed is given below:

The Laplace transformation of any fractional order time function and its inverse Laplace transformation are obtained using Eqs. (6) and (7), respectively:

$$\mathcal{L}\left[t^{\alpha-1}\right] = \int e^{-st} t^{\alpha-1} \mathrm{d}t = \frac{\Gamma(\alpha)}{s^{\alpha}}, 0 < \alpha < 1 \tag{6}$$

$$\mathcal{L}^{-1}\left[\frac{1}{s^{\alpha}}\right] = \frac{1/\,\Gamma(\alpha)}{t^{1-\alpha}} \tag{7}$$

To enhance the process of fitting, the step response of the system is calculated using Eq. (8):

$$\mathcal{L}^{-1}\left[\frac{1}{s^{\alpha+1}}\right] = \frac{\frac{1}{\Gamma(\alpha)}}{\alpha} \cdot t^{\alpha} \tag{8}$$

Now, the above-obtained function (8) is fitted using TLBO method on a fifth order polynomial function (9) which is selected by performing trial experiments.

$$g_s(t) = l_1 e^{-q_1 t} + l_2 e^{-q_2 t} + l_3 e^{-q_3 t} + l_4 e^{-q_4 t} + l_5 e^{-q_5 t} + c \tag{9}$$

The unknown coefficients present here are found using the teacher–learner-based optimization technique. TLBO generates a random population of coefficients depending upon the maximum and minimum value of constraint set for the coefficients, and then, it compares the plot of both functions and tries to minimize the difference using Huber loss function. Here, the unknown coefficients will act as subjects or courses of TLBO. Therefore, a total of 11 subjects will form an 11×1 vector.

The constraints that are set on the vector are as follows. For defining the constraint on the variables, an assumption made is that the variables $l_1, l_2, l_3, l_4, l_5, q_1, q_2, q_3, q_4, q_5, c$ all lie between -1000 and 10,000, i.e.,

$$(-1000 \leq l_1, l_2, l_3, l_4, l_5, q_1, q_2, q_3, q_4, q_5, c \leq 10,000).$$

The population size taken here is 20. Thus, there are 20 possible set of solutions of $[l_1, l_2, l_3, l_4, l_5, q_1, q_2, q_3, q_4, q_5, c]$. So, we have twenty random process models. Then, Eq. (8) is plotted after generating sample values for different values of α where α can take values between 0 and 1. For plotting Eq. (9), random population is generated, and TLBO method is applied to find the coefficients value which fits the polynomial approximately on the generated data set. For minimizing the error, the cost function applied is Huber loss function.

The vector which gives the minimum cost provides the optimized solution, i.e., the optimized values of $[l_1, l_2, l_3, l_4, l_5, q_1, q_2, q_3, q_4, q_5, c]$.

The Laplace transform of function $g_s(t)$ (9) is given by

$$G_s(s) = \frac{l_1}{s+q_1} + \frac{l_2}{s+q_2} + \frac{l_3}{s+q_3} + \frac{l_4}{s+q_4} + \frac{l_5}{s+q_5} + \frac{c}{s} \tag{10}$$

$G_s(s)$ is the step response and thus we can obtain integer order transfer function for $\frac{1}{s^\alpha}$ by multiplying $G_s(s)$ with (s) and eventually obtaining $G(s)$.

$$G_S(s) = \frac{1}{s} G(s) \cong \frac{1}{s^{\alpha+1}} \tag{11}$$

Hence, the integer order's approximation for $\frac{1}{s^\alpha}$ is given as

$$G(s) \cong \frac{1}{s^\alpha} \tag{12}$$

The value of coefficients obtained using TLBO method is substituted in Eq. (10), and an approximated fifth order transfer function, i.e., $G(s)$ is obtained.

6 Results

Table 1 shows the approximated fifth order transfer functions using the TLBO method. Consequently, Figs. 1 and 2 show the step response and error plots for the system $G(s) = 1/s^{0.6}$ obtained using TLBO method, least square curve fitting method and the ideal response obtained using FOMCON toolbox.

Observations show that the results obtained using TLBO method outperform those using curve fitting method. Errors are considerably reduced. The technique has been verified for all fractional orders from -0.1 to -0.9, but results are displayed of

Table 1 Approximation table obtained using TLBO

System	Fifth order approximation
$\frac{1}{s^{0.1}}$	$\frac{0.0001235s^5+0.0375s^4+1382s^3+8850s^2+7159s+226.8}{s^5+71.32s^4+1996s^3+10000s^2+5789s+133.6}$
$\frac{1}{s^{0.2}}$	$\frac{0.0012s^5+6309s^4+5019s^3+9395s^2+4281s+128.6}{s^5+7357s^4+4053s^3+10000s^2+2636s+43.54}$
$\frac{1}{s^{0.3}}$	$\frac{0.0259s^5+2069s^4+3366s^3+10000s^2+3751s+109.3}{s^5+2562s^4+3862s^3+10000s^2+1706s+20.64}$
$\frac{1}{s^{0.4}}$	$\frac{0.269s^5+1875s^4+2877s^3+7603s^2+5566s+227.8}{s^5+3022s^4+2190s^3+10000s^2+2160s+25.31}$
$\frac{1}{s^{0.5}}$	$\frac{0.3632s^5+3732s^4+7542s^3+5882s^2+6779s+341.3}{s^5+9890s^4+3529s^3+10000s^2+2167s+21.51}$
$\frac{1}{s^{0.6}}$	$\frac{10^{-10}s^5+0.1092s^4+10s^3+9.16s^2+6.508s+0.3507}{s^5+8.27s^4+10s^3+10s^2+1.5721s+0.0116}$
$\frac{1}{s^{0.7}}$	$\frac{0.0402s^5-970.8s^4+10000s^3+911.6s^2+10000s+845.9}{s^5+8679s^4+3162s^3+10000s^2+2491s+14.95}$
$\frac{1}{s^{0.8}}$	$\frac{0.3708s^5-133.5s^4+10000s^3-1000s^2+9170s+1933}{s^5+8.403s^4+6222s^3+3269s^2+3772s+17.38}$
$\frac{1}{s^{0.9}}$	$\frac{-0.0881s^5-174s^4+8051s^3+4534s^2+10000s+7313}{s^5+80.76s^4+7593s^3-501.8s^2+10000s+25.85}$

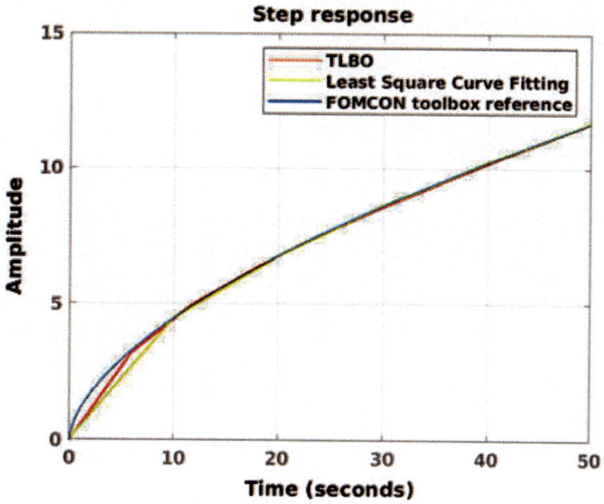

Fig. 1 Comparison of step response of $1/s^{0.6}$

$\alpha = -0.6$. The authors are working on further reducing the errors using other optimization techniques.

7 Conclusion

In this study, TLBO method was successfully applied to fractional order systems to obtain integer order approximated transfer functions. The system obtained using TLBO method has shown promising behavior, especially in initial stages of the response. The error when compared to step response obtained by FOMCON toolbox

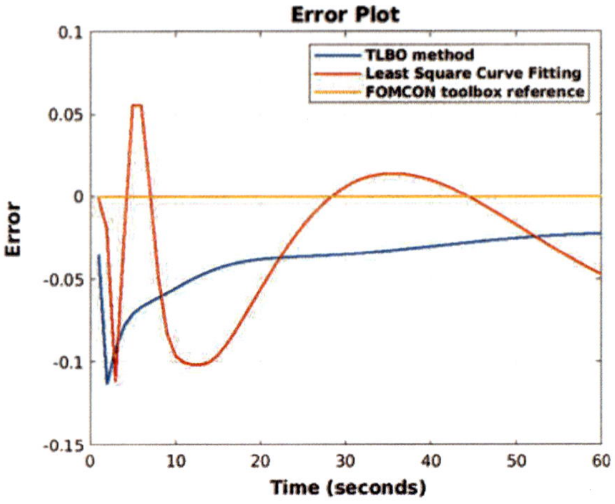

Fig. 2 Error plot

is of the order of 10^{-2}. Thus, the obtained coefficients can be used to design physical realizable systems.

References

1. Oustaloup A, Levron F, Mathieu B, Nanot FM (2000) Frequency band complex noninteger differentiator: characterization and synthesis. IEEE Trans Circuit Syst-I, Fundam Theory Appl 47(1):25–39
2. Carlson GE, Halijak CA (1964) Approximation of fractional capacitors $1/s^{1/n}$ by a regular Proposedton process. IEEE Trans Circuit Theory 11(2):210–213
3. Matsuda K, Fujii H (1993) H∞-optimized wave-absorbing control: analytical and experimental results. J Guid Control Dyn 16(6):1146–1153
4. Rao RV, Savsani VJ, Vakharia DP (2011) Teaching–learning-based optimization: a novel method for constrained mechanical design optimization problems. Comput Aided Des 43(3):303–315
5. Xue D, Chen Y, Atherton DP (2008) Linear feedback control: analysis and design with MATLAB (advances in design and control), 1st edn. Society for Industrial and Applied Mathematics, Philadelphia, PA, USA
6. Chen YQ, Petráš I, Xue D (2009) Fractional order control—a tutorial. In: Proceedings of ACC '09. American control conference, pp 1397–1411
7. Monje CA, Chen Y, Vinagre BM, Xue D, Feliu V (2010) Fractional-order systems and controls: fundamentals and applications. Advances in industrial control. Springer, Berlin
8. Huber PJ (1964) Robust estimation of a location parameter. Ann Stat 53(1):73–101
9. Yüce A, Deniz FN, Tan N (2017) A new integer order approximation table for fractional order derivative operators. IFAC-PapersOnLine 50(1):9736–9741
10. Atherton DP, Tan N, Yüce A (2015) Methods for computing the time response of fractional-order systems. IET Control Theory Appl 9(7):817–830
11. Yüce A, Tan N (2015) Analytical inverse laplace transform for fractional order derivative $(1/s^{\alpha})$. In: International symposium on fractional signals and systems 2015 (FSS 2015), Cluj-Napoca, Romania, pp 70–74

Age-Invariant Facial Expression Classification Method Using Deep Learning

Taurai George Rebanowako, Arvind R. Yadav, and Rahul Joshi

Abstract This paper proposes a method for facial expression classification as a solution to some of the challenges caused by age variations in image classification tasks. The proposed method utilizes deep learning neural networks pre-trained for image classification tasks which are then fine-tuned using a training dataset composed of mixed age facial expression images. The experimental results obtained after testing the model trained on a mixed age dataset, and another model trained on a standard dataset, using new images from a mixed age test dataset, show that the proposed model has better performance at facial expression classification task involving age variations in the input images.

Keywords Facial expression classification · Deep learning · Neural networks · Convolutional neural networks · Age variations

1 Introduction

Facial expression recognition (FER) has become one of the most interesting research areas in the image processing domain in recent years. Annual FER competitions, such as FER2013 and Emotion Recognition in the Wild (EmotiW) [1], have stimulated interest and motivated researchers to participate in the quest to provide solutions to a wide range of challenges presented in these competitions.

T. G. Rebanowako (✉) · R. Joshi
CSE Department, Parul Institute of Engineering and Technology, Parul University, Limda 391760, India
e-mail: tauraigeorge@gmail.com

R. Joshi
e-mail: rahul.joshi@paruluniversity.ac.in

A. R. Yadav
E&C Department, Parul Institute of Engineering and Technology, Parul University, Limda 391760, India
e-mail: arvind.yadav@paruluniversity.ac.in

R. P. Mahapatra et al. (eds.), *Proceedings of 6th International Conference on Recent Trends in Computing*, Lecture Notes in Networks and Systems 177,
https://doi.org/10.1007/978-981-33-4501-0_53

1.1 *Current Trends in Deep Learning*

Deep learning techniques are becoming the most preferred approaches in a wide range of image processing tasks which include facial expression recognition. Their widespread use is mainly due to their automatic feature learning capabilities when given images as input. However, despite their strong automatic feature learning capabilities, a wide range of challenges are still being encountered when applying deep learning techniques to facial expression recognition (FER) tasks. Current trends in deep learning involve the use of a wide range of deep learning techniques to address more and more task-specific issues in FER, so as to achieve performance improvements in deep feature learning and deep feature classification subtasks. The most commonly used deep learning techniques include recurrent neural networks (RNNs) [2], convolutional neural networks (CNNs) [3–6], deep belief networks (DBNs) [7, 8], and deep autoencoders (DAEs) [9–14]. Deep neural networks (particularly CNNs) are employed as feature extraction tools with additional independent classifiers being applied to the extracted representations.

1.2 *Need for FER*

Facial expression recognition, which involves learning the way in which we express our emotions in response to different situations, is important in different application areas as it determines the inference of meaning to the different expressions. It may also determine how we respond to others given the facial expressions displayed in different situations. Some of the facial expression labels that can be identified include angry, disgust, fear, happy, neutral, sad, and surprise. Correct interpretation of these facial expressions can find application in such areas as determination of appropriate intervention strategies by professional counseling organizations, pilot or driver fatigue surveillance by airlines and transport and logistics organizations, respectively, sentiment analysis, study of sign language, and detection of levels of intrinsic motivation among students.

1.3 *Problem of Age Variation*

Age variation has been cited as one of the challenges in FER performance [1, 15] despite the successes achieved to date with deep FER systems. This age variation is probably caused by the composition of the training dataset used, for example, if the dataset contains adult facial expression images only, the performance of the trained model will be affected negatively when it is presented with images from younger age groups. The same facial expression may be expressed differently by someone in the 3–11 years age group, for example, when compared to the same expression

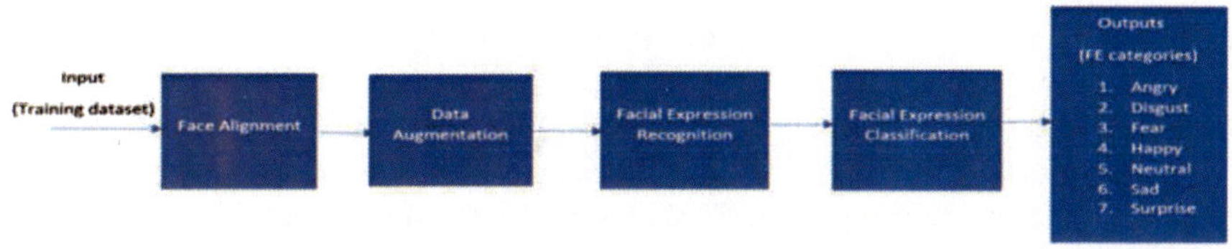

Fig. 1 Basic facial expression recognition process

displayed by someone above 50 years. This paper seeks to propose a model that can be generalized to produce accurate FER classifications regardless of the age of the person in the input FE image.

2 Related Work

A considerable amount of work has been done to date in the area of facial expression recognition aimed at solving different problems in the FER domain. Annual competitions in the field have also been held to stimulate and motivate directed research into more specific FER challenges with winners in different subchallenge categories receiving recognition for their efforts. FER datasets, such as FER2013, have also been made available to support these annual challenges through provision of the required training data.

2.1 *Basic FER*

Facial expression recognition takes place in three basic steps: preprocessing, deep feature learning, and deep feature classification. Preprocessing involves face alignment, data augmentation, and face normalization. This step is followed by deep feature learning in which neural networks are used to learn the different facial features that contribute toward formation of different facial expressions. The final step, deep feature classification, involves assignment of each facial expression into one of the identified expression categories. Figure 1 shows the block schematic diagram representing the basic FER process, while the general pipeline of a deep FER system has been discussed in [1].

2.2 *Machine Learning-Based FER*

Facial expression recognition systems using machine learning have been proposed by a number of researchers. These systems generally make use of support vector machines (SVMs) as classifiers of the identified facial expressions. Kim et al. [16], have proposed a FER system using machine learning in which the FER process extracts histogram of oriented gradients (HOG) features from each facial region with a support vector machine (SVM) used to classify the final facial expression. Their proposed system achieved a total $F1$ score of 0.8759, which was 0.0316 higher than the conventional system (with a total $F1$ score of 0.8443). Other FER techniques have been discussed in [17, 18].

2.3 Deep FER

Deep FER involves use of deep neural networks to build and train models for use in facial expression recognition tasks. The depth of the NN is determined by the number of hidden layers involved which, in turn, determine the performance of the NN (the larger the number of hidden layers the deeper the NN and the better the performance). The deep learning techniques that can be employed have been described in [19, 20].

2.4 Limitations

Some of the limitations of the FER work includes: (1) Lack of sufficient training data in terms of both quantity and quality given that FER is a data-driven task and training of models requires a large amount of training data [1]. (2) Unavailability of large-scale facial expression datasets which include occlusion type and head-pose annotations [1]. (3) Accurately annotating a large volume of image data with large variation and complexity of natural scenarios is an impediment to construction of facial expression datasets [1]. (4) Facial expressions are subject to change with respect to age, as different age groups may present the same facial expression differently [1].

3 Proposed Work

The proposed work involves training a model that is capable of performing well in facial expression classification tasks involving large age variations. The block schematic diagram of the proposed age-invariant FER approach is shown in Fig. 2. It involves extracting facial expression images from mixed age datasets during the data preparation stage. This results in the creation of a mixed age facial expression images training dataset. 75% of this dataset will be used for training and the remaining 25% will be used for validation. The trained model will then be presented with a mixed age test dataset for predictions and its performance will be evaluated from the prediction results.

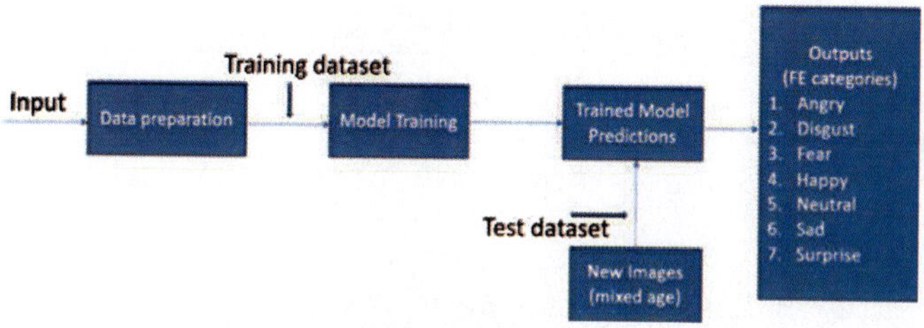

Fig. 2 Proposed age-invariant FE classification block schematic diagram

Facial expression images were extracted from the Large Age Gap (LAG) mixed age dataset to create a training dataset with an average of 100 images in each of the seven facial expression categories. Python will be used to train the CNN model on this mixed age dataset with 75% of the dataset used for training and the remaining 25% used for validation. First, a CNN model will be trained using the Japanese Female Facial Expressions (JAFFE) dataset and then tested using new images from the Large Age Gap (LAG) dataset. Another CNN model will then be trained using mixed age facial expression images and then tested using new images extracted from the LAG dataset. The results of these two models will then be compared to determine if there is any difference in facial expression classification performance emanating from the age variations in the datasets used for training and testing the models.

4 Results and Analysis

4.1 Implementation of Basic FER on JAFFE Dataset and Proposed Approach Using CNN

The CNN model was trained on the Japanese Female Facial Expression (JAFFE) dataset while the proposed age-invariant model was trained on a mixed age dataset (derived from the LAG dataset) using Python, before being tested using facial expression images which included new (mixed age) images. The training dataset was split into two subsets: 75% of the images were used for model training while the remaining 25% was used for model validation. Also, the number of iterations for each training run were adjusted until the best training loss and accuracy results were achieved, and the trained model was saved in the output folder.

Training and Validation Results Summary

The overall performance analysis of the work of this paper is shown in Tables 1 and 2 showing a comparative analysis of the basic FER model trained on the JAFFE dataset and the age-invariant model trained on the LAG dataset.

Table 1 Training loss and accuracy comparative analysis

Model	Train_loss	Train_acc	Val_loss	Val_acc
Basic FER on JAFFE dataset	0.1056	0.9526	0.3768	0.8704
Proposed age-invariant training	0.2497	0.9195	2.3383	0.5909

Table 2 Training performance scores comparative analysis

	Basic FER on JAFFE dataset				Proposed age-invariant prelim training			
	Precision	Recall	*F*1 score	Support	Precision	Recall	*F*1 score	Support
Angry	0.86	0.86	0.86	7	0.67	0.50	0.57	28
Disgust	0.67	0.67	0.67	6	0.48	0.45	0.47	22
Fear	0.89	1.00	0.94	8	0.75	0.60	0.67	30
Happy	0.90	0.82	0.86	11	0.51	0.73	0.60	26
Neutral	0.78	0.88	0.82	8	0.63	0.73	0.68	26
Sad	1.00	0.88	0.93	8	0.48	0.48	0.48	25
Surprise	1.00	1.00	1.00	6	0.67	0.63	0.65	19
Ave/total	0.87	0.87	0.87	54	0.60	0.59	0.59	176

The summary of performance scores also shows that the basic FER model has better training results than the age-invariant FER model. This can also be attributed to the need for a large mixed age training dataset which would provide adequate training leverage on the age-related facial expression variations so as to improve the deep feature learning capability of the CNN FER model.

Sample FE Predictions

Figures 3 and 4 show sample facial expression predictions made by the basic FER and proposed age-invariant models, respectively, on new images.

Fig. 3 Sample FE predictions

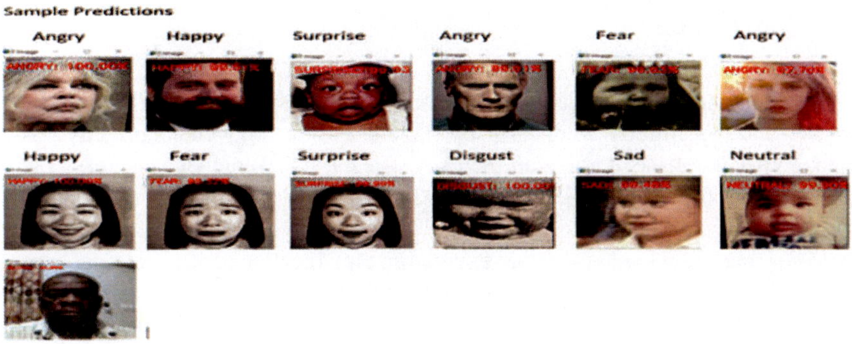

Fig. 4 Model post-training sample predictions

Table 3 Model test results

Parameters	Basic FER model	Proposed model
Total number of images used	2035	2035
Number of images correctly classified	299	1635
Number of images incorrectly classified	1736	400
Model classification accuracy (%)	14.69	80.34

Model Testing Results

A test dataset consisting of 2035 new images extracted from the LAG dataset was used to test the models, and the test results for the basic FER and proposed age-invariant models are summarized in Table 3. However, the sample prediction results show that the proposed age-invariant FER model has much better prediction performance as compared to the basic FER model implementation on the JAFFE dataset. The possible explanation could be the effects of age-related variations in facial expressions, which acted positively in the case of the proposed age-invariant FER model which had exposure to such variations from the mixed age training dataset and acted negatively in the case of the basic FER implementation on the JAFFE dataset which has facial expressions of Japanese females in almost the same age group. It is, therefore, imperative that using a mixed age training dataset for the proposed age-invariant FER approach will yield good results. There is need to further increase the size of the training dataset for improved performance. Further adjustment to the number of iterations per training run may also help improve the trained model's performance.

The trained model test results using 2035 new facial expression images from the LAG dataset produced a model classification accuracy of 80.34% for the proposed model and 14.69% for the basic FER model. However, the trained model's performance in terms of classification accuracy needs to be compared with that of other known models on the same dataset before making the final conclusion.

5 Conclusion

The trained model was tested using 2035 facial expression images from the LAG dataset and 1635 images were correctly classified while 400 were incorrectly classified, giving 80.34% classification accuracy. The basic FER model's test results with 2035 images from the LAG dataset have shown a very low classification accuracy (14.69%) which may be indicative of the effects of age variations across datasets used for training models for facial expression classification tasks. The final step of this phase will involve evaluating the trained model's classification accuracy against other deep learning models on the same dataset before the final conclusion is drawn.

References

1. Li S, Deng W (2018) Deep facial expression recognition: a survey. IEEE
2. Mostafa A, Khalil MI, Abbas H (2018) Emotion recognition by facial features using recurrent neural networks. IEEE
3. Meng D, Peng X, Wang K, Qiao Y (2019) Frame attention networks for facial expression recognition in videos. In: IEEE 2019 international conference on image processing (ICIP 2019), IEEE
4. Georgescu M-I, Ionescu RT, Popescu M (2019) Local learning with deep and handcrafted features for facial expression recognition. IEEE Access 7:64827–64836
5. Zhu Y, Fan H, Yuan K (2019) Explanation of CNN mechanism using traditional facial affective recognition. arXiv: 1904.09737v2 [cs.CV], 24 Apr 2019
6. Touvron H, Vedaldi A, Douze M, Jegou H (2019) Fixing the train-test resolution discrepancy. arXiv: 1906.06423v2 [cs.CV], 19 July 2019
7. Hinton GE (2012) A practical guide to training restricted Boltzmann machines. In: Neural networks: tricks of the trade. Springer, Berlin, pp 599–619
8. Bengio Y, Lamblin P, Popovici D, Larochelle H (2007) Greedy layer-wise training of deep networks. In: Advances in neural information processing systems, pp 153–160
9. Hinton GE, Salakhutdinov RR (2006) Reducing the dimensionality of data with neural networks. Science 313(5786):504–507
10. Vincent P, Larochelle H, Lajoie I, Bengio Y, Manzagol PA (2010) Stacked denoising autoencoders: learning useful representations in a deep network with a local denoising criterion. J Mach Learn Res 11:3371–3408
11. Le QV (2013) Building high-level features using large scale unsupervised learning. In: 2013 IEEE international conference on acoustics, speech and signal processing (ICASSP). IEEE, pp 8595–8598
12. Rifai S, Vincent P, Muller X, Glorot X, Bengio Y (2011) Contractive auto-encoders: explicit invariance during feature extraction. In: Proceedings of the 28th international conference on machine learning. Omnipress, pp 833–840
13. Masci J, Meier U, Ciresan D, Schmidhuber J (2011) Stacked convolutional auto-encoders for hierarchical feature extraction. In: International conference on artificial neural networks. Springer, Berlin, pp 52–59
14. Kingma DP, Welling M (2013) Auto-encoding variational Bayes. arXiv preprint arXiv:1312.6114
15. Pramerdorfer C, Kampel M (2016) Facial expression recognition using convolutional neural networks: state of the art. arXiv: 1612.02903v1 [cs.CV], 9 Dec 2016
16. Kim S, An GH, Kang S-J (2017) Facial expression recognition system using machine learning. IEEE, pp 266–267

17. Kundu T, Saravanan C (2017) Advancements and recent trends in emotion recognition using facial image analysis and machine learning models. In: 2017 international conference on electrical, electronics, communication, computer and optimization techniques (ICEECCOT), IEEE
18. Ashwin TS, Jose J, Raghu G, Ram Mohana Reddy G (2015) An E-learning system with multi-facial emotion recognition using supervised machine learning. In: 2015 IEEE seventh international conference on technology for education. IEEE, pp 23–26
19. Nezami OM, Dras M, Hamey L, Richards D, Wan S, Paris C (2019) Automatic recognition of student engagement using deep learning and facial expression. arXiv: 1808.02324v5 [cs.CV], 8 July 2019
20. Bianco S (2017) Large age-gap face verification by feature injection in deep networks. Pattern Recogn Lett 90:36–42

Crop Recommendation System Using *K*-Nearest Neighbors Algorithm

M. Rekha Sundari, G. Siva Rama Krishna, V. Sai Naveen, and G. Bharathi

Abstract Agriculture is the most important sector of Indian Economy that has ever increasing demand due to tremendous raise in population growth. Lagging in agricultural prosperous is the major reason for Indian economical growth not reaching the expectations. This is due to increasing pressures from climate change, soil erosion, biodiversity, etc. To meet these challenges, advances in the field of agriculture are need of the hour. Sucess of farming veture depends upon selection of crops to be grown. Crop selection depends mainly on weather, soil conditions and past crop yield data. The major setback in Indian Agriculture is that, farmers do not select appropriate crop, they just follow the crop which was sown by their ancestors without understanding the way that yield is fortuitous, conditioning intensely upon the present-day climate, soil conditions. Updated technological traits of machine learning (ML) available in the literature, predicted crop yield production depending on the historical data and had not considered present conditions of weather and soil. The methods only depicted how algorithms can be used in recommendation of the crop but has not validated the efficiency of algorithms in performing the task. Our work concentrates on predicting the crop yield in the future accurately taking into account both past and present conditions by using simple yet strong *k*-nearest neighbor algorithm.

Keywords Crop · Farmers · Prediction · Classification · KNN

M. Rekha Sundari (✉) · G. Siva Rama Krishna · V. Sai Naveen · G. Bharathi
Department of Information Technology, Anil Neerukonda Institute of Technology and Sciences, Visakhapatnam, India
e-mail: mr.sundari@gmail.com

G. Siva Rama Krishna
e-mail: sivaramakrishnagundubilli@gmail.com

V. Sai Naveen
e-mail: vajja.16.it@anits.edu.in

G. Bharathi
e-mail: bharathigadala451@gmail.com

R. P. Mahapatra et al. (eds.), *Proceedings of 6th International Conference on Recent Trends in Computing*, Lecture Notes in Networks and Systems 177,
https://doi.org/10.1007/978-981-33-4501-0_54

1 Introduction

The rapid spread of technology fueled by Internet will lead to the overall development of the country. India is an established nation that has farming as its essential well-spring of pay. With latest technological trends, India is still witnessing a slow growth in the field of agriculture. Agriculture is the backbone for a developing economy like India and there is an enormous need to maintain the agricultural sustainability. Due to this, inventions are been concentrated on developing effective methodologies for improving the productivity of the crops. The customary horticultural practices and procedures are causing a great deal of issues as far as effectiveness, benefits, and asset use. So there is a requirement of new techniques to improve the condition of farmer. The changes in weather will affect the agricultural productivity directly as it is straightly dependent on the climate. The magnitude of the impact depends upon the relationship between the climate and crop yields. Finding out the suitable crops based on the climatic conditions becomes tedious for novice farmers. Climatic conditions of a crop mainly refer to rainfall, soil temperature, soil moisture, pH value of the soil. Accurate models that predict best climate crop relation with good yield are important to increase India's gross domestic productivity.

Over the years, many techniques are evolved in the field of agriculture to increase the standard of living of the farmer. Machine learning techniques play a key role in making the predictions with great accuracy. The three broad categories of ML algorithms belong to supervised learning, unsupervised learning and reinforcement learning. We employ supervised learning that derives relations among the new data learning the rules from the old data. The crops to be yielded by the farmer are recommended by mining large history of data available on crops grown with better profits under different weather conditions over the years. This data is taken into consideration as different crops can be grown in areas with various climatic conditions and it's always not correct to yield only one crop depending on ancestral data. For example, a farmer who yielded a crop under some climatic conditions might have witnessed high yield whereas the other farmer in other climatic conditions might have experienced low productivity. This data when considered while predicting the data for a new field would of great effect in decision making. The main inputs to the system are the diagnosed nutritional features in soil directly from the laboratory test reports. The proposed system with the help of ML algorithms recommends the right crop to the farmer based on the soil test results.

1.1 Literature Survey

Rajak et al. [1] in the proposed work of crop recommendation system considers the soil dataset as for four harvests rice, cotton, sugarcane, and wheat. The dataset is first preprocessed and afterward the group method plays out a basic capacity in the characterization of the four yields. The individual base learners utilized in the ensemble model are random forest, Naive Bayes, and linear support vector machine (SVM). Greater part voting technique has been utilized as the blend strategy to give the best exactness. But the drawback is that it does not check the present soil condition and directly assume the soil type from past data and recommend the crop.

Pudumalar et al. [2] in their work about exactness horticulture, which is an advanced cultivating procedure utilizes research information of soil attributes, soil types, crop yield information, and recommends the ranchers the correct harvest dependent on their site explicit boundaries. This diminishes an inappropriate decision on a yield and expands efficiency. In this paper, this issue is settled by proposing a suggestion framework through an ensemble model with majority voting technique using random forest, *k*-nearest neighbor and Naive Bayes as learning algorithms to suggest a harvest for the site explicit boundaries with high exactness and proficiency.

Kumar et al. [3] states that the farmers are commonly not mindful of the necessities of the yield, for example, the minerals, soil dampness, and other soil necessities. This can cause pain to farmer both psychologically and economically. One more issue that a farmer may largely experience is absence of information about vermin and sicknesses that can influence the harvests they develop. This issue of farmer is addressed in this paper and the authors attempted to understand it with the assistance of a Recommendation System. By the assistance of this model, they anticipated the best reasonable harvest to the rancher, recognized the bug that may influence and proposed the nuisance control strategies. In this paper, they have applied SVM classification algorithm, decision tree calculation, and logistic regression calculation and found that SVM model gives the better precision when contrasted with different calculations.

Gandge [4] in their work used datasets with various variables like soil quality, nitrogen (N), potassium (K), phosphorous (P), and moisture contents over a period of time. They worked on different algorithms on different crops. Multiple linear regression on rice crop. KNN algorithm on wheat and potato crops. Neural networks on corn yield. But there was no algorithm which suits for any crop from this project. Yan et al. [5] considered the relationship between wheat growth and meteorology as their research object. Basically, thirteen features like atmospheric pressure, temperature, light, and precipitation from surface meteorological data are considered, but after discretization, they have selected only three features accumulative temperature, sunshine hours, and temperature. Using Apriori, a model was constructed for selecting farming season with the month as index.

Priya et al. [6] in the proposed work analyzed the data according to seasons. Dataset contain factors like rainfall, soil moisture, temperature, and atmospheric pressure. Map reduce and Naïve bayes classifier were used to predict the crop. The result was depicted and differentiated according to seasons for different crops. This paper recommends us when to sow but not what to sow to get more yield production with specific dataset factors. Pawar and Chillarge [7] in their work used J48 algorithm for classifying fertility and predict the soil toxicity of soil. Features like pH value and nutritional contents (like P, N, Cl, Ca). Accuracy of sensors plays crucial role in this project. It tests the soil and recommends fertilizers to be added to increase the crop production, water level. It does not recommend crop which gives profit for farmers by yielding more production but it predicts the toxicity of soil and its fertility. Kumar et al. [3] in their proposed work considered features like rainfall, pH value, and temperature. SVM, decision tree, and random forest algorithms were used but the accuracies of algorithms were 89.66, 86.8, and 88%, respectively, which did not cross at least 90%. This system was designed using Spyder IDE. All the existing methods including Priya et al. and Priyanka et al. [8, 9] only depicted how algorithms can be used in recommendation of the crop but has not validated the efficiency of algorithms in performing the task.

2 Dataset and Methodology

2.1 Dataset

The datasets comprises the soil specific attributes and the production quantity of the crop that was produced by crop previously yielded in that soil. The attributes that were taken into consideration were:

1. Rainfall
2. Soil temperature
3. pH value of the soil
4. Soil moisture
5. Crop data
6. Crop production.

The parameters above play a major role to extract the quality of the soil.

Soil is a complex, living, changing, and dynamic component of the ecosystem. It is an important variable that can be degraded or can be wisely managed. Learning soil moisture variations help us to figure out and predict changes in surface temperature, precipitation, drought, flood, and the impacts of future climate changes. This is an important variable in the climate system. Intelligence about the soil moisture content and how it changes is a key for the design and management of agricultural system.

Soil temperature also assists in the prediction of plant growth. It affects the plant growth indirectly by influencing the water and nutrient uptake there by affecting root growth. Low temperatures lower the traffic from the root to the shoot and vice versa. Each plant reacts contrastingly to soil temperature, contingent upon its scope of resistance of soil temperatures, and its temperature ideal. The objective is to keep up soil temperatures as near to the yield's ideal temperature as could be expected under the circumstances and to keep variations in soil temperature inside a harvest's scope of resilience.

Soil pH is the major sign of soil health. It perturbs crop yields, crop suitability, plant nutrient availability and is the combined measure of soil acidity and alkalinity. Adding proper amount of nitrogen fertilizer, liming and by following best yielding techniques good pH value that improvises soil health can be attained. Soils with pH more no than 7.0 are viewed as basic soils. Micronutrient insufficiencies, for example, iron inadequacy, are regular in these dirts. Yields developed in soils with pH level lower than 5.5 may show poisonous quality side effects of metals (e.g., iron, manganese) and lacks of different supplements, for example, magnesium. Liming the dirt is generally prescribed in such soils. The perfect soil pH go for most harvests is somewhere in the range of 5.8 and 6.5, a range where most supplements are accessible for the yields to take them in.

Expanded levels of precipitation would build filtering rates in all around depleted soils with high infiltration rates and would cause transitory flooding or water-immersion, bringing about diminished natural issue decay in numerous dirts or depression locales. This may influence a maximum area, particularly the better soils that would give rise to more larger sums and recurrence of overflow on soils in inclining landscape, with sedimentation down slant, and more regrettable downstream.

2.2 *Step by step procedure for crop recommendation system is shown in Fig. 1*

2.2.1 The Parameters of the Soil

The parameter of the soil where the crop is to be predicted are estimated by taking the samples to the laboratory and is supplied as test data to the classifier.

2.2.2 Collecting Dataset

The dataset logs contain historical data about the different crops yielded under various soil parameters, rainfall concentration, and their production rate.

2.2.3 Preprocessing

The dataset collected has some noisy data, so the data is first preprocessed. Preprocessing includes populating of missing data, scaling of values, and encoding of categorical data.

2.2.4 Classifying New Soil Pattern to the Existing Based on KNN Algorithm

The testing dataset is applied on the classifier, and the individual class names are acquired.

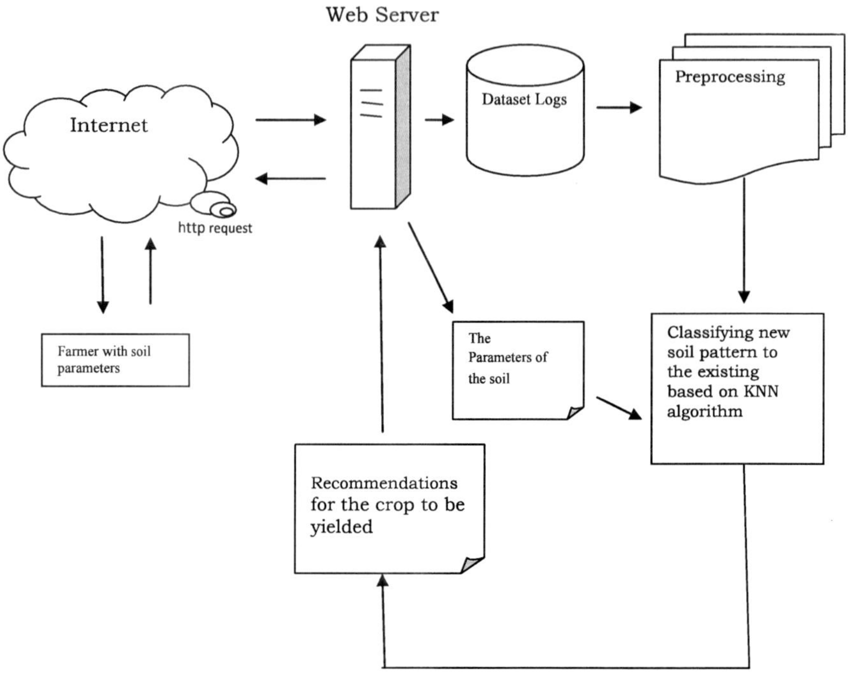

Fig. 1 Crop recommendation system

***K*-Nearest Neighbor (KNN)**:

The k-nearest neighbors (KNN) algorithm is a basic, supervised machine learning algorithm that can be used to solve both classification and regression problems. It is easy to execute and comprehend. The k-nearest neighbor is an instance-based learning which completes its grouping dependent on a likeness measure, like Euclidean, Mahanttan, or Minkowski distance functions. The initial two separation estimates function admirably with continuous variables while the third suits categorical variables. The Euclidean distance measure is used in this study for the KNN classifier. The Euclidean distance (D_{ij}) between two input vectors (X_i, X_j) is given by:

$$D_{1j} = \sqrt{\sum_{k=1}^{n} \left(x_{ik} - x_{jk}\right)^2}, \quad \text{for } k = 1, 2, 3, \ldots$$

For each datum point in the dataset, the Euclidean separation between a new point and current point is determined. These separations are arranged in moving up order and k items with most reduced separations to the new point are chosen. The dominated class among these items is found and the classifier restores the dominated class as the class for the new point.

Table 1 Accuracy of training data at different values of *k*

k-value	5	9	13	17	21	25	21	37
Accuracy	98.10	96.63	95.48	94.40	93.42	92.63	91.92	90.43

2.2.5 The Above Steps Are Applied for Recommendation of the Crop to Be Yielded Using *K*-NN as Follows

The input dataset contains data which contains the details like temperature, rainfall, pH, production, and crop. It has some missing and noisy data, so it was preprocessed. The missing values are filled with the average of remaining rows of same column. Encoding of categorical data is done by replacing all the 56 crops with values.

The noisy free data is now provided as input to the classifier to predict the data and make necessary recommendations to the farmer. The classifier is implemented using Python sklearn package. The dataset is partitioned into 80% learning and 20% testing data and using the confusion matrix, the results are analyzed.

3 Results

As stated in the methodology, the historical data of crops grown under various soil parameters are taken into consideration while predicting the best crop that is to be grown in the present soil parameters. KNN classifier, the simple and easy to interpret classification algorithm is applied on various percentages of training data and test data with different values of *k* to check the accuracy. The algorithm gave an accuracy of 98.1% at $K = 5$ for training data and is given in Table 1. The results associated with Fig. 2a, b here depict the accuracy of KNN with training data and test data.

4 Conclusion

In the proposed work, KNN algorithm is implemented on the data collected from different sources. As our future work, we aim to collect the features of the soil by using IOT devices with sensors that measure moisture and temperature so that we can predict the crop according to the changing features of the crop instantly without facing the difficulty of taking the samples to the lab. As KNN is lazy learning algorithm, we implement the methodology with eager learning algorithms by including other features like electrical conductivity and texture of soil to improve the predictive accuracy of the crop to be yielded.

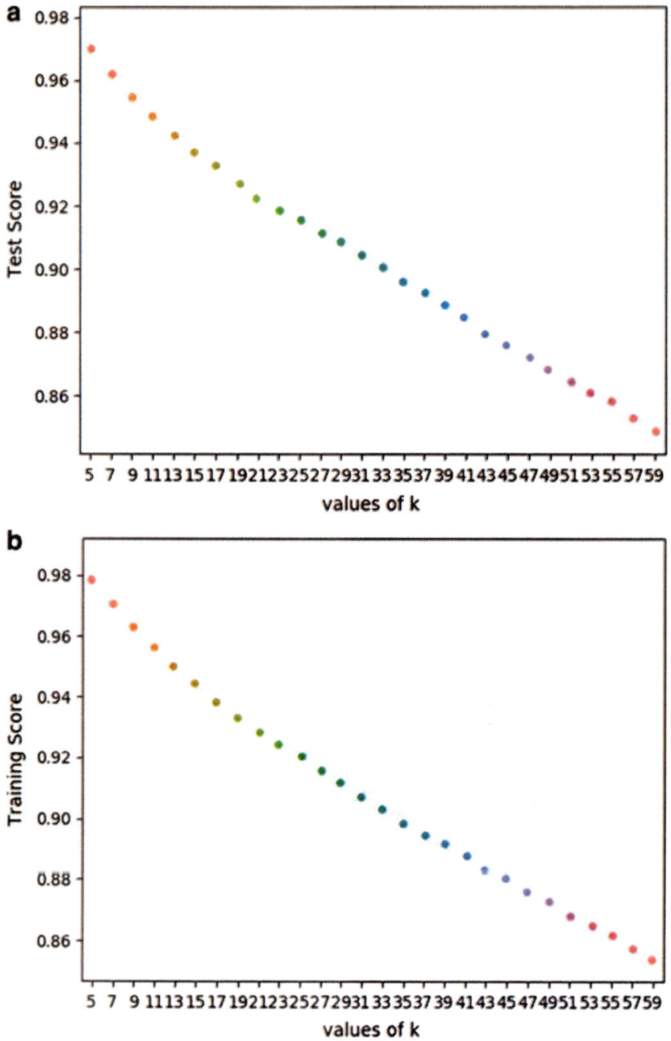

Fig. 2 **a** Graph between test data and *K*-value. **b** Graph between training data and *K*-value

References

1. Rajak RK et al (2017) Crop recommendation system to maximize crop yield using machine learning technique. Int Res J Eng Technol 4(12):950–953
2. Pudumalar S et al (2017) Crop recommendation system for precision agriculture. In: 2016 eighth international conference on advanced computing (ICoAC). IEEE
3. Kumar A, Sarkar S, Pradhan C (2019) Recommendation system for crop identification and pest control technique in agriculture. In: 2019 international conference on communication and signal processing (ICCSP). IEEE

4. Gandge Y (2017) A study on various data mining techniques for crop yield prediction. In: 2017 international conference on electrical, electronics, communication, computer, and optimization techniques (ICEECCOT). IEEE
5. Yan M et al (2018) Research on precision management of farming season based on big data. EURASIP J Wireless Commun Networking 2018(1):143
6. Priya R, Ramesh D, Khosla E (2018) Crop prediction on the region belts of India: a Naïve Bayes MapReduce precision agricultural model. In: 2018 international conference on advances in computing, communications and informatics (ICACCI). IEEE
7. Pawar M, Chillarge G (2018) Soil toxicity prediction and recommendation system using data mining in precision agriculture. In: 2018 3rd international conference for convergence in technology (I2CT). IEEE
8. Priya P, Muthaiah U, Balamurugan M (2018) Predicting yield of the crop using machine learning algorithm. Int J Eng Sci Res Technol 7(1):1–7
9. Priyanka T, Soni P, Malathy C (2018) Agricultural crop yield prediction using artificial intelligence and satellite imagery. Eurasian J Anal Chem 13(SP):6–12. ISSN: 1306-3057

Enhancing Spectrum Utilization in 2G to 5G Cognitive Radio Networks

Bablu Kumar Singh and Rajesh Bhadada

Abstract The wireless communication has expanded rapidly during last few years attributed to its adoption by commercial, consumer, medical, public safety, industrial, and military applications. Concurrent users by these applications puts strain on the availability of radio spectrum, i.e. need more spectrum which results in overload on the current Global System for Mobile Communication (GSM) spectrum allocation. Under the current spectrum allocation rules, nearly all the usable electromagnetic spectrums have already been allocated. The allocated spectrum bands mobile communication with high spectrum occupancy is overcrowded and spectrum band with low occupancy are under-utilized. This paper focused on the scope and method of spectrum sensing and spectrum allocation dynamically using Store Forward Base Transceiver System (SFBTS) model in 2G GSM and proposed this model for 5G which may be used as database spectrum sharing model for enhancing efficiency of spectrum utilization in 2G to 5G cognitive radio network.

Keywords SFBTS · 2G · 5G · Cognitive radio network · Spectrum sensing · Spectrum utilization

1 Introduction

Spectrum is considered as a finite and limited resource; it needs to be carefully managed at the national and international levels. The fixed assignment techniques in current spectrum prevent mobile as well as other wireless devices from utilizing allocated spectrum efficiently, as a results white spaces remain present in spectrum [1]. These white spaces are vital to fulfil the requirements of the fifth-generation (5G) network, and the spectrum range used in 5G will be expanded up to 100 GHz.

B. K. Singh (✉) · R. Bhadada
Department of Electronics and Communication Engineering, M.B.M Engineering College, JNV University, Jodhpur, India
e-mail: bablukumar.singh@gmail.com

R. Bhadada
e-mail: rajesh_bhadada@rediffmail.com

R. P. Mahapatra et al. (eds.), *Proceedings of 6th International Conference on Recent Trends in Computing*, Lecture Notes in Networks and Systems 177,
https://doi.org/10.1007/978-981-33-4501-0_55

The key technologies in 2G for spectrum sharing and utilization will be our proposed SFBTS model to release the current strain on spectrum. The similar technologies can be related to 5G in future and can be proposed in terms of database-based spectrum sharing model. The survey of spectrum usage shown from Federal Communications Commission (FCC) [2] reveals that spectrum have different utilization percentage with time and space dimension due to fixed allocation rules for licensed spectrum. This survey also reveals that some crowded bands are almost busy, while other portion of the spectrum is not used frequently. Various multiplexing technologies such as time division multiple access, frequency division multiple access, code division multiple access, and multiple input multiple output are used to accommodate more users within limited spectrum resources [3], but they are unable to solve the problem of scarcity of spectrum caused by fixed allocation rules. To solve this problem, the concept of Cognitive Radio (CR) was introduced and defined for wireless communication system [4, 5], which can automatically sense the usage of the nearby spectrum band and use the free spectrum band without affecting licensed users over same band in normal communication. The users are either primary users (PUs) referred to as licensed user or the secondary users (SUs) termed as unlicensed users which are also known to be cognitive users. According to CR, SUs should have the ability to sense free spectrum for its usage and free the spectrum channel immediately when the PUs comes back. The spectrum can be sensed using energy sensing method and decision on channel characteristics can be made for availability of spectrum. Once the channel characteristics is known, then by applying various hopping techniques the spectrum efficiency in both 2G and 5G can be enhanced using different model proposed in this paper.

2 Techniques Used for Enhancing Spectrum Utilization

Enhancing spectrum is vital for wireless communication but security is also the main important aspect of it. 2G considered to be most secure communication since it uses hopping techniques which improve the quality of the transmission on the air interface and increases the efficiency of the channel coding and interleaving in multipath fading environment [6]. Frequency hopping can be classified based on channel characteristics as channel unaware and channel aware. In channel unaware, the hopping pattern is selected regardless of the channel behaviour which results into higher error during data transmission while in channel aware, the hopping pattern is selected after determining behaviour of the channel, which results in better performance due to reduced hop set. In channel aware hopping, the busy channels are avoided completely and also provide probabilistic channel usage where all channels are used but busy channels have smaller usage probability [7]. Similar property of hopping can be applied in 5G using database-based spectrum allocation method and dynamic adaptive channel allocation scheme within the database of the stored channel spectrum. CR technology makes it possible to increase valuable spectrum utilization rate solving the problem of low utilization rate in the current fixed allocation rules. The CR uses allocation

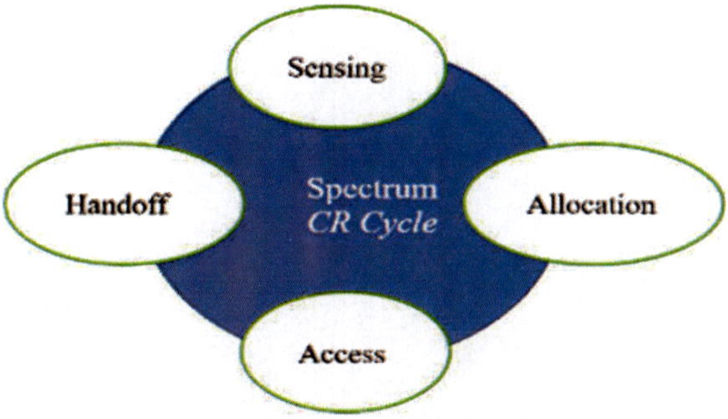

Fig. 1 Spectrum cycle of CR

and sharing of spectrum through dynamic spectrum allocation, in which SUs can use the idle spectrum, and do not interfere with the communication of PUs. CR spectrum cycle can be implemented using methods as shown in Fig. 1.

3 System Description and Algorithm

In 2G, the communication [8, 9] is based on Time Division Multiple Access frame on the radio carrier with channel rate 270.833 kbps with 25 MHz bandwidth and each channel carries 200 kHz band. These channels are very different in LTE, 4G and 5G which has much wider bandwidth and different modulation scheme to increase data rate. As shown in Fig. 2. The spectrum database and SFBTS store the historical spectrum information based on the 2G or 5G network and generate a cluster of available spectrum channel based on the current characteristics of spectrum.

Spectrum database channel holds nearest base stations channels and then the sensing is done to find all available spectrum holes for SU and according to various hopping algorithm [10] based on frame structure of network is used to enhance spectrum utilization. This sensing technique has been applied in a channel aware environment using LabVIEW and Universal Software Defines Radio Peripheral (USRP) [11] for continuously for seven days over individual channel in the test region and recorded spectrum to know the behaviour of channels. A cluster of six channels is made as database and hopping techniques [12, 13] are applied using *Dynamic Adaptive Channel* (*DAC*) hopping for transmission of SUs data making use of clustered channels.

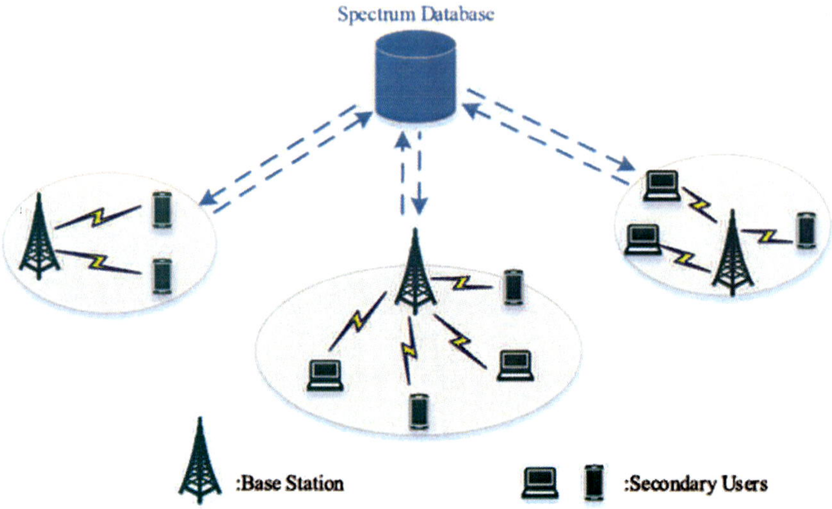

Fig. 2 Proposed SFBTS/database channel model

4 Simulation and Results

The simulation has been done using PC installed MATLAB having windows10 operating system with core i7 processor and 16 GB RAM. The channel frequency selected based on based on the availability of channel around MBM Engineering College Jodhpur. (Longitude 26.2679° N, Latitude 73.0352° E) of the GSM network based on received power level [14]. The SU data set for 15 min, 30 min, 45 min, and 60 min for transmission and reception using frequency hopping techniques and simulated result in Table 2, and Fig. 4, shows comparative analysis of optimised cyclic hopping, random hopping and our developed "DAC hopping" algorithm which may be implemented in current 2G and future 5G technology for enhancing spectrum efficiency. It is evidenced that there is significant increase in channel utilization and DAC hopping performance can be proposed for better channel utilization. The flow diagram for DAC hopping is shown in Fig. 3. Designed for optimum utilization of channel. In order to achieve 100% utilization of the test channel selected from 2G network, overload condition is considered in the SFBTS model. The sharing of unused channel with SU in real time with GMSK modulation [15] having frame duration of 4.615 ms is implemented in simulation over real spectrum of six channel considering all other constraints of the mobile communication network working perfectly and in synchronized manner. The simulated result obtained from sensing clustered channel data in Tables 1 and 2, is a real-time spectrum measurement of the individual channel and average value of spectrum usage is calculated in percentage form has been presented. The channel selection criteria are another important task for spectrum utilization; in this paper, we have selected six and nine channels to make cluster of spectrum database and applied algorithm over one-hour period using different algorithm, i.e.

optimize cyclic hopping, random hopping, and DAC hopping for optimizing the channel utilization [16].

The overload condition is very important in order to achieve the 100% utilization and this can be achieved when all the idle frame of spectrum is utilized by SU. The results shown in Table 3 show overload condition of one-hour spectrum during idle, busy, and any random hours selected for SU with prominent six and nine channels around the test region. The simulated result is encouraging and can be implemented for channel utilization. Further our simulated result is compared with the existing and upcoming technology (LTE and 5G) is shown in Table 4.

5 Conclusion

The spectrum sensing and spectrum allocation using SFBTS/database spectrum sharing model have been proposed for 2G and 5G. The simulated results show that spectrum utilization can be enhanced in both 2G to 5G cognitive radio network. Using CR technology, SU can use the primary licensed channel which results in significate improvement in the spectrum utilization. In Table 3, overload condition shown shows conclusive result that increasing number of channels the overload in utilization can be improved keeping fixed number of SUs, such that all the channel taken in the model should reach up to 100% utilization. Table 4 shows the result of the simulation obtained using channels of GSM and implemented using SFBTS model to enhance spectrum utilization. The similar model, i.e. database model can be implemented in 5G to increase spectrum utilization from 95.4% to nearly 100% without changing any existing hardware of the present existing network.

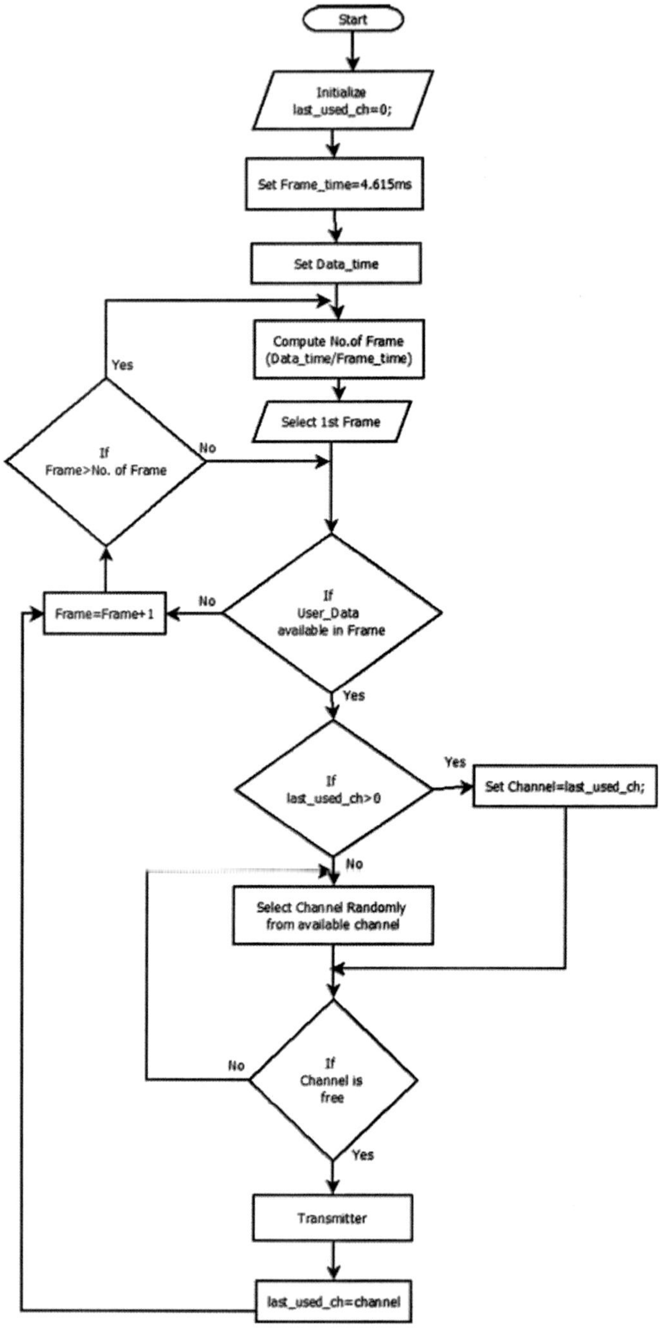

Fig. 3 Flow diagram of DAC hopping algorithm

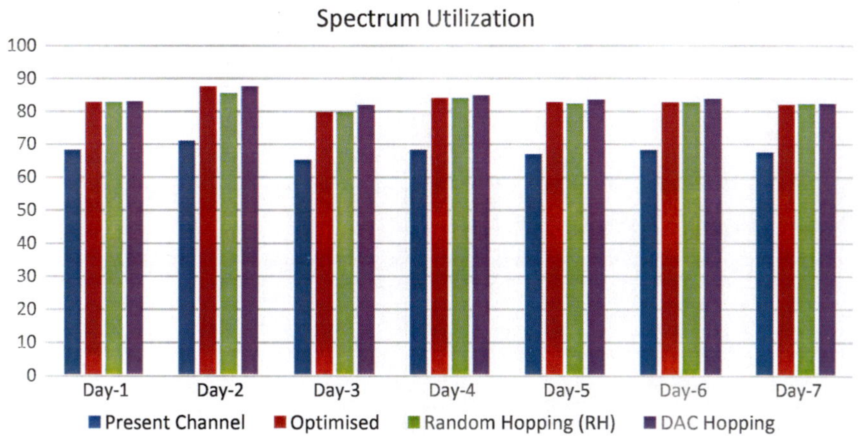

Fig. 4 Results of spectrum utilization compared with OCH, RH, and DAC hopping

Table 1 Comparison of hopping techniques applied over 2G GSM

Week days	Present channel utilization	Optimised cyclic hopping (OCH)	Random hopping (RH)	DAC hopping
Day-1	68.0402	82.6269	82.6236	82.7902
Day-2	70.7699	87.4363	85.3532	87.4362
Day-3	65.041	79.6248	79.6251	81.7076
Day-4	68.0095	83.919	83.9427	84.6719
Day-5	66.7686	82.666	82.2551	83.4245
Day-6	68.0417	82.6251	82.625	83.7469
Day-7	67.3802	81.9636	82.1239	82.1905

Table 2 Results obtained using DAC hopping algorithm with simulation time

Week days	Present utilization	Simulation time (min)			
		15	30	45	60
Day-1	68.0402	71.407	75.433	78.981	82.627
Day-2	70.7699	74.519	78.269	83.269	87.436
Day-3	65.041	68.687	72.332	75.978	79.625
Day-4	68.0095	71.715	75.543	80.457	83.929
Day-5	66.7686	70.518	74.259	79.258	82.666
Day-6	68.0417	71.687	75.333	78.979	82.625
Day-7	67.3802	70.953	74.672	78.318	81.964

Table 3 Overload condition for SU

Channel	Idle hour (%)	Busy hour (%)	Random hour (%)
Nine channel	248	138	165
Six channel	130	86	83

Table 4 Comparison of spectrum utilization LTE, 5G and 2G

Technology	LTE [5]	5G [5]	2G
Bandwidth (MHz)	20	20	25
Spectrum utilization (%)	90	95.4	100

References

1. Steenkiste P, Sicker D et al (2009) Future directions in cognitive radio network research. NSF workshop report, 9–10 Mar 2009
2. Spectrum policy task force report, document ET Docket 02-155, FCC (2002)
3. Wang B, Liu KJR (2011) Advances in cognitive radio networks: a survey. IEEE J Sel Top Signal Process 5(1):5–23
4. Mitola J, Maguire GQ Jr (1999) Cognitive radio: making software radios more personal. IEEE Pers Commun 6(4):13–18
5. Hu F, Chen B, Zhu K (2018) Full spectrum sharing in cognitive radio networks toward 5G: a survey. IEEE Access 6:15754–15776
6. Ericsson Radio System (2000) Ericsson-bts-installation-and-commissioning manual. LNZ 30276 R7B, p 279, 28 Apr 2000
7. Fernández de Gorostiza E, Berzosa J, Mabe J, Cortinas R (2018) A method for dynamically selecting the best frequency hopping technique in industrial wireless sensor network applications. Sensors (Basel) 18(2)
8. Nitin Babu KM, Vinaymurthi KK (2011) GMSK modulator for GSM system, an economical implementation on FPGA. In: International conference on communications and signal processing, Calicut, pp 208–212
9. Barreto AN, Deissner J, Fettweis G (1998) A frequency hopping algorithm for cordless telephone systems. In: ICUPC '98. IEEE international conference on universal personal communications. Conference proceedings, vol 2, Florence, Italy, pp 1273–1277
10. Kaplun DI, Klionskiy DM, Gulvanskiy VV, Bogaevskiy DV, Kupriyanov MS (2017) Development and study of demodulators for frequency-hopping spread spectrum signals. In: Progress in electromagnetics research symposium—Spring (PIERS), St. Petersburg, pp 445–448
11. Subramaniam S, Reyes H, Kaabouch N (2015) Spectrum occupancy measurement: an autocorrelation based scanning technique using USRP. In: IEEE 16th annual wireless and microwave technology conference (WAMICON), pp 1–5
12. Putzke M, Wietfeld C (2011) Self-organizing OFDMA systems by random frequency hopping. In: 2011 IFIP wireless days (WD), Niagara Falls, ON, pp 1–6
13. Verdone R, Zanella A, Zuliani L (2001) Dynamic channel allocation schemes in mobile radio systems with frequency hopping. In: 12th IEEE international symposium on personal, indoor and mobile radio communications. PIMRC 2001. Proceedings (Cat. No.01TH8598), San Diego, CA, USA
14. Perin JK, Roberts I, Kahn JM (2011) Improving the capacity of terrestrial and submarine systems via channel power optimization. In: 2018 20th international conference on transparent optical networks (ICTON), Bucharest, pp 1–4

15. Manjunath RK, Raju KN (2015) Synchronization of GSM TDM network through microwave media and its impact on GSM KPI's. In: 2015 IEEE international conference on electronics, computing and communication technologies (CONECCT), Bangalore, 1–6
16. Singh BK, Bhadada R (2020) Enhancing spectral efficiency of mobile communication by store forward base transceiver system. In: 7th international conference on signal processing and integrated networks (SPIN), Noida, India, 27–28 Feb 2020, pp 781–784

Color Face Recognition Based on Quaternion Moment Vectors Using Polynomial Kernel Extreme Learning Machine

Supriya Anand and Sukhjeet K. Ranade

Abstract The quaternion moments (QMs) have been successfully utilized for numerous image processing and pattern recognition applications due to their ability to capture and represent the color information holistically. In this paper, we propose a method based on the robust quaternion pseudo-Zernike moments (QPZMs) and polynomial kernel extreme learning machine (P-KELM) for color face recognition. A new hybrid color space-X^nSCr is presented that outperforms the accuracies achieved in commonly used color spaces. The comparative performance analysis of the variants of extreme learning machine (ELM) is performed and the experiments exhibit that the proposed method achieves higher recognition accuracies with minimal training time on the benchmark datasets: Faces95 and Georgia Tech.

Keywords Color space · Face recognition · Quaternion moments · Kernel extreme learning machine · Quaternion extreme learning machine

1 Introduction

Face recognition has been continuously addressed by the researchers owing to the broad spectrum of its applications. While the initial research in the field of recognition was primarily based on grayscale images, the contemporary studies depict that color provides discriminative, and complementary information which is crucial for face recognition. The color space enhances the feature extraction and classification performance by providing an optimal way of representing the color images.

The Zernike moments (ZMs) and pseudo-Zernike moments (PZMs) have been considered as powerful descriptors due to their favorable properties, such as low information redundancy, rotational invariance, and higher noise resilience. In recent

S. Anand (✉)
Department of Computer Science and Engineering, Punjabi University, Patiala, India
e-mail: supriya.anand70@gmail.com

S. K. Ranade
Department of Computer Science, Punjabi University, Patiala, India
e-mail: mrsdjjr@gmail.com

R. P. Mahapatra et al. (eds.), *Proceedings of 6th International Conference on Recent Trends in Computing*, Lecture Notes in Networks and Systems 177,
https://doi.org/10.1007/978-981-33-4501-0_56

studies, the quaternion representation of color images has been popular owing to its ability to preserve correlation among the color channels. The quaternion representation of ZMs was introduced by Chen et al. [1]. The quaternion moment-based approaches in the literature for the recognition applications have used the primary (RGB) color channels. Recently, a technique for image splicing detection using quaternion PZMs and YCbCr color space was given by Chen et al. [2]. However, no existing study has laid focus on constructing a suitable color space representation that could enhance the performance of quaternion moments.

A non-iterative algorithm named extreme learning machine (ELM) given by Huang [3] has been widely accepted as it facilitates fast and efficient learning. However, the dimensionality L (number of nodes in the hidden layer) of the ELM space has to be selected empirically for each application. It has been observed that a sufficiently high value of L is required to achieve good results, which consequently leads to an increase in the training time to a great extent. The Quaternionic extension of ELM (QELM) was recently proposed by Minemoto et al. [4] and was utilized for the face recognition application [5]. Although the QELM seems to be the natural choice for quaternion features, it is dependent on parameters like L and the number of training images (N). Another variant of ELM-kernel extreme learning machine (KELM) [6] is widely used as it eliminates the dependency on the parameters L and N. The KELM with polynomial kernel provides one of the best results especially with lower degrees polynomials.

In this study, we have performed a comparative analysis of the machine learning algorithms- ELM, QELM, and KELM for the features extracted using the quaternion moment function. The proposed framework for face recognition in the color domain consists of: (1) a new color space-X^nSCr, (2) quaternion moment vector (QMV)-based feature extraction using Quaternion pseudo-Zernike moment (QPZM) function and (3) Polynomial-based KELM (P-KELM) of degree 2 for classification. The experiments demonstrate superior results achieved by the proposed method on two benchmark face image databases.

2 Background

2.1 Color Space

The conventional alternative color spaces like YIQ, YCbCr, and HSV [7–9] have been explored over the period of time to boost the feature extraction capabilities in the various face recognition methods. Apart from the conventional color spaces, some hybrid color spaces have been formulated for improved representation of color images. A hybrid RQCr color space has been widely applied in color face recognition using various feature extraction techniques [10–13]. The Z^nRG^n color space, composed by the combination of the normalized color spaces (CSN-II) has also been effective as it reduces the correlation and amplifies the discriminating power of the color channels [12].

2.2 Quaternion Moments

A quaternion is a four-element vector with one real and three imaginary units, which was given by the mathematician Hamilton [14]. It can be considered as the generalization of the complex numbers and is defined as follows:

$$\boldsymbol{q} = a + b\boldsymbol{i} + c\boldsymbol{j} + d\boldsymbol{k} \tag{1}$$

where *i*, *j*, *k* are the imaginary unit vectors. The multiplication rules followed by the three imaginary unit vectors are:

$$\boldsymbol{i}^2 = \boldsymbol{j}^2 = \boldsymbol{k}^2 = -1, \boldsymbol{ij} = -\boldsymbol{ji} = \boldsymbol{k}, \boldsymbol{jk} = -\boldsymbol{kj} = \boldsymbol{i}, \boldsymbol{ki} = -\boldsymbol{ik} = \boldsymbol{j}. \tag{2}$$

The digital color image can be denoted using pure quaternion representation wherein the real part $a = 0$ and the imaginary parts encode the three color channels, respectively. For instance, the RGB image in the polar domain $f(r, \theta)$ can be expressed using the quaternion representation as:

$$f(r, \theta) = f_R(r, \theta)\boldsymbol{i} + f_G(r, \theta)\boldsymbol{j} + f_B(r, \theta)\boldsymbol{k}. \tag{3}$$

The quaternion moment, $\mathrm{QM}_{n,m}$ of order '*n*' and repetition '*m*,' for the color image is defined over a unit disk in the polar domain, which is given below:

$$\mathrm{QM}^R_{n,m}(f) = \frac{n+1}{\pi} \int_0^{2\pi} \int_0^1 R_{n,m}(r) f(r, \theta) \mathrm{e}^{-\mu m \theta} r \mathrm{d}r \mathrm{d}\theta \tag{4}$$

where $\mu = \frac{\boldsymbol{i}+\boldsymbol{j}+\boldsymbol{k}}{\sqrt{3}}$. The real-valued radial polynomial, $\mathrm{R}_{n,m}(r)$, for ZMs and PZMs is given as:

$$\mathrm{R}^{\mathrm{ZM}}_{nm}(r) = \sum_{s=0}^{\frac{(n-|m|)}{2}} \frac{(-1)^s (n-s)!(r)^{n-2s}}{s!\left(\frac{(n+|m|)}{2} - s\right)!\left(\frac{(n-|m|)}{2} - s\right)!},$$
$$n \geq 0, |m| \leq n, n - |m| = \text{even}. \tag{5}$$

$$\mathrm{R}^{\mathrm{PZM}}_{nm}(r) = \sum_{s=0}^{(n-|m|)} \frac{(-1)^s (2n+1-s)!(r)^{n-s}}{s!(n+|m|+1-s)!(n-|m|-s)!}, \quad n \geq 0 \text{ and } |m| \leq n. \tag{6}$$

For a discrete color image, QM can be expressed in the Cartesian domain as follows:

$$\mathrm{QM}_{n,m}^{R} = \frac{n+1}{\pi} \sum_{i=0}^{N-1} \sum_{k=0}^{N-1} R_{nm}(r_{ik}) f(x_i, y_k) \mathrm{e}^{-\mu m \theta_{ik}} \Delta x \Delta y \tag{7}$$

such that $x = \Delta y = \frac{2}{D}, r_{ik}^2 = x_i^2 + y_k^2 \leq 1$ and $\Theta_{ik} = \tan^{-1}(\frac{y_k}{x_i})$ and $f(x, y) = f_R(x, y)\boldsymbol{i} + f_G(xy)\boldsymbol{j} + f_B(x, y)\boldsymbol{k}$.

The originally introduced magnitude-based quaternion moments produced single real-valued features as it used modulus operator to achieve rotation invariance. It should be noted that the vital phase information was not retained in these invariants.

2.3 Extreme Learning Machine (ELM)

ELM was presented by Huang [3] as a simple, non-iterative algorithm using a single hidden layer feed forward neural network (SLFN), wherein the tuning of the parameters is not required. The input parameters are assigned using random function, whereas the output weights are determined using the mathematical inverse function. For a given training sample set $(x_i, \boldsymbol{t}_i), i = 1, 2, \ldots, n$ with 'm' number of classes, 'n' number of features each. With $x_i = [x_{i1,} x_{i2}, \ldots, x_{in}]^{\mathrm{T}} \in R^n$ and $\boldsymbol{t}_i = [t_{i1,} t_{i2}, \ldots, t_{im}]^{\mathrm{T}} \in R^m$ as the input and target output vectors, respectively, the output function is given as:

$$f_L(x) = \sum_{j=1}^{L} \beta_j g_j(x_i) = O_i, \quad i = 1, 2, \ldots, n \tag{8}$$

where 'L' is the number of hidden layer nodes and $g(x)$ is the activation function. $\beta = [\beta_1, \ldots, \beta_L]^{\mathrm{T}}$ is the matrix of weights between the hidden layer and the output layer nodes. The objective of the training in ELM is minimizing the error between the computed and the actual target output such that $\sum_{i=1}^{n} \| O_i - t_i \| = 0$ thus, Eq. (8) can be written as:

$$f_L(x) = \sum_{j=1}^{L} \beta_j g_j(x_i) = \boldsymbol{t}_i, \quad i = 1, 2, \ldots, n. \tag{9}$$

The above expression is given using the matrix representation as:

$$H\beta = T. \tag{10}$$

Here, $H = \begin{bmatrix} g_1(x_1) & \dots & g_L(x_1) \\ \vdots & \ddots & \vdots \\ g_1(x_n) & \dots & g_L(x_n) \end{bmatrix}_{n \times L}$ is the output matrix for the hidden layer and T is the target matrix. The output weight matrix β is computed using the minimal norm least-squares solution [3] of the Eq. (10) and is given as:

$$\beta = H^{\dagger} T. \tag{11}$$

Here, $H^{\dagger}$ is the Moore–Penrose generalized inverse of matrix H computed using the orthogonal projection method [15].

Quaternionic Extreme Learning Machine (QELM). The QELM encodes the inputs, outputs, and weights as a quaternion. Consequently, the activation function has to be adapted in the form of split type activation function to accept quaternion [4], which is given as:

$$g(x) = g\left(x^a\right) + g\left(x^b\right)\boldsymbol{i} + g\left(x^c\right)\boldsymbol{j} + g\left(x^d\right)\boldsymbol{k}. \tag{12}$$

In order to compute the Moore–Penrose inverse of the quaternion matrix, the Cayley-Dickson notation is used to represent the quaternion matrix using its complex adjoint matrix [16]. The necessary conditions for the quaternion matrix to be invertible are given in [17]. The notation for a quaternion matrix Q is as given below:

$$Q = \left(Q^a + Q^b i\right) + \left(Q^c + Q^d i\right) j = A + Bj. \tag{13}$$

Here, $A = \left(Q^a + Q^b i\right)$ and $B = \left(Q^c + Q^d i\right)$ are complex matrices. The complex adjoint matrix of Q is given as:

$$Q_c = \begin{pmatrix} A & B \\ -B^* & A^* \end{pmatrix} \tag{14}$$

where * is the complex conjugate. Hence, the complex adjoint matrix βc for the weight matrix β is defined by:

$$\beta_c = \begin{cases} H_c^h \left(H_c H_c^h\right)^{-1} T_c (N \le L) \\ \left(H_c^h H_c\right)^{-1} H_c^h T_c (N > L) \end{cases} \tag{15}$$

where ^{h} denotes the Hermitian transpose.

Kernel Extreme Learning Machine (KELM). The kernelized ELM was proposed [6] to achieve better stability and generalized performance. The kernel function Ω_{ELM} is given as:

$$\Omega_{\mathrm{ELM}} = HH^{\mathrm{T}} = K(x_i, x_j). \tag{16}$$

Hence, the output function of KELM can be defined as:

$$f(x) = \begin{bmatrix} K(x, x_1) \\ \vdots \\ K(x, x_N) \end{bmatrix}^{\mathrm{T}} \left(\frac{I}{C} + \Omega_{\mathrm{ELM}}\right)^{-1} T. \tag{17}$$

Here, C is the regularization parameter and I is the identity matrix.

The essence of KELM is that the feature mapping may not be known; rather, the kernel is specified. Additionally, it is not required to identify the optimum number of hidden layer nodes [18], thus making the results more stable, reliable, and independent from parameters.

3 Proposed Method

The proposed method for color face recognition aims to enhance the performance by refining the three fundamental steps of a face recognition process namely preprocessing, feature extraction, and classification.

3.1 *Preprocessing*

The hybrid color space, $X^{n}SCr$ is presented on the basis of extensive empirical analysis. The new space is composed of X^{n}, S and Cr from the normalized-XYZ [19], HSV and YCbCr color space, respectively. The input RGB images are transformed into the proposed color space and then further represented in the quaternion domain for the subsequent stages.

3.2 *Feature Extraction*

We propose the Quaternion moment vector (QMV) approach where we incorporate the four elements (one real and three imaginary) of the quaternion moments into the feature set without the application of any invariance operations. The QMV approach preserves the critical phase information and relies on the inherent invariant properties of the moments. For robust feature extraction from face images, we have used the QPZM function, and using the QMV approach, we obtain the QPZM vector (QPZMV).

3.3 Classification

For stable, reliable, fast, and parameter-free classification, we propose the use of polynomial KELM (P-KELM). The proposed method uses the polynomial kernel of degree 2. The polynomial kernel function of degree, d is defined as:

$$K(x_i, x_j) = (c + x_i x_j)^d. \tag{18}$$

4 Experiments and Results

For the experimental analysis, feature extraction has been performed in Microsoft's Visual Studio, and classification has been performed in MATLAB using ELM [3] and ELM kernel [18] functions under Windows 10 environment on a system with 8 GB RAM and 2.20 GHz CPU.

The benchmark databases-Faces95 [20] and Georgia Tech (GT) [21] have been used for the experiments. The Faces95 database consists of a total of 1440 images for 72 persons and images are split into equal halves for training and testing, whereas the GT database has 750 images for 50 persons that are split in a ratio of 7:8 between the training and testing sets.

4.1 Selection of Parameter L for ELM and QELM

The parameter order of moments, $n_{\max}$ for the quaternion moment function has been set to 6 [1], leading to 64 and 112 features, respectively, for quaternion Zernike moment vector (QZMV) and quaternion pseudo-Zernike moment vector (QPZMV). The number of hidden layer nodes (L) in the case of ELM and QELM has been set empirically using the following set of experiments on the two databases. Ten trial runs have been carried out for each case. The mean recognition accuracies along with standard deviations and the average training time have been depicted in Figs. 1, 2, 3, and 4.

In the first set of experiments using ELM, it is observed that QPZMV performs better than QZMV as seen in Figs. 1 and 2. The recognition accuracy is computed by varying the value of parameter L from 1000 to 8000 with a fixed step size of 1000. The recognition accuracy achieves an optimal peak at $L = 4000$ after which there is no significant improvement in accuracy whereas the training time increases considerably.

The recognition accuracy using QELM is computed by varying L such that $L \ll N$, that is the number of hidden layer nodes is much less than the number of training images. Therefore, for GT database L is varied between 90 and 240 with a step size

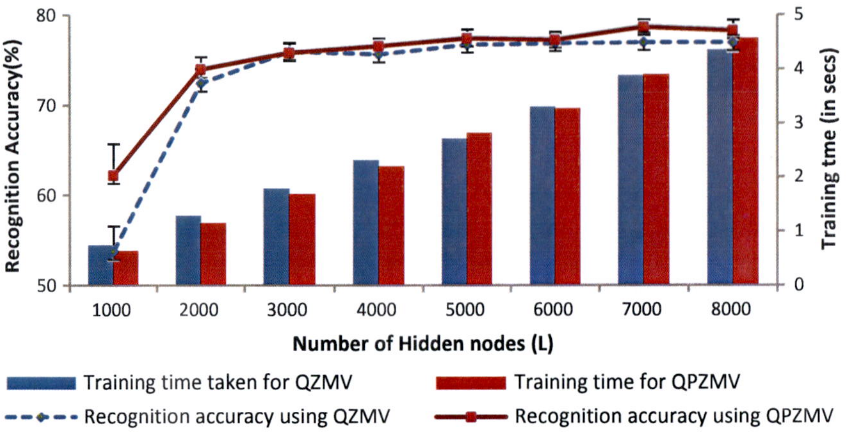

Fig. 1 Recognition accuracy (%) and training time taken for the different values of L in ELM on Georgia Tech Face database

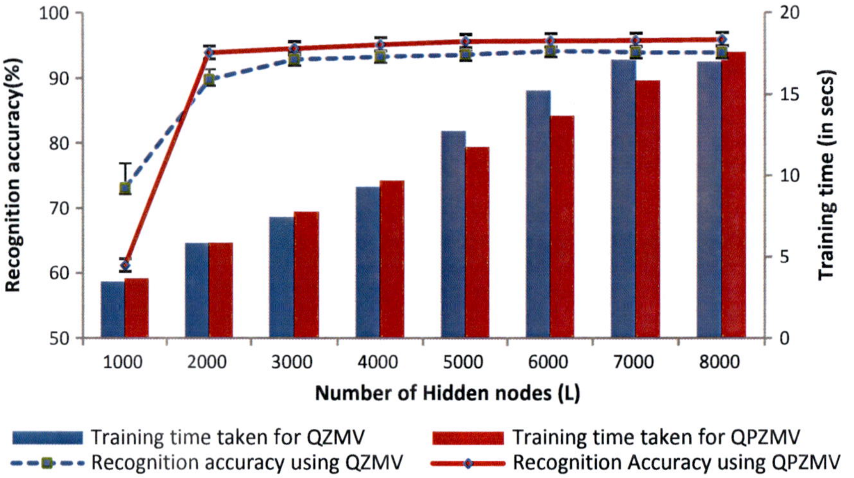

Fig. 2 Recognition accuracy (%) and training time taken for the different values of L in ELM on Faces95 database

of 30, and for the Faces95 database, it is varied from 50 to 500 with a step size of 50. The feature set QPZMV shows better results with QELM as compared to QZMV. From the recognition trends observed in Figs. 3 and 4, the value of L is chosen as 150 and 300 for the GT and Faces95 databases, respectively.

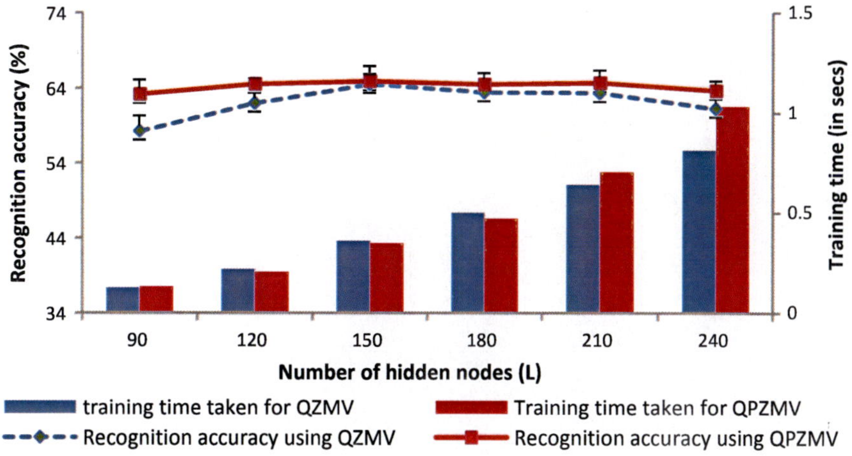

Fig. 3 Recognition accuracy (%) and training time taken for the different values of L in QELM on Georgia Tech database

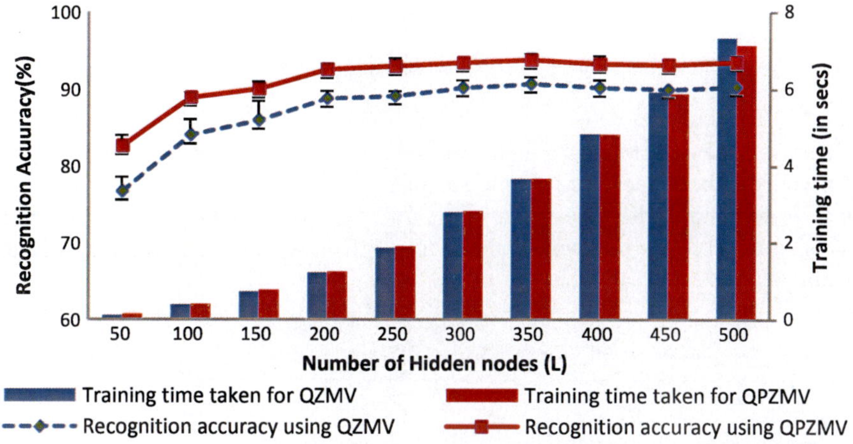

Fig. 4 Recognition accuracy (%) and training time taken for the different values of L in QELM on Faces95 database

4.2 Comparison of Proposed Method with the Various Color Spaces and Classifiers

We compare the performance of three variants of extreme learning machine classifiers- ELM, QELM, and P-KELM in the existing (RGB, YIQ, RQCr, Z^nRG^n) and proposed color space (X^nSCr) using the QPZMV features. In the case of ELM and QELM, 10 trials for each case have been carried out and the mean recognition accuracies, and standard variations achieved have been demonstrated in Table 1. It is

Table 1 Recognition performance comparison of the extreme learning methods in various color spaces using QPZMV

Color space	Faces 95			Georgia Tech		
	ELM	QELM	P-KELM	ELM	QELM	P-KELM
RGB	95.13 ± 0.70	93.42 ± 0.47	92.22	75.65 ± 0.71	64.85 ± 2.19	78
YIQ	95.76 ± 0.59	94.42 ± 0.70	94.58	76.98 ± 0.85	68.98 ± 1.62	77.5
RQCr	95.17 ± 1.12	93.93 ± 1.03	94.31	75.78 ± 1.22	62.78 ± 2.11	80.25
Z^nRG^n	95.29 ± 0.61	94.14 ± 0.83	94.44	78.20 ± 1.37	67.88 ± 1.55	79.75
X^nSCr (Proposed)	94.69 ± 0.61	92.67 ± 0.70	**96.94**	82.33 ± 0.71	66.28 ± 1.96	**86.5**

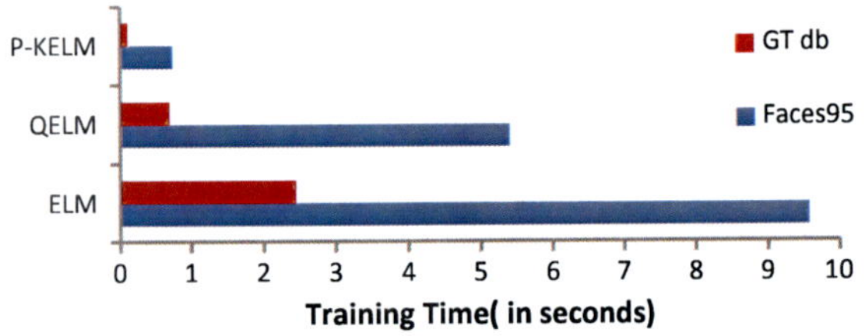

Fig. 5 Training time comparison for the various extreme learning-based classification methods

observed that P-KELM attains higher recognition accuracies than ELM and QELM, especially in the proposed color space-X^nSCr.

Figure 5 shows the time taken in the training phase for the different extreme learning methods for the two databases. The P-KELM takes the minimum time in training for both the databases.

5 Conclusion

In this study, an efficient method for color face recognition has been proposed. A new color space- X^nSCr has been introduced which boosts the feature extraction ability of the quaternion moment descriptors. The performances of the classifiers-ELM and the recently introduced QELM are compared with P-KELM in terms of mean recognition accuracies and time required for training. It is observed that apart from providing classification independent from parameters like the number of hidden nodes and the images in the training set, P-KELM produces superior results. The experiments demonstrate that the proposed method with QPZMV and P-KELM, as feature descriptor and classifier, respectively, achieves higher accuracy within minimum training time in the X^nSCr color space as compared to the state-of-the-art techniques.

References

1. Chen B, Shu H, Zhang H, Chen G, Toumoulin C, Dillensege JL, Luo L (2012) Quaternion Zernike moments and their invariants for color image analysis and object recognition. Signal Process 92:308–318
2. Chen B, Qi X, Sun X, Shi YQ (2017) Quaternion pseudo-Zernike moments combining both RGB information and depth information for color image splicing detection. J Vis Commun Image Represent 49:283–290
3. Huang GB, Zhu QY, Siew CK (2006) Extreme learning machine: theory and applications. Neurocomputing 70(1):489–501
4. Minemoto T, Isokawa T, Nishimura H, Matsui N (2017) Feedforward network with random quaternionic neurons. Signal Process 136:59–68
5. Dad N, En-nahnahi N, El Alaoui Ouatik S (2019) Quaternion Harmonic moments and extreme learning machine for color object recognition. Multimed Tools Appl 78(15):20935–20959
6. Zong W, Zhou H, Huang GB, Lin Z (2011) Face recognition based on kernelized extreme learning machine. In: Kamel M, Karray F, Gueaieb W, Khamis A (eds) Autonomous and intelligent systems. AIS 2011. LNCS, vol 6752. Springer, Berlin, Heidelberg
7. Liu Q (2014) Within-component and between-component discriminant analysis for color face recognition. Optik 125:6366–6374
8. Li J, Sang N, Gao C (2016) Completed local similarity pattern for color image recognition. Neurocomputing 182:111–117
9. Anbarjafari G (2013) Face recognition using color local binary pattern from mutually independent color channels. J Image Video Proc 2013:6
10. Liu Z, Liu C (2010) Fusion of color, local spatial and global frequency information for face recognition. Pattern Recogn 43:2882–2890
11. Choi JY, Ro YM, Plataniotis K (2009) Color face recognition for degraded face images. IEEE Trans Syst Man Cybern 39(5):1217–1230
12. Choi JY, Ro YM, Plataniotis K (2012) Color local texture features for color face recognition. IEEE Trans Image Process 21(3):1366–1380
13. Lee SH, Choi JY, Ro YM, Plataniotis KN (2012) Local color vector binary patterns from multichannel face images for face recognition. IEEE Trans Image Process 21(4):2347–2353
14. Hamilton WR (1866) Elements of quaternions. Longmans, Green and Company, London
15. Rao CR, Mitra SK (1971) Generalized inverse of matrices and its applications. Wiley, New York
16. Zhang F (1997) Quaternions and matrices of quaternions. Linear Algebra Appl 251:21–57
17. Kyrchei I (2013) Explicit representation formulas for the minimum norm least-squares solutions of some quaternion matrix equations. Linear Algebra Appl 438:136–152
18. Huang GB, Zhou H, Ding X, Zhang R (2012) Extreme learning machine for regression and multiclass classification. IEEE Trans Syst Man Cybern 42(2):513–529
19. Yang J, Liu C, Zhang L (2010) Color space normalization: enhancing the discriminating power of color spaces for face recognition. Pattern Recogn 43:1454–1466
20. Description of the collection of facial images. https://cswww.essex.ac.uk/mv/allfaces/index.html
21. Ara Nefian face recognition page. https://www.anefian.com/research/face_reco.htm

Human Activity Recognition Using Long Short-Term Memory

Vibhor Vimal, Anshuman Singh, and Priyanka Meel

Abstract The aim of this project is to perform human activity recognition using recurrent neural networks by long short-term memory. Human activity recognition involves dividing the accelerometer input sequences into well-defined movements. The project uses tensor flow which is a math library that performs programming tasks across a wide range of objectives. Recurrent neural networks are a type of artificial neural network that are designed to identify patterns in data sequences. The project uses deep learning architecture. The problem of long-term dependencies is solved using a long short-term memory mechanism of recurrent neural networks, also known as LSTM. The advantage of using LSTM for the model is that it can learn directly from the time series data. The project also develops an Android application to study the real-life application of the model by directly using data from the accelerometer in the device. Additionally, a GUI was also developed to display the results.

Keywords Activity recognition · Recurrent neural network · LSTM · Accelerometer · Android

1 Introduction

We human beings do not start our thinking from scratch every time. Even when we read a book, we understand each word based on our understanding of previous words. Thus, we can say that our thinking and understanding depend on previous knowledge. The classical neural networks cannot achieve this which becomes a major limitation.

V. Vimal (✉) · A. Singh · P. Meel
Department of Information Technology, Delhi Technological University, New Delhi 110042, India
e-mail: vibhorvimal5598@gmail.com

A. Singh
e-mail: singhanshumansingh5@gmail.com

P. Meel
e-mail: priyankameel86@gmail.com

R. P. Mahapatra et al. (eds.), *Proceedings of 6th International Conference on Recent Trends in Computing*, Lecture Notes in Networks and Systems 177,
https://doi.org/10.1007/978-981-33-4501-0_57

Our aim here is to construct a model that is capable of recognizing several human activities performed by a user using a triaxial accelerometer built in the user's smartphone. The sensor in the triaxial accelerometer gives an estimated acceleration along all the three dimensions x, y and z. Human activity recognition can thus be considered as a supervised classification problem, whose training data can be obtained from the sensors on smartphones worn by a set of subjects who perform a set of activities. We have used LSTM, which is a complex form of recurrent neural network. The model is trained on the dataset and can identify the type of activity being performed. For this project, we have considered six basic activities. However, the project can also be expanded to include the postural transitions that occur between the basic activities as labels.

In Sect. 2, we presented some related work done using different mechanisms to solve the problem of human activity recognition and the technologies used. Then, we discussed the method proposed by us, its mathematical background as well as the tools used for implementation. Finally, the results and relevant conclusions were drawn out of them.

2 Related Work

Human activity recognition involves dividing the accelerometer input sequences into movements which are properly defined. To obtain relevant features and classify the activities are challenging tasks. The problem of detecting and recognizing human activity has been tackled through various approaches in the past.

In the past few years, we have seen a rise in the number of CNN-based methods used to solve the problem. Jiang and Yin [1] proposed a method for constructing a novel activity image for deep CNN. The proposed method aimed at constructing a novel activity image for deep CNN (DCNN), using the linear acceleration and gyroscope signals. Recognition performance of 97.59%, 99.93%, and 97.83% on the benchmark dataset of UCI, SHO, and USC were obtained.

The performance for activity recognition of deep belief networks (DBNs) using different parameter settings was tested by Alsheikh et al. [2]. A hybrid deep learning and hidden Markov model (DL-HMM)-based method were proposed in this paper. The method uses a spectrogram signal from the triaxial accelerometer instead of the raw data. The accuracy obtained was 89.38%, 91.5%, and 98.23% on the Skoda, Daphnet, and WISDM datasets, respectively.

The three types of deep learning models for activity recognition, namely, convolutional networks (CNN), recurrent neural networks (RNN), and deep feedforward networks (DNN) were used by Hammerla et al. [3] on the benchmark datasets of PAMAP2, the Daphnet Gait, and the Opportunity. Human activity recognition was collected using wearable sensors. A novel regularization approach was proposed for recurrent methods. It was found that the DNNs require a significant parameter tuning. It was also found that for activities like walking and running, CNN gives the best performance.

Milenkoski et al. developed a real-time human activity recognition system using LSTM and worked on smartphones data which is easily implementable on mobile applications [4]. Raw data were collected in a laboratory setting and then processed into sequences. LSTM model was trained training set and the developed architecture was also transferred as a mobile application to an Android device. It was observed that for the "walking upstairs", "sitting", "standing" and "walking" classes, LSTM gave a significantly better performance than other methods like J48, straw man and multilayer perceptron.

A major limitation faced by CNN is that they require a huge amount of data to train and often have a high computational cost. DBNs have no recurrent connections between units that pass data from one unit to another. RNNs are not able to model long-term dependencies. Our work proposes a method based on LSTM networks, which directly works with the raw accelerometer data and can handle long-term dependencies.

3 Long Short-Term Memory

LSTM was firstly proposed by Hochreiter [5]. It was later improved by Gers in 2000 [6]. In this section, we define the mathematical model of LSTM. The cell state is the key to LSTMs. It is said that the cell state is similar to a conveyor belt. It performs very minor linear interactions as it runs down the chain. Sometimes, information might just flow through it unchanged. LSTM possesses the ability to remove or add certain information to the cell state through structures called gates. The function of gates is to let information through. They are made up of two things: a pointwise multiplication operation and a sigmoid function.

To control and protect the cell state, LSTM has three such gates. Inside a gate, the sigmoid layer is responsible for producing a number between 0 and 1, telling the quantity of each component to pass through the cell state. The sigmoid layer is also known as the forget gate. The exact quantity of information that should be allowed is controlled by the second gate, also known as the input gate. This gate plays a major factor in determining how much an old memory should impact the new memory. As we proceed, we see a "+" operator. The "+" operator stands for piecewise summation. We merge the present input and the old memory with this operation. This process is known as elementwise summation. Toward the end, we need to produce an output for the LSTM structure. The third gate, also known as output gate, comes into play. The output gate is controlled by three factors: the present input, the previous output, and the fresh new memory. The amount of memory that is to be produced to the next LSTM unit is controlled by the output gate. The input, forget, and output gates are known as i_t, f_t and o_t.

The following equations are used at the interacting layers of the LSTM:

$$i_t = \sigma(W_i[h_{t-1}, x_t])$$

$$
\begin{aligned}
f_t &= \sigma(W_f[h_{t-1}, x_t]) \\
o_t &= \sigma(W_o[h_{t-1}, x_t]) \\
l_t &= \tanh(W_l[h_{t-1},\ x_t]) \\
c_t &= f_t \cdot c_{t-1} + i_t \cdot l_t \\
h_t &= o_t \cdot \tanh(c_t)
\end{aligned} \tag{1}
$$

The computational complexity per time step and weight is $O(1)$ [5].

4 Implementation

The activity recognition problem has been solved by various methods before, in this paper, we focus on achieving the goal using long short-term memory method, which is a complex form of recurrent neural network (RNN).

4.1 Procedure

In this work, we have used LSTM and not plain RNN as LSTM have several advantages over RNN. The major reason for the popularity of RNNs is the fact that they connect previous information to the current task. Many times, we only need current information to accomplish the task. In situations where the gap between the relevant information and the place of need is small, RNNs can themselves learn how to use the previous information. In other words, RNNs have a short-term memory. But in cases where we need context, i.e., where the gap increases, RNNs are unable to connect the information. LSTMs are exclusively developed to handle the long-term dependency problem. They remember information for longer periods.

Figure 1 shows the flow diagram for the entire model. In the beginning, data is acquired and it is preprocessed. The input data is divided into training (70%) and testing (30%) data. After that, the data parameters are decided and modeled appropriately. The LSTM model is trained using the training set. Once the model is trained, the testing data set is passed and results are observed. Relevant accuracy predictions are then made.

4.2 Datasets and Tools Used

For implementation, we used Tensor Flow, which is a Python library [7]. Google Collab was also used. "Activity Recognition Using Smartphones Dataset" which was made available in 2012, is one of the standard datasets for human activity recognition. The dataset was created by Davide Anguita, et al. from the University of Genova,

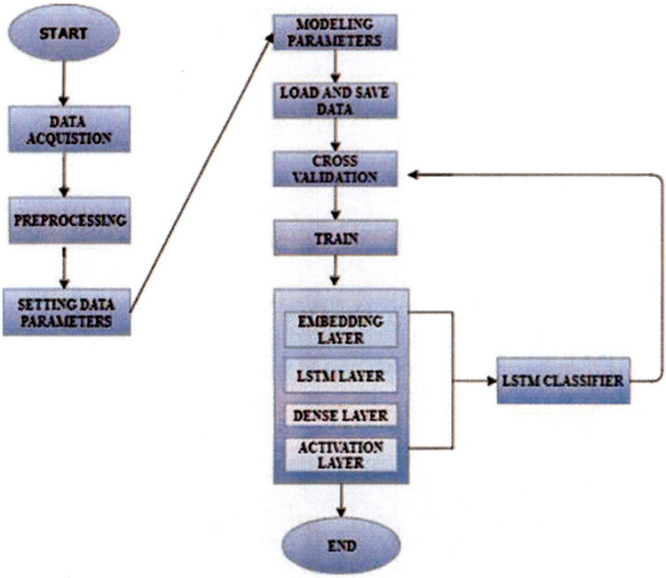

Fig. 1 Model flow diagram

Italy. The dataset is present and can be downloaded from the UCI Machine Learning Repository for free of cost, under the name "Human Activity Recognition Using Smartphones Data Set, UCI Machine Learning Repository".

The dataset contains observations from subjects who performed one of the six basic activities while they were wearing a smartphone across their waist. Every subject performed a set of activities twice, one with the device on the left side, other on the right side. The sensors in the smartphone collected the data using a triaxial accelerometer. The sensor in the triaxial accelerometer gives an estimated acceleration along all the three dimensions x, y and z. Velocity and displacement can also be estimated from the same.

Since the raw data is unavailable, a pre-processed version was made available. The preprocessing of the sensor signals from accelerometer and gyroscope was accomplished by noise filters and sampling in fixed-size sliding windows. The acceleration signal from the sensor consisting of both gravitational and body readings was separated using a low-pass filter. We assume that the gravitational force consists of low-frequency components, so we used a low cutoff frequency.

4.3 Android Application

To test the performance of our model in real life, we developed a mobile-based application [4]. The Android application fetches the data from the smartphone's

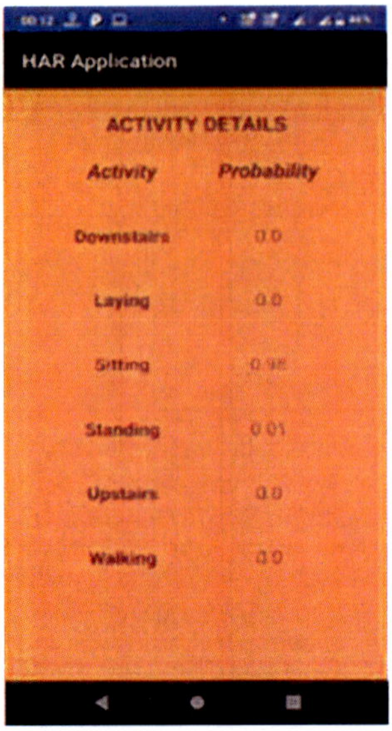

Fig. 2 Android application snapshot for sitting activity

accelerometer after a fixed timestamp. As a result, the application predicts and displays the activity that was performed in the previous 10-s window. Our Tensor Flow model was imported in Android for implementation. The text-to-speech Android API was also used to tell the user the predicted activity. The application serves as a journal/diary that records the user's activity by fetching data from the triaxial accelerometer. Figures 2 and 3 display two of the snapshots of the applications predicting different activities.

4.4 Addition of GUI

As an extra addition to the project, we have provided a GRAPHICAL USER INTERFACE for predicting or showing the results. The user is asked to enter the file location. Once entered, the user can press the "Load Dataset" button. This immediately makes a call to the press function created, which plots the graphs we had obtained just a few cells earlier. Also, we can print several other information regarding accuracy of the classifier, labels of the output data etc., to provide visual insights into the results of the human activity recognition. The basic screen of the GUI is attached below in Fig. 4 and it will have outputs printed same as we obtained earlier. The only difference is that we are creating a separate display view for the user to hide the intricacies of the code, i.e., achieving the ideology of abstraction.

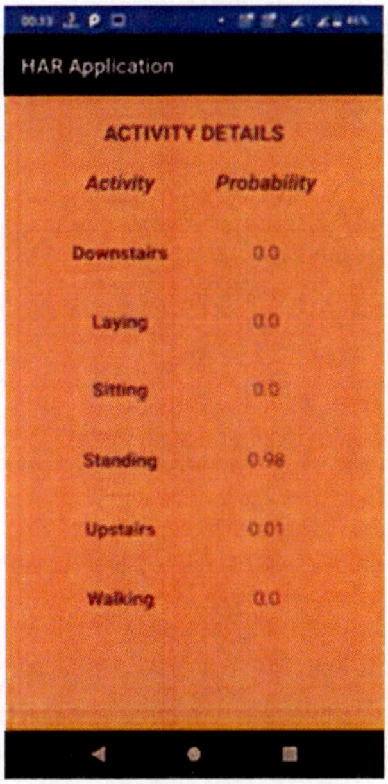

Fig. 3 Android application snapshot for standing activity

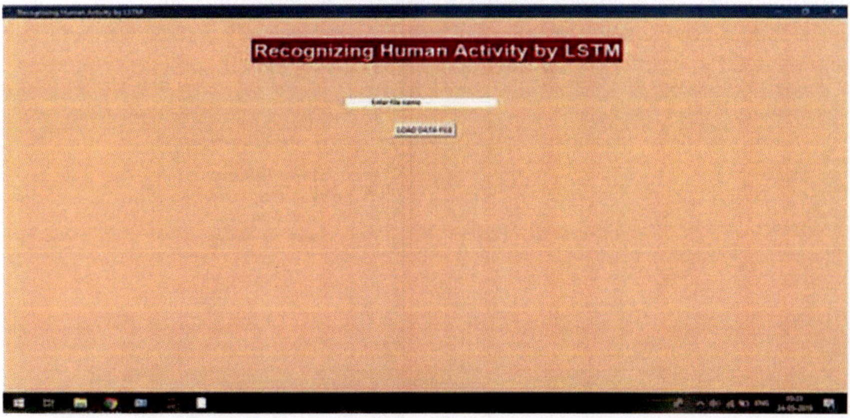

Fig. 4 Display of GUI

5 Results and Discussions

Figure 5 shows a snapshot of the time when the data is being trained in which the training data is divided into batch sizes of 1500 and we iterate unless we reach training iteration, which is 300 times the size of training data count to iterate through

Fig. 5 Snapshot of training

the dataset. For every epoch, the test data's performance is shown in term of batch loss and accuracy and finally, batch loss of 0.62 and accuracy of 0.86 is achieved.

Now to get visual insights matplotlib is used to obtain the graphs of train losses and accuracies represented by blue curve and green curve, respectively, in Fig. 6. The abscissa denotes the training iteration shown on an interval of 500,000 and the ordinate shows the training progress, i.e., the loss/accuracy values are depicted. We observe that upon reaching the end of the iteration, loss decreases and becomes constant while the accuracy starting from 0.12 increases and obtains a value of 0.86 (approx.) in the end iteration. After getting the accuracy results, we pictorially represented the correctly classified data and wrongly classified data by plotting a

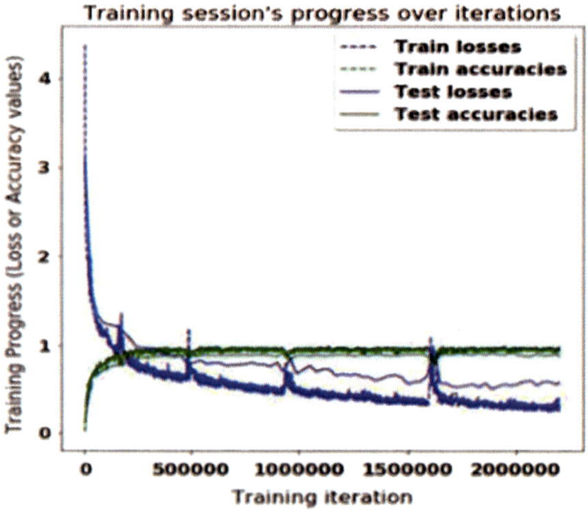

Fig. 6 Plot of training loss and accuracy versus iterations

```
Accuracy of Testing: 86.15541458129883%

Precision: 87.41322797417543%
f1_score: 85.97403228347925%
Recall: 86.15541228367832%
```

Fig. 7 Performance measures of model testing

```
Confusion Matrix is shown below:
[[439  10  47   0   0   0]
 [  5 435  31   0   0   0]
 [  3  34 383   0   0   0]
 [  6  24   0 297 164   0]
 [  5  25   0  27 475   0]
 [  0  27   0   0   0 510]]

NORMALISED CONFUSION MATRIX:
[[14.896504   0.3393281  1.5948422  0.         0.         0.        ]
 [ 0.16966406 14.760775   1.0519172  0.         0.         0.        ]
 [ 0.10179844  1.1537156 12.996266   0.         0.         0.        ]
 [ 0.20359688  0.8143875  0.        10.078046   5.5649815  0.        ]
 [ 0.16966406  0.8483203  0.         0.916186  16.118088   0.        ]
 [ 0.         0.916186   0.         0.         0.        17.305735  ]]
```

Fig. 8 General and normalized confusion matrices

confusion matrix. As represented in Fig. 7 the accuracy value is 86.15% bundled with $f1$ score, precision and recall.

The confusion matrix is shown both in normalized form and original number of data points form in Fig. 8. There is a 6 * 6 matrix corresponding to the six output labels and the entries in the matrix represents the number of data points classified as (i, j) where i and j are belonging to activity list (6 output labels), as shown in Fig. 8.

Figure 9 depicts the confusion matrix where predicted values are along the columns and true values are along the rows.

6 Conclusion

In this paper, a method has been proposed by us to solve the problem of human activity recognition with the use of long short-term memory networks. LSTM, which is a relatively new approach, is capable of handling long-term dependencies and can learn features from completely raw sensor data, thus requiring no expertise in designing features manually. The computation complexity of LSTM is $O(1)$ which makes it suitable to extend to real-world applications through a mobile application.

We demonstrated how LSTM overcomes the limitations of classical approaches and traditional models based on CNN. We successfully categorized the sensor data from smartphones into one of the basic six activities. The model achieved an accuracy of 86.15%.

The Android application correctly classified the activities performed by the user in fixed-time window based on the data collected by the sensor in the device [8]. The

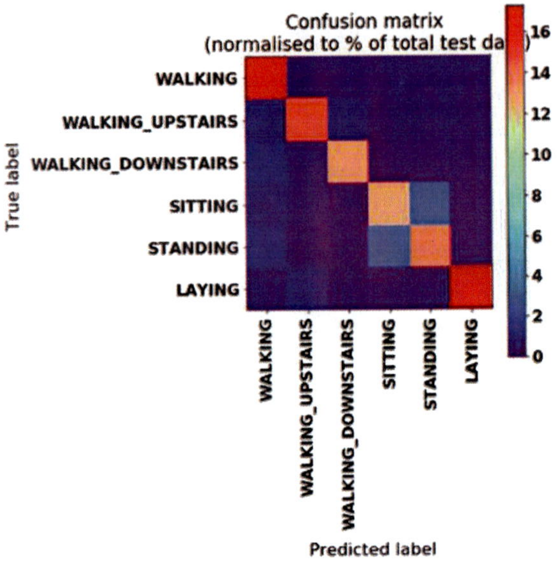

Fig. 9 Graphical Plot of normalized confusion matrix

Graphical user interface works successfully to show the results to the user hiding the unnecessary details, thus establishing abstraction.

7 Future Work

As a future scope of this project, one of the goals can be to overcome the battery draining feature of gyroscope and implementing the model just with accelerometer reading to make it more suitable in real-life scenarios. Study/research on deep LSTM architecture which further improves the accuracy and uses Bidirectional RNN, stacked cell and residual connections, can be done as an extension to this project. The project can be extended to include furthermore activities like postural transitions such as the transition from standing to sitting, sitting to lying, and so on.

References

1. Jiang W, Yin Z (2015) Human activity recognition using wearable sensors by deep convolutional neural networks. In: Proceedings of the 23rd ACM international conference on multimedia (MM '15), ACM, New York, NY, USA, pp 1307–1310
2. Alsheikh MA, Selim A, Niyato D, Doyle L, Lin S, Tan H-P (2016) Deep activity recognition models with triaxial accelerometers. In: Proceedings of the AAAI workshops, Phoenix, Az, USA, Feb 2016

3. Hammerla NY, Halloran S, Plotz T (2016) Deep, convolutional, and recurrent models for human activity recognition using wearables. In: Proceedings of the 25th international joint conference on artificial intelligence (IJCAI '16)
4. Milenkoski M, Trivodaliev K, Kalajdziski S, Jovanov M, Stojkoska BR (2018) Real time human activity recognition on smartphones using LSTM networks. In: MIPRO 2018
5. Hochreiter S, Schmidhuber J (1997) Long short-term memory. Neural Comput 9(8):1735–1780
6. Gers FA, Schmidhuber J, Cummins F (2000) Learning to forget: continual prediction with LSTM. Neural Comput **12**(10):2451–2471
7. Abadi M, Barham P, Chen J, Chen Z, Davis A, Dean J, Devin M et al (2016) TensorFlow: a system for large-scale machine learning. In: OSDI, vol 16, pp 265–283
8. Android Developers (n.d.) Retrieved 24 May 2019, from https://developer.android.com/

An Extended Oddball Technique to Detect Anomaly in Static Attributed Graphs

Prakhyat G. Kulkarni, S. Y. Praneet, R. B. Raghav, Ashwin Ashok, and Bhaskarjyoti Das

Abstract Social media sites connect us to every part of the world and assist in sharing information and enable personal and professional growth. All these positives come with their own pitfalls in the form of invasion of privacy, fraud, and identity thefts. This solicits a need for anomaly detection in real time. Since most of the available data sets are unlabelled and unweighted, we extend an unsupervised Oddball technique so that it works on attributed graphs. We demonstrate that considering both the structural aspects and the attributes yield better results by comparing various similarity measures that can be employed on these attributed graphs.

Keywords Anomaly detection · Attributed graph · Static graph · Social graph · Oddball · Weighted graph · Similarity measures

1 Introduction

Social media is a forum to help people stay connected with friends and colleagues and it helps them to talk, share ideas and interests, or for business purposes. Social media networks hosted online such as Reddit and Facebook have myriads of people. Any such social network can be modelled as a graph structure with users as nodes and edges representing the relations between the users. In such social networks, each

---

P. G. Kulkarni · S. Y. Praneet · R. B. Raghav · A. Ashok (✉) · B. Das
PES University, Bengaluru, India
e-mail: ashwinashok1998@gmail.com

P. G. Kulkarni
e-mail: prakhyat2612@gmail.com

S. Y. Praneet
e-mail: praneetsy@gmail.com

R. B. Raghav
e-mail: rbongole@gmail.com

B. Das
e-mail: bhaskarjyoti01@gmail.com

R. P. Mahapatra et al. (eds.), *Proceedings of 6th International Conference on Recent Trends in Computing*, Lecture Notes in Networks and Systems 177,
https://doi.org/10.1007/978-981-33-4501-0_58

user has attributes linked to them that includes information like the opinions that the user holds and so on. The trouble however with online social networks is that they have no built-in authentication system to verify that someone is indeed who they say they are and no elaborate system to verify that their activities are not malicious. It becomes crucial to detect these non-typical affairs as precisely and as ahead of time as possible to counteract the impending disasters and attacks. In general, any novelty, noise, outlier, exception in a data set can be termed as anomaly that requires special attention.

Anomaly detection initially evolved as a data mining problem on structured data and researchers have created an array of techniques. However, when it comes to graph data, the data points cannot be treated as IID (independent and identically distributed) as the structural information to an extent influences the nodal attributes. This is the distinction that anomaly detection on social graph data has when compared to non-graph data. However, the existing anomaly detection research has mostly come up with methods using either the structural information or attribute information of such graph data. This work attempts to bridge the two ends by proposing an innovative way of combining both for the purpose of anomaly detection.

2 Related Work

Yu et al. [26] and Savage et al. [22] have provided an excellent overview of the field of anomaly detection for social graph data. On the basis of nature of the anomalies, they can be categorized into point and group anomalies. An anomaly is a point anomaly when an isolated data point is unusual and uncommon from the rest of the points. Group anomaly is observed when a subset of the sample data differs appreciably from the rest of the points. If the data points are from a static graph, it is called a static anomaly. On the contrary, if the anomaly analysis is carried out on a dynamic graph with time dimension involved, it is called dynamic anomaly. If the attributes (of nodes and edges) are used in the anomaly analysis, then it is called labelled anomaly and otherwise it is called unlabelled anomaly.

Anomaly detection as a field started as a branch of data mining. From data mining perspective, all three methodologies, i.e. supervised, unsupervised, and semi-supervised have been used for anomaly detection. Among supervised learning methods, Bayesian [1, 11], support vector machine [15–17, 19], and artificial neural networks [3, 18] have been used. However, the supervised approach suffers from the lack of labelled data. Unsupervised approach is used to address this issue and researchers have mostly relied on various clustering methods [4, 20, 21]. Researchers group anomaly detection techniques for graph data into behaviour-based technique and structure-based technique. The former deals with behaviour of the users or their posts. This includes the number of posts, the number of messages sent to friends, the content of these posts and messages, number of likes and shares, etc. Structure-based approach deals with the different characteristics of the graph structure and derived metrics from the graph. Behaviour-based approaches encompass an expansive array of features of social media users that are perplexing and misleading. The structure-based approaches, on the contrary, deal with well-defined metrics.

2.1 Behaviour Based

The prevalent approach here is to pick up features from structure and content of normal users and use those features in various supervised learning methods to predict anomaly [25, 28]. Silva and Willett [23] detect anomalous meeting in a social network by considering the probability of such events. Chu et al. [6] have detected spamming on twitter based on shared URL properties used in a supervised learning method. Viswanath et al. [24] used principle component analysis (PCA) to distinguish regular users from the anomalous ones. Two years of user behaviour were recorded from multiple social networking sites. PCA uses input features (temporal, spatio-temporal, and multiple features) to capture user behaviours in a system. PCA was concluded to be a superior technique than other methods because it was able to make use of complex behavioural data to catch the anomalous users. The space complexity is $O(m + n)$ where n is number of inputs and m is the number of users, hence making it scalable.

2.2 Structure Based

Social network data are considered as graphs which are massive and exhibit complex behaviour. Therefore, to analyse such complex data, it becomes easier to find distinct patterns and then learn anomalies which don't conform to these patterns. This approach [8] depends on structural properties of nodes and substructures like centrality, clustering coefficient, reciprocity etc. Kaur et al. [10] have examined the nature of networks by using power laws. The approach considers plotting curves for linear and power laws considering the PN centrality (centrality that takes into account negative ties of a node) and degree centrality measures. Chen et al. [5] used detection of communities in dynamic networks for anomaly detection. The proposed approach was used on evolutionary networks where anomalies are identified by overlapping communities. Since they were trying to find anomalies in communities, they covered all six possible types of anomalies that are possible in a community. They developed a parameter-free and scalable algorithm which was applied to Food Web and Enron Email data set. Lam and Yeung [12] used social network metrics as features to detect spams. Doostari et al. [7] used a novel hybrid approach that uses fuzzy logic and combines both parametric and non-parametric approaches to spot anomalies. They detect abnormal users in complete sub-graphs (cliques) of users in an online social network. Cliques are employed as the basic building blocks since it is laborious to evaluate the entire social network. Akolglu et al. [2] proposed Oddball algorithm that extracts egonet features and validates the patterns that pairs of such features follow in large social graphs. The methods described here are more inclined towards using graph-related properties commonly used in social network analysis (SNA). These properties relate to notifiable information about nodes and edges like their counts, various centrality-based measures, etc. Recently embedding techniques have been used to represent network structure information in a vector form that can be easily used in a machine learning algorithm. Hu et al. [9] have used network embedding coupled with dimensionality reduction for detection of anomalous nodes in a network. To spot structural anomalies, Yu et al. [27] have come up with a network embedding model.

3 Our Contributions

The cornerstone of our approach is to find anomalies using both the attributes and the structural information of the graph. Due to the lack of labelled graph data set for anomaly detection, we have used a synthetic attributed data set. The generated synthetic graph emulates some essential properties of real-world social graphs. Finally, the receiver operating characteristic curve has been utilized to compare the respective results.

3.1 Modified OddBall Algorithm

The OddBall algorithm [2] is an anomaly detection algorithm defined for a weighted network rather than an attributed network. The OddBall technique performs the anomaly detection task by defining a set of empirical rules describing relationships between pairs of ego network features. This algorithm has been fairly successful in detecting three prevalent types of anomaly in social graphs. These are patterns such as near-clique and stars, dominating link that indicates a stalker, extreme total edge weights for a given set of neighbours of an ego. These carefully chosen features include:

- degree of ego
- number of edges in egonet
- total weight of egonet
- principal eigenvalue of the weighted adjacency matrix of egonet.

The empirical rules are four power laws formed by pairs of the above features, i.e. egonet versus density, number of edges versus total weight, principal eigenvalue of the adjacency matrix versus total weight and rank in terms of edge weight versus weight of destination node of an edge. After plotting these power laws for a given data set, it was observed that anomalous nodes are away from normal nodes. The algorithm proposes an oddball score that are s are calculated with the following oddball formula as described in the paper.

$$\text{Outine}(i) = \frac{\max(y_i, Cx_i^{\Theta})}{\min(yi, Cx_i^{\Theta})} * \log(|y_i - Cx_i^{\Theta}| + 1) \tag{1}$$

where the y-value of a node i is y_i and x_i denote the x-value of node i for a particular feature pair $f(x, y)$. In this work, a method is proposed to convert an attributed network to a weighted network to be eventually used as an input to the OddBall algorithm. Formally, for each edge in the unweighted graph, we assign a weight to the edge by evaluating how similar the two adjacent nodes are. We evaluate similarity between any two adjacent nodes by utilizing various metrics to compare the two nodes' attribute vectors. We then compare which metric performs the best. The intuition behind this method is that, if two adjacent nodes are more similar to each other, they should have a larger weight between them than two nodes which are not as similar.

3.2 Similarity Measures

We have used the following metrics to assign weights between two adjacent nodes i and j. The distance metrics include Euclidean, Manhattan, Chebyshev, and Minkowski with ($p = 4$) and we then assign weights according to the following formula:

$$W_{ij} = \frac{1}{\text{distance}(a_i, a_j) + 1}$$

If the two attribute vectors of adjacent nodes are farther apart in distance, they should receive a lower weight between them. However, for Cosine similarity, the formula becomes:

$$W_{ij} = \text{cosine similarity}(a_i, a_j) + 1$$

This is because, if the two vectors have a higher cosine similarity value, then they are more similar and should have a higher weight assigned.

4 Data set

Due to the scarcity of social graph data sets with true labels for anomalous nodes, we have opted to synthetically insert anomalies in graphs. Largeron et al. [13] have provided a new generator for an attributed network that has communities following real world networks. This generator offers an extensive set of parameters such as number of vertices, maximum within and between community edges for each vertex,

Table 1 Data set

Data set	# of nodes	# of edges	# of attributes
Synthetic	10,000	17,848	2
Amazon product	1418	3695	21

minimum number of edges in the graph, number of communities, number of members in the generated community, etc. Two data sets have been employed for conducting this research as shown in Table 1:

- Using the software developed by Largeron et al. [13], synthetic graph with characteristics of real world social graphs
- Amazon (co-purchased network) data set [14] which is based on "customer who bought this also bought this" information. The tag information and attribute values were extracted for March 2012. A graph is constructed out of this data set. Every node in this extracted data set represents a product and an edge between any two nodes indicate frequent co-purchasing of these products.

4.1 Injecting Structural Anomalies

To introduce structural anomalies, we inject small cliques in the graph. This is considered anomalous in a social graph since it is peculiar that a small number of participants of a social graph are very closely tied to each other. We inject n small clique-structures, each containing m nodes using the following algorithm:

- Randomly pick m nodes and add an edge between each pair among them to form a clique and mark these m nodes as anomalous nodes.
- Repeat this process n times.

Hence, we obtain a graph with $m * n$ structurally anomalous nodes.

4.2 Injecting Attribute Anomalies

We subsequently inject attribute-based anomalies into the graph. We employ the following algorithm:

- Randomly pick $m * n$ nodes at random and mark them as anomalous nodes.
- For each picked node i:
  - Randomly pick k nodes from graph.
  - Among the k nodes picked, obtain node j which has the most divergence (considering the Euclidean distance of attribute vectors) from node i.
  - Duplicate the attribute vector of node j to the attribute vector of node i.

Table 2 Results

ROC-AUC scores		
Methodology	Amazon	Synthetic
Cosine similarity	0.7372	0.4735
Chebyshev distance	0.7399	0.5105
Minkowski distance	0.7401	0.5119
Euclidean distance	0.7405	0.5135
Manhattan distance	0.7443	0.5148

5 Experimental Results

Table 2 shows the methodology used, aspects of the network considered, and the AUC scores of the ROC curve for each method on Amazon and the synthetic data set. We observed that Manhattan distance metric is performing the best.

6 Conclusion

In this paper, we introduce a novel technique to convert graph structures with attributes into weighted matrices for the Oddball technique to effectively consider both nodal attributes and the structural aspects of the graph data. We also experiment with various similarity measures that can be used on graph data to convert it into weighted matrices and found that simple Manhattan similarity works best. The logical next step is to evaluate this approach on a much larger real world data set with millions of nodes and compare the effectiveness of this approach with a network embedding technique that works with deep neighbourhood information.

References

1. Abraham B, Box GE (1979) Bayesian analysis of some outlier problems in time series. Biometrika 66(2):229–236
2. Akoglu L, McGlohon M, Faloutsos C (2010) Oddball: spotting anomalies in weighted graphs. In: Pacific-Asia conference on knowledge discovery and data mining. Springer, Berlin, pp 410–421
3. Augusteijn M, Folkert B (2002) Neural network classification and novelty detection. Int J Remote Sens 23(14):2891–2902
4. Beutel A, Xu W, Guruswami V, Palow C, Faloutsos C (2013) Copycatch: stopping group attacks by spotting lockstep behavior in social networks. In: Proceedings of the 22nd international conference on world wide web, pp 119–130
5. Chen Z, Hendrix W, Samatova NF (2012) Community-based anomaly detection in evolutionary networks. J Intell Inf Syst 39(1):59–85
6. Chu Z, Widjaja I, Wang H (2012) Detecting social spam campaigns on twitter. In: International conference on applied cryptography and network security. Springer, Berlin, pp 455–472

7. Doostari MA, Zeinali R, Lashkari H, Ajamzamani M (2013) Fuzzy node-fuzzy graph
8. Hassanzadeh R, Nayak R, Stebila D (2012) Analyzing the effectiveness of graph metrics for anomaly detection in online social networks. In: International conference on web information systems engineering. Springer, Berlin, pp 624–630
9. Hu R, Aggarwal CC, Ma S, Huai J (2016) An embedding approach to anomaly detection. In: 2016 IEEE 32nd international conference on data engineering (ICDE). IEEE, pp 385–396
10. Kaur R, Kaur M, Singh S (2016) A novel graph centrality based approach to analyze anomalous nodes with negative behavior. Procedia Comput Sci 78:556–562
11. Kruegel C, Mutz D, Robertson W, Valeur F (2003) Bayesian event classification for intrusion detection. In: 19th annual computer security applications conference, 2003. Proceedings. IEEE, pp 14–23
12. Lam HY, Yeung DY (2007) A learning approach to spam detection based on social networks. Ph.D. thesis, Hong Kong University of Science and Technology
13. Largeron C, Mougel PN, Rabbany R, Zaïane OR (2015) Generating attributed networks with communities. PLoS ONE 10(4):e0122777
14. Leskovec J, Adamic LA, Huberman BA (2007) The dynamics of viral marketing. ACM Trans Web (TWEB) 1(1):5-es
15. Li KL, Huang HK, Tian SF, Xu W (2003) Improving one-class svm for anomaly detection. In: Proceedings of the 2003 international conference on machine learning and cybernetics (IEEE Cat. No. 03EX693), vol 5. IEEE, pp 3077–3081
16. Ma J, Perkins S (2003) Online novelty detection on temporal sequences. In: Proceedings of the ninth ACM SIGKDD international conference on knowledge discovery and data mining, pp 613–618
17. Ma J, Perkins S (2003) Time-series novelty detection using one-class support vector machines. In: Proceedings of the international joint conference on neural networks, vol 3. IEEE, pp 1741–1745
18. Moradi M, Zulkernine M (2004) A neural network based system for intrusion detection and classification of attacks. In: Proceedings of the IEEE international conference on advances in intelligent systems-theory and applications, pp 15–18
19. Perdisci R, Ariu D, Fogla P, Giacinto G, Lee W (2009) Mcpad: a multiple classifier system for accurate payload-based anomaly detection. Comput Netw 53(6):864–881
20. Pires A, Santos-Pereira C (2005) Using clustering and robust estimators to detect outliers in multivariate data
21. Portnoy L, Eskin E, Stolfo S (2001) Intrusion detection with unlabeled data using clustering. In: Proceedings of ACM CSS workshop on data mining applied to security (DMSA-2001)
22. Savage D, Zhang X, Yu X, Chou P, Wang Q (2014) Anomaly detection in online social networks. Soc Netw 39:62–70
23. Silva J, Willett R (2008) Detection of anomalous meetings in a social network. In: 2008 42nd annual conference on information sciences and systems. IEEE, pp 636–641
24. Viswanath B, Bashir MA, Crovella M, Guha S, Gummadi KP, Krishnamurthy B, Mislove A (2014) Towards detecting anomalous user behavior in online social networks. In: 23rd {USENIX} security symposium ({USENIX} security 14), pp 223–238
25. Xiao C, Freeman DM, Hwa T (2015) Detecting clusters of fake accounts in online social networks. In: Proceedings of the 8th ACM workshop on artificial intelligence and security, pp 91–101
26. Yu R, Qiu H, Wen Z, Lin C, Liu Y (2016) A survey on social media anomaly detection. ACM SIGKDD Explor Newsl 18(1):1–14
27. Yu W, Cheng W, Aggarwal CC, Zhang K, Chen H, Wang W (2018) Netwalk: a flexible deep embedding approach for anomaly detection in dynamic networks. In: Proceedings of the 24th ACM SIGKDD international conference on knowledge discovery & data mining, pp 2672–2681
28. Zheng X, Zeng Z, Chen Z, Yu Y, Rong C (2015) Detecting spammers on social networks. Neurocomputing 159:27–34

Secure Routing-Based AODV to Prevent Network from Black Hole Attack in MANET

Anant Ram, Jagrati Kulshrestha, and Vishesh Gupta

Abstract Mobile ad hoc networks (MANETs) are the collection of mobile nodes that transmit data through intermediate nodes. MANET does not depend upon any existing infrastructure. Securing MANET is the toughest task because all the nodes are movable in nature. In MANET, every node can perform as a relay and transmit data through intermediate node to the destination node. Just because of its dynamic nature, multiple attacks occur in this network. The most occurring attack in the MANET is black hole attack. In this, the attacker is a malicious node that works as a destructive node. During the transmission of the data packet from the source to the destination, the attacker node becomes the part of route and it does not forward further. To overcome this attack, we proposed an approach that uses the dynamic threshold to detect the attacker nodes and avoid the entire path having the attacker nodes. Identification of the attacker node is done during the path formation. We performed the evaluation of the proposed approach and compared with the performance of existing approach. Outcome shows that the proposed methodology is more effective than existing method in context to throughput and packet delivery ratio.

Keywords Mobile ad hoc network (MANET) · Black hole attack · Ad hoc on-demand distance vector (AODV) · Dynamic threshold

A. Ram · J. Kulshrestha (✉) · V. Gupta
GLA University, Mathura, India
e-mail: jagrati.kulshrestha@gla.ac.in

A. Ram
e-mail: anant.ram@gla.ac.in

V. Gupta
e-mail: vishesh.gupta_mtcs17@gla.ac.in

R. P. Mahapatra et al. (eds.), *Proceedings of 6th International Conference on Recent Trends in Computing*, Lecture Notes in Networks and Systems 177,
https://doi.org/10.1007/978-981-33-4501-0_59

1 Introduction

The ad hoc networks are characterized as a infrastructure based and infrastructure less. MANET is the self-organizing, self-configuring and multi-hope network. It does not require any preexisting infrastructure. MANET plays significant role in various types of application like disaster relief, military operation and vehicle to vehicle communications [1]. As MANET is useful in such applications, so it also requires very secure communication to protect data from data breach. Just because of its dynamic nature, it can have contact with malicious node. There are multiple types of attacks that occur in MANET. One of the well-known attacks that occur in MANET is the black hole attack. It sends spoofed destination sequence and small number of hops to the source that indicate the better route to the destination [2]. This happens at the time of route formation. Once the route is established, it drops all the packets that come to it, without forwarding these packets further towards to the destination node [2]. In this work, we come up with a new dynamic threshold-based approach to overcome our network from malicious node. In MANET, all the nodes are moving in nature. Therefore, static threshold-based methods are not able to reveal the existence of the malicious nodes that contribute in the path formation.

In the proposed approach, when the source node receives the path information in the of route reply (RREP) messages. This message may be sent by the intermediate node or the destination. After that source node calculates the appropriate threshold based on the RREP sequence number (S. No.). If any RREP S. No. is larger than or equal to the calculated threshold, then it considers that the sequence number is coming from malicious node. Therefore, the proposed approach discards that route to deliver the data packets and consider another path that is free from malicious nodes.

The rest of the paper is structured as follows: in Sect. 2, we discussed related work and working of AODV is discussed in Sect. 3. The brief of the proposed work presented in Sect. 4. Section 5 discusses how the dynamic threshold is calculated. Section 6 presents the simulation result and finally we conclude the paper in Sect. 7.

2 Related Work

Numerous techniques have been proposed to find out and inhibit the network from black hole attack by many researchers. In this section, we gave a concise discussion of few approaches that are related to the proposed work.

Lu et al. [3] came up with the solution that is the secure AODV. This proposal is designed to enucleate any attacker node in the network. SAODV scheme in MANETs is designed to endure black hole attacks. Deswal and Singh [4] developed a renovate version of the secure AODV that involve additional password security for every node and every routing table. They are up to date and depend upon timeliness. Ramaswamy et al. [5] came up with a solution for recognizing various malicious nodes. They have shown a "data routing information (DRI)" table by altering AODV protocol. This approach includes reliable nodes to pass the data. Deng et al. [6] developed a method that asks each hope to give information about next hop along with RREPs once a path towards destination has been identified. It can only identify one black hole. It is not able to detect multiple malicious nodes. Raj and Swadas [7] alter the functionality of the AODV and involve a method that checks the sequence number of the incoming RREP. Source node compares the sequence number with the calculated threshold value, if the calculated value is larger then that of path is ignored and deleted. Kumar and Kumar [8] introduced a method that work on arriving RREP to discriminate a malicious node attack. This table is preserved by source node and it saves DSN and ID of replying node. Gambhir and Sharma [9] came up with a solution that uses a prime product method for the deletion of malicious node in MANETs. A prime number is given to every hop. Prime number is unique to every node. When the source node gets RREP, the prime product number is checked by the source node to establish a path from source node to the destination node. Al-Shurman et al. [10] introduced two methods, i.e. in first method, the data is sent through shared hop and in second method the S. No. of last sent and received data is used to identify malicious node. Dokurer et al. [11] came up with an approach to inhibit malicious node in MANET. In this method, the first RREP is ignored. Author assumed that first route reply is coming from attacker side. Sergio et al. [12] developed a method that provide secure route to prevent malicious attack. In this, authors modify the AODV to prevent black hole attack based on a static threshold method. The limitation of this paper is that, due to dynamic nature of nodes, this method gives more false positive. Zhang et al. [13] came up with a solution that use watchdog and path rates to discriminate black hole attack. This watchdog method detects those malicious nodes which are intentionally dropping the packets. Alem and Xuan [14] gave an authentication method for recognizing attacker nodes. An authentication method is built on hash function, MAC and PRF, which is applied for identifying the RREPs. Kshirsagar and Patil [15] came up with a method that depend upon intrusion detection to prevent network from multiple black hole attacks. Scarcity of routing packets cause minimum network overhead and also do a reliable communication.

3 AODV and Black Hole Attack

AODV is most wildly used protocol in mobile ad hoc network. It selects a path between source and destination only when it is needed that is why it is called reactive routing protocol [1]. This protocol is based on sequence number to validate whether the route is updated or not. According to the AODV, larger the sequence number fresher will be the route [7]. This protocol has various advantage as compare to other protocol because it is dynamic in nature and it provides loop-free communication [1, 7].

AODV protocol uses three types of messages, "route request (RREQ), route reply (RREP), route error (RERR)" [1]. When the source node wishes to establish a communication to the destination node, first it checks if there is any other

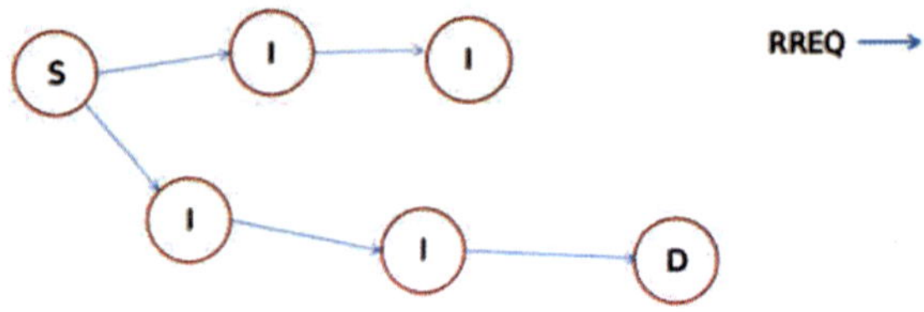

Fig. 1 AODV route discovery process

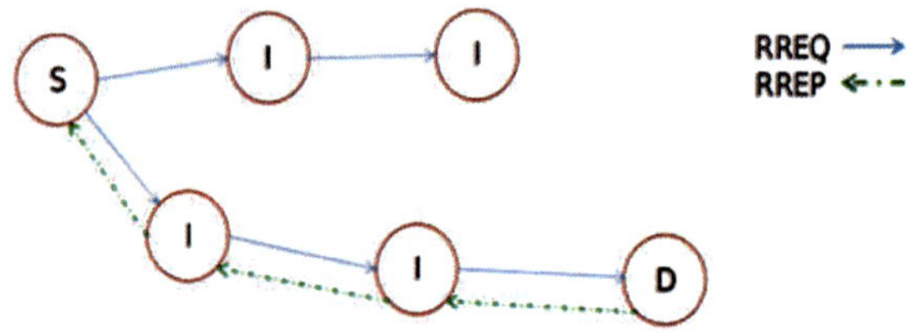

Fig. 2 AODV route reply process

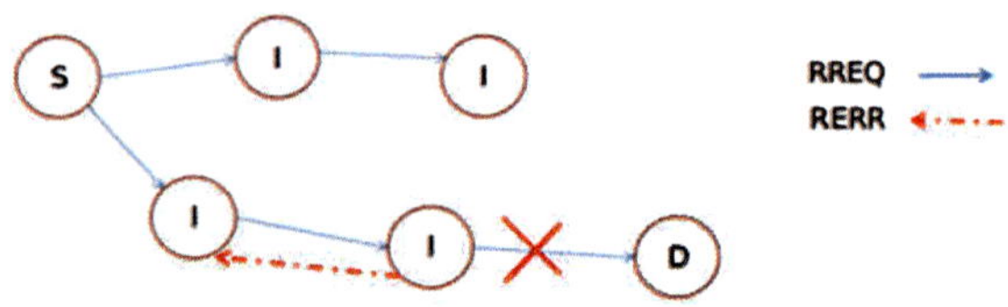

Fig. 3 AODV route error process

path to the destination, if not then it broadcasts a route request (RREQ) message towards the destination. If the intermediate node does not have a path towards the destination, it rebroadcast the RREQ further towards the destination. Figure 1 shows the route-finding process. When the RREQ reached to the destination node, it passes route reply (RREP) back to the source node along the reverse path. Figure 2 shows route reply (RREP) process [7].

In route error (RERR) phase, if there is any link break identified by the node, then it will send a route error message [7], shown in Fig. 3.

Black hole attack is a serious attack in MANET. Black hole attack alters the amount of data traffic during the communication. There is an attacker node that imbibes all the data packets that comes to that node [8]. First, the route request (RREQ) is broadcasted in the network then the source node waits for the RREPs. Attacker node receives a route request than it sends a fake route reply with minimum hop count and large sequence number [2].

Malicious node sends smallest hop count and significantly greater sequence number just to show that the path it has is small and fresh. After receiving the path information the from malicious node, it incorrectly assumes it to be the smallest and updated route to the destination, so it sends data packet through that path. Once data packet reaches to the attacker node it does not pass data packet further and drops all the packets [9].

In Fig. 4, S represents the source node that broadcast a route request (RREQ) packet and after that wait for the route reply (RREP) packet. The destination node

Fig. 4 Black hole attack

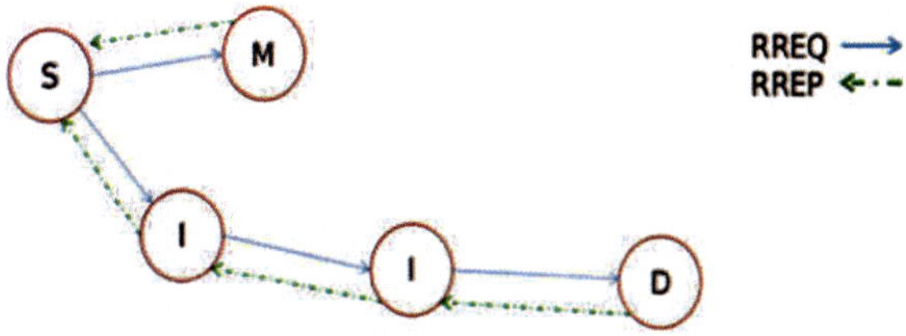

and the malicious node send the RREP packet. In this case, the source node gets two route reply from two different nodes. It becomes difficult for the source node to identify, which is the fake route reply and which is legitimate route reply. So it selects the route reply with higher sequence number (S. No.) and smallest hop count. Since, it assumes that higher sequence number and small hop count are the sign of fresh path [2]. Therefore, it sends the data packet to the malicious node.

4 Proposed Method

In this work, threshold-based method is proposed to prevent the network from black hole attack. The malicious node sends a large sequence number (S. No.) and tiny number of hops in route reply message. To protect the network from this type of attack, the proposed dynamic threshold-based method performs well.

In this method, we check the S. No. in route reply that is greater than or equal to our threshold value, if it is, then it is identified as a malicious node. So the data packets do not send through that route and that path is discarded from the network. One of the difficult issues in this is to calculate the appropriate threshold.

In Fig. 4, when the source node S receives two route reply packets, i.e. one from legitimate node and other from malicious node. The source node identifies the authenticity of route reply by checking the sequence number with the calculated threshold. If it is larger than or equal to the threshold, then it is discarded else it is considered as a normal node so the data packet is sent through that path. Figure 5 shows the black hole node detection process.

In this flowchart, RSN is defined as received sequence number and TH is threshold.

The algorithm for prevention of black hole attack is as follows:

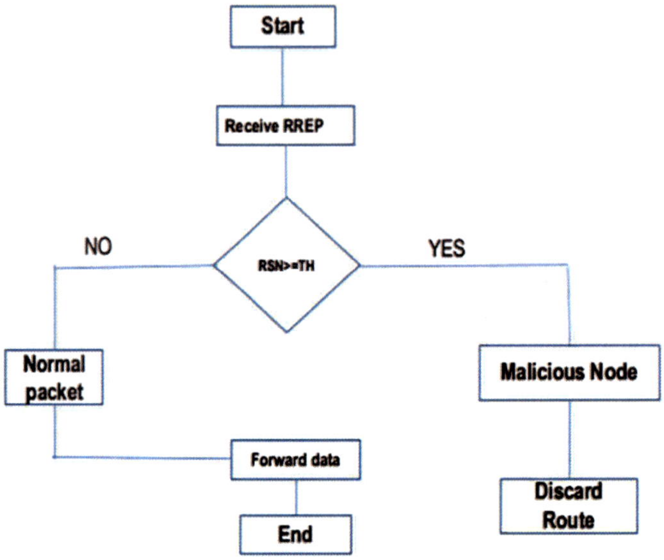

Fig. 5 Black hole node detection process

1. Source node (Sou_SNo) circulate RREQ to all nodes
2. Intermediate node gets RREQ and forwards it until it reaches the Destination node (Des_SNo)
3. Destination node gets RREQ from Source Node or Intermediate node
4. Destination Node gets Sequence Number(S.N) From RREQ and verify S.N in its table

   If S.No. Of RREQ is greater than S.No. Of its table **then**
   Des_SNo. take S.No. Of RREQ and increment by 1
   Else
   Des_No. take S.No. Of its table and increment by 1 and
   Des_No. sends RREP back to Sou_SNo.
   End if
5. **If** Sou_No. receives RREP **then**
   Sou_SNo. checks the RREP
   If D_SNo. is greater than or equal to TH then
   Discard the message
   Else
   Path is established
6. **End if**

5 Threshold Calculation

To get rid of above issue, we developed a new approach in MANET that prevent network from the black hole attack. In this, we designed a new approach that comprises modification of the standard AODV protocol. That is known as secure routing-based AODV to prevent network from black hole attack in MANET. In this approach, source node calculates the dynamic threshold based on the number of route reply.

$$M = \sum_{i=1}^{N} \frac{DSN_i}{N} \tag{1}$$

where DSN_i is the destination sequence number for different paths, N, received by the source. M is the average of the received destination sequence numbers by the source from different paths to the destination.

In Eq. (1), first, we check the how many sequence numbers are coming. Based on this, we calculate the average of all the sequence number.

$$\delta = \sum_{i=1}^{N} \frac{(M - \mathrm{DSN}_i)}{N} \tag{2}$$

In Eq. (2), we calculate the δ as the average of the difference of M and every destination sequence number.

$$\mathrm{TH} = M + \delta \tag{3}$$

This is the final threshold. If the received sequence number is greater than or equal to this threshold value, then it is assumed that this sequence number is coming from malicious node, so that route is discarded. We deem that there is at least one black hole node in network and it always responds to all received route request messages by sending fake RREPs with significantly large destination sequence number. Black hole node sends very large sequence number in order to show itself having the fresh path. So, in this, we reckon that the sequence number in fake reply is ten times larger than the original sequence number.

6 Performance Evaluations

In this section, we evaluated the proposed approach regarding black hole prevention and then compare it with the existing method. For this simulation, we used network simulator (NS 2.35). The simulation parameters are mentioned in Table 1.

Table 1 Simulation parameters

Parameter	Value
Simulation time	200 s
Number of nodes	10, 20, 30, 40, 50
Network area	1000 × 1000 m
Network mobility	Random way point
Traffic type	CBR

6.1 Simulation Model

In our simulation, we assume that when malicious node receives RREQ packet from source node, it always sends spoofed destination sequence number (DSN) and tiny number of hops to validate the freshness of the path to the destination.

6.2 Performance Measurement

We measure the performance of our method by calculating packet delivery ratio (PDR) and throughput and compare with existing method.

6.2.1 Packet Delivery Ratio

Packet delivery ratio (PDR) can be figured out as how many packets are sent by the source node and how many of them are reached to the destination.

$$\text{PDR} = \frac{N_{\text{recv}}}{N_{\text{send}}} * 100 \tag{4}$$

6.2.2 Throughput

The throughput is calculated by using this formula.

$$\text{Throughput} = \frac{\text{Pktsize} * 8 * N_{\text{recv}}}{T} \tag{5}$$

In the above equation, the packet size is the size of data packet. T is the time interval from the time when first route reply is received by any node to the end of the simulation.

6.3 Simulation Result

For packet delivery ratio (Fig. 6), we calculate the PDR on different number of nodes and check with existing method that is secure AODV. In this Fig. 6, we can clearly see that our proposed method performs better with respect to the AODV.

In Fig. 7, we compared the throughput with existing secure AODV method. Our proposed method performs better in terms of throughput.

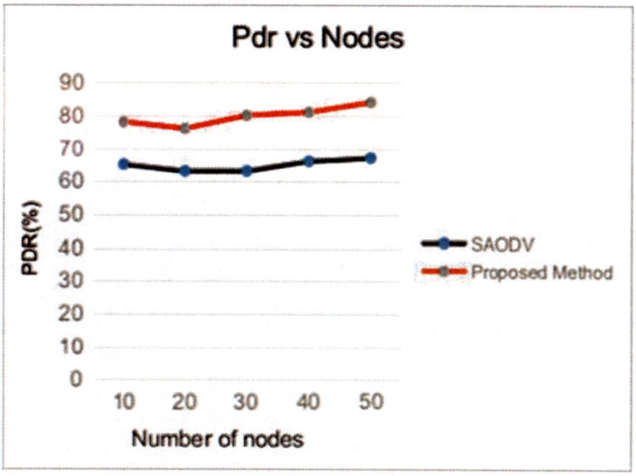

Fig. 6 Packet delivery ratio versus number of nodes

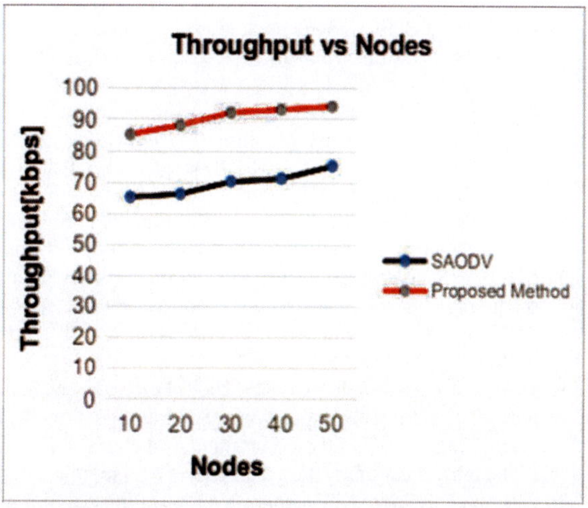

Fig. 7 Throughput versus number of nodes

7 Conclusion

In MANET, base station is not available. Black hole attack is one of the well-liked attacks in MANET. A black hole node is an attacker node that works as a destructive node. When the data packet reaches to that node, then it drops all the packets. To overcome this attack, we proposed a new dynamic threshold method and examine the performance of our technique with existing approach. Our simulation outcome shows that our proposed technique is more effective than existing method in context of throughput and packet delivery rate.

References

1. Tan S, Kim K (2013) Secure route discovery for preventing black hole attacks on AODV-based MANETs. In: 2013 IEEE 10th international conference on high performance computing and communications, 2013 IEEE international conference on embedded and ubiquitous computing. IEEE
2. Noguchi T, Yamamoto T (2017) Black hole attack prevention method using dynamic threshold in mobile ad hoc networks. In: 2017 federated conference on computer science and information systems (FedCSIS). IEEE, pp 797–802
3. Lu S, Li L, Lam K-Y, Jia L (2009) SAODV: a MANET routing protocol that can withstand blackhole attack. In: Proceedings of international conference on computational intelligence and security (CIS '09), Dec 11–14, Beijing, China, pp 421–425
4. Deswal S, Singh S (2010) Implementation of routing security aspects in AODV. Int J Comput Theor Eng 2(1)
5. Ramaswamy S, Fu H, Sreekantaradhya M, Dixon J, Nygard K (2003) Prevention of cooperative black hole attack in wireless ad hoc networks. In: 2003 international conference on wireless networks (ICWN 03), Las Vegas, Nevada, USA
6. Dong H, Li W, Agrawal DP (2002) Routing security in wireless ad hoc networks. IEEE Commun Mag 40(10):70–75. https://doi.org/10.1109/MCOM.2002.1039859
7. Raj PN, Swadas PB (2009) PRAODV: a dynamic learning system against blackhole attack in AODV based MANET. IJCSI Int J Comput Sci 2
8. Kumar V, Kumar R (2015) An adaptive approach for detection of blackhole attack in mobile ad hoc network. Proc Comput Sci 48:472–479
9. Gambhir S, Sharma S (2013) PPN: prime product number based malicious node detection scheme for MANETs. In: 3rd international advance computing conference (IACC), IEEE
10. Al-Shurman M, Yoo S-M, Park S (2004) Black hole attack in mobile ad hoc networks. In: ACMSE04, 23 Apr 2004, p 9697
11. Dokurer S, Erten YM, Acar CE (2007) Performance analysis of ad-hoc networks under black hole attacks. In: IEEE proceedings Southeast conference, 22–25 Mar 2007, pp 148–153
12. Marti S et al (2000) Mitigating routing misbehavior in mobile ad hoc networks. In: Proceedings of the 6th annual international conference on mobile computing and networking. ACM
13. Zhang XY, Sekiya Y, Wakahara Y (2009) Proposal of a method to detect black hole attack in MANET. In: International symposium on autonomous decentralized systems, 2009. ISADS '09, 23–25 Mar 2009, pp 1–6
14. Alem YF, Xuan ZC (2010) Preventing black hole attack in mobile ad-hoc networks using anomaly detection. In: Future computer and communication (ICFCC), 2010
15. Kshirsagar D, Patil A (2013) Blackhole attack detection and prevention by real time monitoring. In: 2013 fourth international conference on computing, communications and networking technologies (ICCCNT). IEEE

Stability Analysis of Fractional-Order Plant with Delay

Nitisha Shrivastava and Pragya Varshney

Abstract The field of fractional calculus has changed the way we model and control real-time phenomena. The fractional calculus-based models are more accurate and closer to reality. The purpose of this paper is mainly to investigate the performance of fractional-order systems with significant delay, in terms of stability. It is well known that introduction of delay in systems results in phase lag. This may lead to instability of the system. In this paper, a fractional-order plant model with delay has been chosen for analysis. The stability of the plant is verified by obtaining Nyquist plot and step response. The simulations have been performed in MATLAB and the results are compared with ideal responses. The effect of increase in delay on the performance of the fractional-order plant has also been studied.

Keywords Fractional-order systems · Time delay · Stability · Steady state error

1 Introduction

Fractional calculus-based mathematical models of the dynamical systems are more real than integer calculus-based models. Application of fractional calculus in science and engineering resulted in the accurate modeling, designing, and control of the physical systems. As fractional-order models are irrational in nature, their realization and implementation are not easy. Hence, proper integer order approximation of fractional-order systems has been obtained using the various approximation techniques [1–7]. Stability is an important factor in control system [8, 9]. The stability of fractional-order systems (FOS) can be carried out by studying the solutions of the differential equations that characterize them. An alternative way is to analyze the transfer function of the system. Since time delay in a fractional-order systems results

N. Shrivastava (✉)
Department of EEE, Inderprastha Engineering College, Delhi-NCR, Ghaziabad 201010, India
e-mail: nitishashrivastav@gmail.com

P. Varshney
Division of ICE, NSUT (Formerly NSIT), New Delhi 110078, India
e-mail: pragya.varshney1@gmail.com

R. P. Mahapatra et al. (eds.), *Proceedings of 6th International Conference on Recent Trends in Computing*, Lecture Notes in Networks and Systems 177,
https://doi.org/10.1007/978-981-33-4501-0_60

in phase lag, such systems tend to become unstable. In this paper, we have performed analysis on a fractional-order plant model with delay. Nyquist stability criterion is used to test the stability, and step responses are also obtained and compared with ideal responses. The procedure adopted in this paper is: an integer approximation of the fractional plant model is obtained using five different approximation techniques and the delay term is substituted by Pade approximation. Since the resulting model is of higher order, the large-scale system is reduced to lower order. The stability is tested by obtaining Nyquist plot and step response. Analysis is performed in MATLAB and simulations and discussions are presented.

2 Approximation Technique

In this paper, a fractional-order plant model with delay has been analyzed. For obtaining integer order approximation of this plant model, five approximation methods have been used. They are Oustaloup method [2], Carlson method [3], Matsuda method [4], Charef method [5], and Modified Oustaloup method [6].

Since these are well-known approaches available in literature, only the parameters chosen for the different methods have been given in this paper as follows:

1. Oustaloup method: The frequency band chosen and the order of the approximation are [0.01, 100] rad/s and 5.
2. Carlson method: Second iterate approximation is obtained using Newton method.
3. Matsuda method: The frequency band chosen is [0.01, 100] rad/s and the number of frequencies within the band is taken as 9.
4. Charef method: The frequency band chosen is [0.01, 100] rad/s and maximum permissible error is 2 dB.
5. Modified Oustaloup method: The frequency band chosen and the order of the approximation are [0.01, 100] rad/s and 7.

Since the approximation techniques usually yield integer models of high orders, these are further reduced using Balanced-truncation method of order reduction. This method is based on eliminating the lower energy states of the higher order model [10].

The delay component is approximated to a 3rd order model using Pade approximation.

3 Illustrative Example

A fractional-order transfer function (FOTF) with input–output delay is considered

$$G_{\text{frac}}(s) = \frac{6.902}{2.686s^{0.73} + 1} e^{-0.5s} \quad (1)$$

Using Oustaloup method, the integer order approximation is

$$G_{\text{frac\_ousta}}(s) = \frac{\left(-6.902s^{8} - 473.9s^{7} + 5590s^{6} + 3.173e4s^{5} - 9.546e5s^{4} + 4.085e6s^{3} + 1.409e7s^{2} + 4.567e6s + 1.911e5\right)}{\left(78.47s^{8} + 3847s^{7} + 7.33e4s^{6} + 7.284e5s^{5} + 3.781e6s^{4} + 8.388e6s^{3} + 5.463e6s^{2} + 9.151e5s + 3.027e4\right)} \quad (2)$$

Since the order of Eq. (2) is 8, it is further reduced to 6th order model Eq. (3).

$$G_{\text{frac\_ousta\_red}}(s) = \frac{\left(-0.08796s^{6} - 4.871s^{5} + 139.8s^{4} - 1434s^{3} + 5145s^{2} + 6078s + 583.8\right)}{\left(s^{6} + 35.7s^{5} + 432.3s^{4} + 2753s^{3} + 4441s^{2} + 1496s + 93.28\right)} \quad (3)$$

The Nyquist plot is obtained for the reduced approximated integer order (AIO) transfer function of the FOS in (1) as shown Fig. 1a. The system is stable for a time delay of 0.5 s as can be seen from the Nyquist plot. Figure 1b shows the step response. The steady-state error and the time-domain specifications are given in Table 1.

Carlson method of approximation is based on Newton iterative technique; therefore, the frequency band for analysis of the system cannot be chosen initially. Since for the other methods, the frequency band chosen is [0.01, 100] rad/s, and we have obtained the Carlson-based approximated transfer functions whose characteristics fit the ideal characteristics in the frequency band [0.01, 100] rad/s. Carlson-based second iterate approximations are considered here since its frequency response curve is close to the ideal characteristics in the frequency band [0.01, 100] rad/s. Also in this method, the numerator of the fractional term must always be 1. Therefore, in order to obtain approximation of $s^{0.73}$, it is written as $s^{0.73} = \frac{s^{0.5} * s^{0.25}}{s^{0.02}}$. The integer approximations of fractional terms $s^{0.5}$, $s^{0.25}$ and $s^{0.02}$ obtained have orders 6, 4, and 52, respectively. Combining all the three will result in very high-order transfer function. So approximation model of $s^{0.02}$ was reduced to 4th-order model using Balanced-truncation method. The delay term is substituted by a third-order Pade approximation. Now, the resultant model which comes out to be is of $O(17)$ which is further reduced to an approximated model of $O(5)$. The higher- and lower-order integer models are Eqs. (4, 5):

$$G_{\text{frac\_carl}}(s) = \frac{\left(\begin{array}{l} -5454s^{17} - 2.604e5s^{16} - 1.943e6s^{15} + 3.197e7s^{14} + 9.976e6s^{13} - 3.915e9s^{12} - 5.438e9s^{11} + 2.051e11s^{10} + 1.339e12s^{9} \\ +3.947e12s^{8} + 6.782e12s^{7} + 7.357e12s^{6} + 5.151e12s^{5} + 2.292e12s^{4} + 6.159e11s^{3} + 9.177e10s^{2} + 6.86e9s + 1.994e8 \end{array}\right)}{\left(\begin{array}{l} 4.974e4s^{17} + 3.036e6s^{16} + 8.322e7s^{15} + 1.345e9s^{14} + 1.409e10s^{13} + 9.914e10s^{12} + 4.737e11s^{11} + 1.543e12s^{10} + 3.456e12s^{9} \\ +5.379e12s^{8} + 5.848e12s^{7} + 4.416e12s^{6} + 2.275e12s^{5} + 7.717e11s^{4} + 1.627e11s^{3} + 1.964e10s^{2} + 1.244e9s + 3.225e7 \end{array}\right)} \quad (4)$$

$$G_{\text{frac\_carl\_red}}(s) = \frac{\left(-0.1096s^{5} - 1.42s^{4} + 68.53s^{3} - 820.4s^{2} + 3454s + 1560\right)}{\left(s^{5} + 25.87s^{4} + 263s^{3} + 1540s^{2} + 1669s + 252.6\right)} \quad (5)$$

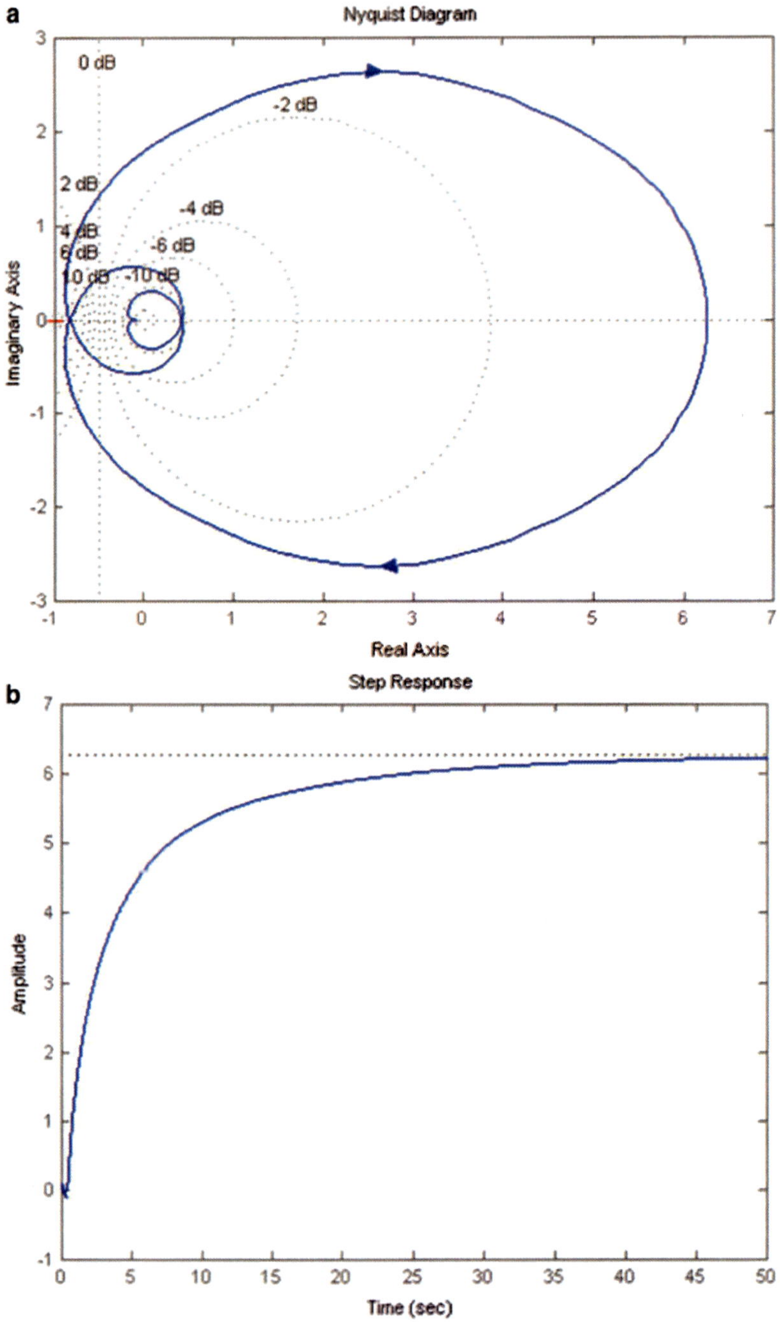

Fig. 1 **a** Nyquist plot of Oustaloup-based FOS with delay. **b** Step response of Oustaloup-based FOS with delay

Table 1 Time-domain specifications of fractional-order plant using approximation methods

Method	Rise time (s)	Settling time (s)	Peak time (s)	Peak	Steady state error
Oustaloup mthod	13.4792	33.4380	58.0016	6.2411	5.2252
Carlson method	**10.2030**	**19.7356**	**35.7706**	**6.1690**	**5.1588**
Matsuda method	20.3037	61.6528	127.5516	6.7477	5.7372
Charef method	18.8181	53.9036	92.7382	6.6920	5.6735
Modified Oustaloup method	19.3474	57.8528	120.9210	6.7457	5.7356

The Nyquist plot is obtained for the reduced approximated integer order transfer function as shown Fig. 2a. It is seen that the system is stable for a time delay of 0.5 s. Figure 2b shows the step response. The steady-state error and the time-domain specifications are listed in Table 1.

In Matsuda method of approximation, from the chosen frequency band, an odd number of frequencies within the band is selected. For our analysis, we have taken the following nine logarithmically spaced frequencies: [0.01, 0.0316, 0.1, 0.3162, 1, 3.1623, 10, 31.6228, 100]. The delay term is substituted by a third-order Pade approximation. Using this method, the integer-order model of the fractional-order system in (1) is of order 7 (Eq. 6), which is further reduced to 6th order model using Balanced-truncation method (Eq. 7).

$$G_{\text{frac\_mat}}(s) = \frac{\left(-6.902s^7 - 2563s^6 + 3.621e4s^5 - 4736s^4 - 3.545e6s^3 + 2.188e7s^2 1.844e7s + 8.343e5\right)}{\left(339.3s^7 + 1.61e4s^6 + 2.871e5s^5 + 2.593e6s^4 + 1.127e7s^3 + 1.51e7s^2 3.752e6s + 1.235e5\right)} \tag{6}$$

$$G_{\text{frac\_mat\_red}}(s) = \frac{\left(-0.02034s^6 - 7.353s^5 + 177.4s^4 - 1721s^3 + 6120s^2 + 5592s + 253.3\right)}{\left(s^6 + 37.75s^5 + 482.2s^4 + 2983s^3 + 4489s^2 + 1139s + 37.48\right)} \tag{7}$$

The Nyquist response and step response of the Matsuda method-based FO plant model are given in Fig. 3a, b, respectively. Again the plots reveal that the system is stable for a time delay of 0.5 s. Table 1 gives the steady-state error and the time domain specifications.

Using Charef method of approximation, the AIO model of the FOS(1) is obtained and it has an $O(13)$ (Eq. 8). Therefore, it is further reduced to 6th-order model (Eq. 9).

$$G_{\text{frac\_char}}(s) = \frac{\begin{pmatrix} -6.902s^{13} - 1.389e6s^{12} - 2.461e10s^{11} - 4.131e13s^{10} - 5.883e15s^9 + 4.588e16s^8 + 8.431e17s^7 - 1.562e19s^6 + \\ 6.524e19s^5 + 1.55e20s^4 + 2.38e19s^3 + 3.421e17s^2 + 4.699e14s + 5.69e10 \end{pmatrix}}{\begin{pmatrix} 2.337e4s^{13} + 8.551e8s^{12} + 2.77e12s^{11} + 9.229e14s^{10} + 4.759e16s^9 + 9.302e17s^8 + 9.33e18s^7 + 4.776e19s^6 + \\ 9.628e19s^5 + 4.391e19s^4 + 4.001e18s^3 + 5.097e16s^2 + 6.844e13s + 8.251e9 \end{pmatrix}} \tag{8}$$

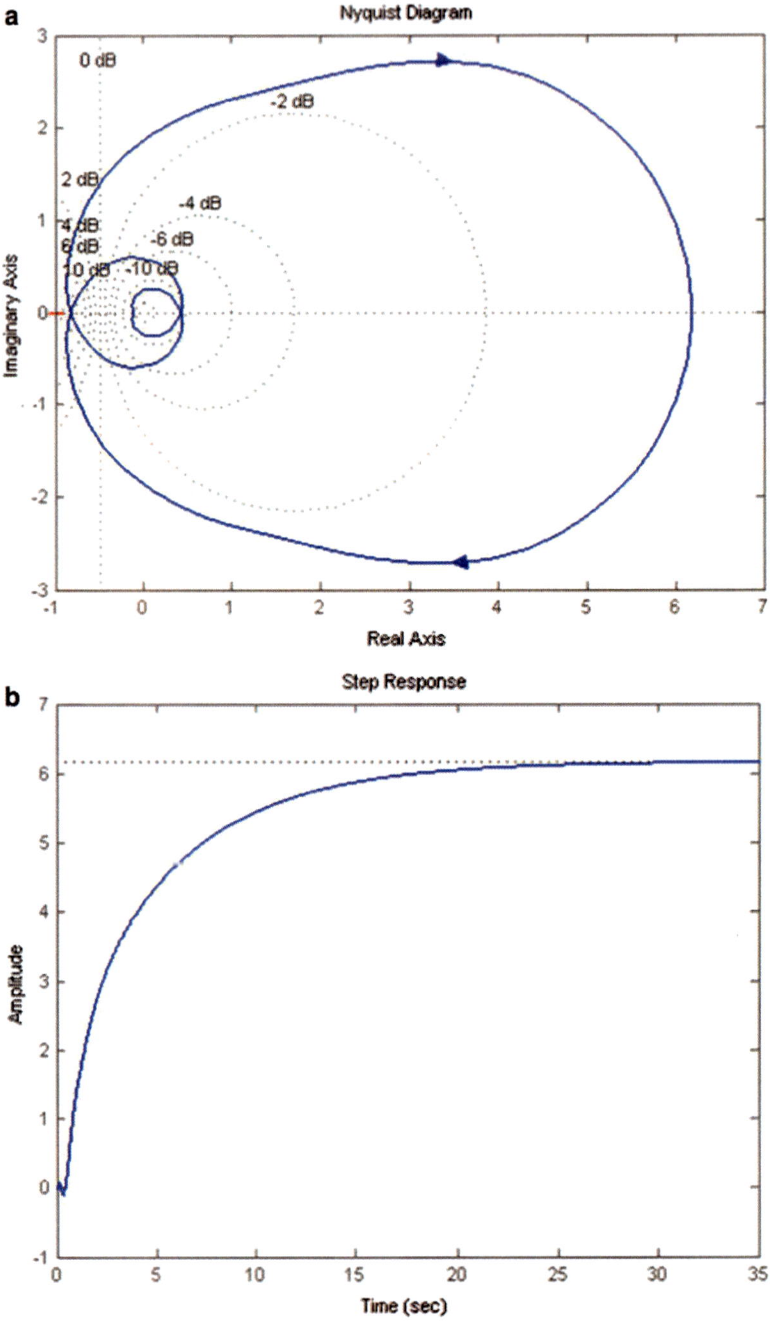

Fig. 2 **a** Nyquist plot of Carlson-based FOS with delay. **b** Step response of Carlson-based FOS with delay

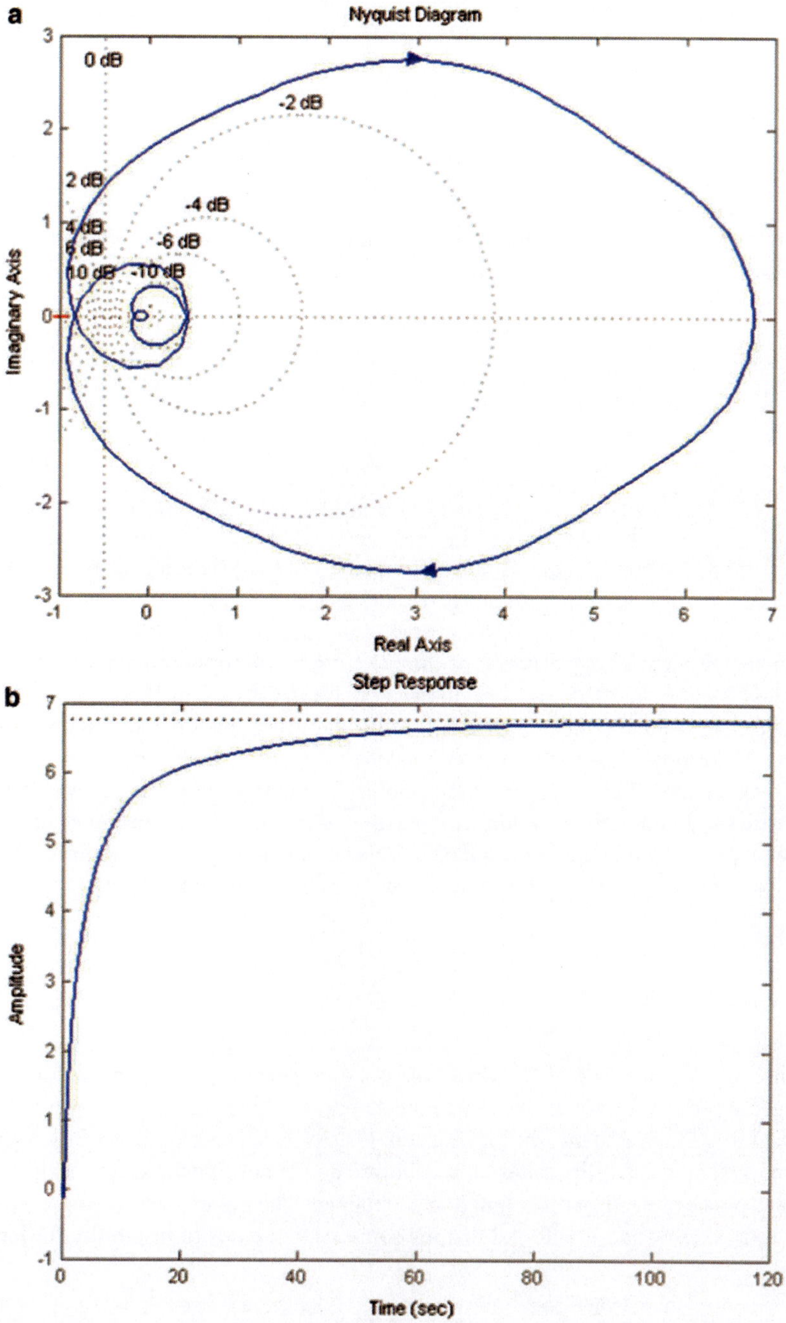

Fig. 3 **a** Nyquist plot of Matsuda-based FOS with delay. **b** Step response of Matsuda-based FOS with delay

$$G_{\text{frac\_char\_red}}(s) = \frac{\left(-0.0002954s^6 - 10.73s^5 + 250s^4 - 2413s^3 + 8708s^2 + 6103s + 314.7\right)}{\left(s^6 + 56.04s^5 + 591.4s^4 + 4157s^3 + 5288s^2 + 1268s + 46.85\right)} \tag{9}$$

The Nyquist and step responses for this method are shown in Fig. 4a, b, respectively, and specifications listed in Table 1.

The modified Oustaloup approximation method is a refined version of the Oustaloup method. This method yields an integer order model of $O(10)$ (Eq. 10) and then reduced to $O(6)$ (Eq. 11).

$$G_{\text{frac\_moust}}(s) = \frac{\begin{pmatrix} -16.77s^{10} - 8203s^9 - 5.215e5s^8 + 5.886e6s^7 + 4.232e7s^6 - 1.126e9s^5 - 4.697e9s^4 + 1.666e10s^3 \\ +5.356e9s^2 + 2.408e8s + 1.356e6 \end{pmatrix}}{\begin{pmatrix} 755.4s^{10} + 1.224e5s^9 + 5.001e6s^8 + 9.004e7s^7 + 8.709e8s^6 + 4.46e9s^5 + 9.88e9s^4 + 6.382e9s^3 \\ +1.056e9s^2 + 3.757e7s + 1.964e5 \end{pmatrix}} \tag{10}$$

$$G_{\text{frac\_moust\_red}}(s) = \frac{\left(-0.0222s^6 - 7.553s^5 + 183.2s^4 - 1795s^3 + 6456s^2 + 4679s + 219.7\right)}{\left(s^6 + 41.16s^5 + 469.4s^4 + 3101s^3 + 4053s^2 + 952.7s + 35.52\right)} \tag{11}$$

Figure 5a, b shows Nyquist and step responses of the Modified Oustaloup-based FO plant and the specifications are given in Table 1.

Figure 6a shows the comparison of step responses of AIO obtained using the five methods with ideal step response of the FOS. From the time-domain specification given in Table 1 and Fig. 6a, it is observed that the Carlson method-based integer model is closest to the ideal response. It is also observed that the input–output delay of the AIO models is less than 0.5 s as shown in Fig. 6b, which is desirable.

To check the effect of delay on the stability of the system, the delay was further increased and simulations obtained. It was found that the system becomes unstable for an input–output delay beyond 0.65 s. This shows that for a larger value of delay, the system tends to become unstable due increase in the lagging angle.

4 Conclusion

In this paper, analysis of FOS with time delay is performed. A comparison of its integer approximation with respect to the stability is presented. Time delay affects the stability of the system since it results in phase lag. To check the stability of such systems, Nyquist stability criterion is followed and to observe the steady-state error, step response of each transfer function is obtained. The system is also simulated for increase in delay and it is found that the same system tends to instability for larger delay.

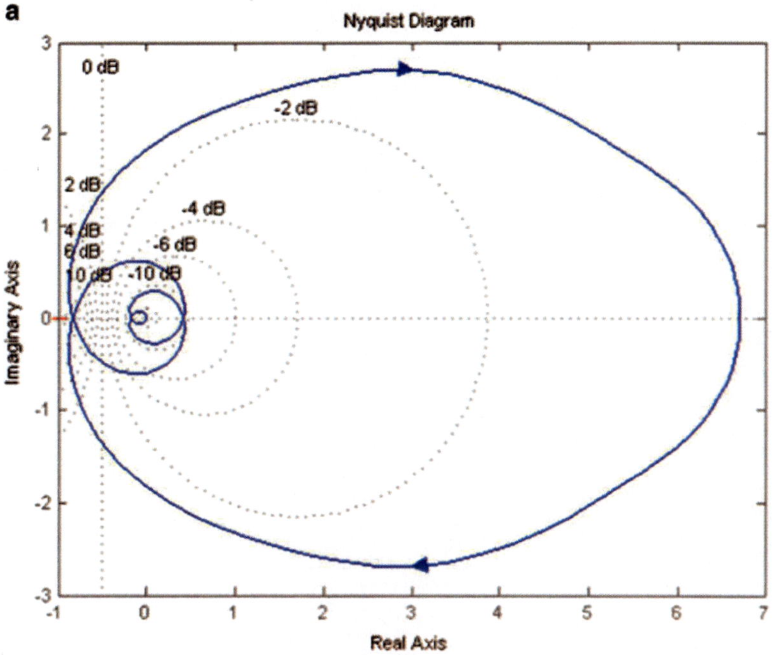

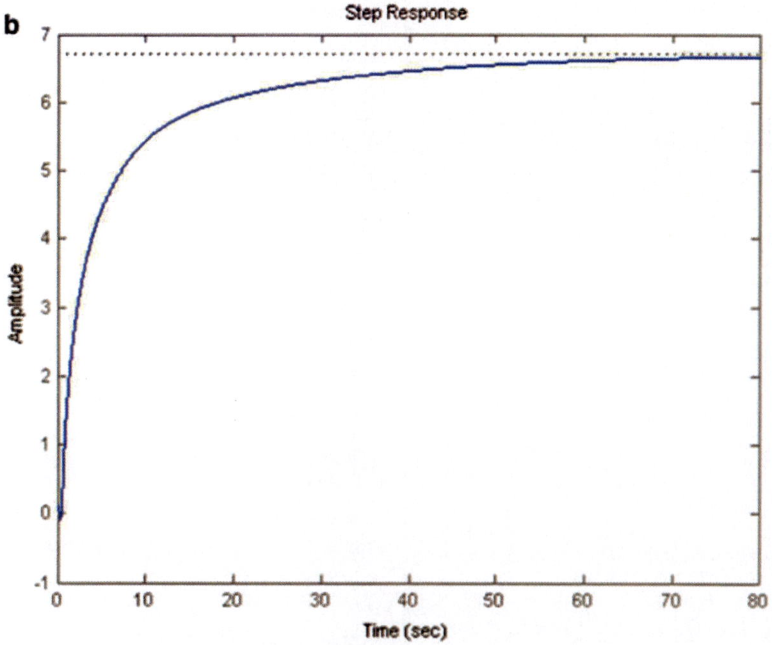

Fig. 4 **a** Nyquist plot of Charef-based FOS with delay. **b** Step response of Charef-based FOS with delay

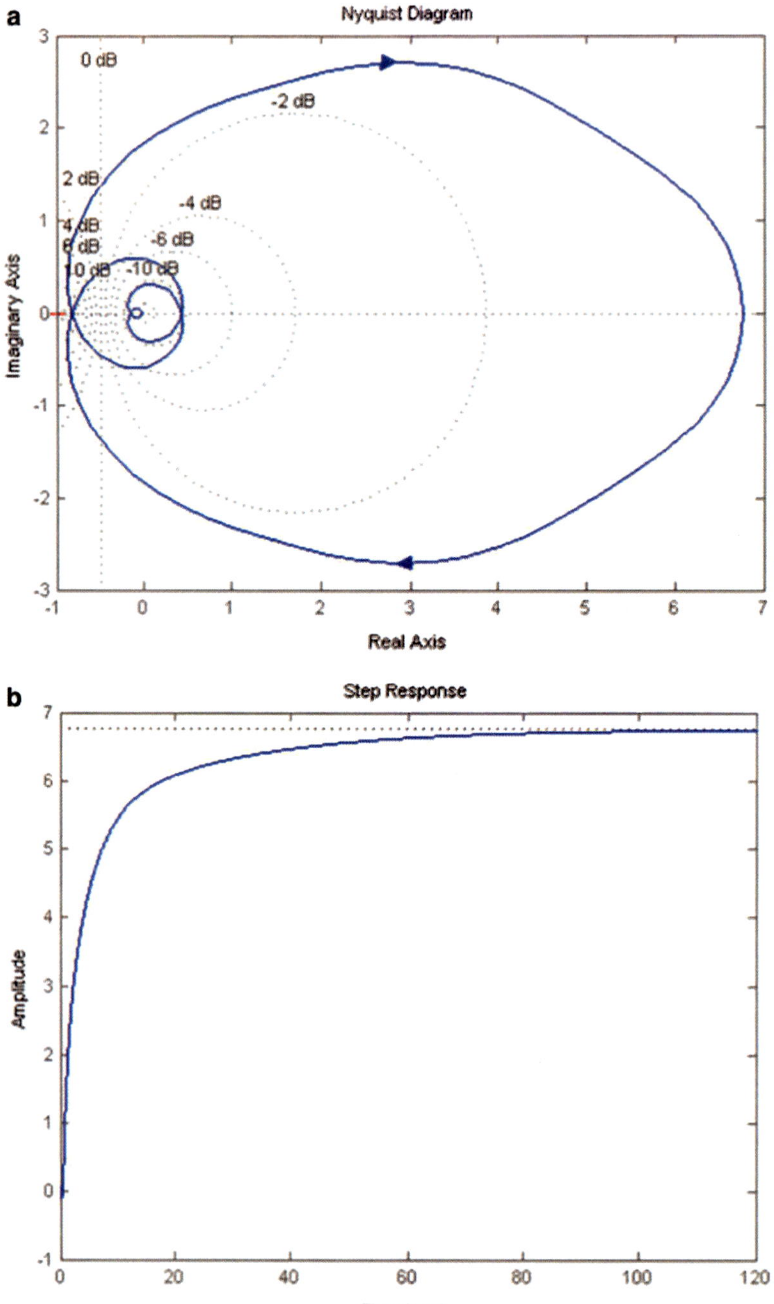

Fig. 5 **a** Nyquist plot of modified Oustaloup-based FOS with delay. **b** Step response of modified Oustaloup-based FOS with delay

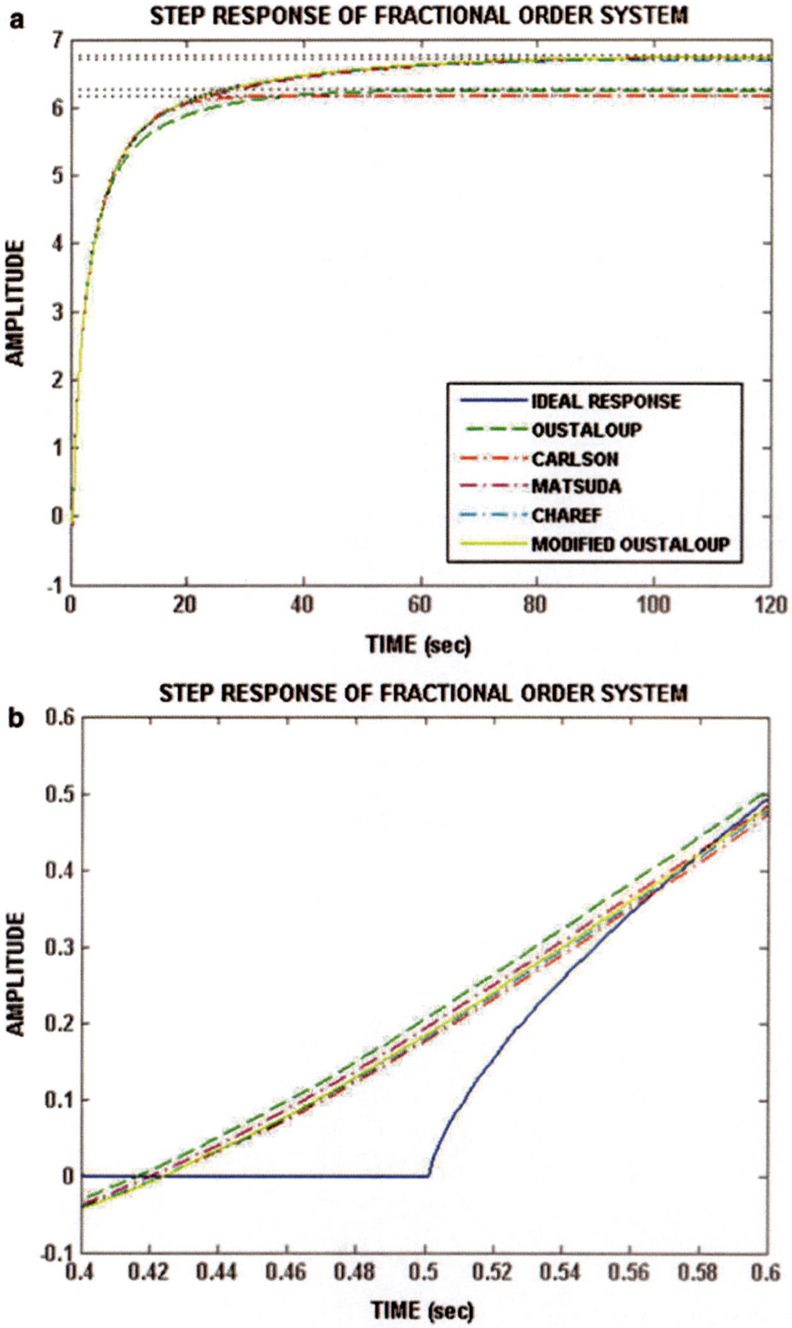

Fig. 6 **a** Comparison of step responses of the fractional-order plant with delay using five different approximation methods. **b** Zoom of step response of fractional-order plant of part a showing reduction in delay

References

1. Vinagre BM, Podlubny I, Hernandez A, Feliu V (2000) Some approximations of fractional order operators used in control theory and applications. Fractional Calculus Appl Anal 3(3):239–248
2. Oustaloup A, Levron F, Mathieu F, Nanot FM (2000) Frequency-band complex non-integer differentiator: characterization and synthesis. IEEE Trans Circ Syst I Fundam Theor Appl 47:25–39
3. Shrivastava N, Varshney P (2019) Application of Newton iterative process in approximation of fractional differentiator and integrator. In: Proceedings of 2019 2nd international conference on intelligent computing, instrumentation and control technologies, Kannur, Kerala, India, pp 589–592
4. Matsuda K, Fuji H (1993) H_{∞}-optimized wave absorbing control: analytical and experimental results. J Guidance Control Dyn 16(62):1146–1153
5. Charef A, Sun H, Tsao Y, Onaral B (1992) Fractal systems as represented by singularity function. IEEE Trans Autom Control 37:1465–1470
6. Xue DY, Zhao C, Chen YQ (2006) A modified approximation method of fractional order system. In: Proceedings of IEEE international conference on mechatronics and automation, pp 1043–1048
7. Khanra M, Pal J., Biswas K (2013) Use of squared-magnitude function in approximation and hardware implementation of SISO fractional order system. J Franklin Inst 350:1753–1767
8. Jiang H, Liu Y (2016) Disturbance rejection for fractional-order time delay systems. Math Probl Eng. ArticleID1316046
9. Petras I (2011) Stability test procedure for a certain class of the fractional-order systems. In: 12th international Carpathian control conference (ICCC), pp 303–307
10. Safonov MG, Chiang RY (1989) A Schur method for balanced model reduction. IEEE Trans Autom Control 34(7):729–733

Cluster-Based Approaches for Energy Efficiency in Wireless Sensor Network

Shweta Sinha, Anjali Dhamiwal, and Poonam Sharma

Abstract In today's time, throughout the world, the science and technology domain are looking for the improvements in existing solutions. These improvements can be in terms of efficiency, time, or cost. Similar scenario exists in WSN domain. The development of tiny sensor nodes capable of communicating with each other in wireless environment have paved their existence in many critical application areas related to disaster management and health care. These tiny nodes have limited battery life and hence are short lived. Continuous efforts are being made to increase the network life expectancy by decreasing the energy consumption of the nodes. Besides highlighting the challenges related to WSN, the present paper outlines the efforts made for reduction of energy consumption. Some of the optimized approaches for cluster creation have been studied and performance of LEACH and LEACH-FL is evaluated in simulated environment.

Keywords Wireless sensor network · Routing protocols · Clustering · LEACH · Fuzzy LEACH

1 Introduction

Wireless sensor network consist of spatially distributed nodes or sensor node that is suitable for building a temporary network. It acts as a distributed network of nodes which can receive and send data to the neighbour nodes. All nodes can move freely in the network and can form either heterogeneous or a homogenous network. Each sensor node has unique capabilities that may range from simple nodes that

S. Sinha (✉) · A. Dhamiwal · P. Sharma
Amity University, Gurgaon, Haryana, India
e-mail: meshweta_7@rediffmail.com

A. Dhamiwal
e-mail: duggu.sharma17@gmail.com

P. Sharma
e-mail: poonamsharma.2289@gmail.com

R. P. Mahapatra et al. (eds.), *Proceedings of 6th International Conference on Recent Trends in Computing*, Lecture Notes in Networks and Systems 177,
https://doi.org/10.1007/978-981-33-4501-0_61

are capable of sensing simple and single physical phenomena to more diversified nodes with advance capabilities to sense. Several existing options can help to decide for the topology of connection between these nodes. The basic aim of this network structure is to offer a sustainable combination of computing and communication in a distributed sensing environment that utilizes the power of the entire network even though the individual capabilities of nodes are limited. The choice of sensors is influenced by the physical property to be sensed or monitored. Their operational capabilities make them exclusive as compared to the already existing communication systems. Some of the unique characteristics of WSN are:

- Limited energy capacity leading to limited storage and computation power of the nodes.
- Self-configurable network; no central authority to manage network arrangements for communication.
- Sensor location; mainly placed in an open area that makes it prone to damage and malicious reading.
- Scalability of the network; provides easy inclusion of nodes into the network.
- Robustness of the network; data is redundantly placed at multiple sensors and hence guarantee data availability even in case of failure.

These unique characteristics also create some limitations on the network. However, in the current scenario, they are being deployed at an accelerated pace to enable a massive amount of applications ranging from precision agriculture, environmental monitoring, healthcare, supply chain management that are influencing the quality of life and economy. Also, recently, the healthcare domain has witnessed the deployment of an enormous number of sensors that are playing an active role in patient monitoring and care [1]. The demand for this self-configurable network is increasing day by day. There is a need for continuous efforts to overcome the limitations of the sensor network for its horizon to expand to day-to-day activities.

Energy consumption is one of the major limitations of WSN. Most of the energy is consumed during communication. Choice of routing protocols and formation of clusters with the deployed node can contribute to the reduction of consumed energy. There are many approaches for the maintenance of the energy of the nodes in the cluster of the wireless sensor network, such as management of topology, cost reduction, change of routing or optimal routing and clustering. This paper discusses the major issues and challenges associated with the WSN and presents a review of some of the energy efficiency approaches used in the cluster-based WSN. Their comparative performance analysis for a network of 100 nodes is presented here. Some of the recent techniques based on artificial intelligence approaches for energy preservation is also discussed in this paper.

2 Background and Literature Survey

Wireless communication and the evolution of MEMS have shaped WSN as a new class of communication network. The use of several micromachining techniques and fabrication process [2] has paved the way to obtain integrated low-cost, low-powered tiny sensor nodes with sensing capabilities. The communication technology has moved through significant advances related to antenna technology, synchronization, and routing methods. But as the sensor networks are infrastructure-less and the nodes communicate through lossy lines, this poses additional challenges related to energy. Apart from communication rules, other factors such as network architecture, management of topology can also affect energy consumption. This section discusses the network architectures possible for wireless communication and briefly review the energy-efficient approaches used under the most promising network structure.

2.1 Network Architecture

A typical communication network is composed of nodes that sense data and send to the sink nodes in the network, which is located nearby or inside the sensing region. These sinks serve as a gateway to the network which connects with the outside network. The connection between the sink and the sensor nodes can be established through a direct link which has to be a distant connection. Since the energy consumption of sensor nodes has to be reduced this direct connection needs to be avoided and replaced by a multiple hop connection. Multi-hop connections enable short distance communication. Several intermediate nodes exist between the sink and the sensor nodes and use less power during communication. Two possible structure exists for establishing a multi-hop connection: The flat and the hierarchical structure. Figure 1 represents the two structures.

The flat architecture obeys a data-centric approach and flooding technique is used for data transmission. Even though data is received by each node of the network, only the intended node responds back. In the case of hierarchical architecture, the nodes are divided into clusters and each cluster is represented by a head node (CH). Nodes communicate with the CH and the CH then aggregates, process, and transmit data to the sink node. In flat architecture, all nodes are equally capable and are designated with equal responsibility, whereas in hierarchical structure, the nodes with less energy are the sensing nodes and the one with high energy is designated as CH. Undoubtedly, the latter is more suitable for energy-efficient ways of communication.

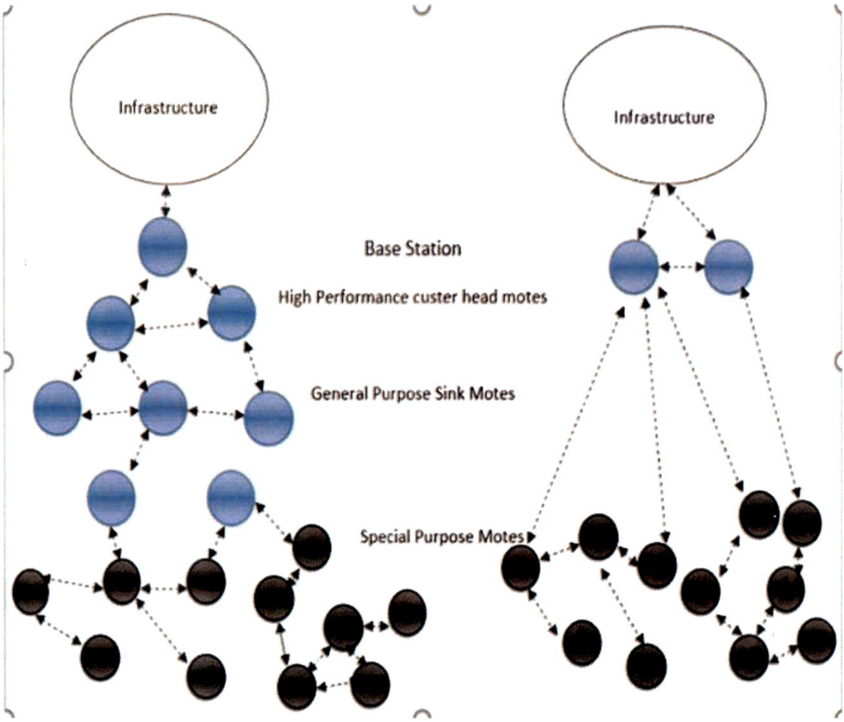

Fig. 1 Hierarchical and flat architecture of WSN

2.2 *Clustering in WSN*

The cluster formation supported by the hierarchical structure of network possesses multiple benefits:

- Cluster formation prevents redundancy of message exchange and can maintain communication bandwidth.
- Due to cluster formation, node interacts only with the CH reducing communication overhead and hence the network topology can be stabilized at the sensor level itself.
- Optimized management strategies can easily be implemented in a clustered network.

Two approaches are used for clustering in WSN: the distributed approach for cluster formation and the centralized clustering techniques. Furthermore, based on the cluster creation rules and the parameters used for CH election, the distributed cluster formation approach is divided into four subcategories: Neighbourhood information based, a probabilistic approach based, the identity-based and iterative approach based. The authors of [3] proposed linked clustering algorithm (LCA), and identity-based clustering method that takes a unique node identifier as a key to select a cluster head. Highest connectivity cluster algorithm identifies a node with the highest number of neighbours at a 1-hop distance as a cluster head. MECH [4] also uses neighbourhood information and gives the liberty of not claiming to be a CH even if conditions are satisfied. On the principle of non-periodic invocation of itself in case of topology, reconfiguration is obeyed by weighted clustering algorithm [5]. Grid clustering routing protocol [6] uses the concept of multiple sinks while designating the responsibility of dynamic selection of cluster head to one node called the primary sink.

Protocols based on probabilistic approach rely on prior assigned probability values. Low-energy adaptive clustering hierarchical protocol [7] randomly rotates the responsibility of cluster head to provide a balance of energy usage, while maintaining uniform load distribution. The dynamic approach of LEACH leads to energy dissipation during rotation of responsibilities and does not guarantee good CH distribution. Iterative clustering algorithm, SPAN [8] uses a localized decision making based on its own energy and other sensor nodes energy in the network for random selection of cluster head. As LEACH follows a proactive approach, threshold sensitive energy efficient sensor network protocol (TEEN) [9] uses reactive approach and reduced the energy consumption during transmission by applying the mechanism to transmit sensed data only when a threshold is reached thereby reducing the frequency of data transmission.

2.3 *Routing Protocols*

In addition to the clustering, the energy consumption also depends upon the routing decisions made for sending the sensed data. Efficient routing protocols can considerably decrease the energy consumption and, consequently, increase the network lifetime. Therefore, effective design of routing protocol is needed, which should take into consideration the parameters such as energy consumption constraints, node deployment scenario, node capabilities and responsibilities, data aggregation and fault tolerance. Table 1 highlights some of the routing protocols along with their merits and limitations.

Clustering and routing algorithms both work for achieving several objectives such as stability of network topology, scalability, energy consumption reduction, data aggregation, load balancing, fault tolerance, increase in connectivity, maximizing network lifetime, decrease in delay, and reducing the number of control messages. There is always a threshold while trying to achieve these and not all can be achieved in several scenarios. Hence, some of these objectives such as scalability, fault tolerance, data aggregation, and load balancing may be considered as the main objectives to be achieved while maximizing the network lifetime.

Table 1 Routing protocols with their merits and limitations

Category	Name of protocol	Merits	Limitations
Location-aided protocols	GAF (Geographic Adaptive Fidelity) [10]	Energy conservation	Processing overhead
	GEAR (Geographic and Energy Aware Routing) [11]	Recursive data dissemination instead of flooding	Lifetime depletion
Data centric protocol	SPIN (Sensor Protocols for Information Negotiation) [12]	Energy aware and resource aware	Does not apply to mobile links, works only for lossless network
	Cougar [13]	Provides in-network data aggregation saves energy efficiently	Need extra storage and synchronization
Layered and in network	LEACH	Efficient resource utilization, uniform load distribution	Energy dissipation does not guarantee good CH distribution
	APTEEN [14]	Works for hybrid network, flexible	Additional overhead and complexity associated with threshold
Multi-path based	Disjoint paths [15]	Low scalability, failure on one path leads to selection of alternate paths	More resilient to sensor failures, less energy efficient
Mobility based	TTDD (Two-tier Data Dissemination Model) [16]	Scalable, handles multiple mobile sinks	Can not handle mobile sensor nodes
QoS based	SPEED [17]	Congestion avoidance, less energy consumption	
Heterogeneity based	CADR (Constrained anisotropic diffusion routing) [18]	Information gain, minimized latency	

3 Working of LEACH and LEACH-FL Protocol

LEACH is a hierarchical protocol which works on the principle of communication through the cluster heads. It obeys a proactive approach. The selection of nodes as CH is periodic in nature and each period consists of two stages. In the first stage, i.e. the construction of clusters, the CH is decided based upon the probability assigned to them in advance to achieve an optimized cluster. Once the cluster head is decided, the clusters are formed under them and in the second phase, CH prepares a TDMA schedule for the sensor nodes giving the slot details in which the cluster nodes can send data and then the CH aggregates and communicate with sink by transferring data. The drawback of LEACH is associated with its principle of the probability model. A cluster head may be selected in very close vicinity of the other CH or, can be located at the WSN edge. Use of probability model does not guarantee good CH distribution.

LEACH-FL [19] is a protocol based on the adaptive approach of cluster formation followed by LEACH algorithm, but the selection of CH is done using fuzzy logic. The decision of CH is based upon three factors: the residual energy or the battery life, distance from the sink, and node density that helps in calculating the probability of becoming CH [20]. The distributed process employed by LEACH-FL for cluster formation and CH selection follows the same mechanism as that of LEACH but uses the fuzzy output to take the final decision. The defuzzification module helps to convert the output in the range and on the basis of decided threshold CH is selected. This model works only when the sensor nodes can get their coordinates.

In the next section, we evaluate the performance of both the network under a simulated environment.

4 Performance Evaluation and Results

The two protocols discussed in the previous section are evaluated under a simulated environment. The comparison parameters being:

- Energy dissipation in the simulated environment.
- The number of dead nodes in the two algorithms is computed in different rounds to provide the measure of network lifetime.

4.1 Simulation Parameters

The system is modelled using NS2 and MATLAB simulator and the simulation environment is described below:

- A huge number of dynamic nodes is placed in a 100 by 100 square region uniformly distributed throughout.
- Initially, the nodes dissipate their information by sending "Hello" packet to all the other sensor nodes placed in the network.
- Initially, the cluster size is optimized based on the deployed nodes, and the cluster number is kept at optimal. As the nodes die, re-clustering is done during which the number of clusters may merge to create fewer clusters. Initial cluster size is kept at 10.
- It is assumed that the sink of the network is a node without any energy limitations. They are placed in the centre of the region of interest.
- The consumption of energy for sending m bit data is varies proportionally to the square of the distance d among the nodes.
- The energy is depleted when packets are transferred in the cluster.

Table 2 Parameters used during simulation

Parameters	Value	Parameters	Value
Network span	100 × 100 (static nodes)	Number of nodes	100
Node distribution	Uniform throughout the area	Node energy coefficient (a_i)	Random number between (0, 5)
Node initial essential Energy	0.75 J	E_{elec}	50 nJ/bit
BS position	(50, 50)	E_{frs}	10 nJ/bit/m^2
Data packet size	500 bytes	E_{amp}	0.0013 pJ/bit/m^4
Rounds	4000	EDA	5 nJ/bit/signal

where E_{elec}: Consumed energy per bit during running state of emitter components
E_{frs}: Consumed energy in free space by the unit power
E_{amp}: Consumed energy in the double path propagation model of amplifier

- The energy used is the resultant of average rate of power consumption times the duration of transfer. Power consumption depends upon signal acquisition and processing (Table 2).

5 Results and Analysis

Figures 2 and 3 represent the network lifetime measured in terms of the overall energy of the network in different rounds and the number of dead nodes in different rounds of simulation of both the protocols respectively. The cumulative energy is again an indirect measure of network lifespan and also highlights the energy conservation aspect of each algorithm.

From the figure, it is clear that the energy consumption in the fuzzy logic-based LEACH is less as compared to the classical LEACH algorithm. Even if the two starts

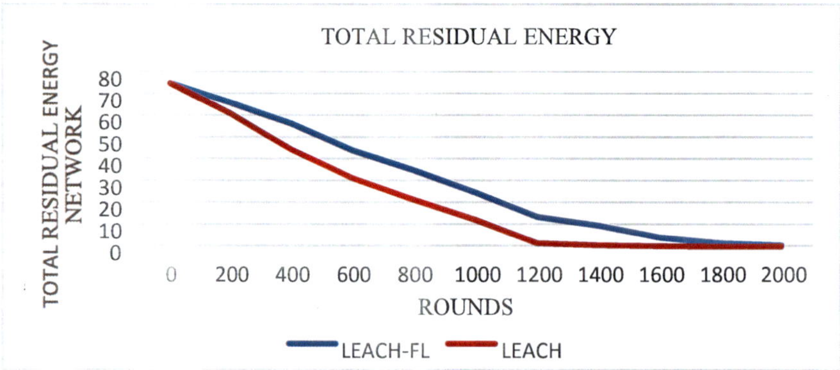

Fig. 2 Comparison of system energy consumption

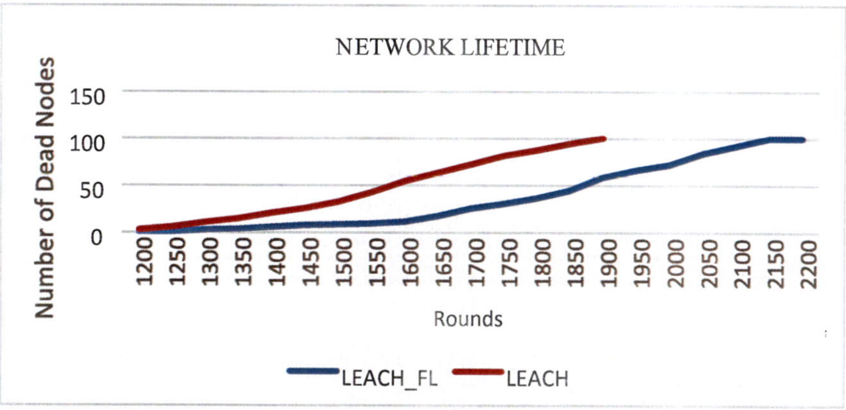

Fig. 3 Comparison of network lifetime of two algorithms

with the same energy level of the network, the LEACH protocols consume more energy and this signifies that the CH selection in the classical approach requires much computation and frequent processing that consumes energy at a faster rate. Figure 3 highlights that even at the start the number of nodes in both the network is the same and remains constant for a considerable time span, in the classical LEACH the nodes start dying early giving the lifespan of the network using LEACH-FL much more compared to the former one.

6 Conclusion and Future Work

The review of the wireless sensor network highlights that energy conservation is the most important issue in the design and deployment of the network. The paper describes the importance of clustering as a mechanism for energy conservation to increase the network lifetime. Several routing protocols that support the clustered approach of the network have been reviewed in the paper. With the increase in the application of artificial intelligent approaches in other domains, the paper evaluates the performance of classical LEACH and fuzzy logic-based LEACH algorithm under a simulated environment. The results obtained are in line to the literature findings. The promising outcome of AI-based approach motivates us to work further in this direction and apply AI/machine learning approaches for cluster formation to further increase the lifespan of the network.

References

1. Jafari R, Encarnacao A, Zahoory A, Dabiri F, Noshadi H, Sarrafzadeh M (2005) Wireless sensor networks for health monitoring. In: The second annual international conference on mobile and ubiquitous systems: networking and services, pp 479–481
2. Senturia SD (2007) Microsystem design. Springer Science & Business Media, Berlin
3. Kumarawadu P, Dechene DJ, Luccini M, Sauer A (2008) Algorithms for node clustering in wireless sensor networks: a survey. In: 4th international conference on information and automation for sustainability, pp 295–300
4. Chang RS, Kuo CJ (2006) An energy efficient routing mechanism for wireless sensor networks. In: 20th international conference on advanced information networking and applications, vol 2, pp 5–9
5. Chatterjee M, Das SK, Turgut D (2002) WCA: a weighted clustering algorithm for mobile ad hoc networks. Cluster Comput 5(2):193–204
6. Yu L, Neng W, Wei Z, Chunlei Z (2006) GROUP: a grid-clustering routing protocol for wireless sensor networks. In: Proceedings of IEEE conference on wireless communications, networking and mobile computing (WiCOM)
7. Mehmood A, Mauri JL, Noman M, Song H (2015) Improvement of the wireless sensor network lifetime using LEACH with vice-cluster head. Ad Hoc Sens Wireless Netw 28(1–2):1–17
8. Chen B, Jamieson K, Balakrishnan H, Morris R (2002) Span: an energy-efficient coordination algorithm for topology maintenance in ad hoc wireless networks. Wireless Netw 8(5):481–494
9. Manjeshwar A, Agrawal DP (2001) TEEN: a routing protocol for enhanced efficiency in wireless sensor networks. In: ipdps, vol 1, pp 189
10. Xu Y, Heidemann J, Estrin D (2001) Geography-informed energy conservation for ad hoc routing. In: Proceedings of the 7th annual international conference on mobile computing and networking, pp 70–84
11. Yu Y, Govindan R, Estrin D (2001) Geographical and energy aware routing: a recursive data dissemination protocol for wireless sensor networks
12. Heinzelman WR, Kulik J, Balakrishnan H (1999) Adaptive protocols for information dissemination in wireless sensor networks. In: Proceedings of the 5th annual ACM/IEEE international conference on mobile computing and networking, pp 174–185
13. Yao Y, Gehrke J (2002) The cougar approach to in-network query processing in sensor networks. ACM Sigmod Rec 31(3):9–18
14. Manjeshwar A, Agrawal DP (2002) APTEEN: a hybrid protocol for efficient routing and comprehensive information retrieval in wireless sensor networks. In: ipdps, p 0195b
15. Ganesan D, Govindan R, Shenker S, Estrin D (2004) Highly-resilient, energy-efficient multipath routing in wireless sensor networks. ACM SIGMOBILE Mob Comput Commun Rev 5(4):11–25
16. Ye F, Luo H, Cheng J, Lu S, Zhang L (2002) A two-tier data dissemination model for large-scale wireless sensor networks. In: Proceedings of the 8th annual international conference on mobile computing and networking, pp 148–159
17. Akkaya K, Younis M (2005) A survey on routing protocols for wireless sensor networks. Ad Hoc Netw 3(3):325–349
18. Chu M, Haussecker H, Zhao F (2002) Scalable information-driven sensor querying and routing for ad hoc heterogeneous sensor networks. Int J High Perform Comput Appl 16(3):293–313
19. Ran G, Zhang H, Gong S (2010) Improving on LEACH protocol of wireless sensor networks using fuzzy logic. J Inf Comput Sci 7(3):767–775
20. Parwekar P (2014) Comparative study of hierarchical based routing protocols: wireless sensor networks. In: ICT and critical infrastructure: proceedings of the 48th annual convention of computer society of India, vol I, pp 277–285

Optimal Tuned Linear Active Disturbance Rejection Control Applied to a 2-DOF Robotic Arm Manipulator

Suhail Ahmad Suhail, Mohammad Abid Bazaz, and Shoeb Hussain

Abstract In this paper, a novel control strategy called the active disturbance rejection control (ADRC) is used for trajectory tracking control of a 2-DOF robotic arm manipulator. The manipulator system is a highly nonlinear, dynamically coupled and time-varying system. ADRC is a model-independent control strategy requiring knowledge of only the relative order and the output of the system. The conventional PID controller does not work well in case of highly coupled systems and the sliding mode control shows good dynamic control performance but requires an accurate model of the system. In contrast, ADRC treats coupling and nonlinearity as an external disturbance and provides better results. A second-order optimal tuned linear ADRC with reduced-order observer has been used in this paper for the trajectory tracking of the manipulator system.

Keywords ADRC · Robotic arm · Linear quadratic regulator (LQR) · Extended state observer (ESO)

1 Introduction

Robotic manipulators have become a major component of manufacturing industry because of their advantage of repeatability, high speed and precision. Two-link robotic arm manipulator is a very basic, simple and classical example of a robot. Robotic manipulators are highly nonlinear, dynamically coupled and time-varying

S. A. Suhail (✉) · M. A. Bazaz
Department of Electrical Engineering, National Institute of Technology Srinagar, Srinagar, India
e-mail: suhailahmad@nitsri.net.com

M. A. Bazaz
e-mail: abid@nitsri.net.com

S. Hussain
Department of Electrical Engineering, University of Kashmir, Srinagar, India
e-mail: shoeb@uok.edu.in

R. P. Mahapatra et al. (eds.), *Proceedings of 6th International Conference on Recent Trends in Computing*, Lecture Notes in Networks and Systems 177,
https://doi.org/10.1007/978-981-33-4501-0_62

systems which are extensively used in industries. The robotic manipulators are generally subjected to uncertainties. Because of these uncertainties and nonlinear behavior, it is a challenging task to control the motion of robot manipulator at accurate position. For the accurate work, any robot has to follow predefined trajectories as close as possible, which makes trajectory tracking control a most important task in control of robotic manipulator. ADRC which is model independent and has high disturbance rejection was first introduced in its nonlinear form by Han [1] in 1998. To make its tuning and frequency domain analysis easy, Gao [2] introduced the linear active disturbance rejection control (LADRC) which uses a linear observer and linear state feedback error. Initially, a nonlinear tuning procedure was proposed for the ADRC but a large number of parameters made tuning an art. Gao [2] proposed a linear ADRC (LADRC) and parameterized the tuning procedure into a few gains. The tuning procedure is actually a pole placement technique and the location of the poles of the controller and the extended state observer (ESO) helps in achieving the desired performance indirectly. The final location of these poles, however, becomes a trial and error strategy which can be difficult to understand fully and to efficiently apply to real-world systems. Linear quadratic regulator (LQR) is a well-known design technique and has been widely used in many applications. The desired performance is achieved directly by minimizing a quadratic cost function of the control input and the states, contrary to the pole placement technique. The rest of this paper is arranged as: In Sect. 2, dynamic model of 2-DOF robotic manipulator is introduced. Section 3 describes the design of a second-order ADRC and its tuning method for trajectory tracking control. Section 4 includes the simulated results. Section 5 includes some conclusions.

2 Dynamic Model of a 2-DOF Robotic Manipulator

The dynamic equation for a n-link rigid robotic manipulator can be expressed as:

$$\mathbf{N}(\Theta)\ddot{\boldsymbol{\Theta}} + \mathbf{V}(\boldsymbol{\Theta}, \dot{\boldsymbol{\Theta}})\dot{\boldsymbol{\Theta}} + \mathbf{G}(\boldsymbol{\Theta}) = \tau \tag{1}$$

where $\boldsymbol{\Theta} \in R^n$ is a matrix of joint displacements, $\tau \in R^n$ is a matrix of applied torques. $\mathbf{N}(\boldsymbol{\Theta})$ is the symmetric positive definite inertia matrix, $\mathbf{V}(\boldsymbol{\Theta}, \dot{\boldsymbol{\Theta}})\dot{\boldsymbol{\Theta}}$ is the $n \times 1$ matrix of centrifugal and Coriolis forces and $\mathbf{G}(\boldsymbol{\Theta})$ is the gravity matrix. Friction has been neglected. For the two-link robotic arm manipulator shown in Fig. 1, the dynamic Eq. (1) can be written in the state-space form as:

$$\begin{bmatrix} N_{11}(\boldsymbol{\Theta}) & N_{12}(\boldsymbol{\Theta}) \\ N_{21}(\boldsymbol{\Theta}) & N_{22}(\boldsymbol{\Theta}) \end{bmatrix} \begin{bmatrix} \ddot{\Theta}_1 \\ \ddot{\Theta}_2 \end{bmatrix} + \begin{bmatrix} \mathbf{V}_{11}(\boldsymbol{\Theta}, \dot{\boldsymbol{\Theta}}) \\ \mathbf{V}_{21}(\boldsymbol{\Theta}, \dot{\boldsymbol{\Theta}}) \end{bmatrix} \begin{bmatrix} \dot{\Theta}_1 \\ \dot{\Theta}_2 \end{bmatrix} + \begin{bmatrix} \mathbf{G}_1(\boldsymbol{\Theta}) \\ \mathbf{G}_2(\boldsymbol{\Theta}) \end{bmatrix} = \begin{bmatrix} \tau_1 \\ \tau_2 \end{bmatrix} \tag{2}$$

where

Fig. 1 Two-link robotic arm

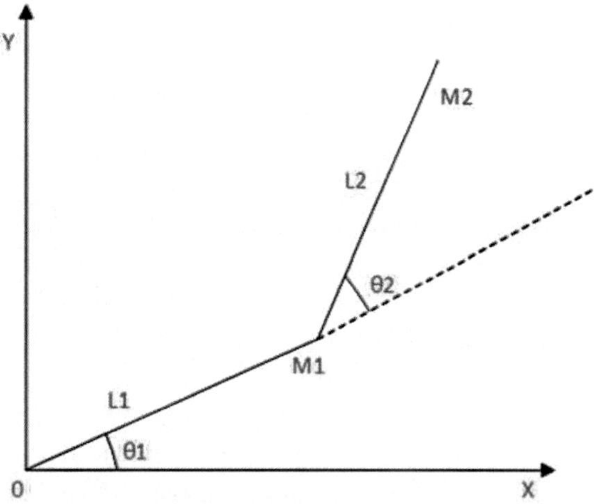

$$N_{11}(\Theta) = (M_1 + M_2)L_1^2 + M_2 L_2^2 + 2M_2 L_1 L_2 \cos\theta_2$$
$$N_{12}(\Theta) = M_2 L_2^2 + 2M_2 L_1 L_2 \cos\theta_2,\ N_{21}(\Theta) = M_2 L_2^2 + 2M_2 L_1 L_2 \cos\theta_2$$
$$N_{22}(\Theta) = M_2 L_2^2$$

and

$$V_{11}(\Theta, \dot{\Theta}) = -M_2 L_1 L_2 (2\dot{\theta}_1 \dot{\theta}_2 + \dot{\theta}_2^2) \sin\theta_2,\ V_{21}(\Theta, \dot{\Theta}) = M_2 L_1 L_2 \dot{\theta}_2^2 \sin\theta_2$$

and

$$G_1(\Theta) = (M_1 + M_2) g L_1 \cos\theta_1 + M_2 g L_2 \cos(\theta_1 + \theta_2),\ G_2(\Theta) = M_2 g L_2 \cos(\theta_1 + \theta_2)$$

This is a special form of state model called "Brunovsky canonical form". M_1 and M_2 are the link masses assumed to be concentrated at the ends of the links of lengths L_1 and L_2, respectively. Torques τ_1 and τ_2 are applied by the actuators to control the angles θ_1 of first link and θ_2 of second link, respectively.

3 Active Disturbance Rejection Control

A general ADRC proposed by Han [3] comprises of three parts: (i) A tracking differentiator (TD) constructed to generate a transient profile for the plant to follow the desired system response; (ii) an ESO for estimate the disturbances and the states; (iii) a control law. The original ADRC as proposed by Han had nonlinear gains and too many parameters to be tuned. To make its tuning and frequency domain analysis

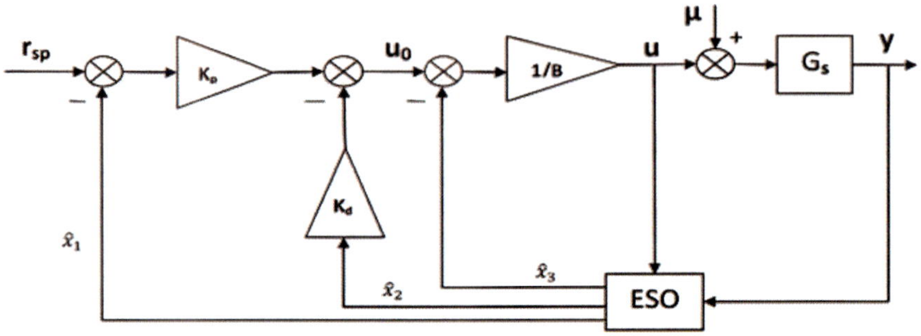

Fig. 2 Structure of second-order ADRC

easy, Gao [2] introduced the LADRC which uses a linear observer and linear state feedback error. The second-order LADRC structure is shown in Fig. 2.

G_s is the process which is to be controlled. The plant output y and the control signal u are the inputs to the ESO. The ESO is used to estimate the states as well as the parameter uncertainties and external disturbances u in real time. The output taken from the ESO are the state estimates $\hat{x}_1$, $\hat{x}_2$ and $\hat{x}_3$.

The ESO in second-order LADRC is

$$\begin{aligned} \dot{\hat{x}}_1 &= x_2 + l_1(y - \hat{x}_1), \dot{\hat{x}}_2 = x_3 + l_2(y - \hat{x}_1) \\ \dot{\hat{x}}_3 &= l_3(y - \hat{x}_1) \end{aligned} \tag{3}$$

where l_1, l_2 and l_3 are the gains of the ESO.

If G_s is somehow approximated by a second-order system model structure:

$$\ddot{y} = f(y, \dot{y}, t, \mu) + bu \tag{4}$$

and $\hat{x}_1$, $\hat{x}_2$, $\hat{x}_3$ will track y, $\dot{y}$ and f, respectively, if the ESO (3) is well tuned [4]. f is the generalized disturbance and is the comprehensive effect of external disturbances μ and internal dynamics. ADRC being model independent, the explicit mathematical characterization of f is not required in model (4).

A second-order ADRC generally uses a control law as follows:

$$u_0 = K_p(r_{sp} - \hat{x}_1) - K_d\hat{x}_2 \tag{5}$$

$$u = \frac{u_0 - \hat{x}_3}{b} \tag{6}$$

where K_p, K_d are the controller parameters. With $\hat{x}_3 \approx f$ obtained from the ESO, (6) reduces (4) to an approximate double integrator:

$$\ddot{y} = f + u_0 - \hat{x}_3 \approx u_0 \tag{7}$$

which can be easily controlled.

3.1 Optimal PID-ADRC Design

The actual system/plant is approximated by a local plant as:

$$\ddot{y} = f(y, \dot{y}, t, \mu) + u \tag{8}$$

The model (8) is only used for controller design purpose. The designed controller is then applied to the actual process/plant. The extended state-space model of (8) for

$$x = \begin{bmatrix} x_1 \ x_2 \ x_3 \end{bmatrix}^{\mathrm{T}} = \begin{bmatrix} y \ \dot{y} \ f \end{bmatrix}^{\mathrm{T}}$$

is given as

$$\begin{bmatrix} \dot{x}_1 \\ \dot{x}_2 \\ \dot{x}_3 \end{bmatrix} = \begin{bmatrix} 0 & 1 & 0 \\ 0 & 0 & 1 \\ 0 & 0 & 0 \end{bmatrix} \begin{bmatrix} x_1 \\ x_2 \\ x_3 \end{bmatrix} + \begin{bmatrix} 0 \\ 1 \\ 0 \end{bmatrix} u + \begin{bmatrix} 0 \\ 0 \\ 1 \end{bmatrix} \dot{f} \tag{9}$$

When the output y is available for feedback, there is no need to estimate it using the ESO. In that case, a second-order reduced ESO (RESO) (Tian 2007; Zheng et al. 2011, 2012) can be used to estimate the state variables x_2 and x_3 as

$$\begin{bmatrix} \dot{\hat{z}}_1 \\ \dot{\hat{z}}_2 \end{bmatrix} = \begin{bmatrix} -l_1 & 1 \\ -l_2 & 0 \end{bmatrix} \begin{bmatrix} \hat{z}_1 \\ \hat{z}_2 \end{bmatrix} + \begin{bmatrix} 1 \\ 0 \end{bmatrix} u + \begin{bmatrix} l_1 \\ l_2 \end{bmatrix} \dot{y}$$

$$f = \begin{bmatrix} 0 & 1 \end{bmatrix} \begin{bmatrix} \hat{z}_1 \\ \hat{z}_2 \end{bmatrix} \quad \text{where} \begin{bmatrix} \hat{z}_1 \ \hat{z}_2 \end{bmatrix}^{\mathrm{T}} = \begin{bmatrix} \dot{y} \ f \end{bmatrix}^{\mathrm{T}}$$

To avoid noise due to numerical differentiation in the second-order RESO, Teppa-Garran and Garcia [5] suggested the following;

$$\begin{bmatrix} \dot{\bar{z}}_1 \\ \dot{\bar{z}}_2 \end{bmatrix} = \begin{bmatrix} -l_1 & 1 \\ -l_2 & 0 \end{bmatrix} \begin{bmatrix} \bar{z}_1 \\ \bar{z}_2 \end{bmatrix} + \begin{bmatrix} 1 \\ 0 \end{bmatrix} u + \begin{bmatrix} l_2 - l_1^2 \\ -l_1 l_2 \end{bmatrix} y \tag{10}$$

with

$$\begin{bmatrix} \hat{z}_1 \\ \hat{z}_2 \end{bmatrix} = \begin{bmatrix} 1 & 0 \\ 0 & 1 \end{bmatrix} \begin{bmatrix} \bar{z}_1 \\ \bar{z}_2 \end{bmatrix} + \begin{bmatrix} l_1 \\ l_2 \end{bmatrix} y.$$

Taking the control signal of the PID type

$$u_0 = K_i \int e\mathrm{d}t + K_p e + K_d \dot{e}$$

where $e = r_{sp} - y$ is the tracking error.

In the case, when the reference r_{sp} is set, (??) can be written as

$$\dddot{e} = -\dddot{y} \tag{11}$$

Putting $u = u_0 - \hat{z}_2$ in (8) gives

$$\ddot{y} = u_0 \tag{12}$$

Taking the derivative of (12) and using (11) gives

$$\dddot{e} = -\dot{u}_0 \tag{13}$$

Let $b_1 = e$, $b_2 = \dot{e}$ and $b_3 = \ddot{e}$ then (13) can be written in state-space as

$$\dot{\mathbf{b}} = \mathbf{F}\mathbf{b} + \mathbf{G}\dot{u}_0 = \begin{bmatrix} 0 & 1 & 0 \\ 0 & 0 & 1 \\ 0 & 0 & 0 \end{bmatrix} \mathbf{b} + \begin{bmatrix} 0 \\ 0 \\ -1 \end{bmatrix} \dot{u}_0 \tag{14}$$

To get an LQR formulation for the ADRC problem, the quadratic cost considered is given as [5];

$$J = \int_0^\infty (\mathbf{b}^{\mathrm{T}}\mathbf{Q}\mathbf{b} + \rho \dot{u}_0^2)\mathrm{d}t \tag{15}$$

where $\mathbf{Q}$ is a positive semi-definite matrix and $\rho > 0$. The minimization of (15) yields the state feedback control given as

$$\dot{u}_0 = -\mathbf{K}\mathbf{b} = -K_1 e - K_2 \dot{e} - K_3 \ddot{e} \tag{16}$$

where $\mathbf{K} = \rho^{-1}\mathbf{G}^{\mathrm{T}}\mathbf{P}$. $\mathbf{P}$ is the symmetric positive definite solution of the continuous algebraic Riccati equation (CARE) given by

$$\mathbf{F}^{\mathrm{T}}\mathbf{P} + \mathbf{P}\mathbf{F} + \mathbf{Q} - \rho^{-1}\mathbf{P}\mathbf{G}\mathbf{G}^{\mathrm{T}}\mathbf{P} = 0 \tag{17}$$

Integrating (16) on both sides results in the optimal PID-ADRC

$$u_0 = K_i \int e + K_p e + K_d \dot{e} \tag{18}$$

with

$$\begin{bmatrix} K_i & K_p & K_d \end{bmatrix} = \begin{bmatrix} -K_1 & -K_2 & -K_3 \end{bmatrix}$$

The expressions for K_i, K_p and K_d are obtained as: Taking $\mathbf{Q} = \text{diag}(q_{11}\ q_{22}\ q_{33})$, the symmetric matrix

$$\mathbf{P} = \begin{bmatrix} P_{11} & P_{12} & P_{13} \\ P_{21} & P_{22} & P_{23} \\ P_{31} & P_{32} & P_{33} \end{bmatrix}$$

and substituting in (17), we get the positive definite solution given as [5]

$$P_{33} = \rho(\lambda + 2)\zeta\omega_n,\ P_{23} = \rho(2\lambda\zeta^2 + 1)\omega_n^2,\ P_{13} = \rho\lambda\zeta\omega_n^3 \quad (19)$$

where $\lambda \gg \zeta\omega_n$ for guaranteed dominant pole placement (GDPP).

Also, we have

$$P_{11} = \frac{P_{13}P_{23}}{\rho},\ P_{12} = \frac{P_{13}P_{33}}{\rho},\ P_{22} = \frac{P_{23}P_{33}}{\rho} - P_{13} \quad (20)$$

Also, the optimal state feedback gain is given as

$$\mathbf{K} = \rho^{-1}\mathbf{G}^T\mathbf{P} = \begin{bmatrix} \frac{-P_{13}}{\rho} & \frac{-P_{23}}{\rho} & \frac{-P_{33}}{\rho} \end{bmatrix} \text{i.e.}$$

$$\begin{bmatrix} K_i & K_p & K_d \end{bmatrix} = \begin{bmatrix} \frac{P_{13}}{\rho} & \frac{P_{23}}{\rho} & \frac{P_{33}}{\rho} \end{bmatrix}$$

4 Results

In this section, simulated results of linear ADRC applied to a two-link arm manipulator are shown. The controller is tested for two desired trajectories shown in the simulated results. Simulated results are also shown in the presence of noise and external disturbances. The simulated results of link 1 and link 2 for the trajectory 1 are shown in Figs. 3 and 4, respectively.

The simulated results of link 1 and link 2 for the trajectory 2 are shown in Figs. 5 and 6, respectively.

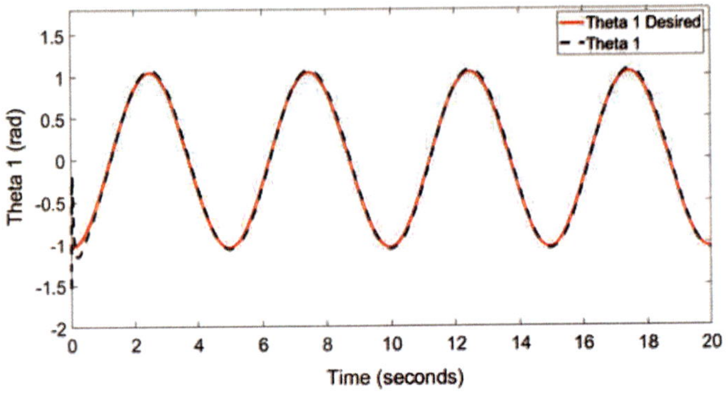

Fig. 3 Trajectory tracking of link 1

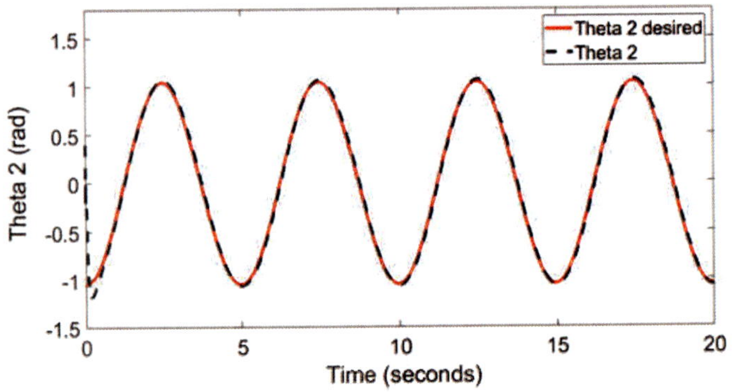

Fig. 4 Trajectory tracking of link 2

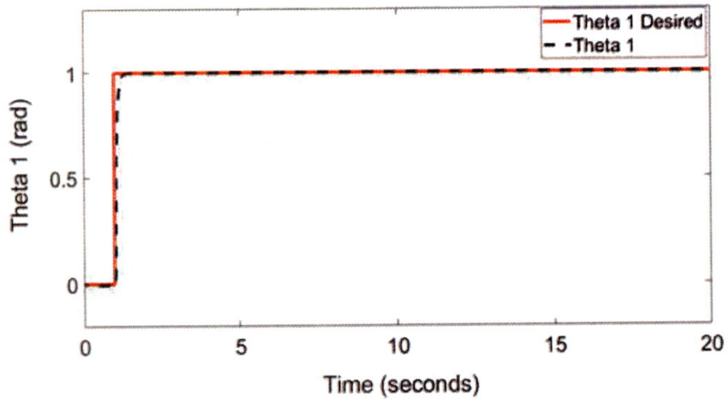

Fig. 5 Trajectory tracking of link 1

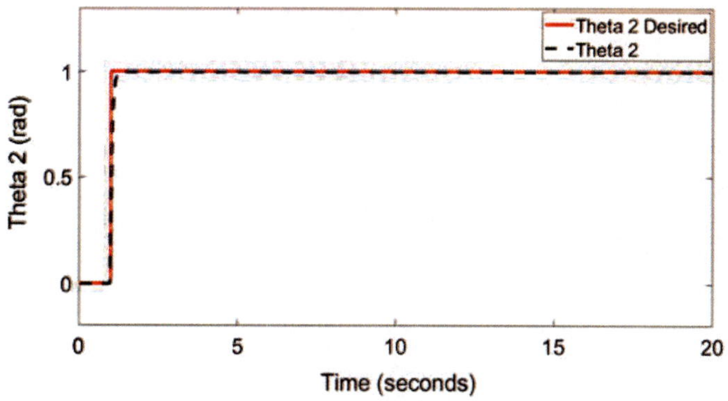

Fig. 6 Trajectory tracking of link 2

4.1 Simulated Results in the Presence of Noise and Disturbances

Since the process and measurement noise can be present in the system, a band-limited white noise signal with noise power = 0.00001 and sample time = 0.01 is added input. A manipulator system is generally prone to external disturbances like change in load, parameter variation, etc. To simulate these disturbances, an external signal is applied. The simulated results for both the links for trajectory 1 in the presence of noise and external disturbances are shown in Figs. 7 and 8.

The simulated results for both the links for trajectory 2 in the presence of noise and external disturbances are shown in Figs. 9 and 10.

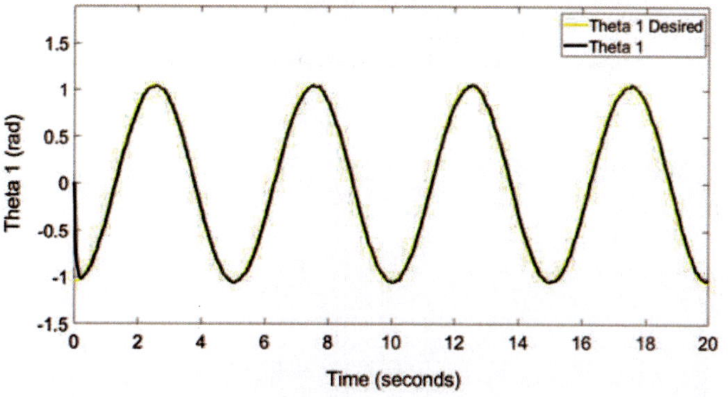

Fig. 7 θ_1 in the presence of noise and external disturbances

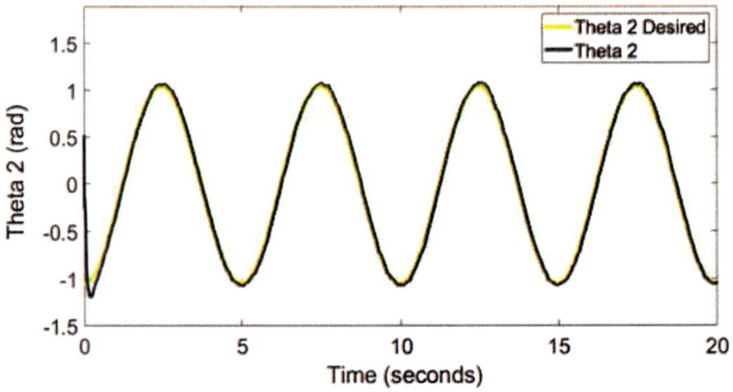

Fig. 8 θ_2 in the presence of noise and external disturbances

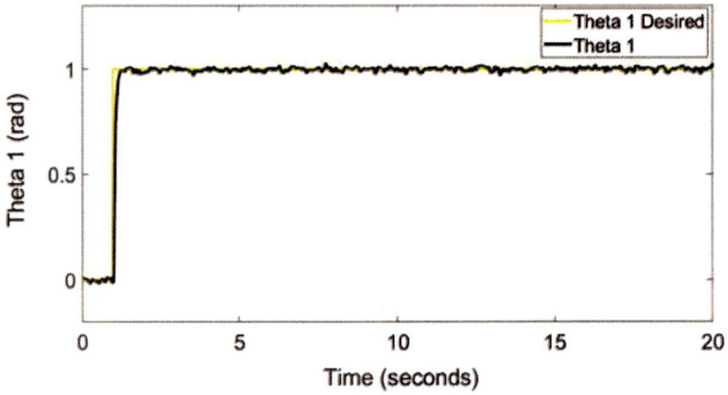

Fig. 9 θ_1 in the presence of noise and external disturbances

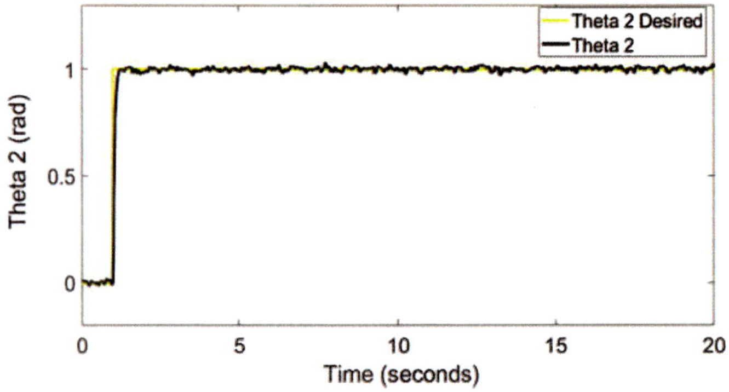

Fig. 10 θ_2 in the presence of noise and external disturbances

5 Conclusions

This papers presents the implementation of ADRC control strategy to a nonlinear, coupled and MIMO system. In order to understand the behavior of physical system, MATLAB simulation has been carried out. Simulation results show that ADRC can be a very effective tool. A second-order linear ADRC with a reduced-order observer was used in this paper. ADRC is able to track two different trajectories without changing the tuning parameters for either desired trajectory. To emphasize the effectiveness of control strategy, the system was subject to noise and different external disturbances. Although under noise, the system response is somewhat debatable, ADRC still effectively tracks the reference value. Simulation results under external disturbances show that ADRC has a good anti-disturbance rejection capability.

References

1. Han JQ (1998) Auto-disturbance-rejection controller and it's applications. Control Decis 13(1):19–23
2. Gao ZQ (2003) Scaling and bandwidth-parameterization based controller tuning. In: Proceedings of the 2003 American control conference, 2003, pp 4989–4996
3. Han JQ (2009) From PID to active disturbance rejection control. IEEE Trans Ind Electron 56(3):1–7
4. Chen X, Li D, Gao Z, Wang C (2011) Tuning method for second-order active disturbance rejection controller. In: Proceedings of the 30th Chinese control conference, July 22–24, 2011, Yantai, China
5. Teppa-Garran P, Garcia G (2013) Optimal tuning of PI/PID/PID (n-1) controllers in active disturbance rejection control. CEAI 15(4):26–36
6. Sun B, Gao Z (2005) A DSP-based active disturbance rejection control design for a 1-kW H-bridge DC-DC power converter. IEEE Trans Ind Electron 52:1271–1277
7. Su YX, Zheng CH, Duan BY (2005) Automatic disturbances rejection controller for precise motion control of permanent-magnet synchronous motors. IEEE Trans Ind Electron 52:814–823
8. Vincent J, Morris D, Usher N, Gao Z, Zhao S, Nicoletti A, Zheng Q (2013) On active disturbance rejection based control design for superconducting RF cavities. Nuclear Instrum Methods Phys Res A 643:11–16. (ADRC) as a control tool for practitioners. Electronics 2:246–279
9. Huang C, Li J, Mu S, Yan H (2017) Linear active disturbance rejection control approach for load frequency control of two-area interconnected power system. Trans Inst Meas Control
10. Ming C, Sun R, Wang X (2018) Velocity control based on active disturbance rejection for air-breathing supersonic vehicles. Complexity 2018:11 p

Reproducible Analysis and Intelligent Scientific Criteria in Engineering Papers Classification Using Data Science

Linga Reddy Nagulapalli, Syed Umar, Tariku Birhanu Yadesa, Tadele Debisa Deressa, and Kamal Mohammed Jimalo

Abstract In this paper, we discuss, any work must be reproducible in order to influence research and contribute to our profession's knowledge. Nevertheless, studies show that 70% of university laboratory work cannot reproduce. Reproducible work with not always available data sets or procedures not clearly specified is uncommon in software engineering and complex specifications engineering. The lack of reproducible research prevents development, which means that researchers must replicate a scratch study. The RE researcher will read conference papers, find empirical articles and then review data that can (if available) be replicated in the empirical report. This paper deals with two aspects of the problem and discusses in RE articles, RE documents and theoretical documents. In learning and development of an automatic classification in RE and empirical document identification, recent conference papers and RE documents have been utilized. They use the ERRC approach for performing supervised lecture classifications using natural language therapy and machine learning. Our software is equated with a fundamental keyword approach. We study the paper collections from the engineering conference sponsored by IEEE on Requirements and the International Software Testing and Evaluation Symposium sponsored by IEEE in order to test our process. In all but a few cases, we examine that the procedure of ERRC worked superior than the reference procedure.

Keywords Machine learning · Information retrieval · Empirical research · Statistical analysis · Reproducible research · Supervised classification learning text classification · Requirements engineering

L. R. Nagulapalli (✉)
Department of Computer Science & Engineering, Himalayan University, Itanagar, India
e-mail: nagulapalli.lingareddy@gmail.com

S. Umar · T. B. Yadesa · T. D. Deressa
Department of Computer Science, Wollega University, Nekemte, Ethiopia
e-mail: umar332@gmail.com

K. M. Jimalo
Department of Information Technology, Wollega University, Nekemte, Ethiopia

R. P. Mahapatra et al. (eds.), *Proceedings of 6th International Conference on Recent Trends in Computing*, Lecture Notes in Networks and Systems 177,
https://doi.org/10.1007/978-981-33-4501-0_63

1 Introduction

Innovative work is important in the field of engineering for the development of modern, faster and better technology. Although work is being performed on the "greenfield," reproductive research declines regardless of whether the reprocessed sample is a greenfield or not. Evidences have shown that up to 70% of academic research cannot be replicated, which represents tremendous waste of money and time [1]. There is no accurate way to determine whether the analysis is true without replication of work. A research system called Trace Lab [3] has recently been developed by the National Science Foundation. Trace Lab is intended to "provide researchers with the ability to design and perform experiments" [3]. Trace Lab lets scientists reproduce studies easily, but initial of all we should know whether the research can be even reproduced in a search article. Although our research is primarily aimed at determining quickly whether a book is repeated in an experiment or test, for this reason this paper answers previous questions. The initial step in our entire procedure is to decide if a paper is suitable for requirement engineering (RE), since we concentrate on demand replication of engineering. Initially, we will see if the empirical paper is RE. We describe a paper that examines the observations and draws conclusions based on the phenomenon observed and measured.

They tend to use the procedures of natural language processing (NLP). After we have extracted and checked the text of the study paper, we use two procedures to determine the characteristics. The initial solution is a norm of certain key words. A supervised learning process is the other form, the empirical research classifier (ERRC). All procedures were introduced as components of Trace Lab.

We also developed a training course for several years with labeling various one not. Still, for each model we employed the classification procedures: When evaluating the ERRC and reference processes, we have utilized accuracy, extraction and f measurement and calculation precision.

2 Procedureology for Reproducible Empirical Requirements

Figure 1 clearly illustrates paperwork automatically for reproducible empirical requirements. The shaded blocks are within this text's reach. We have created the ERRC, a tool to recognize empirical engineering records, to achieve our goal. The figure shows that each conference is initially made up of a directory each year and the file is saved for each article. That paper has been manually graded to promote model creation as a non-ERornon-empirical RE. That paper was initially scanned. The newline characters and return characters were then deleted. For example, all phrases spanning several lines can be read in their entirety. Preprocessing took place: All but apostrophes and dashboards have been replaced with spaces to improve utilized a filter to exclude words from delay. Word stops are common words with

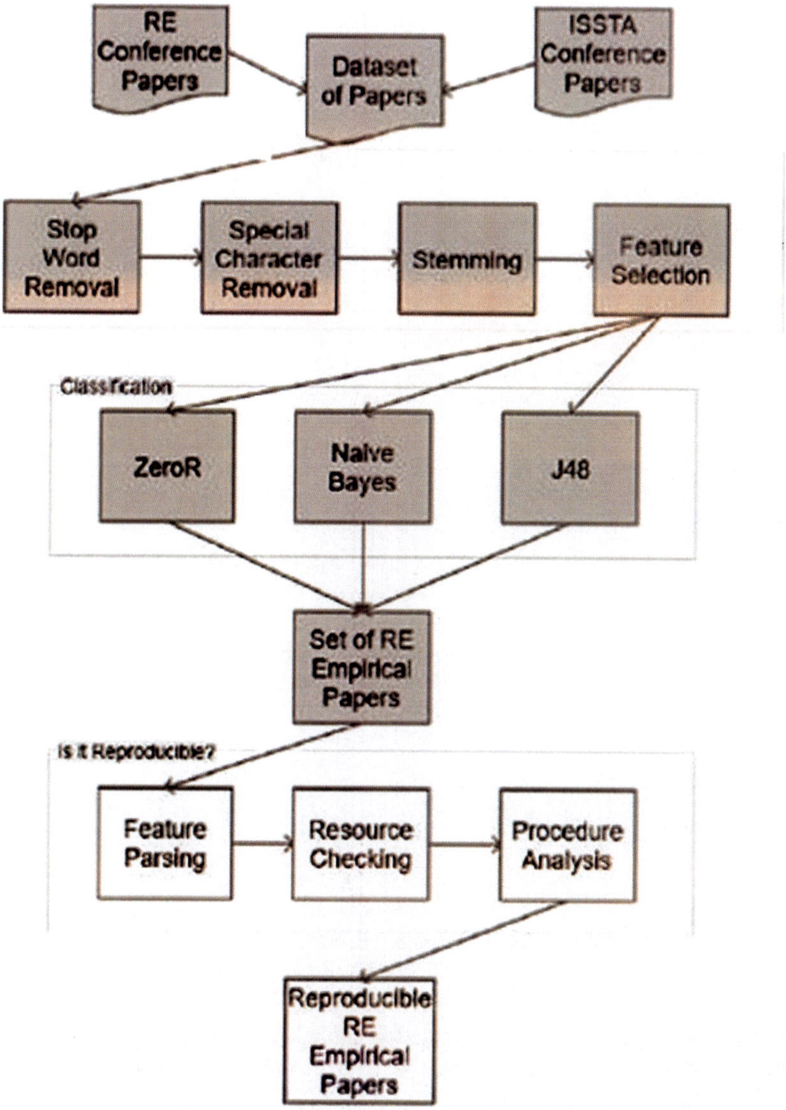

Fig. 1 Description of the high-level approach to recognition of engineering papers of empirical requirements for reproducible

no context (e.g., "the,") eventually removed from the document. After reading the text, we collected a trunk data e-mailing word (e.g., "need," "need," "required") and included in the trunks list (in this case, if trunk number was already established). Stemming words to their morphological origins increase the likelihood that words or phrases suit. Because collection of data is complete, and we have utilized two various selection procedures for classification models.

3 Basic Approach

The model creation guide approach uses a simple term frequency count. In previous RE and theoretical papers, we have established key words. In total, more than 200 publications have been included in the top five key words utilized in the latest RE and ISSTA conferences. Key words selected were then utilized to construct the basic model.

4 ERRC Procedure

The ERRC approach for paper assessment is more general. The ERRC protocol indicates the frequency of all the stemmed word utilized in the text, compared with the basic process, which uses only the key terms. When a complete list was made, the conditions were sorted from the most common to the least. The top ten words were then listed as the only features of the text. The outcome that could be converted into a classification method.

5 Model Analysis

Weka's results identified the efficacy of model strategies and ERRC approaches. For every system type, method, zero, Naive Bayes and J48 were applied. The models have been tested with validation crosses 10, 20, 30 and 40 subsets during cross-validation. Just k will be utilized. The benefit of this procedure is that every element in the training set is set once and once exactly. Though cross-validation was performed randomly with another seed ten times, we utilized the average results in the analyses of all races.

6 Empirical Studies

We conducted a case study to find out whether empirical problems of research were discussed. RQ1: Could NLP features be utilized for characterization of analytical and/or RE documents? RQ2: Could ERRC predict paper classifications better than the basic procedure? Determine whether a model for RE empirical documents can be established or not. In response to this question, we must move our long-term research project to determine if documents can be identified as reproducible.

H0: AERRC = AB and a various hypothesis of H1: AERRC > AB, where A is process accuracy, procedure and B the basis of a system.

7 Objects of Analytical Purposes

For practical purposes, IEEE Requirement Technology Evaluation Symposium has chosen conference papers. The conference ensures that research in the area of requirements engineering is based on papers. The ISSTA conference ensures that work in engineering is not needed. Such two conferences were chosen.

The description of the materials utilized is provided in Table 1. We decided to separate the years 2000, 2005 and 2015 in the final years of the assembly, for the same reason and for ISSTA. The articles ranged from 5 to 10 pages, most of them 10 pages. When we obtained a suitable subset of publications, we categorized all papers manually. One of the that categorized or did not classify them as RE whether they were empirical or not. Remember, documents are balanced, as shown in the table below.

The independent and dependent variables of our analysis are explained in this section.

7.1 *Variables Independent*

We had two various variables available. Additionally, we have updated the procedures of choosing applications using the approach of standard fundamental structure as to the ERRC system evaluation. Furthermore, in our work we utilized three ranking strategies: utilized the weak data mining software [6] to implement these procedures.

Table 1 Conference papers variables and behaviors

Year	Empirical	Non-Empirical	RE	Non-RE	Total
IEEE RE					
2000	7	6	12	1	13
2005	21	23	41	3	44
2015	29	18	43	4	47
IEEE ISSTA					
2000	12	9	1	20	21
2004	11	17	1	27	28
2015	21	21	0	42	42
Total	101 (52%)	94 (48%)	98 (50%)	97 (50%)	195 (100%)

Zero R is one of the easiest procedures of classification. Based on the data goal alone, there are no predictors. It cannot predict anything other than the principal class, but it is useful to obtain a comparative result with other methods. The classification is based on various theories for classification predictions. J48 is an open-source Apache C4.5 algorithm implementation. It creates decision-making trees from the training session.

7.2 Dependent Variables

As dependent variables, we selected precision, recall, accuracy and calculation.

Precision is tested for the proportion documents. Test the found ZeroR documents are RE documents?) are important.

On the other side, remember to calculate the actual percentage of instance found (Did ZeroR find all the RE documents?) and F-accuracy tool measurement.

8 Procedure of Evaluation

We have utilized the Apache Lucent and Solar libraries [12] to build and select the system. These Java-based libraries allow text to be indexed, scanned, evaluated and tokenized. These libraries are utilized to delete stop words, to eliminate other characters and numbers, to stop words and to count their frequencies. To speed up the development and evaluation process and allow others to easily replicate our experiment, we have implemented the study as a number of components in Trace Lab.

9 Authenticity Risks Data Sets Were the Main Threat to External Authenticity

Some sets of data can be larger or have various frequencies of terms. A larger data set creates various classification types.

We have also reduced the risk by having a model that uses two various materials in each meeting area, using papers worth many years, variations and gaps. We cannot permit the extension of our findings to other data sets. Naming the training system by eye was the principal danger to corporate reputation. For this chance, we had a co-author of the labeling. Such marks were then reviewed by a co-author. All the co-authors were separate. According to an agreement, the co-authors addressed contradictory issues. We have utilized this and validated procedures for research (i.e., Weka). The primary concern for building validity was the dependent variables utilized to answer questions. In order to combat this hazard, we have employed normal and well agreed exact reminder, accurate and f-measurement steps. To minimize the risk of uncertainty, we have evaluated the findings statistically.

10 Conclusion

The number of non-repeated studies has also risen research also with number publications. This even measured whether paper is RE the reproductively of academic research paper is an analysis paper. We also gathered information and data for training sets from the IEEE Requirement Software IEEE International on research in this context. We have developed a key word-based basic procedure, and our ERRC procedure has employed numerous typing methods to model academic research. Our results show that than that of the basic procedure for empirical paper classification and 12% better in the RE paper classification.

References

1. Maalej W, Nabil H (2015) Bug report, feature request, or simply praise? On Myself categorize app commentary. In: 2015 IEEE 23rd international needfulness engineering convention (RE), IEEE
2. Hall et al (2009) The WEKA referred fossorial software: an up-to-date. ACM SIGKDD Reconnoiter News Letter 11(1):10–18
3. Lobo LMRJ, Aher SB (2012) Comparative study of classification algorithms. Int J Information Technol 5(2):239–243
4. Cahoy E. What is empirical research? Penn State University Libraries. https://psu.libguides.com/emp
5. About Trace Lab, COEST. https://www.coest.org/index.php/tracelab/about-tracelab
6. Apache Lucent. https://lucene.apache.org/
7. Sayad S. "ZeroR" An introduction to data mining. https://www.saedsayad.com/zeror.htm
8. Sayad S. "Naive Bayesian" An introduction to data mining. https://www.saedsayad.com/naive_bayesian.htm
9. Schneider J. Cross validation. https://www.cs.cmu.edu/~schneide/tut5/node42.html

Effect of Beta-Frequency Binaural Beats on Cognitive Control in Healthy Adults

Ritika Mahajan, Ronnie V. Daniel, Akash K. Rao, Vishal Pandey, Rishi Pal Chauhan, and Sushil Chandra

Abstract Cognitive control encompasses some core cognitive qualities like taking the appropriate time to think before acting on a decision, efficiently dealing with novel and unanticipated challenges, resisting temptations, and staying focused. Response inhibition, conflict control, working memory, and cognitive flexibility are some of the sub-cognitive processes which have a detrimental effect on the cognitive control of an individual. There has been a considerable amount of research on different methods/interventions through which cognitive control could be enhanced through time. But the effect of beta-frequency binaural beats on cognitive control is lacking and much needed in the literature. In this research, our main objective is to evaluate repercussions of beta range binaural beats on cognitive control of healthy adults. Subjects were divided into two groups— experimental and sham. Individuals in the experimental condition were exposed to beta-frequency binaural beats (25 Hz) for 20 min. Both the groups then executed a computerized version of the Stroop task. Behavioral measures (Accuracy, reaction time) and various event-related potentials (ERP) features like P300 and N400 were extracted and considered for analysis. Data analysis revealed that the participants in the experimental group recorded higher accuracy and slower reaction times in the Stroop task compared to the sham condition. The behavioral results were also complemented by higher averaged P300 amplitudes and lower N400 amplitudes recorded for the experimental condition compared to the sham condition. We intend to use these results as substantial evidence to imply beta-frequency binaural beats as an effective means of enhancement of cognitive control in healthy adults over sham controlled, longitudinal studies.

Keywords Cognitive control · Binaural beats · Event-related potentials

---

R. Mahajan (✉) · R. P. Chauhan
National Institute of Technology Kurukshetra, Kurukshetra, India
e-mail: ritikamahajan2202@gmail.com

R. V. Daniel · V. Pandey · S. Chandra
Institute of Nuclear Medicine and Allied Sciences, Delhi, India

A. K. Rao
Indian Institute of Technology Mandi, Mandi, India

R. P. Mahapatra et al. (eds.), *Proceedings of 6th International Conference on Recent Trends in Computing*, Lecture Notes in Networks and Systems 177,
https://doi.org/10.1007/978-981-33-4501-0_64

1 Introduction

The effect of cognitive control is what makes the cognitive system of human beings far different from others [1]. It comprises of certain capabilities like goal maintenance, resource allocation, and conflict control. Stroop task is one of the widely used methods to evaluate cognitive control and cognitive interference [1]. In color-word version of Stroop task (CWST), the participants are presented with words written with dissimilar colors which are directed to make a riposte to the colors swiftly and efficiently while disregarding words in itself [1]. In the conventional CSWT, the incongruent condition wherein both word and color are different (e.g., "red" written with "yellow" color), which requires conflict resolution and cognitive control process [2]. Both these processes cause delay in response time in comparison to their congruent version, wherein the word and color are same [2]. This lag in the response time in case of incongruent stimuli as compared to the congruent stimuli is known as the Stroop effect [2].

Research investigating cognitive control with the Stroop task through different modalities like fMRI, PET, and EEG has obtained significant activation in the anterior cingulate cortex (ACC) while executing the tasks [2–5]. Early studies suggest that ACC is primarily responsible for conflict detection, monitoring, and resolution that encompasses cognitive control [6, 7]. According to earlier EEG studies, N400 component of ERP has been found to be manipulated by congruency effect in CSWT. This N400 component is a non-negative component occurring approximately 280 to 580 ms after presentation of the stimulus with more negative amplitude in incongruent condition. N400 was more localized over frontal and left centro-parietal region of human brain [8, 9]. The source reconstruction methods of electroencephalogram point the origin of N400 in ACC [10]. However, if N400 is related to the conflict detection or resolution processes, then it still holds debate [9]. EEG studies carried out so far failed to comment on the actual mechanism underlying conflict resolution or detection involved with N400 [11–14], but concludes to a point that N400 is responsible for conflict resolution and detection.

Recently, a new form of meditation, binaural beats has been recognized. With its origin in 1839 [15], its use as an intervention was not explored till twentieth century. Binaural beats are a form of sound wave fluctuations perceived as oscillating beats. When two sine waves with minute difference in frequency are presented to either ear, beats perceived are called binaural beat. For example, if a pure tone of 300 Hz is given to the right ear and a pure tone of 320 Hz is given to the left ear, then binaural beat of 310 Hz frequency modulating with an amplitude10 Hz is perceived [16]. Unmixed tones from 210 Hz to less than 1000 Hz are proven to be highly effective in producing these beats. Also, binaural beats have been found to be maximally effective around 400 Hz. The nerve signals from nuclei of superior and medial olivary colliculus seemed to be involved in binaural beat perception [17].

One of the studies [18] with binaural beats found that when theta beats with pink noise was presented for twenty minutes, there was an increase in relaxation scores of individuals in comparison to those presented with only pink noise. Thus, theta beats introduced a state of relaxation. When binaural beats were tested on military personnel [19], its potential in lowering stress was observed in both work as well as in home environment. Here, the military personnel when presented with delta and theta beats experienced decline in stress. Reedijk et al. [20] observed that with gamma beats, there was a reduction in attentional blink [21], implying higher cognitive flexibility. Hence, high-frequency binaural beats can an improve individual's performance in task related to memory retrieval. Thus, binaural beats have the potential to cause cognitive enhancement. But the behavioral and cognitive effects of using beta-frequency binaural beats for enhancing cognitive control have not been documented and hence are much needed in the literature.

In this experiment, a study was conducted to evaluate repercussions of beta-frequency binaural beats on cognitive control of healthy adults. The study hypothesized that the participants who underwent the beta-frequency binaural beats intervention would perform relatively better in the Stroop task compared to the control group (i.e., the participants who did not undergo any intervention).

2 Materials and Methods

2.1 *Participants*

A total of 40 male participants (20.5 $\pm$ 1.4) took part in the experiment. The participants were grouped randomly to either control or experimental group. Both the control and experimental group executed the standardized, computerized version of Stroop task. The participants were reported to be dextro-manual with normal or corrected to normal vision. No participant had any history of neurological diseases or was under the influence of any medications. Participants had provided both verbal and written informed consent participate in the experiment. All the participants were from either one of science, technology, engineering, or mathematics background.

2.2 *Experiment Design*

The participants were assigned randomly to control or experimental group. The experimental group took the standardized version of the Stroop test designed in Open Sesame [22] twice, both before and after the single-session binaural beat intervention. The experimental group was exposed to bb in the frequency of 25 Hz (250 and 275 Hz for left and right ear, respectively). Binaural beats were designed using the audacity speech recording and editing software.

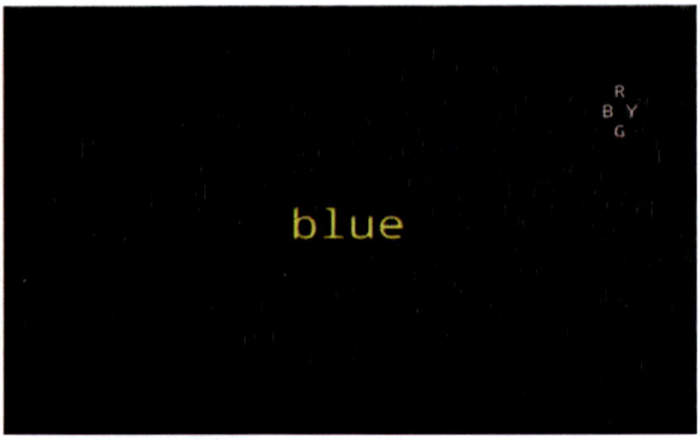

Fig. 1 Computerized version of the Stroop task

2.3 *Procedure*

The individuals were seated in a comfortable chair at a distance of approximated 1 m from the stimulus screen. The stimuli were displayed on a 24-inch LCD screen. The words were printed with Times New Roman font of size 70-point size. As shown in Fig. 1, the individuals were requested to respond to the color of the word irrespective of the word being displayed. The participants responded to the color red by pressing the UP arrow key, the color blue with the LEFT arrow key, the color green with DOWN arrow key, and to the color yellow with the RIGHT arrow key. These instructions were also displayed on along with the stimuli on the upper right corner of the screen. The words were displayed on a black background. A total of 200, excluding the 10 practice trials, trials were presented per participant; out of which 80 were congruent, and 120 were incongruent in randomized order. The Stroop task consisted of a fixation point for 250 ms at the center of screen, followed by the stimulus for 2000 ms or till a response is made. Once the participant has responded or the timeout has reached, a black screen was shown for 1000 ms followed by the next trial. During the initial briefing, the participants were instructed to respond to the stimuli as swiftly and efficiently as possible.

2.4 *EEG Acquisition*

We acquired electroencephalographic data using Ag/AgCl electrodes from sixty-four channel location conforming to 10–20 electrode placement system. The recorded data was sampled continuously at 2048 Hz with an input impedance below 5 KΩ. The data was acquired using an eego™ mylab EEG system (ANT Neuro, Enschede, Netherlands). Figure 2 shows the EEG data acquisition of a subject while executing the Stroop task.

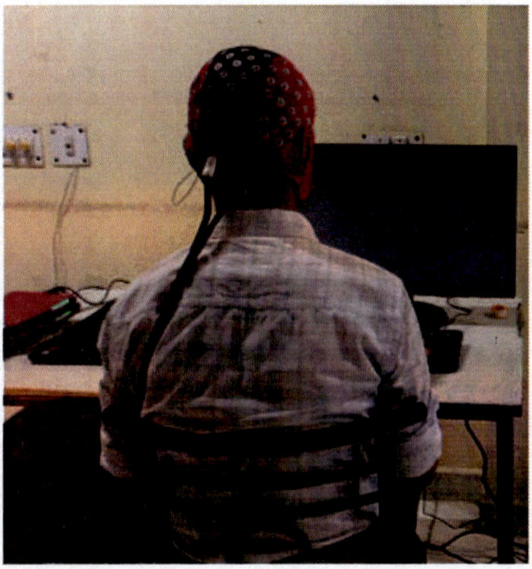

Fig. 2 EEG data acquisition of a participant while executing the Stroop task

2.5 *Analysis of EEG Data*

The offline EEG analysis was performed using custom scripts using MATLAB 2015b (MathWorks). The signals were zero-phase filtered using FIR bandpass filter with cut-off frequencies at 0.1 and 45 Hz. Those segments with high amplitude and variance were marked to be removed. The data was then re-referenced to the average electrode. Single-trial EEG segments (−200 to 1000 ms) were then epoched from the continuous data. Baseline corrections were then performed using the 200 ms before the stimulus onset. The single-trial EEG epochs were then averaged to form the ERP waveforms. The ERP operation can mathematically be represented as:

$$\overline{x}(t) = \frac{1}{N_T} \sum_{k=1}^{N_T} x(t, k) \tag{1}$$

where k is the trial number, and t is the time that has transpired after k-th trial, N_T is the number of trials, and $x(t, k)$ is the recorded trial. The recorded trials can again be represented as the sum of the signal of interest s(t) and the noise $\eta(t, k)$ as:

$$x(t, k) = s(t) + \eta(t, k) \tag{2}$$

Henceforth, Eq. (1) can then be re-written as:

$$\overline{x}(t) = s(t) + \sum_{k=1}^{N_T} \eta(t, k) \tag{3}$$

3 Results

A one-way ANOVA was carried out to compare the main effects of high frequency, single-session binaural beats on the different behavioral, and ERP measures mentioned above.

3.1 *Behavioral Measures*

As shown in Fig. 3a, the accuracy in the Stroop task (congruent condition) was high in experimental group than to the sham group (Experimental: $\mu = 98.33\%$, $\sigma = 1.35\%$ > Control: $\mu = 97.91\%$, $\sigma = 2.14\%$; $F\ (1, 31) = 0.46$, $p = 0.5$, $\eta_p{}^2 = 0.97$). But the difference did not yield statistical significance. Also, as shown in Fig. 3b, the response time in the Stroop task (congruent condition) was low in the experimental

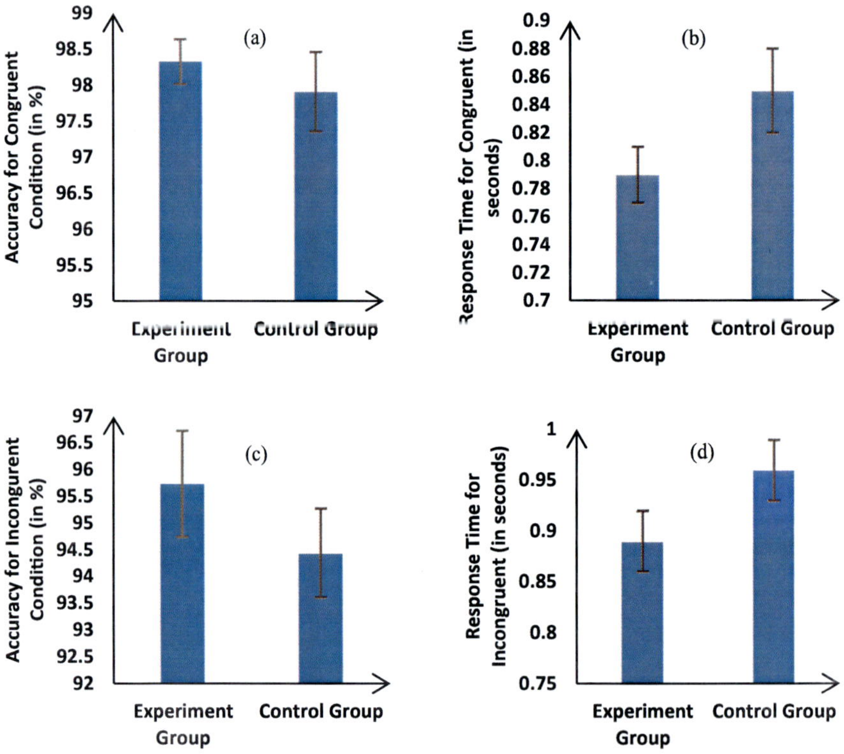

Fig. 3 The **a** The average accuracy in the congruent condition across both groups **b** The average response time in the congruent condition across both groups **c** The average accuracy in the incongruent condition across both groups **d** The average response time in the incongruent condition across both groups. The error bars show standard errors around point estimates

condition compared to the sham condition (Experimental: $\mu = 0.79$ s, $\sigma = 0.11$ s < Control: $\mu = 0.85$ s, $\sigma = 0.13$ s; F (1, 31) = 1.75, $p = 0.1$, $\eta_p{}^2 = 0.92$). This difference though did not yield statistical significance.

As shown in Fig. 3c, the accuracy in the Stroop task (incongruent condition) was higher in the experimental condition compared to sham condition (Experimental: $\mu = 95.74\%, \sigma = 4.22\% >$ Control: $\mu = 94.44\%, \sigma = 3.23\%; F\,(1, 31) = 0.94, p = 0.3$, $\eta_p{}^2 = 0.88$). But the difference did not reach statistical significance. However, the response time in Stroop task (incongruent condition) was significantly lower in the experimental condition compared to the sham condition (Experimental: $\mu = 0.89$ s, $\sigma = 0.13$ s < Control: $\mu = 0.96$ s, $\sigma = 0.15$ s; F (1, 31) = 5.44, $p < 0.05$, $\eta_p{}^2 = 0.95$) as shown in Fig. 3d.

3.2 ERP Measures

The effect of high-frequency binaural beats on P300 and N400 amplitudes extracted out of the ERP data was also analyzed. Table 1 shows the differences between the P300 amplitudes on different electrodes which reached statistical significance in the congruent condition. Similarly, Table 2 shows the differences between the N400 amplitudes on different electrodes which reached statistical significance in the incongruent condition. Figures 4, 5, 6, and 7 show the averaged ERPs from −200 ms (before stimulus) to 1000 ms (after stimulus) indicating the P300 and N400 amplitudes for all the significant electrodes in the congruent/incongruent conditions.

3.3 Correlation Analysis

A two-tailed Pearson correlation analysis was conducted to examine the strength of the relationship between different dependent variables taken as a part of the experiment.

Table 3 shows the correlations (only significant ones) obtained in the congruent condition, and Table 4 shows the correlations (only significant ones) obtained in the incongruent condition.

4 Discussion and Conclusion

The present work evaluated the ramifications of beta-frequency, single-session binaural beats on cognitive control in healthy adults. The participants were randomly divided into groups called experimental and sham (control). The experimental group undertook a 20-min intervention session of beta-frequency binaural beats before they executed the Stroop task; the control group directly executed the Stroop task without

Table 1 Means, standard deviations of the P300 amplitudes and the corresponding F-ratio, p-value and the effect size obtained for the significantly different electrodes in the congruent and incongruent condition

Electrodes	Mean and SD Experimental and control group		F-ratio (F)	p-value (p)	Effect size ($\eta_p{}^2$)
Congruent					
AF7	Exp	Control	5.06	$p < 0.05$	0.91
	$\mu = 1.78\ \mu V$ $\sigma = 0.88\ \mu V$	$\mu = 1.78\ \mu V$ $\sigma = 0.94\ \mu V$			
CP1	$\mu = 0.93\ \mu V$ $\sigma = 0.90\ \mu V$	$\mu = 0.39\ \mu V$ $\sigma = 0.56\ \mu V$	3.99	$p < 0.05$	0.94
CP2	$\mu = 1.31\ \mu V$ $\sigma = 0.82\ \mu V$	$\mu = 0.86\ \mu V$ $\sigma = 0.5\ \mu V$	3.41	$p < 0.05$	0.83
CP3	$\mu = 0.94\ \mu V$ $\sigma = 0.94\ \mu V$	$\mu = 0.44\ \mu V$ $\sigma = 0.59\ \mu V$	3.12	$p < 0.05$	0.79
Incongruent					
AF7	$\mu = 1.89\ \mu V$ $\sigma = 0.58\ \mu V$	$\mu = 1.34\ \mu V$ $\sigma = 1.12\ \mu V$	8.92	$p < 0.05$	0.61

any intervention. As hypothesized, the experimental group recorded relatively higher accuracy and relatively lower reaction times in the Stroop task compared to the control group as shown in Fig. 2. This difference, however, did not yield statistical significance. A statistically significant difference in the performance parameters of the experimental, and the control group would probably be reached if the beta-frequency binaural beats intervention was administered longitudinally (i.e., several instances of intervention) rather than single-session intervention.

As shown in Figs. 4, 5, 6, and 7, and Tables 1 and 2, the ERP measures like P300 and N400 also yielded significant results which validated our hypothesis. As shown in Figs. 4 and 6, the averaged P300 amplitude was significantly higher in experimental condition compared to the sham condition. These results were consistent with [10], where it was articulated that binaural beats intervention leads to a gradual increase in the excitatory potentials and a gradual decrease in the inhibitory potentials of the neurons, which ultimately leads to better synaptic information transmission across the brain. These effects were particularly evident in the frontal electrodes and the parietal electrodes, which, according to [5], are the premier decision-making and

Table 2 Means, standard deviations of the N400 amplitudes and the corresponding F-ratio, p-value and the effect size obtained for the significantly different electrodes in the congruent and incongruent condition

Electrodes	Mean and SD Experimental and control group		F-ratio (F)	p-value (p)	Effect size ($\eta_p{}^2$)
Congruent					
F4	Exp	Control	4.19	p < 0.05	0.82
	μ = −0.59 μV σ = 0.70 μV	μ = −1.22 μV σ = 1.04 μV			
F6	μ = −0.39 μV σ = 0.62 μV	μ = −0.93 μV σ = 0.73 μV	5.25	p < 0.05	0.91
TP8	μ = −0.97 μV σ = 1.07 μV	μ = −2.01 μV σ = 1.31 μV	6.27	p < 0.05	0.84
T8	μ = −1.05 μV σ = 1.22 μV	μ = −1.76 μV σ = 1.09 μV	3.08	p < 0.05	0.76
CP6	μ = −0.39 μV σ = 0.46 μV	μ = −0.88 μV σ = 0.64 μV	5.25	p < 0.05	0.44
Incongruent					
F5	μ = −0.41 μV σ = 0.73 μV	μ = −0.84 μV σ = 0.63 μV	3.1	p < 0.05	0.98
F6	μ = −0.40 μV σ = 0.69 μV	μ = −0.83 μV σ = 0.76 μV	2.78	p < 0.05	0.93

cognitive control centers of the brain. In addition, as shown in Figs. 5 and 7, the averaged N400 amplitudes were significantly lower in the experimental condition compared to the control condition. These results were consistent with [11], which had empirically and qualitatively shown that N400 was an involuntary neurological signature of decrease in cognitive interference and a gradual drop in N400 indicated the drop in cognitive interference, leading to better clarity in the decision-making centers of the brain. Our results were well complemented by the correlation analyses carried out between the behavioral and the cognitive descriptors (as shown in Table

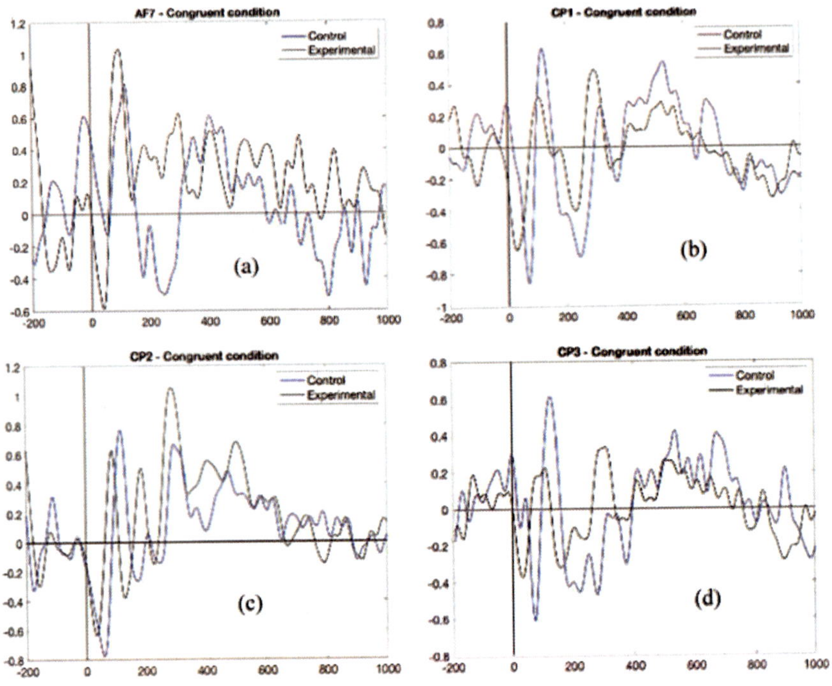

Fig. 4 Averaged ERPs from (−200 ms [before stimulus] to 1000 ms [after stimulus] for different electrodes indicating P300 amplitudes (~250 ms to ~ 350 ms) across both groups in the congruent condition: **a** For electrode AF7 **b** For electrode CP1 **c** For electrode CP2 **d** For electrode CP3 **e** For electrode F6 **f** For electrode condition across both groups. All the averaged differences in P300 amplitudes across these electrodes yielded statistical significance

3 and Table 4), with the averaged P300 amplitudes and the accuracy indicating a significant positive correlation and the averaged P300 amplitudes and the response time indicating a significant negative correlation.

In future, the authors intend to administer beta-frequency binaural beats as an intervention through sham controlled, longitudinal studies. Longitudinal studies, where the intervention is administered through several sessions, would possibly lead to significantly better performance at transfer. In addition, beta-frequency binaural beats can be a reliable intervention technique alongside other modern cortical modulation techniques like transcranial direct current stimulation (tDCS) and neurofeedback in personalized performance enhancement frameworks.

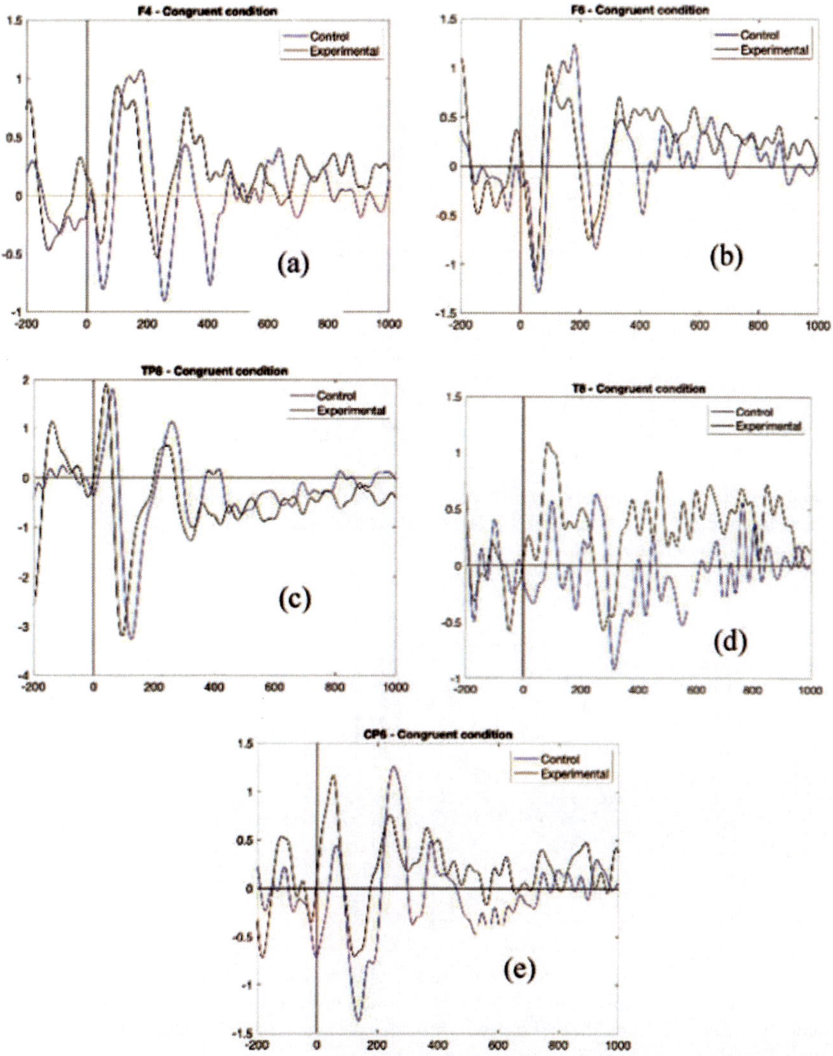

Fig. 5 Averaged ERPs from (-200 ms [before stimulus] to 1000 ms [after stimulus] for different electrodes indicating N400 amplitudes (~350 ms to ~ 475 ms) across both groups in the congruent condition: (a) For electrode F4 (b) For electrode F6 (c) For electrode TP8 (d) For electrode T8 (e) For electrode CP6. All the averaged differences in P300 amplitudes across these electrodes yielded statistical significance

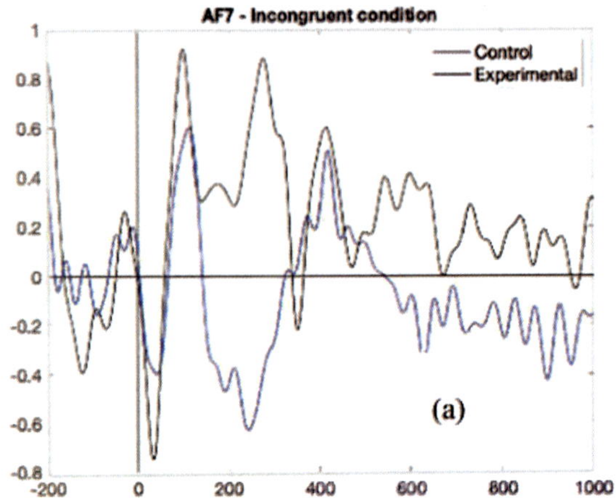

Fig. 6 Averaged ERPs from (-200 ms [before stimulus] to 1000 ms [after stimulus] for different electrodes indicating P300 amplitudes (~250 ms to ~ 350 ms) across both groups in the incongruent condition: (a) For electrode AF7

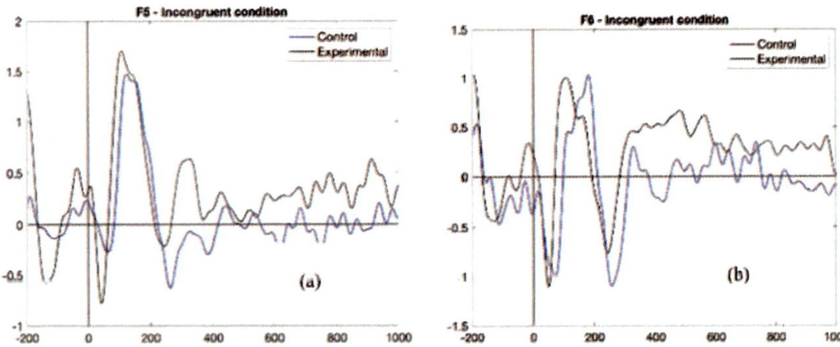

Fig. 7 Averaged ERPs from (-200 ms [before stimulus] to 1000 ms [after stimulus] for different electrodes indicating N400 amplitudes (~350 ms to ~ 475 ms) across both groups in the incongruent condition: (a) For electrode F5 (b) For electrode F6. All the averaged differences in N400 amplitudes across these electrodes yielded statistical significance

Table 3 Correlation between different dependent variables in the experiment (congruent condition)

Dependent variable 1	Dependent variable 2	Extent of correlation (r)	Significance
P300 amplitude—electrode AF7	Accuracy	0.67	$p < 0.05$
P300 amplitude—electrode AF7	Response time	−0.95	$p < 0.01$
P300 amplitude—electrode CP2	Response time	−0.55	$p < 0.05$
N400 amplitude—electrode F4	Response time	0.44	$p < 0.05$

Table 4 Correlation between different dependent variables in the experiment (incongruent condition)

Dependent variable 1	Dependent variable 2	Extent of correlation (r)	Significance
P300 amplitude — electrode CP5	Accuracy	0.84	$p < 0.01$

Acknowledgements The authors thank the Defense Research and Development Organization (DRDO) for supporting our research from the project titled "Vision Research in Cognitive Neuroscience" (ST-14/DIPR-734) to Dr. Sushil Chandra.

References

1. Stroop JR (1935) Studies of interference in serial verbal reactions. Journal of Experimental Psychology 18:643–662
2. Peterson BS et al (1999) An fMRI study of stroop word-color interference: evidence for cingulate subregions subserving multiple distributed attentional systems. Biol Psychiat 45(10):1237–1258
3. Taylor SF et al (1997) Isolation of Specific Interference Processing in the Stroop Task: PET Activation Studies. NeuroImage 6(2):81–92
4. Aine CJ, Harter MR (1984) Hemispheric Differences in Event-Related Potentials to Stroop Stimuli. Ann N Y Acad Sci 425(1):154–156
5. Markela-Lerenc J et al (2004) Prefrontal-cingulate activation during executive control: which comes first? Cognitive Brain Research 18(3):278–287
6. Melcher T, Gruber O (2009) Decomposing interference during Stroop performance into different conflict factors: An event-related fMRI study. Cortex 45(2):189–200
7. Peterson BS et al (2002) An event-related functional MRI study comparing interference effects in the Simon and Stroop tasks. Cognitive Brain Research 13(3):427–440
8. Appelbaum LG, Meyerhoff KL, Woldorff MG (2009) Priming and Backward Influences in the Human Brain: Processing Interactions during the Stroop Interference Effect. Cereb Cortex 19(11):2508–2521
9. Larson MJ, Kaufman DAS, Perlstein WM (2009) Neural time course of conflict adaptation effects on the Stroop task. Neuropsychologia 47(3):663–670
10. Liotti M et al (2000) An ERP study of the temporal course of the Stroop color-word interference effect. Neuropsychologia 38(5):701–711
11. Hanslmayr S et al (2007) The Electrophysiological Dynamics of Interference during the Stroop Task. Journal of Cognitive Neuroscience 20(2):215–225
12. West R (2003) Neural correlates of cognitive control and conflict detection in the Stroop and digit-location tasks. Neuropsychologia 41(8):1122–1135
13. Chen S, Melara RD (2009) Sequential effects in the Simon task: Conflict adaptation or feature integration? Brain Res 1297:89–100
14. West R et al (2005) Neural correlates of conflict processing. Exp Brain Res 167(1):38–48
15. Oster G (1973) Auditory beats in the brain. Sci Am 229(4):94–102
16. Licklider JCR, Webster JC, Hedlun JM (1950) On the frequency limits of binaural beats. J Acoust Soc Am 22:468–473
17. Wahbeh H et al (2007) Binaural Beat Technology in Humans: A Pilot Study to Assess Neuropsychologic, Physiologic, And Electroencephalographic Effects. the Journal of Alternative and Complementary Medicine 13(2):199–206

18. McConnell PA et al (2014) Auditory driving of the autonomic nervous system: Listening to theta-frequency binaural beats post-exercise increases parasympathetic activation and sympathetic withdrawal. Frontiers in Psychology 5:1248
19. Gantt MA et al (2017) The Effect of Binaural Beat Technology on the Cardiovascular Stress Response in Military Service Members With Postdeployment Stress. J Nurs Scholarsh 49(4):411–420
20. Reedijk SA et al (2015) Eliminating the Attentional Blink through Binaural Beats: A Case for Tailored Cognitive Enhancement. Frontiers in Psychiatry 6:82
21. Olivers CNL, Nieuwenhuis S (2006) The beneficial effects of additional task load, positive affect, and instruction on the attentional blink. J Exp Psychol Hum Percept Perform 32(2):364–379
22. Mathôt S, Schreij D, Theeuwes J (2012) OpenSesame: An open-source, graphical experiment builder for the social sciences. Behav Res Methods 44(2):314–324

Enterprise IoT Core Platform

Anil Joseph D'Costa, Siddhartha Sankar Biswas, Md Tabrez Nafis, M. Afshar Alam, and Sherin Zafar

Abstract The term IoT is used very liberally. As of now, there are no specific deployment guidelines laid down by TEC, Govt. of India. The Indian industry needs automation. However, they do not have a clear direction in which they should move ahead since different IoT solutions exist in the market. This paper is an attempt to prescribe a solution which will facilitate an integrated approach to a wholistic IoT network design. This all-encompassing solution will drive India towards a networked society by connecting all stakeholders together. Value engineering brought about by this approach will have a drastic improvement in efficiency and productivity for industry and society with the ultimate gain being in an increased GDP for the country.

Keywords Enterprise · LTE · IoT core · IoT broker · Subscriber service · Signaling service · Payload service · EPC

1 Introduction

The growth of wireless networks in India is poised to become a launch pad for enabling automation across Enterprises/Industries. The reliability of LTE or 5G (expected to be deployed in India during the year 2021 time frame) as a technology

A. J. D'Costa · S. S. Biswas (✉) · M. T. Nafis · M. A. Alam · S. Zafar
Department of Computer Science and Engineering, School of Engineering Sciences and Technology, Jamia Hamdard, New Delhi 110062, India
e-mail: ssbiswas1984@gmail.com

A. J. D'Costa
e-mail: anil_dcosta@yahoo.co.in

M. T. Nafis
e-mail: tabrez.nafis@jamiahamdard.ac.in

M. A. Alam
e-mail: aalam@jamiahamdard.ac.in

S. Zafar
e-mail: sherin.zafar@jamiahamdard.ac.in

R. P. Mahapatra et al. (eds.), *Proceedings of 6th International Conference on Recent Trends in Computing*, Lecture Notes in Networks and Systems 177,
https://doi.org/10.1007/978-981-33-4501-0_65

will force many enterprises to re-think their strategy in terms of automation, be it the traditional M2M or IIoT, in their quest to be a part of the Industry 4.0 revolution. This need will eventually give rise to private LTE networks in India, wherein more and more enterprises will bid for RF spectrum exclusively for their captive use. A typical example of this is Indian Railways which has been allotted 5 MHz bandwidth by WPC, Govt. of India in the cherished 700 MHz band based on recommendations from TRAI. This spectrum is essentially for a private LTE network and is targeted at public safety [1–3]. In addition to voice communication, this will enable automatic updation of public displays indicating status/arrival/departure of trains, security services, e.g., IoT-based video surveillance, asset monitoring, etc.

2 Enterprise Packet Core

A typical enterprise packet core network architecture which can be deployed as part of a private LTE network and its integration with the operator's network is as depicted in the Fig. 1.

This enterprise packet core can be built by infrastructure providers who can in turn lease out network capacity to enterprises (e.g., Automobile companies like Maruti Udyog Ltd, Tata Motors) for a guaranteed SLA/KPI. Alternatively, enterprises can build the core network themselves in cases where they have the necessary expertise [4–6]. The private EPC (Carrier Grade) can be hosted On premise or on a private cloud/public cloud and can connect to the operator's EPC through the access network.

This virtualized solution should contain 3GPP compliant network functions that can connect with WiFi (unlicensed band) and with LTE, 5G (licensed band) networks.

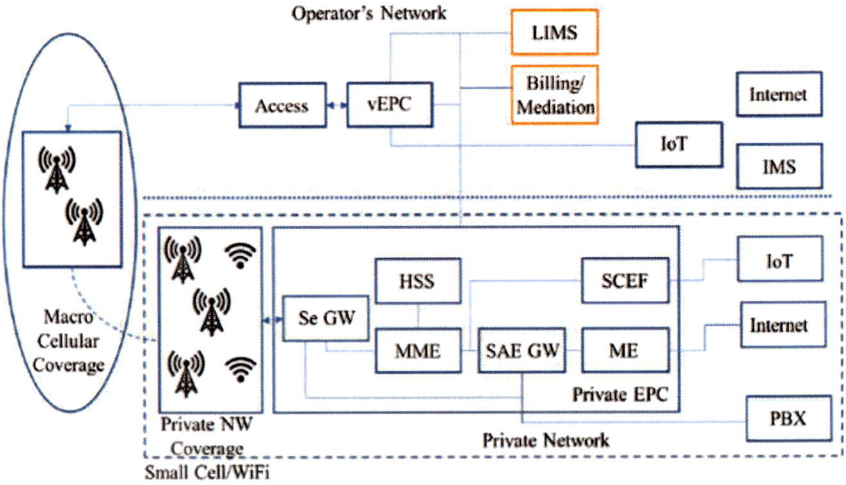

Fig. 1 Enterprise packet core network architecture

The core must support VoLTE and Internet. Additional key features should include mobility support, SRVCC support, vCGNAT capability, vFirewall capability, TCP optimization capability, video optimization capability, and traffic shaping capability. The solution should be capable of scaling up quickly. An evolution path toward future technologies should exist [7–9]. The infrastructure provider or the enterprise (as the case may be) must have the flexibility to add/remove components based on use case so that CAPEX/OPEX remains in check and remains optimized at any instant of time. IoT aspects are described in the sections below.

3 Overview of the IoT Architecture

Telecom operators should strive to achieve an IoT connectivity, as shown in Fig. 2, which will help result in the following:

- Simplified production with a reduced amount of vendors and integration needs
- Faster time-to-market for new functionalities
- Reduced production costs per SIM (user) per year and transferred GB
- Higher degree of automation for testing, software deployment, and operation
- Reduction of efforts and duration for customer on-boarding
- Separation of connectivity production and business support systems (BSS) platform.

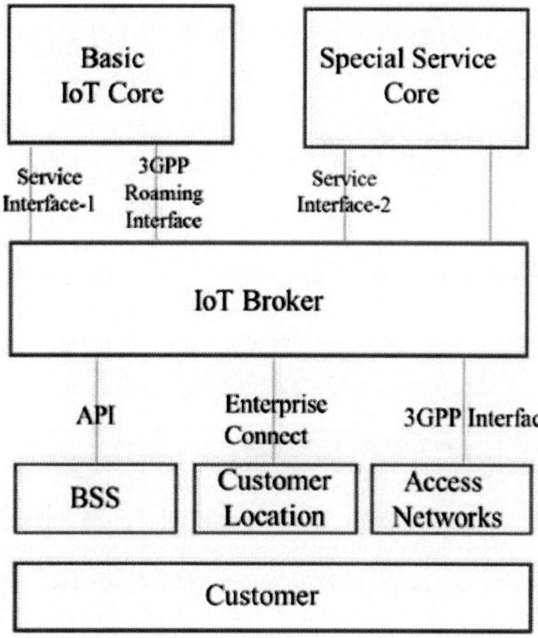

Fig. 2 IoT architecture overview

The IoT network will be technology agnostic, i.e., NB-IoT, LoRaWAN, Sigfox, Zigbee, 6 LowPAN, WiFi protocol stacks will be able to interconnect to the IoT Core Network seamlessly as against the present day networks where each technology needs to connect to its core network.

Elements in the above block diagram in Fig. 2 are described in the sub-sections below.

3.1 IoT Broker

The IoT Broker will be the interface and integrating element of the overall production platform to all surrounding systems. Hence, it will serve as an anchor which combines, aggregates and abstracts the internal structure toward external systems/network elements.

This includes:

- Integration of Mobile Networks
- Integration of Roaming Networks
- Customer Integration
- Integration to Operator Business Support Systems

The IoT Broker provides three main services:

- Subscriber Service
- Signaling Service
- Payload Service

3.2 Basic IoT Core

The Basic IoT Core is meant to provide all network capabilities for 3GPP compliant M2M/IoT Data and SMS. The solution should support 2G/3G voice services and must be capable of integrating with the operator's existing MSC, MGW, and MSS. Additionally, the Basic IoT Core should provide the needed functions to integrate a third party IMS Core for VoLTE.

3.3 Special Service Core

The special service core is designed at providing services which are not typical for a basic M2M/IoT connectivity service but are necessary to cover special customer demands, e.g., an IMS core for M2M, VoLTE.

3.4 BSS

The BSS system (as normally mandated) should include CRM, order management and billing. It could offer an API and/or a GUI to customers of an operator which can used for ordering/managing SIM cards and related services.

4 IoT Broker

This section describes elements, functions, and interfaces of this building block. Figure 3 depicts the elements of an IoT Broker.

- Subscriber Service
- Signaling Service
- Payload Service.

4.1 Subscriber Service

The Subscriber Service enables distribution of network related subscriber and service data to a single IoT Core or to multiple, multi-vendor IoT Cores, as shown in Fig. 4. The Service interacts with customer focused business support systems on one side and the subscriber profile repository of the HSS/HLR of the Basic IoT Core [7–9]. It also transfers and synchronizes data across subscriber profile repositories of different Basic IoT Core instances.

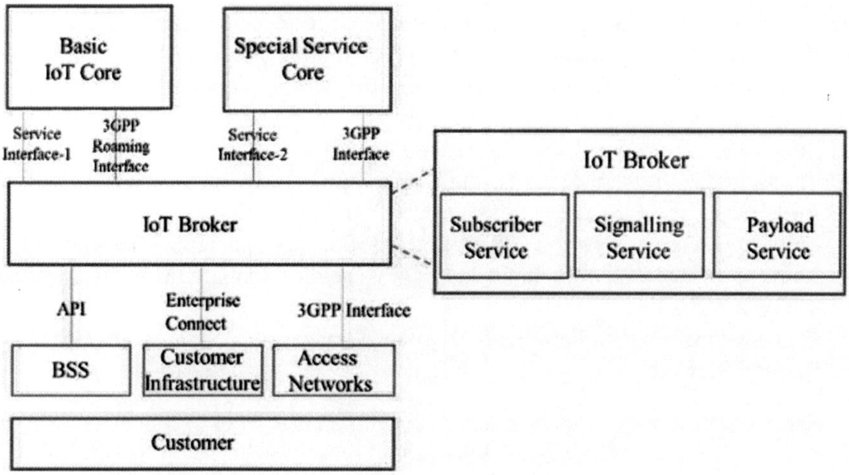

Fig. 3 IoT broker-building blocks

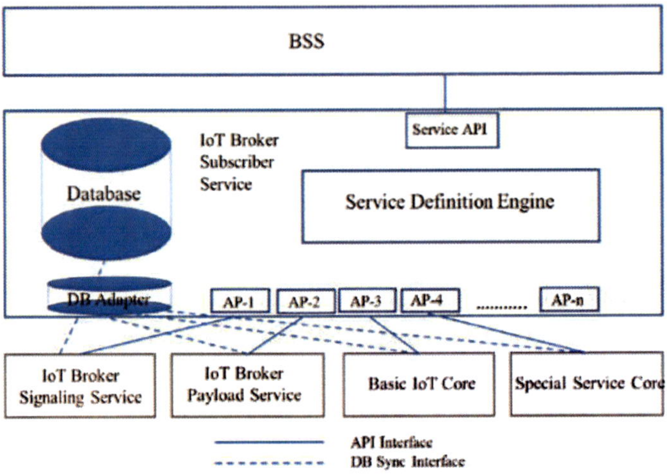

Fig. 4 Subscriber service overview

The Subscriber Service should have versatile functions as mentioned below:

APIs: The Subscriber Service should offer a service API to business support systems. This API should be capable of offering different services to the BSS as mentioned below:

- Service to order, provision, change, modify and delete subscriber, and their services
- Service to order, pre-provision, activate a SIM
- Service to order an APN realization including IP Sec connectivity toward customers
- Service to order a SMS short code including definition of SMS forwarding method
- Service to order a volume limit for a SIM. Additionally, this API should offer capabilities to provide information back to the BSS, if requested, e.g.
- Notifications that a volume limit is reached
- Information about SIM/subscriber status
- Through specific service APIs, the Subscriber Service should enable the IoT Broker to
- Provision different services of the Basic IoT Core (or Special Service Core).
- Retrieve information from the IoT Core (e.g., session status, SIM status, location information)
- Exchange information between IoT Broker Services, i.e., Signaling Service and Payload Service.

Service Definition Engine: All service orders received from BSS shall be checked for consistency and correctness. These pre-processed orders shall be pushed to a database which is part of the Subscriber Service.

API calls containing a service order for SIM activation should be enriched by IMSI, MSISDN or ICC-ID information. These number resources will be managed by the Subscriber Service. For an enrichment, it should be possible to use information pre-configured as templates and link these to API calls. This will allow a BSS to omit parameters on the API required for an activation of a SIM. These parameters should be managed and added by the Subscriber Service as and when needed. The templates shall provide parameter values, which are used by several Subscriber/SIMs, e.g., QoS information as needed for a subscriber profile repository [10–12].

The service definition engine should be able to translate an API (e.g., Service API) into different APIs which can be offered by other service blocks such as the payload service, the signaling service or the Basic IoT Core. The Subscriber Service is a kind of mediator and should enable and simplify migration of SIMs from one Basic IoT Core to an alternate second Basic IoT Core as required information is already available within the Subscriber Service Database. Instead of the above described API translation, it should also be able to synchronize database information between the Subscriber Service Database and other databases which are part of other service blocks and a Basic IoT Core (Subscriber Profile Database). For these tasks, the Subscriber Service needs to provide an API translation as well as data replication functionalities of mainstream databases, e.g., noSQL, MySQL, MongoDB.

Process Automation Handling: The Subscriber Service shall provide workflow mechanisms so that a single service order, received through the service API can be split into several sub tasks to be executed in a defined sequence. For example, it shall be possible to request the initiation of a private APN via the service API. The Subscriber Service needs to split the APN initiation request into a trigger to the Basic IoT Core and to the Payload Service.

4.2 Signaling Service

Signaling Service incorporates.

Diameter Agent (DEA/DRA)
Signaling Transfer Point (STP)
Signaling Firewall

The Signaling Service interfaces to the local access network, the Basic IoT Core as well as roaming networks via an international interconnection carrier, as shown in Fig. 5. The Signaling Service network elements will route signaling traffic to the Basic IoT core instance. The Signaling Service should provide a well-documented API and a GUI to enable service deployment and configuration. For example, it should be possible to change via API, the SS7/SIGTRAN and diameter routing easily. It should be possible to define on a per SIM basis (i.e., per IMSI/MSISDN) to which HLR/HSS/SMSC/MSC/MSS, corresponding traffic shall be routed to.

The per SIM routing is required to enable a customer by customer migration of SIMs from one HLR/HSS and/or SMSC (e.g., from an existing M2M production) to

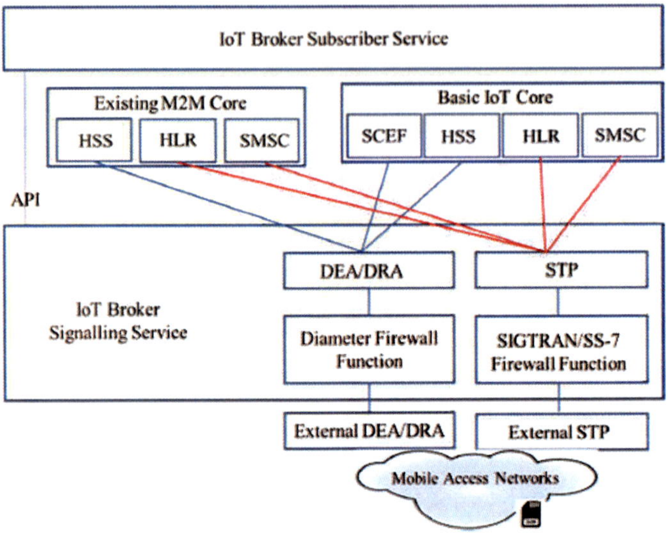

Fig. 5 Signaling service overview

a second HLR/HSS and/or SMSC. Besides a pure signaling routing, it is required that signaling security functionalities are provided as part of the Signaling Service. Hence, SIGTRAN and diameter firewall functionalities are provided. These functionalities need to be configured and managed via an API and a GUI.

4.3 *Payload Service*

The Payload Service should incorporate the following functions:

- Gateway for Connectivity
- GTP Load Balancer
- IP Firewall
- NAT Functionality

The Basic IoT Core network and the specialized core network will be de-coupled from the access network by this service, as shown in Fig. 6. Migration of the basic core network to a different core network will be supported by this service without a change requested by the customer. The GTP load balancer routes GTP payload traffic to the Basic IoT Core. It shall have the function to route GTP payload traffic to different IoT cores according to a set of criteria provided by the operator. These criteria could be based on, e.g., IMSI range, RAT type.

The GTP load balancer shall support high availability cloud architectures by distributing GTP payload traffic to different instances of the Basic IoT Core. It will also support non-disruptive service re-location during maintenance updates. The

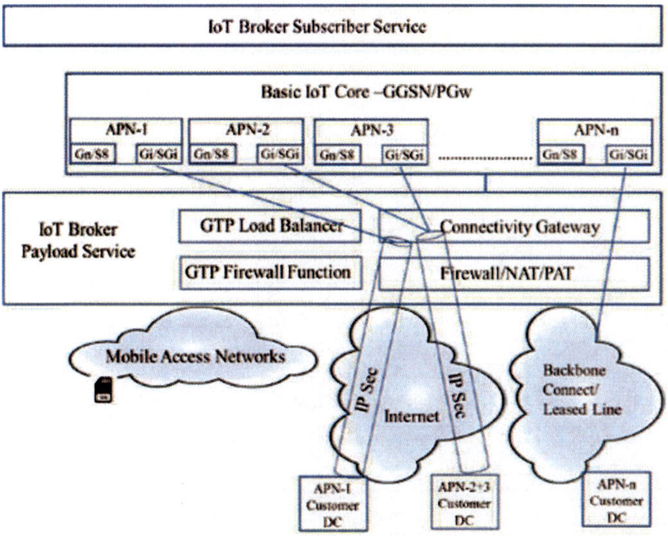

Fig. 6 Payload service overview

GTP load balancer will interface to the Basic IoT Core P-GW/GGSN elements, the roaming network via international interconnection carrier as well as the operator access network (SGSN, S-GW). For all GTP traffic, a firewall functionality will be provided in addition to a load balancer.

The connectivity gateway connects the Gi/SGi interface of the Basic IoT Core to the customer infrastructure/data center and provides a secure payload data transfer between them. For this task, IP Sec and leased line termination should be supported.

Additionally, the Payload Service should provide IP firewall and NAT functions to support a secure connectivity to the internet.

For the Payload Service, a well-documented API and a GUI to enable service deployment and configuration should be provided. The following procedure will be supported via the API and the GUI.

- Provision, configure and setup an SGi/Gi interface toward a new configured APN within the Basic IoT Core.
- Provision, configure and setup an IP Sec connectivity toward a customer data center
- Provision, configure and setup rules for the user payload routing toward the Internet and/or leased line and/or backbone connection and/or IP Sec Connectivity
- Provision, configure and setup GTP load balancer configuration to enable a country/geographical zone specific routing of GTP sessions. It should be possible to define based on SIM and APN which GGSN/P-GW will be used for the user data session.

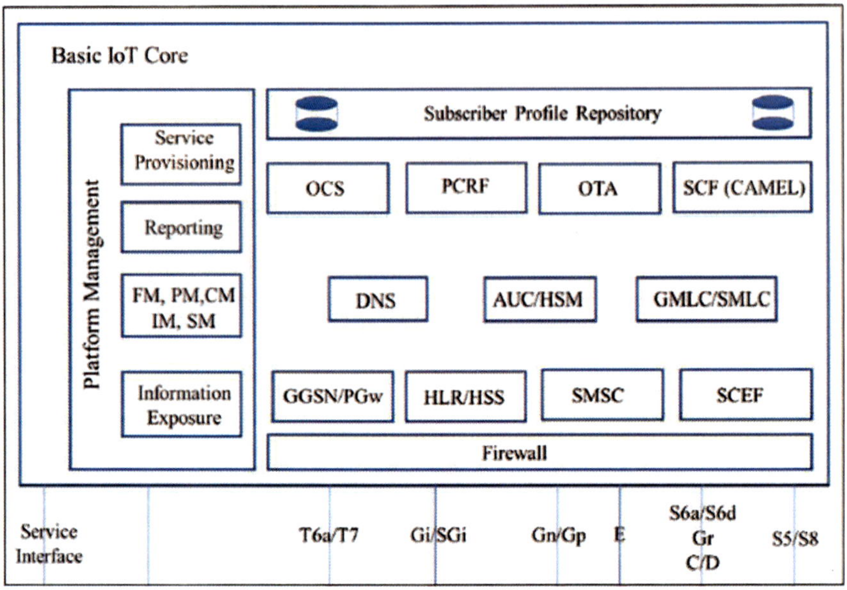

Fig. 7 IoT basic core—functional elements

5 Basic IoT Core Overview

The following Fig. 7 provides an overview about the standardized functional elements which are needed to fulfill the requirements of a Basic IoT Core.

From the Fig. 7, it becomes clear that the Basic IoT Core will include all functions of a technical MVNO platform. Within the new Basic IoT Core, internal interfaces between the functional elements do not necessarily need to be implemented according to 3GPP Standards as long as they do not need to be exposed outside the Basic IoT Core.

The interfaces toward external partner network elements, e.g., SGSN/S-GW, MSC or MME should comply to the corresponding 3GPP standards. These interfaces are used in case of roaming.

Overall, the Basic IoT Core should be seen as a kind of black box, offering the required interfaces to external partner systems while the internal implementation is not restricted as long as the functional requirements are met.

6 Conclusion

The Government of India aims to make India a global economic power house by targeting to position the country into the five trillion dollar economy club within the next five years. Industrial automation will play a key role in driving this growth through various solutions (e.g., deploying robots) that will improve efficiency and reduce costs. Govt. of India should go ahead, reserve a designated frequency band/bandwidth and auction spectrum to the enterprise segment for an exclusive industrial use. IoT through licensed frequency spectrum is the most reliable means for enabling automation. In the near future, the enterprise segment should consider this no risk proposition to boost efficiency/productivity and thereby exports which will earn valuable foreign exchange for the country.

References

1. ETSI TS 124 302 V9.3.0 (2010–06) Technical Specification Universal Mobile Telecommunications System (UMTS); LTE; Access to the 3GPP Evolved Packet Core (EPC) via non-3GPP access networks; Stage 3 (3GPP TS 24.302 version 9.3.0 Release 9)
2. ETSI TR 103 591 V1.1.1 (2019–10), Technical Report- SmartM2M; Privacy study report; Standards Landscape and best practices
3. Nafis MT, Biswas R (2019) A secure technique for unstructured big data using clustering method. Int J Information Technol. Springer. ISSN 2511–2112. https://doi.org/10.1007/s41870-019-00278-x
4. Nafis MT, Biswas R (2018) (N,α)-means algorithm for clustering big data. Int J Eng Technol 7(2.27):50–55 [Scopus Indexed] [S.l.], June 2018. ISSN 2227–524X
5. Garg R, Nafis M, Nafis T, Garg B (2018) Ordered weighted averaging operator used to enhance the accuracy of fuzzy predictor based on genetic algorithm. Int J Intelligent Syst Technol Appl 17(1–2), 229–253 [Scopus Indexed, ACM Digital Library]. ISSN: 1740–8873
6. Nafis MT, Biswas R (2018) A technique for classifying unstructured big data files. Int J Comput Sci Eng (IJCSE) 6(6):198–200 [UGC Approved],ISSN: 2347-2693
7. Khan A, Nafis MT (2018) Neoteric breast cancer detection through machine learning algorithms. Int J Comput Appl 179(49):8–12
8. Biswas SS (2019) Neutrosophic Shortest Path Problem (NSPP) in a directed multigraph, In: Neutrosophic sets and systems, vol 29. ISBN 978-1-59973-620-4
9. , Memory diagram for quantum computers: a proposal. International Information Institute (Tokyo). Information; Koganei, vol 21, issue 2 (Feb 2018): 765–775
10. Biswas SS, Alam B, Doja MN. Intuitionistic fuzzy shortest path in a multigraph, data science and analytics. REDSET 2017. In: Communications in computer and information science, vol 799. Springer, Singapore. https://doi.org/10.1007/978-981-10-8527-7_44. Print ISBN : 978-981-10-8526-0, Online ISBN : 978-981-10-8527-7
11. Ahad MA, Tripathi G, Zafar S, Doja F. IoT data management—security aspects of information linkage in IoT systems, principles of Internet of Things (IoT) ecosystem: insight paradigm, pp 439–464, January 2020. https://doi.org/10.1007/978-3-030-33596-0_18
12. Batra NSU, Zafar S. A neoteric swarm intelligence stationed IOT–IWD algorithm for revolutionizing pharmaceutical industry leading to digital health. https://doi.org/10.1016/B978-0-12-819593-2.00001-7

Optimization of Web Applications Using Eager Loading

M. Deva Priya, B. Sabarinathan, and M. Surya

Abstract Web applications are widely recognized as the building blocks of service-based applications. Performance of an application mainly depends on the components of web applications. All legal services like appointment, court cases, hearings, to-do list, notification, etc., are available online. When we fetch all the records and data related to those records in web applications, the number of queries increases, thereby increasing the query processing time and reducing the performance of the application. This project mainly focuses on optimizing the performance of web application using eager loading technique on a key-value database. One way to improve the performance is to reduce the number of SQL queries by implementing a cache layer in-between the web application and the database server. In eager loading, a query loads the associated entities as part of the query. This is achieved by using the include method. Most Object Relational Mappings (ORMs) are lazy loading enabled by default. Queries are issued for the parent record, followed by one query for each child record. Eager loading technique is introduced to deal with a greater number of queries, and key-value database is used to support caching. By making use of the previous logs, many web applications can be optimized. As a result, the performance of the web applications can be improved, and the users can get rapid responses to their queries.

Keywords Eager loading · Lazy loading · Query optimization · Caching · Redis

M. D. Priya (✉) · B. Sabarinathan · M. Surya
Department of Computer Science & Engineering, Sri Krishna College of Technology, Coimbatore, Tamil Nadu, India
e-mail: m.devapriya@skct.edu.in

B. Sabarinathan
e-mail: 16tucs203@skct.edu.in

M. Surya
e-mail: 16tucs237@skct.edu.in

R. P. Mahapatra et al. (eds.), *Proceedings of 6th International Conference on Recent Trends in Computing*, Lecture Notes in Networks and Systems 177,
https://doi.org/10.1007/978-981-33-4501-0_66

1 Introduction

There are various areas of research like optimized web mining [1–3, 6] and pattern mining in text and images [4, 5]. The main focus of this work is to optimize web applications and increase the performance (speed) using eager loading technique for key-value databases. Building an effective product relies on two fundamental parameters, namely functionality and performance. Functionality is what the application lets its users achieve, and performance is the system's capacity to complete transactions and provide information swiftly and precisely, despite increased multi-user collaboration or restricted hardware resources [7].

Currently, users are fond of fast applications, as slow applications annoy them and lead to loss in revenue. In this paper, diverse ways to improve performance of web applications are dealt with. Active records and Object Relational Mapping (ORM) are powerful tools in ruby on rails, but only if it's utility is understood and used. Initially, a task can be performed in several ways in ruby on rails, but the costs of using one over another should be analyzed. The "N + 1" problem occurs when the first query populates the primary object, and the second query populates all the children objects for every unique primary object returned.

Eager loading associations are very useful and provide better performance, but can also cause serious harm in case of increased nested associations. There are few essential optimizations for leveraging Redis as a cache in rails infrastructure. Most of these optimizations are still helpful even if a hosted solution is used. Unlike Memcached which is multi-threaded, Redis runs only a single thread per process. Considering the speed of Redis, a single process seems to be heavy in many workloads. When the platform traffic starts increasing, the background jobs are fired continuously, pub/sub channels relay thousands of payloads over the network, and the cache is hit continuously. Each request to Redis is blocking, as the background jobs consume more time and is an absolute bottleneck for a set of load balanced servers.

Databases like Redis are optimized for storing a lot of small values (such as links and their short URLs) as they are loaded into memory (RAM). When a call is made to Redis, it reads data from RAM (faster) instead of the hard drive (slower). Model level caching is often overlooked even by experienced researchers. It is due to the delusion that when views are cached, it is not essential to cache at the lower levels. Though much challenge in the rails lies in the view layer, it is not always true.

2 Related Works

This section discusses about the works done by various researchers related to query optimization.

Yadav et al. [7] have propounded an algorithm for organizing the cache memory. They have introduced the concepts of cache and the architecture of cache memory organization which comprises of cache manager and memory, private file, and shared file caches. In addition, they have dealt with varied kinds of cache algorithms, their merits and demerits. The proposed optimization algorithm primarily offers better hit ratio and involves low latency.

Kurpanik and Pańkowska [8] have done an extensive research to deal with the related NoSQL schemes, defining them and indicating the popular works done. Not only SQL (NoSQL) mechanisms are discussed, and the NoSQL issues are discussed. They have used mind maps which aid in the visualization of concept relations. Initially, the map is produced for one concept using the mind map maker which is linked directly to an added concept, and additional ideas originate from that.

NoSQL systems offer big data services on the cloud. Nevertheless, the prevailing NoSQL systems depend on skilled administrators to organize and alter the extensive variety of configurable parameters so as to attain better performance. Bao et al. [9] have presented a policy-based configuration management system called Policy-driven Configuration Management (PCM) which identifies workload-based configuration parameters and identifies the altered parameters for various workloads as configuration strategies. PCM can be used to examine a variety of configuration parameters that may influence the runtime performance of NoSQL systems in terms of read and write workloads. PCM is used to optimize the NoSQL systems such as HBase for both individual node and cluster in the cloud. The PCM configuration outdoes the default configuration on a variety of workloads.

NoSQL databases are widely used with some merits like scalability and flexibility. Chevalier et al. [10] have investigated the utilization of NoSQL systems for implementing Online Analytical Processing (OLAP) systems. More exactly, OLAP systems are instantiated from the conceptual to the logical level and later to an aggregation lattice (optimization). A set of rules are defined to map the star schemas into two NoSQL models namely, column and document-oriented models. The variances between the NoSQL systems are considered. Column-oriented HBase is faster than document-based MongoDB in terms of loading time.

Dart is an open source, structured programming language developed for the web. Its syntax resembles Java, JavaScript, and C#. It can be applied to both the server and client environment, which helps in building a bridge between back-end and front-end. MongoDB is a modern database solution. Though it may not substitute relational databases, it has merits over traditional databases which enable it to work impeccably over distributed storages. Chonkar et al. [11] have propounded a dart coupled with MongoDB that can be a flawless formula for increasing demand of quicker and more scalable applications.

Handling huge quantity of data that is sent to the Robot Operating System (ROS) network seems challenging. In the former days, this was handled by storing the messages in a flat file that acts as storage for the future. Unfortunately, this system is not appropriate for tasks like interrogating messages sent during a certain period to find points in time at which the robot enters or leaves particular states. Fiannaca and Huang [12] have dealt with many databases with changing storage models so as to find the finest data store for an application which incorporates the present rosbag system. They have analyzed the throughput of MongoDB, PostgreSQL, and SQLite3 databases by applying stress testing on every database with differing messages, message sizes, and number of topics. Moreover, each database is verified against exemplar queries that are beneficial for roboticists, and runtimes are conveyed. Based on these studies, endorsements are made for the growth of upcoming ROS message logging systems.

NoSQL storage systems are scalable and easy to use. Ron et al. [13] have examined the security measures to be taken for NoSQL databases, concentrating on query and access schemes. Due to the advent of novel query formats, the former SQL injection techniques may appear to be inappropriate, but NoSQL databases are not resistant to injection. Injections and Cross-Site Request Forgery (CSRF) are some attacks against NoSQL databases. The authors have examined the source of vulnerabilities and have presented procedures to alleviate these attacks. They have expounded that the current technology lacks the security measures and alertness developed over the years.

Wassan [14] have focused on educational modeling based on big data methods. With the progression in Internet worldwide, the claim for online education is increasing. Many institutions offer different kinds of online courses. The analytical models in the field of data mining and computer science heuristics aid in data investigation and visualization, forecasting student performance, producing endorsements for students and teachers, providing feedback, categorizing associated courses, e-content, finding objectionable student behaviors, etc. The data access patterns of students including their logs in online systems are explored to find associations. Massive amount of data is to be explored as the number of students and courses increase progressively. Big data platforms and parallel programming models may be employed to hasten analysis of exploding data and finding the competence of computational patterns.

Owing to large volume of data in websites, web innovative services are needed to ensure scalability and availability. Huge volumes of data are to be considered, and large-scale disseminated computations are expressed using MapReduce methods. To improve the service quality of customers and resolve the issues of massive amount of data in websites like Facebook, Google, and Twitter, the relational database technology is progressively substituted with the NoSQL database to enhance the performance and expansion elasticity. Wu et al. [15] have compared both NoSQL MongoDB and MS-SQL databases and have discussed the efficiency of inquiries. Further, relational database cluster systems demand better server competence and volume with increased cost. Instead, NoSQL databases enable expanding the capacity without any additional costs. Through experiments, it is seen that NoSQL MongoDB is nearly ten times effective for reading and writing when compared to MS-SQL database. NoSQL databases will find a dominant place in the future.

Cache plays an important role in computer architecture. It provides data and instruction at a faster rate to increase the utilization of CPU. Gautama and Singh [16] have focussed on concepts related to the memory hierarchy, cache memory, its organization, and several methods used to enhance cache to increase the performance of Central Processing Unit (CPU). Memory hierarchy deals with low-memory bandwidth and latency of main memory access, followed by compiler-based cache optimization.

NoSQL databases like MongoDB, Redis, and Cassandra are progressively general as they are easy, flexible, and lightweight. Applications using these databases change over time, occasionally demanding a change to data organization. Saur et al. [17] have dealt with the progress of high-availability applications and NoSQL data online without extreme delays, even in the presence of backward mismatched data format variations. They have proposed KVolve, an enhancement to the existing Redis NoSQL database. It permits a designer to upgrade specification that defines how to transform present data to the latest version. This transformation is slowly applied as applications interact with databases, thus circumventing extended pause times. KVolve is sensitive to sustain practical updates, incorporating format changes to a Redis-backed File System (RedisFS), while imposing no overhead with negligible pause time during updates.

With the introduction of collecting real-time information, the demand for NoSQL databases has risen steeply. No research is done on watermarked defense of such schema-less databases. Outdated watermarking methods cannot be applied as NoSQL databases dynamically increase and often have anomalies in schema. Khanduja [18] have propounded a novel viewpoint of implanting watermark into databases, wherein watermarks act as a signature for each tuple separately. The proposal deals with the issues by leveraging the flexible schema features provided by NoSQL database. A new feature is dynamically inserted into the tuple before adding it to the database. Small modifications in tuples lead to more database changes, thus enabling instant recognition with localization in the tuple level.

Many database-sponsored web applications are based on ORM structures. Though these works simplify application enhancement by abstracting constant information, they involve presentation costs. Currently, Yan et al. [19] have considered 27 open-source applications based on mainstream ruby on rails ORM structure, with an aim to understand database-related execution of inefficient features in applications. The unwanted aspects range from challenges related to physical structure to how questions are communicated in the application code. They have applied static program examination to identify and enumerate pervasiveness. The database inquiry and the site reaction times are reduced.

Multicore processors are used in development, research, and in real-time as they offer better processing leading to improved performance. This has involved a number of researchers as the processors are fixed on a chip, and the performance is amplified by saving space, dropping power, and reducing delay. One method of optimization is cache handling, which offers better results in terms of power, performance, and area of the processor. The main challenge with the cache is the consistency which needs to be applied to plan for an appropriate and effective technique. Lalwani and Chawla [20] have propounded a novel concept from which a cache controller is designed using the Least Recently Used (LRU) cache replacement.

Web developers face the challenge of designing informative web pages while preserving the page load time. This task is more challenging as web contents are generated by processing amassed volumes of user data stored in back-end databases. It is tedious to comprehend the cost of producing web page elements, discover, and select the web design with the trade-off between performance and functionality.

Yang et al. [21] have presented panorama, a view-centric and database-aware development environment for web developers. Using database-aware program analysis and novel Integrated Development Environment (IDE) design, it provides developers with instinctive information about cost and performance improving options, and suggests various global code refactorings that permit designers to effortlessly discover an extensive spectrum of performance and functionality trade-offs.

3 Proposed System

In eager loading, a query for a type of entity loads associated entities as part of the query. It is achieved by using the include method. For instance, consider queries written for load blogs and posts connected to each blog. In the former works, accessing these services is seen to be a challenge, as a greater number of queries are to be loaded based on the details and associations. But in this paper, association is used for fetching details from the database, and eager loading technique is incorporated to pair queries, thus improving the performance of the web applications. After predicting a smaller number of queries, the user can easily access the services effectively. The user can access the web page with improved performance and memory caching.

With entity framework, diverse patterns can be used to load related entities. Entity types define navigation properties that signify relations in the data model, and these properties can be used to load entities that are associated with the entity returned by the defined association. When entities are produced based on the data model, navigation properties are produced for entities at both ends of an association.

These properties provide either a reference on the "one" end of one-to-one or many-to-one relationships or a collection on the "many" ends of a one-to-many or many-to-many relationships. Eager loading is applied when the data in associated entities is too expensive to be loaded at the cost of queries being made to the database. The data can be fetched in a single draw along with the key entity using eager loading. Regrettably, when focusing on a volume of friend related information, a lazy loading strategy can force the framework to send an added query for each friend. For a list of "n" friends, lazy loading demands "N + 1" queries. Lazy loading is suitable, but possibly expensive in some circumstances.

3.1 *Dataset Description*

Following datasets are considered for analyzing the performance of early and lazy loading.

- **Firm user table** contains the users' personal details like name, email address, contact number and type of subscription

- **Subscription table** contains the details of user preferred types of subscription like basic, supreme, and plus
- **Appointment table** has an appointment title, date and time of applicant
- **CourtCase table** holds the record of case types, case id, case date, court locations
- **ProfessionalCourt table** contains the details of lawyer names, court id, court locations

Table 1 gives the list of queries and the description of each query. Table 2 shows the time taken for each query applied to the dataset.

4 Implementation of the Proposed System

Preparing queries before applying eager loading strategy is shown in Fig. 1. Query handling by lazy loading method with the time length of 227 ms is shown in Fig. 1. The screenshot after the execution of eager loading technique is displayed in Fig. 2. A similar query prepared by lazy loading system is executed for a period length of 80 ms by using eager loading strategy (Fig. 2).

Eager loading supports extraction of related data in addition to relevant data in a single call. It reinforces a wide assortment of information types. Redis offers a bit of flexibility to continue extracting, if necessary. It is reasonable and limits the beginning uptime of the application. Application expends less memory due to on-request stacking. Pointless database SQL execution is avoided.

For instance, consider a query that stacks blogs of web journals and the data related to each blog. More queries are stacked based on subtleties and associations. Association aids in generating subtleties from the database and combining queries.

5 Work Flow

Queries processed by lazy loading technique are shown in Fig. 3. Consider an example of fetching the user's book details from the database. If lazy loading is used, then the organization table should be hit each time to fetch the user of a specific organization based on Org_ID in user table, and again hit the user table to fetch the details of book preferred by a particular user based on User_ID in the book table.

Eager loading causes stacking all the required elements at once; i.e., all the substances are stacked in a single database call. This is accomplished by utilizing the include strategy which restores the related substances as a piece of query, and huge information is stacked in a single database call. When Lazy loading is used for retrieving book details of multiple users, the database should be hit multiple times. When eager loading technique is used, book details of each and every user can be retrieved in a single database call as shown in Fig. 4.

Table 1 Query Details

<table>
<tr><th>Dataset</th><th>Description</th><th>Technique</th><th>Query</th></tr>
<tr><td rowspan="2">Firm user table</td><td rowspan="2">Query for fetching all the first names and email addresses of users from user database</td><td>Lazy loading</td><td>@user = User.all.map{|a| a.firm_users.pluck(:first_name,:email)}</td></tr>
<tr><td>Eager loading</td><td>@user = User.includes(:firm_users).all.map{|a| a.firm_users.pluck(:first_name,:email)}</td></tr>
<tr><td rowspan="2">Subscription table</td><td rowspan="2">Query for fetching all the first names of users from subscription database</td><td>Lazy loading</td><td>firm_name = Subscription.all.map{|a| a.firm_user.first_name}</td></tr>
<tr><td>Eager loading</td><td>firm_name = Subscription.includes(firm_user: [user::professional]).map{|a| a.firm_user.first_name}</td></tr>
<tr><td rowspan="2">Appointment table</td><td rowspan="2">Query for fetching all the appointment titles from appointment database</td><td>Lazy loading</td><td>appointment = Appointment.all.map{|a| a.applicant.title}</td></tr>
<tr><td>Eager loading</td><td>appointment = Appointment.includes(applicant::applicantable).all.map{|a| a.applicant.title}</td></tr>
<tr><td rowspan="2">CourtCase table</td><td rowspan="2">Query for fetching all the case types from CourtCase database</td><td>Lazy loading</td><td>case_type = CourtCase.all.map{|a| a.try(:case_type).try(:name)}</td></tr>
<tr><td>Eager loading</td><td>case_type = CourtCase.includes(:case_type).all.map{|a| a.try(:case_type).try(:name)}</td></tr>
<tr><td rowspan="2">ProfessionalCourt table</td><td rowspan="2">Query for fetching the names of all the court lawyers from ProfessionalCourt database</td><td>Lazy loading</td><td>prof_crt = ProfessionalCourt.all.map{|a| a.court_record_lawyer_names.pluck(:name)}.uniq</td></tr>
<tr><td>Eager loading</td><td>prof_crt = ProfessionalCourt.includes(:court_record_lawyer_names,:court_record).map{|a| a.court_record_lawyer_names.pluck(:name)}.uniq</td></tr>
</table>

Table 2 Time Taken for Retrieving Records

Datasets	Lazy loading (ms)	Early loading (ms)
Firm user table	1220	170
Subscription table	1878	271
Appointment table	841	51
CourtCase table	20,348	743
ProfessionalCourt table	882	91

Fig. 1 Screenshot of Lazy Loading Technique (Query: case_type = CourtCase.all.map{|a| a.try(:case_type).try(:name)})

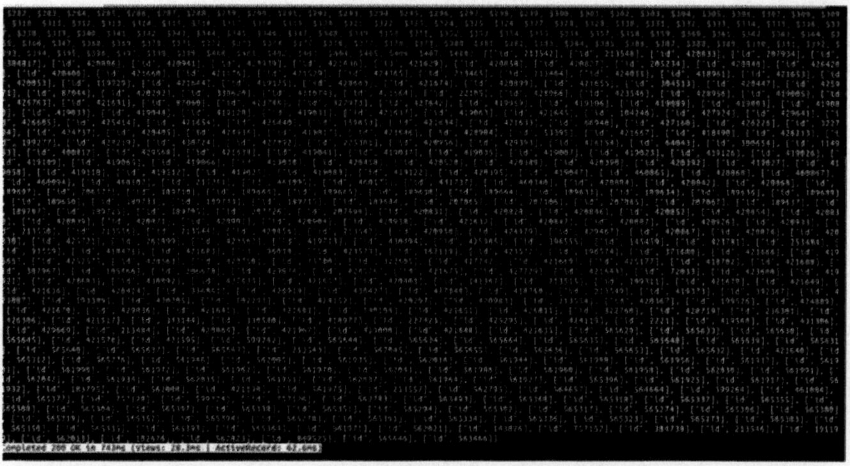

Fig. 2 Screenshot of Eager Loading Technique (Query: case_type = CourtCase.includes(:case_type).all.map{|a| a.try(:case_type).try(:name)})

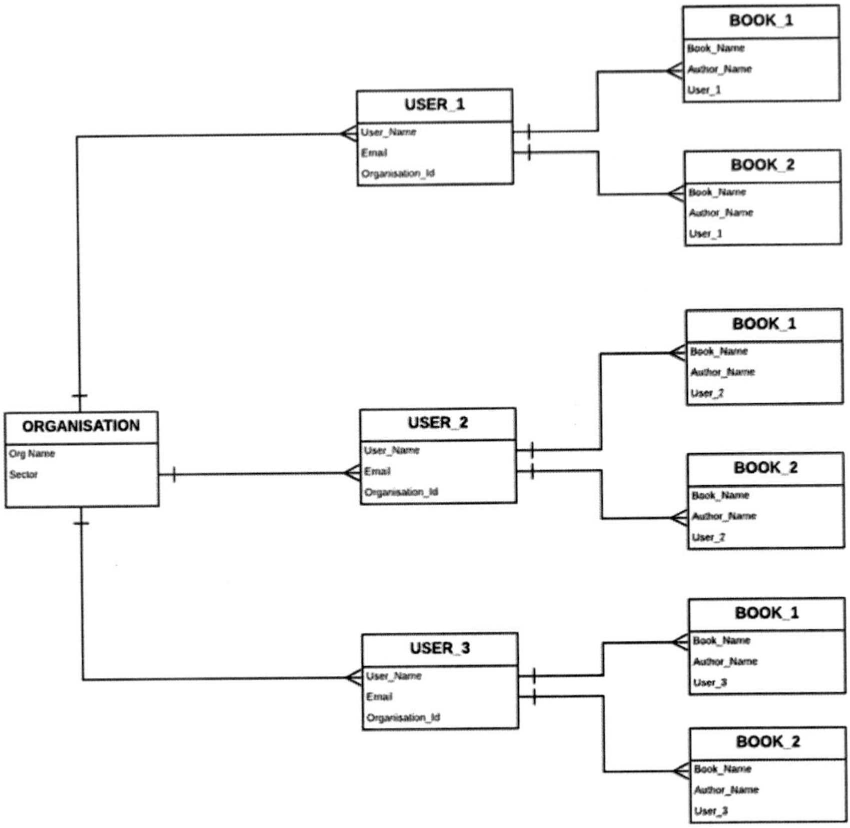

Fig. 3 Query processing by lazy loading technique

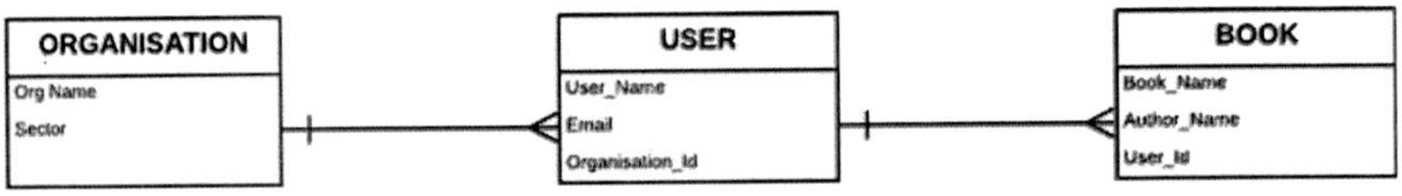

Fig. 4 Query processing by eager loading

With Lazy loading, only less amount of information can be recovered in a single query. When more information is needed with the underlying data, additional queries are to be given to the database. This implies that there are a few full circle trips between the application server and the database server. As a rule, full circle trips are significant bottlenecks in many applications. Lesser the trips to and fro, better the exhibition, wherein Lazy loading is not applicable. Lazy loading leads to increased delay as it is challenging to estimate the sizes of query results and/or the frequency distributions of values in them.

6 Conclusion

In this Paper, a novel scheme is developed for web application optimization using eager loading technique for key-value databases. The eager loading technique is implemented to improve the performance of web applications, whereby the users need not wait for a response to their requests. From the experimental results, it is evident that eager loading for key-value databases provides easy access to the web applications. The execution time of each action performed in the back-end is analyzed. It gives a fair idea on which parts of the application lead to increased delay and demand optimization by using Redis. In the future, it is intended to extend the performance of the web applications based on the user requirements and working on "Dynamic Conditions on Preloaded Records." All plans are similar in terms of terminal output but vary in the quantum of time taken to run. Query optimizations are undeniably essential in database management systems.

References

1. Thilagu M, Nadarajan R (2012) Efficiently mining of effective web traversal patterns with average utility. Procedia Technology. 6:444–451
2. Bama S, Ahmed MI, Saravanan A (2013) Improved PageRank Algorithm for Web Structure Mining. International Journal of Computers & Technology. 10(9):1969–1976
3. Sathya Bama, S., Irfan Ahmed, M. S., Saravanan, A.: A Mathematical Approach For Improving The Performance of the Search Engine through Web Content Mining. Journal of Theoretical & Applied Information Technology. 60(2) (2014)
4. Sreeja NK, Sankar A (2015) Pattern matching based classification using ant colony optimization based feature selection. Applied Soft Computing. 31:91–102
5. Sreeja NK, Sankar A (2015) Ant colony optimization based binary search for efficient point pattern matching in images. Eur J Oper Res 246(1):154–169
6. Nivedha, S., Gokulan, S., Karthik, C., Gopinath, R., Gowshik, R.: Improving Phishing URL Detection Using Fuzzy Association Mining. International Journal of Engineering and Science (IJES). (2017) 21–31
7. Yadav P, Sharma V, Yadav P (2014) Cache Memory-Various Algorithm. International Journal of Computer Science and Mobile Computing. 3(9):838–840
8. Kurpanik J, Pańkowska M (2015) NOSQL Problem Literature Review. Studia Ekonomiczne. 234:80–100
9. Bao, X., Liu, L., Xiao, N., Zhou, Y., Zhang, Q.: Policy-driven configuration management for NoSQL. In: Proceedings of the 8th IEEE International Conference on Cloud Computing. 2015 245–252
10. Chevalier, M., El Malki, M., Kopliku, A., Teste, O., Tournier, R.: Implementing multidimensional data warehouses into NoSQL (2015)
11. Chonkar, S., Suman, S., Sawant, S., Moondra, S.: Supercharging web applications with Dart and MongoDB. International Journal of Current Engineering and Technology. Valley Village, USA. 5(1) 2015 409–412
12. Fiannaca AJ, Huang J (2015) Benchmarking of relational and nosql databases to determine constraints for querying robot execution logs. University of Washington, USA, Computer Science & Engineering, pp 1–8
13. Ron, A., Shulman-Peleg, A., Bronshtein, E.: No sql, no injection? examining nosql security. arXiv preprint arXiv:1506.04082 (2015)

14. Wassan JT (2015) Discovering big data modelling for educational world. Procedia-Social and Behavioral Sciences. 176:642–649
15. Wu CM, Huang YF, Lee J (2015) Comparisons between mongodb and ms-sql databases on the twc website. American Journal of Software Engineering and Applications. 4(2):35–41
16. Gautama, V. K., Singh, P. K.: An overview of compiler-based cache optimization techniques. Int. J. of Scientific Research and Education, 4(3) (2016)
17. Saur, K., Dumitraş, T., Hicks, M.: Evolving nosql databases without downtime. In: Proceedings of IEEE International Conference on Software Maintenance and Evolution (ICSME). (2016, October) 166–176
18. Khanduja, V.: Dynamic Watermark Injection in NoSQL Databases. Journal of Computer Science Applications and Information Technology. Symbiosis. (2016) 1–5
19. Yan, C., Cheung, A., Yang, J., Lu, S. Understanding database performance inefficiencies in real-world web applications. In: Proceedings of the ACM Conference on Information and Knowledge Management. (2017, November) 1299–1308
20. Lalwani K, Chawla M (2017) FPGA Implementation of High Throughput Cache Controller using VHDL & its FPGA Implementation. International Journal of Engineering Sciences & Research Technology. 6(10):346–353
21. Yang, J., Yan, C., Wan, C., Lu, S., Cheung, A. View-centric performance optimization for database-backed web applications. In: Proceedings of 41st IEEE/ACM International Conference on Software Engineering (ICSE) (2019, May) 994–1004

Identifying Criminal Suspects by Human Gestures Using Deep Learning

P. Anantha Prabha, S. A. Vighneshbalaji, C. S. Surya, and K. Vijaya Kumar

Abstract When it comes to investigations, suspects are well aware of the questions asked to them and know every possible way to answer them without getting caught But in the aspect of behaviour and attitude, each person is unique and it can be identified through body language, and hand gestures. Gestures are used in criminology departments to identify criminals based on body language, eye contact and hand gestures. Since the above process is manually done by the investigators, they begin to lose the ability to identify the suspects by investigating throughout the day. To assist police personnel during the investigation, a system is proposed which automatically identifies the suspects with deep learning techniques based on gestures. Gestures are found to be appealing as they are a part of how humans communicate. The proposed system follows the standards of Federal Bureau of Investigation (FBI) former agent which helps to read body languages based on different types of positive and negative hand gestures. Convolutional Neural Networks (CNNs) utilized in the proposed system is capable of learning and classifying multifarious hand gestures at lower error levels. With the dataset of negative and positive gestures, the networks are well trained and tested, and the model gives an accuracy level of 98.96%.

Keywords Deep learning · Convolutional Neural Networks · Human gesture

P. A. Prabha (✉) · S. A. Vighneshbalaji · C. S. Surya · K. V. Kumar
Department of Computer Science and Engineering, Sri Krishna College of Technology, Coimbatore, Tamilnadu, India
e-mail: ap.prabha@gmail.com

S. A. Vighneshbalaji
e-mail: savighneshbalaji@gmail.com

C. S. Surya
e-mail: suryatips@gmail.com

K. V. Kumar
e-mail: vijayrigid@gmail.com

R. P. Mahapatra et al. (eds.), *Proceedings of 6th International Conference on Recent Trends in Computing*, Lecture Notes in Networks and Systems 177,
https://doi.org/10.1007/978-981-33-4501-0_67

1 Introduction

Gestures are a major form of human communication. The gesture could be an image of physical behaviour and emotional expression which include both body and hand gestures. Hand gesture recognition is used in several applications like the robotic arm to control the mini robot, human action recognition, augmented reality, sign language interpreters for the disabled, and so on. Since there is no gesture recognition application in the field of police investigations, a system is proposed which automatically identifies the suspects using deep learning techniques based on gestures. Gestures are classified into positive and negative gestures. The positive gestures include direct eye contact with the investigator and keeping hands stable or calm without any movements. Negative gestures include holding their palms, making movements often, and raising their hands up and down. In every case, the final step to identify the criminals would be investigation through direct eye contact. Humans can be acquainted with body and sign language. This is attainable with the combination of vision and synaptic relations that were shaped along with brain development. The disadvantage in human investigations is that, if the same investigator investigates throughout the day, they can get exhausted and begin to lose the ability to identify them. The other disadvantage is that not all the investigators will have the knowledge to identify the criminals based on hand gestures. Hence, to apply this talent in computers and to circumvent these drawbacks, a new medium of human–machine communication is developed wherein, devices can capture human hand gestures. In this paper, an effective methodology is proposed for hand gesture recognition. To identify hand gestures, CNNs are used in the proposed system. CNN is computationally efficient and it is extensively used for image classification, beholding, and detection. For example, given many pictures, it learns distinctive features for each class by itself which makes the identification rate of gestures faster. The procedure involves segmentation during preprocessing, in which they contribute an improved feature extraction. It automatically detects the important features without any human supervision. In this methodology, it is possible to attain good results for suspect identification using hand gesture analysis. The next section includes a brief description of the techniques, the proposed methodology and, also the experiments which were done. The next sections of this work shows the results obtained, followed by conclusion and future scope.

2 Literature Survey

There are numerous publications and research work done on the concepts of gesture recognition using a deep learning approach. The work by Wei et al. [1] is on electromyography-based gesture recognition by multiview deep learning with the CNN. Wei et al. [1]'s research is about the identification of gesture based on electromyography by applying deep learning methods using CNN. To achieve better results, they incorporated surface electromyography along with inertial measurement units. Lu et al. [2] have utilized a deep CNN with spatial–temporal separable 3D convolution to recognize hand gestures. The system generated is robust and accurate. Liu et al. [3] have detailed on the integration made for both humans and robots using gesture recognition while replacing the workers in the workplace. This would help the workplace jobs to be more efficient and automated. The identification of American sign language numbers using deep CNNs is carried out by Bheda et al. [4]. The depth-sensing technology is used instead of the conventional camera technology. The work of Tzirakis et al. [5] demonstrates the ways of recognizing emotions using Deep Neural Networks (DNNs) and used CNNs as well as Long Short-Term Memory (LSTM) to deal with anomalies or outliers. Using CNN's deep learning, Hussain et al. [6] have introduced hand gesture recognition where various functionalities of operating devices use those gestures to operate. Ronao et al. [7] have discussed the recognition of human activity through smartphone sensors with 1D time-series signals. Deep Learning based hand gesture recognition techniques are discussed in [8, 9]. The methods are effective, efficient and data-adaptive. Pattern matching and edgelet-based human detection are discussed in [10–12]. Various gestures [13], body language [14] and pattern recognition [15] and related techniques are discussed in [16–18]. Although a significant number of researches are done on gesture recognition of voluntary actions, this study aims to focus on the deliberate and unconstrained hand actions by the suspects which help in the prediction of criminal of an incident. The proposed study aims to solve the problem of the suspect diverting the investigators away from the truthiness of the incident by capturing the negative gestures.

3 Dataset

The datasets for our system were first initiated with the help of American investigators and former FBI agents [11] who describe how positive and negative gestures will be. Since the gestures do not have any proper datasets, the datasets were generated by our own with reference to various gestures classified according to standards of security agencies globally. Based on those documentaries, the dataset was recorded and created. The datasets were recorded using canon EOS high-resolution DSLR with ISO: 1006400 sensitivity range for better clarity. The dataset is organized into three classes. There are two negative classes and one positive class and one non-manual class. There are 9159 images which consist of 4938 negative gesture images, 1897 positive gesture images and 2324 non-manual combined gesture images.

Fig. 1 Positive getures

3.1 *Positive Class Gesture*

The positive class gestures consist of images of normal human behaviour as shown in Fig. 1.

3.2 *Negative Class Gesture*

The negative gesture consists of movements that make the investigator lose focus on the suspect like shaking hands, wiggling hands, making movements using hands often towards the face, raising their hands up and down and wringing hands as shown in Figs. 2 and 3. Two kinds of negative gestures are taken for the training purpose of the network mentioned below as type 1 and type 2.

Type 1 The dataset consists of 2708 negative gesture images with each image of size 1.5 MB approximately with dimension 384 × 216.

Type 2 The dataset consists of 2230 negative gesture images with each image of size 3 MB approximately with dimension 384 × 216.

Fig. 2 Negative class gestures—type 1

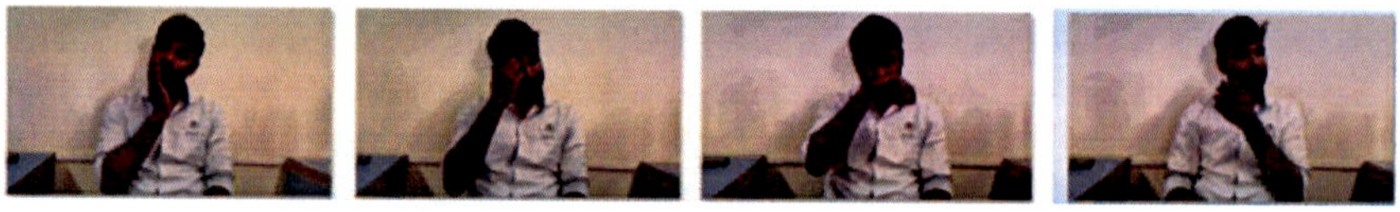

Fig. 3 Negative class gestures—type 2

3.3 *Non-manual*

The non-manual gestures category consists of pictures that embody head drooping and shaking, shoulder shrugging, and facial features changes, etc. as shown in Fig. 4. The dataset consists of 2324 negative gesture images with each image of size 3 MB approximately with dimension 384×216.

4 Preprocessing

Preprocessing involves the process of converting raw data into usable means and it is shown in Fig. 5. The captured video of the suspect is converted into multiple frames. The lossy data in the dataset is removed. Then, the images are converted from RGB format to the greyscale image. Further, the size of the image is resized with the dimension of size 50×50. The above steps are done to speed up the training process.

5 Methodology

The proposed model is built to identify negative gestures exhibited by the suspect and predict the criminality level. The emotions and gestures of the suspect are captured by the high-resolution camera. The captured video is converted into frames and then

Fig. 4 Other gestures

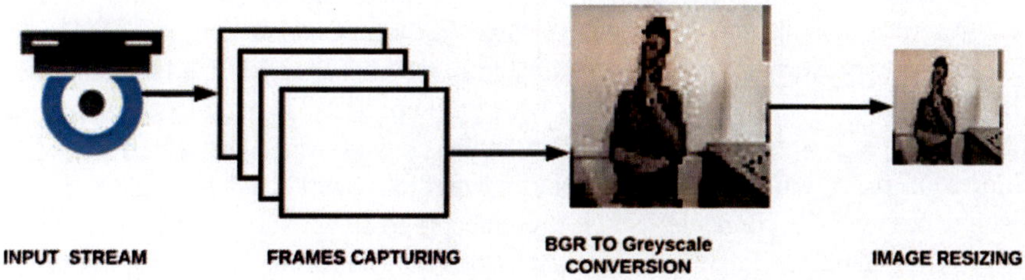

Fig. 5 Steps in preprocessing

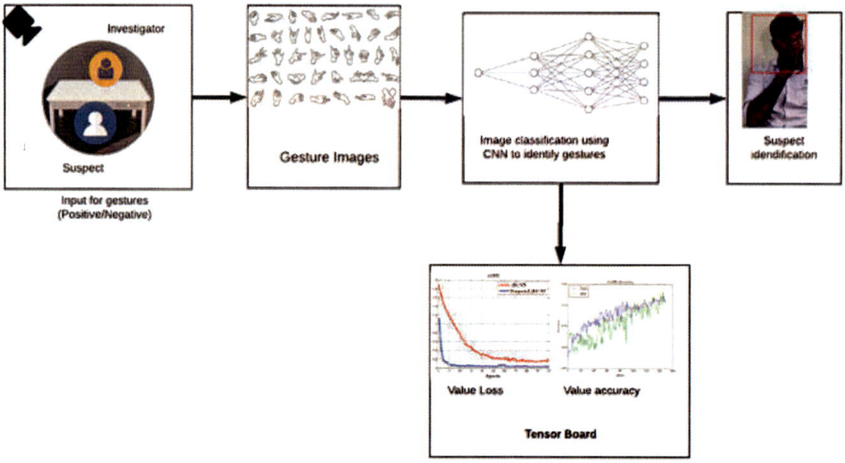

Fig. 6 System architecture

preprocessed. The frames generated are converted from the RGB to the greyscale. The architecture of the proposed model is shown in Fig. 6.

CNN is used to classify the images. CNN is a specially designed multilayer perceptron to spot two-dimensional shapes. Therefore, dimensional data preserved in wave shape points are effectively used by CNN. CNN model has the characteristic of reconciling feature extraction, so it's applied for image and gesture recognition. The output with the probability value of criminality of the suspect is shown. In parallel, the performance metrics with value accuracy, value loss, train accuracy and train loss are graphically generated and visualized.

5.1 CNN Model

In the input layer, image data are fed via the convolution layer to the convolution network. The layered architecture is depicted in Fig. 7a.

The convolution layer performs filtering of the input data to get the most relevant features from the data that propagates through the network's convolution layers. The convolution kernels are used and the kernel weights are initialized in the model. The ReLu layer traits to the properties which are nonlinear in the decision mapping function. The pooling layer is added after the convolution layer which performs dimension reduction. Usually, max-pooling methods work well for image inputs. Each neuron is fully connected and it is connected to all activations in the preceding layer of the network. Each layer simply carries out the role of classification, while the previous layers do the learning of features. Depending on the degree of precision required, the whole dataset is passed both forward and in reverse through the neural network as per the number of epoch values. As the segment describes the CNN, the

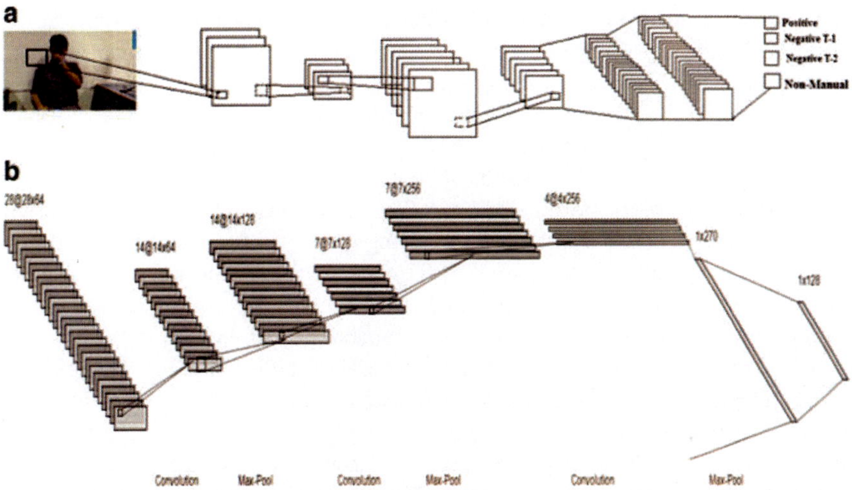

Fig. 7 **a** CNN architecture, **b** Internal CNN architecture

layers perform all extraction and classification functions. As shown in Fig. 7b, the model has four layers; they are Network layer 1, Network layer 2, Network layer 3 and Network layer 4. In the first convolution network layer, a kernel of size 3 $\times$ 3 with 64 feature maps of size 28*28 with a stride value of 1 are used for the convolution operation. For the sub-sampling process, a pooling window of size 2 and a stride value of 2 are also used. In the second convolution network, a kernel of size 3 $\times$ 3 with 128 feature maps of size 21*21 with a stride value of 1 are used for the convolution operation. For the sub-sampling process, a pooling window of size 2 and a stride value of 2 are also used. In the third convolution network, a kernel of size 3 $\times$ 3 with 256 feature maps of size 14*14 with a stride value of 1 are used for the convolution operation. For the sub-sampling process, a pooling window of size 2 and a stride value of 2 are also used. In the fourth convolution network, a kernel of size 3 $\times$ 3 with 256 feature maps of size 4*4 with a stride value of 1 are used for the convolution operation. For the sub-sampling process, a pooling window of size 2 and a stride value of 2 are also used.

6 Experimental Results

The system of gesture recognition is implemented using Intel Core i5-6200U 6th gen with RAM 16 GB and NVIDIA GeForce 940mx 2 GB GPU accelerator. The process is done with the help of the Jupyter Notebook environment which is an open-source Python, geulia and R-distribution is used for implementing various machine learning and statistical methods. Preprocessing of data involves the process of converting raw data into usable means. The captured video of the suspect is converted into multiple frames using OpenCV, and then the images are converted from RGB format to greyscale image. The primary data are stored into a NumPy array in the form of a 2D matrix. Keras and TensorFlow are used to visualize feature maps of the CNN model to show the details of the feature extraction process on several images. TensorBoard is used to visualize the results in the graphical format with train and test accuracy with various hyper-parameters. The results of the convolutional network will be displayed with a percentage of the criminality level. The probability value of the suspect will be determined based on the gestures indicating the level of criminality. The results of train accuracy and train Loss are shown in Fig. 8a and the results of test accuracy and test loss are shown in Fig. 8b. Table 8a represent the results obtained using TensorFlow. The experimental results show an accuracy of 98.96%.

The graph shown in Fig. 8a represents the train accuracy and loss of the model with batch size = 64, learning rate = 0.01 with 50 epochs. The accuracy increases gradually with a decrease in loss value. The graph shown in Fig. 8b represents the test accuracy and loss of the model with batch size = 64, learning rate = 0.01 with 50 epochs. The accuracy increases gradually with a decrease in loss value.

Table 1 indicates the results obtained using TensorFlow with batch size = 64, learning rate = 0.01 with 50 epochs. The model gives a good classification accuracy of 98.96% and the loss is 0.62%.

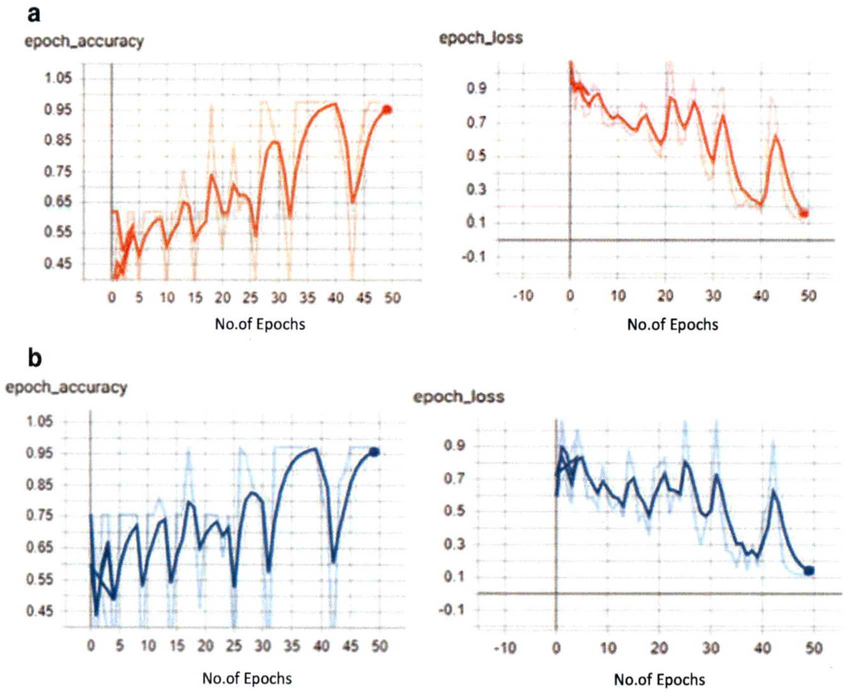

Fig. 8 **a** Train accuracy and loss, **b** test accuracy and loss

Table 1 Results

Epoch value	Batch size	Learning rate	VAL_Accuracy %	VAL_Loss %
50	64	0.01	98.96	0.62

7 Conclusion

A system to automatically identify the suspects during the process of investigation is designed. It requires only a web camera to capture the input image. This will be a new generation of HCI within which no physical contact with the device is required. The experimental results indicate that the proposed method offers better performance and is suitable for applications in the real world. In the future, the project can be up-scaled to simultaneous prediction of multiple suspects. Other features like body temperature sensors can be included. When a criminal is under investigation, the body temperature and blood pressures typically increase when they have committed a crime.

References

1. Wei W, Dai Q, Wong Y, Hu Y, Kankanhalli M, Geng W (2019) Surface-electromyography-based gesture recognition by multi-view deep learning. IEEE Trans Biomed Eng 66(10):2964–2973
2. Lu Z, Qin S, Li L, Zhang D, Xu K, Hu Z (2019) One-shot learning hand gesture recognition based on lightweight 3D convolutional neural networks for portable applications on mobile systems. IEEE Access 7:131732–131748
3. Liu H, Wang L (2018) Gesture recognition for human-robot collaboration: a review. Int J Ind Ergon 68:355–367
4. Bheda V, Radpour D (2017) Using deep convolutional networks for gesture recognition in American sign language. arXiv preprint arXiv:1710.06836
5. Tzirakis P, Trigeorgis G, Nicolaou MA, Schuller BW, Zafeiriou S (2017) End-to-end multi-modal emotion recognition using deep neural networks. IEEE J Selected Topics Signal Process 11(8):1301–1309
6. Hussain S, Saxena R, Han X, Khan JA, Shin (2017) Hand gesture recognition using deep learning. In: 2017 International SoC Design Conference (ISOCC), IEEE, pp 48–49
7. Ronao CA, Cho SB (2016) Human activity recognition with smartphone sensors using deep learning neural networks. Expert Syst Appl 59:235–244
8. Strezoski G, Stojanovski D, Dimitrovski I, Madjarov G (2016) Hand gesture recognition using deep convolutional neural networks. In: International conference on ICT innovations. Springer, Cham, pp 49–58
9. Oyedotun OK, Khashman A (2017) Deep learning in vision-based static hand gesture recognition. Neural Comput Appl 28(12):3941–3951
10. Sreeja NK, Sankar A (2015) Pattern matching based classification using ant colony optimization based feature selection. Appl Soft Comput 31:91–102
11. Bhuvaneswari K, Rauf HA (2009) Edgelet based human detection and tracking by combined segmentation and soft decision. In: 2009 international conference on control, automation, communication and energy conservation
12. Shelvy PT, Palanisamy V, Purusothaman T (2011) Performance analysis of clustering algorithms in brain tumor detection of MR images. Eur J Sci Res 62(3):321–330
13. Pinto RF, Borges CDB, Almeida AMA, Paula TC (2019) Static hand gesture recognition based on convolutional neural networks, vol 2019

14. Navarro J (2018) The dictionary of body language
15. Dey D, Chatterjee B, Dalai S, Munshi S, Chakravorti S (2017) A deep learning framework using convolution neural network for classification of impulse fault patterns in transformers with increased accuracy. IEEE Trans Dielectr Electr Insul 24(6):3894–3897
16. Deva Priya M, Balamurugan A, Dhivyaprabha E, Vignesh R, Vishnu Varthan LR (2019) Intelligent navigation system for emergency vehicles in the 4th international conference on Smart City Applications (SCA 2019) at Hassania School of Public Works EHTP-Casablanca in partnership with Mediterranean Association of Sciences and Technologies (Medi-AST). Casablanca. Morocco during 02nd to 04th October 2019. The paper is published in ACM Digital Library. https://doi.org/10.1145/3368756.3369085
17. Deva Priya M, Christy Jeba Malar A, Lavanya G, Vishnu Varthan LR, Balamurugan A (2019) Smart life saving navigation system for emergency vehicles. In: The 3rd international conference on Smart City Applications, SCA 2019: Innovations in Smart Cities Applications Edition 3, Lecture Notes in Intelligent Transportation and Infrastructure. Book Series. Springer, Cham, pp 921–932. ISBN: 978-3-030-37628-4
18. Christy Jeba Malar A, Deva Priya M, Sengathir J, Kiruthiga N, Anitha R, Sangeetha T (2019) An ıntelligent multi-floor ındoor positioning system for cloud based environment. Int J Comput Appl. Taylor & Francis, (Elsevier). ISSN: 1925-7074. https://doi.org/10.1080/1206212X.2019.1696447

Batch Signature-Based Verification of Data Computation in Cloud Applications

M. Deva Priya, J. Muhammad Haaris, C. Y. Poovizhi Selvan, R. Ruban, and J. Sengathir

Abstract Ensuring security in a cloud environment is an uphill task. In this paper, a scheme for ensuring cloud security called Batch Signature-based Enhanced Data Verification Technique (BSEDVT) is presented for identifying its significance in verifying the amount of application data stored on the cloud. BSEDVT supports faster batch verification comprising of multiple digital signatures when compared to the verification time of the benchmarking schemes in the literature. It utilizes the process of signature tree-based approach for dynamic integrity checking with the inclusion setup, challenge-prove and update phases. It uses component-characterized relations derived from cloud applications that are to be highly processed for constructing the component graph. It also uses a component ranking algorithm for estimating the significant values of cloud components. The proposed BSEVT is used for identifying most of the significant parameters as they represent the potentiality of each component.

Keywords Batch signature · Data verification · Signature tree · Digital signatures · Cloud storage privacy

M. D. Priya (✉) · J. M. Haaris · C. Y. P. Selvan · R. Ruban
Department of Computer Science & Engineering, Sri Krishna College of Technology, Coimbatore, Tamil Nadu, India
e-mail: m.devapriya@skct.edu.in

J. M. Haaris
e-mail: 16tucs122@skct.edu.in

C. Y. P. Selvan
e-mail: 16tucs132@skct.edu.in

R. Ruban
e-mail: 16tucs156@skct.edu.in

J. Sengathir
Department of Information Technology, CVR College of Engineering, Mangalpally, Vastunagar, Hyderabad, Telangana, India
e-mail: j.sengathir@gmail.com

R. P. Mahapatra et al. (eds.), *Proceedings of 6th International Conference on Recent Trends in Computing*, Lecture Notes in Networks and Systems 177,
https://doi.org/10.1007/978-981-33-4501-0_68

1 Introduction

Cloud computing plays a predominant role in many applications due to its capability to offer huge data storage [1–6]. It has founded its way from the mainframe computers which exist from 1950. Cloud computing involves the process of accessing services which includes storage environments, applications and servers involving Internet [7]. It is formed by making use of the existing resources of various organizations as remote services for payment. This processing enables an organization to function by owning fewer amounts of computer resources and by storing or accessing big data and programs virtually on a cloud [8, 9].

Mainframe computers are central processing systems that have high level configurations and can be used by many users through dummy terminals [10]. The costs of deployment are too high and are not feasible for many organizations to buy and maintain. This critical situation can be handled by providing shared access to a resource which is also an economically feasible solution for an organization. Later in 1970s, Virtual Machines (VMs) which enable the execution of one or more Operating Systems (OSs) concurrently in a remote environment were created [11]. The VM concept proved that the complete virtual system could be implemented in a physical hardware which in turn can run a wholly different OS. Virtualization concept is an important catalyst in the communication and information evolution.

In 1990s, the development of Virtualized Private Network (VPN) connections offered by the telecommunication services seemed to be a significant milestone in the development of communication and information era [12]. In VPN, the Quality of Service (QoS) is as good as those of Point-to-Point (P2P) dedicated services at a reduced cost. The evolution of VPN also flagged the way for deploying shared Accessing Points (APs) based on single physical infrastructure for the users.

Another important development in the communication and information era is the design of grid computing environment. It handles the major issues of sharing resources by means of parallel computing. Utility computing enables sharing of computing resources as metered or paid service. Further, the sharing of resources is performed through network referred as Software-as-a-Service (SaaS). Cloud computing evolved by providing resource sharing not only in terms of software but also in terms of information technology infrastructure. It serves as a platform for the growth of various applications. It enables anytime, anywhere access to the computing resources in a more flexible and dynamic way.

There are three different types of delivery service models namely, Software-as-a-Service (SaaS), Infrastructure-as-a-Service (IaaS) and Platform-as-a-Service (PaaS) [13]. The SaaS delivery model enables the user to access software services over the internet. The IaaS delivery model provides virtual infrastructure for large organizations. The PaaS delivery model provides the user with the complete platform for the development of their customized applications.

The implementation/deployment models of clouds are categorized into four types namely, public cloud, private cloud, community cloud and hybrid cloud [14]. Public cloud can offer services to anybody and anywhere through Internet for fee by means of pay-per-usage method. The main advantage of public cloud is that it is simple and easier to implement since the cost is based on the usage of the application, hardware and bandwidth. A private cloud is a registered computing architecture which provides services to restricted number of users. It is also referred to as internal or business cloud. The data security in the private cloud is an important issue faced by an organization or a business deploy private cloud environment in order to exercise more control over the critical data. In case of community cloud, the necessary cloud services are shared between two or more organizations which require the same type of services. The accessing of services is restricted in this type which opens the challenge for securing the data. The combination of private, public or community clouds forms a hybrid cloud.

1.1 Cloud Security

In cloud computing, massively scalable IT-enabled capabilities are delivered 'as-a-service' to external customers using Internet. It offers several beneficiary aspects to the users such as fast deployment of services in the user's environment, enabling users to access services and pay only for the usage which offers cost effectiveness [15].

Though the cloud offers many benefits, the main challenge is the risk involved in rightly understanding the security aspects. In case of any crisis, the responsibility will be shouldered on either the Cloud Service Provider (CSP) or the user [16]. Apart from these issues, the cloud business also has the challenge of addressing privacy issues raised in the way of computing. The application software and data of an organization is transferred to their data centers, which do not guarantee 100% reliability of data and service management. Transferring an application to the cloud imposes new security issues in the environment. Some of the issues include: accessing the data of an organization by an unauthorized user, attacks based on virtualization, attacks through web application, for example, Structured Query Language (SQL) injection attack and cross-site scripting attack, privacy and other control issues raised by the owner of the application software and data, identification-based issues in handling data, various issues related to data verification, interfering and integrity, message confidentiality, loss or hack of data, challenges related to authentication of an authorized user and data or message spoofing [17]. Though cloud computing provides better service in sharing of resources, it suffers from the high level of security risks.

In the CSP, the user's data will be transmitted and stored in the provider's data center wherein all types of operations are carried out. Users need not deploy a separate data center model but share the resources. However, the security aspects related to data transfer and storage remain crucial in this type of implementation. The organization applies some standard classifications on the data and also follows some cryptographic strategies which need to be accommodated in the cloud-based processing also.

Among the challenges in cloud computing, security is considered to be one of the crucial factors in the dynamic resource sharing and computing environment. The security challenge is a crucial factor since computing is dynamic and data location is transparently maintained. Location transparency is one of the important elasticity for cloud computing, which is a security hazard at the same time—without knowing the location of data storage, data protection of a number of regions might be severely affected and violated. The users' personal data security is therefore a vital concern which needs high level of concentration. As far as the customer data or business data security is concerned, the confinement of strategic policies plays a major role as the technical security alone is not sufficient to handle the problem. Trust establishment in cloud service is another issue which is of more concern. Trust establishment plays a major role which is directly related to the trustworthiness and legitimacy of the CSPs. The success of cloud computing is in the process of establishing trust. Hence, it demands the establishment of the trust model for the all stakeholders. The factors like automation, management, processes and policies are the key factors to be considered for establishing trust [18]. All the different vulnerabilities related to computer networks are also applicable for cloud-based services. Some threats include man-in-the-middle attack, phishing, eavesdropping, sniffing, etc. Distributed Denial of Service (DDoS) attack is one common but major attack and is a potential problem in cloud computing.

2 Related Work

Wang et al. [19] have proposed an information re-appropriating scheme based on the concept of information neighborhood weights. In this approach, the clients never establish infrastructure ownership due to the inclusion of appropriate information that makes testing a major challenge. It provides major review potentiality for the cloud information which is considered as the important goal that aids in evaluating the objective of information that could be shared based on the demands of the user and data. It is also phenomenal in introducing and investigating different demands of the multi-client established for the purpose of maintaining different errands in a clustered manner. Chen et al. [20] have contributed a cloud security scheme based on algebraic signature. This algebraic signature-based cloud security scheme aids in enhancing the execution and efficiency of the security provisioning for servicing demands of hundreds of megabytes per unit second. It supports high degree of verification without the essentiality of the challenger to compare it with the original data. It is developed as a challenge-response scheme which is capable of transmitting a constant and minimal amount of data. The algebraic signatures generated by the chaotic property make it feasible for converting this cloud security approach into a significant one. A privacy preserving public auditing scheme using the concept of 'think imagination' with the benefits of watermarking is proposed by Paigude and Chavan [21] for achieving cloud storage security. This privacy scheme includes the Information of Element (IoE) for facilitating the clients to carry out diversified tasks using information that includes erase, refresh and addition similar to cluster inspection. It performs group investigation involving trusted party which handles them for different clients based on the parallel demands that minimize calculation overhead and correspondence. It also incorporates the basics of bilinear marking for establishing cluster inspection process based on the benefits of Merkle hash tree.

Further, a cluster evaluating scheme for cloud security is proposed by Shin and Kwon [22] for clarifying the remote information investigation plan. It completely concentrates on the process of plan evaluation that specifically focuses on provable information execution plans that enhances the possibility of group data and auditing. It includes seven categories of execution plans aiming at the process of auditing productive cluster. It is determined to be highly imperative with respect to the process of information re-appropriating. Then, Shen et al. [23] have contributed a secure examination for cloud privacy with the view to inspect the process of clustering. This scheme proves the significant ideal properties of the cloud secure approach based on the execution and generation time incurred during the process of facilitating security with maximized authentication. In addition, a signature scheme termed Zero Short Signature (ZSS) scheme is proposed by Zhu [24] for verifying the process of information trust that includes the degree of ensuring security and closely evaluating confidence from the point of the outsider. This ZSS scheme minimizes the degree of hash work overhead based on the process of mark procedure that handles the issue of Cumulative Distribution Hash (CDH) property. It is proved to be highly significant in attaining maximized security and effectiveness under different situations of reproduction that aids in checking uprightness information.

3 Proposed System

An enhanced verification technique known as Batch Signature-based Enhanced Data Verification Technique (BSEDVT) is formulated and presented for understanding its role in verifying data stored in the cloud applications. The key steps in verification of the potential features in the implementation are studied and evaluated using simulation-based investigations.

3.1 Batch Signature-Based Enhanced Data Verification Technique (BSEDVT)

The cloud users use resources based on their handheld devices such as laptop, tablet and the like. The potential service is provided by the CSPs. When user registration is complete, the user uses the storage space for storing messages to which the index of the user is returned by the CSP. In the proposed Batch Signature-based Enhanced Data Verification Technique (BSEDVT), the batch verification process speedily considers multiple digital signatures in contrast to the verification times of the existing techniques. In this verification, the signature tree-based approach proposed for dynamic integrity checking includes three phases: setup, challenge-prove and update.

Phase 1-Intial setup

The setup phase is a series of operations at the client side which involves initializing the system including following functions: (a) GenKey(I_k) → $\{s_k, p_k\}$ is a probabilistic key generation algorithm taking security parameter 'k' as input. The outputs include the public key 'p_k' and a private key 's_k.' The client publishes 'p_k' and keeps 's_k' as a secret. GenerateTag(s_k,p_k,M) → {ϕ, metadata} takes the client's file 'M,' 'p_k' and 's_k'as inputs and the outputs M's tag 'ϕ' and the metadata which are stored locally by the client. File 'M' and the corresponding tag 'ϕ' will be sent by clients and remotely stored on the server side.

Phase 2-Challenge-response

The challenge-response phase is an interaction procedure between the client and the server. The client computes challenges and sends them to the server. The server computes the corresponding responses and sends the output to the client. In this phase, initially, GenChallenge(s) → {chal-client} takes the client's private parameter 's' as input and outputs challenge chal-client for the future query. Then, GenProof(g_s, M, ϕ, chal) → {P} takes the user's public parameter 'g_s,' file 'M,' metadata 'ϕ' and challenge 'chal' as its inputs and outputs the evidence 'P' for the user to verify that the server possesses the outsourced file correctly. Finally, CheckProof(pk, chal, metadata, P) procedure is applied to verify the server's possession of the target files. It takes the user's public key p_k,' challenge 'chal' generated by the client, metadata and evidence 'P' as inputs. If 'P' passes the verification, the function outputs 'Accept'; Otherwise, CheckProof outputs 'Reject.'

Phase 3-Update the verified results

In the final phase, the verification process supports dynamic data operation. The verification procedure is run for the clients to check the correctness of the dynamic operation results at the server side. The verification is initiated by updating the metadata and inputs. If the server verifies the updates correctly, it outputs 'Accept' and retains the new metadata locally; Otherwise, the procedure returns 'Reject.'

The potential entities are identified from the components of the cloud based on two ranking techniques and a scheme is implemented for choosing an optimal fault tolerance technique suitable for each cloud component. Hence, the architecture consists of ranking and optimal fault tolerance. The steps of ranking include:

1. In the first step, component-characterized relations are derived from the cloud applications and then processed for constructing the component graph.
2. Component ranking algorithm is applied over the constructed component graph for estimating the significant values of the cloud components.
3. The ranking results are used for identifying most of the significant parameters as they represent the potentiality of each component.

4. The investigation on the possible fault tolerant approaches is performed, and based on the estimated significant parameters, an efficient and effective fault tolerant scheme is chosen.
5. Reliable cloud services are enabled based on the chosen fault tolerant approach and the results of component ranking are then fed to the system designer for triggering maximum cloud security.

4 Result Evaluation and Inferences

In Experiment 1, the performance of the proposed Batch Signature-based Enhanced Data Verification Technique (BSEDVT) is examined in terms of query processing time for varying file sizes and verification thresholds. The plots of query processing times for different verification thresholds are shown in Figs. 1 and 2. Figure 1 proves that query times of BSEDVT, PBDVT and RKDVT systematically increase with increase in the file size. This increase in file size induces extra amount of file processing time and hence the query processing time. BSEDVT involves 12%-16% and 14%-17% reduced query processing times in contrast to PBDVT and RKDVT

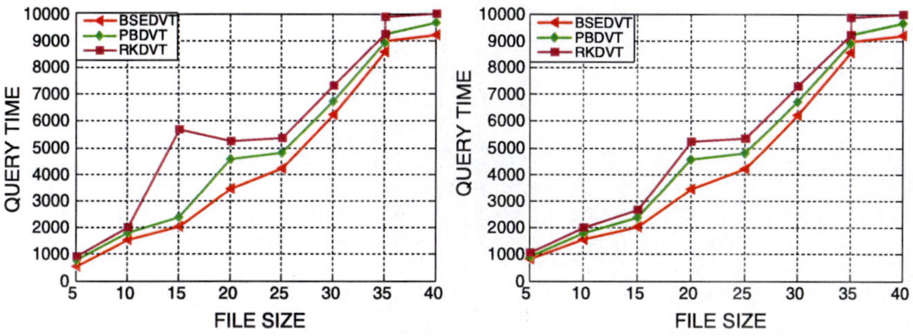

Fig. 1 Query Processing Times based on File Size (Verification Thresholds = 0.2 and 0.4)

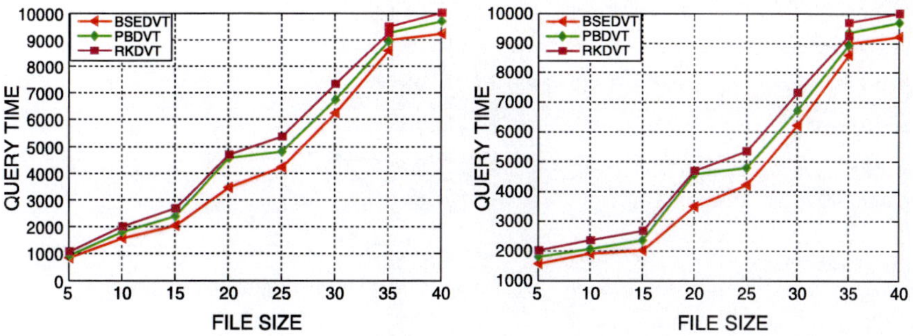

Fig. 2 Query Processing Times based on File Size (Verification Thresholds = 0.6 and 0.8)

respectively for a threshold of 0.2. Figure 1 also highlights the query processing times of BSEDVT, PBDVT and RKDVT for a threshold of 0.4 and BSEDVT is proved to minimize the query times by 11%-13% and 16%-19% in contrast to PBDVT and RKDVT, respectively.

Figure 2 depicts the query processing times of BSEDVT, PBDVT and RKDVT for the threshold of 0.6. BSEDVT involves 17%–19% and 20%–23% reduced query processing times in contrast to PBDVT and RKDVT, respectively. Similarly, Fig. 2 highlights the query processing times of BSEDVT, PBDVT and RKDVT for a threshold of 0.8, and it is seen that it transparently minimizes the query times by 7%–10% and 13%–17% in contrast to PBDVT and RKDVT, respectively.

In Experiment 2, the performance of BSEDVT is investigated to examine the system failure probabilities for varying number of tasks and verification thresholds. The plots of system failure probabilities for different verification thresholds are shown in Figs. 3 and 4.

Figure 3 confirms that the system failure probabilities of BSEDVT, PBDVT and RKDVT systematically increase with increase in the number of tasks. The increase

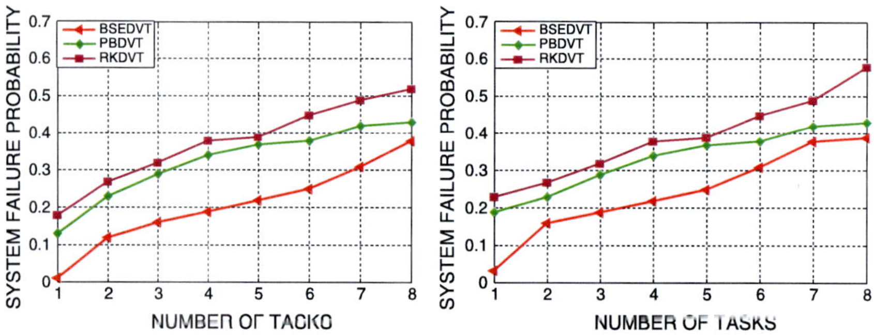

Fig. 3 System Failure Probabilities based on Number of Tasks (Verification Thresholds = 0.2 and 0.4)

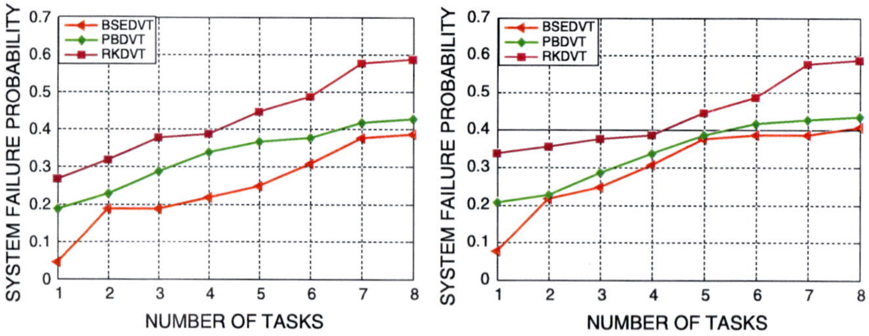

Fig. 4 System Failure Probabilities based on Number of Tasks (Verification Thresholds = 0.6 and 0.8)

in the number of tasks incurs an extra time for processing tasks and BSEDVT is determined to enforce a rapid task processing rates of 22% in contrast to the classical approaches of PBDVT and RKDVT. It is transparent that for a threshold of 0.2, BSEDVT offers 10%–13% and 13%–16% reduced system failure probabilities in contrast to PBDVT and RKDVT respectively. Figure 3 also highlights the system failure probabilities of BSEDVT, PBDVT and RKDVT for a threshold of 0.4. BSEDVT is proved to minimize the system failure probabilities by 8%–10% and 11%–14% in contrast to PBDVT and RKDVT, respectively. Figure 4 shows the system failure probabilities of BSEDVT, PBDVT and RKDVT for a threshold of 0.6. BSEDVT reduces the system failure probabilities by 7%–10% and 11%–15% in contrast to PBDVT and RKDVT, respectively. Similarly, Fig. 4 highlights the system failure probabilities of BSEDVT, PBDVT and RKDVT for a threshold of 0.8. It is transparent that BSEDVT minimizes the system failure probabilities by 5%–8% and 10%–13% in contrast to PBDVT and RKDVT, respectively.

In Experiment 3, the performance of BSEDVT is investigated in terms of computation overhead by varying the number of users and verification thresholds. The plots of computation overhead for different verification thresholds are shown in Figs. 5 and 6.

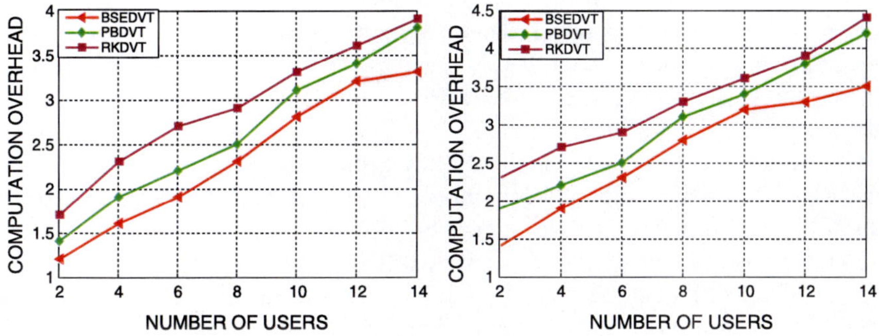

Fig. 5 Computation Overheads based on Number of Users (Verification threshold = 0.2 and 0.4)

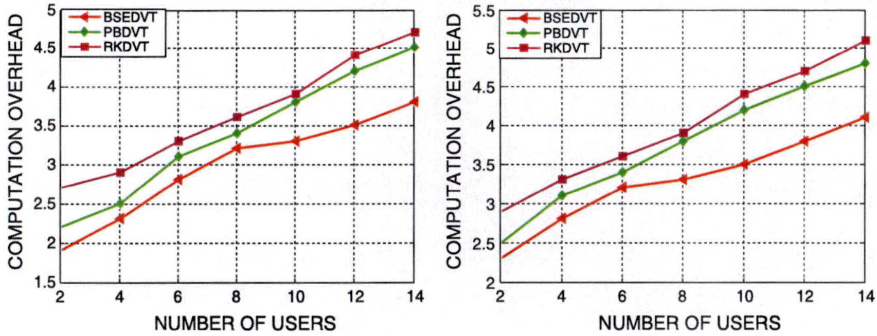

Fig. 6 Computation Overheads based on Number of Users (Verification threshold = 0.6 and 0.8)

Figure 5 proves that the computation overheads of BSEDVT, PBDVT and RKDVT systematically increase with increase in the number of users. This increase in the number of users incurs additional time for processing users' requests and BSEDVT handles this context by enforcing a rapid user response rate of 15% in contrast to the classical approaches of PBDVT and RKDVT. It is transparent that for a threshold of 0.2, BSEDVT decreases the computation overheads by 12%–14% and 17%–19% in contrast to PBDVT and RKDVT, respectively. Figure 5 also portrays the computation overheads of BSEDVT, PBDVT and RKDVT for a threshold of 0.4. This increase in threshold reduces the additional time for processing the tasks by ensuring a rapid user response rate of 16% greater than the considered traditional approaches like PBDVT and RKDVT. It is determined that BSEDVT decreases the computation overheads by 10%–15% and 18%–21% in contrast to PBDVT and RKDVT, respectively.

Figure 6 presents the computation overheads of BSEDVT, PBDVT and RKDVT for a threshold of 0.6. BSEDVT reduces the computation overheads by 9%–13% and 16%–19% when compared to PBDVT and RKDVT, respectively. Similarly, Fig. 6 also highlights the computation overheads of BSEDVT, PBDVT and RKDVT for a threshold of 0.8. It is transparent that BSEDVT minimizes the computation overheads by 13%–16% and 17%–19% in contrast to PBDVT and RKDVT, respectively.

In Experiment 4, the performance of BSEDVT is examined in terms of response times for varying number of tasks and verification thresholds. The plots of response times for different verification thresholds are shown in Figs. 7 and 8. It is evident that the response times of BSEDVT, PBDVT and RKDVT systematically decrease with increase in the number of tasks. This increase in the number of tasks incurs an additional time for processing each task and BSEDVT handles this context by enforcing a rapid task processing rate by 22% in contrast to the classical approaches of PBDVT and RKDVT. It is transparent that for a threshold of 0.2, BSEDVT involves 14%–16% and 19%–21% response times in contrast to PBDVT and RKDVT, respectively.

Figure 7 shows the response times of BSEDVT, PBDVT and RKDVT for a threshold of 0.4. This increase in threshold reduces an additional time for processing the tasks by ensuring a rapid task processing rate of 20% greater than the considered traditional approaches like PBDVT and RKDVT. It is determined that BSEDVT

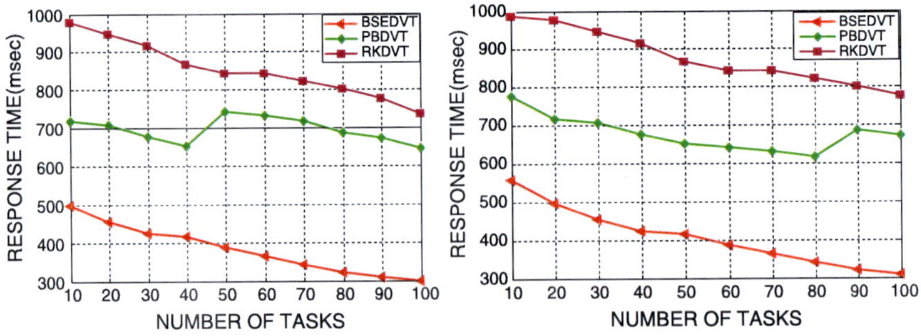

Fig. 7 Response Times (Verification Threshold = 0.2 and 0.4)

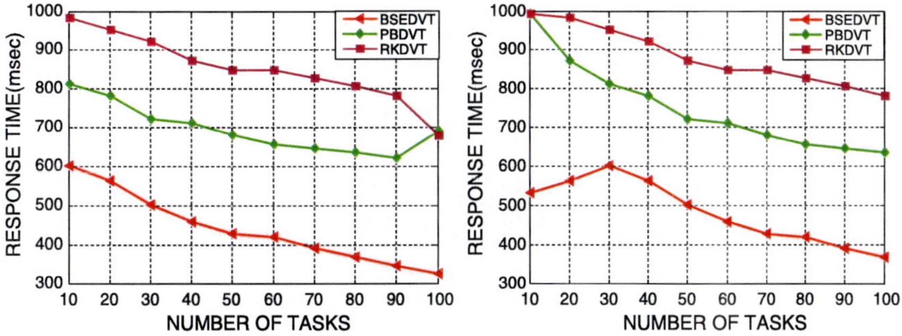

Fig. 8 Response Times (Verification Threshold = 0.6 and 0.8)

enhances the response time by 12%–18% and 21%–24% when compared to PBDVT and RKDVT, respectively. Figure 8 presents the response times of BSEDVT, PBDVT and RKDVT for a threshold of 0.6. This increase in ranking threshold reduces the additional time for processing the tasks by increasing the task processing rate by 26% in contrast to the considered traditional approaches like PBDVT and RKDVT. BSEDVT improves the response time by 11%–16% and 19%–23% when compared to PBDVT and RKDVT, respectively. Similarly, Fig. 8 highlights the response times of BSEDVT, PBDVT and RKDVT for a threshold of 0.8. This increase in threshold reduces the additional time for processing the tasks by ensuring a rapid task processing rate of 28% in contrast to the considered traditional approaches such as PBDVT and RKDVT. It is transparent that BSEDVT offers 15%–18% and 23%–26% reduced response times in contrast to PBDVT and RKDVT, respectively.

In Experiment 5, the performance of BSEVT is analyzed in terms of percentage decrease in system failure probabilities, percentage increase in throughputs, decrease in cost functions and increase in response times for varying verification thresholds. Figures 9 and 10 present the decrease in system failure probabilities and increase in throughput of BSEVT, PBDVT and RKDVT for varying verification thresholds. From the results, it is confirmed that the proposed BSEVT involves 4%–6% and

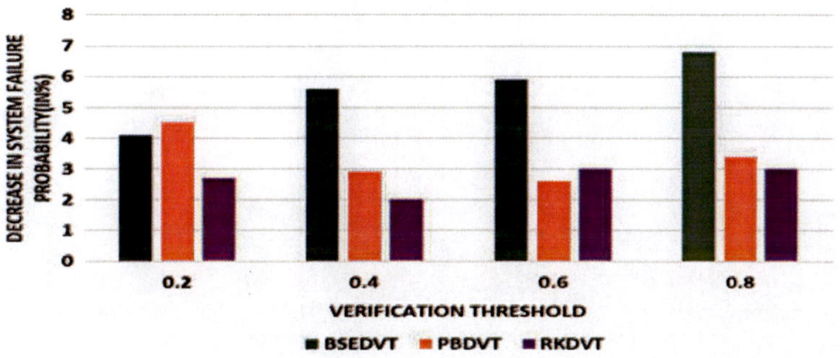

Fig. 9 Decrease in System Failure Probabilities

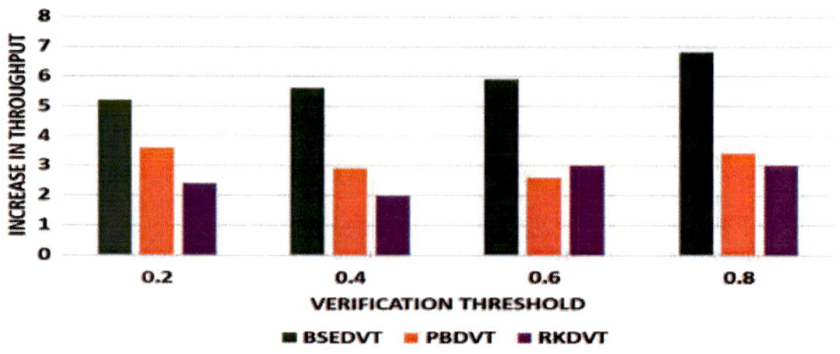

Fig. 10 Increase in Throughput

10%–13% reduced system failure probabilities in contrast to PBDVT and RKDVT, respectively. BSEVT offers 3%–5% and 11%–14% better throughput in contrast to PBDVT and RKDVT.

Figures 11 and 12 highlight the performance of BSEVT, PBDVT and RKDVT in terms of percentage decrease in cost function and percentage decrease in response times for different verification thresholds. Figure 11 highlights the decrease in cost function of BSEVT, PBDVT and RKDVT for varying verification thresholds. BSEVT

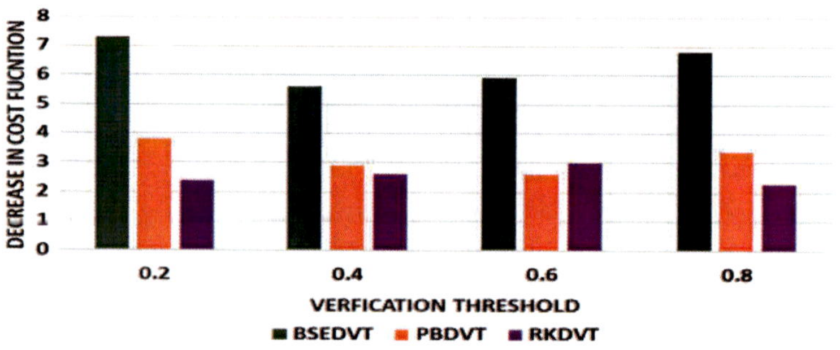

Fig. 11 Decrease in Cost Functions

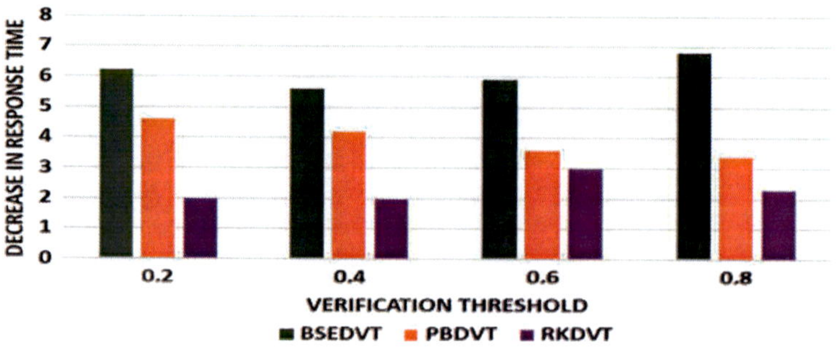

Fig. 12 Increase in Response Times

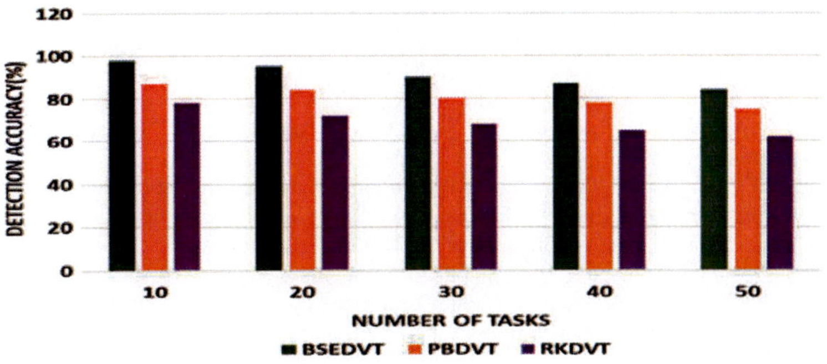

Fig. 13 Improvement in Verification Accuracy

decreases the rate of cost function by 4%–7% and 12%–16% in contrast to PBDVT and RKDVT, respectively. In addition, Fig. 12 presents the percentage increase in response times of BSEVT, PBDVT and RKDVT for varying verification thresholds and BSEVT is found to minimize the response times by 9%–11% and 16%–19% when compared to PBDVT and RKDVT, respectively.

Figure 13 presents the improvement in verification accuracies of BSEVT, PBDVT and RKDVT for a verification threshold of 0.8. BSEVT enhances the verification accuracy by 3%–6% and 6%–9% in contrast to PBDVT and RKDVT, respectively.

5 Conclusion

This paper has presented a Batch Signature-based Enhanced Data Verification Technique (BSEDVT) for improving the verification accuracy rate of the data stored on the cloud. The performance of BSEDVT is found to be predominant under different investigations conducted for testing its potential. The results infer that the verification accuracies of BSEVT, PBDVT and RKDVT for different verification thresholds seem to enhance the verification accuracies by 3%–7% and 9%–12% in contrast to PBDVT and RKDVT, respectively. Similarly, the decrease in response times of BSEVT when compared to PBDVT and RKDVT for different verification thresholds is confirmed to enhance the verification accuracy by 3%–6% and 6%–9% when compared to PBDVT and RKDVT, respectively.

References

1. Sathya Bama S, Irfan Ahmed MS, Saravanan A (2014) A mathematical approach for improving the performance of the search engine through web content mining. J Theor Appl Information Technol 60(2)
2. Kumar TG, Vijayan VP (2007) A multi-agent optimal path planning approach to robotics environment. In: Proceedings of the IEEE International Conference on Computational Intelligence and Multimedia Applications (ICCIMA 2007) 1 (2007, December), pp 400–404
3. Poornaselvan KJ, Kumar TG, Vijayan VP (2008) Agent based ground flight control using type-2 fuzzy logic and hybrid ant colony optimization to a dynamic environment. In: Proceedings of the 1st IEEE international conference on emerging trends in engineering and technology, (2008, July), pp 343–348
4. Venkatesh V, Parthasarathi P (2013) Enhanced audit services for the correctness of outsourced data in cloud storage. Int J Adv Res Comput Eng Technol (IJARCET) 2
5. Aswin MR, Kavitha M (2012) Cloud intelligent track-risk analysis and privacy data management in the cloud computing. In: Proceedings of the IEEE international conference on recent trends in Information Technology (2012, April), pp 222–227
6. Anto S, Pandian PA (2013) An optimal solution of resource provisioning cost in cloud computing environments. Int J Comput Appl Information Technol 2(2):35–40
7. Islam S, Ouedraogo M, Kalloniatis C, Mouratidis H, Gritzalis S (2015) Assurance of security and privacy requirements for cloud deployment models. IEEE Trans Cloud Comput 6(2):387–400
8. Sengathir J, Deva Priya M, Christy Jeba Malar A (2020) Improved privacy multi-keyword based secure retrieval scheme for cloud data. Int J Sci Technol Res 9(02):909–914, ISSN 2277–8616
9. Christy Jeba Malar A, Deva Priya M, Sengathir J, Kiruthiga N, Anitha R, Sangeetha T (2019) An intelligent multi-floor indoor positioning system for cloud based environment. Int J Comput Appl. https://doi.org/10.1080/1206212X.2019.1696447. ISSN: 1925–7074, Taylor & Francis (Elsevier)
10. Zhang H, Zhou Z, Ye L, Du X (2015) Towards privacy preserving publishing of set-valued data on hybrid cloud. IEEE Trans Cloud Comput 6(2):316–329
11. Esposito C, Palmieri F, Choo KKR (2018) Cloud message queueing and notification: challenges and opportunities. IEEE Cloud Comput 5(2):11–16
12. Chenthara S, Ahmed K, Wang H, Whittaker F (2019) Security and privacy-preserving challenges of e-Health solutions in cloud computing. IEEE Access 7:74361–74382
13. Du M, Wang Q, He M, Weng J (2018) Privacy-preserving indexing and query processing for secure dynamic cloud storage. IEEE Trans Inf Forensics Secur 13(9):2320–2332
14. Cheng H, Rong C, Qian M, Wang W (2018) Accountable privacy-preserving mechanism for cloud computing based on identity-based encryption. IEEE Access 6:37869–37882
15. Jiang Q, Ma J, Wei F (2016) On the security of a privacy-aware authentication scheme for distributed mobile cloud computing services. IEEE Syst J 12(2):2039–2042
16. Ahmad W, Wang S, Ullah A, Mahmood Z (2018) Reputation-aware trust and privacy-preservation for mobile cloud computing. IEEE Access 6:46363–46381
17. Wang T, Zhou J, Chen X, Wang G, Liu A, Liu Y (2018) A three-layer privacy preserving cloud storage scheme based on computational intelligence in fog computing. IEEE Trans Emerging Topics Comput Intell 2(1):3–12
18. Xue K, Li S, Hong J, Xue Y, Yu N, Hong P (2017) Two-cloud secure database for numeric-related SQL range queries with privacy preserving. IEEE Trans Inf Forensics Secur 12(7):1596–1608
19. Wang C, Wang Q, Ren K, Lou W (2010) Privacy-preserving public auditing for data storage security in cloud computing. In: Proceedings of IEEE Infocom (2010, March), pp 1–9
20. Chen L (2013) Using algebraic signatures to check data possession in cloud storage. Future Gener Comput Syst 29(7):1709–1715
21. Paigude T, Chavan TA (2013) A survey on privacy preserving public auditing for data storage security. Int J Comput Trends Technol (IJCTT). 4(3):412–418

22. Shin S, Kwon T (2015) A survey of public provable data possession schemes with batch verification in cloud storage. J Internet Serv Inf Secur 5(3):37–47
23. Shen J, Shen J, Chen X, Huang X, Susilo W (2017) An efficient public auditing protocol with novel dynamic structure for cloud data. IEEE Trans Inf Forensics Secur 12(10):2402–2415
24. Zhu H, Yuan Y, Chen Y, Zha Y, Xi W, Jia B, Xin Y (2019) A secure and efficient data integrity verification scheme for cloud-IoT based on short signature. IEEE Access 7:90036–90044

A Cellular Automata-Based Bilateral Filter for De-speckling in Ultrasound Imaging

Deboshree Seal, A. P. Shukla, and Ankur Bhardwaj

Abstract The ultrasound imaging is much more recommended in the medical field for the clinical recognition than any other imaging mode, as it is of adaptable, befitting, safer, and viable nature. The images which are captured thereafter per-processed; as the process plays an important aspect in medical diagnosis. The accession of the images can incorporate different kinds of noise. The much the degraded the image is, the more is the problem arise in for the better disentangling the required features, recognition, studying clarity of edges. In this paper, the comparison is done in between the various filters like Gaussian, kuan, lee, etc., which were discussed in previous paper with the bilateral imposed together with the cellular automaton filter, both together has been imposed. The comparative result has been depicted in the table format as a result.

Keywords MSE · PSNR · SNR · Bilateral filter · Cellular automata

1 Introduction

Ultrasound known as ultrasonography, which works in disclosing the body part to the high-frequency ultrasound waves in order to fabricate interior body images. These images are always captured in the real-time; it can very well depict the motion of the blood flow of interior body organs along via blood vessels of any living being. Apart from X-ray, the imaging with ultrasound contains no ionizing radiation linked to it. The medical test uses ultrasound scan that makes the use of the high-frequency sound waves for getting the live images of the internal body. This is done so as to identify

D. Seal (✉) · A. P. Shukla · A. Bhardwaj
Computer Science and Engineering Department, KIET Group of Institutions, Ghaziabad, India
e-mail: deboshreesheal@gmail.com

A. P. Shukla
e-mail: ap.shukla@kiet.edu

A. Bhardwaj
e-mail: ankurbhardwaj2002@gmail.com

R. P. Mahapatra et al. (eds.), *Proceedings of 6th International Conference on Recent Trends in Computing*, Lecture Notes in Networks and Systems 177,
https://doi.org/10.1007/978-981-33-4501-0_69

and becoming aware of the reason of pain, lump, and infection that may come in internal organs of the body and also to examine women with embryo in womb, who so ever is expecting and the brain and elbows, feet in an infants. Speckle noise is entitled cumulative or coarse-grained noise. This can result in supple synthetic aperture radar (SAR) and particularly in imaging via ultrasound. The resultant of this kind of noise is due to trespassed waves that seem to be returning back. In imaging via ultrasound, speckle noise is a critical problem as the difficulty of whose simplification can be seen in the image. As in SAR is a kind of cumulative type [1], in radar applications, the speckle can create problem. When the speckle is present in SAR images results in the degradation of their quality. The ultrasound imaging's value decreases by the presence of noise called speckle that is considered to be of cumulative type. Speckle noise is a type of an underlying quality in ultrasound imaging in medical field, like in charge for the resolution of captured image and variations that causes depletion, which results in the error of the value of diagnosis of the computing the modality [2]. So, it tends to become very important to decrease speckle noise in the very first step, i.e., pre-processing, so that these ultrasound imaging can be used in the medical purpose. Hence, de-speckling the image becomes chief feature, and the removal of noise is done of the image [1–6], without wrecking the needed features. The resultant noise creates signal disruption in those returning from the objects which tend to be smaller than a single image-processing element. This creates a difficulty in picturing the image due to the noise present. The statistical property analysis for the speckle noise was given by Goodman [7], which helped to much extent in denying, accepting, and considering the laser speckle. A method was provided by Goodman so as in conquering the speckle noise via, aid of various filtering (linear) method that helped in improving the image quality. This statistical attribute that arises due to noise helped a lot in seeking difference, between the different types of tissues [8, 9]. The resultant of [7] was done with up gradation in [8].

2 Methodology

2.1 *Bilateral Filter*

A **bilateral filter** is a kind of irregular approach ability that provides reduction of noise by applying the smoothing filter over the images while maintaining edge preservation. The job of this filter is blurring the captured or pictures to work with but with the edges preserved strongly. It has great capability so as to break down an image into various parameters without generating post-modification haloes—this has made it universal in computational photography implementation such as tone aligning, style shift, renewing, and DE noising [4]. It changes the magnitude of each pixel with a weighted average of the magnitude of adjacent pixels values, where the weight can depend on a Gaussian distribution. This feature is dominant because it makes it simpler to capture instinct about its behavior, to make its requirement applicable to its application-specific need, and to implement it. More importantly, the weights depend on the radiometric differences (like on the range differences, e.g., in the color intensity, depth, or distance, etc.) as opposed to Euclidean distance of pixels. This is helpful in preserving the sharp edges [10].

2.2 *Cellular Automata*

The cellular automata at very first coined by Ulam [1] is an individual dynamical system. Discrete system can be termed as the parameters of space and that of time. Here, the lattice (cells) contains the fixed number of states that are organized in some particular kind of the pattern. The dimensions of the grid can be fixed with any number. The state of a particular lattice at any point of time depends only on itself and to the neighboring cells at one step ahead. The state of cell is constructed by some kind of transformation rule which can be a kind of mathematical function and changes accordingly and hence updated simultaneously.

3 Filters Used for Comparative Study

(i). **Anisotropic diffusion**: This filter focuses in reducing the noise that is present in the image without disturbing or losing the chief details of the content of the image, mainly the edges, lining, or the fine details that seem to be important for understanding an image keenly [6, 11, 12]. It is similar to the processing and making a multi-scale single representation; therefore, an image constructed with the collection of many indistinct pictures that depends on diffusion procedure.

(ii). **Gaussian smoothing filter**: It is well known in taking the edge off (softening) and squeezing the present noise [12] anywhere that has been captured inside the picture. The filtering applies a 2D intricacy operator that works by multiplying the components of (an average) by factors to take account of their importance which is embraced in well-balanced way in sculpturing of Gaussian function [13].

(iii). **Kuan filter**: It works good for de-speckling the SAR image [14] that is worked for in the image processing. In spite of the other local statistics filters, this is pretty much well-known DE speckling algorithm. And works effectively within the alike area of SAR images in comparison to the boundary is imposed with retribution [15].

(iv). **Kuwahara filter**: This filter is a type of non-linear smoothening filter which is considered for the adaptive noise reduction. This filter works for the smoothening the image while protecting the image quality. In order to access images' pixel information, this filter is made to slide over the image.

(v). **Lee filter**: This filter tends to smoothen the noise [16] present in the image which has dominance related to the scene of the image and hence has an additional/cumulative part. The function of the Lee filter works with the standard deviation of those images that have been filtered depending on the statistics which is calculated within the individual windows of the filter.

4 Quality Parameters for Images

(i). **Mean Square Error (MSE)**: Here, this looks for any kind of dissimilarity in grade of a captured image (*R*) to that of processed (noise free) one, (*S*) of size $M \times M$ [2]. It is given as

$$\mathrm{MSE} = \frac{1}{m^2} \sum_{i,j=0}^{m-1} \left(R_{i,j} - S_{i,j}\right)^2$$

The scaling in the image standard is done by using MSE, but cannot work by itself, because it cannot correlate enough strongly with those non-cognitive standards. This must hence be compared to various different quality metrics and perceptible approach [17].

(ii). **Signal-to-Noise Ratio (SNR)**: This recognizes the amount of required signal with the noise in the background. As the proportion increases, the background noise is lesser prominent [18]. It is measured in terms of decibels (dB) as

$$\mathrm{SNR} = 10\log_{10}\left(\frac{\sigma^2}{\sigma_e^2}\right)$$

where σ^2 is the initial image variance [1] and σ_e^2 be error variance (Dissimilarity between the captured and image which is error free which is $|R - S|$).

(iii). **Peak Signal-to-Noise Ratio (PSNR)**: This can be [19] given as

$$\mathrm{PSNR} = 10\log_{10}\left(\frac{L^2}{\mathrm{MSE}}\right)$$

In this, L taken as maximal intensity of the captured (original) picture. As the value of PSNR increases, image hence considered to be of better quality and vice-versa. Here, the evaluation is done for the image accuracy, so as to check that how keenly the transmuted image is close to the initial image [2]. It is the ratio that is taken in between the reference signal to the deform signal of an image, termed in decibels [18]. The higher the PSNR, the closer is distorted image is to the original.

5 Proposed Algorithm

The input that is fed in MATLAB is the ultrasound image that is captured [2]. Thereafter the bilateral filter with the proposed CA is imposed over the speckled image in order to receive DE noised image [2] with clarity. When the error free images are processed out by applying the filter over, the error metrics is hence taken into account. Then, after the image quality analysis is calculated of the error-free image, so as to check the quality issues with that of original image [20]. The numbers of quality parameters are like PSNR, SNR MSE functions are called in the program [2]. Thereafter the error, i.e., (Error = Captured Image — Resulted Image) is calculated. The results are here compared with other [3] filters with the new one. A proposed algorithmic approach which is to be applied is shown below.

Input: Images from ultrasound **Output**: Ultrasound images that are error free.

- Ultrasound image is called in the program
- Various image parameters functions are called in the program
- DE noised image, i.e., the filtered image is hence resulted.
- Error present in the image is calculated by using the given formula (Error = Captured Figure—Reconstructed Figure).

Various parameters to consider image such as MSE, SNR, PSNR are took into account for examining the various aspects that have been resulted (Figs. 1, 2, 3, and 4).

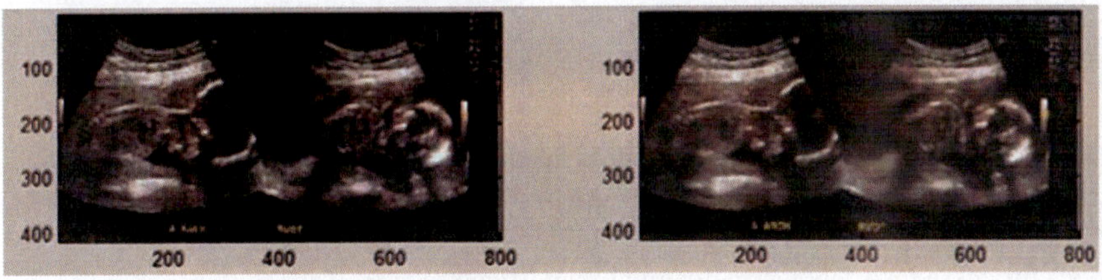

Fig. 1 A ARCH the original (speckled) and processed image

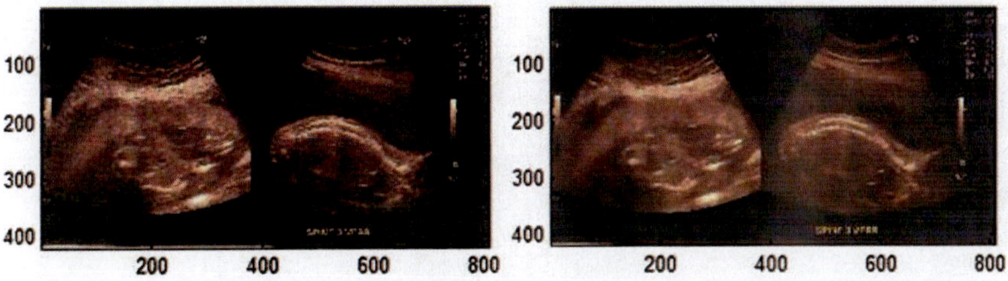

Fig. 2 SPINE the original (speckled) and processed image

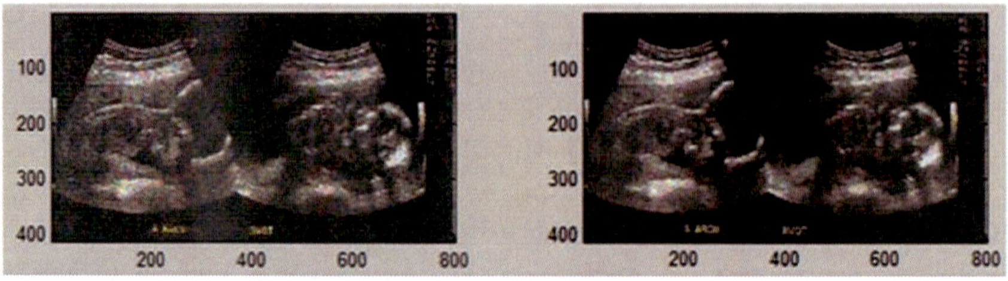

Fig. 3 A ARCH the original (speckled) and processed image

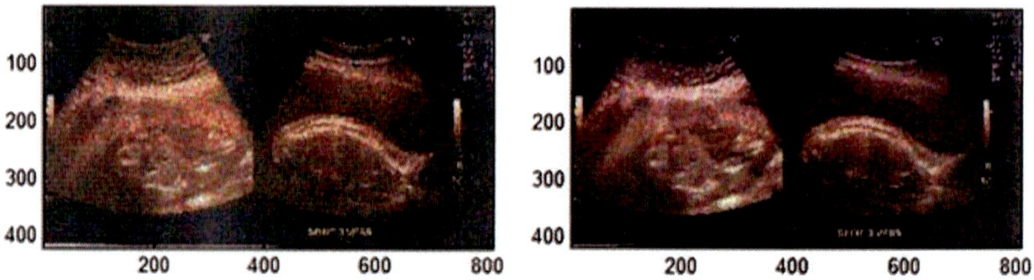

Fig. 4 SPINE the original and processed image

6 Result and Discussion

The dataset that which we have taken into consideration is approx. 200 in number, over on that the processing is done. The ample amount of data which has been processed out; the results of some of them are shown below. The comparison of some filter with the CA based on bilateral filter has been depicted in the table over some of the ultrasound images that have been processed (Tables 1, 2, 3, and 4).

The calculated results on basis of the above parameters used are made available of the data processed, and hence, the tables contain the details of the data that has been processed, so that we may conclude with the results of the tables that are shown above. That can be taken into account depending on the different parameters for the quality measurement known to us; among which the parameters MSE, PSNR,

Table 1 Outcome of filters in terms of quality parameters on A ARCH while de-speckling 0.01 speckle value

Filters	MSE	PSNR	SNR
Aniso_diff	87.9013	31.7011	22.9865
Gaussian	34.5515	25.8218	45.5999
Kuan	320.5666	26.0819	17.2967
Kuwahara	118.7747	30.3939	21.6868
Lee	36.18	26.83	36.18
CA-based bilateral	35.8646	30.53	24.358

Table 2 Outcome of filters in terms of quality parameters on SPINE while de-speckling 0.01 speckle value

Filter	MSE	PSNR	SNR
Aniso_diff	81.2057	32.0452	9.0114
Gaussian	51.4549	34.0268	7.1732
Kuan	319.2779	26.0994	17.8684
Kuwahara	110.7335	30.6983	10.523
Lee	32.2779	36.0825	5.6615
CA-based bilateral	31.2407	31.9556	64.0364

Table 3 Outcome of filters in terms of quality parameters on A ARCH while de-speckling 0.05 speckle value

Filter	MSE	PSNR	SNR
Aniso_diff	87.9013	31.7011	22.9865
Gaussian	45.5999	34.5515	25.8218
Kuan	320.567	26.082	17.297
Kuwahara	118.7747	30.3939	21.6868
Lee	36.1761	35.5569	26.831
CA-based bilateral	203.45	25.0802	50.1605

Table 4 Outcome of filters in terms of quality parameters on SPINE while de-speckling 0.05 speckle value

Filter	MSE	PSNR	SNR
Aniso_diff	81.2057	32.0452	24.2425
Gaussian	51.4549	34.0268	26.196
Kuan	319.2779	26.0994	18.2148
Kuwahara	110.7335	30.6983	22.8988
Lee	32.2779	36.0825	28.2584
CA-based bilateral	157.86	26.19	52.3638

and SNR have been discussed. And the various filters that have been taken into the account gives the various values that have to be compared to each other, so as to check for the filter that has to be compared with each other.

We can see while comparing the values of these parameters the bilateral filter along with CA imposed over the pictures so as to provide us the better result. In case of MSE when taken into account, if the values of the filtered images are if closer to 0, then the image quality tends to be good. By which, it can be said that the bilateral filter along with the CA imposed gives the better result. When we talk about the SNR values, it is taken into account with the range. SNR value if tends to be < than 24, then the image quality tends to be the poor, $24 \leq \text{SNR} \leq 36$ images are considered to be of better quality, and if SNR > 36, then magnificent. So for this kind, all the resulted images show the poorest quality, but the CA with bilateral again gives the good result.

The third parameter that has been taken into consideration is PSNR. In this parameter, the more is the result of the processed picture, the better is the image quality. And hence, this can be seen that again the bilateral with CA resulted in the better image quality at the pixel level. The processed work of the paper is highlighted in the table. These all tends to help in minimizing the speckle values with varied iterations applied over the images.

7 Conclusion

On inspecting the resulted data, it can be stated that when we increase the number of iterations on using different filters, their execution on considered image parameters fluctuates. On using the various kind of quality measuring with the different parameters and different filters resulted in different results. None such filters gave the best results in the terms of various parameters that have been taken into account in all the various cases on the bases of quality of the images. The result can be considered that non such filter can be made in a manner which gives all the best results in comparing all the grade estimating specification for the multiplicative speckle noise. In future, the researches can be able to analyze about the effects of multiple iteration of various filters based on considered parameters.

References

1. Ulam S (1972) Some ideas and prospects in biomathematics. In: Annual review of biophysics and bioengineering, vol 1, no 1, pp 277–292. Godman JW (1976) Some fundamental properties of speckle. J Opt Soc Am 66(11):1145–1149
2. Gupta N, Shukla AP, Agarwal S (2016) Despeckling of medical ultrasound images: a technical review. Int J Information Eng Electronic Bus
3. Yongjian Y, Acton ST (2002) Speckle reducing anisotropic diffusion. IEEE Trans Image Processing 11(11):1260–1270S. Ulam (1972) Some ideas and prospects in biomathematics. In: Annual review of biophysics and bioengineering, vol 1, no 1, pp 277–292 (1972)
4. https://www.ijarest.com/papers/finished_papers/150312133754.pdf
5. Gupta M, Taneja H, Chand L (2018) Performance enhancement and analysis of filters in ultrasound image denoising. In: Procedia computer science
6. Perona P, Malik J (1990) Scale-space and edge detection using anisotropic diffusion. IEEE Trans Pattern Anal Mach Intell 12(7):629–639. https://doi.org/10.1109/34.56205
7. Prager RW, Gee AH, Treece GM, Berman L (2002) Speckle detection in ultrasound images using first order statistics, GUED/ F-INFENG/TR 415, University of Cambridge, Dept. of Engineering, July, 2002, pp 1–17
8. Goodman JW (1975) Statistical properties of laser speckle patterns. Laser speckle and related phenomena. Springer, Heidelberg, pp 9–75
9. Wagner F, Smith SW, Sandrik JW, Lopez H (1983) Statistics of speckle in ultrasound B-scans. IEEE Trans Sonic Ultrason 30:156–163

10. Mengqi L (2011) Research on image de-noising enhancement. Savonia University of Applied Sciences, Bachelor's thesis, September 2011
11. Perona P, Malik J (1987) Scale-space and edge detection using anisotropic diffusion. In: Proceedings of IEEE computer society workshop on computer vision, pp 16–22
12. Sapiro G (2001) Geometric partial differential equations and image analysis. Cambridge University Press. p. 223. ISBN 978-0-521-79075-8. Kadam C, Borse SB (2017) An improved image denoising using spatial adaptive mask filter for medical images. In: 2017 International Conference on Computing, Communication, Control And Automation, (ICCUBAE), 2017 Publication
13. Pei YH, Shin SC, Feng CH (2007) Generic 2D Gaussian smoothing filter for noisy image processing. In: IEEE Region 10 Conference(TENCON)
14. www.ijoart.org Internet Source
15. Information Systems Design and Intelligent Applications. Springer Nature, 2016 Publication
16. Kadam C, Borse SB (2017) An improved image denoising using spatial adaptive mask filter for medical images. In: 2017 International Conference on Computing, Communication, Control and Automation, (ICCUBAE), 2017 Publication
17. Godman JW (1976) Some fundamental properties of Speckle. J Opt Soc Am 66(11):1145–1149
18. Burckhardt CB (1978) Speckle in ultrasound B mode scans. IEEE Trans Sonics Ultrasonics 25:1–6
19. Seal D, Shukla AP, Bhardwaj A (2019) Effect of iterative variations on de-speckling filters in ultrasound imaging. In: 2019 International Conference on Issues and Challenges in Intelligent Computing Techniques (ICICT), GHAZIABAD, India, pp 1–6
20. Sehgal (1993) Quantitative relationship between tissue composition and scattering o of ultrasound. J Acoust Soc Amer 94:1944–1952

National Time Synchronization Augmenting Cybersecurity

Amutha Arunachalam, K. Seetharaman, and Ashish Agarwal

Abstract The essential matter is to get the time synchronization of country-linked communication infrastructures, computer systems and devices to the IST (Indian Standard Time). The competence to demonstrate its sequence of traceability to Coordinated Universal Time (UTC) originated by BIPM, with adequate lawful/governing support. This has acute share in real-time applications in critical service sector and to national security. Given the significance of cybersecurity and cybercrime angle developing the formation of Distributed Timing Infrastructures to several critical sectors and customer-centric timing delivery methods is the requirement of time. The International Standards ISO/IEC 18014 under a generally used title of Information technology—Security techniques—Time stamping services consisting of three parts, including a general notion, time stamping service model, data structures as well as protocols in its part 1(2008), 2(2008) & 3(2009) and Part 4(2015) for traceability of time sources has been adopted in many countries. It gives a speedy thought to have a precise Nationwide Timing strategy that shall address time delivery services that could be chosen by the wide range of users with acceptable level of uncertainties.

Keywords Cybersecurity · Time synchronization · Indian Standard time and time stamp

A. Arunachalam (✉) · A. Agarwal
Time & Frequency Metrology, National Physical Laboratory, New Delhi, India
e-mail: amutha@csir.res.in

A. Agarwal
e-mail: ashish.npl@nic.in

K. Seetharaman
Department of Computer & Information Sciences, Annamalai University, Chidambaram, India
e-mail: kseethadde@yahoo.com

R. P. Mahapatra et al. (eds.), *Proceedings of 6th International Conference on Recent Trends in Computing*, Lecture Notes in Networks and Systems 177,
https://doi.org/10.1007/978-981-33-4501-0_70

1 Introduction

The government has a special accountability to a citizen, to businesses and organizations working in Indian sovereign and to our worldwide partners and associates. A surety must be provided to them assuring that all efforts are made to solidify the safe infrastructures and to safeguard the network and the related data from attack or intrusion. Setting ourselves with the highest standards of cybersecurity and follow them will make the groundwork for national security of India as well as the well-being economically, along with a precedence for many to adapt [1].

The Internet dependence is characteristically not secure and will include attempts for exploiting certain weaknesses for launching the cyberattacks. Absolute eradication of threat though difficult, the risk needs to be reduced expressively up to some level which enables the society for continuous displaying and benefitting from higher prospects which is brought in by the digital technologies. The government administration, country's economic conditions as well as establishment of services trust, cyberspace reliability as well as on data, infrastructure and the system that help to strengthen it. Loss in integrity and trust will have an impact on the advantages of scientific and technical revolution.

The SI second is outlined because the length of nine, 192,631,770 radiation periods comparable with present transition among 2 thin levels of a bottom state of a Cs 133 atom. Such an explanation was adopted by Bureau International des Poids et Measures (BIPM) in 1967. By 1997, an amendment was introduced to the SI definition of second as "on second is the time referred for 9,192,631,770 oscillators between doubly spotted hyperfine ground state of 133 cs atoms at 0 degree kelvin temperature, which is possible to achieve in an atomic function." The length of SI second is expected to be time invariant and unaffected by the environmental changes. More the 400 industrial high-performance Cesium clocks operated in natural measurement institutes across the world contribute to the generation of international time scale called Coordinated Universal Time.

The Indian NMI National Physical Laboratory (CSIR NPL) is considered for local realization of UTC in India, using the Cesium clocks and disseminate "time" and "frequency" to fulfil needs of the country. It is desired that all the networks in the country should have uniform time that is synchronized accurately to UTC. Even if the network clocks are accurately set initially, they would still have some difference after some time because of the drift in the clock, which can cause several problems. Hence, all the network clocks need to be time synchronized to an accurate primary reference clock.

Traceability associate exceedingly in a very scientific discipline term may be the parameter for measuring where the outcomes may be associated with the help of reference using the unbroken calibration chains, each one of the contributing for measuring uncertainties [1]. UTC is used to refer time as per the International Bureau of Weights and Measures, and hence, all clocks used in legal procedures must be tracked as per this time. As per this time of reference, the traceability provided is direct by comparing the CCTF-K001 (Fig. 1).

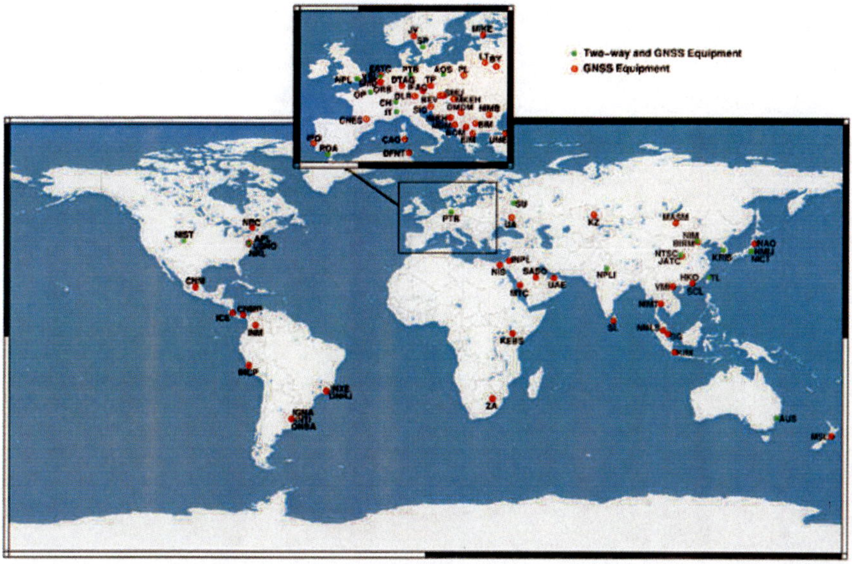

Fig. 1 Geographical distribution of the laboratories that contribute to TAI and time transfer equipment [3] (December 2018)

UTC key in which the NMIs have to send its measurement of time and the data of time transfer in to BIPM [2]. After every month, the BIPM processes its measurement of time from the previous month and results a UTC in the form of a highly stable and smooth reference to the international scale of time. There is a variation on the deviations in every part time of laboratory unit which is printed every month by the International Bureau of Weights and Measures in the Circular T. Users for obtaining their traces by comparing local clock to that of other participating NMIs in CCTF-K001.UTC. The time traceability can be achieved by remotely comparing, number of technical aspects used for (a) transferring time codes in 2-time labs or (b) the way in which a comparison can be made regarding their native clocks to those or another relevant institute area. Such kinds of technical methods are Global Navigation System (GNSS), and Two-Way Satellite Time and Frequency Transfer (TWSTFT). International network of atomic clocks participating in generation of UTC using international links (Fig. 2).

2 Dissemination of Indian Standard Time

CSIR NPL is country's time keeper and helps in measuring the highest levels of frequency and time estimations in India according to the internationally set standards. A primary time scale called UTC(NPLI) was maintained using hydrogen maser and 5 Cesium (Cs) atomic clocks. Every 5 metallic element clocks and therefore the

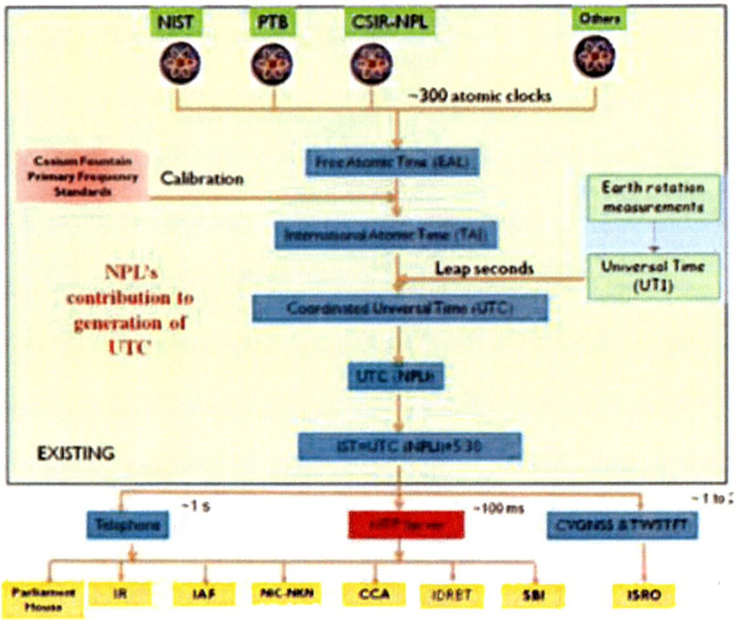

Fig. 2 NPL's contribution to generation of UTC [5]

chemical element amplifier area unit conducive to International Atomic Time by precise satellite links, automated, regulated, and precise inter-comparison of clocks, along with daily automated data upload into International Bureau of Weights and Measures [5].

UTC(NPLI) + 5:30 known as the IST. Indian Standard Time is complete with 7.2 ns of uncertainty. Exploitation of time transfer to the time links which are ultra-stable is carried out by CSIR-NPL for contributing to the set standards of TAI and maintaining national duration. Adding the leap seconds from time to time to the Indian Standard Time is also done by CSIR-NPL Dissemination of Time and frequency carried out with the help of NTP (Network Time Protocol) service. Recently, CSIR-NPL launched a web-based time dissemination service (Fig. 3) [6].

3 Methodology Used

NTP elaborated as Network Time Protocol is a standard Internet protocol used for maintaining and organizing a complete set of servers for time and their respective paths or transmission in a synchronized subset. Network Time Protocol was developed on the basis of UDP: User Datagram Protocol and IP: Internet protocol that helps in providing a mechanism of connectionless transfer. However, this mechanism is highly adaptable to every protocol suites of Internet. This was evolved by an ICMP

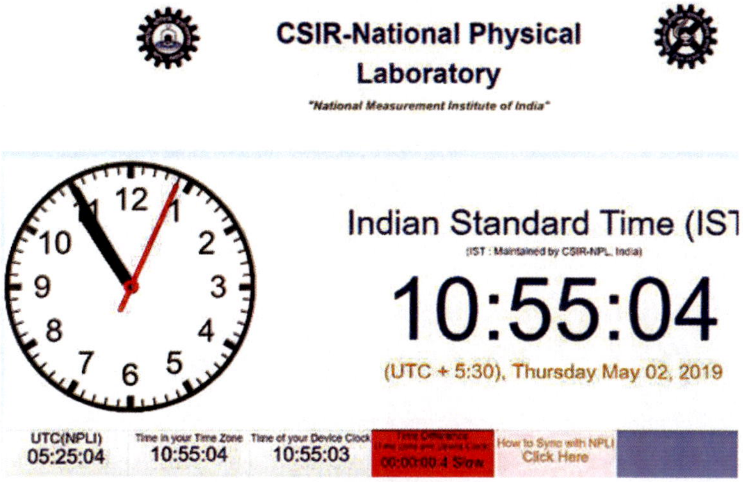

Fig. 3 Web Time Display of Indian Standard Time [6]

time stamp message and time protocol and is designed specifically for maintaining the reliability and accuracy, when applied on typical paths of Internet which involves multiple unreliable networks and gateways. There exists no provision for acquisition peer discovery, or even configuration in the process of Network Time Protocol, but many implementations also comprise of such features. Data integrity is provided by UDP and IP checksums. No facilities of retransmission, circuit management or duplicate detection are necessary or provided [4].

This protocol could be operated in many different modes which are applicable to varying scenarios involving network configurations, private workstations as well as public servers. Variable mechanisms of poll-interval and dynamic reachability under a lightweight capability of association management are used for managing the information regarding the state and also help in reducing the requirements of resources. The option features comprise of authentication of messages depending on the provisions and crypto-checksums for monitoring and remote control. The implementation of this protocol is quite easy as only a single format for NTP message is used and are applicable to many different operating systems as well as networking environments. In Network Time Protocol, one or more than one primary server is directly synchronized to an external source of reference like in timecode receivers. Figure 4 illustrates the overall Network Time Protocol organization of a timeserver model.

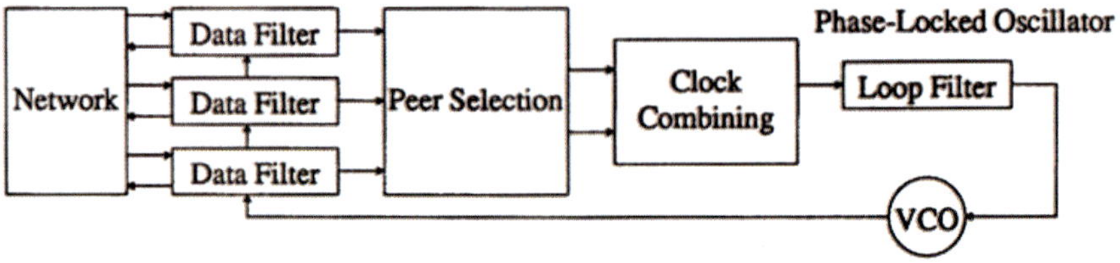

Fig. 4 Network Time Protocol

3.1 Establishing Secure and Resilient Time in Networks, Systems and Devices

Security: Diminishing a chance of entering into a cyber-physical infrastructure by threats which could be both natural or manmade. **Resilience**: An ability of adapting and preparing for the changing conditions as well as recovering and withstanding rapidly from any disruptions.

Necessities for resilient as well as secure time also exist in every layer of network till application from physical layer. Time is a physical quantity, its perception in the networks as well as complex system of information transforms the security into physical and cyber concerns. Hence, time has an impact on both the architectures, physical and cybersecurity. Timing could be unintentional or vulnerable-interference, can be impacted due to space weather, abnormalities in the network, along with intentional threats like spoofing and jamming simulating because of cyberattack or injection of RF signal. Developing other techniques for the distribution of national-level time reference for augment and backup the IRNSS in case of failure should be proactively worked upon. The methods of timing distribution in the communication sector, like the distribution protocols on a dedicated optical network or in combination of PTP and SyncE, could act as an alternative national reference time source. Distribution of network timing controls the NTP- or PTP-type packet-based protocol for distributing the time information on a number of receivers working in a hierarchy. The hierarchy has timing source working on the top of this hierarchy helps in deriving a national reference time which is traceable through the constellation of a satellite like IRNSS or from some other source of time transfer. Methods of securing network time distribution are comprised of authentication assurance too refer to the traceable time, time stamp integrity, and other exchanged metadata in the form of synchronized packets, as well as availability of diverse and redundant paths. Another main requirement of securing time in the networks is the detection ability of any intrusions or any kind of anomalies present in the network before the threat cases any effect on the time of the network.

Due to the shortage of secured temporal arrangement sources globally offered nowadays, an affordable approach to securing time is to confirm systems to keep up temporal arrangement among the application tolerance of its compromised timing duration [7].

4 Impact of Timing in the National Critical Infrastructures

Timing has significant position in national critical infrastructures and to distribute secure timing surges evident encounters regarding to security and resilience. Critical Infrastructure Defined: "Assets, systems, and networks, whether or not or not physical or virtual, so vital to the country that their incapacitation or destruction would have a debilitating result on security, national economic security, national public health or safety, or any combination therefrom." For securing the temporal order signal needs each physical signal to be secured and also the information related to it. The timing signal data must be secured, and this is equivalent to numerous cybersecurity issues. A secure physical signal results in many phases distinctive to timing. Usually, the users are distant from the timing signal source demonstrating the specific timescale system. The physical characteristics of time are on probabilities of the mechanism of the system, important to core problems in critical infrastructures. Networking, software, computer hardware and data systems have been enhanced by intellectualizing away the properties of time in the physical layer. Such systems segregate all the processes of timing, which enables the processing of data with extreme competence due in asynchrony part. Where the real-time control, latency measurement, time stamping of events and harmonization of processes were enhanced and enabled with a robust timing sense. Locating and timing are powerfully interconnected. Critical Infrastructures are mixture of cyber and physical components having acquaintances of data processing and networking systems to the systems live in the basic physics laws. Such mixture of systems in the timing systems resonant upgrades or changes in software or hardware might result in a need of re-calibration for the whole system.

Three different timing signals are in synchronization: Time, frequency and phase. Precise amount of frequency could be supplies using a system of discrete clock on a Cesium standard, which drives the practical usage of oscillators requiring the dynamic and calibration signals of reference. For divergence, the time and phase synchronization would require signal transport and possibly data. Unlike the data transfer, the transfer of phase and time needs compensation for delay of transmission of the timing signals to achieve accuracy in synchronization. Taking the case of GNSS distributes locating by sending synchronized signals of time space and its known locations. Delay in transmission is on the order of 70 ms. For providing an accurate range of 1 m, the true delay should be removed by 3 ns, which is a factor or nearly 1 part in 20 million.

Often the data supplements a physical signal of timing, where synchronization of phase might not be required. The simplest data timing is for time, occasionally known as the time-of-day, where a signal specified the accuracy of time, but the accurate time-of-day and date of that particular signal of time might be transferred in the form of data. In such cases, the time signals are known as on-time marker. Data of time is transferrable with significant latency as well as noise, as long as it becomes accurate to the time where an on-time marker is referred with respect to the data. For applications, many other types of data are linked to the time signals. For instance, a source clock quality level is required often with the timing data. Synchronization over networks generally comprise broadcast the data and the time markers by applying the two-way time protocol for cancelling the delay by the network. Common protocols are Precise Time Protocol and Network Time Protocol. Requirements of time could be detailed in terms of the time interval between noteworthy events. The idea of specifying time intervals indicates the system supporting a timescale at which the intervals are measured. Characterization of time scale is done using two features: (a) time zero or the origin called: the epoch, (b) definition of a second: advancement rate of time [2].

In an absence, the time-aware architecture plugs suitable time on the components at which the applications are built in critical Infrastructures such as smart systems (transportation systems, buildings, cities, grid, telecom), system based on location, entertainment, environmental monitoring, and medical devices, are rapidly rolled under various restrictions because of lower accessibility of precise time [3]. Using firewalls on the network might isolate time in critical Infrastructures defense from the network external to the system. These firewalls allow unified control of boundary security of data added as assurance provided to the user. Nonetheless, with isolation of time, a clock drift might occur in both external and internal networks which results in degradation of performance and in some instances failure on many levels. On the similar manner, when the sources of time reference from networks or GNSS are compromised which results in errors in synchronization which attempts to restore or normalize the time services able of running high discontinuities risk of timing along with other issues of alignment.

For reliably and safely operating in the present environment threat, National Critical Infrastructures should implement as many elements of secure timing as possible. Preferably, every critical infrastructure in applications which are critical to safety must have a legally traceable, assured, independent and multiple time sources with predictable and safe modes of failure which must be denied or detectably handled. Here the mixture of insecure and secure sources of timing is available and includes a common standard for time existing among them, an insecure source of timing should be validated against the secure sources of timings. Secured signals of time and its related measurement must be assured to critical infrastructures with well-defined metrics performance which includes switchover time, traceability, mean detection time, holdover capability, frequency stability and phase accuracy [7].

5 Time Synchronization Significance in National Security

Time is more necessary aspect of our lives these days, and therefore, the interconnectivity degree for these networks depicts that everything and something would be exposed, where everything from our basic human rights to national security could be compromised. Thus, the government is ironed to mirror policies which supports continuous technology growth for sophistication in technology security and access and as an important opening move, to adopt a national temporal order strategy.

Though increase in access to Internet and highly mature developments in the technology are correlated to the cybersecurity improvements on a global level, which is certainly not adopted in India as developing economy due to lower development levels in technology in the use of National Legal Time practices in legislation of national cybercrimes in terms of misusing computer systems, system and data interception or interference, and unauthorized access, toward digital forensic investigation.

In the present, laptop networks' time synchronization is essential as a result of each side of debugging, planning, securing and managing a network involves crucial once events happening in terms of cybersecurity. Correlation of log files among the devices is difficult with no time synchronization. The following are some important necessities of time synchronization:

(a) Network usage, tracking security breaches related issues which effects various components could be impossible in case of inaccurate time stamps in logs.
(b) Time is an important aspect which enables the events on a single node of network to be mapped on another event.
For reducing confusion for the file systems shared over the network, the modification times should be importantly consistent, irrespective of the machine with all the file systems. Services used for billing and related applications must accurately apprehend time.
(c) The financial services need accurate timekeeping as per the law. Indian forces require time accuracy in the time of war.

6 Regulations Specific to Sectors in India

6.1 Securities and Exchange Board of India

Circular SEBI/HO/MRD/DP/CIR/P/2017/08 dated January 20, 2017, to other stock exchanges stated for fair along with transparent access to the information feeds of stock exchanges directed them vide point 4. Additionally, SEBI vide circular dated March 30, 2013, had entomb alia instructed stock exchanges to harmonize their system clocks with all the timepiece prior to the beginning of marketplace so that their clocks hold the accuracy of more than 1 microsecond as well as precision with a minimum of 1 ms. By the way, the securities sector should ensure that almost all the clocks of servers & various connected methods have been synchronous. Stock exchanges might follow a good mechanism to confirm such harmonization of the system clocks.

6.2 Controller of Certifying Authorities, Department of Electronics and Information Technology Ministry of Communications and Information Technology

Recommendations for setup of time stamping services by the CA, i.e., to give provision to the IT Act of 2000, the India's Government established CCA. The time stamping system offered by the certifying Authorities should be physically and logically break away the CA methods. Nevertheless, CA uses a corresponding physical resource as well as infrastructure. The time stamping service's audit should be enclosed to the audit of the CA amenities. In India, clause 4.2 regarding time stamping services clock—time values in time stamping amenities utilized in time stamp token shall be noticeable to a Standardized Time Source. The Time stamping amenities clocks shall be shielded against risks which may end in associate degree undiscovered amendment to the clock which takes it outside the standardization of it.

Instances of risks incorporate meddling by unofficial personnel, stereo or maybe electric shocks. The CA (Controller Authority) shall offer a power to identify time stamping provider's clock actually being away reliability specified in the suggestions. Once the time stamping provider's clock is recognized as getting from the precision specified in the suggestions, the occasion shall be audited as well as time stamp tokens shan't be given. Moreover, this particular non-assurance should be examined. The NPLI (National Physical Laboratory, India) is liable for the upkeep as well as advancement of the Indian Standard Time (IST). NPLI keeps time scope of Indian Standard Time (IST) by the assistance of a professional Cesium atomic timepiece. The time scope looked after by NPL is specified as UTC [11].

6.3 The Legal Metrology Act, 2009 an Act

To create as well as implement requirements of measures and weights, control commerce and swap in weights, methods along with other foods that are offered or maybe sent out by body weight, quantity or measure as well as for issues connected incidental thereto or therewith. Ministry of customer Affairs, Public Distribution along with Food, Department of Consumer Affairs. Under THE FOURTH Schedule (Rule 15) Units Permitted to be used with base, Supplementary or Derived Units Permitted units of time (1) The permitted units in relation to time shall be as follows, namely: "(i) the minute, equal to 60 s (Symbol: min), (ii) the hour, equal to 3600 s or 60 min (Symbol: h), and (iii) the day, equal to 86,000 s or 24 h (Symbol: d) The week, month and year shall correspond to the Saka Calendar or the Gregorian calendar" [12].

7 International Standards / IEC 18014 Practiced for Time stamping and Time Traceability

ISO (the International Organization for Standardization) along with IEC (the International Electro Technical Commission) develop the special program for global standardization. ISO/IEC 18014 involves the subsequent components, under the overall label Information technology-Security techniques-Time stamping services. It is composed of three components, including the common thought, designs for a time stamping program, information components, and protocols in its part 1(2008), 2(2008) and 3(2009). Later added with Part 4(2015) for the traceability of time sources.

The first part of ISO/IEC 18014 gives framework for time stamping techniques with a) the certification of the goals of a time authority, b) the explanation associated with a broad design on what moment stamping services are based, c) the characterization of time stamping services and d) the characterization of the fundamental protocols of time stamping as well as the disclaimers of the protocols in between the integrated entities. Second part of ISO/IEC 18014 gives mechanisms creating impartial tokens. A time stamping assistance provides proof that an information product been around before a particular point of time slot. Time stamp tokens are produced by time stamp tokens that are information constructions that contains a verifiable cryptographic binding in between an information item's representation and a time value. This component of ISO/IEC 18014 describes time stamping systems which produce freelance tokens that could be confirmed individually.

In its third part of ISO/IEC 18014, mechanisms creating related tokens (a) describe a basic style for time stamping service solutions creating related tokens, (b) describes the standard elements utilized to create a time stamping program of this particular type, (c) defines the information constructions utilized to have interaction with a time stamping service of this type, (d) describes certain situations of such time stamping service solutions. In its third part of ISO/IEC 18014, traceability of time sources (a) describe the performance of the precious time evaluation expert (TAA), (b) describe a general structure for giving the time period in order to the time stamping power (TSA) and also in order to assure the correctness of it throughout the usage of c), and the TAA gives specialized standards for any TAA to supply, also to offer assurance [13].

8 Results and Discussion

For national security and critical infrastructures, the IST needs to declared as the "legal Time" of the country. The role of legal time in different sectors is discussed below:

8.1 Telecommunication Sector

In the field of communications, networks timing could be as complex or as easy as timing in video systems and digital audio. In communications networks, clock precision is made close to a four-level structured hierarchy with probably of the maximum correct clock in the very first phase and also minimum correct at the fourth level. The electronic community of the 1960s, followed as well as standardized by ANSI, ITU, along with other standards forming systems, the method is described as stratum one. In the start, there's just one stratum one clock in functioning for the whole community at frequent time period. This particular clock was sent out to local running markets and moving over centers and also used to time and also synchronize community components in that amount in the hierarchy respectively, to hometown central workplaces in the bottom part of the hierarchy. Every clock phase be governed by the above one for synchronization. In case it mislaid the given reference signal from above, the nearby clock was allowed to run inside a selection broader compared to the bigger guide until the higher-level guide was again offered. The clock stays inside and keeps the frequency of its of functioning for a certain time and, in the curiosity of general balance, not swap again to the better hierarchy form of given reference clock while waiting after several time of feasible functioning, and in a case likely, switch superbly, at times known as hitless moving over. Main guide timepiece (PRC) products compliant with ITU and ANSI requirements can be found from many resources. These timepiece resources are able to operate independently, or maybe they may be referenced as well as locked to various other sources noticeable to everyone. Standard of timing for time-of-day is known as UTC (Coordinated Universal Time) [8].

To distribute the PRC to each of the components digital network infrastructure with developments in decreasing clock reference price of merchandise and also the accessibility of the world wide ranking system (Internet-based references and GPS) operating under NTP (network time protocol) have significantly reduced price on all fronts and enhanced precision along with reliability of the energy sources as well as the references of theirs.

Fundamentally, indicators produced from multiplexing minimal-speed digital bit streams to higher-order cumulative little channels, several of which offer private line, along with others which offer packet-switched providers and circuit cell have to be synchronized as well as timed precisely how video signals and digital audio are timed as well as synchronized to bit-level precision inside a frame. Telecom electronic indicators are not subject to changing transitions like a cross-fade or split-screen, and also telecom, as well as electronic system articles, operate on completely diverse time period bases.

No matter the kind of what the transmission media (fibre wire), or radio, bits are mailed & bits are obtained. Each final exchange in the system has to treat with a new serial bit stream which has payload bits and clock. Clocking and information, likewise known as payload, have the timing of theirs and phasing associations started in the purpose of development.

On the manner, multiplexed serial bit stream might be with extra serial bit streams, multiple carriers by cross-connection, or maybe exchanged at the circuit, cellular, or packet layer. Sorting out clock information from the payload, or even mucking around with time connection between signal transitions quantities to errors. In case for irrespective of cause, a system component will lose the synchronization reference of it as well as wanders outside keeping everything, anything, and limits turning it into a synchronization reference is out of time together with the bigger community, altering important expertise to invaluable e-garbage.

Data recovery as well as clocking actually comes as a critical feature. Thinking about the period notion after the perspective of a receiver exchange holding a system component or even the getting surface of any transmission course offers the freedom to look again to the cause plus advanced toward various other community components and amenities influenced by the clock for correct shipping and functioning of the connected payload.

The Telecommunication Centre in the whitish papers of theirs promptly synchronization in IP networks concludes that time synchronization among the disparate networks will be the demand of the hour. Treatments are obtainable on the market but call for regulatory enforcement to wonder all of the service providers to absolutely apply time synchronization results in the networks of theirs. The fundamental timing infrastructure up to stratum-1 this time server will be constructed by the government and get all of the service providers as well as ISP's to create the own infrastructure of theirs for stratum-2-time servers onward that will gain the indicators of theirs coming from the stratum-1 server. This way a coordinated nationwide implementation may be ensured for scientific discipline network time synchronization [9].

8.2 Energy Sector

The energy and power sector necessitate timing in methods supplying regular dimensions applicable on the system condition and identifying the location of faults around a transmission type in an application program like the Phasor Measurement Unit (PMU). Network automated safeguard of methods (Wide Area Measurement Systems/Wide Area Control Systems) is utilizing Phasor Measurements Units (PMUs) source of supply of timing and also synchronized record information for network monitoring (present use) along with automatic protection (future use). A lot of redundancy and accuracy in the PMU fitness level is demanded by automatic protection. PMUs are deployed throughout remote places of the energy system (nodes), with inner period references it should be a legally traceable timing system. The Indian Power Systems comes as one of the major synchronous interconnections over the world.

A large interconnection caters to various class of shoppers. A slim in operation varies for power grid frequency considerably lowers usage in electric devices so will increase the lifetime of theirs. Additionally, with the increased in proportion of advanced customer lots including procedure industries, traction locomotives as well

Table 1 Permissible deviation adopted in various countries

Country/interconnection	Frequency band (Hz)	Nominal frequency permissible	Permissible deviation (%)
India	49.5–50.2	50	−1%/ + 0.4%
Other SAARC	49.5–50.5	50	−1%
Europe	49.8–50.2	50	+ //0.4

as silicon a load the hope of greater energy quality from the power grid is additionally rising. The big selection of permissible frequency on purpose has economic furthermore as security issues in an exceedingly giant grid. In this particular context, it the allowable deviation comes through the nominal frequency current in various other places was analyzed, and it is shown in Table 1.

Time Dimensions of Power—There square measure if truth be told 2 key roles time data plays in operation an influence grid. It is usually to ascertain source, another to make certain a sense of balance between need as well as supply. In order to ascertain source, power grid operators monitor the cycle period of alternating electrical present—nominally 50 cycles/second. In case cycle period begins to drift lower or higher, they may signify that voltage is absurdly very high and very low. A difference in cycle period may also harm generators since turbines are properly healthy to run in that particular frequency. Not merely does strength have to get universally synchronized to 50 cycles/second—but those cycles have been saved in stage.

Otherwise, power transmission from one generator would possibly really work in contradiction to power flow from alternative generators. With the speedy enlargement within the installation infrastructure in Bharat, a sturdy communication infrastructure is additionally needed for installation operation and management. The main problem in this regard is the immediate requirement of a special regulation on transmission process for Power Sector Standards.

for Power Frequency in India. According to the Indian Electricity Rules 1956 (amended as many as 25th Nov 2000), the allowable selection for grid frequency was ±3% of nominal, i.e., 48.5–51.5 Hz. The allowable frequency limits (by manufacturers) for functioning of different makes of vapour turbines. The nominal frequency of functioning in the Indian power grid is 50.0 Hz, and also the allowable frequency band identified is 49.5–50.2 Hz, i.e., May 3, 2010, by Indian Electricity Grid Code (IEGC) [10].

8.3 *Finance Sector*

The financial trading industry requires highly accurate, reliable and documentable clock synchronization for business logic, competitive advantage and regulatory compliance. Without reliable clock synchronization, trading records can only be identified by time stamps that are off by seconds or more. The finance subsegment utilizes timing methods to time stamp monetary transactions, enabling a person to trace causal associations and also synchronize monetary computer systems. The uses in finance usually are stock exchanges as well as banks. Financial services have confidence terribly effective IT methods as well as networks needing a top degree of availableness, safety measures and dependability. Legally traceable timing systems for time stamping and synchronization capabilities are should to log quotes or functions in a chronologic fashion. There's prevalent usage of transfer protocols as NTP/PTP to disperse some time (an NTP Primary Server is linked to about 1500/2000 NTP clients).

Internationally, the regulations such as MiFID II, FINRA, CAT NSM, SEC Rule 613, ESMA and MiFIR regulations in Europe and the USA are similar to emerging regulations around the world and call for clock traceability (to UTC, NIST, or other national atomic clock standards), reliable time sync, and transaction time stamping accuracy performance in applications, as well as time sync data analytics and auditable records for proof-of-compliance coverage over multiple years. MiFID II RTS twenty-five presently needs 100 μs accuracy in high-frequency applications as a "minimum" demand for participants in CAT 613 NMS, while industry members are currently required to meet the 50 ms accuracy standard. CAT's time stamp coverage specification needs accuracy in milliseconds; however, finer time stamp increases up to nanoseconds for order handling or execution systems should be recorded and reported to the CAT furthermore. Meeting the standard is not sufficient; however, the same must be able to document compliance [14].

8.4 *Need of Time for General Consumers*

Currently in India, the common consumers adopt the time available by various sources of networks—for example, GPS using mobile, Internet time using Internet, television and radio broadcasting services; however, these sources are not synchronized and traceable to IST leaving a gap in its legal backing for the correlation of events or to establish the authenticity of accuracy and traceability.

9 Conclusion

Having discussed the importance of time, its accuracy, trusted and legal traceability with its criticality issues in the critical infrastructures and available Indian regulations of various sectors raise the missing uniformity.

(a) In the current digitized governance, there is a need to address the Synchronized National Legal Timing that shall help the Indian critical information infrastructure and consumer consumption.

(b) It can be implemented by having a national approach with effective regulation for use of Indian Standard time.
(c) The use of International Standards ISO/IEC 18014 by defining the national framework mandating the role-playing organizations with the specific roles and responsibilities as followed in countries like Japan, Europe, USA, etc.
(d) The national architecture for trusted time infrastructure shall assure reliability in cyberspace
(e) This shall help critical infrastructures, systems, devices and data to handle cybersecurity issues with legally backed precise time and time stamp.

Acknowledgements We acknowledge our thanks for support by DG CSIR, Council of Scientific and Industrial Research, New Delhi, DIRECTOR CSIR NPL, National Physical Laboratory and colleagues of the CSIR NPL IST Division to this work.

References

1. https://www.bipm.org/en/bipm-services/calibrations/traceability.html
2. https://kcdb.bipm.org/AppendixB/AppBResults/CCTF-K001.UTC/2018/CCTF-K001.UTC_1810.pdf
3. https://www.bipm.org/utils/en/pdf/time_ann_rep/Time_annual_report_2016/8_tt_equip_TAR16.pdf (Figure 1) URL as found on 12/06/2019
4. CCTF-K001.UTC—At its 17th meeting in September 2006, the CCTF decided to assign the identifier "CCTF-K001.UTC" to MNI s key comparison
5. https://www.nplindia.in/csir-npl-time-keeper-nation (Source-Figure 2) url as on 12/06/2019
6. https://www.nplindia.in/clockcode/html/index.php (Source Figure-3)
7. https://nvlpubs.nist.gov/nistpubs/SpecialPublications/NIST.SP.1500-202.pdf
8. https://simple-telecom.blogspot.com/2010/11/clocking-timing-and-synchronization.html
9. https://tec.gov.in/pdf/Studypaper/Paper-3-Time%20Synchronization.pdf
10. https://www.cercind.gov.in/2011/WhatsNew/AGENDA_NOTE_FOR_15TH_CAC_MEETINGHI.pdf
11. https://www.cca.gov.in/cca/sites/default/files/files/Guidelines/CCA-TSG.pdf
12. https://upload.indiacode.nic.in/showfile?actid=AC_CEN_21_44_00007_201001_1517807327712&type=rule&filename=lm_natl_standards_rules_2011.pdf
13. https://en.wikipedia.org/wiki/ISO/IEC_18014
14. EURAMET Countries' Legal Time Regulations and Practices April 2011 BY Rado Lapuh Metrology Institute of the Republic of Slovenia Tkalska 15, 3000 Celje, Slovenia rado.lapuh@gov.si

An Approach to Detect Alopecia Areata Hair Disease Using Deep Learning

Gaurav Rai, Naveen, Shubham Sharma, Akbar Ansari, and Namit Khanduja

Abstract Hair is one of the defining characteristics of human beings. Hair is a protein filament that grows from follicles found in the dermis. Keratin is the source of hairs and nails. There are different kinds of hair diseases in humans like Alopecia Areata, Congenital Hypotrichosis, Telogen Effluvium, Scarring Alopecia, Anagen Effluvium, etc., in which Alopecia Areata is the most common. It is the subject of research for early identification. People typically lose so many hairs per day. Hair loss is a very common problem nowadays seen in more or less in every age group, Alopecia is similar to hair loss. The reason behind hair loss is some vitamin deficiency that helps to build the sources like keratin. At the age of 32–36 near 40% people are facing the problem of hair loss. In the dataset there are 609 images in which 310 images of normal hair and 299 images which are infected from Alopecia Areata. In this study, we used pre-trained models (VGG-16, VGG-19, SqueezeNet and Inception-V3) for feature extraction. The dataset is splitted into the ratio of 7:3 in which 70% and 30% used as training data and testing data, respectively. Then, apply the proposed algorithms like ANN, SVM, logistic regression, Naive Bayes and get the maximum 98.3% accuracy.

G. Rai · Naveen (✉)
IIIT Delhi, Delhi, India
e-mail: ynaveen569@gmail.com

G. Rai
e-mail: gauravrai15011998@gmail.com

S. Sharma
Quantum School of Technology, Roorkee, India
e-mail: shubhams6123@gmail.com

A. Ansari
IIIT Allahabad, Allahabad, Uttar Pradesh, India
e-mail: akbar815315@gmail.com

N. Khanduja
Gurukul Kangri Vishwavidyalaya, Haridwar, India
e-mail: namit.khandujal@gmail.com

R. P. Mahapatra et al. (eds.), *Proceedings of 6th International Conference on Recent Trends in Computing*, Lecture Notes in Networks and Systems 177,
https://doi.org/10.1007/978-981-33-4501-0_71

Keywords Deep learning · ANN · Feature extraction · SVM · Logistic regression · Naive Bayes

1 Introduction

Uncertain hair deficit trouble arises due to Alopecia Areata which is a similar autoimmune disarray. The symptoms of Alopecia Areata are patchy hair loss. Hair begins to fall out in the coin sized patches. In some cases, it can cause total hair deficit on the head scalp and in ultimate cases, to the all body. The situation takes place when different colours of blood cells mainly like white blood cells assail the cells in hair follicles, dramatically slowing devour hair yield and causing them to shrink [1]. Causes of baldness can be many such as stress, hormonal changes and medical conditions, radiation therapy to the head, certain hairstyles and discourse. "Alopecia Areata" is a mutual autoimmune disarray that gives results in uncertain hair deficit [2] which means that the body's immune system starts acting against its own hair follicles, resulting in hair loss. It is a polygenic disorder and can be genetic [3]. Alopecia Areata becomes one of the most mutual reasons that causes the body's resistant structure to point hair follicles in the similar manner. In the many research, a conclusion bring out that Alopecia Areata turns into the family history of autoimmune disarrays like atopy, disorder identified using propensity hyperallergic, thyroiditis and vitiligo. Hair disease classification is a field which has a lot of potential for research. In recent years, some work has been done in this field [4, 5]. Due to unavailability of standard dataset, we collect 609 images of hairs in which 310 is normal hairs and 299 images of hair infected from Alopecia Areata. We have used different scenarios to classify the hairs. We used deep learning approach such as Inception-V3, VGG-19, VGG-16 and SqueezeNet for extracting the features. The dataset is split as 70% for training purpose and 30% as testing. Among the different scenarios, we got the highest 98.3% accuracy with Inception-V3 on logistic regression classifier and with SqueezeNet on ANN and SVM classifier algorithm.

2 Dataset Description

The collection of dataset is one of the most important and challenging parts in any machine learning problem. In this problem statement, there is two classes one is normal hair and other is infected hair from Alopecia Areata. There is no open-source dataset available, we collect some good quality images of hair both normal and infected from Alopecia Areata as given in Fig. 1.

In the dataset, there are 609 images in which 310 images of normal hair and 299 images which are infected from Alopecia Areata that is mentioned in Table 1. So we divide the dataset as 70% for training data and 30% for testing data.

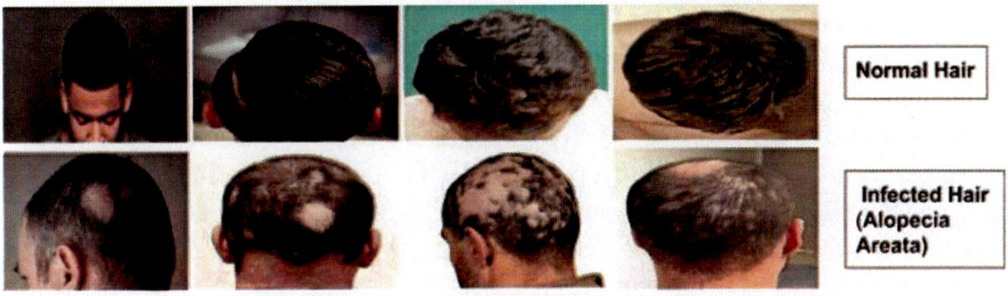

Fig. 1 Sample images of normal hair and infected hair (Alopecia Areata)

Table 1 Specification of images

Types of hair	Number of images
Normal hair images	310
Hair infected form Alopecia Areata	299

3 Method

We have run this algorithm on a machine with configuration 8 GB RAM, 2 GB Nvidia GPU, i5 processor of 8th generation. The machine includes 8 cores and 4 threads. In this problem, we have used deep convolutional neural network (VGG-16, VGG-19, Inception-V3 and SqueezeNet) for feature extraction as shown in Fig. 2 [6, 7].VGG-16 has 13 layers for finding features by applying the filters is known as convolution, 3 layers is used to add fully connected layer is called dense layers and 5 layer is used for preserving the feature, reduce overfitting and reduce size is known as max pooling layers. VGG-16 architecture is like 2 convolution layer

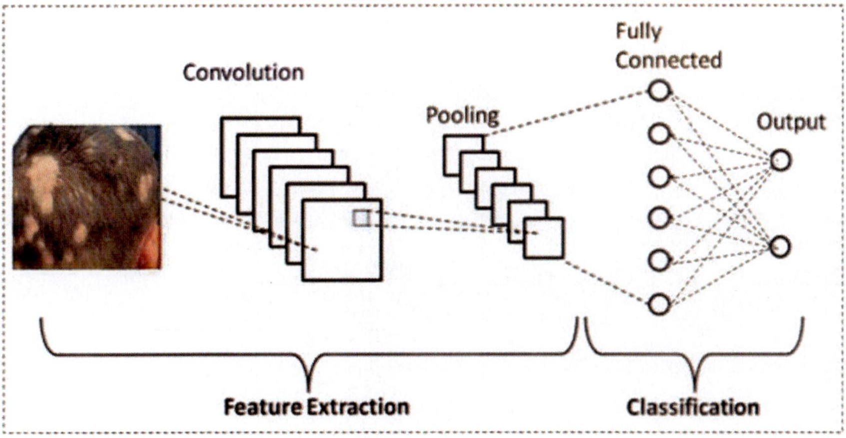

Fig. 2 Working of convolutional neural network

(C1.1, C1.2) then 1 max pooling layer again 2 convolution layer (C2.1, C2.2) then 1 max pooling layer again 3 convolution layer (C3.1, C3.2, C3.3) then 1 max pooling layer again 3 convolution layer (C4.1, C4.2, C4.3) then 1 max pooling layer again 3 convolution layer (C5.1, C5.2, C5.3) then 1 max pooling layer after it 3 dense layer is used to add fully connected layers. VGG-19 has 19 layers for finding features by applying the filters is known as convolution, 5 layer is used for preserving the feature, reduce overfitting and reduce size is known as max pooling layers, 3 fully connected layer and 1 softmax as activation function [6, 7]. Inception-V3 has two parts with fully connected and softmax layers is feature extraction and classification. There are three versions of inception that are V1, V2 and V3. It has 48 deep learning network layers and is similar to GoogleNet architecture. SqueezeNet is a convolutional neural network that strained on more than one million images from the ImageNet database. SqueezeNet is 3 times faster and 500 times smaller than AlexNet architecture. We split the dataset 70% as training and rest 30% for testing. We used state of the art algorithms ANN, SVM, NB and logistic regression [8, 9]. We got 98.3% accuracy on SqueezeNet using ANN and SVM classifiers and on Inception-V3 using logistic regression classifier [10, 11].

The mathematics behind the algorithm such as ANN and SVM which we have used in this machine learning problem is given below:

3.1 ANN

In ANN that is given by the equation below where parameters like "*a*" is the input,"*g*" is the output of the neurons, "*a*" and "w" are biased and weight [11]. Matrices denote all the parameters that depicts as:

$$g = f(\text{net}) = f(n) = f\left(w^T \cdot a + d\right) = f\left(\sum_{i=1}^{R} w_R^T \cdot a_d + d\right)$$

$$a = [a_1, a_2, \ldots a_R,], w = [w_1, w_2, \ldots, w_R]$$

In the process to train the function properly, we must draw out and iterate the procedure of evaluating and enhancing. The peak of wi, di(s) will result in decreasing mean square error (MSE) value. This process will persist until we obtain the prescribed precision [12]. In the given process, biases and weights will switch each time that the process is rerun. The evaluation process for wi, di(s) is:

$$w_{i,j}^{l+1} = w_{i,j}^{l} - \alpha \frac{\partial e(w, d)}{\partial w_{i,j}^{l}}$$

$$d_{i,j}^{l+1} = d_{i,j}^{l} - \alpha \frac{\partial e(w, d)}{\partial d_{i,j}^{l}}$$

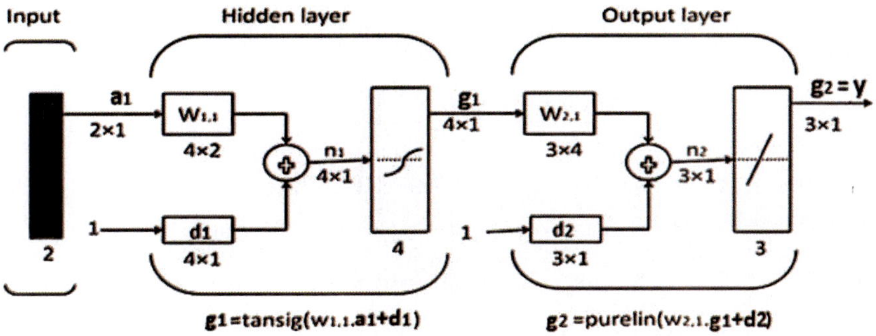

Fig. 3 Working of ANN

$$\frac{\partial e(w,d)}{\partial w_{i,j}^{l}} = \left[\frac{1}{m}\sum_{i=1}^{m}\frac{\partial}{\partial w_{i,j}^{l}}e\left(w,d;x^{(i)},y^{(i)}\right)\right] + \alpha\partial w_{i,j}^{l}$$

$$\frac{\partial e(w,d)}{\partial d_{i,j}^{l}} = \left[\frac{1}{m}\sum_{i=1}^{m}\frac{\partial}{\partial d_{i,j}^{l}}e\left(w,d;x^{(i)},y^{(i)}\right)\right] + \alpha\partial d_{i,j}^{l}$$

where α denotes rate of learning. Figure 3 depicts that the previous layers output works as input for the upcoming layer and the output produced is compared to target value.

$$\text{Mean Square Error} = \frac{1}{m}\sum_{i=1}^{m}e^2 = \frac{1}{m}\sum_{i=1}^{m}(t - g_i)^2$$

where target value is ti and output is g_i.

3.2 *SVM*

SVM is one of the most powerful out-of-the-box supervised machine learning algorithms [10]. The length of a vector x is called its norms that is represented as $||x||$ [13, 14]. To calculate the norm of a vector $x = (x_1, x_2,\ldots,x_n)$, the Euclidean norm formula is:

$$\|X\| = \sqrt{x_1^2 + x_2^2 + \cdots + x_n^2}$$

The direction of a vector $x = (x_1, x_2)$ is written as w and is defined as:

$$w = \left(\frac{x_1}{||x||}, \frac{x_2}{||x||}\right)$$

We can see that $\cos(\theta) = \frac{x_1}{||x||}$ and $\cos(\alpha) = \frac{x_2}{||x||}$. Thus, the direction vector w can also be written as:

$$\mathrm{w} = (\cos(\theta), \cos(\alpha))$$

Two vectors x and y and the angle θ between them. The geometric formula of dot product is defined as:

$$x \cdot y = ||x||\,||y||\cos(\theta)$$

By looking at Fig. 4, we can see $\theta = \beta - \alpha$. Then, we can get:

$$\begin{aligned}\cos(\theta) &= \cos(\beta - \alpha)\\ &= \cos\beta\cos\alpha + \sin\beta\sin\alpha\\ &= \frac{x_1}{||x||}\frac{y_1}{||y||} + \frac{x_2}{||x||}\frac{y_2}{||y||}\\ &= \frac{x_1 y_1 + x_2 y_2}{||x||||y||}\end{aligned}$$

We substitute this into the geometric dot product formula, we get:

$$x \cdot y = ||x||\,||y||\frac{x_1 y_1 + x_2 y_2}{||x||||y||} = x_1 y_1 + x_2 y_2$$

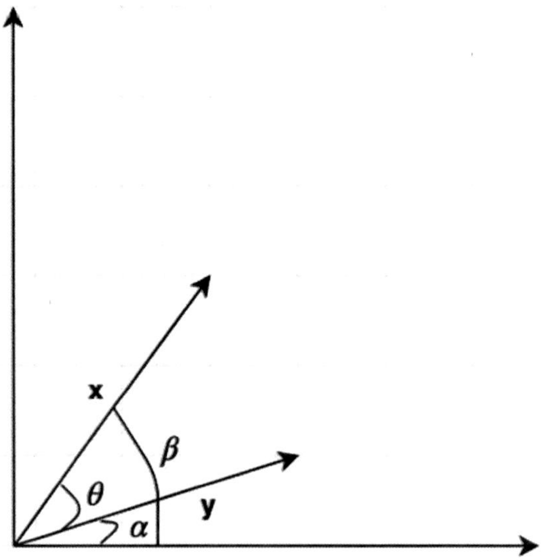

Fig. 4 Graphical representation of SVM

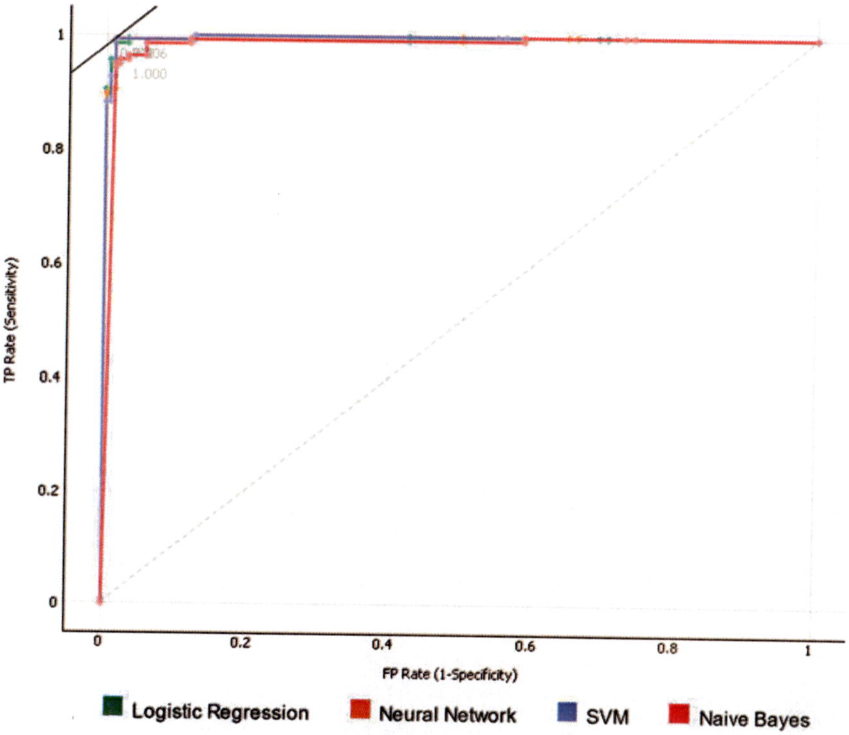

Fig. 5 ROC curve of hair disease (Alopecia Areata)

This is the algebraic formula of dot product [15]. In general, the dot product can be computed as the following for two n-dimensional vectors:

$$X \cdot y = \sum_{i=1}^{n} x_i y_i$$

We evaluate the ROC curve for the calculation of performance on our framework. On X-axis "Specificity" is defined and on Y-axis "Sensitivity" is mentioned. Figure 5 represents the ROC curve on hair disease (Alopecia Areata) images.

4 Result and Discussion

In this section, we discuss the result by using the four different approaches. We used pre-trained models (VGG-16, VGG-19, SqueezeNet and Inception-V3) for feature extraction [6, 7]. The dataset is splitted into 70% and 30% as training data and testing data, respectively. In scenario 1, we used VGG-16 for feature extraction and

Table 2 Accuracy assessment

Feature extraction	Model	Accuracy	AUC	Precision	Recall
VGG-16	Neural network	96.0	99.1	96.2	96.0
	SVM	90.6	98.4	91.3	90.6
	Naive Bayes	96.3	96.9	96.5	96.3
	Logistic regression	97.0	99.4	97.0	97.0
Inception-V3	Neural network	96.7	99.6	96.8	96.7
	SVM	96.0	99.3	96.0	96.0
	Naive Bayes	94.6	96.1	94.8	94.6
	Logistic regression	98.3	99.7	98.3	98.3
VGG-19	Neural network	95.7	98.8	95.8	95.7
	SVM	86.0	96.6	87.1	86.0
	Naive Bayes	95.3	96.2	95.4	95.3
	Logistic regression	95.3	99.3	95.3	95.3
SqueezeNet	Neural network	98.3	99.4	98.3	98.3
	SVM	98.3	99.8	98.3	98.3
	Naive Bayes	96.7	97.3	96.7	96.7
	Logistic regression	97.3	99.5	97.3	97.3

got 97.0% accuracy on logistic regression. In scenario 2, we used Inception-V3 for feature extraction and got 98.3% accuracy on logistic regression [8]. In scenario 3, we used VGG-19 for feature extraction and got 95.7% accuracy on the artificial neural network [11]. In scenario 4, we used SqueezeNet for feature extraction and got 98.3% accuracy on ANN and SVM [10, 12]. The accuracy assessment is given in Table 2.

5 Conclusion

This paper initiates the hair disease classification which used the deep convolutional neural network by using VGG-16, VGG-19, Inception-V3 and SqueezeNet for feature extraction. The proposed method of categorization of the Alopecia Areata hair disease gives an accuracy of 98.3%. This paper primarily aims to improve the medical adeptness at an early time. Our study facilitates the early diagnosis of Alopecia Areata to prevent adverse consequences in such remote areas. This model can help in the future for skin cancer detection, skin disease, nail disease and other types of hair disease. We plan to contrive our algorithm further by implementing image magnification techniques to counter for the lost feature in images.

Acknowledgements We would like to express our special thanks to Dr. Ankush Mittal (300+ publications) and Dr. Rahul Nijhawan for guiding us and motivating us to do some good work in the field of research in Machine Learning.

References

1. Srivastava AK (2015) Hair disorders, treatment and care: an overview. In: Srivastava AK, Srivastava SC, Srivastava N (eds) Research Centre For Human Mycoses And Allergy 250, Keshav Nagar, Sitapur Road, Lucknow-226020, Up India. Journal of Engineering and Applied Science
2. Gordon K, Tosti A (2011) Alopecia: evaluation and treatment. Clinical Cosmetic Investigational Dermatol 4:101–106. https://doi.org/10.2147/CCID.S10182
3. Darwin E et al (2018) Alopecia Areata: review of epidemiology, clinical features, pathogenesis, and new treatment options. Int J Trichol 10(2):51–60. https://doi.org/10.4103/ijt.ijt_99_17
4. Lloyd GR, Ahmad S, Wasim M, Brereton RG. Pattern recognition of inductively coupled plasma atomic emission spectroscopy of human scalp hair for discriminating between healthy and Hepatitis C patients. Analytica Chimica Acta
5. Linch C, Whiting D, Holland M (2001) Human hair histogenesis for the mitochondrial DNA forensic scientist. J Forensic Sci 46(4):844–853. https://doi.org/10.1520/JFS15056J.ISSN0022-1198
6. O'Shea K, Nash R (2015) An introduction to convolutional neural networks. ArXiv e-prints
7. Aishwarya, Goel A, Nijhawan R (2019) A deep learning approach for classification of onychomycosis nail disease. https://doi.org/10.1007/978-3-030-30577-2_98
8. Peng J, Lee K, Ingersoll G (2002) An introduction to logistic regression analysis and reporting. J Educ Res 96:3–14. https://doi.org/10.1080/00220670209598786
9. Kaviani P, Dhotre S (2017) Short survey on Naive Bayes algorithm. Int J Adv Res Comput Sci Manage 04
10. Evgeniou T, Pontil M (2001) Support vector machines: theory and applications 2049:249–257. https://doi.org/10.1007/3-540-44673-7_12
11. Mishra M, Srivastava M (2014) A view of artificial neural network. In: 2014 International Conference on Advances in Engineering & Technology Research (ICAETR - 2014), Unnao, pp 1–3
12. Gao N (2010) Artificial neural network management system. In: The 2nd international conference on Information Science and Engineering, Hangzhou, pp 1891–1894
13. Yang Y, Li J, Yang Y (2015) The research of the fast SVM classifier method. In: 2015 12th International Computer Conference on Wavelet Active Media Technology and Information Processing (ICCWAMTIP), Chengdu, 2015, pp 121–124
14. Duan H, Liu N (2013) A greedy search algorithm for resolving the lowermost C Threshold in SVM classification. In: 2013 ninth international conference on computational intelligence and security, Leshan, pp 190–193
15. Keerthi SS, Lin C (2003) Asymptotic behaviors of support vector machines with Gaussian Kernel. Neural Comput 15(7):1667–1689

Heart Anomaly Detection by Analyzing Heartbeat Sounds Using Deep Learning

P. Anantha Prabha, S. Deepak, A. Deva Surya, and R. Jaswanth Aakash

Abstract Heart disease is the main source of death all around, bringing about more individuals passing on consistently because of Cardio Vascular Disease (CVDs) compared with some other reason for death. Rural mortality rates have exceeded urban mortality rates as 75 percent of essential provincial consideration is given by inadequate experts because of intense specialist deficiencies. There are two compelling methods for following one's heart condition: Electrocardiogram (ECG) and Echocardiogram. All strategies, in any case, are moderately costly for mass assessment and require specialized involvement with their usage. The main idea is to develop a reliable, fast, and low-cost system that can be used by anyone with internet access for an expert diagnosis. This will also assist in the early diagnosis of CVDs which will drastically reduce the potential risk factors of such deaths. We take stethoscope sounds and even waveforms recorded using a cell phone's microphone as input for this work and apply substantial figuring out how to interpret cardio-vascular auscultation in the computerized task, for example, perceiving heart sound irregularities.

Keywords Heart disease · Diagnosis · Stethoscope sounds · Classification · CNN

P. A. Prabha (✉) · S. Deepak · A. D. Surya · R. J. Aakash
Sri Krishna College of Technology, Coimbatore, Tamilnadu, India
e-mail: p.ananthaprabha@skct.edu.in

S. Deepak
e-mail: 16tucs029@skct.edu.in

A. D. Surya
e-mail: 16tucs030@skct.edu.in

R. J. Aakash
e-mail: 16tucs048@skct.edu.in

R. P. Mahapatra et al. (eds.), *Proceedings of 6th International Conference on Recent Trends in Computing*, Lecture Notes in Networks and Systems 177,
https://doi.org/10.1007/978-981-33-4501-0_72

1 Introduction

An Electrocardiogram (ECG) is a complete portrayal of the heart's electrical action on the human body's surface and is broadly applied in the clinical finding of heart disease [1]; it tends to be utilized dependably as a measure to screen the cardiovascular framework's capacity. Because of its effortlessness and non-obtrusive, ECG signals have been broadly utilized for the location of heart infections. ECG signal attributes can be registered from ECG tests and removed utilizing some techniques. For instance, millions experience the ill effects of sporadic pulses that can at times be deadly. Exact, minimal effort analysis of arrhythmic pulses is accordingly profoundly attractive [2]. Numerous examinations have created arrangement approaches for arrhythmia which depend on ECG signals utilizing programmed investigation and indicative frameworks. The most significant variables for the heart illness examination and analysis are the extraction highlights and the order of pulsates. As of late, various procedures have been proposed for grouping ECG signs, and great outcomes have been accomplished [3–5]. The efficiency of the ECG pattern classification is highly dependent on the characterization power of the features drawn from the nature of the ECG signal and classifier (classification model). Many investigators have previously reported an automated classification of heartbeats utilizing an assortment of highlights to represent the ECG and various techniques for arrangement. By and large, heartbeat attributes incorporate an ECG morphology, heartbeat interim attributes (transient qualities), beat connections, pinnacle esteems [6]. Huge information investigation assumes a crucial job in dealing with the colossal measure of social insurance information and improving the nature of the medicinal services benefits that patients are advertised. One of the difficulties right now is information arrangement, which depends on productively dispersed preparing stages, propelled information mining, and AI strategies. Thus, right now it is utilized to address the difficulties of characterizing the ECG beats. As of late, numerous organizations have utilized profound learning methods, including Facebook, Google, IBM, Microsoft, NEC, Netflix, NVIDIA [7–9], and in a wide scope of uses, for example, client agitates forecast in telecom organizations. A tale, profound learning approach for the arrangement of ECG beats is presented in this paper.

2 Related Works

There are numerous works identified with the order of the ECG without the utilization of huge information instruments when the dataset size isn't enormous. Then again, a few examinations depend on big data methods. In [10], Indonesia has a high cardiovascular death rate. To limit mortality, a Tele-ECG framework for early location and checking of coronary illness was proposed utilizing the Hadoop system to deal with huge information preparing. Utilizing a choice tree (DT) and arbitrary backwoods

(RF), the framework can group the ECG information; it was the main genuine heartbeat order framework utilizing enormous information instruments. The framework was based on 4 hub bunch PC. The server oversaw 60 demands at the same time. The exactness for choice tree and irregular woodland was 97.14%, and 98.92%, individually. Neural systems and measurement decrease strategy were utilized in [10], and the methodology was tried on the database of the Massachusetts Institute of Technology for Arrhythmia. The characterization execution on a test set of only 18 ECG records of 30 min each accomplished 96.97% exactness. In [1], numerous heartbeat types have been separated and ordered, and characterization technique is utilized to arrange autonomous sort (3 records for each kind); each kind of heartbeat has its own model (typical heartbeat grouping model, type 1 heartbeat arrangement model, etc.). Neural system and SVM were applied, and results were profoundly precise (more than 90%), yet there was no bound together model without a moment's delay to group all multi-types. In [9], two patterns of heartbeats ventricular ectopic beats (VEBs) and supra-VEBs (SVEBs) are recognized by feed-forward and fully connected artificial neural networks that are assisted by particle swarm optimization technique. Identification of significant features and data mining techniques in predicting heart disease are dealt in [11]. Tuning parameters of the proposed method improved classification accuracy to 96% compared to the same method with the parameter default value. Not all features (morphological and temporal only) were used, and they selected a total of 83,648 beats for training and testing. These are the various works carried out previously which are limited to their capabilities of using the algorithms that could be applied for the normal numeric dataset and that could work efficiently in the neural networks.

3 Proposed Framework

Figure 1 depicts the overall workflow of the proposed system. The process is as follows. First the audio dataset has been collected from the sources.

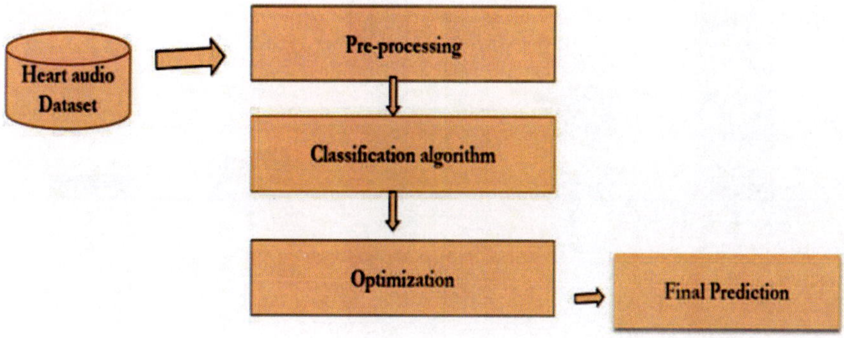

Fig. 1 Overall architecture

Next the collected data is preprocessed which forms as the input for the model. The input given to the model is processed using the algorithm defined. The optimization process is also done along with classification algorithm and the final results are predicted.

3.1 Dataset

This dataset was originally used for Mr. Peter Bentley's classification of heartbeat sounds in a machine learning challenge [2]. The dataset is divided into 2 sets. Set A (set a.csv) data is collected from hospitals as a clinical trial using the Digital Stethoscope. DigiScope was collected from the general public via the iStethoscope Pro iPhone application and which is Set B (set b.csv).

There are 5 classes of heartbeat sounds in this dataset:

(a) Normal: healthy heart sounds
(b) Murmur: extra sounds that occur when blood flow turbulence causes the extra vibrations that can be heard
(c) Artifact: It is usually no discernable heart sound
(d) Extrahls: Heartbeats with an additional sound
(e) Extrasystoles: Extrasystoles are additional heartbeats that occur outside the physiological heart rhythm and may cause unpleasant symptoms.

Figure 2 gives the information about the five different types of the heart sounds that are present in the dataset which is taken as the input for the model.

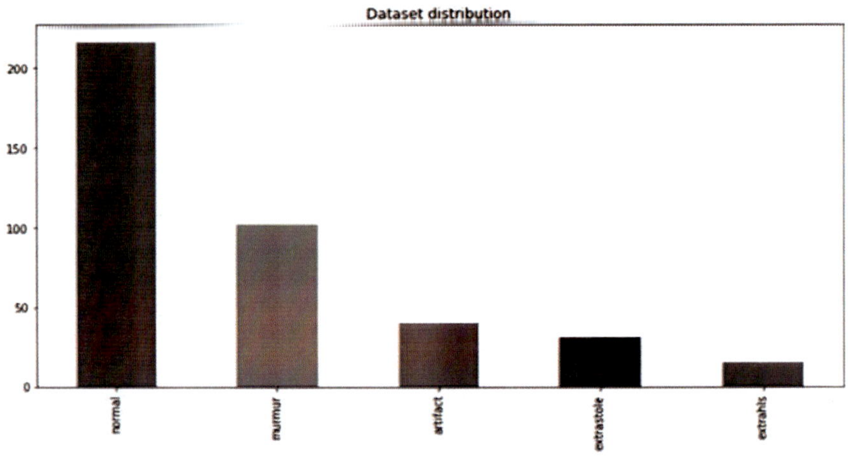

Fig. 2 Dataset distribution

3.2 *Preprocessing*

Background noise will be present when the heart beats are recorded from digital stethoscope from the microphone of and mobile phone of the persons using it. The preprocessing of heart sounds is an important and crucial step toward the automated recording of heartbeats. The audio files that have duration of less than 3 s is not needed because they do not contain enough data points to classify heartbeats accurately. Before training, the recordings have to be converted to a certain fixed length, so the heart sounds are sliced into 3 s long, fixed-length segments. To expand the size of the dataset, we isolate huge documents into different littler records while as yet holding their unique name (i.e., normal or abnormal). The audio signal is also trimmed about half a second from the beginning and end of all audio files because of the noise which is due to the microphone's contact with the body.

In the time domain, the data in the form of raw audio used in the.wav type (Waveform Audio File Format) was in the amplitude versus time form which is transformed into one-dimensional time series signal and then into a two-dimensional heat map capturing the signal's time–frequency distribution which is shown in Fig. 3. This representation of the frequency spectrum of a signal is called a spectrogram, as it varies with time. Since the data was collected using various instruments (i.e., digital stethoscopes and mobile phone microphone) resulting in varying ranges of amplitude, converting the data into the frequency domain results in more accuracy.

From the librosa library in python, we used the built-in function as follows: The preprocessed sound files are transformed into spectrograms of magnitude and then mapped to the Mel-scale using librosa's built-in feature method to get the mel-scaled spectrogram.

On the Mel-Spectrum, a cepstral analysis is performed to obtain Mel-Frequency Cepstral Coefficients (MFCC) by passing the log-power Melspectogram to the MFCC function as an argument. So the audio file is now represented as Cepstral vector sequence. At that point, the first one-dimensional time series is changed into

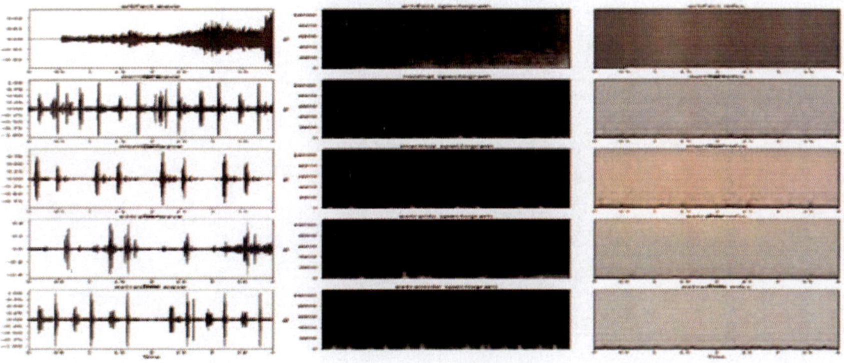

Fig. 3 Spectrogram waves into image

a two-dimensional portrayal of the time recurrence (e.g., spectrogram) that permits the picture preparing of every heart sound section. This spectrogram is provided as the input for heart anomaly detection model.

4 CNN Model

One of the neural system models explicitly utilized for picture characterization is the convolutional neural network (CNN). Likewise with different techniques for neural systems administration, CNN is additionally roused by human cerebrum tissue. The Convolutional neural system comprises principally of two parts, extraction of highlights, and arrangement. A convolutional neural system's design acknowledges a solitary channel 40 albeit130 MFCC heat map as information and yields a twofold characterization, anticipating whether the information portion is an ordinary or strange heart sound. Right now, convolutional neural network utilizes 4 convolution layers, 4 max pooling layers, 4 dropout layers, 1 worldwide normal pooling layer, and a thick layer finally. In convolution layers, the initiation work utilized is rectifier linear unit (ReLU). The ReLU is the most usually utilized enactment work in convolutional neural systems. It is shown in Fig. 4.

Convolutional neural networks (CNNs) are one of the variants of neural networks used heavily in the field of computer vision. The hidden layers of a CNN typically consist of convolutional layers, pooling layers, fully connected layers, etc. A CNN model can be thought as a combination of two components are feature extraction part and the classification part. The convolution and pooling layers perform feature extraction. The fully connected layers then act as a classifier on top of these features and assign a probability to provide final output. The main advantage of CNN compared to its predecessors is that it automatically detects the important features without any human supervision. Efficient model performs automatic feature extraction to achieve highest accuracy. *Pooling* performs action to reduce the dimensionality. This enables us to reduce the number of parameters, which both shortens the training time and combats overfitting.

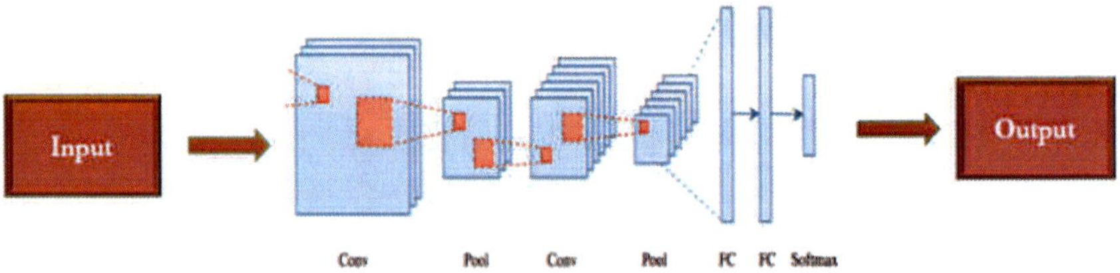

Fig. 4 CNN model for heart anomaly detection

Convolution Layer: In this layer, filters are applied to the original image to extract features in the image. The number of filters and the size of the filters are most important parameters.

Max Pooling: Max Pooling aims at down-sampling a representation of inputs. It helps to reduce the dimensionality and soften extraction of features. It reduces the cost of the computation, reducing the number of parameters to learn.

Global Average Pooling Layer: It calculates the average output of every feature map in previous layer. This layer reduces the data significantly and prepares the model for the final classification layer. It is just like max pooling without trainable parameters.

Dense layer: Every input here is connected by weight to each output. And after this layer, we use softmax as the nonlinear activation function. Adam technique is utilized to refresh the weight on the convolutional neural network for streamlining process. This strategy has proficient computation (memory and time), invariant to inclination scaling and proper when applied to enormous information or parameters.

The dropout layer: The term "dropout" refers to units (both hidden and visible) being dropped in a neural network. It is a very efficient way of using neural networks to perform model averages. Model averaging is a natural response to uncertainty about the model. The dropout layer allows for regularization through training by automatically setting certain neurons to zero in previous layers.

The model developed for the heart anomaly detection uses the convolutional neural network for the whole process of predicting the results. The model uses 4 hidden layers to process the input and 1 output layer for producing the output. The model takes the input in the form of the images which contains the neurons and then the process to be carried is done in the hidden layer. Finally in the output layer, the optimizer is also used which reduces the model and time and gives the final output. The model developed using the CNN works well for the data that is been provided which contains the various forms of the heart beats, and it classifies the various types provided in an easy way. Thus, the model developed could predict the anomalies in the better and cost-efficient way when the model is trained with the necessary information.

5 Experimental Results

Google Colab is the tool used for developing the model. The dataset is collected from its source which contains the five different types of the heartbeat sounds along with the normal. The collected audio dataset is then preprocessed to a ceptral vector sequence which is again converted into the spectrograms of type heat maps. This process is carried out using the librosa package that is available in the python. This final spectrograms is taken as the input for the model. The model developed uses the convolutional neural network since it is best algorithm for the image classification.

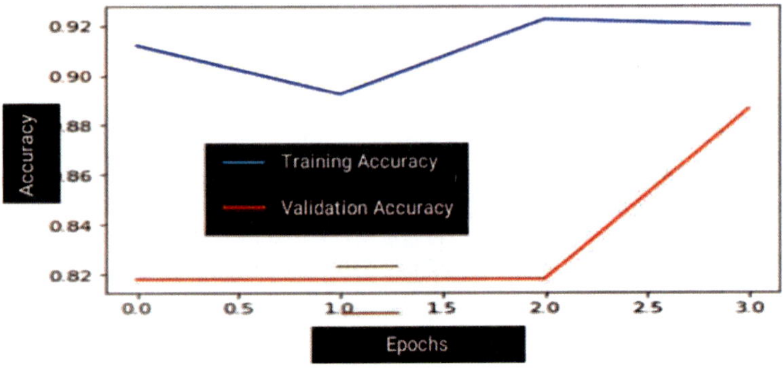

Fig. 5 Accuracy of CNN model

The model contains the hidden layers and the output layer for presenting the final output.

The output is represented in the form of graph which is given in Fig. 5. The accuracy of about 92% is obtained. This shows that the model developed gives highest accuracy for provided input. It represents the accuracy obtained by the models in the time they have been processed the data.

6 Conclusion

This study was intended to distinguish a strange heartbeat sound recorded through a cell phone's amplifier and utilizing stethoscope sounds and pulses. The heartbeat sounds were characterized utilizing a convolutional neural network. There is no other time succession-based neural networks; for example, RNNs are utilized since the heartbeat's transient conduct was rehashed inside the perception window and distinctive consecutive examples were not required to be educated. The model introduced here is one of the first to apply profound convolutional neural systems to the assignment of arranging heartbeat sound consequently recorded through a stethoscope. We have built up a novel model utilizing CNN calculation that initially changes the contribution of a one-dimensional time-arrangement into a two-dimensional Mel time-recurrence spectrogram. Those spectrograms are taken as the contribution for the model. The prepared model consequently separates among typical and irregular heartbeat inputs. The future extent of the examination is that proposed model could be applied for a bigger dataset and that could be executed in the centers for examining the patients.

References

1. Delima AJP (2020) An enhanced K-nearest neighbor predictive model through metaheuristic optimization. (IJACSA) Int J Adv Comput Sci Appl 11(1)
2. Gandhi M, Singh SN (2020) Predictions in heart disease using techniques of data mining. In: 2015 international conference on futuristic trends on computational analysis and knowledge management (ABLAZE). IEEE, pp 520–525
3. Chen AH, Huang SY, Hong PS, Cheng CH, Lin EJ (2011) HDPS: Heart disease prediction system. In: 2011 computing in cardiology. IEEE, pp 557–560
4. Aldallal A, Al-Moosa AAA (2018) Using data mining techniques to predict diabetes and heart diseases. In: 2018 4th International Conference on Frontiers of Signal Processing (ICFSP). IEEE, pp 150–154
5. Alarsan FI, Younes M (2019) Analysis and classification of heart diseases using heartbeat features and machine learning algorithms. J Big Data 6(1):1–15
6. Potes C, Parvaneh S, Rahman A, Conroy B (2016) Ensemble of feature-based and deep learning-based classifiers for detection of abnormal heart sounds. In: 2016 Computing in Cardiology Conference (CinC). IEEE, pp 621–624
7. Latif S, Usman M, Rana R, Qadir J (2018) Phonocardiographic sensing using deep learning for abnormal heartbeat detection. IEEE Sens J 18(22):9393–9400
8. Jayaraman V, Sultana HP (2019) Artificial gravitational cuckoo search algorithm along with particle bee optimized associative memory neural network for feature selection in heart disease classification. J Ambient Intell Humanized Comput, 1–10
9. Dwivedi AK (2018) Performance evaluation of different machine learning techniques for prediction of heart disease. Neural Comput Appl 29(10):685–693
10. Hassan A, Bilal HM, Khan MA, Khan MF, Hassan R, Farooq MS (2018) Enhanced fuzzy resolution appliance for identification of heart disease in teenagers. In: International conference on intelligent technologies and applications. Springer, Singapore, pp 28–37
11. Aminc MS, Chiama YK, Varathan KD (2019) Identification of significant features and data mining techniques in predicting heart disease, IEEE

Improved ResNet-Based Image Classification Technique for Malaria Detection

P. Anantha Prabha, M. Deva Priya, A. Christy Jeba Malar, S. Karthik, G. Dakshin, and S. Dinesh Kumar

Abstract Malaria is a communicable disease instigated by mosquitoes and is life-threatening. As per the World Health Organization (WHO), there are 216 million cases of malaria spread across 91 countries. According to the report dated December 2019, there are 228 million people affected by malaria in 2018 in contrast to 231 million ones in 2017. The traditional way of diagnosing malaria is by visually inspecting blood samples of Red Blood Cells (RBCs) infected by the parasite under a microscope. However, this method is time-consuming as the diagnosis is entirely manual and is based on the expertise of the examiner. Other alternate diagnostic methods include Automatic Image Recognition (AIR) technologies based on Machine Learning (ML). These methods when applied to blood smear yield non-optimal results. To overcome the drawbacks associated with these techniques, the proposed system uses a Residual Neural Network (ResNet) architecture for detecting malaria from blood sample images. Classification is performed on the augmented image dataset to achieve more accuracy. The model offers 93.25% diagnostic accuracy on the training data and 87.35% accuracy on the validation data.

P. A. Prabha · M. D. Priya (✉) · G. Dakshin · S. D. Kumar
Department of Computer Science and Engineering, Sri Krishna College of Technology, Coimbatore, Tamilnadu, India
e-mail: m.devapriya@skct.edu.in

P. A. Prabha
e-mail: p.ananthaprabha@skct.edu.in

G. Dakshin
e-mail: 16tucs026@skct.edu.in

S. D. Kumar
e-mail: 16tucs033@skct.edu.in

A. C. Jeba Malar
Department of Information Technology, Sri Krishna College of Technology, Coimbatore, Tamilnadu, India
e-mail: a.christyjebamalar@skct.edu.in

S. Karthik
SRM Institute of Science and Technology, Delhi-NCR Campus, Ghaziabad, India
e-mail: karthick.usilai@gmail.com

R. P. Mahapatra et al. (eds.), *Proceedings of 6th International Conference on Recent Trends in Computing*, Lecture Notes in Networks and Systems 177,
https://doi.org/10.1007/978-981-33-4501-0_73

Keywords Malaria · Infectious disease · ResNet · Image dataset · Deep learning

1 Introduction

Malaria is a deadly disease instigated by parasites communicated through female anopheles mosquito. This deadly disease causes fever, vomiting, headache and fatigue; in some cases, it can lead to coma or even demise of the affected. The disease is generally transmitted by female anopheles mosquitoes. The mosquito bite injects the parasite into the affected person's blood which then travels to the liver, where it matures and reproduces. There are five species of parasites. The most dominant among those is the Plasmodium falciparum. There is an increase in demise of the affected people—405,000 in 2018 and 416,000 in 2017. Children under the age of 5 years are more susceptible to be affected. As per the 2018 records, they contribute to 67% (272,000) of overall malaria deaths.

Microscopic examination is less costly than polymerase chain reaction-based diagnosis. Though it is extensively used, microscopic diagnosis has many shortcomings. As malaria is normally related to deficiency and happens typically in less economic countries, there is a lacking of normal testing facilities. Furthermore, the efficiency of diagnosis is based on the expertise of the person inspecting the blood film and level of parasites present thereon. The global shortage of pathologists in general has a serious impact on the healthcare system of developing countries. Without proper health care and diagnosis, this can escalate to epidemic extents. Manual investigation is taxing for large-scale analyses in endemic provinces. This may result in deprived quality and needless medication, causing severe economic influence on the individual health programme. There are different types of diagnosis methods available for malaria testing such as clinical and microscopic diagnosis, antigen identification, molecular analysis, serology and drug resistance test. Nevertheless, these diagnosis methods are subject to several inefficiencies.

To speed up malaria testing, scientists and researchers have designed antigen tests for Rapid Diagnosis Testing (RDT). Automated malaria screening using ML techniques and image classification methods are the best inspective aids. In this paper, the efficacy of using deep learning technique for malaria diagnostics is explored. A subclass of the Malaria dataset from National Institute of Health (NIH) is taken which is classified into parasitized and uninfected classes. Classification of medical imaging using deep learning is developed and ResNet method is applied to increase the accuracy.

2 Related Works

Many authors have proposed intelligent diagnostic systems based on ML techniques to help physicians.

The algorithm proposed by Anggraini et al. [1] offers better results for a dataset with 450 malaria images. The model does not perform well when applied wholly to blood images in the infection phase. Purwar et al. [2] have propounded an unsupervised deep learning-based scheme to detect the malaria affected cells found in microscopic images. The proposed method is based on a 16-layer Convolutional Neural Network (CNN) that outperforms the transfer learning-based model using the AlexNet architecture pre-trained on the CIFAR-100 dataset. He et al. [3] have propounded a different approach based on deep residual learning for image recognition and have come up with a pattern recognition that includes pre-trained Probabilistic Neural Networks (PNN), Learning Vector Quantization (LVQ) neural networks and Backpropagation Artificial Neural Networks (BP-ANN).

Zhang et al. [4] have dealt with detecting unlabelled malaria parasites in RBCs. The challenges in automated examination of microscopy images include overlying and strangely shaped cells. This scheme forms strong templates that are capable of classifying infected and uninfected cells. The Histogram of Oriented Gradient (HOG) features are mined and involved in training the classifier. Viola–Jones object identification framework is used to identify the affected and unaffected cells along with the image background. Designing algorithms for unstained parasites is an uphill task. This method permits analysis of parasite development in living cells with diverse drug treatments. Mishra et al. [5] have developed a Computer-Aided Diagnosis (CAD) system for timely identification of leukaemia. The existing schemes involve progressive technologies like medical image processing. Suitable feature extraction scheme is needed for classifying the disease. Discrete Cosine Transform (DCT) is used in image processing and ML. The propounded method provides a chance to perfectly adjust the accuracy of detection.

Ran et al. [6] have dealt with the microscopic hyperspectral images obtained by linking the microscope and the imager, followed by classification. A spatial-spectral classification scheme for hyperspectral images is designed by enhancing Extreme Learning Machine (ELM). The spatial context is integrated into the process of Markov Random Field (MRF) model-based classification. This method offers better results when compared to pixel-based methods (ELM, SVM) in terms of precision and accuracy in localization. Another approach based on deep learning proposed by Chollet [7] involves forming depth-wise separable convolutions with 8 convolution layers using pattern recognition techniques. It also involves segmentation process. The scheme offers better accuracy but the percentage of false negatives is high for medical images.

Yang et al. [8] have designed a transportable image-based cytometer for malaria identification and quantification. It is a portable android version of image classification which offers lesser accuracy. The prevailing methods for automatic slide processing are assessed on comparatively small image sets. Though the stated consequences are hopeful, these methods are not robust and scalable [9]. Though deep Convolutional Neural Network (CNN) and conventional ML schemes support classification, they are incapable of exploiting the long-term dependency association between the key features and labels. To deal with this issue, Liang et al. [10] have included Recurrent Neural Networks (RNNs) and have designed CNN-RNN framework that is capable of extending the details of image content, learning the features and commencing training of big data in medical image analysis. They have used the transfer learning scheme to transfer the weight parameters that were trained to the CNN section. A custom loss function is designed to support training and faster converge with more precise weight parameters.

Huang et al. [11] have used CNN and Gabor filter to exploit deep characteristics from Medical Hyper-Spectral Image (MHSI). In the proposed framework, the CNN kernels capture distinct features. Deep learning for MHSI classification is an improved research aspect. In contrast to the RGB image, HSI has bands to get fullness information. To deal with this, a kernel fusion scheme based on CNN with Gabor kernel is designed. Gabor kernels are built with dot product in the first layer. Based on backpropagation, it is anticipated to capture distinct features to aid CNN in improving the classification accuracy. Wang et al. [12] have dealt with the labelling and classification of microscopic images using diverse comparatively independent framework methods to mine features from microscopic images of coronary artery atherosclerosis plaque. By using unvarying quadrilateral method and medical microscopic image classification algorithm based on several characteristics, the accuracy is improved. ML is used to detect parasitaemia in Giemsa-stained blood smears [13]. This method is based on the ResNet-50 and incorporates transfer learning-based classification of the infected cells. The correlation between automatic and manual detection of parasitaemia is examined, but both the infected and uninfected Red Blood Cells (RBCs) are not identified. Qin et al. [14] have proposed a data-driven approach called Evolutionary Convolutional Deep Network (ECDN) to identify malaria parasites. Optimized evolutionary algorithms are used to generate Deep Neural Networks (DNNs). An optimal network structure is automatically generated without the necessity for any previous knowledge of building a neural network as associated with a traditional Artificial Convolution Network (ACN). Diverse irregularities in White Blood Cells (WBCs) are connected to various diseases. Conventional method of classification deals with cell segmentation, feature extraction and classification. Inadequate or unstable samples reduce the classification accuracy. Ma et al. [15] have dealt with these issues by proposing a classification framework based on Deep Convolutional Generative Adversarial Network (DC-GAN) and ResNet. They have introduced a loss function to deal with the discriminative proficiency of the features.

3 Methodology

In this section, the different phases of the proposed methodology are discussed. The phases include data preparation, data augmentation, data processing and Residual neural Network-50 (ResNet-50) Algorithm.

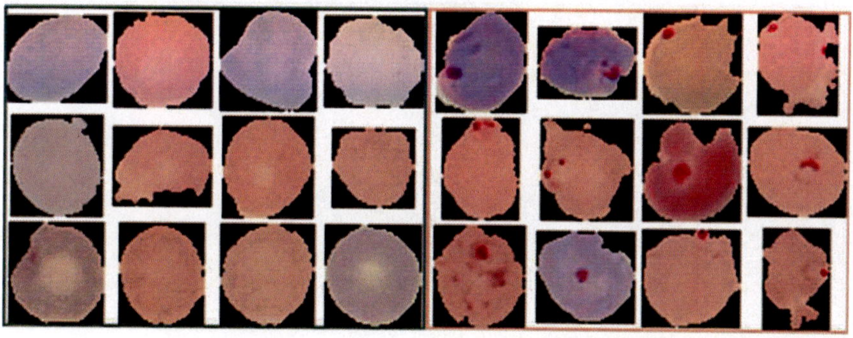

Fig. 1 Sample Malaria Dataset

3.1 Data Preparation

Malaria dataset provided by the National Institute of Health (NIH) is used. A deep learning model is developed for classification of medical imaging using Python, OpenCV and Keras [16]. A sample malaria dataset is shown in Fig. 1. The dataset includes 27,588 images belonging to two distinct classes.

- **Parasitized:** Indicating the presence of malaria
- **Uninfected:** Indicating the absence of malaria.

The patches of segmented RBCs are of 3 channels (RGB) with sizes varying from 110–150 pixels which are later re-sampled to 200×200 output dimensions, a channel depth of 3 and 32-bit Floating Point precision (FP32) to suit the input requirements of different classification algorithms used in this work. Different pre-processing techniques are also applied to achieve faster convergence.

3.2 Data Augmentation

Data augmentation deals with increasing the size of input data by applying some basic transformation algorithms to make the model classify data more easily. The following augmentation algorithms are applied to get the augmented data: Flip, Rotation, Scale, Crop, Translation and Gaussian Noise. The resulting dataset containing 47,640 images is used for training. This offers improved accuracy. Figure 2 shows how data is augmented and input into the model.

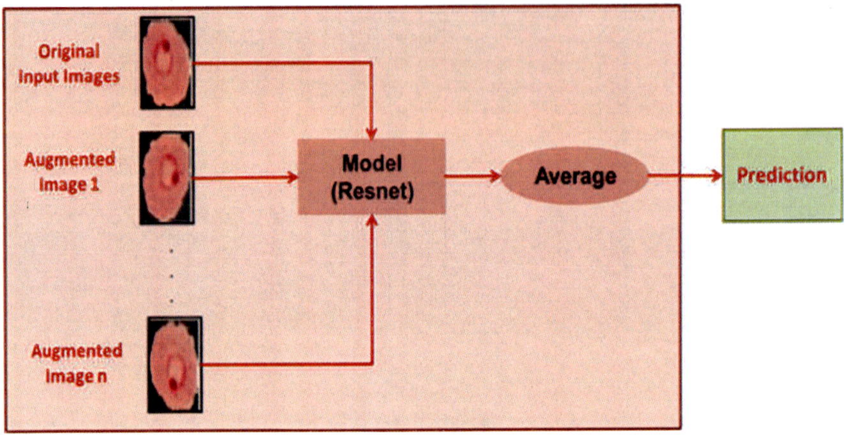

Fig. 2 Augmentation Phase

3.3 *Data Processing*

The augmented data is fed into the ResNet algorithm and the model is trained. The resulting model is the required ResNet model capable of classifying the given blood sample as infected or uninfected.

3.4 *Proposed System*

The proposed model uses ResNet-50 architecture. It deals with the degradation and the fading gradient problems of the existing models by using building block and bottleneck structures. The remaining network involves a simple network structure that is eight times larger than VGGNet. ResNet-50, a Deep Residual Network (DRN) is a subclass of CNN and includes 50 layers. ResNet is most popularly used for image classification. The main feature of ResNet is the skip connection. Without changes, deep networks often suffer from vanishing gradients, i.e. as the model backpropagates, the gradient gets reduced. Tiny gradients make learning inflexible. Figure 3 shows the block diagram of the ResNet architecture.

There are 12 cases, and the images are of $64 \times 64 \times 3$ size. ResNet performs (3, 4, 6) stacking involving (64, 128, 256, 512) convolution layers. Preceding reduction of spatial dimensions, the first layer includes 64 filters. Then, three sets of the remaining modules are stacked. The three convolution layers in each module learn 32, 32 and 128 filters correspondingly. The spatial dimensions are then reduced. Four sets of the remaining modules are stacked, where each layer has 64, 64 and 256 filters respectively. Spatial dimensions are reduced once again. Lastly, six sets of the remaining modules are stacked, where every layer learns 128, 128 and 512 filters,

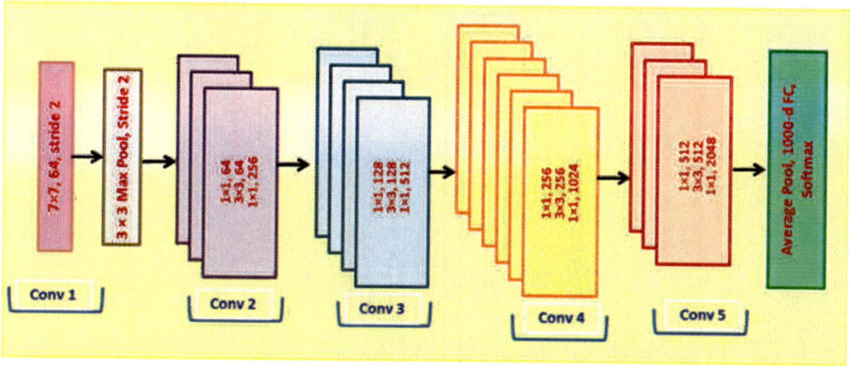

Fig. 3 ResNet Architecture

Table 1 Precision, Recall and F1-Score

	Precision	Recall	F1-Score
Parasitized (%)	77	95	88
Uninfected (%)	96	70	80

respectively. Reduction of spatial dimensions is carried out before average pooling and classification using Softmax classifier.

4 Results

The proposed malaria classifier model offers better training and testing accuracies involving less loss. The model does not show any sign of over or under fitting. The model is developed with ResNet50 architecture using Keras. At the end of the 50th epoch, an accuracy of 93.25% is obtained on the training data and 87.35% on the validation data. Table 1 shows the Precision, Recall and F1-Score, and Fig. 4 shows the training and testing accuracies and the loss plot.

5 Conclusion

The proposed malaria classifier model is a smaller variant of ResNet, wherein the model is 17.7 MB in size. An accuracy of 87.35% which is far better than most of the existing models is obtained. The accuracy is high enough that the classifier can be used for medical examinations. In the future, this model can be tuned to be applied for more real-word applications.

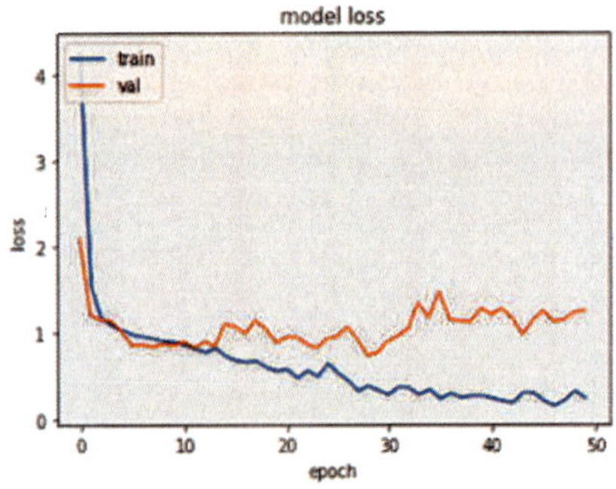

(a) Loss

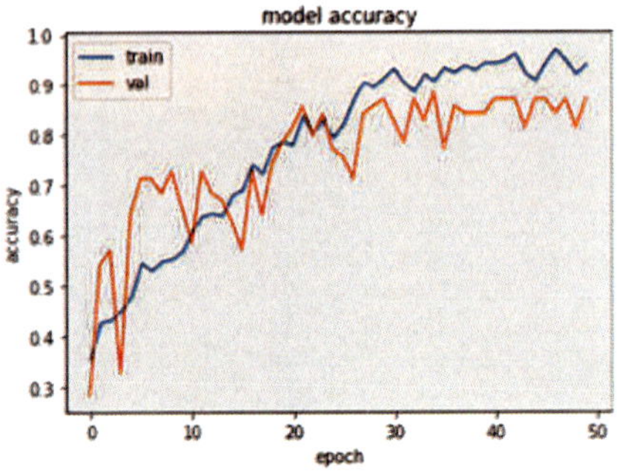

(b) Accuracy

Fig. 4 **a** Loss, **b** Accuracy

References

1. Anggraini D, Nugroho AS, Pratama C, Rozi IE, Pragesjvara V, Gunawan M (2011) Automated status identification of microscopic images obtained from malaria thin blood smears using Bayes decision: a study case in *Plasmodium falciparum*. In: Proceedings of the IEEE international conference on advanced computer science and information systems, pp 347–352
2. Purwar Y, Shah SL, Clarke G, Almugairi A, Muehlenbachs A (2011) Automated and unsupervised detection of malarial parasites in microscopic images. Malaria J 10(1):364
3. He K, Zhang X, Ren S, Sun J (2016) Deep residual learning for image recognition. In: Proceedings of the IEEE conference on computer vision and pattern recognition, pp 770–778
4. Zhang Z, Ong LS, Fang K, Matthew A, Dauwels J, Dao M, Asada H (2016) Image classification of unlabeled malaria parasites in red blood cells. In: Proceedings of the 38th Annual IEEE international conference of the IEEE Engineering in Medicine and Biology Society, pp 3981–3984
5. Mishra S, Sharma L, Majhi B, Sa PK (2017) Microscopic image classification using DCT for the detection of acute lymphoblastic leukemia (all). In: Proceedings of the international conference on computer vision and image processing. Springer, Singapore, pp 171–180
6. Ran Q, Chang L, Li W, Xu X (2017) Spatial-spectral blood cell classification with microscopic hyperspectral imagery. In: AOPC 2017: optical spectroscopy and imaging international society for optics and photonics, vol 10461, p 1046102
7. Chollet F (2017) Xception: deep learning with depthwise separable convolutions. In: Proceedings of the IEEE conference on computer vision and pattern recognition, pp 1251–1258
8. Yang D, Subramanian G, Duan J, Gao S, Bai L, Chandramohanadas R, Ai Y (2017) A portable image-based cytometer for rapid malaria detection and quantification. PLoS One 12(6)
9. Poostchi M, Silamut K, Maude RJ, Jaeger S, Thoma G (2018) Image analysis and machine learning for detecting malaria. Transl Res 194:36–55
10. Liang G, Hong H, Xie W, Zheng L (2018) Combining convolutional neural network with recursive neural network for blood cell image classification. IEEE Access 6:36188–36197
11. Huang Q, Li W, Xie X (2018) Convolutional neural network for medical hyperspectral image classification with kernel fusion. In: Proceedings of the international conference on biological information and biomedical engineering, pp 1–4
12. Wang Y, Liu S, Cui F, Sun H, Ru M, Yang W, Wang G, Wang Z (2018) Examination of medical microscopic image classification algorithm and its application in the evaluation of coronary artery atherosclerosis plaque. Acta Microsc 28(6)

13. Reddy ASB, Juliet DS (2019) Transfer learning with ResNet-50 for malaria cell-image classification. In: Proceedings of the IEEE international conference on communication and signal processing, pp 0945–0949
14. Qin B, Wu Y, Wang Z, Zheng H (2019) Malaria cell detection using evolutionary convolutional deep networks. In: Proceedings of the IEEE international conference on computing, communications and IoT applications, pp 333–336
15. Ma L, Shuai R, Ran X, Liu W, Ye C (2020) Combining DC-GAN with ResNet for blood cell image classification. Med Biol Eng Comput 1–14
16. Jaeger S Malaria Datasets. nih.gov. https://ceb.nlm.nih.gov/repositories/malariadatasets/

Automated Water Flow Control System in Overhead Tanks Using Internet of Things and Mobile Application

Sujatha Karimisetty, Prasanna Kumar Lakineni, M. Lakshmi Kamalanjali, and Pritee Parwekar

Abstract Basically, automation in systems would help to reduce wastage of precious natural resources in various ways, wherein prevention of wastage of water resource to control the water tanks from overflow is crucial. In this automated world, the handling of any kind of devices with comfort is made to be the easiest factor everywhere. However, the overflow and underflow sensors and mobile application should be reliable so that the consumer can rely on this system. Automated water flow control system in overhead tanks using Internet of Things and mobile application uses reliable sensors and mobile application. The application is developed after testing various sensors and finding the best basing on the cost and work efficiency. The mobile application includes features that are not existing such as automated control of tank based on water level selected by user. For this case, three levels are proposed minimum, medium, high. The consumer can change his preference any time. This reduces the water stagnancy for long time. Hence, the proposed methodology has advantages over existing system in saving more water.

Keywords Overflow and underflow sensors · Automated water flow control system in overhead tanks (AWOT) · Internet of Things · Mobile application

---

S. Karimisetty (✉) · P. Kumar Lakineni · M. L. Kamalanjali
Department of CSE, Dadi Institute of Engineering and Technology, Visakhapatnam, India
e-mail: sujathakota29@gmail.com

P. Kumar Lakineni
e-mail: lpk.lakineni@gmail.com

M. L. Kamalanjali
e-mail: kamalanjalimetta3198@gmail.com

P. Parwekar
Department of CSE, SRMIST, NCR Campus, Delhi-NCR, Ghaziabad, Uttar Pradesh, India
e-mail: priteep@srmist.edu.in

R. P. Mahapatra et al. (eds.), *Proceedings of 6th International Conference on Recent Trends in Computing*, Lecture Notes in Networks and Systems 177,
https://doi.org/10.1007/978-981-33-4501-0_74

1 Introduction

A huge amount of water is being wasted because of uncontrolled use and exploitation of water resource. Some other automated water level monitoring systems are also present, but so far most of the methods have some shortcomings in practice. Many times people depend on borewells or municipal water which is usually diverted to a sump. Here, the consumer is worried about switching on the motor when he is at work or needs to go out of station. This issue multiplies for apartment maintenance team [1, 2].

Usually, this controlling requires continuous man intervention and monitoring. Power is wasted if motor is running continuously though tank is overflowing. Maintenance cost is very high. Hence, the operation is difficult. Nowadays, most of the people in urban areas are working, and they have less time to think about switching on/off the pumps. Hence, an automated system is required to continuously monitor the water levels of overhead tank and switch on automatically. This helps not only the public but also the health of motor and working condition. The system should be complex and should be protected from attacks. This should be reliable as the people's dependency increases and should not lose faith on the system [3].

There are other issues like power problems which are to be checked by user before switching on the motor that are common in summer. Here, user, either for individual house, group house, or apartment water tank controlling is a major issue. Many times the user need to go to roof top to check water levels. Small tools and sensors are used for controlling overhead tank level, and alarm systems to indicate overflow conditions are all feasible only if the consumer is at location. Water tank controlling in offices, institutes, and industries are also crucial. Hence, a system is required for effectively using this water and avoid wastage and stagnancy [4, 5].

2 Automated Water Flow Control System in Overhead Tanks (AWOT)

AWOT brings in technology into the house by mechanizing of household requirements, which make the work of water storage system much easier and convenient over handling. The overall system has a great impact in the prevention of water wastage more effective than the manual work. The objectives of the AWOT are listed as follows.

- Switching on or off the motor from anywhere over mobile.
- Check the overflow and underflow levels of water and automatically switch on/off the motor.
- Periodically update the tank water level and user can view a graphical image of overhead tank with water level in his mobile.
- Set the levels in the tank from 1 to 4, where user can set maximum level depending on usage in order to avoid water wastage.

- Generate the monthly report utilization of motor, water, and power consumption.

2.1 Architecture of AWOT

The main components of AWOT are mobile app, IOT kit, server. These components can be fitted to the regular water tank setup, and control can be automated without any additional cost.

1. Mobile app—Mobile app is developed for user to check periodic notifications. The best feature of AWOT compared to existing systems is user which can set the water level basing on his requirement for the particular day. This allows the user to use fresh ground water all the time.
2. IOT kit—Various sensors are tested, and finally, magnetic sensors are used as they are good in detecting the water level to the best. The sensors are fixed at various levels as per the choice of the user. Here, this is fixed to 30%, 60%, and 100%.
3. Server—Here, server stores all the data, and monthly or daily usage reports can be generated using the information obtained from server.

The architecture diagram is as shown in Fig. 1.

The pseudocode for the functioning of AWOT is as follows

1. The system periodically checks if water tank is empty. Then, notification is given to consumer, and motor is switched on automatically.
2. If the consumer is out of station and he doesn't want tank to be filled, then he can switch off setting timer. Consumer can set three water levels depending on

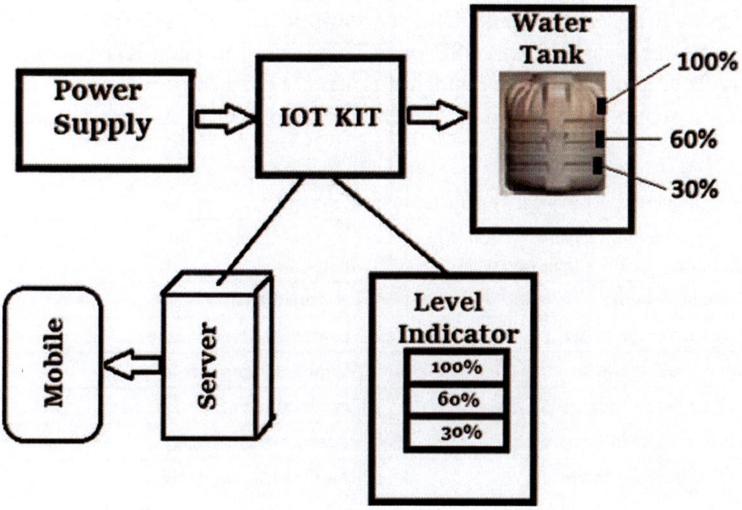

Fig. 1 Architecture diagram of AWOT

usage which indicates 30, 60, 100% of tank. This can be changed whenever user requirement changes.
3. If the water reaches the overflow condition, then the motor is switched off automatically. This is the water level chosen by the user.
4. Send notification to consumer directly by the app on any failure conditions.

The most important merits of the system are:

- The conservation of water resource.
- Manual effort reduction.
- Set water level basing on utilization.
- Time to time alters over the status water storage.

3 Results and Discussion

The AWOT system can be used in any areas from individual houses to huge apartment as an automated water controlling system. This can be used in water controlling basing on utilization that saves both power and water. The system is tested and is found reliable (Table 1).

3.1 *Experimental Setup of AWOT*

The components required in developing AWOT are NodeMCU, relay, resistors, voltage regulator, diode, floating switch, submersible pump. The main hardware components used in developing AWOT prototype are shown in Fig. 2.

NodeMCU has wi-fi module, USB, headers, power pins, reset, and flash buttons. It needs power of 5 V by the USB port. This is used to build IOT-based projects. Relay acts as switch, voltage regulator, diode are used for supporting NodeMCU. Floating switch is used to find the water levels in overhead tanks and send the data

Table 1 Comparison of manual system and AWOT system

Should run under human supervision. Requires a lot of manpower for huge systems	Fully automated system needs very less human intervention
Maintenance is high	Low maintenance
Power consumption is more	Power consumption is less
Motor wear and tear is more	Motor wear and tear is less
Continuous monitoring is required	Less monitoring is required
Overfilling results in water wastage	Water wastage is less
Does not depend on Internet	Requires Internet facility
Motor control and tank filling are difficult	Automated motor control and tank filling

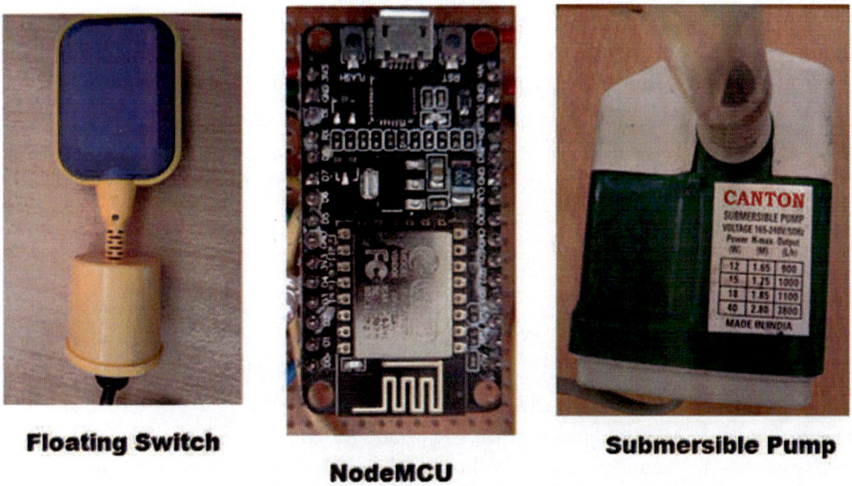

Fig. 2 Main hardware components of AWOT prototype

to Node MCU. Submersible pump is switched on/off basing on specified conditions of water levels. The prototype of AWOT is shown in Fig. 3.

The green bucket is used to represent borewell and blue buckets to represent overhead tanks. The floating switch is placed in green bucket to find the water underflow

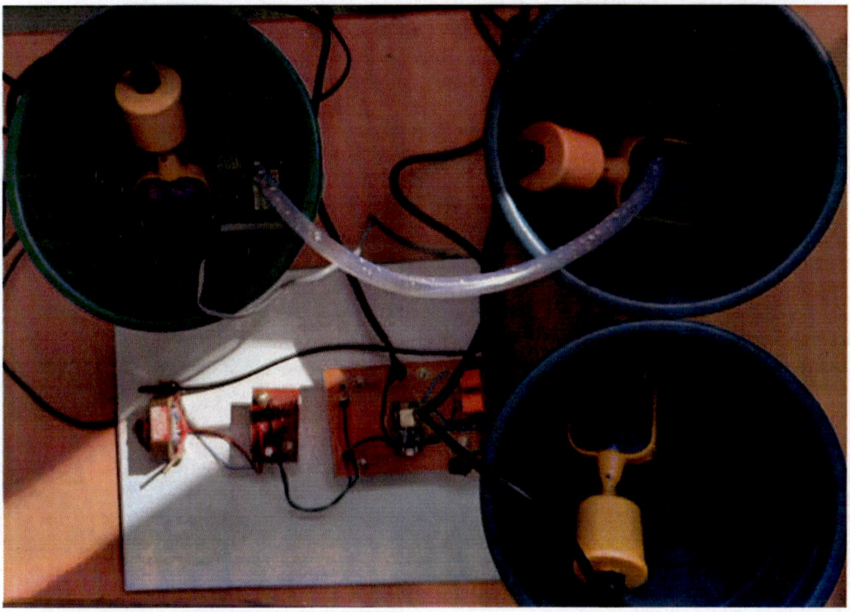

Fig. 3 Prototype of AWOT

condition and switch off motor automatically. The floating switches placed in blue buckets check the medium and high conditions and switch of automatically when the tank is in overflow condition. The level of water in the tank is displayed on the mobile app and the user can control the pump from anywhere basing on the utilization. This reduces water wastage and stagnance. The same concept can be implemented to check the water levels in borewell which increases the life of pump [6].

4 Conclusion

The AWOT system can be used in any areas from individual houses to huge apartment complex as an automated water controlling system. This is economical and can be used by any category of people as it is budget friendly. The system is utilized by numerous applications related to home, agriculture, industries. This can be used in water controlling basing on utilization that saves both power and water. The future scope of study involves quality testing of water and water management in industries.

Acknowledgements The authors would like to express their deepest gratitude to Sri Dadi Ratnakar, Chairman of Dadi Institute of Engineering and Technology for his encouragement and support in executing this work successfully and B-Technologies for sponsoring for the idea collected from requirement of existing customer.

References

1. Getu BN, Attia HA (2016) Automatic water level sensor and controller system. In: 2016, 5th International Conference on Electronic Devices, Systems and Applications (ICEDSA), Ras Al Khaimah, pp 1–4
2. Chen H, Chen Z, Su J (2007) Design of a sliding mode controller for a water tank liquid level control system. In: Second International Conference on Innovative Computing, Information and Control (ICICIC 2007), Kumamoto, p 335
3. Chen H, Chen Z (2008) Implement of a cascade integral sliding mode controller for a water tank level control system. In: 2008 3rd international conference on innovative computing information and control, Dalian, Liaoning, p 162
4. Zuo X (2010) Liquid level control of water tank system based on improved polyclonal selection algorithm and RBF network. In: 2010 2nd international conference on computer engineering and technology, Chengdu, pp V2-528–V2-532
5. Brito N, Ribeiro P, Soares F, Monteiro C, Carvalho V, Vasconcelos R (2009) A remote system for water tank level monitoring and contro—a collaborative case-study. In: 2009 3rd IEEE International Conference on E-Learning in Industrial Electronics (ICELIE), Porto, pp 19–23
6. Karimisetty S, Rugada VR, Harshitha D (2021) IOT-based borewell water-level detection and auto-control of submersible pumps. In: Satapathy S, Bhateja V, Janakiramaiah B, Chen YW (eds) Intelligent system design. Advances in Intelligent Systems and Computing, vol 1171. Springer, Singapore. https://doi.org/10.1007/978-981-15-5400-1_6

Student Result Prediction Before Attempting Exams Using Machine Learning Algorithm

Sujatha Karimisetty, Surendra Talari, Baratam Renuka, and Baratam Sailaja

Abstract In present scenario, students results are creating drastic problems if they are not as expected. Many parents are upset, and some students are even losing their lives as they feel that they could not face their parents with such results. Faculty usually guess the results based on performance of student which helps in predicting student result and inform parents. However, there is no proper mechanism for communicating this as both faculty and parents have their own busy schedule. Hence, this student result prediction before attempting exams using machine learning algorithm is a well-trained system with the training set adapted by faculty in guessing the student failure. This is usually predicted by the system by attendance and internal assessment of the student. Then, a suitable mechanism is adapted for intimating this to students, parents, faculty, mentors, heads of department. This will help in taking extra care before exams instead of becoming aggressive after exams. The algorithm used for this prediction is Naïve Bayes algorithm. Then, the accuracy is being tested on existing dataset of students.

Keywords Student result prediction · Machine learning algorithm · Naïve Bayes algorithm

S. Karimisetty (✉) · B. Renuka
Department of CSE, Dadi Institute of Engineering and Technology, Visakhapatnam, India
e-mail: sujathak@diet.edu.in

B. Renuka
e-mail: renukab@diet.edu.in

S. Talari
Department of Mathematics, GITAM Institute of Science, Visakhapatnam, India
e-mail: stalari@gitam.edu

B. Sailaja
Department of CSE, Vignan Institute of Engineering for Women, Visakhapatnam, India
e-mail: sailu.gld@gmail.com

R. P. Mahapatra et al. (eds.), *Proceedings of 6th International Conference on Recent Trends in Computing*, Lecture Notes in Networks and Systems 177,
https://doi.org/10.1007/978-981-33-4501-0_75

1 Introduction

Though the present examination system is criticized for years but still taking exams and result analysis has become a routine. The parents are bothered about their wards attempting exams and further worried if the results are not as expected. Earlier in schools and colleges, many parents were attending parents meetings organized by educational institution and used to enquire about their child's performance. However, nowadays the parents are busy with their job and business, and hence, they are unable to attend meeting organized by schools and colleges. But they are upset after viewing the result of their children. Many children are also worried to show their results to their parents and few have lost their lives for this silly reason. Many parents are comparing the grades with the classmates which are hurting the student's feelings. However, if the results are expected prior to writing the exam, they can take extra care on their ward and motivate him to excel in education.

The student many times fails due to his over expectation that he will pass basing on some assumptions. However, many times these assumptions go wrong. Hence, if the student is informed about the probability of failure, then he may work hard and try to pass the exam. Similarly, the parents and faculty would have taken more concern before the child attempts the exam. Sometimes if extra care is required for a child to pass in some subject that can be provided by taking it into concern. Some system is needed replacing the manual system which is paper based, slow and inefficient and facing difficulty in report generation, eradicating improvement in student academic performance.

2 Literature Review

Rumsey in his paper of innovative approach discussed the approach of teaching undergraduate students about an application [1]. Bloomfield discussed approach for evolution of a perfect exam grading system [2]. Shahiri depicted a system for predicting performance of the students using data mining techniques [3]. Ibrahim has discussed about predicting academic performance by using ANN and other techniques [4]. Sujatha has proposed and encrypted question paper distribution system in exam cell [5]. Transformations in undergraduate and engineering education need to be done for bringing reformations in creating good graduates [6, 7]. However, predicting if a student will pass or fail is still needed to be worked out. Farrell discussed about the reforms required in transforming education and role of teaching to support these transformations [8]. Karimisetty have proposed a system of developing mobile application for examcell for effective communication to parents [9].

3 Student Result Prediction Before Attempting Exams Using Machine Learning Algorithm

Student result prediction before attempting exams using machine learning algorithm aims to predict that the student will pass/fail in the examination before the student writes the exam. Results prediction is based on some conditions like attendance, assignments and internal marks.

The objectives of this student result prediction are

- To uphold entire student, attendance and marks database.
- Predict, if a student will pass or fail based on Naïve Bayes algorithm.
- SMS/Notify predicted results to students and parents.
- Keep up the expectation of parents on the institute.
- Inform student about his calibre.

3.1 *Naïve Bayes Algorithm*

Naïve algorithm is used for the result prediction which is used to measure the performance of the student during that year. Student academics will be improved with the help of result prediction as prior care can be taken by parents and faculty for avoiding failure. This is Bayes theorem-based classification technique which is easy to develop and can be used on huge datasets. This classifier is based on assumption that one feature of a class is not related to existence of any other feature [6].

Bayes' theorem is mathematically represented by Eq. (1) where m and n are events

$$P(m|n) = (P(n|m) \times P(m))/P(n) \tag{1}$$

Here first, the probability of m is calculated when n is true. $P(m)$ is the probability of occurrence of m before event n occurs. $P(n|m)$ is the probability of occurrence of n. The probability is calculated as shown in Eq. (2)

$$P(m|n) = P(n_1|m) \times P(n_2|m) \times \ldots P(n_i|m) \times P(m) \tag{2}$$

3.2 *Student Result Prediction Using Naïve Bayes Algorithm*

The student dataset is considered with attendance and internal assessment with failures. The student dataset is taken from exam cell of educational institutions. Initially, the system needs to be trained with dataset that predicts if a student passed or failed based on existing training set which is the dataset of students who have taken exam and result announced. Once the system is trained, the test set can be applied to predict

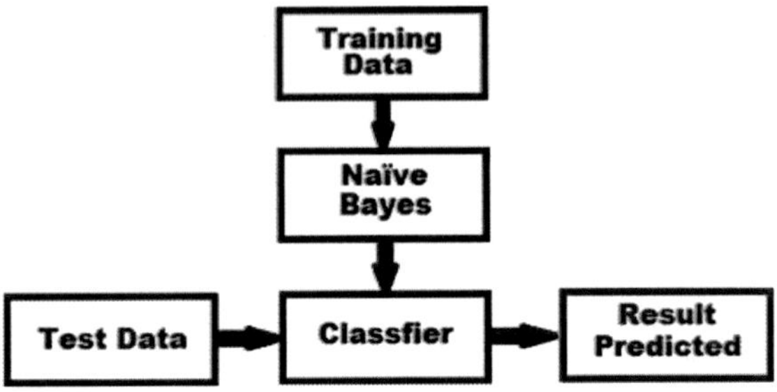

Fig. 1 Student result prediction using Naïve Bayes

if a student will pass or fail. The process is depicted in Fig. 1.The classifier is used to classify the trained data into specific classes and uses this when test data is given for prediction.

4 Results and Discussion

For predicting the student failure, the student database is set by which the faculty comes to assumption and informs parents that there is a probability that the student may fail. The sample dataset regarding guessing if the students pass/fail is tabulated as in Table 1. The fields used in the table are SNo indicating the serial number. Rno is used to indicate the registered number that is the unique identification number alloted to student in the university. For representing student name, name is used. Att and Asg are the attendance percentage and assignment marks of the student. Mid1 and Mid2 are the total of internal assessment tests conducted by the college at two periodical intervals. SF indicates the number of subjects in which the student failed and Res is the result of the student predicted manually.

The predicted result can be used to inform parent over SMS or app/email notifications. Parent can take extra care and counsel student and give him confidence to attempt the exam properly. The result is informed prior to writing exam to the parent,

Table 1 Sample student dataset

SNo	Rno	Name	Att	Asg	Mid1	Mid2	SF	Res
0	16U41A0501	Riya	80	25	165	170	0	Pass
1	16U41A0502	John	45	15	90	80	4	Fail
2	16U41A0503	Peter	90	28	175	Ab	0	Pass
3	16U41A0504	Bobby	69	23	147	158	0	Pass
4	16U41A0505	Robin	35	10	54	75	5	Fail
5	16U41A0506	Diya	75	19	154	162	0	Pass
6	16U41A0507	Karan	82	20	70	66	5	Fail
7	16U41A0509	Razia	96	22	140	133	1	Pass

the parent or student will not get upset by the exam failure. Extra care can be taken by all the persons related to student.

There are three modules in the proposed system. First, the student can login with his account and check his performance and also run the result prediction module. He can collect suggestions from concerned faculty and try to work smart to pass the exam. He will get an idea on which he has to concentrate more. He can also get the percentile of the class and can analyse the status of his grade in the class.

Second, module is related to parents informing the status of their ward. Here, parent can come to know the status of student and will try to mentor them in proper direction. Third module is related to faculty where concerned subject handling faculty can verify the students with below average performance and try to take extra care. This increases pass percentage of faculty which is considered to be important in many institutions. The student also will show gratitude towards faculty as his issue is resolved before viewing bad results.

5 Conclusion

The student result prediction system can generate the output as whether the student will pass or fail in the examination, so that the students academics can be improved by extra care taken by student themselves, parents and concerned faculty. This system can be used by all the colleges who autonomously take exams in their colleges for improving the students' career. The proposed system is developed, and mobile applications are shared to parents, students and faculty. The test results have proved that the results are accurate and lie as motivation to develop such applications useful for society and education. In this competitive environment, students are not only worried about failure, but their concern is about the grades which they obtain. The future scope can be extended to expect the grade that will be obtained by the student based on past performance.

Acknowledgements The authors would like to express their deepest gratitude to Sri Dadi Ratnakar, Chairman of Dadi Institute of Engineering and Technology for his encouragement and support in executing this work successfully.

References

1. Rumsey N, Russell J, Tarhini K (2010, December) Innovative approach to teaching undergraduate reinforced concrete design. University of Virginia, IEEE
2. Bloomfield A (2010, December) Evolution of a digital paper exam grading system. University of Virginia, IEEE
3. Shahiri AM, Husain W, Rashid NA (2015) A review on predicting student's performance using data mining techniques. Elsevier 72:414–422

4. Ibrahim Z, Rusli D (2007) Predicting students academic performance: comparing artificial neural network, decision tree and linear regression. In: 21st Annual SAS Malaysia Forum, 5 Sep 2007. Romero C, Ventura S Educational data mining: a review
5. Sujatha K, Rao PVN, Rao AA (2016) Encrypted examination paper distribution system for preventing paper leakage. Int J Adv Res Sci Eng 5(1):394–400
6. Huang Y, Li L (2011) Naive Bayes classification algorithm based on small sample set. In: International Conference on Cloud Computing and Intelligence Systems. IEEE
7. Transforming undergraduate education in engineering. In: Phase I: synthesizing and integrating industry perspectives, Workshop Report, 9–10 May 2013. National Science Foundation (2013)
8. Farrell S, Hoyer HJ, Fraser DMK (2014) Transformations in engineering education globally. In: Proceedings of the international conference on transformations in engineering education. Springer, pp 3–7
9. Karimisetty S, Thulam S, Talari S (2021) Institute examcell automation with mobile application interface. In: Satapathy S, Bhateja V, Janakiramaiah B, Chen YW (eds) Intelligent system design. Advances in intelligent systems and computing, vol 1171. Springer, Singapore. https://doi.org/10.1007/978-981-15-5400-1_7

Analysis of Threshold Voltage and Subthreshold Behaviour for 400 NM Gate of Ultrathin Nanoscale FD SOI MOSFET

Chandra Shakher Tyagi, R. L. Sharma, and Prashant Mani

Abstract In the current paper, we analyze the electrical characteristics of single-gate fully depleted SOI MOSFET. This paper was focused on following electrical characteristics that are threshold voltage and subthreshold voltage. The performance of the device is analyzed for the 400 nm gate length. The device is also modeled for threshold voltage and subthreshold voltage. The device characterization of proposed device is taken in account by making other device parameter constants—short channel effect, substrate current, heating effect, doping effect, kink effect of single-gate FDSOI MOSFET. The device structure and characteristics were constructed examined and simulated using Silvaco Atlas.

Keywords SOI technology · Fully depleted SOI MOSFET · Silvaco software

1 Introduction

Over the previous decades, semiconductors have been downsized in size to expand execution and lessen power utilization, prompting better hardware gadgets for advanced application. Truly it has been seen from the last numerous years that scaling, for example, decrease of physical elements of the semiconductor structure, has consistently improved the presentation of gadgets, and furthermore its thickness scaling likewise prompts increments of the circuit speed and decline of the force utilization and zone involved by circuit [1] (Fig. 1).

C. Shakher Tyagi · R. L. Sharma (✉)
Department of Electronics and Communication Engineering, Noida International University, Noida, U.P., India
e-mail: rlsharmadigudr@yahoo.co.in

C. Shakher Tyagi
e-mail: tyagi_shekhar@yahoo.co.in

P. Mani (✉)
Department of Electronics and Communication Engineering, SRM Institute of Science and Technology, Ghaziabad, U.P., India
e-mail: prashanm@srmist.edu.in

R. P. Mahapatra et al. (eds.), *Proceedings of 6th International Conference on Recent Trends in Computing*, Lecture Notes in Networks and Systems 177,
https://doi.org/10.1007/978-981-33-4501-0_76

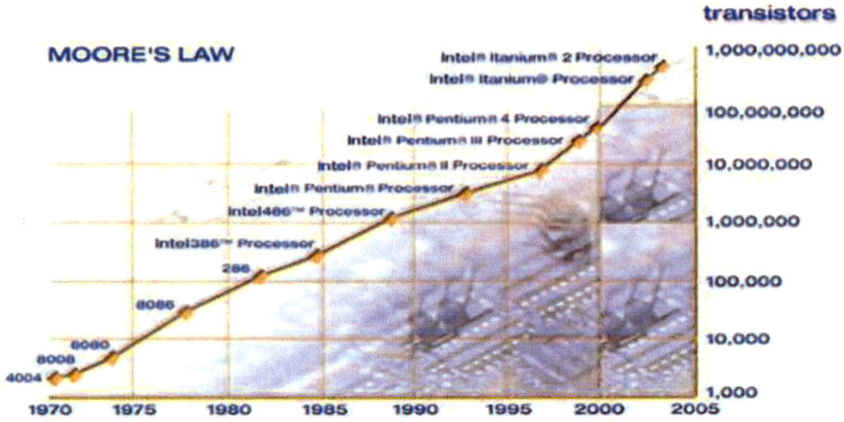

Fig. 1 Moore's law

Scaling of MOSFET gadget is required to speed up and bundling thickness, yet it debases the gadget execution. So as to keep on conveying better, while holding the spillage under control mass silicon semiconductors have become considerably increasingly complex including extra assembling advances. The mass gadgets in the microelectronics business having same unfortunate impacts like lock up and parasitic capacitance among source and channel and substrate, both of result from absence of electrical seclusion between the dynamic gadget and the substrate. SOI empowers expanded chip usefulness without the expense of significant procedure gear changes (e.g. higher goal and lithography tools). The focal points of IC gadgets based on SOI wafers (for the most part higher circuit execution and lower working voltages) have delivered a ground breaking flood in the exhibition of rapid system and new structure for hand held registering and specialized gadgets with longer battery life.

The MOSFETs being used as switch need an outside control which can portray the on what's more, off nature. DC voltage is used as the outside control in MOSFET contraption and therefore may be called as voltage-controlled device. This DC voltage expected to turn on the MOSFET is called edge voltage at which a reasonable channel current is accomplished. In the language of material science, it is the voltage at which the station gets changed and depends upon the contraption estimations and materials utilized while creating [2, 3] and [4]. A similar meaning of material science is being applied here to discover the limit voltage articulation in subsegment. Clearly, limit voltage will rely upon gadget measurements what's more, boundaries picked and accordingly the variety in limit voltage is being concentrated in the following subscaling of MOSFET gadget is required to speed up and bundling thickness; however, it corrupts the gadget execution. So as to keep on conveying better, while holding the spillage under control mass silicon semiconductors have become significantly progressively complex including extra assembling advances. The mass gadgets in the microelectronics business having same unwanted impacts like hook up and parasitic capacitance among source and channel and substrate, both of result from absence of electrical disengagement between the dynamic gadget and the substrate. SOI empowers expanded chip usefulness without the expense of significant procedure hardware changes (e.g. higher goal and lithography tools). The focal points of IC gadgets based on SOI wafers (principally higher circuit execution and lower working voltages) have created an incredible flood in the presentation of fast organize and new structure for handheld registering and specialized gadgets with longer battery life.

2 The SOI MOSFET

The SOI gadgets are smidgen not quite the same as mass CMOS. Silicon on Insulator (SOI) wafers comprise of three layers: a meagre (200 A to a few microns, contingent upon the application) layer of single gem silicon on a thick (1000–4000 A) silicon dioxide layer that is attached to a customary to "handle" wafer. The whole semiconductor is situated in slim top layer of silicon and electrically secluded from mass wafer by the covered oxide (BOX) layer. Fig. 2.

In a MOS semiconductor shaped by on a SOI wafer, the whole semiconductor is in (generally under 0.2 μm thick) silicon layer protector from the main part of the silicon "handle" wafer by a thick (normally 0.1 to 0.4 μm) oxide [5]. The littler volume of silicon that is drained during switching of a SOI semiconductor speed up signal preparing a permit activity at lower drive voltages. As appeared in Fig. 2 the change from miniaturized scale innovation to nanotechnology has just occurred in 2002, also as anticipated by Moore the downsizing procedure speed up while decreasing the

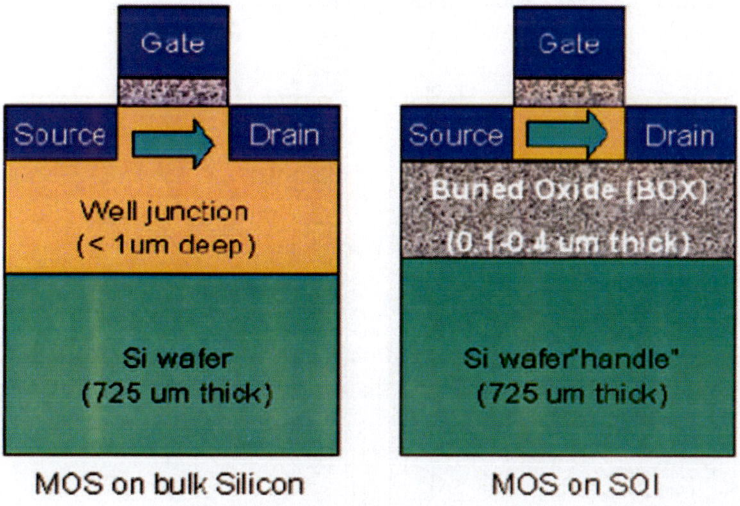

Fig. 2 Sketch of metal oxide semiconductor transistor on a bulk silicon wafer (left) a on-Insulator (SOI) wafer (right)

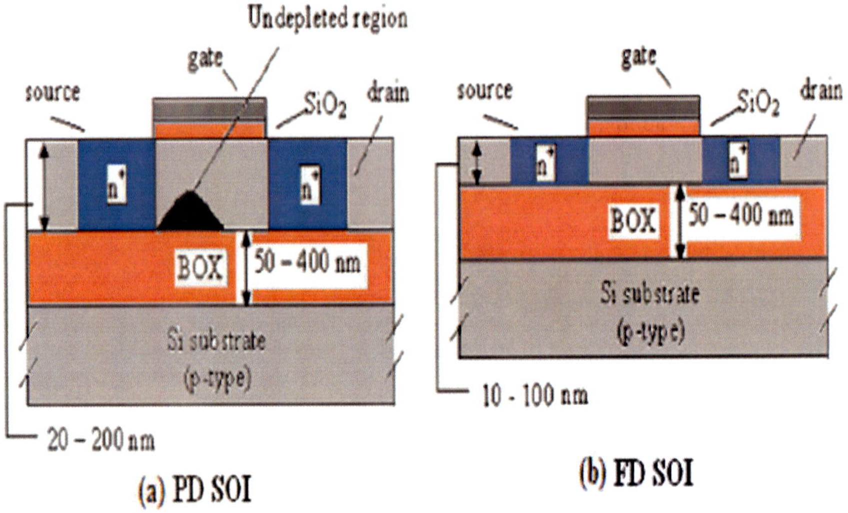

Fig. 3 A cross-view for **a** partially depleted and **b** fully depleted SOI MOSFET

expense definitely, safeguarding the semiconductor execution with downsizing was the key of microelectronics innovation development and its prosperity [6] (Figs. 3 and 4).

The fundamental issue in PDSOI is the Floating Body Effect (FBE) since film isn't associated with the any of the supplies. FDSOI MOSFETS has the upside of lower parasitic capacitance and better subedge swing, decreasing short channel sway (SCE) and freed from crease impact [7]. The affiliated institutions are to be listed directly below the names of the authors. Multiple affiliations should be marked with superscript arabic numbers, and they should each start on a new line as shown in this document. In addition to the name of your affiliation, we would ask you to give the town and the country in which it is situated. Do not include the entire postal address. E-mail addresses should start on a new line and should be grouped per affiliation.

3 Device Structure and Simulation

Simulation of single-gate fully depleted SOI n-MOSFET (silicon film thickness = 0.02 μm) was performed by the Silvaco Atlas tool. Normal estimations of different semiconductor boundaries utilized in this recreation are appeared in the beneath given Table 1 [8].

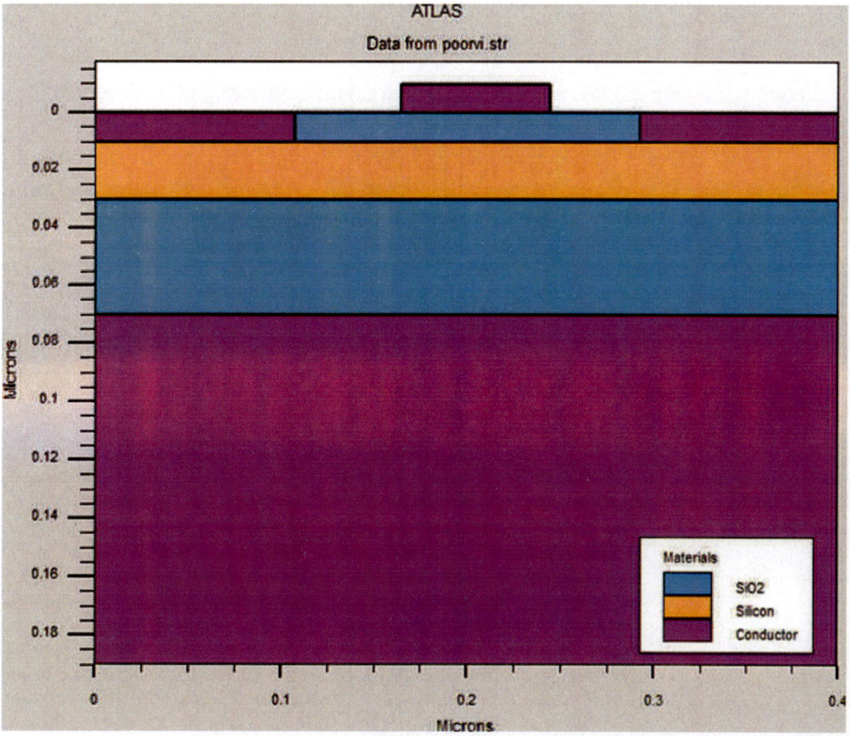

Fig. 4 Device structure of the single-gate FDSOI n-MOSFET with compelling channel length 0.4 μm channel doping is 1E18 cm^{-3} and channel and source doping fixation is 1E20 cm^{-3}; door oxide thickness is 0.01 μm

Table 1 Parameters of single-gate SOI *n* MOSFET transistors

Symbol	Description	Value
LD,LS,LG	Drain length, source length, gate length	0.11 μm, 0.11 μm, 0.10 μm
L	Channel length	0.10 μm
TOX	Gate oxide thickness	0.01 μm
TSI	Silicon film thickness	0.02 μm
TBOX	Buried oxide thickness	0.04 μm
	Substrate thickness	0.120 μm
NA	Substrate concentration	1×10^{18} cm^{-3}
ND	Drain and source concentration	1×10^{20} cm^{-3}

3.1 Threshold Voltage

The edge voltage of a FD SOI MOSFET with significantly doped channel could be depicted as the applied passage voltage at any rate of the surface potential becomes $2\Phi_F$ [2], where Φ_F is the Fermi capability of the channel territory. In any case [9], considering summed up case for channel doping the accompanying meaning of limit voltage has been applied for the recessed gadget in this work [3, 10].

$$\varphi(x = 0/t_{\mathrm{si}}, y)|\min|_{V_{\mathrm{g}}=V_{\mathrm{thf}}/V_{\mathrm{thb}}} = 2\Phi_F^* = \begin{cases} 2\Phi_F, & \text{for } N_a > n_T \\ \Phi_F + \frac{kT}{q}\ln\left(\frac{n_T}{n_i}\right), & \text{for } N_a < n_T \end{cases} \tag{1}$$

where V_{thf} and V_{thb} are front surface controlled limit voltages and back surface controlled limit voltage individually. What's more, n_T is a basic grouping of electron in channel to turn on the gadget [3, 4]. The boundary relies upon gadget boundaries which have been decided after the strategy for Chen et al. [4].

The front surface minimum potential can be obtained from equation

$$\varphi(x = 0, y)|_{\min} = V_{sf} + 2 \cdot \sqrt{\alpha\beta} \cdot \mathrm{e}^{-\lambda L/2} \tag{2}$$

where

$$\alpha = \frac{\left[(v_{\mathrm{bi}} - v_{\mathrm{sf}}) - (v_{\mathrm{ds}} + v_{\mathrm{bi}} - v_{\mathrm{sf}})\mathrm{e}^{-\lambda L}\right]}{1 - \mathrm{e}^{-2\lambda L}} \tag{3}$$

$$\beta = \frac{\left[(V_{\mathrm{ds}} + V_{\mathrm{bi}} - V_{\mathrm{sf}}) - (V_{\mathrm{bi}} - V_{\mathrm{sf}})\mathrm{e}^{-\lambda L}\right]}{1 - \mathrm{e}^{-2\lambda L}} \tag{4}$$

And back surface minimum potential is obtained from equation

$$\varphi(x = t_{\mathrm{si}}, y)|_{\min} = V_{\mathrm{sb}} + 2 \cdot \sqrt{\alpha'\beta'} \cdot \mathrm{e}^{-\lambda L/2} \tag{5}$$

where

$$\alpha' = \frac{\left[(V_{\mathrm{bi}} - V_{\mathrm{sb}}) - (V_{\mathrm{ds}} + V_{\mathrm{bi}} - V_{\mathrm{sb}})\mathrm{e}^{-\lambda L}\right]}{1 - \mathrm{e}^{-2\lambda L}} \tag{6}$$

$$\beta' = \frac{\left[(V_{\mathrm{ds}} + V_{\mathrm{bi}} - V_{\mathrm{sb}}) - (V_{\mathrm{bi}} - V_{\mathrm{sb}})\mathrm{e}^{-\lambda L}\right]}{1 - \mathrm{e}^{-2\lambda L}} \tag{7}$$

$$V_{\mathrm{sft}} + 2 \cdot \sqrt{\alpha\beta} \cdot \mathrm{e}^{-\lambda L/2} = 2\Phi_F^* \tag{8}$$

where

$$V_{\text{sft}} = V_{\text{sf}}|_{V_g = V_{\text{thf}}} \tag{9}$$

Further Eq. (8) can be simplified as explicit equation as given below by second-order polynomial.

$$A_1 \cdot V_{\text{sft}}^2 + A_2 \cdot V_{\text{sft}} + A_3 = 0 \tag{10}$$

where

$$A_1 = 4\mathrm{e}^{-\lambda L}\left(1 - \mathrm{e}^{-\lambda L}\right)^2 - \left(1 - \mathrm{e}^{-2\lambda L}\right)^2 \tag{11}$$

$$A_2 = 4\Phi_F^*\left(1 - \mathrm{e}^{-2\lambda L}\right)^2 - 4(V_{\text{ds}} + 2V_{\text{bi}})\mathrm{e}^{-\lambda L}\left(1 - \mathrm{e}^{-\lambda L}\right)^2 \tag{12}$$

$$A_3 = 4V_{\text{bi}}(V_{\text{ds}} + V_{\text{bi}})\left(\mathrm{e}^{-\lambda L}\right)\left(1 + \mathrm{e}^{-2\lambda L}\right) - 4\left\{(V_{\text{bi}})^2 + (V_{\text{ds}} + V_{\text{bi}})^2\right\}\mathrm{e}^{-2\lambda L} - 4\Phi_F^{*2}\left(1 - \mathrm{e}^{-2\lambda L}\right)^2 \tag{13}$$

Equation (10) is a quadratic equation, and its root could be easily obtained as

$$V_{\text{sft}} = \frac{-A_2 + \sqrt{A_2^2 - 4A_1A_3}}{2A_1} \tag{14}$$

Now, at threshold condition ($V_g = V_{\text{f)}}$ V_{sf} becomes V_{sft} and hence equation can be modified as

$$V_{\text{sft}} = \frac{C_{\text{ox}}(2C_r + C_b + C_{\text{si}})}{C_{\text{ox}}C_{\text{si}} + (c_{\text{si}} + C_{\text{ox}})(2c_r + C_b)}(V_{\text{thf}} - V_{\text{fb1}}) - \frac{C_{\text{ox}}c_{\text{si}}}{c_{\text{ox}}c_{\text{si}} + (c_{\text{si}} + c_{\text{ox}})(2c_r + c_b)} \left[\frac{qN_a t_{\text{si}}}{c_{\text{ox}}}\left(1 + \frac{2c_r + c_b}{2c_{\text{si}}}\right) - \frac{c_r}{c_{\text{ox}}}(V_d + V_s - 2V_{\text{fb2}}) - \frac{c_b}{c_{\text{ox}}}(V_{\text{sub}} - V_{\text{fb3}} + \Delta V)\right] \tag{15}$$

After rearranging the terms of Eq. (15), following equation of front channel threshold voltage could be obtained

$$V_{\text{thf}} = V_{\text{fb1}} + V_{\text{sft}} \cdot \frac{c_{\text{si}}}{2c_r + c_b + c_{\text{si}}} \cdot \left[1 + \left(\frac{c_{\text{si}} + c_{\text{ox}}}{c_{\text{ox}}}\right)\left(\frac{2c_r + c_b}{c_{\text{si}}}\right)\right] + \frac{c_{\text{si}}}{2c_r + c_b + c_{\text{si}}} \left[\frac{qN_a t_{\text{si}}}{c_{\text{ox}}}\left(1 + \frac{2C_r + c_b}{2c_{\text{si}}}\right) - \frac{c_r}{c_{\text{ox}}}(V_d + V_s - 2V_{\text{fb2}}) - \frac{c_b}{c_{\text{ox}}}(V_{\text{sub}} - V_{\text{fb3}} + \Delta V)\right] \tag{16}$$

$$V_{\text{thb}} = V_{\text{fb1}} + V_{\text{sbt}} \cdot \frac{C_{\text{si}} + C_{0x}}{C_{\text{ox}}} \cdot \left[\left(\frac{c_{\text{ox}}}{c_{\text{si}} + c_{\text{ox}}}\right) + \left(\frac{2C_r + C_b}{c_{\text{si}}}\right)\right] + \frac{C_{\text{si}} + c_{0x}}{c_{\text{ox}}}$$

$$\left[\frac{qN_a t_{\rm si}}{2c_{\rm si}}\left(\frac{2c_{\rm si}+c_{\rm ox}}{c_{\rm si}+c_{\rm ox}}\right)-\frac{c_r}{C_{\rm si}}(V_d+V_s-2V_{\rm fb2})-\frac{c_b}{c_{\rm si}}(V_{\rm sub}-V_{\rm fb3}+\Delta V)\right] \tag{17}$$

where $V_{\rm sbt} = V_{\rm sb}|V_g = V_{\rm thb}$ is found from quadratic equation given below

$$A_1 \cdot V_{\rm sbt}^2 + A_2 \cdot V_{\rm sbt} + A_3 = 0 \tag{18}$$

Since $V_{\rm thf}$ and $V_{\rm thb}$ are two limit voltages model, really the physical gadget becomes on if adequate current can move through either front or back surface of channel. The surface with higher potential first beginnings leading and guarantees the edge of physical re S/D SOI MOSFET. In like manner, the edge voltage $V_{\rm th}$ of a recessed source/channel SOI MOSFET could be dictated by front surface edge voltage as follows (25).

$$V_{\rm th} = \begin{cases} V_{\rm thf} \text{ for, } \varphi(x=0,y)|_{\min} > \varphi(x=t_{\rm si},y)|_{\min} \\ V_{\rm thb} \text{ for, } \varphi(x=0,y)|_{\min} < \varphi(x=t_{\rm si},y)|_{\min} \end{cases} V_{\rm thf} \tag{19}$$

3.2 Subthreshold Voltage

Subedge qualities of a MOS gadget are viewed as significant boundary for both simple and computerized applications. It is straightforward that though MOS is a DC voltage controlled switch yet channel current increments bit by bit with door voltage rather than step increment. The extent of channel current underneath limit voltage is viewed as subedge current. It is additionally named as subedge leakage, since it is not on current rather moving through channel without applying required door inclination. Sublimit current is fundamentally feeble reversal current and for the most part debase computerized circuits made utilizing MOS device. However, it can be valuable in simple circuits to create various capacities.

The subedge incline is a proportion of pace of increment of channel current with door voltage sublimit area of activity.

3.3 Subthreshold Current and Subthreshold Slope Model

The subthreshold current density may be expressed as

$$J_n(x) = qD_n\frac{n_{\min}(x)}{L_e}\left(1-e^{\frac{V_{\rm ds}}{V_T}}\right) \tag{20}$$

where D_n is electron diffusion constant, $n_{\min}(x)$ is the minimum free electron density at $(x, y_{\min})$ coordinate in channel, L_e is the effective length of channel and V_T is thermal voltage. The expression of $n_{\min}(x)$ is given by

$$n_{\min}(x) = \frac{n_i^2}{N_a}\left(e^{\frac{\varphi(x, y_{\min})}{V_T}}\right) \tag{21}$$

Here, $\varphi(x, y_{\min})$ is the base channel potential along channel length.

As far as possible inclination is portrayed as the regressive of pace of augmentation of channel current with respect to gateway voltage in the subedge locale of action. It has been earlier shown that in the current model at low doping back channel controls the device action while at high back channel controls the device movement while a high doping front surface is prevailing. Following this finding subedge grade of Re S/D SOI MOSFET may be described by the going with condition

$$S = \begin{cases} \frac{\ln(10)\cdot kT}{q} \cdot \frac{\partial\varphi(x, y_{\min})}{\partial V_g}|_{x=0} \text{ for, } \varphi(0, y_{\min}) > \varphi(t_{si}, y_{\min}) \\ \frac{\ln(10)\cdot kT}{q} \cdot \frac{\partial\varphi(x, y_{\min})}{\partial V_g}|_{x=t_{si}} \text{ for, } \varphi(0, y_{\min}) < \varphi(t_{si}, y_{\min}) \end{cases} \tag{22}$$

The base of surface potential along the channel length $\varphi(x, y_{\min})$ can be found from equation being separated and likened to zero to discover $y_{\min}$. Further replacement of $y_{\min}$ results in the accompanying articulation

$$\varphi(x, y_{\min}) = V(x) + U(x, y_{\min}) \tag{23}$$

and

$$U(x, y_{\min}) = 2\sqrt{QR}\cdot[\cos(\lambda x) + P\cdot\sin(\lambda x)] \tag{24}$$

where P, Q, R and λ are taken from equations

In this way, sublimit current per unit channel width can be acquiring utilizing condition (23) into Eq. (20) and resulting joining as (Figs. 5, 6, 7 and 8)

$$I_{ds} = \int_0^{t_{si}} J_n(x)dx \tag{25}$$

Since, $\varphi(x, y_{\min})$ is an element of ΔV and ΔV is itself an element of V_g henceforth thorough numerical inference is required to discover $\left(d\varphi(x, y_{\min})/d\,V_g\right)$in unequivocal form. However, the equivalent can be fathomed utilizing the accompanying can be obtained

$$\frac{dV(x)}{dV_g} = \left\{\frac{C_{ox}C_{si}}{C_{ox}C_{si} + (C_{si} + C_{ox})(2c_r + C_b)} + \frac{c_b(C_{si} + C_{ox})}{C_{ox}C_{si} + (C_{si} + C_{ox})(2c_r + C_b)}\frac{d\Delta V}{dV_g}\right\}\frac{x}{t_s}$$

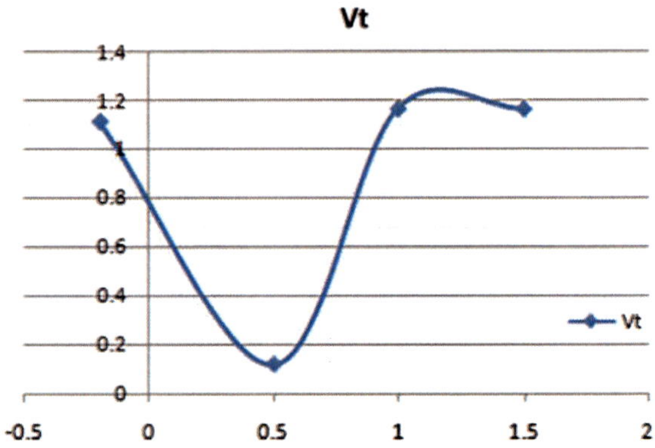

Fig. 5 Device behaviour under the constant drain voltage = 0.02

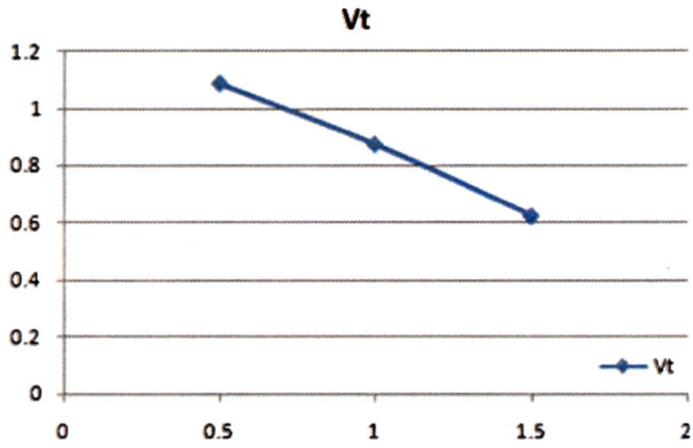

Fig. 6 Device behaviour under the constant gate voltage = −0.2

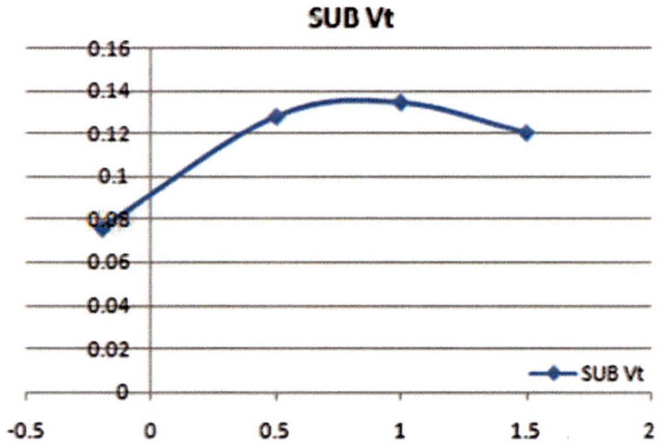

Fig. 7 Device behaviour under the constant drain voltage = 0.02

$$+\left\{\frac{c_{\text{ox}}(2C_r + C_b + C_{\text{si}})}{C_{\text{ox}}C_{\text{si}} + (C_{\text{si}} + C_{\text{ox}})(2c_r + C_b)} + \frac{c_b c_{\text{si}}}{C_{\text{ox}}c_{\text{si}} + (c_{\text{si}} + C_{\text{ox}})(2c_r + c_b)}\frac{d\Delta v}{dV_g}\right\}\left(1 - \frac{x}{t_s}\right) \tag{26}$$

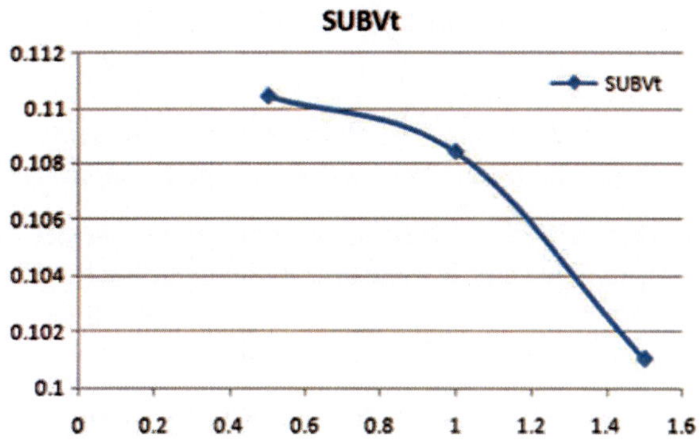

Fig. 8 Device behaviour under the constant gate voltage = −0.2

and

$$\frac{dU(x, y_{\min})}{dV_g} = \frac{1}{e^{\lambda L}-e^{-\lambda L}}\left[\sqrt{\frac{R}{Q}}\left(e^{-\lambda L-1}\right) + \sqrt{\frac{R}{Q}}\left(1 - e^{\lambda L}\right)\right]\frac{dV(x)}{dV_g} \quad (27)$$

Thus, results $\left(d\varphi(x, y_{\min})/dV_g\right)$ in Eq. (22) are given below

$$\frac{\partial\varphi(x, y_{\min})}{\partial V_g} = \frac{dV(x)}{dV_g} + \frac{dU(x, y_{\min})}{dV_g} \quad (28)$$

4 Results and Discussions

In Fig. 5 the proposed device behaviour under the constant drain voltage = 0.02, the figure represents the variation in threshold voltage at applied

gate voltage = −0.2, the threshold voltage we get $V_t = 1.1167$, Sub $V_t = 0.07584$
gate voltage = 0.5, the threshold voltage we get $V_t = 0.1167$, Sub Sub$V_t = 0.1280$
gate voltage = 1.0, the threshold voltage we get $V_t = 1.1670$, Sub $V_t = 0.1347$
gate voltage = 1.5, the threshold voltage we get $V_t = 1.1670$, Sub Sub$V_t = 0.1250$

As the continuous gate voltages increases the beyond 0.5 V, the threshold voltage increases. The continuous variation in threshold voltage stabilize after the gate voltage increases beyond 1.0 v.

In Fig. 6, the proposed device behaviour under the constant gate voltage = −0.2, the figure represents the variation in threshold voltage at applied

drain voltage = 0.5, the threshold voltage we get $V_t = 1.0866$, Sub$V_t = 0.1104$
drain voltage = 1.0, the threshold voltage we get $V_t = 0.8748$, Sub$V_t = 0.1084$
drain voltage = 1.5, the threshold voltage we get $V_t = 0.6250$, Sub$V_t = 0.1010$

It can be concluded that as the increase in drain voltage, the drain side rise in electric field attracts more electron from source side thus continuously lower V_t. DIBL also improves as barrier lowering.

In Fig. 7, the proposed device behaviour under the constant drain voltage = 0.02, the figure represents the variation in subthreshold voltage at applied

gate voltage = −0.2, the threshold voltage we get $V_t = 1.1167$, Sub $V_t = 0.07584$
gate voltage = 0.5, the hreshold voltage we get $V_t = 0.1167$, Sub$V_t = 0.1280$
gate voltage = 1.0, the threshold voltage we get $V_t = 1.1670$, Sub $V_t = 0.1347$
gate voltage = 1.5, the threshold voltage we get $V_t = 1.1670$, Sub$V_t = 0.1250$

From above Fig. 7, it is clear that in off state subthreshold slope reduces as the gate voltages. In off state device will become in sleep mode but as applied gate voltage increases the device will on after V_t. In case of subthreshold voltage slope is improving as the device operate in off mode.

In Fig. 8, the proposed device behaviour under the constant gate voltage = −0.2, the figure represents the variation in subthreshold voltage at applied

drain voltage = 0.5, the hreshold voltage we get $V_t = 1.0866$, Sub $V_t = 0.1104$
drain voltage = 1.0, the threshold voltage we get $V_t = 0.8748$, Sub $V_t = 0.1084$
drain voltage = 1.5, the threshold voltage we get $V_t = 0.6250$, Sub$V_t = 0.1010$

In this figure as the drain side voltage increases, the value of Sub V_t continuously start decreasing. As drain voltage increases beyond 1.2 v then subthreshold voltage lowers. The device performance improves as Sub V_t slope reduces.

5 Conclusion

In this paper, we successfully obtained the electrical characteristics of narrow channel single-gate fully depleted MOSFET. The characteristics of FDSOI MOSFET are obtained using Silvaco Atlas software. Here we see that the electrical characteristics of FDSOI MOSFET are much better than PDSOI MOSFET. As the device shrinks, the short channel effect is very important in conventional MOSFET, and the short channel effects (SCE) of FDSOI MOSFET are much better than conventional and PDSOI MOSFET. In the above analysis, we obtained improved electrical characteristics of FDSOI MOSFET (like lattice heating effect, net doping effect, kink effect) as compared to PDSOI MOSFET. The major drawback of in PDSOI n-MOSFET is kink effect, which is eliminated in FDSOI n-MOSFET.

References

1. Mani P, Sharma S (2015) Analysis of narrow channel single gate fully depleted SOI MOSFET. Int J Eng Res Manag Technol 2(2)

2. Choi C-H, Goo J-S, Oh T-Y, Yu Z, Dutton RW, Bayoumi A, Cao M, Voorde PV, Vook D, Diaz CH (1999) MOS C-V characterization of ultra-thin gate oxide thickness (1.3–1.8 nm). IEEE Electron Device Lett 20:292–294
3. Wu Y, Lee Y-M, Lucovsky G (2000) 1.6 nm oxide equivalent gate dielectrics using nitride/oxide (N/O) composites prepared by RPECVD/oxidation process. IEEE Electron Device Lett 21:116–118
4. Choi C-H, Wu Y, Goo J-S, Yu Z, Dutton RW (2000) Capacitance reconstruction from measured C-V in high leakage, nitride/oxide MOS. IEEE Trans Electron Devices 47:1843–1850
5. The International technology Roadmap for semiconductor (2009) Emerging Research Devices
6. Colinge JP, Institute TN (2008) The new generation of SOI MOSET. Romanian J Inf Sci Technol 1(1):3–15
7. Rahou FZ, Guan-Bouazza A, Rahou M (2013) Electrical characteristics comparison between fully-depleted and SOI MOSFET and partially-SOI MOSFET using Silvaco software. Glob J Res Eng Electr Electron Eng (0975-5861) 13(1)
8. Doganis K, Scharfetter DL (1983) General optimization and extraction of IC device model parameters. IEEE Trans CAD 30:1219–1228
9. Choi C (2002, April) Modelıng of nanoscale MOSFETS
10. Vogel EM, Henson WK, Richter CA, Suehle JS (2000) Limitations of conductance to the measurement of the ınterface state density of MOS capacitors with tunneling gate dielectrics. IEEE Trans Electron Devices 47:601–608

An Investigation on Route Identification Through the Dynamic Source Routing Protocol in Multi-hop Wireless Mobile Ad-Hoc Networks

Ezhilarasu Umadevi Palani, R. P. Mahapatra, P. Karthik Kumar, and S. Karthick

Abstract Mobile Ad hoc Networks (MANET) is a wireless network that has varying mobility for the collection of nodes. The communication path is identified by routing protocol that has a predefined procedure to determine the efficient path. It is primarily classified into three types based on the method adapted for the protocol activation. They are (1) proactive, (2) reactive, and (3) hybrid. The Dynamic Source Routing protocol (DSR) is a reactive protocol, and this work investigates the advantages and drawbacks of DSR protocol. The reactive nature of DSR is analyzed for multi-hop wireless mobile ad hoc network scenario. The two main aspects of routing protocol are related with the route. Those are discovery through recording cum matching and maintenance. The overhead of maintaining routing table is eliminated since it is a reactive protocol. The loop formation is also eliminated using host matching with route record. The complete route information of intermediate nodes between source node and target nodes is well known in advance. The two route messages, Route Request message (RREQ) along with Route Reply message (RREP) are used in discovery of route and maintenance. Finally, the performance of the DSR protocol is compared with the well-advanced reactive protocol, Ad hoc On-demand Distance Vector (AODV).

Keywords AODV · DSR · Loop elimination · Route discovery · Route maintenance · Routing protocol

---

E. U. Palani (✉) · R. P. Mahapatra · S. Karthick
Computer Science and Engineering, SRMIST, Delhi-NCR Campus, New Delhi, India
e-mail: prof.p.ezhilarasu@gmail.com

R. P. Mahapatra
e-mail: mahapatra.rp@gmail.com

S. Karthick
e-mail: karthick.usilai@gmail.com

P. K. Kumar
Information and Communication Engineering, Anna University, Chennai, India
e-mail: karthikkumarphd@gmail.com

R. P. Mahapatra et al. (eds.), *Proceedings of 6th International Conference on Recent Trends in Computing*, Lecture Notes in Networks and Systems 177,
https://doi.org/10.1007/978-981-33-4501-0_77

1 Introduction

The computer network establishes the communication between the nodes. This can be facilitated with the support of physical medium directly or indirectly. In wireless networks, the communication takes place without any physical medium, and hence, the communication is indirect. MANET is one form of wireless networks [1]. The increase in the use of desktop, laptop, and the mobile phones are the major source for growth of MANET. MANET performs communication between the collection of nodes without the support of the network infrastructure. One of the main functionality of MANET is the route establishment. The dynamic mobility of every node in the network makes the route discovery a challenging one in MANET [2].

2 Routing Protocol

The nodes are communicating through the route established statically or dynamically. The route formation is implemented through the routing protocol. The various forms taken by the node are originating node, final destination node, and intermediate node (router). There exist three types of protocol based on their activeness and are proactive, hybrid, and reactive protocols [3]. The proactive protocols are basically table driven. The reactive protocol uses instant request (on demand). The hybrid protocol utilizes the best aspects of both the protocols. The DSR protocol comes under reactive protocol, and it doesn't use table for route discovery as it is on-demand routing protocol [4]. The active routes are maintained for further usage to minimize the delay in new route discovery. The route cache contains active route. The packet header has full route details, and the route is discovered dynamically, and the starting node is source. These factors are main reasons for naming this routing protocol as dynamic source routing protocol [5].

3 Related Works

DSR is an instant request routing protocol used to find the route when an initiating node wants to send a data packet. The two primary operations performed by the DSR are route discovery and maintenance. It uses soft state routing and supports high data rate [6].

Performance of AODV, DSDV, and DSR protocol are analyzed by network simulator NS-3. The analysis implemented by packet delivery ratio, throughput, delay, and the overhead occurring in the routing. The simulation results prove that overall DSR is better than AODV and DSDV [7]. Analysis of AODV, Ad hoc On-demand Multiple path Distance Vector (AOMDV), and DSR using NS-3 proves that DSR is more efficient [8].

DSR-ant is a variation of conventional DSR routing protocol. It works based on the routes adapted by ants for finding the food. The ant uses the foot prints and pheromones of the previously traversed ants for finalizing the shortest path. The simulation result proves that DSR-ant performs better than the DSR protocol.

Efficient Power Routing DSR (EPRDSR) is a scheme designed by modifying the DSR protocol considering the power constraints into account. Route discovery is carefully made considering the energy consumption. This results in the network life time increase and reduced consumption of power. EPRDSR is simulated using network simulator NS-3, and the simulated results prove that the network life time is increased 60–65% when compared with DSR and Minimum Total Transmission Power Routing (MTPR) [9].

AODV and DSR implemented and evaluated through NS-3. The averaging delay and the throughput are used as the performance metrics. The simulation is run by varying the node density. From simulation results, it is clear that DSR protocol outperforms AODV based on the discussed performance metrics [10].

DSDV, AODV along with DSR are simulated using NS-3 and the performance is evaluated. The performance metrics considered for evaluation are throughput, packet delivery ratio, and the delay. The simulation results show that DSR performs better than DSDV and AODV in terms of the throughput and the packet delivery ratio. The DSDV shows improved performance for delay [11].

4 Scenario

The multi-hop wireless mobile ad hoc network scenario for DSR [6] is shown in Fig. 1. The assumed network has totally sixteen nodes with the nodes "E" and "L" as starting and destination nodes. The intermediate nodes are "Z," "H," "I," "U," "M," "A," "D," "V," "N," "O," "P," "B," "K," and "R" nodes. The nodes "E" are directly connected with the nodes "V," "Z," "U." This direct link is known as one hop distance. The other twelve nodes are connected with the node "E" indirectly. In other term, it is known as more than one hop distance nodes.

5 Route Discovery

The node E needs route to the node L. It checks whether any unexpired path to node L exist in its route cache. If there is unavailability of node, then the starting node "E" performs the discovery of route initially by broadcasting the message RREQ to its neighbor nodes as shown in Fig. 2. Each RREQ has sender address for identifying the originator node, destination address for identifying the target node, unique ID for eliminating the duplication, route record for identifying the path sequence. Initially, the nodes "V," "Z," and "U" receive the RREQ from the node E. The RREQ has four fields with the values E, L, 10, E. The intermediate nodes V, Z, and U matched with

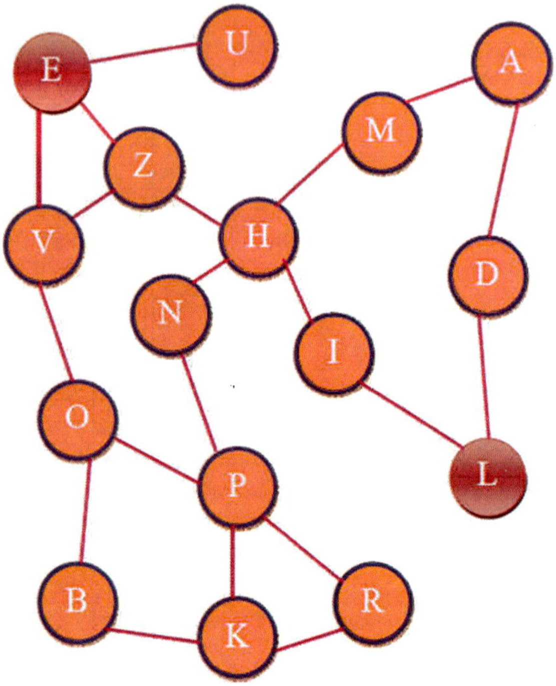

Fig. 1 Scenario for DSR protocol with originating node E and final destination node L

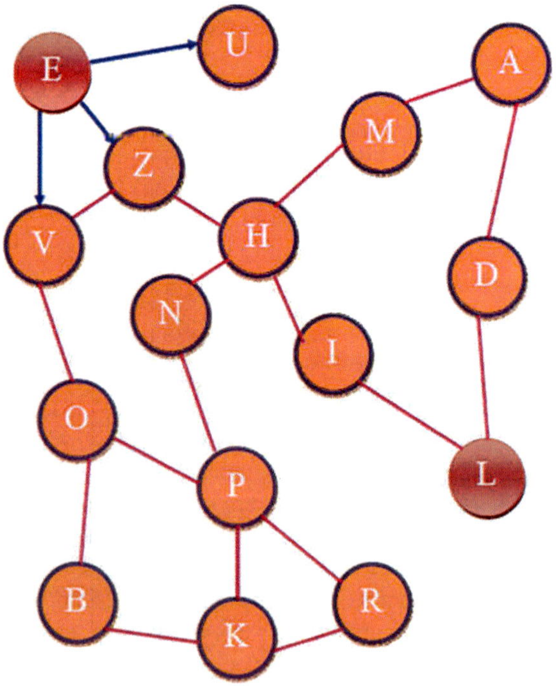

Fig. 2 Flooding of RREQ to neighbor nodes ("V node," "Z node," "U node") from the originating node E

the destination node U. As mismatch occurs for three RREQ, the flooding continues for their neighbor node. Since the node U is having only one connectivity with the node E, RREQ received by the node U checks its route record value E. As neighbor node value and route record value both matches the node U won't send the RREQ to the node E. The node V has three neighbor nodes E, Z, and O. Because of matching of neighbor node E and route record value E, the node V will not send the RREQ to its neighbor node E. The node V sends the RREQ with the value E, L, 10, and EV to its neighbor node Z and O. The node Z has three neighbor nodes E, V, and H. Because of matching of neighbor node E and route record value E, the node V will not send the RREQ to its neighbor node E. The node Z sends the RREQ with the value E, L, 10, and EZ to its neighbor node V and H.

This process continues until the destination address matches with the flooding nodes any one of the neighbor node. In the hop4, the destination address L matches with the neighbor node (H, L) of the flooding node I. Hence, flooding of RREQ stopped at the node L as shown in Fig. 3. The consolidated hop value for each node is shown in Table 1 [12].

The route reply (RREP) sends the route discovery information to the starting node "E" as shown in Fig. 4.

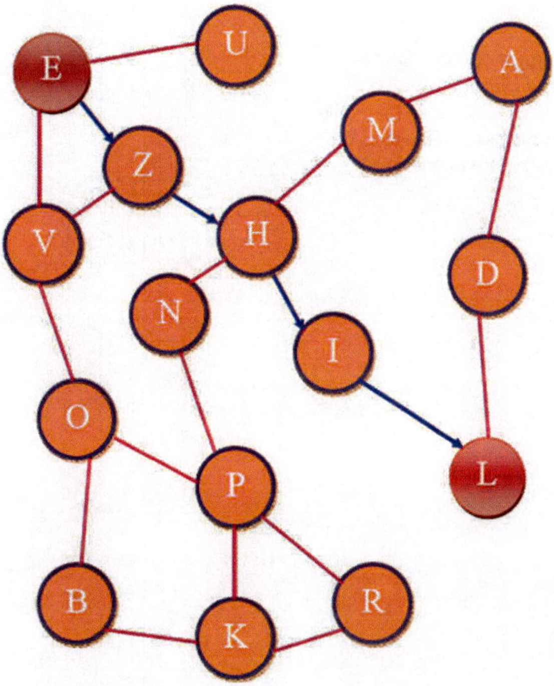

Fig. 3 Route discovery between originating node E and the final destination node L using RREQ message

Table 1 Route record value for each hop at each node

S. no.	NODES	HOP1	HOP2	HOP3	HOP4
1	"E node"	–	–	–	–
2	"Z node"	E	E,V	–	–
3	"H node"	–	E,Z	E,V,Z	–
4	"I node"	–	–	E,Z,H	"E node," V,Z,H
5	"L node"	–	–	–	"E node," Z,H,I
6	"U node"	E	–	–	–
7	"M node"	–	–	E,Z,H	"E node," V,Z,H
8	"A node"	–	–	–	"E node," Z,H,M
9	"D node"	–	–	–	–
10	"V node"	E	E,Z	–	–
11	"N node"	–	–	E,Z,H	"E node," V,Z,H and "E node," V,O,P
12	"O node"	–	E,V	E,Z,V	–
13	"P node"	–	–	E,V,O	"E node," Z,H,N and "E node," Z,V,O
14	"B node"	–	–	E,V,O	"E node," Z,V,O
15	"K node"	–	–	–	"E node," V,O,B and "E node," V,O,P
16	"R node"	–	–	–	"E node," V,O,P

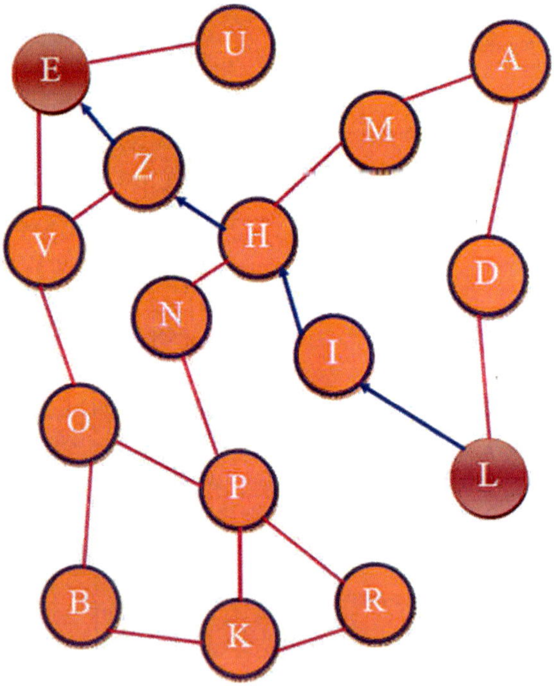

Fig. 4 Route discovery information between originating node E and the final destination node L using RREP message

6 Loop and Duplication Elimination

The DSR protocol eliminates the loop by checking the route record with its neighbor node. If matching occurs, then RREQ is not sent. In the hop2, RREQ from the node Z is not sent to its neighbor node E, because the route record E matches with the neighbor node E. Hence, the loop formation EZE is eliminated. The duplication of RREQ message is eliminated through the unique id value. If node Z receives the RREQ as E, L, 10, E two times, then the second RREQ is eliminated. If the unique ID value differs, then the route is updated [13].

7 Route Maintenance

Route maintenance is used to support new route discovery within a short time. If we need the route from the node V to L, then we can use the route cache stored at the node Z (LIHZ). The final route will be VZHIL. The route cache stored at the node V will be LIZHV. The route maintenance used to reduce the route discovery time [14].

8 Link Failure Management

The link failures between the two nodes are more when there is increase in the mobility of the nodes. There is a need to manage this link failure to identify the route. In the mentioned scenario, if the link between the nodes "L" and "I" fails, then the node "I" sends the message RERR (Route Error) to the node "E" in the route IHZE. Then, based on this RERR message, the nodes update the route cache by removing the link I-L from their route record. The node I cannot find the new route from E to L because now the node I is connected with the node H only. The route E-Z-H-I-H will result in the loop formation. Hence, the node I is removed from the route record. Then, the next previous node H finds the new route using the RREQ message. Now, the new route to the destination node L will be E-Z-H-M-A-D-L [15].

9 Comparison Between DSR and AODV

The important characteristic differences are compared for AODV and DSR in Table 2 [16].

Table 2 Comparison between DSR and AODV

S.No	ASPECTS	AODV	DSR
1	Periodic activity	Through hello message	None
2	Routing overhead	High	Less
3	Routing information	Lesser amount	Greater amount
4	Route record	Not maintained	Stored in cache
5	Number of routes	One	Multiple
6	Timer	Yes	No
7	Mobility support	More	Less
8	Link failure	New route is discovered	Alternate route is used
9	Performance	Better in high mobility	Better in low mobility
10	Routing method	Hop by hop	Source

10 Conclusion

If the link failure is less and the mobility of the nodes is very less, then DSR is preferred for finding the route. Through the route cache, the route is obtained. The time taken to determine the route is less, if the route already exists between the source and destination. It also reduces the broadcasting of RREQ and RREP messages between various nodes. The sequence number field eliminates the duplication of the route hops. Initially, the route identification will take more time. After finding more routes, the route identification time will be lesser when compared to DSR protocol. The reason is DSR protocol uses the reusability concept through route cache. The routing identification suffers heavily because of more link failure. If nodes are making more amount of mobility, then routing between the source and destination also suffers. The table-driven approach is best suited when the route identification is more, whereas the on-demand approach is best suited for less number of route identification. There is a need to modify the route cache whenever link failure or node mobility occurs.

References

1. Karimi M, Pan D (2009) Challenges for Quality of Service (QoS) in Mobile Ad-hoc Networks (MANETs). In: IEEE 10th Annual Wireless and Microwave Technology Conference, Clearwater, FL, pp 1–5
2. Corson, Macker (2010) Mobile Ad hoc Networking (MANET): routing protocol performance issues and evaluation considerations (No. RFC 2501)
3. Quy VK, Ban NT, Nam VH, Tuan DM, Han ND (2019) Survey of recent routing metrics and protocols for mobile Ad-Hoc networks. J Commun 14(2):110–120
4. Abolhasan M, Wysocki T, Dutkiewicz E (2004) A review of routing protocols for mobile ad hoc networks. Ad Hoc Netw 2(1):1–22
5. Tarique M, Tepe KE, Adibi S, Erfani S (2009) Survey of multipath routing protocols for mobile ad hoc networks. J Netw Comput Appl 32(6):1125–1143
6. Johnson D, Hu YC, Maltz D (2007) The Dynamic Source Routing protocol (DSR) for mobile ad hoc networks for IPv4 (No. RFC 4728)
7. Tuteja A, Gujral R, Thalia S (2010) Comparative performance analysis of DSDV, AODV and DSR routing protocols in MANET using NS2. In: IEEE international conference on Advances in Computer Engineering (ACE), pp 330–333
8. Araghi TK, Zamani M, Mnaf ABA (2013) Performance analysis in reactive routing protocols in wireless mobile ad hoc networks using DSR, AODV and AOMDV. In: IEEE International Conference on Informatics and Creative Multimedia (ICICM), pp 81–84
9. Shankar S, Varaprasad G, Suresh HN (2014) Importance of on-demand modified power aware dynamic source routing protocol in mobile ad-hoc networks. IET Microw Antennas Propag 8(7):459–464
10. Sahu PK, Acharya B, Panda N (2018) QoS based performance analysis of AODV and DSR routing protocols in MANET. In: 2nd International Conference on Data Science and Business Analytics (ICDSBA 2018), pp 221–225
11. Sureshbhai TH, Mahajan M, Rai MK (2018) An investigational analysis of DSDV, AODV and DSR routing protocols in mobile ad hoc networks. In: International Conference on Intelligent Circuits and Systems (ICICS 2018), pp 281–285
12. Devi TR, Rameswari B, Vikram K, Abhishek J (2013) Implementation of Dynamic Source Routing (DSR) in Mobile Ad Hoc Network (MANET). Int J Res Eng Technol 2(11):339–345
13. Yufei C, Egemen KC, James PGS (2012) Dynamic Source Routing (DSR) protocol implementation in ns-3. In: Proceedings of the 5th international ICST conference on simulation tools and techniques, pp 367–374
14. Liang Q, Thomas K (2002) Pro-active route maintenance in DSR. ACM SIGMOBILE Mobile Comput Commun Rev 6(3):79–89
15. Naseer AH, Osman G, Suhaidi H, Mohammed MK (2012) Redirect link failure protocol based on dynamic source routing for MANET. In: International conference on networked digital technologies. Springer, Berlin, pp 577–591
16. Mamoun HM (2007) Important characteristic of differences between DSR and AODV routing protocol. In: MCN 2007 conference, pp 7–13